1 11 0832912 8
551.305/A44
AF782469
RSITY OF SUNDERLAND
RMAT ON SERVICES
AY
SUNDERLAND
POLYTECHNIC
Polytechnic Library

DEVELOPMENTS IN SEDIMENTOLOGY 30A

SEDIMENTARY STRUCTURES

THEIR CHARACTER AND PHYSICAL BASIS

VOLUME I

FURTHER TITLES IN THIS SERIES
VOLUMES 1, 2, 3, 5, 8 and 9 are out of print

4 *F.G. TICKELL*
THE TECHNIQUES OF SEDIMENTARY MINERALOGY
6 *L. VAN DER PLAS*
THE IDENTIFICATION OF DETRITAL FELDSPARS
7 *S. DZULYNSKI* and *E.K. WALTON*
SEDIMENTARY FEATURES OF FLYSCH AND GREYWACKES
10 *P.McL.D. DUFF, A. HALLAM* and *E.K. WALTON*
CYCLIC SEDIMENTATION
11 *C.C. REEVES Jr.*
INTRODUCTION TO PALEOLIMNOLOGY
12 *R.G.C. BATHURST*
CARBONATE SEDIMENTS AND THEIR DIAGENESIS
13 *A.A. MANTEN*
SILURIAN REEFS OF GOTLAND
14 *K.W. GLENNIE*
DESERT SEDIMENTARY ENVIRONMENTS
15 *C.E. WEAVER* and *L.D. POLLARD*
THE CHEMISTRY OF CLAY MINERALS
16 *H.H. RIEKE III* and *G.V. CHILINGARIAN*
COMPACTION OF ARGILLACEOUS SEDIMENTS
17 *M.D. PICARD* and *L.R. HIGH Jr.*
SEDIMENTARY STRUCTURES OF EPHEMERAL STREAMS
18 *G.V. CHILINGARIAN* and *K.H. WOLF*
COMPACTION OF COARSE-GRAINED SEDIMENTS
19 *W. SCHWARZACHER*
SEDIMENTATION MODELS AND QUANTITATIVE STRATIGRAPHY
20 *M.R. WALTER, Editor*
STROMATOLITES
21 *B. VELDE*
CLAYS AND CLAY MINERALS IN NATURAL AND SYNTHETIC SYSTEMS
22 *C.E. WEAVER* and *K.C. BECK*
MIOCENE OF THE SOUTHEASTERN UNITED STATES
23 *B.C. HEEZEN, Editor*
INFLUENCE OF ABYSSAL CIRCULATION ON SEDIMENTARY
ACCUMULATIONS IN SPACE AND TIME
24 *R.E. GRIM* and *N. GÜVEN*
BENTONITES
25A *G. LARSEN* and *G.V. CHILINGARIAN, Editors*
DIAGENESIS IN SEDIMENTS AND SEDIMENTARY ROCKS
26 *T. SUDO* and *S. SHIMODA, Editors*
CLAYS AND CLAY MINERALS OF JAPAN
27 *M.M. MORTLAND* and *V.C. FARMER*
INTERNATIONAL CLAY CONFERENCE 1978
28 *A. NISSENBAUM, Editor*
HYPERSALINE BRINES AND EVAPORITIC ENVIRONMENTS
29 *P. TURNER*
CONTINENTAL RED BEDS
31 *T. SUDO, S. SHIMODA, H. YOTSUMOTO* and *S. AITA*
ELECTRON MICROGRAPHS OF CLAY MINERALS
32 *C.A. NITTROUER, Editor*
SEDIMENTARY DYNAMICS OF CONTINENTAL SHELVES

DEVELOPMENTS IN SEDIMENTOLOGY 30A

SEDIMENTARY STRUCTURES

THEIR CHARACTER AND PHYSICAL BASIS

VOLUME I

BY

JOHN R.L. ALLEN, *F.R.S.*

Professor of Geology, University of Reading, England

UNIVERSITY OF SUNDERLAND
WITHDRAWN

ELSEVIER SCIENTIFIC PUBLISHING COMPANY
Amsterdam — Oxford — New York 1982

ELSEVIER SCIENTIFIC PUBLISHING COMPANY
Molenwerf 1,
P.O. Box 211, 1000 AE Amsterdam, The Netherlands

Distributors for the United States and Canada:

ELSEVIER/NORTH-HOLLAND INC.
52, Vanderbilt Avenue
New York, N.Y. 10017

SUNDERLAND POLYTECHNIC
Priestman Library
Class No. 551.305 A44
Checked KT
Access No. 822912
Order No. K5987
UNIVERSITY OF SUNDERLAND
WITHDRAWN

Library of Congress Cataloging in Publication Data

Allen, John R. L.
Sedimentary structures.

(Developments in sedimentology ; 30A-30B)
Includes bibliographies and index.
1. Sedimentary structures. I. Title. II. Series.
QE472.A44 551.3'05 81-12561
ISBN 0-444-41935-7 (v. 30A) AACR2
ISBN 0-444-41945-4 (v. 30B)

ISBN 0-444-41935-7 (Vol. 30A)
ISBN 0-444-41238-7 (Series)
ISBN 0-444-41946-2 (Set)

© Elsevier Scientific Publishing Company, 1982
All rights reserved. No part of this publication may be reproduced, stored in a retrieval system or transmitted in any form or by any means, electronic, mechanical, photocopying, recording or otherwise, without the prior written permission of the publisher, Elsevier Scientific Publishing Company, P.O. Box 330, 1000 AH Amsterdam, The Netherlands

Printed in The Netherlands

To the genius of Henry Clifton Sorby who, combining keen powers of observation with a taste for experiment and quantitative analysis, pointed out the way.

GENERAL PREFACE AND INTRODUCTION TO VOLUME I

Sedimentary structures arise in immediate or close association with the transport of sedimentary materials. Some form where erosion predominates, others as net deposition prevails, and yet further kinds in the brief interval of time between sediment deposition and significant lithification. Many sedimentary structures are ordered features visible on sedimentation (bedding) surfaces, whereas others, often related to surface forms, are expressed as compositional and/or textural patterns (stratification) within sedimentary deposits. Sedimentary structures can be created by chemical and biological as well as by physical agencies, but this book is about those structures wholly or predominantly shaped by physical mechanisms. The latter are much the most important and have continued to attract attention since the earliest days of the earth sciences in their modern form.

The work of describing sedimentary structures, both from rocks (stratigraphic record) and modern sediments, began early in the last century and rightly continues to the present day. Several excellent atlases of structures have been published in the past twenty-five years, of which perhaps the most valuable is that by C.E.B. Conybeare and K.A.W. Crook (1968, Manual of Sedimentary Structures. *Bull. Bur. Miner. Resour. Geol. Geophys.,* No. 102, Canberra, Australia). Most early investigators were concerned with sedimentary structures as contributing to historical geology based on uniformitarian principles. Out of these studies have sprung two divergent uses of sedimentary structures. Particularly since the end of the last century, they have on the one hand been employed by structural geologists and in geological mapping as criteria for the way-up (attitude) of deformed rocks (e.g. R.R. Shrock, 1948, *Sequence in Layered Rocks.* McGraw-Hill, New York). The important syntheses by A.W. Grabau (1913, *Principles of Stratigraphy,* Seiler, New York) and W.H. Twenhofel (1926, *Treatise on Sedimentation,* Williams and Wilkins, New York) foreshadow the other major use of sedimentary structures, namely, as criteria contributing to the environmental interpretation of the stratigraphic record by comparative methods. This type of application of sedimentary structures is of major practical as well as academic importance, and is now an essential element in what we may call *historical sedimentology,* with its emphasis on vertical sequence, spatial patterns and temporal change, and processes on a broad scale. Two excellent recent books —*Sand and Sandstone* (F.J. Pettijohn, P.E. Potter and R. Siever, 1972, Springer-Verlag, Berlin) and *Sedimentary Environments and Facies* (H.G. Reading, 1978, Blackwell, Oxford)— exemplify in different ways this particular flowering.

But there is another and more widely ranging conception of physically-based sedimentary structures available to us, for these features are worthy of study in their own right, as expressions of what in detail happens during and/or shortly after the erosion, transport and deposition of sedimentary materials. Under this view sedimentary structures belong to *dynamical sedimentology,* which operates on much smaller temporal and spatial scales than is typical of historical sedimentology, and which, looking toward the fundamental sciences, seeks to account (ideally in quantitative terms) for sedimentary features of every kind in terms of forces and mechanisms. The beginnings of this kind of understanding of physically-based sedimentary structures are to be found in the work of Henry Clifton Sorby, a Sheffield ironmaster who combined great intellectual gifts with the independent means to indulge his scientific interests. In papers of 1859 (*The Geologist*, 2:137–147) and 1908 (*Q. J. Geol. Soc. London,* 64:171–233), partly based on his own experiments, he described many sedimentary structures and gave a tentative account of the mechanisms and hydraulics of several bedforms. J.S. Owens (1908, *Geogr. J.,* 31:415–420) was perhaps the first to recognize that there existed a definite sequence of bedforms in relation to increasing current strength, a result soon confirmed in detail by the extensive flume experiments of G.K. Gilbert (1914, *U.S. Geol. Surv., Prof. Pap.,* 86). The many subsequent marriages between the descriptive and experimental approaches have been very successful in enlarging our understanding of the origin of sedimentary structures, as can be seen from the influential synthesis by Å. Sundborg (1956, *Geogr. Ann,* 38:217–316) and, particularly, in the milestone of papers (one by a group of hydraulic engineers) compiled by G.V. Middleton (1965, *Primary Sedimentary Structures and their Hydrodynamic Interpretation,* Soc. Econ. Paleontol. Mineral., Spec. Publ., No. 12). In the most successful of these marriages, the ideas and data of the traditional disciplines—geology, geomorphology, engineering in its several forms, and fluid dynamics—are blended together to form, as far as our knowledge and techniques will allow, new explanatory syntheses.

The present book is written in this spirit. It is offered as a hopefully comprehensive summary and review of what we know (or think we know) of physically-based sedimentary structures and, specifically, is a provisional attempt (1) to describe the most important of these structures as they occur at the present day and in the stratigraphic record, and (2) to offer for them an explanation (wherever possible quantitative) in terms of general principles, or at least to suggest by juxtaposition amongst which set of principles their explanation should be found to lie. This study is therefore not a text-book, but rather a handbook or source work, intended for a wide readership (no one traditional discipline is implied), and as much to show where we remain ignorant and require further studies as to indicate where the truth would seem to rest. I venture to think that it will interest those who, like myself, believe that sedimentary structures are worth studying for their own sake, but should also prove useful to historical sedimentologists in their task of environmental reconstruction, to geomorphologists and oceanographers concerned with understanding sea and land forms, to engineers involved with sediment control in deserts, rivers and seas or with offshore structures, and to applied mathematicians and fluid dynamicists looking in the natural environment for examples of phenomena studied mathematically or experimentally, or for the special challenges of two-phase flow phenomena. The wide scope of this book (merely encyclopaedic or bibliographic approaches were never intended) arises from the richness and diversity of material that demands consideration as soon as one examines sedimentary structures for their own sake and not from some narrower and essentially technological (science of a technique) standpoint. It seemed necessary to summarize rather fully the individual contributions made by the several traditional disciplines, for perhaps the most important difficulty, and ultimately the most limiting, faced by anyone who tries to study sedimentary structures in the above way, is that of discovering what relevant work exists. Here a catholic taste in scientific literature seems essential, if the right analogies are to be made and connections drawn. It is true that one reads in order ultimately to reject, but at this comparatively early stage in our understanding of sedimentary structures, it would be dangerously presumptuous to leave final decisions about this other than largely to the reader.

The material is presented in two volumes, and organized as far as possible according to broad physical ideas, aside from the first two chapters in Volume I, which are introductory and preparatory. Volume I is concerned with sedimentary structures in relatively simple physical settings, and Volume II with the more complex situations, in some of which it is necessary to consider groups or hierarchies of structures. To an extent, however, the material of the first volume shades into that of the second. Because the book as a whole is organized according to broad physical ideas, a loose dynamical classification of sedimentary structures emerges, though I have otherwise set aside the thorny and often rather barren problems purely of classification and nomenclature, except in a few cases in which action seemed unavoidable.

Volume I begins with a sketch of environmental fluid dynamics and an introduction to sediment transport. Perhaps the simplest of all sedimentary structures as regards setting are those (some types of grading, packing, fabric) related to the motion of sedimentary particles through, or their emplacement from, various media. Turbulent boundary layers and their transitional states include small-scale flow configurations to which a number of sedimentary structures seem to be related. It is natural to progress from there to the bedforms and internal structures related to sand transport by unidirectional currents. These have been extensively studied, theoretically, in the laboratory, and in the field in modern environments as well as from the stratigraphic record. It seemed appropriate also to include in this volume some account of bedforms and internal structures related to sand transport by oscillatory currents representing waves and tides, despite the relative complexity and ill-understood nature of the mechanisms involved. Research in this hitherto rather neglected area is rapidly gathering pace, and further important developments are to be expected over the next decade. Volume I finishes with an account of sandy bedforms beneath currents subject to spatial and/or temporal change, where the non-uniformity or unsteadiness is not the fundamental cause of the structures. Such changeable currents are, of course, the norm in the real world, yet they have been insufficiently modelled either theoretically or in the laboratory. There is much scope here for future research.

Having written this book I am conscious of my very great debt and lasting obligation to other people. My family and friends have been a source of encouragement and strength, supporting me wholeheartedly and bearing patiently for more years than they should with my preoccupation. It is a delight to acknowledge the help I have received over a period of numerous years from very many individuals, by no means all geologists, who have unstingingly answered my queries, supplied data, illustrative material or references, or given their time to comment on sections of this book. Errors of commission or omission are of course my own responsibility. It is also a delight to thank Milly Oates, who typed and retyped the manuscript more times than I am sure she would care to remember and who helped in many other ways; to Jim Watkins for photographic work over many years; to Gordon Smith and Denys Hutchings who helped with experimental work; and to Alan Cross who made several of the drawings.

I owe a special debt of gratitude to all those geologists, geomorphologists, engineers, and fluid dynamicists whose work is incorporated in various ways in this book. No book is entirely the product of its author, and whatever novel features mine may be thought to possess, will inevitably have depended on their findings and insights.

Reading, May 1980 J.R.L. ALLEN

CONTENTS

Synopsis of Volume II

Chapter 1

ENVIRONMENTAL FLUID DYNAMICS

INTRODUCTION

The *mechanical structures* of sedimentary deposits are ordered shapes on bedding surfaces, and three-dimensional patterns of textural and/or mineralogical layering internal to strata, that were created solely by mechanical forces. *Primary* structures are made directly by an agent of sediment transport, whereas *secondary* structures arise during the short interval between sediment deposition and the noticeable start of lithification. Primary structures far outweigh the secondary in number and variety.

Most mechanical structures arise at the surface of the Earth, and therefore in milieux dominated by the sediment-transporting agents of the *atmosphere* and *hydrosphere,* the planet's outermost fluid shells. A minority of structures, however, form as igneous magmas cool and crystallize in chambers buried deep within the Earth's crust, under physical conditions very different than at the surface. Sedimentary structures record the action of forces associated with the motion of just a few natural fluids—air, water, and magma—or of quasi-fluid materials, such as sediments themselves can temporarily become.

NATURAL FLUIDS AND SOLIDS

Atmospheric air is a mixture of gases, principally nitrogen and oxygen with minor amounts of carbon dioxide, the noble gases, and water vapour. Its density at sea level and 15°C is 12.2 kg m^{-3}, while the dynamic viscosity, the degree of stickiness or cohesiveness, is 1.78×10^{-5} Ns m^{-2} (or Pascal second, Pa s). The viscosity is effectively independent of pressure but increases gradually with temperature (W. Sutherland, 1893; Goldstein, 1965). Since the atmospheric temperature declines upward at about 0.0065°C m^{-1} from the Earth's surface to the top of the troposphere, at a height of roughly 10 km, the dynamic viscosity has decreased by this height to about 80% of its sea-level value. Assuming constant atmospheric temperature, good enough for present purposes, the pressure must decline exponentially upward (Prandtl and Tietjens, 1957) whence, applying the gas laws, the density of air at the top of the troposphere is a mere 25% of its sea-level value. Hence the kinematic viscosity of air—the dynamic viscosity divided by the density—increases upward through the troposphere over an approximately three-fold range. This increase exerts an important control on the aerodynamic drag on sufficiently small bodies moving through the atmosphere.

The temperature generally increases with height in the stratosphere overlying the troposphere. Since the pressure continues to decline approximately exponentially, the air in all but the lowermost parts of the stratosphere is extremely rarified, and the continuum hypothesis which underpins classical fluid dynamics is invalid in this region. For example, under hypersonic conditions, when a body travels at many times the speed of sound (335 m s^{-1} at sea level), we cannot assume as is done for lower speeds a condition of no slip between body and air. At extreme heights, the air is so thin that the motion of a body through it is governed primarily by the frequency of impacts with individual gas molecules, and a "corpuscular" rather than "continuum" approach is demanded.

Whereas the atmosphere, theoretically of infinite depth, envelops the whole Earth, the hydrosphere is today but a few kilometres thick on average and fails to cover some 30% of the surface. The hydrosphere is conveniently divided between fresh and salt water. Fresh water, found in rivers and lakes, can be regarded as pure for present purposes and has a practical density of 1000 kg m^{-3}. The density is very weakly dependent on temperature, reaching a maximum at 4°C (Pounder, 1965). The dynamic viscosity of pure water is 0.00179 Ns m^{-2} at 0°C and decreases markedly with increasing temperature (Goldstein, 1965), a fact of practical importance (e.g., dynamic viscosity at 25°C is 0.000894 Ns m^{-2}). Normal salt or seawater holds about 34.5 parts per thousand of dissolved salts, chiefly sodium chloride, and has a density of approximately 1025 kg m^{-3} (Neumann and Pierson, 1966). The freezing point gradually falls with increasing salinity, but the density shows no temperature-controlled maximum for salinities in excess of 24.7 parts per thousand (Neumann and Pierson, 1966).

The magmas that rise into the Earth's crust consist of mixtures of silicate minerals (chiefly silica, feldspars, micas, amphiboles, pyroxenes, olivines) with dissolved water and gases at high temperatures and pressures. They range in density from about 2700 kg m^{-3} for silica-rich granitic melts to about 3100 kg m^{-3} for basaltic magmas rich in iron-bearing pyroxenes and olivines. Their dynamic viscosity is not known directly, unless one counts measurements from artificial melts, but can be estimated given magma composition and temperature (Bottinga and Weill, 1972; H.R. Shaw, 1972; Scarfe, 1973). Viscosity is strongly affected by composition, that of a basaltic and a granitic melt at 1300°C being respectively 10 and 10^7 Ns m^{-2}. The presence of water lowers magma viscosity (H.R. Shaw, 1963, 1972), the addition of some 8% by weight reducing the value under anhydrous conditions by about two-thirds. Increase of temperature above the chilling point causes a rapid fall in viscosity. Magmas are not dissimilar from water in density, but are very much more viscous than either water or air.

The detrital solids contributing to non-magmatic sedimentary structures originate directly or indirectly in rocks weathered at the Earth's surface, and consist of one or a combination of the following minerals (listed with their densities): quartz (2650 kg m^{-3}), feldspars (2570–2770 kg m^{-3}), micas

(2800–3400 kg m^{-3}), amphiboles (3000–3400 kg m^{-3}), pyroxenes (3100–3900 kg m^{-3}), olivines (3220–4390 kg m^{-3}), calcite (2710 kg m^{-3}), aragonite (2930 kg m^{-3}), and the clay minerals (2600–2900 kg m^{-3}). Many other species occur in sediments, but usually in only minor amounts. Of the minerals listed, quartz and feldspar, the carbonates calcite and aragonite, and the clay minerals are overwhelmingly predominant, mainly in the form of monomineralic grains. Their density is in the order of 2650 kg m^{-3}, it being customary in accounts of sediment transport to describe this value as *mineral density* and the grains concerned as *mineral-density solids.* Magmatic structures are shaped from crystals of silicate minerals, chiefly the feldspars, pyroxenes and olivines, that grew in the melts that created the forms. Wadsworth (1973) gives a short account of magmatic sediments. Comparing transported solids with transporting media, it will be noticed that there is a very close agreement in density in the case of the crystallized silicates and magmas, a moderate agreement between the common minerals and water, and a difference in density of two orders of magnitude in the case of air. Substantial differences in physical behaviour may therefore be expected between these three systems.

The three natural fluids so far considered agree in displaying *Newtonian* behaviour when steadily sheared under laminar conditions between parallel plates (Fig. 1-1). The velocity gradient dU/dy, or rate of shear, within the fluid is then linearly related to the shearing force per unit area τ, or shear stress, by the simple flow-law:

$$\tau = \eta \frac{dU}{dy} \tag{1.1}$$

in which the constant of proportionality η is the afore-mentioned dynamic

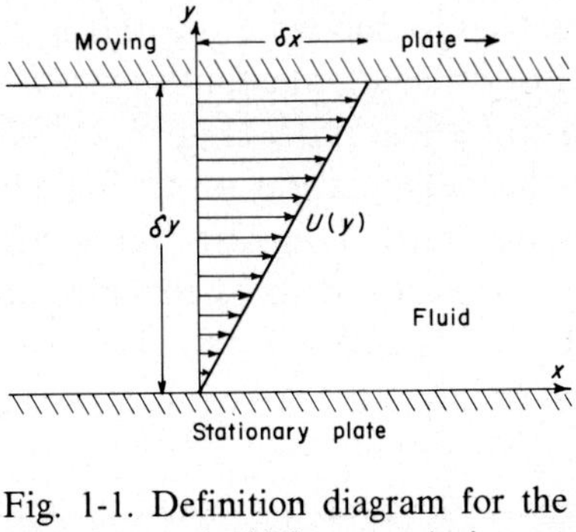

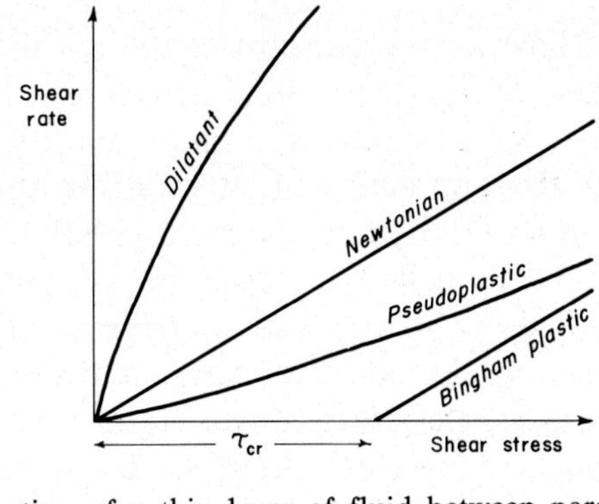

Fig. 1-1. Definition diagram for the deformation of a thin layer of fluid between parallel plates, one of which moves in its own plane.

Fig. 1-2. Schematic relationships between shear rate (strain) and shear stress (stress) in various kinds of fluid.

viscosity. The dynamic viscosity of these fluids, while varying with temperature and pressure, is independent of shear rate. But there is a category of fluid and quasi-fluid materials typified by flow-laws different from eq. (1.1), the description *non-Newtonian* being appropriate. These laws, reviewed by Reiner (1959, 1969), Skelland (1967), Van Wezer et al. (1963), and W.L. Wilkinson (1960), show the "viscosity" to depend not only on temperature and pressure but also on such factors as shear rate and deformational history. Granular solids mixed at high concentration with a Newtonian fluid can exhibit non-Newtonian behaviour, and there are many natural examples of similar mixtures (e.g. mud flows, magmas with dispersed crystals or gas bubbles). Glacier ice is an especially important natural material that cannot be treated as a Newtonian fluid, despite early attempts in this direction.

The deformation of many naturally occurring non-Newtonian materials can be satisfactorily represented by flow-laws in which the shear rate is a function only of the stress (A.M. Johnson, 1970). A *Bingham plastic* has a linear shear stress-shear rate graph with a positive intercept τ_{cr} on the stress axis, in contrast to the Newtonian fluid yielding a graph through the origin (Fig. 1-2). The quantity τ_{cr} is the *yield stress,* which must be exceeded before flow will begin. The flow law therefore is:

$$\tau = \tau_{cr} + \eta_a \frac{dU}{dy} \tag{1.2}$$

in which η_a is the apparent (plastic) viscosity of the material. The flow laws of *pseudoplastic* and *dilatant* fluids are:

$$\tau = k\left(\frac{dU}{dy}\right)^n \tag{1.3}$$

$$\eta_a = k\left(\frac{dU}{dy}\right)^{n-1} \tag{1.4}$$

in which k is a measure of the consistency, n is an exponent denoting the degree of non-Newtonian behaviour ($n = 1$ for Newtonian liquids), and η_a is the apparent viscosity, a function of dU/dy. For a pseudoplastic fluid $n < 1$ and the apparent viscosity falls with increasing shear rate (Fig. 1.2). As Pounder (1965) and Paterson (1969) explain, the deformation of glacier ice can be modelled in terms of the pseudoplastic fluid. Dilatant fluids are typified by $n > 1$, and increase in apparent viscosity with shear rate, a behavioural mode of concentrated sand (O. Reynolds, 1885, 1886). The flow curves are therefore convex-up in Fig. 1-2. Power laws such as eq. (1.2) have created many difficulties, as discussed by Reiner (1969), one of the more obvious being that k lacks unique dimensions, but retain practical appeal and are capable of considerable further generalization (Nutting, 1921; Scott Blair, 1965).

There are yet other classes of non-Newtonian fluid, but of lesser importance naturally. In one class, that of *thixotropic* and *rheopectic* fluids, resistance to deformation depends on the duration of shear, reflecting respectively the breakdown and build-up of structure within the deforming material. Dispersions of certain clay minerals in water belong to this general class. The *viscoelastic* fluids, for example, bitumen and "bouncing putty", possess both elastic and viscous properties, and which property manifests itself depends on the time-rate of change of shear stress.

SEDIMENTATION: ENVIRONMENTS, AGENTS AND PRODUCTS

The grains composing the sediments in which we mainly see mechanical structures came together after having been derived either directly by the weathering of rocks exposed at the Earth's surface, or indirectly through the mediation of organisms that precipitated the minerals from natural waters

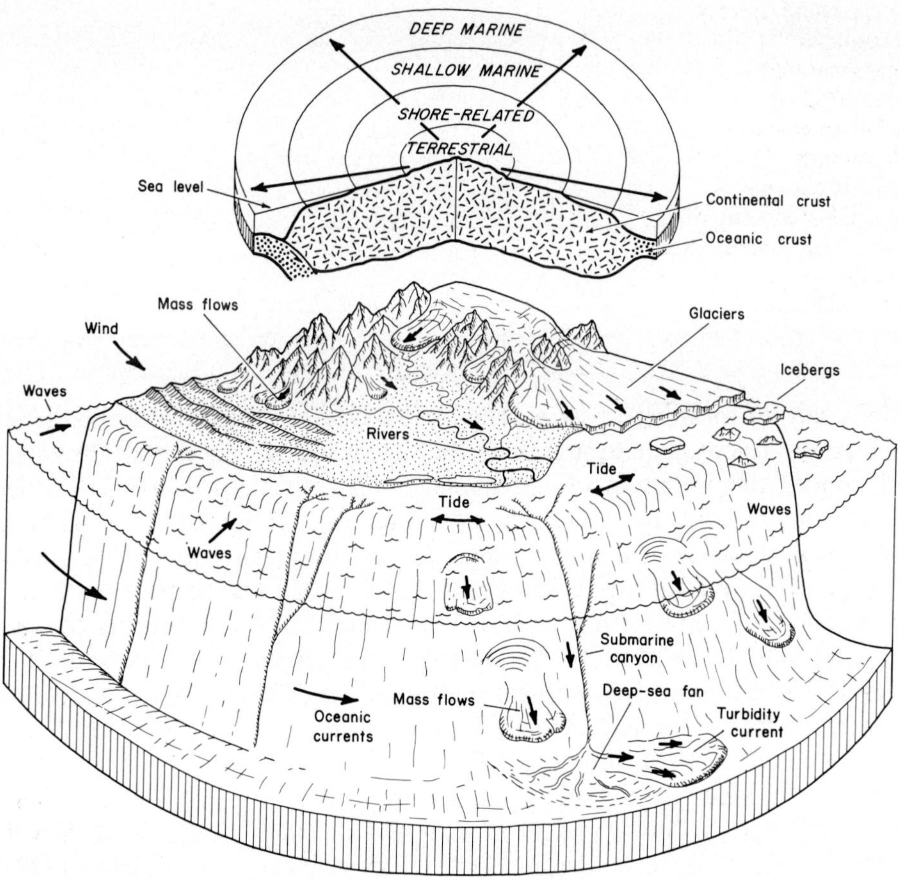

Fig. 1-3. Sedimentary environments, sedimentary agents, and sediment transport paths (source-sink relationship) for an ideal continent and bordering ocean.

TABLE 1-I

Summary of the main transporting agents and their typical speeds

Transporting agency	Typical speed ($m\ s^{-1}$)
Mass flows	
creeping soils	1×10^{-9}–3×10^{-8}
rock slides/avalanches	$\lesssim 10.0$
debris flows	$\lesssim 1.0$
Rivers	
normal flow stages	0.5–1.0
flood stages	1.0–5.0
Wind	
breeze	1.0
gale/hurricane	10.0–20.0
Wind-generated waves	
maximum orbital velocity	$\lesssim 1.0$
Tidal currents	
open ocean	0.01–0.1
offshore in restricted sea	0.5
estuaries	$\lesssim 3.0$
Oceanic currents	
thermohaline	0.1–1.0
wind-generated	0.1–0.5
storm surge	$\lesssim 1.0$
Turbidity currents	
small, low concentration	0.1–1.0
large, high concentration	1.0–10.0
Glaciers	
at equilibrium	3×10^{-7}–3×10^{-6}

(chiefly salt) to form hard tissues. The typical weathered grain travelled from a *source* on a continent to a *sink* at the ocean bed on an overall downhill path, under the influence of a sequence of agents that exchanged either thermal or potential for kinetic energy (Fig. 1-3; Table 1-I). The particle traversed a sequence of major sedimentary environments—*terrestrial, shore-related, shallow-marine,* and *deep-marine*—each characterized by a particular combination of transporting agents. Each combination, acting in concert with environmental factors such as salinity and temperature, and under the general constraints of geographical boundary conditions, created a distinctive sort of deposit, or sedimentary facies (Reineck and Singh, 1973; Friedman and Sanders, 1978; Reading, 1978).

Prominent amongst transporting agents in the terrestrial environment (Fig. 1-3; Table 1-I) are a variety of *mass flows* (Sharpe, 1938; W.H. Ward, 1945; Varnes, 1958; Scheidegger, 1975; Ives et al., 1976; Voight, 1978). These are gravity-driven overland surges or slower and more continuous streams of concentrated dry, moist, saturated or frozen debris. The mantle of soil and rock-waste is involved in slow downhill flow (*creeping regolith*) (Carson and Kirkby, 1972), distinguished as *solifluction* when saturation is more or less permanent, but as *gelifluction* when movement above frozen ground and freezing are involved (Washburn, 1973). In cold climates such flows may become majestic *rock glaciers* (P.G. Johnson, 1978). Weathered fragments as they tumble off crags and cliffs create *rockfalls* (Rapp, 1960a, 1960b; Crandell and Fahenstock, 1965; Luckman, 1976; Gordon et al., 1978). More spectacular because larger and less frequent are *rock slides* and *rock avalanches* (Crandell and Fahnestock, 1965; Hoyer, 1971; Plafker and Ericksen, 1978; Voight and Pariseau, 1978), in which the failed rocks retain much of their original arrangement (slides) or become thoroughly broken up and jumbled (avalanches). *Mud flows* or *debris flows* (A.M. Johnson, 1970; Prior and Stephens, 1972; Morton and Campbell, 1974; Pilgrim and Conacher, 1974; R.H. Campbell, 1975; Hsu, 1975; Blong and Dunkerley, 1976; Statham, 1976), a mixture of coarse debris in a muddy matrix, are formed usually after heavy rain, either from saturated ground or from debris-charged rivers that lose water to their beds. The variety called a *lahar* arises from the loose ash accumulated on the slopes of volcanoes (Scrivenor, 1929; Crandell and Waldron, 1956; Ulate and Corrales, 1966; Waldron, 1967; Hyde, 1975), often as the result of heavy rain accompanying eruption.

Mass flows ultimately lead to *rivers* (Fig. 1-3; Table 1-I), the second major transporting agent of the terrestrial environment (Leopold et al., 1964; Morisawa, 1968; Schumm, 1977). These are channelized flows of water found in three main setting, on alluvial fans, which are flat, cone-shaped bodies of sediment formed where streams debouch from uplands abruptly on to plains, on deltas formed where receptive water-bodies such as a lake or the ocean are encountered (C.C. Bates, 1953; Wright and Coleman, 1973, 1974; Broussard, 1975; Coleman and Wright, 1975; Audley-Charles et al., 1977), and in alluviated valleys with floodplains (Fisk, 1944, 1947). By lateral coalescence, alluvial fans create piedmont aprons, and valley fills make coastal plains of alluviation (Bernard and LeBlanc, 1965).

The *wind* is a potent transporting agent (Bagnold, 1954b) in the hot deserts (Cooke and Warren, 1973; Glennie, 1970; Mabbutt, 1977) and the cold deserts (Antevs, 1928; Fristrup, 1952; Sundborg, 1955) of the terrestrial realm (Fig. 1-3; Table 1-I), where precipitation is low and vegetation sparse. Sand and dust storms typify these regions (Coles, 1938; Lunson, 1950; Ashwell, 1966, 1972, 1973; Péwé, 1975; Nickling, 1978), from which there is substantial transport of dust direct to the oceans (Folger, 1970; Prospero et al., 1970; Chester and Johnson, 1971a, 1971b; Chester, 1972; Prospero and Carlson, 1972; Chester et al., 1972; Windom and Chamberlain, 1978; Tullett, 1980).

The shore-related environment (Fig. 1-3; Table 1-I) is complex (Guilcher, 1958a; Zenkovitch, 1967; C.A.M. King, 1972; Komar, 1976; R.A. Davis, 1978), consisting of some combination of tidal river channels, estuaries, tidal flats, lagoons and bays, sandy barrier islands, and fresh to salt water marshes. Here sediment is transported by rivers emerging from their mouths as jet-like flows, by wind-generated *surface waves,* and by *tidal currents,* particularly in estuaries (Ippen, 1966; Barnes and Green, 1972; Cronin, 1975; Dyer, 1973; Officer, 1976), by thermohaline and wind-generated *oceanic currents,* and by the wind itself on intertidal and supratidal parts. The wind, acting through the waves and storm tides that it generates, becomes a potent modifier of the coast when raised to storm force (Morgan et al., 1958; McKee, 1959; Stoddart, 1962; Niddrie, 1964; Hayes, 1967; Perkins and Enos, 1968; Andrews, 1970; McGowen and Scott, 1975).

The shallow-marine environment (Fig. 1-3; Table 1-I) fringes the continental masses, or forms outlying platforms, and extends to a depth of 100–200 m, where there is a significant steepening of the sea bed at the edge of the continental shelf. Dominant here are wave-related currents, tidal currents, wind-driven flows, and thermohaline circulations (Swift et al., 1972; Stanley and Swift, 1976), the patterns of water and sediment movement being complex and often seasonally dependent (e.g. Bumpus, 1973; Harlett and Kulm, 1973; Carlson, 1974; Komar et al., 1974; Carter and Heath, 1975; Kulm et al., 1975). The long-term effect, however, is for sediment to disperse and diffuse outward from the shore, where it was introduced by rivers and spread laterally by tidal and wave-generated longshore currents. Wave and tidal currents can ordinarily mobilize the coarser grades of sediment only in the shallows near shore, but during storms wave-related flows can entrain sand at depths of 100–200 m. The other currents, generally relatively sluggish, are chiefly involved in transporting suspended silt and clay, and any sand dispersed upward by wave-action.

The deep-marine (Fig.1-3) and terrestrial environments are in many ways morphologically comparable (M.N. Hill, 1963; Shepard, 1963). On the ocean floor at a depth of 4–6 km lie extensive depositional plains (abyssal plains) above which tower mountain ranges (mid-oceanic ridges) larger and more extensive than any on the land. The continental borders are marked by long slopes (continental slope and rise) of substantial inclination (about 5° maximum) crossed by deep valleys (submarine canyons) fed by tributaries some of which head in the shallow-marine realm. At the foot of the continental rise, the canyons pass into systems of distributaries that meander toward the abyssal plains across gently sloping aprons of sediment called deep-sea fans or cones, analogous to terrestrial alluvial fans (Normark, 1970a, 1970b, 1978).

Correspondingly, the transporting agents (Fig. 1-3; Table 1-I) resemble those of the terrestrial realm. Various mass flows occur in the oceans, some on slopes considerably less than 1°, with the larger involving in the order of 10 km^3 of sediment (D.G. Moore, 1961; Dott, 1963; Morgenstern, 1967; Middleton and Hampton, 1973, 1976; Lowe, 1976a, 1976b; Embley, 1980). Slides (integrity of stratification largely retained) and slumps (strata disrupted and jumbled) are common, particularly on the relatively steep, sediment-fed continental slope and rise (D.G. Moore, 1961, 1978; Heezen and Drake, 1964; Morgenstern, 1967; Uchupi, 1967; Stanley and Silverberg, 1969; K.B. Lewis, 1971; Coleman et al., 1974; Herzer, 1975; Le Pichon et al., 1975; Jacobi, 1976; McGregor and Bennett, 1977, 1979; Almagor and Garfunkel, 1979; Prior and Coleman, 1980a, 1980b). The other types of mass flow are *grain flows,* composed predominantly of sand, and mud-rich debris flows similar in character to terrestrial examples (Shepard, 1951; Middleton and Hampton, 1973, 1976; Embley, 1976; Lowe, 1976a, 1976b; Stanley and Taylor, 1977). Slumps and debris flows, through prolonged dilution with seawater, apparently can change into *turbidity currents* (Van der Knaap and Eijpe, 1968; Allen, 1971b; Hampton, 1972), which are vigorous river or surge-like flows sustained by an excess of density imparted by dispersed sediment. These currents are regarded as primarily responsible for submarine canyons and channeled deep-sea fans and cones. The deep-marine environment possesses "winds" in the form of wind-driven flows, confined to a relatively thin (50–500 m) warm surface layer above the thermocline, and thermohaline currents that operate in deeper and cooler aqueous strata (Neumann and Pierson, 1966; Neumann, 1968; Heezen and Hollister, 1971; Webster, 1971; Gould, 1978). These deeper currents (e.g. Stander et al., 1969; Reid and Nowlin, 1971; Webster, 1971; H.B. Zimmerman, 1971; Connary and Ewing, 1972; Eittreim et al., 1972; Betzer et al., 1974; Tucholke and Eittreim, 1974), strongest along the western boundaries of oceans, arise through the sinking at high latitudes of seawater made relatively dense through cooling and partial freezing. Locally, as in the Mediterranean Sea (Lacombe, 1965; Heezen and Johnson, 1969), an increase of density due to evaporation leads to undercurrents.

Glaciers are gravity-driven streams of ice of importance as transporting agents (Fig. 1-3; Table 1-I) at high altitudes in the terrestrial environment and at high latitudes in the terrestrial, shallow-marine and deep-marine realms (Lliboutry, 1964, 1965; Paterson, 1969; Embleton and King, 1975; Sugden and John, 1976). Land-based glaciers are customarily divided between temperate, when some slip is possible between the ice and its meltwater-lubricated bed, and cold, when the ice is so chilled that it firmly adheres to the substrate. Stagnating glaciers generally include systems of dendritic internal drainage tunnels that collect and transmit meltwater to their downslope margins.

BOUNDARY LAYERS ON A ROTATING EARTH

General

A vital consequence of viscosity is that where a fluid and rigid surface are in relative motion, the fluid in a layer adjoining the surface experiences a retardation, that directly at the surface being brought to rest, except under special circumstances. This zone of retardation is a *boundary layer,* the flow in it being either *laminar,* when fluid particles follow smooth parallel streamlines, or *turbulent,* when there is an irregular eddying motion of relatively high velocity. The flow is *two-dimensional* when the same profile of velocity is observed over the extent of the layer, but *three-dimensional* when the velocity must be resolved between three orthogonal components. Boundary layers can, furthermore, be *steady,* when the flow remains constant in time at a station, or *unsteady*, when conditions change with time. Some boundary layers are *uniform,* maintaining a constant character over their extent, whereas others are *non-uniform,* varying spatially. The boundary layers of interest in environmental fluid dynamics all occur in the context of a rotating spherical Earth, and rotational effects are negligible only for those of a sufficiently small scale.

Rotating flows

Two fictitious forces on fluid elements must be accounted for when using the rotating Earth as a frame of reference for boundary-layer study (Craig, 1973). The first is the *centrifugal force* $mr\sigma^2$, where m is the particle mass, r its radial distance from the axis of rotation, and σ the angular velocity ($=7.3 \times 10^{-5}$ rad s^{-1}). The second is the *Coriolis force,* which is zero for particles that are stationary with respect to the Earth, but non-zero when the particles are moving. The magnitude of its horizontal component is $2m\sigma U \sin\theta$, where U is the particle velocity and θ the latitude of the motion. This component acts perpendicularly to the right of the motion in the Northern Hemisphere and to the left in the Southern Hemisphere; its value is zero at the Equator and a maximum at the Poles.

The boundary layer formed on a flat disc rotating uniformly about an axis perpendicular to its plane in an infinitely extensive Newtonian fluid otherwise at rest illustrates well the influence of centrifugal force and the nature of three-dimensional boundary layers (Schlichting, 1960). Fluid adjacent to the disc is transported with the disc because of viscous drag, but on account of the centrifugal force is thrown outward, a compensating axial flow toward the disc being induced (Fig. 1-4).

Recognizing that, under steady conditions, the radial shear-stress component is balanced by the centrifugal force, while the circumferential component is balanced by the circumferential velocity gradient in the fluid at the surface of the disc, the boundary-layer thickness can be shown to be of order $\delta = (\nu/\sigma)^{1/2}$, in which $\nu = \eta/\rho$ is the fluid kinematic viscosity where ρ is the fluid density. Viewed from a stationary frame of reference in the undisturbed fluid, and compared to the radial velocity of the disc, the radial velocity component U reaches a maximum at about the outer edge of the boundary layer, while the circumferential component W is a maximum at the surface

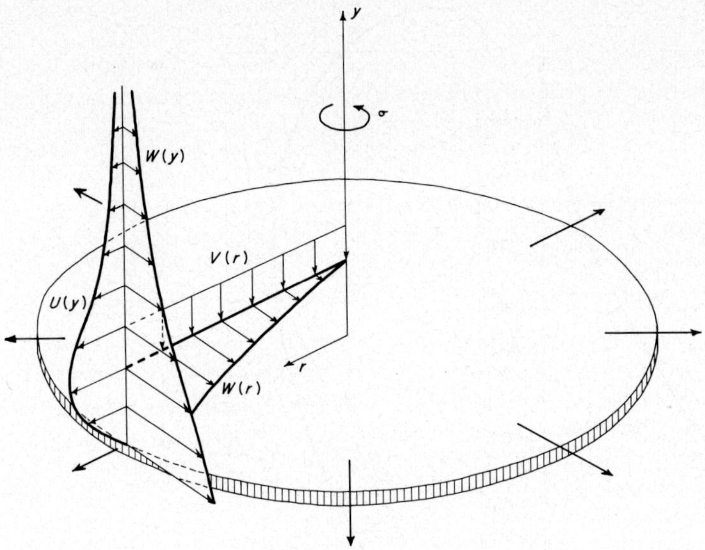

Fig. 1-4. Schematic radial, axial, and tangential profiles of velocity generated within the boundary layer on a smooth, circular disc rotating in an otherwise still fluid. Velocity measured relative to stationary coordinates.

of the disc and declines upward. The axial component V increases upward from zero at the surface. The direction of the shear stress on the surface of the disc therefore deviates considerably from the radial direction.

A related motion illustrating the role of Coriolis force arises when a disc in rigid-body rotation with an infinitely extensive Newtonian fluid is speeded up slightly (Greenspan, 1968). The changed motion of the disc is communicated to the fluid through a boundary layer again with thickness of order $(\nu/\sigma)^{1/2}$, the Coriolis force in this layer being balanced by the viscous shear stress on the disc. The resulting steady flow in the boundary layer closely resembles that in Fig. 1-4 and, viewed from a rotating coordinate system, is described by:

$$U = W_{\mathrm{ref}} \exp\left[-y(\nu/\sigma)^{-1/2}\right] \sin y(\nu/\sigma)^{-1/2} \tag{1.5}$$

$$V = \left(\frac{\nu}{\sigma}\right)^{1/2}\left\{-1 + \exp\left[-y(\nu/\sigma)^{-1/2}\right]\left[\sin y(\nu/\sigma)^{-1/2} + \cos y(\nu/\sigma)^{-1/2}\right]\right\} \tag{1.6}$$

$$W = W_{\mathrm{ref}} \exp\left[-y(\nu/\sigma)^{-1/2}\right] \cos y(\nu/\sigma)^{-1/2} \tag{1.7}$$

where U, V and W are the radial, axial and circumferential velocity components as before, W_{ref} is a reference circumferential velocity, and y is distance measured upward parallel with the axis of rotation (Fig. 1-5). Equations (1.5) and (1.7) represent the components of a vector. The tips of the vectors they yield, when plotted in the U–W plane for increasing values of $y(\nu/\sigma)^{-1/2}$,

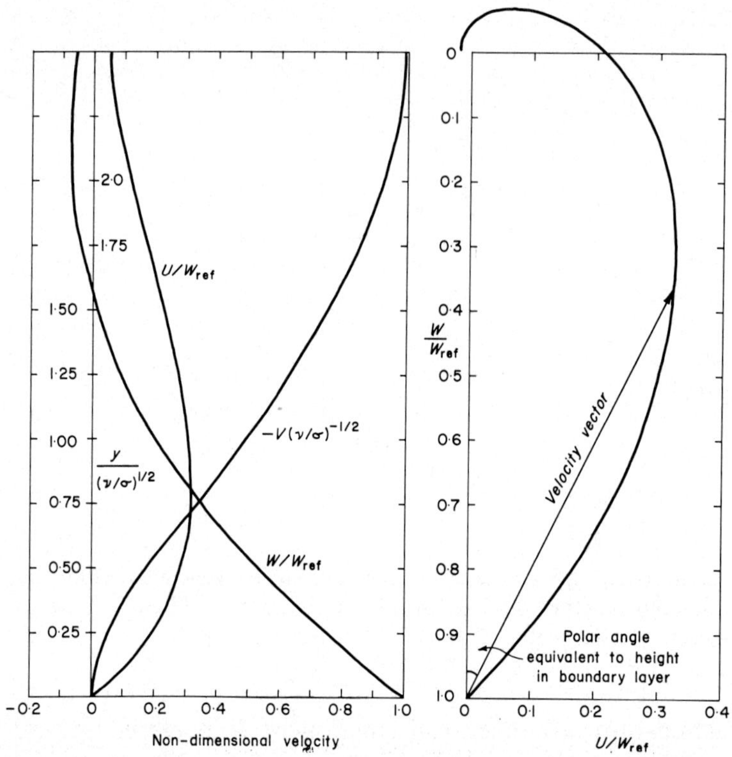

Fig. 1-5. Calculated velocity profiles in three mutually perpendicular directions within the three-dimensional boundary layer developed on a disc perturbed from a state of rigid-body rotation with a fluid. Height within the boundary layer is given non-dimensionally in terms of the boundary-layer thickness $(\nu/\sigma)^{1/2}$. The right-hand graph shows the greater part of the so-called Ekman spiral, that is, the curve defined by the tips of the velocity vector parallel with surface of the disc.

describe a logarithmic spiral called the Ekman spiral, after the discoverer of this type of motion in wind-driven oceanic currents (Ekman, 1905). Applied to motions on the Earth, the angular velocity present in eqs. (1.5)–(1.7) appears in latitude-adjusted form, $\sigma \sin \theta$, as in the definition of the Coriolis force ($\theta = 90°$ for the disc).

Flat plate

Coriolis force can safely be discounted for non-rotating boundary-layer flows on laboratory and similar scales. Two-dimensional flows are then obtainable. Consider the flow of a Newtonian fluid of uniform velocity U_∞ parallel with a smooth, thin flat plate (Fig. 1-6). The fluid on each side is retarded by viscous friction to form a boundary layer (Prandtl, 1952; Schlichting, 1960; Rosenhead, 1963). The flow is laminar over an initial streamwise distance, the continuous transfer or diffusion of momentum from

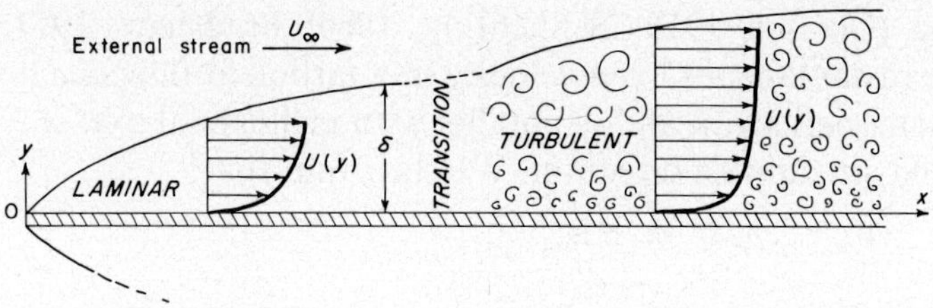

Fig. 1-6. Schematic representation of the boundary layers developed on a flat plate in parallel flow.

faster to slower levels in the fluid taking place only at a molecular scale. The boundary-layer thickness can be shown to grow with distance x as:

$$\delta = 5.64\left(\frac{\nu x}{U_\infty}\right)^{1/2} \tag{1.8}$$

while the boundary shear stress falls as:

$$\tau_0 = 0.332\rho U_\infty\left(\frac{\nu}{U_\infty x}\right)^{1/2} \tag{1.9}$$

in which the bracketed term is the inverse of the *Reynolds number*, $Re = U_\infty x/\nu$, a non-dimensional parameter measuring the ratio of inertial to viscous forces acting in the layer. The local velocity increases very gradually upward within a laminar boundary layer, whence the shear stress proves to be very small for water and air (but not necessarily for magma), having regard to the viscosities and densities quoted above. At $Re_x \approx 10^5$ the boundary layer undergoes transition to turbulence. Momentum transfer from faster to slower layers now occurs on a macromolecular scale, by virtue of the transverse movement of eddying parcels of fluid, as well as at the molecular level. Equation (1.1) written for this case includes an "eddy viscosity" additional to and ordinarily very much larger than the dynamic viscosity. The turbulent boundary layer can be shown to grow in thickness as:

$$\delta = 0.37x\left(\frac{U_\infty x}{\nu}\right) \tag{1.10}$$

the boundary shear stress being given by:

$$\tau_0 = 0.0225\ \rho U_\infty^2\left(\frac{\nu}{U_\infty \delta}\right)^{1/4} \tag{1.11}$$

The velocity profile is much steeper in turbulent than laminar flow and the shear stresses are larger. Most natural currents in water and air are turbulent. Turbulent flow in magma, however, arises only with low-viscosity melts at a large scale.

The structure of turbulent flow and its bearing on other flow properties requires brief mention (Clauser, 1956; Schlichting, 1960; Bradshaw, 1971; Townsend, 1976). The instantaneous local velocity in a turbulent flow can be resolved into three components u, v and w parallel with orthogonal axes x, y and z, where x is in the streamwise direction. We then find that:

$$u = U + u', \qquad v = V + v', \qquad w = W + w',$$
$$V = 0, \qquad W = 0 \tag{1.12}$$

where U, V and W are the velocities averaged over a time large compared with the turbulent fluctuations, and u', v' and w' represent the deviations of the instantaneous components from the time-averaged values. The fluctuating components, some positive and others negative, average to zero and are

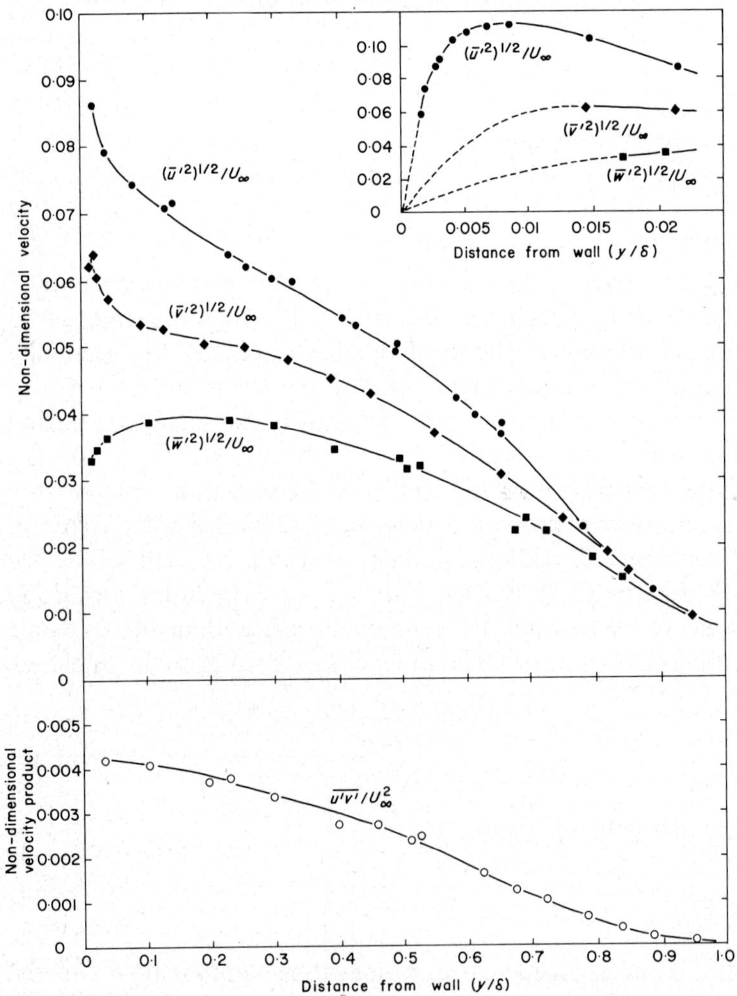

Fig. 1-7. Variation with non-dimensional distance normal to wall of turbulence intensities (root-mean-square values of fluctuating velocity components divided by velocity of external stream) and Reynolds stress (normalized by velocity of external stream). Data of Klebanoff (1955).

distributed in an approximately Gaussian manner. The root-mean-square quantities $(\overline{u'^2})^{1/2}$, $(\overline{v'^2})^{1/2}$ and $(\overline{w'^2})^{1/2}$, where the bar denotes a time-average, are therefore measures of turbulence intensity. The product $\rho \overline{u'v'}$, called a Reynolds stress, represents a rate of change of momentum, and is an effective shearing stress within the flow. Laufer's (1951, 1954) and Klebanoff's (1955) experiments show that the Reynolds stress and root-mean-square quantities attain maxima within a broad zone contained in the lower one-quarter of the boundary layer (Fig. 1-7). The fact that fluid in the turbulent eddies is moving haphazardly across the line of the general stream, in places upward and in others downward, while at the same time these eddies are transported with the flow, means that a fixed point on the flow boundary experiences significant pressure fluctuations on a time-scale that reflects the scale of the eddies (Willmarth and Woolridge, 1962; M.K. Bull, 1967; Blake, 1970; Willmarth and Yang, 1970; Willmarth, 1975).

A turbulent boundary layer is conveniently divided into three layers. In the *viscous sublayer,* an extremely thin zone next to the boundary, turbulence is absent and the velocity profile is practically linear, whence:

$$\tau = \eta \frac{dU}{dy} = \text{constant.} \tag{1.13}$$

Defining $U_*^2 = \tau_0/\rho$, in which U_* is called the shear velocity (another measure of intensity of turbulence), eq. (1.13) becomes:

$$U_*^2 = \nu \frac{dU}{dy} \tag{1.14}$$

which on integration gives:

$$\frac{U}{U_*} = \frac{U_* y}{\nu} \tag{1.15}$$

under the boundary conditions $U = 0$ at $y = 0$. This equation, in which the right-hand side is a Reynolds number, describes the linear increase in velocity within the viscous sublayer, which proves experimentally to have a thickness $\delta_{sub} = 10\,\nu/U_*$. Equation (1.15) is valid over the range $0 < U_* y/\nu \leqslant 10$. In the next thin zone, $10 < U_* y/\nu \leqslant 30$–$70$, called the *buffer layer,* the viscous and turbulent stresses are comparable. In the outer zone, that of *fully developed turbulent flow,* the turbulent stresses greatly predominate.

There is no single function that will satisfactorily represent the velocity profile of such complex flows (see Willis, 1972). The most successful is the *universal velocity-defect law* (Clauser, 1956; Schlichting, 1960), given by:

$$\frac{U_\infty - U}{U_*} = -5.6 \log \frac{y}{\delta} + 2.5 \qquad y/\delta \leqslant 0.15 \tag{1.17}$$

$$\frac{U_\infty - U}{U_*} = -8.6 \log \frac{y}{\delta} \qquad y/\delta > 0.15 \tag{1.18}$$

and valid only outside the viscous sublayer. The defect law successfully describes the velocity profile in all three kinds of flow distinguished by Nikuradse (1933) on the basis of the degree of roughness of the flow boundary, that is, *smooth flow,* in which the boundary irregularities such as sand grains lie wholly within the viscous sublayer, *transitional flow*, in which they penetrate the viscous sublayer but none extend beyond the buffer layer, and *rough flow,* in which all irregularities project into the fully developed flow. These regimes can be distinguished either using Reynolds numbers of the form of $U_* k_s / \nu$, in which k_s is a measure of the size of irregularity, or by values of k_s / δ_{sub}. A rather general relation, useful for laminar as well as turbulent flow, is the *quadratic stress law* (alternatively a definition):

$$\tau_0 = \frac{f}{4} \rho \frac{U_m^2}{2} \tag{1.19}$$

or $U_* / U_m = (f/8)^{1/2}$, in which f is the Darcy-Weisbach friction coefficient, and U_m is the local flow velocity averaged over the boundary layer. The Darcy-Weisbach f is calculable under restrictive conditions, but for other than smooth turbulent flows is best determined from empirical graphs (e.g. Coulson and Richardson, 1965). It decreases with Reynolds number but increases with boundary roughness.

Oscillatory flows

Waves and tides create oscillatory boundary layers. Schlichting (1960) summarizes early work on the boundary layer created on a plate oscillating sinusoidally in its own plane in a stationary fluid, and Knight (1978) reviews much of the literature relating to the corresponding case of wave and tidal boundary layers. In these cases the boundary-layer thickness is of order $(2\nu/\sigma)^{1/2}$, where $\sigma = 2\pi/T$ and T is the wave period, a suitable eddy viscosity being chosen when the flow is turbulent. Lamb (1932) gives the instantaneous velocity profile in the boundary layer created on a smooth plane bed by a deep sinusoidally oscillating fluid as:

$$U = U_{max} \left\{ \sin \frac{2\pi t}{T} - \exp\left[-y \left(\frac{2\nu}{\sigma} \right)^{-1/2} \right] \sin\left[\frac{2\pi t}{T} - y \left(\frac{2\nu}{\sigma} \right)^{-1/2} \right] \right\} \tag{1.20}$$

in which U_{max} is the amplitude of the velocity outside the boundary layer, t is time, and y is distance vertically upward. Figure 1-8 shows velocity profiles calculated at intervals of one-tenth of a period during one-half of an oscillation, rotation of the graph about the ordinate yielding profiles for the other half. Equation (1.20) has excellent experimental support (Knight, 1978), and indicates qualitatively the form of profile in turbulent flow (Kalkanis, 1964; Kajiura, 1968; Sleath, 1970).

Transition to turbulence in oscillatory boundary layers, investigated widely (Li, 1954; Manohar, 1955; Vincent, 1958; Collins, 1963; Kalkanis, 1964;

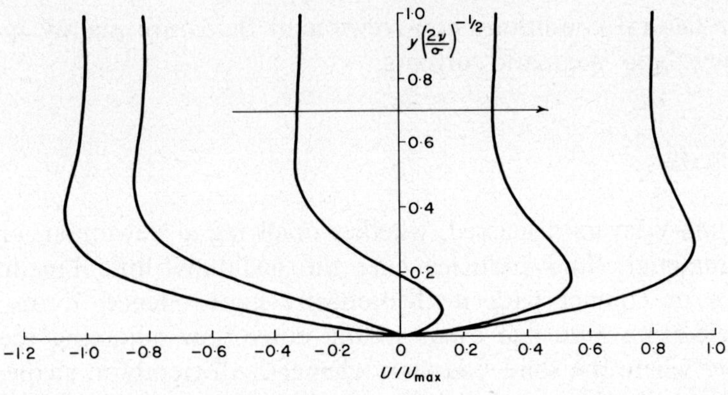

Fig. 1-8. Profiles of non-dimensional velocity (relative to plate) within the boundary layer generated on a flat, smooth, plate oscillating in its own plane within an otherwise still fluid. Distance from the plate is made non-dimensional using the boundary-layer thickness, $(2\nu/\sigma)^{1/2}$.

Obremski and Fejer, 1967; Horikawa and Watanabe, 1967, 1968; Sleath, 1974a, 1975a; Merkli and Thomann, 1975), begins with the appearance of transient turbulent spots (e.g. Miche, 1958) and extends over a very wide and imprecise range of conditions. For smooth boundaries a critical Reynolds number $U_{max}/(\nu\sigma)^{1/2}$ in the general order of 400 is suggested experimentally (Merkli and Thomann, 1975). Transition on rough and rippled beds, however, is influenced by bed irregularities (Sleath, 1974a, 1975a, 1976; Knight, 1978). Knight (1975, 1978) emphasizes that the maximum bed shear stress in oscillatory flow is always greater than that of the corresponding steady flow. The increase can be very substantial under laminar conditions, but may not be marked in turbulent flows. Jonsson (1967), Kamphuis (1975), Jonsson and Carlsen (1976), Saunders (1977) and Vitale (1979) show how the Darcy-Weisbach friction coefficient can be estimated for oscillatory boundary layers. As in steady flows, it varies with Reynolds number and bed roughness.

Non-Newtonian fluids

Wilkinson (1960) and Skelland (1967) give useful introductions to boundary-layer flow in non-Newtonian fluids. This problem can be treated theoretically in ways corresponding to Newtonian fluids, though several effects peculiar to the non-Newtonian condition emerge, for example, *plug flow* (motion as a solid) in those regions of a streaming Bingham plastic where $\tau < \tau_{cr}$. Theoretical and empirical correlations are available for the friction coefficient in a variety of non-Newtonian fluids, and there are criteria for the onset of turbulence, in the form of critical Reynolds numbers involving as may be appropriate the yield stress and exponent in eqs. (1.2) and (1.3). Under natural conditions non-Newtonian fluids are mainly involved in mass flows and magmatic currents.

SEPARATION OF FLOW

In all the boundary layers discussed, whether involving a Newtonian or non-Newtonian material, fluid particles near the bed have little kinetic energy and those in contact with it theoretically none. Hence, if any retarding or accelerating influence exists in the outer flow adjoining the boundary layer, or where the solid boundary changes sufficiently in shape, those particles closest to the bed will be particularly strongly affected. A retarding influence, or a sufficiently abrupt downward turn in the bed, will bring them to a standstill and even reverse their motion. The resultant piled-up fluid forces the main flow away from the boundary, while itself developing a backward flow. This process, called *flow separation* (Prandtl, 1952; Chang, 1970), is widespread and important. It occurs when a current flows toward a region of high pressure, or encounters abrupt changes in the shape of its boundaries, for example, at steps up or down, bluff bodies, or sharp bends. Separation is suppressed, however, where a fluid is accelerated. If the flow is already separated, then the opposite process, *flow attachment,* may eventually be induced.

Figure 1-9 illustrates schematically the processes of two-dimensional flow separation and reattachment. In each case there is a streamline that divides on the solid boundary, giving branches that point upstream and downstream. At these points of division $dU/dy = 0$ and $\tau_0 = 0$, where S is a separation point and A is an attachment point. The streamline originating at S is a *separation streamline,* whereas that joining the bed at A is an *attachment streamline.* Beneath each streamline is a region of sluggish recirculatory flow. Motions similar to these are present during the formation of many sedimentary structures.

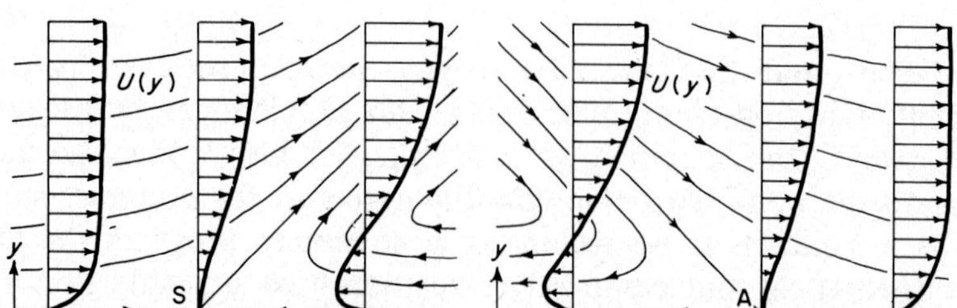

Fig. 1-9. Schematic two-dimensional (a) flow separation, and (b) flow reattachment.

No attempt at being comprehensive will be made, and we shall here restrict discussion to the creep of regolith, debris flows, and the gravity flow of cohesionless grains.

Creeping soils and rock-wastes are the most voluminous but slowest moving of all terrestrial mass flows; Carson and Kirkby (1972) and A. Young (1972) thoroughly review them. Creep is perceptible on slopes in excess of a few degrees and usually affects only the topmost 0.5 m of the ground, though movements down to 10 m are known (Kojan, 1967). The observed creep is composed of continuous *rheological creep,* controlled by the mineralogy and texture of the regolith and by the climatically determined moisture content, and the *"seasonal" creep,* reflecting the action of some combination of wetting and drying, freeze-thaw, thermal contraction and expansion, and churning by organisms. Particles affected by the first three seasonal processes are heaved upward perpendicularly to the surface, but descend along a more vertical path, so moving downslope on zig-zag trajectories. Each type of creep falls in magnitude with increasing depth below the surface, as the weight of overburden increases, and the temperature and moisture-content become less variable.

It is as if the regolith had an apparent viscosity increasing exponentially downward. Consider a uniform regolith beneath a surface of slope β (Fig. 1-10). Equation (1.1) becomes:

$$\tau = \eta_{a(0)} \exp(ky)(dU/dy) \tag{1.21}$$

in which y is distance perpendicularly beneath the surface, $\eta_{a(0)}$ is the apparent viscosity at the surface, and k measures the rate of downward viscosity change. Now the driving stress is:

$$\tau = \gamma g y \sin \beta \tag{1.22}$$

at any level y beneath the surface, where γ is the regolith bulk density, and g is the acceleration due to gravity. Eliminating the stress between eqs. (1.21)

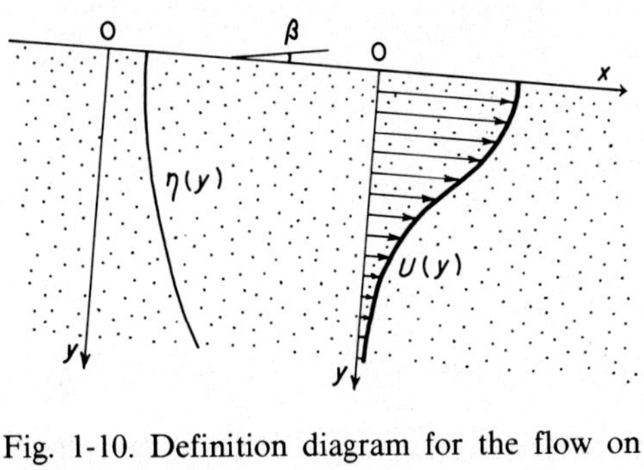

Fig. 1-10. Definition diagram for the flow on a slope of soil with a downward-increasing effective viscosity.

and (1.22), we obtain:

$$-\frac{\mathrm{d}U}{\mathrm{d}y}=\frac{\gamma g \sin\beta}{\eta_{\mathrm{a}(0)}}\exp(-ky)y \tag{1.23}$$

where the minus sign is introduced because the velocity decreases downward. Integrating under the boundary condition $U=0$ at $y=\infty$, we find that:

$$U=\frac{\gamma g \sin\beta}{\eta_{\mathrm{a}(0)}}\frac{(1+ky)}{k^2}\exp(-ky) \tag{1.24}$$

as sketched in Fig. 1-10. This result resembles Kirkby's (1967), though reached from a different starting point, and describes observed profiles quite well (e.g. Kirkby, 1967; Kojan, 1967).

Debris flows are frequent on land and almost certainly exist in the marine environment. Consider an infinitely extensive debris flow of uniform thickness h and bulk density γ flowing steadily over a smooth bed of slope β beneath a stationary medium of density ρ (Fig. 1-11). As R.M. Carter (1975) has indicated, shearing will occur at the upper as well as the lower flow boundary. But if the flow speed is small, and the debris flow greatly exceeds the medium in viscosity, we may ignore shear at the upper boundary. The stress at a perpendicular depth y beneath the surface then is:

$$\tau=(\gamma-\rho)gy\sin\beta \tag{1.25}$$

Writing:

$$\tau=\tau_{\mathrm{cr}}+\eta_{\mathrm{a}}\frac{\mathrm{d}U}{\mathrm{d}y} \tag{1.2}$$

since the Bingham plastic is a good rheological model at least for terrestrial debris flows (A.M. Johnson, 1970), and eliminating the stress between this and the preceding equation, the velocity gradient becomes:

$$-\frac{\mathrm{d}U}{\mathrm{d}y}=\frac{1}{\eta_{\mathrm{a}}}[(\gamma-\rho)gy\sin\beta-\tau_{\mathrm{cr}}] \tag{1.26}$$

where the minus sign appears because the velocity decreases downward. Now

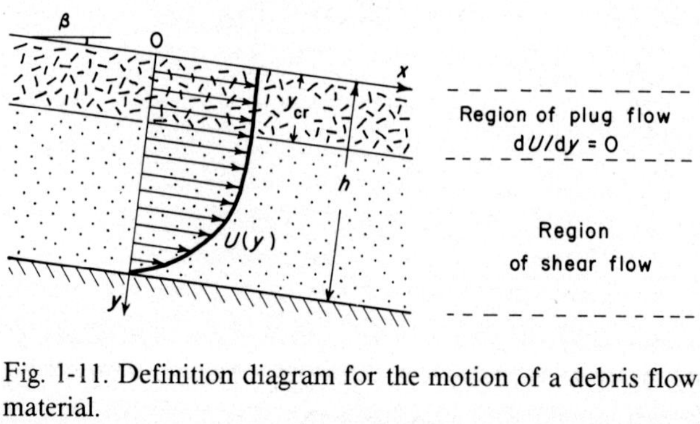

Fig. 1-11. Definition diagram for the motion of a debris flow composed of a Bingham plastic material.

for sufficiently small depths, the stress given by eq. (1.25) is less than the yield stress, whence only plug flow is possible. This critical depth proves to be $y_{cr} = \tau_{cr}/(\gamma - \rho)g \sin \beta$, on setting $dU/dy = 0$ in eq. (1.26). Integration of eq. (1.26) yields the velocity profile:

$$U = \frac{1}{\eta_a}\left[\frac{(h^2 - y^2)(\gamma - \rho)g \sin \beta}{2} - \tau_{cr}(h - y)\right] \qquad y \geqslant y_{cr} \tag{1.27}$$

observing the boundary condition $U = 0$ at $y = h$ (Fig. 1-10). We therefore have plug flow for $0 < y \leqslant y_{cr}$ and viscous flow for $y_{cr} < y < h$. Clearly ρ may be ignored for subatmospheric flows.

Real debris flows are generally restricted laterally, occurring either in valleys or channels, or as tongue-like flows with convex-upward tops on open slopes. A.M. Johnson's (1970) analysis suggests that in the latter case, plug flow will occur in a narrow zone along each flow-margin, creating so-called lateral deposits or levees (e.g. Statham, 1976), as well as in a restricted axial zone. Lateral deposits are also formed by channelized debris flows, provided that the channels depart significantly from a parabolic cross-section.

The flow of cohesionless granular solids, to which we now turn, is of practical importance as well as of significance in the natural environment, for example, in the avalanching of sand on dunes. Although ingenious theories of granular flow have been advanced (e.g. Goodman and Cowin, 1971, 1972; Savage, 1979), the most accessible and appealing is due to Bagnold (1954a, 1956, 1966), who examined theoretically and experimentally the shearing of granular solids dispersed in Newtonian liquids. Bagnold (1954a) reasoned, and confirmed using a rotating drum-apparatus, that when grains were sheared together, there arose an intergranular force divisible between a tangential shear stress T and a normal repulsive or dispersive pressure P. The ratio $T/P = \tan \alpha$ is the coefficient of dynamic solid friction. *Viscous, transitional and inertial* granular shear regimes were recognized, analogous to the laminar, transitional, and turbulent flow regimes of a Newtonian fluid. For smooth, waxy, neutrally buoyant spheres uniformly dispersed in Newtonian liquids, he found empirically for other than very large grain concentrations that:

$$T = 2.25\lambda^{3/2}\eta \frac{dU}{dy} \tag{1.28}$$

under viscous conditions, and:

$$T = 0.0252\sigma\lambda^2 D^2 \left(\frac{dU}{dy}\right)^2 \tag{1.29}$$

for the inertial regime. Here U is the mean grain velocity, y is distance normal to a stationary boundary, and σ and D are respectively the grain density and diameter. Bagnold's quantity λ is the linear grain concentration, defined as the particle diameter divided by the mean free separation distance, or given by:

$$\lambda = \frac{1}{(C_{max}/C)^{1/3} - 1} \tag{1.30}$$

where C is the fractional particle volumetric concentration, and C_{max} is the maximum possible concentration. For the experimental particles, $\tan\alpha$ ranged from a constant value of 0.32 (revised to 0.375 in 1966) in the viscous regime, to another constant but larger value of 0.75 under fully inertial conditions. It will be seen that T is independent of grain size under viscous conditions, but independent of the dynamic viscosity in the inertial regime. The Bagnold number:

$$Ba = \frac{D(\sigma T)^{1/2}}{\lambda^{1/2}\eta} = \frac{D(T/\lambda\sigma)^{1/2}}{\eta/\sigma} \tag{1.31}$$

may be used to specify the grain-shear regime, transition occurring between $10 < Ba < 55$. It will be noticed that this non-dimensional parameter has the form of a Reynolds number founded on a quantity analogous to the fluid shear-velocity.

Southard (1967) doubted Bagnold's intergranular forces but has so far advanced no contrary evidence. Savage (1978, 1979), also using a drum-apparatus but polystyrene pellets in salt water, obtained results similar to Bagnold's (1954a), with the chief exception of a variable dependence of T upon D. Lowe (1976a) indicated that a gravity-driven flow of grains (in a vacuum) can be steady and uniform only if $\tan\beta = \tan\alpha$, where β is the bed slope. Hence a grain mass disturbed on a slope $\tan\beta < \tan\alpha$ must immediately refreeze, whereas one set in motion on a slope of $\tan\beta > \tan\alpha$ must flow and dilate indefinitely. Flows of well-sorted grains do not occur naturally on slopes of less than about $\tan\beta = 0.58$. From this last fact we may infer that $\tan\alpha$ for natural grains (non-spherical, angular, rough-textured) is significantly larger than for Bagnold's waxy spheres. Lowe suggests that uniform steady flow may be possible for $\tan\beta > \tan\alpha$ if there is an intergranular fluid to be dragged along by the grains, since then there can be an added fluid shear on the bed.

WATER IN RIVERS AND ICE TUNNELS

The rivers of such significance as terrestrial transporting agents are wandering, gravity-driven aqueous flows channelled into erodible substrates. The ideal river is a steady, uniform stream of water contained in a straight, rigid, open channel of uniform cross-section, slope, and surface roughness. In calculating this flow, we may safely ignore drag at the free surface, since

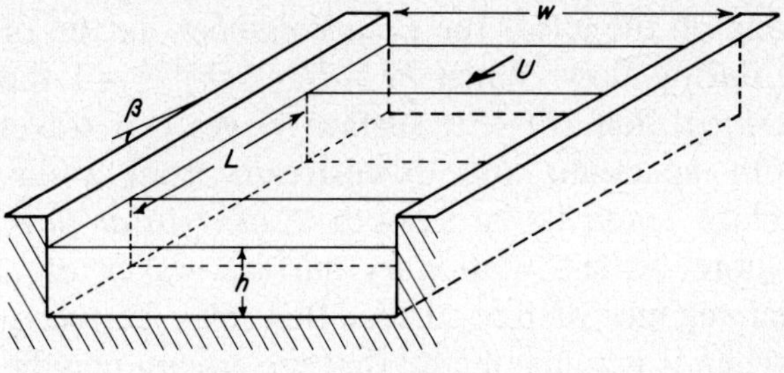

Fig. 1-12. Definition diagram for fluid flow in an open channel.

the air is so much the smaller in density and viscosity. Surveys of flow in idealized channels are given by Chow (1959), Henderson (1966), Rouse (1961) and Sellin (1969).

Consider the uniform steady flow of water of density ρ contained in a rigid open channel of uniform slope β, uniform rectangular cross-section, where w is the flow width and h the flow depth, and uniform bed roughness (Fig. 1-12). The water in a segment of length L is acted on by a body force of magnitude $\rho ghwL \sin\beta$. The opposing drag, exerted over bed and banks, is of magnitude $\tau_0 L(2h+w)$, where τ_0 is the boundary shear stress averaged over the wetted area. Since the motion is uniform and steady, these forces are equal and opposite, and:

$$\tau_0 = \rho g\left(\frac{hw}{2h+w}\right)\sin\beta \tag{1.32}$$

Now introducing as a definition (see eq. 1.19):

$$\tau_0 = \frac{f}{4}\rho\frac{U_m^2}{2} \tag{1.33}$$

where U_m is the velocity averaged over the cross-section, and eliminating τ_0 between the equations, we find that:

$$U_m^2 = \frac{8g}{f}\left(\frac{hw}{2h+w}\right)\sin\beta \tag{1.34}$$

The quantity in brackets is the ratio of cross-sectional area to wetted perimeter, called the hydraulic radius r. Equation (1.34) therefore reduces to:

$$U_m = \left(\frac{8g}{f}rS\right)^{1/2} \tag{1.35}$$

where the slope S has been substituted for $\sin\beta$, since β for real rivers is a small angle. When channel width is large compared to depth, we may substitute h for r in eq. (1.35) without undue error.

On substituting as suggested, eq. (1.35) may be rearranged as:

$$\frac{U_m}{(gh)^{1/2}} = \left(\frac{8S}{f}\right)^{1/2} \tag{1.36}$$

where the non-dimensional group on the left is the *Froude number,* a ratio of inertial to gravitational forces during flow. When $Fr = U_m/(gh)^{1/2} = 1$ the flow is called *critical.* In *subcritical flow,* $Fr < 1$, surface waves can travel (and so become damped) both upstream and downstream from a disturbance, whence the free surface tends to be smooth. *Supercritical flow,* $Fr > 1$, is marked by bold, more or less stationary surface waves of a sustained amplitude. Two other regimes of free-surface flow may be recognized, making a total of four, when it is remembered that the stream may be either laminar or turbulent (Fig. 1-13) (Robertson and Rouse, 1941; Sundborg, 1956). The Reynolds number, $Re = U_m h/\nu = 500$, bounding the turbulent regimes is the lowest at which transition to turbulence occurs in a smooth channel. Real rivers fall mostly in these regimes, but overland sheet-flows, and thin-film flows on exposed rock surfaces, are mainly laminar and frequently supercritical.

Real rivers have deformable beds and banks, and are seen to wander laterally, yet their channels are of a determinate cross-sectional scale which increases with river size as measured by aqueous discharge. Hence at least locally on real rivers there must be cross-sections that are neutral, that is, neither widening nor deepening in the long term as the result of sedimentary processes. If our ideal channel were deformable, all cross-sections could be

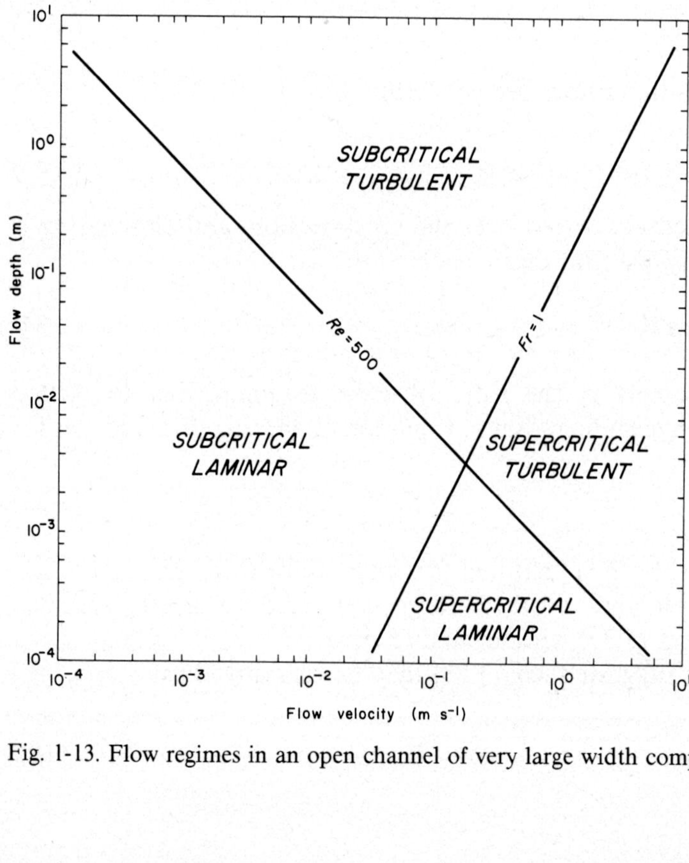

Fig. 1-13. Flow regimes in an open channel of very large width compared to depth.

neutral, provided that an erosion criterion imposed by the character of the surrounding bed and bank material was met by the flow. An appropriate kinematic condition is that $U_m = U_{cr}$, where U_{cr} is the threshold erosion velocity of this material. Writing Q as the discharge and w_n and h_n as respectively the width and depth of the neutral channel, and putting the continuity equation $Q = U_{cr} w_n h_n$ into eq. (1.34) with U_m replaced by U_{cr} and $\sin\beta$ replaced by S, we find two quadratic equations with solutions:

$$w_n = \frac{8gQS + \left[(8gQS)^2 - 8f^2 U_{cr}^5 Q\right]^{1/2}}{2fU_{cr}^3} \tag{1.37}$$

and:

$$h_n = \frac{8gQS - \left[(8gQS)^2 - 8f^2 U_{cr}^5 Q\right]^{1/2}}{4fU_{cr}^3} \tag{1.38}$$

No channel of any kind can exist when the terms in square brackets are smaller than zero, whence the neutral channel cannot be narrower than $w_n/h_n = 2$ for a uniform Darcy-Weisbach coefficient.

Real rivers, the largest discharging from 1×10^3 to 1×10^5 m^3 s^{-1} (Leopold, 1962; Potter, 1978), depart from the previous ideal by exhibiting unsteady flow. For the hydrograph, or discharge–time curve, obtained at a river station shows peaks related to individual rain-storms and other short-term events, superimposed on a seasonally controlled increase and decrease of flow (Beckinsale, 1969; Pardé, 1947; Rodda, 1969). Since the discharge varies with time, the flow width, depth and mean velocity will also vary. The character of the discharge–time curve defines the river regime, of which three broad classes are recognized. Rivers of *microthermal regime,* found at high latitudes and/or altitudes (e.g. Colville River, Arnborg et al., 1966), are fed chiefly from melting snow during late spring and summer, and show a short and intense flood season followed by a long interval of very low flows. Significant diurnal variations of discharge may also be discernible (e.g. Fahnestock, 1963; Church, 1972). Somewhat similar in flow pattern are many rivers of *macrothermal regime,* dependent on monsoon rains (e.g. Brahmaputra River, J.M. Coleman, 1969). However, other macrothermal streams, for example, the Congo (Zaire) River (Peters, 1971), have two flood seasons each year and show only a moderate difference between the wet-season and dry-season discharges. Rivers of *mesothermal regime*, located in temperate climates, are also relatively steady in discharge, showing but a single annual flood.

In a second way real rivers depart from the ideal presented, for their channels are non-uniform. The channel viewed in plan is typically not straight but either turns from side to side (*meandering streams*) or consists of joined curved segments (*braided streams*) (F.A. Melton, 1936; Blench, 1957,

1969a; Simons, 1971; Galay et al., 1973). One consequence of curvature is that large-scale *secondary currents*—corkscrew movements of streamwise axis superimposed on the general downstream flow—are created in each bend, where viscous and centrifugal forces fail to balance (Bagnold, 1960; Leopold and Wolman, 1960; Rozovskii, 1961). Only locally does the channel prove to be symmetrical in cross-sectional shape and velocity distribution (Leopold and Wolman, 1960; Rozovskii, 1961; Morisawa, 1968; Paulissen, 1973). Typically, the channel at a bend is triangular in cross-section, the velocity maximum having been shifted to the deep outer side. Approximately rectangular and also shallower cross-sections can be found only near the turn of one bend into another. Hence the maximum flow depth varies more or less periodically with distance downstream.

Unsteady and non-uniform flow in rivers is extremely difficult to model. Accounts of the underlying theory and traditional methods appear in Chow (1959), Henderson (1966), Sellin (1969) and J.R.D. Francis (1975). Newer techniques are illustrated by the work of Gunaratnam and Perkins (1970), Cunge and Simons (1975), Stevens et al. (1975), and K.S. Richards (1978). One major problem —combined flow in a channel and an adjacent floodplain (e.g. Velikanova and Yarnikh, 1972)—has so far been approached mainly empirically (Sellin, 1964; Toebes and Sooky, 1967; Ghosh and Jena, 1971; Ghosh and Mehta, 1974; Ghosh and Kar, 1975; Myers and Elsawy, 1975; Petryk and Grant, 1978; Rajaratnam and Ahmadi, 1979). This topic underpins an understanding of floodplain construction, for during combined flow there is significant transfer of momentum and sediment transversely away from the main channel.

Meltwater flow through tunnels within glaciers, or that of groundwater in submerged cave shafts and conduits, can be likened to fluid motion in completely filled pipes (Rouse, 1961; J.R.D. Francis, 1975). Consider a small length δx of a long, straight, circular pipe of radius a in which a uniform,

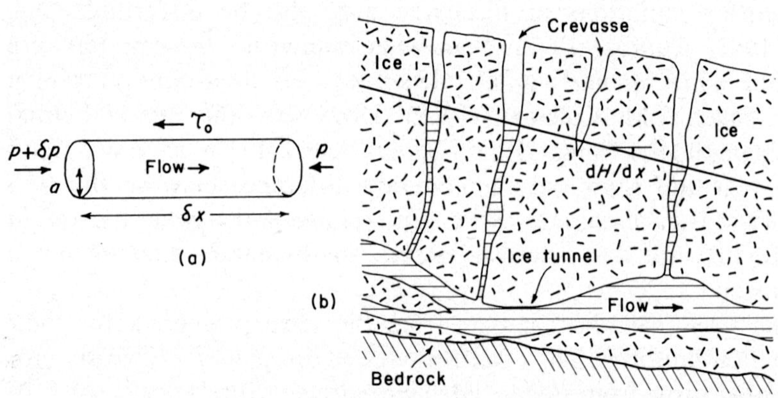

Fig. 1-14. Flow in pipes and conduits. a. General definition diagram. b. Schematic representation of meltwater flow in a glacier tunnel under the action of a hydraulic gradient, $\mathrm{d}H/\mathrm{d}x$.

steady flow has been established (Fig. 1-14a). In this case the driving force is the pressure difference δp across the ends of the length δx, the total pressure force being $\pi a^2 \cdot \delta p$. Opposing this force is a drag distributed over the pipe wall and amounting to $\tau_0(2\pi a) \cdot \delta x$, where τ_0 is the unit boundary shear stress. Equating the forces:

$$\tau_0 = \frac{a}{2} \frac{\mathrm{d}p}{\mathrm{d}x} \tag{1.39}$$

and, introducing the definition eq. (1.33), we find for the mean flow velocity:

$$U_\mathrm{m}^2 = \frac{4a}{f\rho} \frac{\mathrm{d}p}{\mathrm{d}x} \tag{1.40}$$

where f, obtainable as indicated, depends on Reynolds number and wall roughness. Since a free water-surface, the water-table, overlies the submerged parts of cave systems, and above englacial meltwater tunnels (Fig. 1-14b), it is useful to substitute for the pressure p the head of water H causing the pressure, for $p = \rho g H$. Equation (1.40) then reads:

$$U_\mathrm{m}^2 = \frac{4ga}{f} \frac{\mathrm{d}H}{\mathrm{d}x} \tag{1.41}$$

where $\mathrm{d}H/\mathrm{d}x$, the slope of this free surface, is the hydraulic gradient. The velocity in natural conduits should therefore increase with the radius and hydraulic gradient but decrease with increasing wall roughness. Departure from the ideal may be expected where a conduit abruptly changes size, shape or direction, for entry, exit and centrifugal effects not accounted for in the model can then be expected.

THE ATMOSPHERE IN MOTION

The atmosphere is in motion on several scales, basically because of the uneven distribution of solar radiation over the Earth (S. Hess, 1959; Willett and Sanders, 1959; Pick, 1962; Chang, 1972). The Earth's surface receives more heat energy in the hot equatorial than cold polar regions, yet the temperature in each remains sensibly constant over the moderate term. In order for the resulting temperature gradient to persist, heat must be continually transported from equatorial toward polar regions. The required heat transport is provided by convection and advection in the atmosphere and ocean, which operate coupled together as a single system (Neumann and Pierson, 1966). The resulting atmospheric circulation on the rotating Earth takes the form of three girdle-like cells within the troposphere covering each Hemisphere, each cell contributing a particular zonal surface wind (Fig. 1-15). The most striking of these are the vigorous, moist westerlies of the middle latitudes. Significantly gentler are the dry easterly Trade Winds of low latitudes, and the circumpolar east winds.

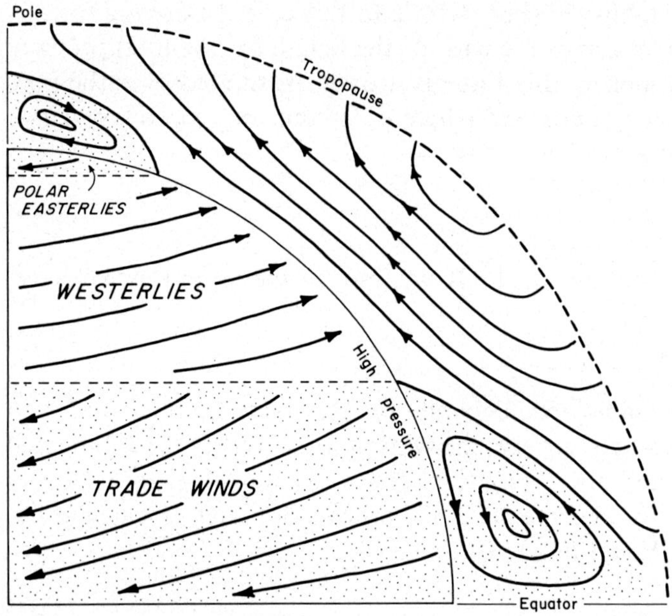

Fig. 1-15. Zonal circulation in the Earth's lower atmosphere (thickness very greatly exaggerated) depicted for one-quarter of the Northern Hemisphere. The circulation is completed in the upper atmosphere (not shown).

The atmospheric circulations of intermediate scale are associated with cyclones (depressions) and anticyclones, which are centres respectively of low and high pressure typified by closed isobars (Fig. 1-16). In the Northern Hemisphere, the wind blows anticlockwise round a cyclone and clockwise round an anticyclone. These perturbations to the zonal winds have either a

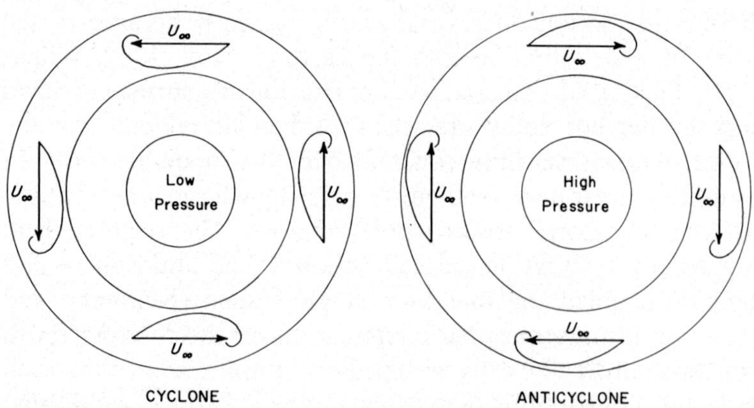

Fig. 1-16. Schematic isobars for a cyclone and anticyclone in the Northern Hemisphere, together with the geostrophic wind outside the atmospheric boundary layer (U_∞) and the Ekman velocity spiral within.

thermal or dynamical origin (Willett and Sanders, 1959; Houghton, 1977). Those of thermal origin reflect the creation by heating or cooling of density differences between very large air masses, and are exemplified by tropical monsoons and storm depressions. The dynamical perturbations express a large-scale atmospheric instability related to the horizontal temperature gradient. They appear within the mid-latitude westerlies as cyclones and anticyclones at a characteristic wavelength in the order of 4000 km. Cyclones bring with them gales and storms, whereas anticyclones yield fair and settled weather with light winds.

The wind is turbulent (J.C. Houbolt, 1973) and the atmospheric boundary layer in the order of 1 km thick (Roll, 1965; Plate, 1971). At greater heights the wind is substantially unaffected by ground friction, and flows roughly parallel with the isobars (Houghton, 1977), as sketched in Fig. 1-15. This is the *geostrophic wind.* Within the boundary layer, however, where friction is significant, Coriolis force causes the wind to deviate from the geostrophic direction, by up to 45° at the ground. The deviation for a cyclone is inward, implying a continual centripetal leading to upward air-flow in such a region. Near-surface winds speeds associated with cyclonic gales and storms exceed 15–20 m s^{-1}. An outward deviation typifies an anticyclone, in which there is a continual downward followed by outward air-flow. For a stationary observer, the velocity profile of the wind forming the atmospheric boundary layer is given by (Houghton, 1977):

$$U = U_\infty \left\{ 1 - \exp\left[-y\left(\frac{2\nu}{\sigma}\right)^{-1/2} \right] \cos y \left(\frac{2\nu}{\sigma}\right)^{-1/2} \right\} \tag{1.42}$$

$$W = U_\infty \exp\left[-y\left(\frac{2\nu}{\sigma}\right)^{-1/2} \right] \sin y \left(\frac{2\nu}{\sigma}\right)^{-1/2} \tag{1.43}$$

where ν is a turbulent eddy kinematic viscosity, and U and W are the components respectively parallel and transverse to the geostrophic wind of velocity U_∞. Observed wind components follow the Ekman spiral only approximately, because the atmospheric kinematic viscosity is not constant, but a complex function of height (Roll, 1965; Plate, 1971). The Ekman spirals sketched in Fig. 1-16 are idealized.

The smallest atmospheric motions, setting aside haphazard turbulence, accompany thermals and waves formed either in the lee of such as a mountain range, or at the interface between two layers of air in relative motion (Scorer, 1958, 1978). Thermals are discrete masses of air, in the order of 100–1000 m across, that rise buoyantly and with internal recirculation because they have been heated by the ground and so made less dense than their immediate surroundings. In moderate winds, thermals become streaked out within the atmospheric boundary layer into large, parallel corkscrew vortices, sometimes called *convection rolls* (e.g. Kuettner and Soules, 1966), another example of secondary flow. The waves produced between layers of air of contrasted velocity (e.g. Browning, 1972) record Kelvin-Helmholtz instability, peculiar to shear layers. Obstacle-related *lee waves* (Scorer, 1978) will be discussed later in connection with oceanic phenomena.

SURFACE AND INTERNAL WAVES

General

Natural wave phenomena are many and varied, but those involved in sediment transport and the production of sedimentary structures are restricted to fluid interfaces. These waves may be classified according to: (1) the nature of the interface, (2) the forces tending to distort or restore the interface, (3) the wave period, and (4) the scale relative to the fluid layer(s).

Surface waves occur where the water of the oceans, seas, lakes and rivers meets the immiscible air above; they exist on a sharp interface across which density and viscosity change abruptly. Two principal distorting forces are involved. The combined drag and push-pull of the turbulent wind (Phillips, 1977) creates *wind waves*, with a period less than about 20 s, while the gravitational pull of the Moon and Sun makes the oceanic *tidal wave,* with either a semidiurnal or diurnal period. The Earth's gravity provides the restoring force for the tidal wave and the longer-period wind waves. For *capillary waves*, of period less than about 0.1 s, surface tension is the chief restoring force. Surface waves can be classified according to the ratio of the wavelength L_0 in deep water to the water depth h, as follows: *deep-water* or *short waves* ($h/L_0 > 0.25$), *intermediate waves* ($0.25 \geqslant h/L_0 > 0.025$), and *shallow-water* or *long waves* ($0.025 \geqslant h/L_0$). Examples of short waves are wind waves generated on a deep ocean far from land. The tidal wave is a good instance of a long wave.

Most *internal waves* occur between water masses in the ocean or in lakes, that is, at "interfaces" that are zones more or less thin across which the fluid density changes continuously (e.g. the thermocline). The internal waves called *lee waves*, also to be found in the atmosphere, are found in thick fluid layers of gradual vertically changing density. Evidently internal waves involve miscible fluids, so that gravity is the only possible restoring force. As a class, internal waves differ considerably in scale from surface waves. The period of internal waves is generally measured in minutes or hours, while the height ranges up to the order of 100 m, and the wavelength can measure hundreds of kilometres (Krauss, 1966; Briscoe, 1975; J. Roberts, 1975; Garrett and Munk, 1979). These waves have many causes and are poorly understood compared to surface waves (J. Roberts, 1975; Phillips, 1977).

Surface and internal waves may be either *progressive* or *standing*. Progressive waves are those whose crests propagate in a single direction, whereas standing waves are formed by the combination of oppositely moving progressive forms. If the waves of the two trains have the same period and height, the water-surface is found to range within an envelope that displays stationary nodes alternating with antinodes.

Real waves

Real surface waves vary from ripplets less than a millimetre high and with a period of a fraction of a second, to storm giants measuring 10–20 s in period and 20–25 m in height. Invariably, real waves are expressed as spectra of wavelength, height and period, which demand statistical treatment (Longuet-Higgins, 1952; M.J. Tucker, 1963; Wiegel, 1964; Kinsman, 1965; D.L. Harris, 1971; LeBlond and Mysak, 1979). Two of the most useful statistics that describe real waves are the *significant wave height* $H_{1/3}$ and the *significant wave period* $T_{1/3}$. The former is the average height of the highest one-third of the waves measured from a suitably long time-series. On averaging the periods of these highest one-third we obtain the significant wave period. These statistics are called significant because they typify the spectrum as regards the ability of the waves to apply forces to the bed or coast. Wave energy is proportional to the square of the height, whence $H_{1/3}^2$ better measures the modal energy than the square of the average height.

In order to comprehend the influence of waves in a sedimentary environment, as well as from other practical standpoints, we must know something of the *wave climate* prevailing there, that is, the character and seasonal changes shown by the wave spectrum. Wave records are difficult to make, especially during bad weather, and long series of data are available for comparatively few areas, notably British waters (Darbyshire, 1962; Draper, 1967a; Draper and Driver, 1971; Sellard and Draper, 1975), the Atlantic Ocean (Brooks et al., 1958; Walden, 1964; Draper and Whitaker, 1965; Draper, 1967b; Snider and Chakrabarti, 1973), and the coastal waters of North America (Todd and Wiegel, 1952; Powers et al., 1969; D.L. Harris, 1972a, 1972b; Thompson and Harris, 1972; E.F. Thompson, 1977). Earle and Malahoff (1979) summarize recent work on ocean wave climate. J.W. Johnson (1948) describes wave conditions in lakes and protected bays, and Dattatri (1973) reports observations from the west coast of India. Several workers describe storm wave conditions (Draper, 1971; Thom, 1973; Earle, 1975). Calculations based on measured spectra suggest that storm waves frequently stir up coarse sediments at depths of many tens of metres (Hadley, 1964; Draper, 1967a; Silvester and Mogridge, 1971; J.A. Ewing, 1973).

Where suitable records are unavailable, an attempt can be made to predict the wave spectrum or climate from known meteorological and geographical conditions. Wind-wave generation involves energy transfer from the wind to the waves. The mechanisms remain incompletely understood (Kinsman, 1965; Phillips, 1977), but the character of a spectrum of wind-waves may be

TABLE 1-II

Expected wave height in the open sea as a function of Beaufort wind force or wind speed at height of 10 m above the sea surface. After Frost (1966)

Mean wind speed at of 10 m ($m\ s^{-1}$)	Beaufort wind force	Descriptive term	Probable height of waves in open sea * (m)
0	0	calm	0
1.5	1	light air	0.1
3.3	2	light breeze	0.2
5.3	3	gentle breeze	0.6
7.5	4	moderate breeze	1.0
9.6	5	fresh breeze	2.0
11.9	6	strong breeze	3.0
14.3	7	near gale	4.0
16.7	8	gale	5.5
19.1	9	strong gale	7.0
21.7	10	storm	9.0
24.1	11	violent storm	11.5
26.8	12	hurricane	$\geqslant$14.0

* Heights less than tabulated values in inshore waters.

expected to depend on: (1) wind speed, (2) the distance or fetch over which the wind operates, (3) the wind duration, or the elapsed time since the wind ceased to act, and (4) the water depth. Generally speaking, the stronger the wind, and the greater its fetch and duration, the larger are the waves. Waves decay once the wind ceases to act, those of largest period and wavelength tending to remain longest in the spectrum. Cornish (1904, 1934) early attempted to develop a method of wave-forecasting. Modern methods, all more or less empirical, are described by Bretschneider (1952a, 1952b, 1966), Darbyshire and Draper (1963), Frost (1966), and Silvester and Vongvises-somjai (1971). Table 1-II, based on Frost's work, gives a greatly simplified impression of the wind-wave relationship.

Ideal waves

Real waves and their attendant fluid motions are so complex and hard to observe that there is no hope of fully understanding them by purely empirical means. A primarily mathematical approach, perforce employing idealized waves, but checked by observation in the laboratory and field, has been inevitable. Numerous thorough surveys of the problem and its voluminous literature are available (Lamb, 1932; Keulegan, 1950; J.J. Stoker, 1957; Wehausen and Laitone, 1960; Wiegel, 1964; Kinsman, 1965; Dean and Eagleson, 1966; Eagleson and Dean, 1966; Neumann and Pierson, 1966;

Barnett and Kenyon, 1975; J. Roberts, 1975; USACERC, 1975; Le Méhauté, 1976; Komar, 1976; Phillips, 1977; LeBlond and Mysak, 1978; Lighthill, 1978), the treatments by Kinsman, Le Méhauté, and Wiegel being particularly fine, and we need do no more here than sketch those results of greatest relevance to sedimentary structures. Three kinds of idealized wave have mainly been studied: (1) those of sinusoidal profile and in principle infinitely small amplitude on an inviscid fluid (*Airy waves*), (2) waves of trochoidal profile and finite amplitude on an inviscid fluid (*Stokes waves*), and (3) *solitary waves.* The solitary wave, a limiting case of the cnoidal wave, a shallow-water type, is a wave of translation. Airy and Stokes waves are oscillatory waves, for fluid elements beneath the surface are found to move back and forth.

Each type of ideal wave has an appropriate application. The solitary wave is a good model for real waves in very shallow water and near breakers. The Airy wave is useful over a wide range of depths and has the advantage that spectra as well as pure waves can be examined theoretically. The Stokes wave is closest to reality but is awkward to handle mathematically.

The celerity c of a two-dimensional Airy wave of wavelength L advancing over water of uniform depth h can be shown to be:

$$c^2 = \frac{g}{k} \tanh(kh) \tag{1.44}$$

where $k = 2\pi/L$ is the radian wave number (Fig. 1-17). Making use of the bounds on the hyperbolic tangent, sketched in Fig. 1-18 (the corresponding sine and cosine functions also appear), we can write as in the top row of Table 1-III the celerity in deep and shallow water, with the subscript distinguishing the deep-water value. For tanh (kh) is approximately equal to unity when $h \gg L$, but approximately equal to kh when $h \ll L$. The quantity $\sigma = 2\pi/T$ is the radian wave frequency. Also listed in Table 1-III are relations for wavelength and wave height, given in terms of deep-water conditions, the wave period being the only invariant property. The symbol n

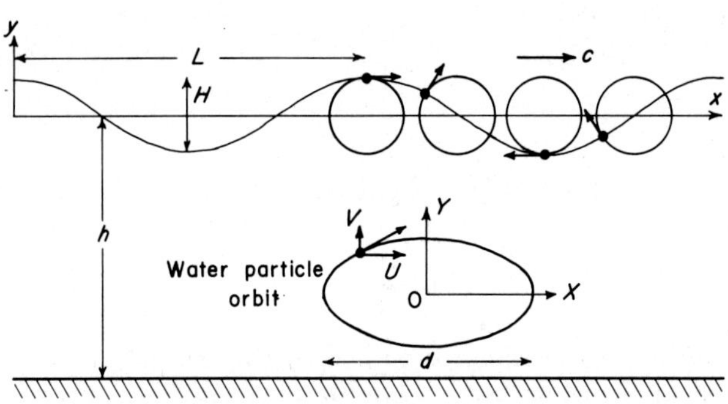

Fig. 1-17. Definition diagram for progressive waves at a fluid interface and for the motion of water-particles.

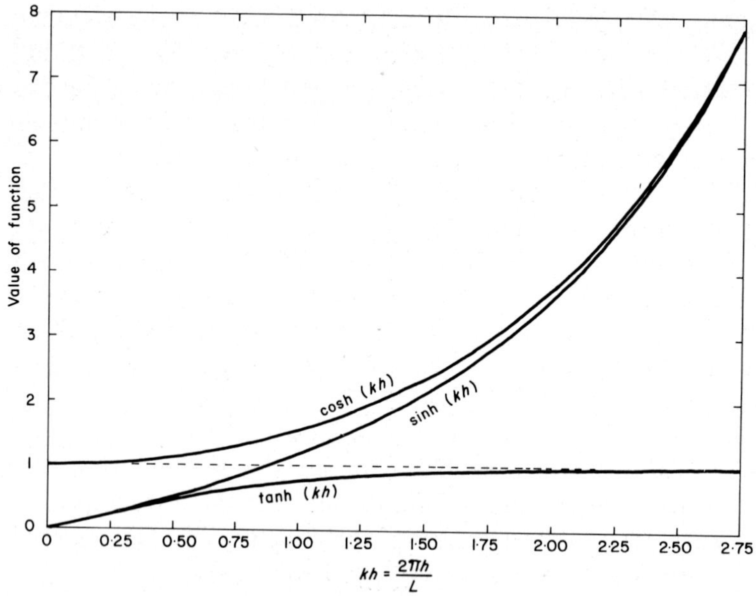

Fig. 1-18. Hyperbolic sine, cosine and tangent as a function of $kh = 2\pi h/L$.

stands for a numerical coefficient ranging between one-half in deep water and one in shallow water. Wiegel (1964) extensively tabulates this and many other functions useful in evaluating wave properties. Inspection of Table 1-III shows that wavelength and celerity decline from deep to shallow water, while the wave-height shows a general increase.

The theory of two-dimensional Airy waves reveals that water particles beneath the surface are in motion over elliptical orbits lying in the vertical plane. Referring to Fig. 1-17 for the particle displacements and velocity components, it can be shown that:

$$X = -\frac{H \cosh k(y+h) \sin(kx - \sigma t)}{2 \sinh(kh)} \tag{1.45}$$

$$Y = \frac{H \sinh k(y+h) \cos(kx - \sigma t)}{2 \sinh(kh)} \tag{1.46}$$

$$U = \frac{\pi H \cosh k(y+h) \cos(kx - \sigma t)}{T \sinh(kh)} \tag{1.47}$$

$$V = \frac{\pi H \sinh k(y+h) \sin(kx - \sigma t)}{T \sinh(kh)} \tag{1.48}$$

where t is time, and x and y refer to the mean position of the water particle. The orbits in the general case of intermediate depth (Fig. 1-19b) are ellipses increasing in eccentricity but decreasing in diameter downward beneath the water surface. Making use of the bounds sketched in Fig. 1-18, the deep-water

TABLE 1-III

Summary of the principal functions describing progressive surface waves

Function	Deep water (short waves)	Intermediate depths	Shallow water (long waves	Near breakers (solitary waves)
Celerity, c	$c_0=\left(\frac{g}{k}\right)^{1/2}=\frac{g}{\sigma}$	$\left[\frac{g}{k}\tanh(kh)\right]^{1/2}$	$(gh)^{1/2}$	$[g(h+H)]^{1/2}$
Wavelength, L	$L_0=\frac{gT}{\sigma}$	$\frac{gT}{\sigma}[\tanh(k_0h)]^{1/2}$		cT
Wave height, H	H_0	$H_0\left(\frac{c_0}{2cn}\right)^{1/2}$	$\frac{H_0}{(4k_0h)^{1/4}}$	$0.32\ H_0\left(\frac{L_0}{H_0}\right)^{1/3}$ approximately
Orbital diameter, d	$d=H_0$ at surface $d=H_0\exp(ky)$ $d=0$ at bottom	$\frac{H\cosh k(h+y)}{\sinh(kh)}$	$\frac{H}{kh}$	approximately $3H/2$ at bottom
Orbital velocity, U_{max}	$\frac{\pi d}{T}$	$\frac{\pi H}{T\sinh(kh)}$	$\frac{Hc}{2h}$	approximately $Hc/3h$ at bottom

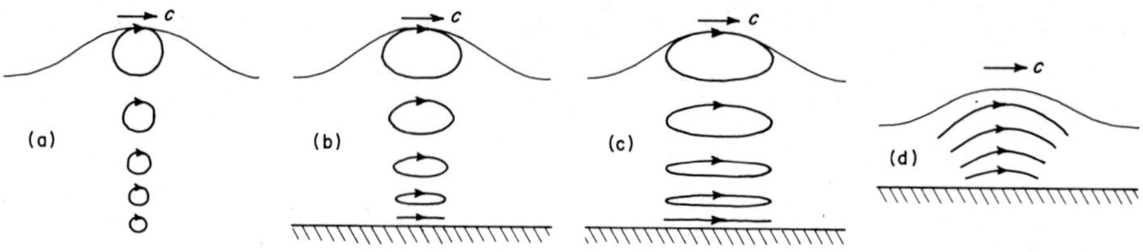

Fig. 1-19. Water-particle orbits beneath progressive surface waves of different types. a. Deep-water waves. b. Waves on water of intermediate depth. c. Shallow-water waves. d. Solitary wave.

orbits (Fig. 1-19a) are evidently circles diminishing exponentially downward, whereas in shallow water (Fig. 1-19c) the diameter remains uniform with depth and only the eccentricity changes. The orbital velocity components behave similarly to the particle displacements. Of particular relevance to sediment transport is the maximum horizontal orbital velocity U_{max} near the bottom, that is, just outside the boundary layer described. The general formula is:

$$U_{max} = \frac{\pi H}{T \sinh(kh)} \tag{1.49}$$

the limiting values appearing in Table 1-III. The orbital diameter and maximum orbital velocity reduce rapidly with depth and, at a depth of $L/2$, often referred to as *wave base,* take only about 4% of their value at the free surface. Water particles beneath solitary waves experience only a forward translation (Fig. 1-19d).

The theories of Airy and Stokes waves have been extended to three-dimensional progressive forms, that is, to waves with crests of finite length (H. Jeffreys, 1924; Fuchs, 1952; Silvester, 1972; Hsu et al., 1979). Water-particle motions are then very complex. Except at certain positions beneath the waves, the plane of the orbit is tilted, so that a fluid element has an additional velocity component normal to the direction of wave propagation.

The Airy theory has also been successfully applied to internal waves (Kinsman, 1965; Neumann and Pierson, 1966), but with one important difference as compared to surface waves. In the case of surface waves, it is assumed that the density of the second medium (air) can be neglected in comparison with that of water. This assumption is not made in the general case, and would certainly be invalid for internal waves, which may occur between water or air masses differing in density by only a few parts per thousand. If the layers are thick compared to the wavelength of the disturbance, the celerity-wave number equation reads:

$$c^2 = \frac{g}{k}\left(\frac{\rho_1 \pm \rho_2}{\rho_1 + \rho_2}\right) \tag{1.50}$$

where ρ_1 and ρ_2 represent the density in respectively the lower and upper layers. The assumption made for surface waves is evidently a good one, since the density of water is two orders of magnitude larger than that of air. We can also see that there are two possible solutions to eq. (1.50), accordingly as the positive or negative sign is taken in the numerator. Hence internal waves of two contrasted modes could arise at a given interface. On taking the positive sign, it can be shown that the internal waves are in phase with waves on a free surface above, but are decreased in amplitude proportionately to depth. Under the negative sign, a wave motion of much longer period is found, but with a much larger amplitude and a phase-difference of π rad from a surface wave.

Wave-tank studies show that the theory of Airy waves satisfactorily describes the main features of water-particle orbits beneath surface waves, except in the friction-dominated boundary layer. Morison and Crooke (1953) provide confirmatory data for progressive waves, and Wallet and Ruellan (1950) describe studies on waves ranging between progressive and fully standing. Daily and Stephan (1953) confirmed in the laboratory the particle motions predicted for solitary waves and, in the field, Inman and Nasu

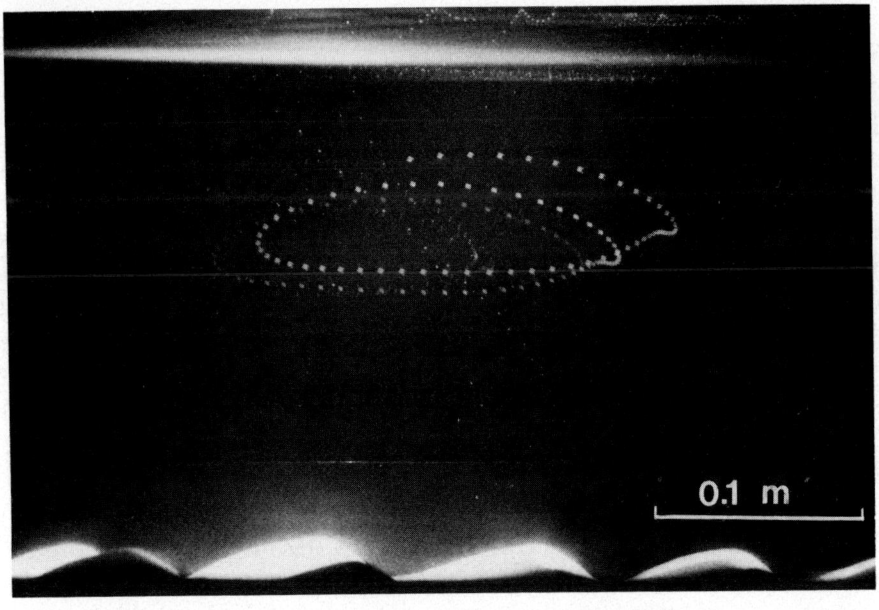

Fig. 1-20. Mass-transport due to wave motion, illustrated by a stroboscopic photograph of a neutrally buoyant particle as it follows slightly distorted elliptical orbits induced by waves travelling from left to right overhead. Notice that the drift of the particle (at mid depth) is "seaward", i.e. toward the wave maker. On the bed are strongly asymmetrical wave-related ripple marks facing in the direction of wave propagation. Water depth = 0.165 m, wave period = 2.9 s; wave height = 0.087 m. Photograph courtesy of K.-W. Tietze (see Tietze, 1978).

(1956) found good agreement between observation and solitary-wave theory. Little is known of orbital motions beneath random progressive waves at sea (Shonting, 1968; Seitz, 1971; Thornton and Krapohl, 1974), though observed motion-spectra broadly agree with theoretical calculations assuming spectra of waves. Laboratory experiments (Lee and Masch, 1971) suggest that a certain amount of wave-related turbulence is also to be expected, but confined to a relatively thin surface layer. The orbital motions accompanying internal waves are also ill-known, but velocity components upward of 1 m s^{-1} are possible theoretically and are observed (J. Roberts, 1975).

Although broadly confirming the theory of Airy waves, the experiments of Morison and Crooke (1953), of Wallet and Ruellan (1950), and Tietze (1978) show that the water particles follow circular or elliptical orbits that are not quite closed (i.e. they are curtate cycloids) (Fig. 1-20), a fact that is not predicted by this particular model. Hence the motion beneath real waves consists of at least two additive parts, a symmetrical oscillatory motion, and a steady streaming of fluid related to the extent to which particle-orbits fail to close. This streaming is potentially of great importance in sediment transport, for the oscillating motion, of itself, cannot yield other than zero net transport, although it may mobilize much sediment. We shall now look more closely at this streaming.

Mass-transport in surface and internal waves

Directly analogous streamings or *mass-transports* are created when: (1) a cylinder oscillates parallel with a diameter in an otherwise still fluid, (2) a fluid oscillates parallel with a stationary wavy wall, and (3) progressive or standing waves occur at an interface. Basically, the mass-transport arises because of the influence of the Reynolds stresses associated with the oscillatory motion (Longuet-Higgins, 1953, 1958). The theory of mass-transport presents considerable difficulties, as does the question of verification in the field and laboratory, yet much progress has been made, particularly in recent years, as may be glimpsed by comparing in sequence the reviews of Ursell (1953), and N. Riley (1967, 1975). Particular success has attended the application of boundary-layer principles to the problem (Longuet-Higgins, 1953), and notably the double boundary-layer concept of Stuart (1963, 1966) and Riley (1965, 1967) (Dore, 1974, 1975, 1976a, 1976b). This concept sees the mass-transport confined to a possible maximum of two boundary layers adjacent to each relevant boundary, an inner viscous layer with thickness of order $(\nu/\sigma)^{1/2}$, where ν and σ are the kinematic viscosity and radian frequency as before, and an outer layer with thickness in the order $d(\nu\sigma)^{1/2}/U$, where d and U are respectively a characteristic length and velocity.

Stokes (1847) developed a theory of finite-amplitude progressive waves on an inviscid fluid, according to which water particles prescribe orbits that are not quite closed, there being a slow drift of fluid in the direction of wave propagation. In water of infinite depth, this mass-transport declines exponentially downward:

$$U_{mt} = (H^2/4)\sigma k \exp(2ky) \tag{1.51}$$

with a gradient at the free surface of:

$$\left(\frac{\partial U_{mt}}{\partial y}\right)_{y=0} = \frac{H^2}{2}\sigma k^2 \coth(kh) \tag{1.52}$$

where U_{mt} is the mass-transport velocity averaged over one wave-period and the other quantities are as before (Fig. 1-21a). Fuchs (1952) extended this analysis to three-dimensional waves, and several authors have applied it to random waves on deep water (Bye, 1967; M.S. Chang, 1969; K.E. Kenyon, 1969, 1970; Huang, 1971; Wang and Liang, 1975). Hsu et al. (1980) have recently described a detailed study of mass-transport beneath short-crested waves. Extended to waves approaching a coast over water of finite depth, eq. (1.49) predicts a *seaward* mass-transport near the bottom (Fig. 1-21b), for continuity demands zero net fluid discharge through a cross-section normal to flow, yet experiment yields a quite different result (Caligny, 1878; Mitchim, 1940; M.A. Mason, 1941; Bagnold, 1947; Vincent and Ruellan, 1957; Russell and Osorio, 1958; Allen and Gibson, 1959; Brebner and Collins, 1961; Brebner et al., 1967; Rubatta, 1971; Bijker et al., 1975), the bottom flow generally being *shoreward* and opposite to that predicted (Fig. 1-21c). Furthermore, experiment shows that the surface gradient of U_{mt} significantly exceeds the value given by eq. (1.52) (Longuet-Higgins, 1960). Evidently the theory of Stokes is in some degree deficient.

The inclusion of viscous effects leads to results agreeing better with experiment. Harrison (1908) analysed progressive surface waves on a viscous fluid but, through an algebraic slip, regained eq. (1.52) for the inviscid case.

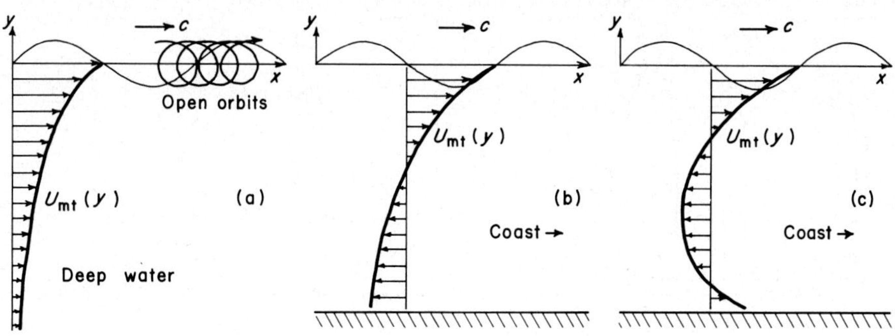

Fig. 1-21. Schematic representation of mass-transport induced by progressive waves. a. Mass-transport (Stokes) beneath waves on deep water. b. Mass-transport (Stokes) beneath waves approaching a coast over water of finite depth. c. Mass-transport (Longuet-Higgins) beneath waves approaching a coast over water of finite depth.

On correcting his result (Longuet-Higgins, 1960), we find:

$$\left(\frac{\partial U_{mt}}{\partial y}\right)_{y=0} = H^2 \sigma k^2 \coth(kh) \tag{1.53}$$

exactly twice the value with viscosity neglected, and in good agreement with observation. Longuet-Higgins (1953), in a now classic contribution, postulated the existence of boundary layers at the free surface and bed to obtain the mass-transport beneath progressive and standing surface waves in water of finite depth. He derived eq. (1.53) for the mass-transport velocity-gradient at the free surface, and the following relations:

$$U_{mt} = \frac{5}{16} \frac{H^2 \sigma k}{\sinh^2(kh)} \tag{1.54}$$

$$U_{mt} = -\frac{3}{16} \frac{H^2 \sigma k \sin(2kx)}{\sinh^2(kh)} \tag{1.55}$$

for the mass-transport velocity just outside the bottom boundary-layer in respectively the progressive and standing wave cases. Equation (1.54) has good experimental support (Vincent and Ruellan, 1957; Russell and Osorio, 1958; Allen and Gibson, 1959; Rubatta, 1971; Bijker et al., 1975). Many investigators have extended the work on progressive waves (Chang, 1969; Dore, 1970, 1977; Ünlüata and Mei, 1970; Liu and Davis, 1977), and attacks on cnoidal (Le Méhauté, 1968; Spielvogel and Spielvogel, 1974; Isaacson, 1976a, 1976b, 1978), interfering (Dore, 1974), edge (Dore, 1975), and internal (Dore, 1973; Al-Zanaidi and Dore, 1976) waves have been attempted.

It will be noticed that eq. (1.55) for standing surface waves due to Longuet-Higgins (1953) is periodic in x, the distance in the direction of wave propagation. This is because the mass-transport beneath standing waves

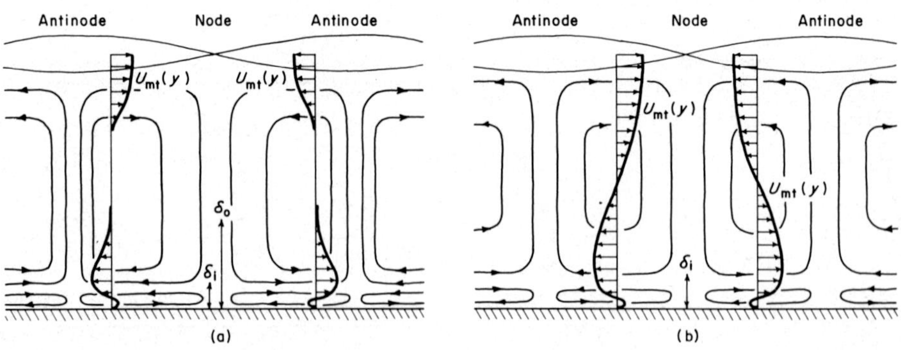

Fig. 1-22. Mass-transport beneath pure standing waves when (a) the mass-transport Reynolds number is large, and (b) when the Reynolds number is small. The outer and inner boundary-layers are denoted respectively by δ_o and δ_i. Modified after Liu and Davis (1977), by permission of the Cambridge University Press.

occurs as a series of cell-like circulations with boundaries at a spacing of one-quarter the wavelength of the surface waves. This result was early obtained by Rayleigh (1883) and has been confirmed by others (Noda, 1969; Mei et al., 1972; Sleath, 1973; Dore, 1976a; Liu and Davis, 1977). The vertical profile of mass-transport velocity within the bottom boundary-layer also shows a position at which the streaming changes direction, as verified by Noda (1969), and the circulation predicted by Lord Rayleigh actually consists of two vertically piled recirculating currents. Figure 1-22 shows these patterns for contrasting values of the mass-transport Reynolds number $Re = \sigma k^2 h^2 H^2 / 4\nu$. At small Reynolds numbers, the inner boundary layer is quite thick, and there is no outer layer. The boundary layers are relatively thin at large Reynolds numbers, so that the fluid in the interior appears to behave inviscidly. In each case the bottom flow is from antinodal toward nodal positions at the free surface. Jet-like flows, formed by the collision of the outer boundary layers, rise upward into the interior from the antinodal positions (Mei et al., 1972; Dore, 1976a; Liu and Davis, 1977). Dore (1973, 1976b) has extended the analysis to standing internal waves.

The results summarized above were derived within the general framework of laminar-flow. Longuet-Higgins (1958), however, has shown that eq. (1.54) for the mass-transport velocity beneath a progressive wave is independent of the viscosity, and is therefore valid for turbulent flow, in which a coefficient of eddy viscosity replaces the kinematic viscosity η/ρ. Johns (1970a) confirmed this result, but recently, in a more refined analysis (Johns, 1977), has obtained a substantially smaller maximum mass-transport velocity. The maximum mass-transport velocity just outside a turbulent bottom boundary-layer is significantly less than in the laminar case, but the profile of mass-transport remains similar (Johns, 1970a).

The above treatment of mass-transport due to surface waves assumes: (1) a perfectly clear air–water interface, (2) no drag between water and air, (3) no influence from the Coriolis force, and (4) the absence of currents other than those due to the wave-motion. Porter and Dore (1974) found that surface contamination decreased or increased the near-surface mass-transport according to circumstances, but had an insubstantial effect on the near-bed flow. The second assumption is widely implicit in studies of mass-transport, and is tantamount to treating the waves as if in a vacuum. Dore (1976a, 1978a, 1978b), however, has shown that drag at the air–water interface significantly affects the mass-transport due even to the short-period waves customary in laboratory experiments. The near-surface drift is increased when free-surface drag is included, though the bottom drift is little affected. Madsen (1978) has analyzed the effects of Coriolis force on the mass-transport associated with deep-water waves, and Huang (1979) explored the influence of wind-driven currents. To a good approximation, mass-transport and shear currents can be combined algebraically (Collins, 1964). Thus the combined current exceeds the mass-transport due to waves when waves and a shear

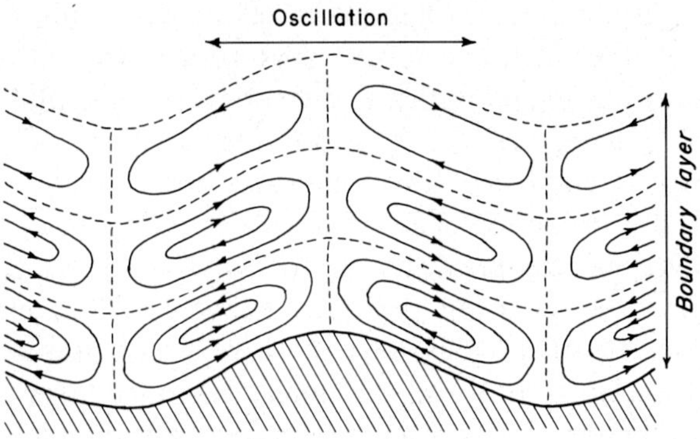

Fig. 1-23. Mass-transport above a transversely wavy bed affected by an oscillatory boundary-layer flow. The motion depicted is that at a large Reynolds number, when the characteristic wall wavelength is small compared to the diameter of a near-bed water-particle orbit. Modified after Lyne (1971), by permission of the Cambridge University Press.

current travel in the same direction, whereas an opposite current may neutralize the wave-induced streaming. Katz (1964) affords a mathematical treatment.

The analogous mass-transport at oscillating cylinders and at wavy walls beneath oscillating flows has been less studied, yet is germane to the origin of ripples and sand waves shaped by wave and tidal currents. As in the case of standing waves, the mass-transport at such wavy walls occurs in cellular regions, containing sometimes stacks of recirculating flows, with streamwise boundaries generally at a spacing of one-half of the wavelength of the wall perturbations (Lyne, 1971; Hall, 1974; Sleath, 1974a, 1974b, 1975b, 1976; Uda and Hino, 1975; Duck, 1978; Kaneko and Honji, 1979a). The flow next to the wall is then from the troughs toward the crests of the wall perturbations. Figure 1-23 shows the pattern predicted by Lyne (1971) for the case when the Reynolds number is large and the orbital diameter of a fluid particle far from the wall is large compared with the wall wavelength. In the outer boundary layer on an oscillating cylinder, the streaming is directed away from the cylinder and in the direction of oscillation (Schlichting, 1932; Stuart, 1963, 1966; Riley, 1965, 1967; Davidson and Riley, 1972; Bertelsen, 1974). It is jet-like at large Reynolds numbers, as in the case of the standing wave. In the inner layer, the streaming is directed toward the cylinder and represents a recirculation in the opposite sense.

Waves close to shore

As surface waves approach land from deep water, their wavelength decreases and height increases (for the most part), until eventually they break near to the beach (e.g. Eckart, 1952; Eagleson, 1956; Koh and Le Méhauté, 1966; Bryant, 1973; Komar, 1976). Waves in deep water have a limiting steepness of approximately $H/L = 0.142$ (Michell, 1893); breaking in shoal water occurs roughly when $H/h = 0.78$. Changes in their profile occur as waves approach the breakers through increasingly shallow water (Wiegel, 1950; Wiegel and Fuchs, 1955; Madsen and Mei, 1970). They may become more pointed, or asymmetrical toward the land, or develop subsidiary crests spaced between the main ones. Subsidiary crests can also arise when waves cross shallowly submerged bars (McNair and Sorensen, 1971; Chandler and Sorensen, 1973; Jeffrey and Tin, 1973). Another change developed close to a shore receiving waves is in the mean slope of the water surface, called *set-down* and *set-up* (Longuet-Higgins and Stewart, 1963, 1964; Bowen et al., 1968). Set-down is a slight downward slope toward the land found in the mean water surface seaward of breakers, whereas set-up is a slight upward slope encountered on the landward side.

Waves break when the forward orbital velocity of a water-particle at the crest exceeds the wave celerity. This topic has been extensively investigated both theoretically and experimentally (Havelock, 1918; Stoker, 1949; Suquet, 1950; Biesel, 1952; Ippen and Kulin, 1955; Laitone, 1963; Divoky et al., 1970; Longuet-Higgins, 1973a; Banner and Phillips, 1974), and was recently reviewed by Cokelet (1977). Galvin (1968) identified in experiments three types of breaking wave: (1) *plunging*, when the crest curls forward, over and down to form an air-filled tunnel (beloved of surf-riders), (2) *spilling*, when agitated "white" water flows like a bore (practically stationary with respect to the wave crest) down the wave front, and (3) *surging*, in which the wave collapses downward, to give on a beach an uprush of water. Plunging breakers in shallow water create large components of velocity directed vertically upward (Iversen, 1952; Miller and Zeigler, 1965; Adeyemo, 1970), and so may be effective scouring agents. Gaughan (1975) has suggested how breaker height and type can be predicted from a knowledge of wave conditions far from land. Internal waves are known from laboratory experiments to break and entrain sediment (e.g. Southard and Cacchione, 1972) in a similar manner to surface waves when in "shoal" water above a submerged slope, fluid from the light layer above being forced down the slope beneath the heavier so that the two partly mix (Cacchione, 1970; Emery and Gunnerson, 1973; Rumer, 1973; Cacchione and Wunsch, 1974). This mechanism could yield significant currents on the outer continental shelf and continental slope, where internal waves at the thermocline encounter the sloping sea-bed.

The water from breaking waves flows over the beach (or submerged sea-bed) as discrete swash and backwash. These surges or bores travel inland up and over the beach surface until their kinetic energy is exhausted, reaching heights significantly above still-water level. The run-up and dynamics of swash (e.g. Amein, 1966; Miller, 1968; Webber and Bullock, 1971; Sachdev and Seshadri, 1976; Kamphuis and Mohamed, 1978; Stoa, 1978) strongly influence sedimentary structures produced on beaches.

Waves are modified in the presence of currents (Biesel, 1950; Yu, 1952; Francis and Dudgeon, 1967; Velthuizen and Van Wijngaarden, 1969; Huang, 1972; Dalrymple, 1974; Vincent and Smith, 1976). An assisting current lowers wave height, whereas an opposing stream increases height, so that, when the current velocity is about one-quarter of the wave celerity, the critical steepness is reached and breaking occurs. Waves change direction where they cross currents (J.W. Johnson, 1947), and the bed shear stress exerted by waves with a current is always greater than could be applied by the waves alone (Bijker, 1967a, 1967b).

Edge waves

A fluid in an open container can engage in various free oscillations. If a part of the container is a sloping surface, for example, a beach marginal to the ocean or a lake, the free oscillations of small period are *edge waves* trapped to the shore. These waves have crests normal to shore and rapidly decrease in height away from land. They may be either progressive or standing, and in natural environments are generally excited by wind-waves. Edge waves were for long neglected, but are now recognised as important to the explanation of many features of beaches and the near-shore.

The theory of edge waves on an inviscid fluid is summarised by Lamb (1932) and Ursell (1951, 1952). Provided that $\beta < \pi/2$, where β is the slope of the fixed surface measured downward from the horizontal, edge waves decline exponentially in height away from shore, assuming a negligible height at outward distances greater than $L \cos \beta$, where L is the wavelength. Their celerity is:

$$c^2 = \frac{g}{k} \sin(2n+1)\beta \tag{1.56}$$

where $n = 0, 1, 2,$ etc. is an integer describing the mode excited and $k = 2\pi/L$. When $n = 0$ we have the lowest mode, the so-called Stokes edge wave. The highest mode is given by the largest integer contained in $(\frac{1}{2} + \pi/4\beta)$. Evidently for any angle $\beta < \pi/2$, at least the zeroth and first modes can exist. Putting $\beta = 10°$, a value typifying sand beaches on exposed shores, any of the first three modes could theoretically be excited. The theory of edge waves has since been considerably extended, some recent developments appearing in Grimshaw (1974), Guza and Bowen (1976), Huntley (1976), and Minzoni and Whitham (1977).

The tide and tidal currents

The tide at a fixed station is the rhythmical rise and fall of the ocean surface attributable to the changing gravitational pull of the Moon and Sun on the total water mass (Russell and MacMillan, 1952; Defant, 1958; R.G.

Dean, 1966; Hendershott and Munk, 1970; Platzman, 1971; Godin, 1972). This fluctuation occurs on several distinct periods simultaneously, related to the spinning of the Earth, the rotation of the Moon around the Earth, and the relative attitude of the Earth's axis. The *semidiurnal* and *diurnal* tides are the shortest oscillations. The semidiurnal tide, with a period of approximately 12.42 hrs, predominates in low and intermediate latitudes. In high latitudes, and at some stations of lower latitude, the tide can be diurnal, with a principal period of 24.83 hrs. The maximum level reached by the tide during any one oscillation is called *high-water level,* and the lowest position *low-water level.* Their difference is the *tidal range* for that particular oscillation. The tidal range in the open ocean is typically between 0.5 and 1.0 m, but in shallow and restricted waters is characteristically several metres, and in narrow estuaries and bays can exceed 15 m. The range varies on a period of approximately 13.66 days, being least (*neap tides*) when the Moon pulls at right-angles to the Sun (first and last lunar quarters), and greatest (*spring tides*) when the Moon and Sun pull on the same line (new and full moons). The range at spring tides is typically 50–100% more than at neaps. The spring tide itself varies in range on a period of one lunar month, approximately 27.32 days. The larger are observed at new moon, when the Moon lies between the Earth and the Sun, and the smaller when the Earth takes the intermediate position (full moon). Because the Earth's axis of rotation is tilted from the ecliptic, the tidal range also varies on a period of approximately 6 months. The smallest springs of all are observed near each Winter and Summer Solstice, whereas the largest of all arise close to the Spring and Autumn Equinox. Even longer tidal periods are known (e.g. De Rop, 1971).

The daily tide is a shallow-water wave. Its celerity is in the order of 20–200 m s^{-1} for depths ranging from the inner shelf to the deep ocean, while the wavelength is extremely large. Water-particle orbits take a diameter in the order of 10 km in a shallow restricted sea. In such a sea, the tidal current is usually rotary, when its value at a fixed station can be described by (Proudman, 1925):

$$U = U_{max} \cos(\sigma t - kx) \quad (1.57)$$

$$V = V_{max} \sin(\sigma t - kx) \quad (1.58)$$

and:

$$\left(\frac{U}{U_{max}}\right)^2 + \left(\frac{V}{V_{max}}\right)^2 = 1 \quad (1.59)$$

in which U and V are the velocity components parallel and transverse to the propagation direction of the tidal wave, U_{max} and V_{max} are the maxima of U and V, and functions of position, σ and k are the radian frequency and wave number as before, t is time, and x is distance measured in the propagation

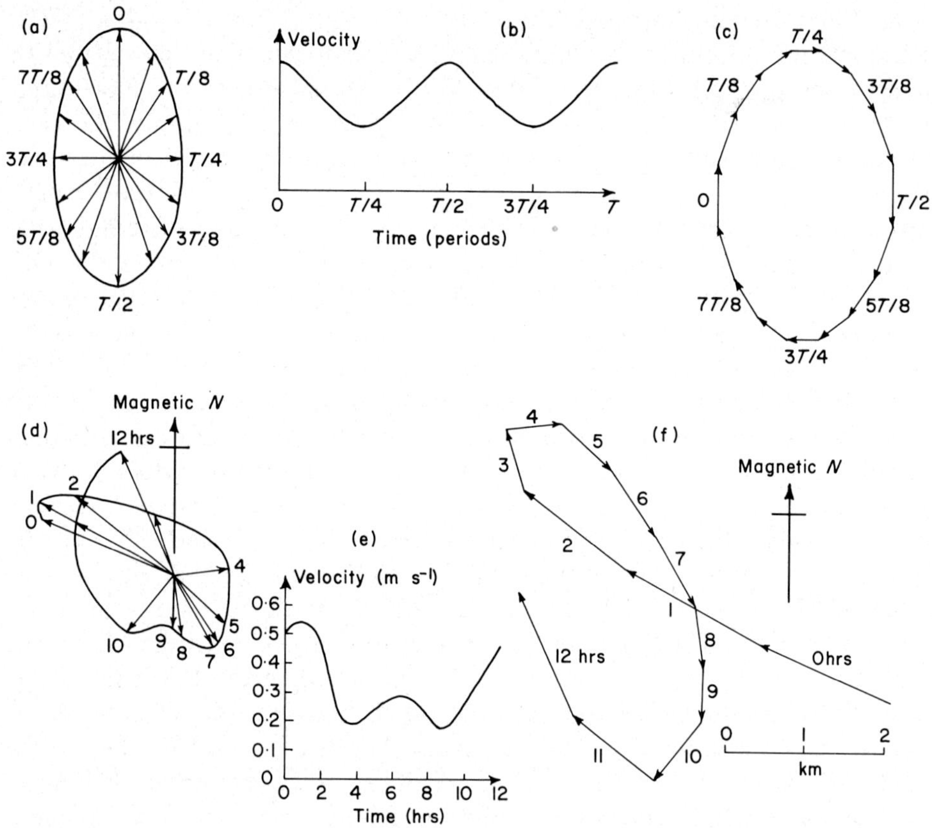

Fig. 1-24. Features of idealized (a–c) and actual (d–f) tidal currents. (a, c) Eulerian and Lagrangian representation of an idealized rotary tidal current over one tidal period, and (b) the time-velocity pattern. (d, f) Eulerian and Lagrangian representations of the tidal current measured and (e) velocity pattern at a depth of 30.5 m (total water depth more than 45 m) at Nantucket Shoals Light Vessel, east coast of U.S.A. 8–10 August, 1923. Partly after Le Lacheur (1924), by permission of the American Geographical Society.

direction. Equation (1.59) describes an ellipse, the *tidal-current ellipse*, with a major semi-axis U_{max} and minor semi-axis V_{max}, which connects the tips of the vectors whose components are given by eqs. (1.57) and (1.58) (Fig. 1-24a).

This ellipse is an *Eulerian* specification of the tide, showing the velocity as it would be observed at a fixed place over one cycle, as also is the velocity–time graph of Fig. 1-24b. We can alternatively follow a single fluid particle over a tidal period, so specifying the motion in a *Lagrangian* manner (Fig. 1-24c). Of course as V_{max} declines relative to U_{max}, the motion becomes increasingly back-and-forth, and is purely oscillatory when V_{max} is zero, as can be observed in very narrow channels.

Tidal currents have been measured or estimated in numerous estuaries and restricted seas (Marmer, 1923; Fleming, 1938; DHIH, 1963; Daboll, 1969; Farrell, 1970; Göhren, 1971a, 1971b; McCave, 1971a; Klein and

Whaley, 1972; Walton and Goodell, 1972; Bhattacharya, 1973; Huthnance, 1973; Ludwick, 1973; Boothroyd and Hubbard, 1974, 1975; Sager and Sammler, 1975), on continental shelves (e.g. Le Lacheur, 1924; Scott and Csanady, 1976), and in straits (e.g. Keller and Richards, 1967; Kranck, 1972). Typically, maximum velocities are in the order of 1 m s^{-1} in such environments. In ocean deeps, however, tidal velocities measure only in the order of 0.01–0.1 m s^{-1} (e.g. Fliegel and Nowroozi, 1970; Keller et al., 1975). Polar diagrams like those in Fig. 1-24a can be constructed where the current direction as well as magnitude are known. All depart from the ideal tide described by eqs. (1.57)–(1.59) in showing some distortion or asymmetry of the tidal ellipse and the velocity–time plot, as well as an open orbit (Figs. 1-24d–f). The latter feature points, as with short-period waves, to a flow that combines oscillatory motion with a residual current. Two sources contribute to the latter.

A predominant fraction of the residual tidal current observed in shallow waters is ultimately attributable to a seeming instability of the tidal flow. Ebb and flood streams become separated horizontally in association with the building of large sand shoals, between which arise channels whose bottom contours close in the direction of the dominant current (Van Veen, 1936). In extreme cases, the flood discharge in such a channel exceeds by a few times that during the ebb, and vice versa. Averaged over all the channels in an area, however, the net tidal discharge amounts to zero.

Mass-transport contributes the remainder of the residual tidal current. Hunt and Johns (1963) and Yamagata (1978) studied a progressive tidal wave creating a rotary current in an inviscid fluid of uniform depth. They found mass-transport in the direction of wave propagation, but of a smaller magnitude than that due to oscillatory wind-waves, owing to the rotary motion induced by the Coriolis force. Johns (1967, 1968, 1970b, 1973) and Johns and Dyke (1972) extended this work, to include the effects of viscosity and non-uniform boundary conditions. Tidal currents are turbulent in varying degrees (Bowden and Fairbairn, 1956; Bowden et al., 1959; Bowden, 1962, 1964, 1970; McGregor, 1972), but are well described by a depth-dependent eddy viscosity (Johns, 1966, 1969). One of the effects of a non-uniform boundary is a reduction of the mass-transport (and even a reversal) in a converging estuary as compared with a uniform channel.

GRAVITY CURRENTS

Character and occurrence

A gravity current, also called a density current, is a surge-like flow that arises when a fluid of one density is juxtaposed against one or more fluids of a different density. The current has a well-defined head that involves a

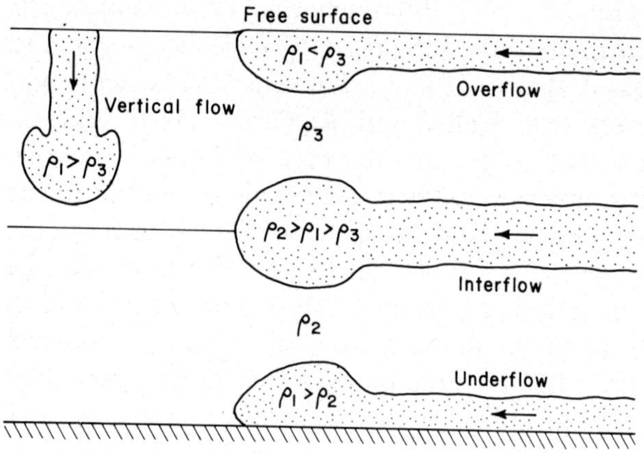

Fig. 1-25. Principal types of gravity current.

mixing region or wake, followed by a more or less long body, in which the motion may approach uniform and steady conditions. The body terminates in a tail. There are four types of gravity current, involving either miscible or immiscible fluids (Fig. 1-25): (1) *underflows,* when the density of the current (ρ_1) exceeds that of the ambient medium, (2) *overflows,* when the ambient medium has the greater density, (3) *interflows*, when the ambient medium is density-stratified and the current is of intermediate density, and (4) *vertical flows,* when the motion is unconstrained. The sources of density contrast are: (1) a difference of fluid composition, (2) a difference of temperature, (3) a difference in the concentration of dissolved salts, and (4) the presence of dispersed solids in the current.

Examples of underflowing gravity currents abound in the atmosphere. Probably the largest, filling the atmospheric boundary layer, are sea-breeze fronts and other small cold fronts (Berson, 1958; Wallington, 1959; R.H. Clarke, 1961, 1965; J.E. Simpson, 1964, 1967, 1969; Steedman and Ashour, 1976; J.E. Simpson et al., 1977). Sea-breezes are masses of air, cooled and moistened over the sea, that spread inland for distances of tens or hundreds of kilometres along well-defined fronts, often marked by distinctive clouds, and in some instances by dust or by concentrations of radar-reflecting birds preying off trapped insect swarms. Considerably smaller in scale are the radially spreading downwashes of cold air originating from thunderstorms (Browning and Ludlam, 1962; Saunders, 1962; Goff, 1976; Gurka, 1976; Hall et al., 1976). In regions of limited vegetation these surges entrain fine sediment, when their well-defined fronts become visible as towering walls of dust-laden air, as in the *haboobs* of the Sahel and elsewhere (Bell, 1942; J.E. Simpson, 1969; Lawson, 1971; Idso et al., 1972; Barth, 1978) or the *chubasio* of western Mexico (Idso, 1973, 1976). The speed of sea-breeze fronts and thunderstorm downwashes is generally less than 15 m s^{-1}. They are very gusty internally and give rise, along the advancing edge, to whirlwinds (Golden, 1974; Idso, 1974). Atmospheric underflows driven by dispersed sediment are represented in mountainous areas by powder-snow avalanches (Allix, 1924; Rohrer, 1954, 1955; Fraser, 1966; Bonington, 1976; Mellor, 1978), and by the *nuées ardentes* (Anderson and Flett, 1903; Perret, 1935; G.A. Taylor, 1958; Moore and Melson, 1969) and base surges (Moore et al., 1966; J.G. Moore, 1967) associated with volcanic eruptions. *Nuées ardentes* are violent surges of hot ash-laden gases that result when magma explodes through the flanks of a volcano; base surges are relatively cool mixtures of gases, water droplets, and ash that follow eruption into an aqueous environment. Having a relatively large excess density, these volcanogenic gravity currents travel much faster than sea-breeze fronts and thunderstorm downwashes.

The best examples of gravity-current overflows in the hydrosphere are the plumes of fresh or freshened water that spread for considerable distances at low speeds over the sea off river and estuary mouths (Bell, 1942; Scruton and Moore, 1953; Knauss, 1957; Duxbury and McGary, 1967; Amos et al., 1972; Grancini and Cescon, 1973; Carlson and McCulloch, 1974; Garvine, 1974a, 1974b, 1977; Garvine and Monk, 1974; Ingram, 1976). Their well-defined steep fronts are marked by an abrupt change of water colour, and by concentrations of foam and floating debris. The small amounts of dispersed mud present are not enough significantly to depress the salinity-related density contrast.

Where a river enters the fresh waters of a lake or reservoir, the muddy waters spread outward from the mouth for a short distance, but eventually plunge downward over the bottom (Singh and Shah, 1971). Such bottom-hugging and often long-sustained sediment-laden currents, called turbidity currents, were long ago inferred by Forel (1885) to occur in Lake Geneva, and have since been detected in other lakes and reservoirs (Grover and Howard, 1938; Gould, 1951; Raynaud, 1951; Nizery and Bonnin, 1953; Lambert et al., 1976). Their speed is in the order of 0.1 m s^{-1}, a magnitude consistent with the small density differences involved. The dumping of particulate wastes into lakes or the sea appears also to give rise to turbidity currents (Jenkins, 1970; Normark and Dickson, 1976). Turbidity currents of a grand scale are thought to occur in the oceans, though they have not been directly observed. Telling amongst the evidence for them is the sudden and unexpected breaking of telegraph cables, in descending sequence, that cross particular slopes on the ocean floor (Heezen and Ewing, 1952, 1955; Kuenen, 1952; Heezen et al., 1954; Heezen, 1956; Houtz and Wellman, 1962; Ryan and Heezen, 1965; Krause et al., 1970). Such breakages followed shortly after earthquakes or river floods, pointing to the involvement of sediments rendered unstable, and thus capable of flowing bodily or of mixing further with sea water to make a vigorous current. Whatever their precise nature, these flows measured tens of metres per second in velocity and covered distances often of many hundreds of kilometres.

The extent to which interflows occur naturally is not clear. They are, however, a well-defined experimental phenomenon (Holyer and Huppert, 1980).

Vertical gravity-current flows can be induced by streams of falling particles, as when wastes are dumped from barges (Arons et al., 1951; Bradley, 1965). Sediment-driven interflows may occur between aqueous strata in the oceans (Stetson and Smith, 1937), though none have so far been reported.

Models for turbidity currents

The observations summarised above, combined with the results of many small-scale laboratory experiments, provide a consistent physical picture of the gravity-current underflow on a stationary bed beneath a miscible ambient medium (Fig. 1-26). The head has a sharp and moderately to steeply sloping front that overhangs the bed toward the base, as can be seen in the laboratory (Schmidt, 1910, 1911; Middleton, 1966a, J.E. Simpson, 1969; Britter and Linden, 1980; Irvine, 1980; Simpson and Britter, 1980), on base surges (J.G. Moore, 1967), *haboobs* (J.E. Simpson, 1969; Barth, 1978), and on powder-snow avalanches (Mellor, 1978). Particularly where it overhangs, the head is marked transversely by a shifting pattern of alternate lobes and clefts (Fig. 1-27). These are clearly visible on laboratory currents (Michon et al., 1955; Juignet et al., 1965; J.E. Simpson, 1969) and are readily seen on the much larger atmospheric underflows (G.A. Taylor, 1958; J.G. Moore, 1967; Moore and Melson, 1969; J.E. Simpson, 1969; Bonington, 1976; Barth, 1978). Using salt solutions flowing through clear water, J.E. Simpson (1969, 1972) obtained shadowgraphs showing that the clefts lead back into the head as tall, narrow tunnels which after a certain penetration lift off the bed and break up (Fig. 1-28). Through these clefts some of the ambient medium is ingested into the head and mixed with the current (Allen, 1971b). With the head for convenience at rest, Allen concluded (see also Irvine, 1980; Simpson and Britter, 1980) that the pattern of streamlines in a vertical streamwise section through a lobe is quite different from that through a cleft (Fig. 1-29).

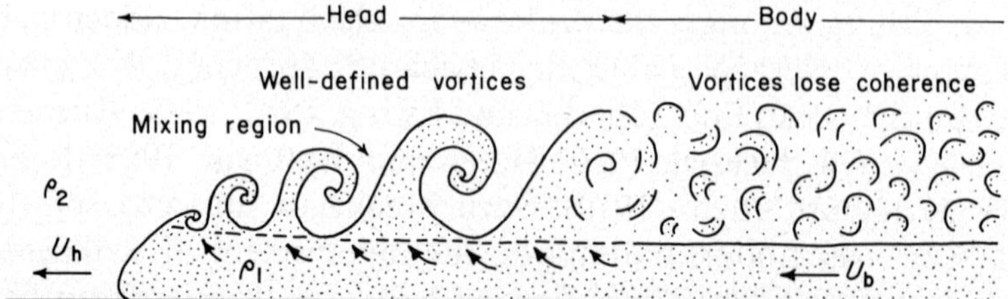

Fig. 1-26. General features of underflowing gravity currents.

Fig. 1-27. Saline gravity current, made visible with milk, flowing from left to right through still water in a horizontal laboratory tank of rectangular cross-section. Note the discrete nature of the head, and the lobes and clefts along its front. Density difference = 10 kg m^{-3}. Scale marked in centimetres. Photograph courtesy of J.E. Simpson (see Simpson, 1969); reproduced by permission of the Royal Meteorological Society.

Most of the ambient medium confronted by the current is lifted up and accelerated over the head (Fig. 1-29). Within the head strong upward velocity components are seen (Schmidt, 1910, 1911; Defant, 1921; Middleton, 1966b; J.E. Simpson, 1972; Britter and Simpson, 1978), while in front of it there may develop vortices with axes steeply inclined to the bed (Golden, 1974; Idso, 1974). Above and backward from the overhang, the smoothly curved leading surface of an experimental current rolls up into transverse vortices or billows which travel backward relative to the head, increasing in diameter, but decreasing in coherence, as their age increases (Fig. 1-30). J.E. Simpson (1969) points out that Atlas (1960) obtained radar echoes from sea-breezes consistent with the occurrence in the field of similar large vortices. Such billows point to a mixing of fluid from the head into the ambient medium, the continual loss being made good by supply to the head from the body of the current, which must therefore have a larger mean forward velocity than the head. The concentrations of debris along the fronts of river plumes are a further proof of this relationship, for they imply a convergence which can arise only if the current flowing into the head travels faster than the head itself.

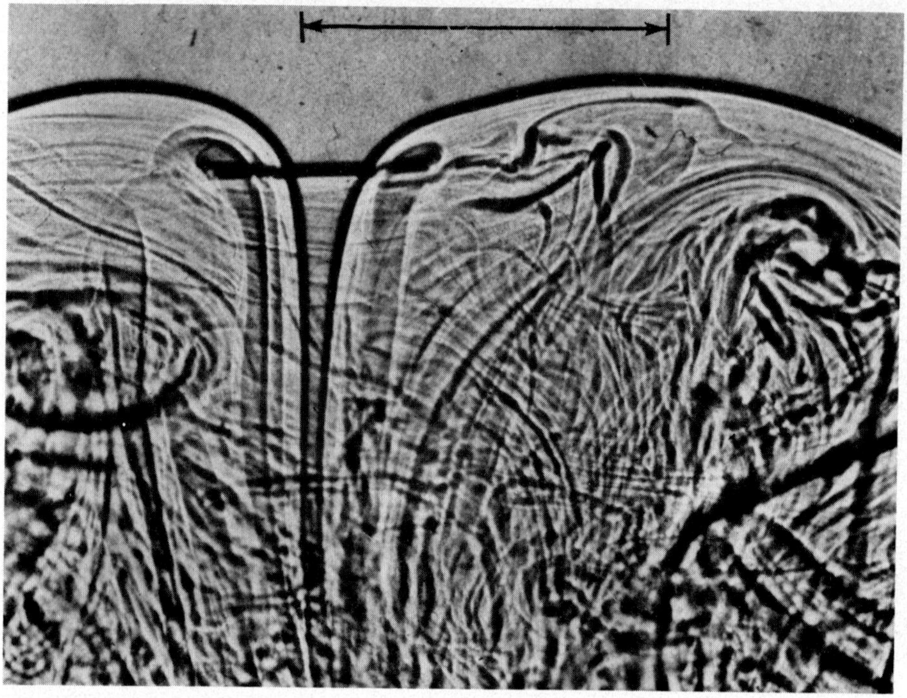

Fig. 1-28. Shadowgraph made by vertical flash lighting of a cleft at the head of a small saline gravity current similar to that in Fig. 1-27 and advancing from the bottom toward the top of the picture. The arrow is 0.01 m long. Photograph courtesy of J.E. Simpson (see Simpson, 1969); reproduced by permission of the Royal Meteorological Society.

Whereas a considerable amount is known that contributes to this physical model of gravity current, it is fair to say that no general theory of such flows, in which the motions of head and body are properly matched, has yet been devised. Attention has so far been concentrated on particular features, either

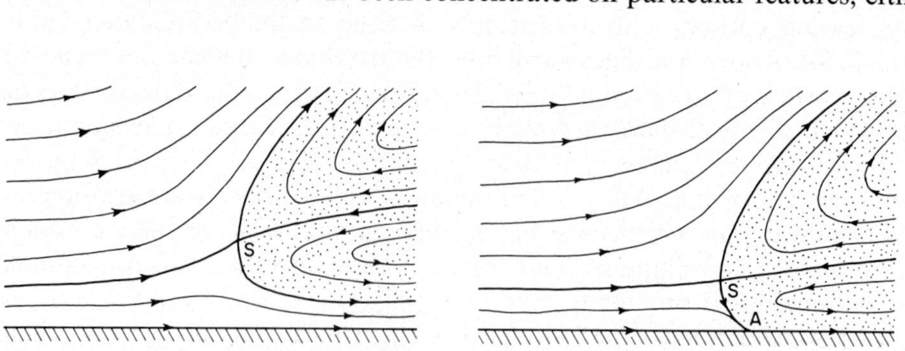

Fig. 1-29. Inferred transverse variation in flow pattern associated with head of underflowing gravity current, relative to the stationary head. a. Vertical flow-parallel plane through a cleft in the head. b. Vertical flow-parallel plane through a lobe.

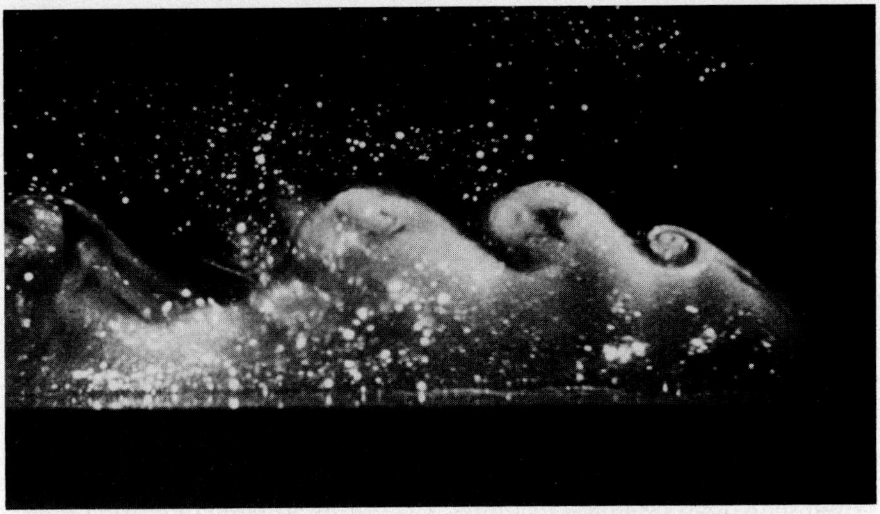

Fig. 1-30. Instability at the head of a gravity current advancing from left to right, made visible by fluorescein and vertical slit lighting in the plane of flow. The current has a density excess of about 10 kg m^{-3} and its head is approximately 0.03 m high. Photograph courtesy of J.E. Simpson (see Simpson, 1969); reproduced by permission of the Royal Meteorological Society.

the surge-like motion of the head, or the notionally steady and uniform flow far back in the body. It appears that the current will pass through a series of regimes, each characterized by specific controls, during its history (Huppert and Simpson, 1980).

The surge-like motion of the head of a gravity-current underflow somewhat resembles that of the wave resulting from the sudden bursting of a dam. Neglecting viscosity and turbulence, and on the supposition that the air is of negligible density, Ritter (1892) obtained for the celerity of a two-dimensional dam-break wave:

$$c = 2\sqrt{gh} \tag{1.60}$$

in which h is the depth of water behind the dam. Later students of the dam-break wave have attempted to include the effects of viscosity and turbulence (Dressler, 1952a, 1954; M.B. Abbott, 1961; Tinney and Bassett, 1961; Wessels and Strelkov, 1968; Sakkas and Strelkoff, 1973; Buckmaster, 1977; Rajar, 1978).

Another hydraulic approach to the two-dimensional spread of one inviscid, immiscible fluid into another under the influence of hydrostatic pressure was initiated by Von Kármán (1940), who concluded that the speed of the current was proportional to the square root of the density difference and the square root of the flow thickness behind the head. He also deduced that the slope from the horizontal of the front of the head was $\pi/3$, defining a shape remarkably similar to that observed, except for the overhang which is a viscous effect. Benjamin (1968) showed that Von Kármán's results were

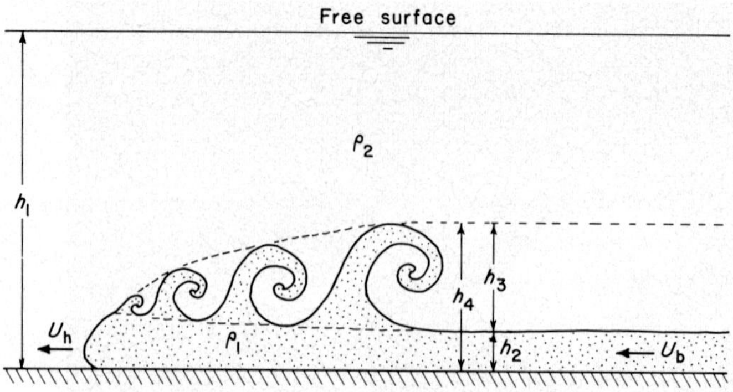

Fig. 1-31. Definition diagram for an underflowing gravity current.

correct but depended on false reasoning. According to Benjamin, the velocity U_h of the head of an inviscid, immiscible, two-dimensional gravity current (Fig. 1-31) of density ρ_1 advancing beneath a still fluid of thickness h_1 and density ρ_2 is:

$$U_h = k\left[\frac{(\rho_1 - \rho_2)g}{\rho_2} h_2\right]^{1/2} \tag{1.61}$$

where h_2 is the thickness of the body of the flow, and:

$$k = \left[\frac{(h_1 - h_2)(2h_1 - h_2)}{h_1(h_1 + h_2)}\right]^{1/2} \tag{1.62}$$

varying with relative depth, ranges from $1/\sqrt{2}$ when $h_1/h_2 = 0.5$ to $\sqrt{2}$ for great depth. Equation (1.61) predicts an infinitely large velocity when $\rho_2 = 0$, but is consistent with observation at small density differences and large Reynolds numbers.

A voluminous literature describes experimental and further theoretical studies of gravity currents. Because of its relevance to oil spillages at sea, the spread of immiscible fluids is receiving increasing attention (Bata and Bogich, 1953; Bata, 1959; M.B. Abbott, 1961; Fay, 1969; Fannelop and Waldman, 1972; Hoult, 1972; Buckmaster, 1973; Bache, 1976). Powder-snow avalanches are analyzed by Shen and Roper (1969) and by Hopfinger and Tochon-Danguy (1977). Much theoretical work has been directed toward the two-dimensional flow of salt-water undercurrents and sediment-driven turbidity currents (Brooks and Berggren, 1943; Duquennois, 1951; Menard and Ludwick, 1951; Schijf and Schönfeld, 1953; Hinze, 1960; Plapp and Mitchell, 1960; M.A. Johnson, 1962, 1964; Hurley, 1964; Komar, 1969, 1971a, 1972, 1973a; Kao, 1977; Pantin, 1979). Allied numerical studies are reported by several investigators (Daly and Pracht, 1968; R.H. Clarke, 1973; Charba, 1974; Xanthopoulos and Kontitas, 1976; Komar, 1977a; Mitchell and Hovermale, 1977; Kao et al., 1978). Corresponding experimental work is extensively reported (Schmidt, 1910, 1911; O'Brien and Cherno, 1934; Ghatage, 1936; Bell, 1942; Knapp, 1943; Kuenen, 1950, 1951; Kuenen and Migliorini, 1950; Kuenen and Menard, 1952; Blanchet and Villatte, 1954; Michon et al., 1955; Keulegan, 1957, 1958; Bonnefille and Goddet, 1959; Fan, 1960; Barr, 1963, 1967; Barr and Hassan, 1963; Wood, 1965, 1967, 1970; Yih, 1965; Middleton, 1966a, 1966b, 1967a; Riddell, 1969, 1970; J.E. Simpson, 1969, 1972; Tesakar, 1969a, 1969b; Wilkinson, 1970; Halliwell and O'Dell, 1971; J.J. Sharp, 1971; Wilkinson and Wood, 1972; Kersey and Hsu, 1976). Some theoretical and experimental work exists on the motion of three-dimensional gravity currents, notably those due to the collapse of a fluid column (Penny and Thornhill, 1951; Martin and Moyse, 1952a, 1952b; Swift, 1962; G.A. Young, 1965; Fietz, 1966; Fietz and Wood, 1967; Sparks and Wilson, 1976; Sparks et al., 1978). Van Andel and Komar (1969) discuss the "sloshing" of a turbidity current within a restricted basin, and Woodcock (1976) its travel over an obliquely sloping surface.

One important outcome of much of this work is that the uniform, steady motion of the body of a gravity current on a slope can be described by a relationship of the form of eq. (1.35) proposed for a river (Middleton, 1966b). With the notation of Fig. 1-31, we can use the principles previously explained to rewrite eq. (1.35) as:

$$U_b = \left[\frac{8g(\rho_1 - \rho_2)h_2 S}{\rho_1(f_0 + f_i)} \right]^{1/2} \tag{1.63}$$

where the flow is assumed to be much wider than deep. Here f_0 and f_i are Darcy-Weisbach friction coefficients respectively at the bed and fluid interface. These coefficients take values of the same general order, but whereas f_0 varies with Reynolds number and bed roughness, f_i is a function of Reynolds number and the density-adjusted Froude number (Middleton, 1966b).

Another and widely quoted outcome is an empirical formula for the velocity of the head of a gravity current in a miscible fluid at large Reynolds numbers (Keulegan, 1957, 1958; Middleton, 1966a):

$$U_h = 0.75 \left[\frac{(\rho_1 - \rho_2)g}{\rho_1} h_4 \right]^{1/2} \tag{1.64}$$

in which $h_4 = (h_2 + h_3)$ is the head thickness. This equation is identical in form to Benjamin's (1968) eq. (1.61), but differs in the presence of h_4 instead of h_2 in the numerator, and in the use of ρ_1 rather than ρ_2 in the denominator. Apparently, the two equations are remarkably similar in the numerical coefficient. The similarity derives, however, from constraints related to the lock-exchange experiments underlying eq. (1.64). This mode of experimentation yields gravity-current heads with only one value for the ratio of the head to body thickness ($h_4 \approx 2h_2$ or $h_3 \approx h_2$). In the less constrained experiments of Britter and Simpson (1978), h_3/h_2 ranged widely, from about 1 for $h_2/h_1 = 0.25$ to approximately 10 for great depths. It would appear from this work that uniform and steady flow for the head jointly with the body of a gravity current involving miscible fluids in very deep water is given by:

$$(U_b - U_h) = \left[\frac{(\rho_1 - \rho_2)g}{\rho_1} h_2\right]^{1/2} \tag{1.65}$$

where $U_b/U_h \approx 1.25$. This ratio gradually declines as h_2/h_1 increases, and eq. (1.65) must be adjusted accordingly with the help of Britter and Simpson's empirical data.

SUMMARY

Mechanical sedimentary structures are attributable mainly to the transporting agents—mass flows, rivers, the wind, wind-wave currents, tidal and oceanic currents, turbidity currents, and glaciers—involved in the movement of detrital particles from source to sink over the Earth's surface. A minor proportion originate in magma chambers within the Earth's crust, where crystallized silicate minerals are available for transport and deposition. A combination of simplified theoretical approaches with field and laboratory observations gives insight into the character and dynamics of these agents. Certain agents, such as many mass flows and magma currents, operate at a low Reynolds number, for they involve highly viscous fluids that flow in a laminar manner. The flow of rivers, turbidity currents, and the wind is fast and turbulent, however, and the Reynolds numbers correspondingly large. Wind-waves and the tide give rise to the most complex boundary layers known, in which oscillatory currents are combined with steady mass-transports. The tide in shallow water flows as fast as rivers, but the currents due to wind-waves span the laminar-turbulent range, and are vigorous only during storms.

Chapter 2

ENTRAINMENT AND TRANSPORT OF SEDIMENTARY PARTICLES

INTRODUCTION

The natural agents of sediment transport were examined in the preceding chapter from the standpoint of their mechanics and the environments in which they operate. Sedimentary structures can form, however, only when these agents can acquire sedimentary materials to shift from place to place. We shall therefore in this chapter look more closely at the nature of sedimentary particles, at their entrainment from cohesionless and cohesive beds, at their settling in still fluids, and at their transport by wind and water.

An understanding of sediment transport is particularly important if the origin of sedimentary structures is properly to be grasped. Many structures are direct expressions of the transport of debris; some cannot have arisen, and none can have been permanently preserved, unless the rate of transport by natural agencies were subject to change in time and/or space. Although much remains to be learned about the mechanics of sediment transport under steady-state equilibrium conditions, our existing knowledge is inevitably the basis for what understanding of structures we presently possess.

SHAPE AND SIZE OF SEDIMENTARY PARTICLES

Particle shape depends primarily on particle origin, whether by: (1) crystallization from magma or aqueous solution, (2) volcanism, (3) rock weathering, or (4) organic activity. The other controls on shape are the mode and duration of subsequent sediment transport, affecting chiefly the degree of rounding of particle edges and corners (Krumbein, 1941a, 1941b). Shape strongly affects the behaviour of particles in bulk, as when flowing *en masse* in a gravity-driven avalanche, or when entrained from a bed and impelled along by a fluid stream, and also individually, as when settling in a fluid.

The most regular sedimentary particles are those that crystallized in a fluid. The orthorhombic olivines form either stubby crystals or irregular grains. The feldspars (monoclinic, triclinic) yield chiefly columnar to tabular crystals. Those of the pyroxenes (orthorhombic, monoclinic) are stubby to columnar, and columnar crystals are also given by the common amphiboles (monoclinic). The clay minerals yield tabular, bladed or acicular crystals (Beutelspacher and Van der Marel, 1968); these in many natural environments become loosely aggregated into large, irregular, porous floccules through the action of electrochemical surface forces (Van Olphen, 1963;

Gillott, 1968; Zabawa, 1978). The evaporation of seawater in hot climates leads to the precipitation of calcium carbonate in the form of acicular aragonite crystals (Cloud, 1962), which settle to the bottom forming carbonate mud.

Fig. 2-1. Some examples of sedimentary particles from recent environments. a. Ooids from an oolite shoal (×3). b. Pumice from the coarse fraction of a pyroclastic fall (×1.4). c. Fragments of the carbonate-secreting alga *Halimeda* (×1.4). d. Fragments of the carbonate-secreting alga *Lithothamnium* (×1.4). e. A sand composed of whole to broken gastropod shells and platy fragments of broken bivalves (×1.4). f. Separated valves of the common cockle, *Cerastoderma edule* (×0.4).

Most sedimentary particles, a variety of which appear in Fig. 2-1, are irregular but can be represented approximately by regular solids. There are extensive accounts of pyroclastic debris (Heiken, 1972, 1974; Walker and Croasdale, 1972), and profuse details of quartz particles appear in Krinsley and Doornkamp (1973). Many workers further illustrate the variety of grain shapes to be found in modern carbonate environments, where organisms produce much of the debris furnished to currents (Newell et al., 1951, 1959; Illing, 1954; Kornicker and Purdy, 1957; Newell and Rigby, 1957; Purdy, 1963a, 1963b; Folk and Robles, 1964; Matthews, 1966; Hoskin and Nelson, 1969; M.S. Lewis, 1969; Wass et al., 1970; Harms et al., 1974; Flood et al., 1978; McLean and Stoddart, 1978).

Irregular grains approximate to a wide range of regular solids. The sphere, the traditional ideal, is approached by comparatively few real particles: some exceptionally well-rounded quartz sand grains, many ooliths, and globular foraminifera. The triaxial ellipsoid is a more acceptable ideal (N.C. Flemming, 1965), particularly for the debris yielded by rock-weathering. Its special cases, the disc-like oblate spheroid, and roller-like prolate spheroid, are especially valuable models. The prolate spheroid is also useful in representing much faecal debris and many ooliths. The right-circular cylinder is a good model for some faecal debris, for fragments from dendritic corals, bryozoa and coraline algae and, of course, vegetable matter other than leaves. Echinoderm spines, some foraminifera, and fragments from certain calcareous algae are modelled by the circular disc. Many bivalves and other shells when sufficiently comminuted, together with mica flakes, can be idealised as rectangular plates. Certain bryozoa and other marine organisms afford fragments that resemble sieves, that is, regularly perforated plates. Individual valves from brachiopods and bivalve molluscs can be represented by spherical or ellipsoidal caps and lenses. Gastropod shells are occasionally important as sedimentary particles, and are best modelled as circular pyramids. An extremely irregular surface, abounding in re-entrants, is typical of vesicular pyroclasts, the smaller particles amongst which approach a tetrahedron with dished faces.

The size of real sedimentary particles is difficult to specify. Only the perfectly spherical grain has a unique size, for no one diameter is different from any other, of course making this shape appealing as an ideal. As a matter of practice, however, particle size generally means one of the following three measures: (1) the length of either the short, intermediate or long grain axis (or the mean of any two or all of these), (2) the diameter of the sphere having the same volume, or (3) the diameter of the smooth sphere of the same density having the same steady falling velocity under specified conditions. Having obtained a measure of size, it is necessary for discussion and comparison to classify and name the particle with respect to that value. Wentworth's (1922) scheme is the one most commonly employed in the earth sciences (Table 2-I). The third listed measure of size (equivalent fall diame

TABLE 2-I

The Wentworth size classification of sedimentary particles

Particle diameter ($\times 10^3$ m)	Name of size class	
>256	Gravel	boulder
64–256		cobble
4– 64		pebble
2–4	Sand	granule
1–2		very coarse sand
0.5–1		coarse sand
0.25–0.5		medium sand
0.125–0.25		fine sand
0.0625–0.125		very fine sand
0.00256–0.0625	Silt	
<0.00256	Clay	

ter) has long been used for silt and clay particles (Krumbein and Pettijohn, 1938), but also has appeal for larger grains (Brezina, 1963; Channon, 1971; Reed et al., 1975; Middleton, 1977).

SEDIMENTARY PARTICLES IN BULK

The ideal sediment consists of particles of a single size, shape, and density immersed in a homogeneous intergranular fluid. Such a mixture is a *monodisperse system.* Real sediments are *polydisperse,* composed of grains of a range of size, shape and, usually, density. It is particularly important to know how size is distributed in the mixtures, and many techniques are available for the purpose of measurement and analysis (Krumbein and Pettijohn, 1938). Two quantities are especially useful in describing these grain-size frequency distributions: (1) the mean or median diameter, which measures central tendency, and (2) the standard deviation of size, which measures the range of sizes represented, that is, the quality of sorting. Most sands deposited in water or by wind yield a size-frequency distribution curve having a single, bell-shaped hump. Most deposits coarser than sand, for example, stream gravels, glacial tills, and debris-flows, are ill-sorted and bimodal or polymodal. Little is known of the distribution curves of muddy sediments at deposition, for such sediments are almost invariably analyzed after dispersion in strong peptizing electrolytes.

It is helpful to divide deposited sediments between *cohesionless* and *cohesive.* The former consist of particles that are neither significantly attracted nor repelled by each other, as is the case with gravels, sands, and the coarser silts. Grains finer than these, however, yield cohesive beds, in which the particles, particularly the clay minerals, are united by electrochemical surface forces, the strength of which varies with grain size, mineralogy, and electrolyte.

An intriguing and important property of cohesionless sediments is their ability, when placed on a sufficiently steep slope, or when supported from below by a sufficiently fast upward intergranular stream, to flow under gravity like a liquid with a free surface. Precisely what is this "sufficiently steep slope", loosely called the angle of rest or repose, and how it relates to the friction angles of soil mechanics, are matters of some confusion and continuing debate (Metcalf, 1966; Statham, 1974, 1977; Carson, 1977). As Van Burkalow (1945) recognised, two now clearly distinguished angles are involved (Bagnold, 1967; Allen, 1969e, 1970b, 1970c), as is easily proved by rolling slowly across a table a cylindrical jar half-filled with dry sand. The surface of the mass must be tilted comparatively steeply, to the *angle of initial yield* ϕ_i, before flow occurs, after which the grains come to rest piled at a lower slope, called the *residual angle after shearing* ϕ_r. Neither angle is unique, however, but varies with grain size, degree of sorting, and closeness of packing, and also with grain shape and surface texture (Van Burkalow,

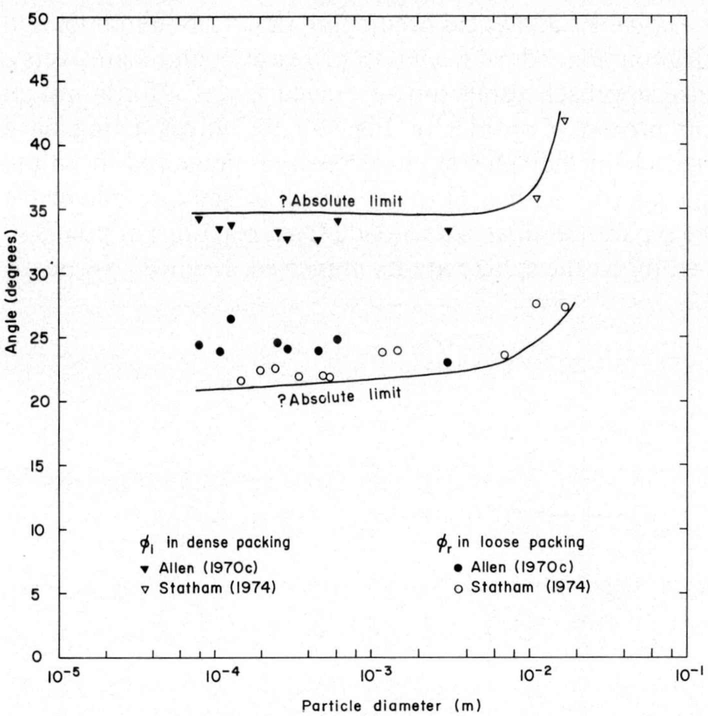

Fig. 2-2. Experimental variation of the angle of initial yield (ϕ_i) and residual angle after shearing (ϕ_r) with particle diameter for polished glass spheres. The angles for natural sediments are in each case approximately 10° larger, depending on particle size, shape and surface roughness.

1945; Fowler and Chodziesner, 1959; Allen, 1965b, 1969b, 1970b, 1970c; Miller and Byrne, 1966; Carrigy, 1967, 1970; Statham, 1974; Mantz, 1977a). Both angles increase markedly with increasing grain roughness and angularity and with increasing closeness of packing. The angles tend to increase with particle size at a fixed closeness of packing, but the difference between them remains constant at 5–20° (Fig. 2-2). A worsening in the quality of sorting increases both angles, probably because of further close packing. Natural gravels and sands yield residual angles after shearing generally in the range 30°–35°; the experiments of Carrigy (1967, 1970) and Allen (1970c) reveal no consistent dependence on the intergranular fluid, a result contrary to popular opinion.

PARTICLE ENTRAINMENT FROM COHESIONLESS BEDS

Steady flows

As the aqueous discharge is gradually increased over a planed sand bed, a fairly definite flow condition is reached when grains begin to be entrained. This condition, marking the start of sediment transport, is called the *plane-bed threshold of particle motion.* Its analytical prediction, practical definition, and empirical characterization are enduring sources of research and controversy.

A simple analytical approach using time-averaged forces affords insight into the entrainment process. Consider in Fig. 2-3 the forces acting on a smooth, spherical particle of diameter D_1 and density σ immersed in a fluid of density ρ flowing parallel with a horizontal bed of smooth spheres of diameter D_0 and free separation distance s. Aside from reactions at points of contact, the forces acting on the sphere are its immersed weight F_G vertically

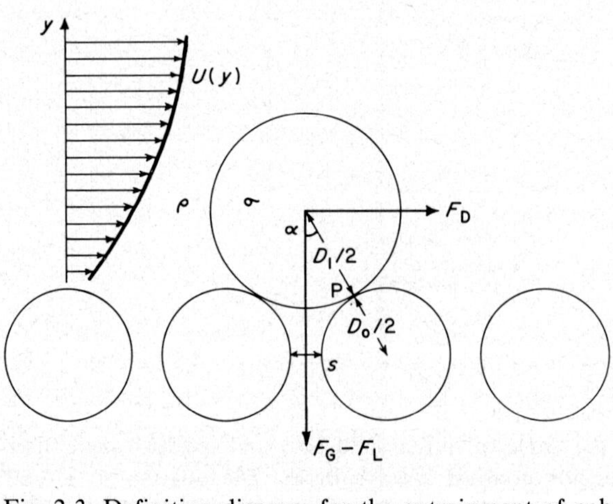

Fig. 2-3. Definition diagram for the entrainment of cohesionless grains from a cohesionless bed acted on by a steady fluid flow.

downward (positive), a fluid-applied lift F_L vertically upward, and a fluid-applied drag force F_D horizontally. Assuming that these forces act through the particle centre, by taking moments above the pivotal point P, the entrainment condition is found to be:

$$F_D \cos\alpha = (F_G - F_L)\sin\alpha \tag{2.1}$$

Now the nominal drag force is:

$$F_D = \frac{\pi}{4}\tau_{0(cr)} D_1^2 \tag{2.2}$$

and:

$$F_G = \frac{\pi}{6}(\sigma - \rho) g D_1^3 \tag{2.3}$$

where $\tau_{0(cr)}$ is the mean boundary shear stress at the threshold of movement, and g is the acceleration due to gravity. If $F_L = kF_D$, where k is a coefficient varying in a known way with the flow conditions controlling F_D, then:

$$\tau_{0(cr)} = \frac{2}{3}(\sigma - \rho) g D_1 \frac{\sin\alpha}{(\cos\alpha + k\sin\alpha)} \tag{2.4}$$

or:

$$\frac{\tau_{0(cr)}}{(\sigma - \rho) g D_1} = \frac{2}{3} \frac{\frac{1}{2}(D_0 + s)}{\left[\frac{D_1^2}{4} + \frac{D_1 D_0}{2} + \frac{D_0 s}{2} + \frac{s^2}{4}\right]^{1/2} + \frac{k}{2}(D_0 + s)} \tag{2.5}$$

in non-dimensional terms and with α expressed using D_1, D_0 and s.

Equations (2.4) and (2.5) reveal that the threshold stress increases with increasing density difference, particle diameter D_1, the diameter ratio D_0/D_1, and the free separation distance. The non-dimensional boundary shear stress appearing on the left in eq. (2.5) is effectively a constant for a given diameter ratio, provided that k is a constant and s is small compared with D_0 and D_1. Several workers give analyses on similar lines to the above, in some instances for horizontal beds (C.M. White, 1940; Kalinske, 1947; Bagnold, 1954b), and in others generalized for a sloping surface (Sundborg, 1956; LeFeuvre et al., 1970; Chen and Carstens, 1973; Everts, 1973; Howard, 1977). H.A. Einstein (1950) indicates the importance of lift forces in the movement of sediment; Helley (1969) and Inokuchi and Takayama (1973) explicitly include lift in their analyses of threshold conditions.

The motion-threshold can be specified otherwise than by the critical mean boundary shear stress. Introducing eq. (1.19), we can rewrite eq. (2.4) in terms of the critical depth-averaged flow velocity. Alternatively, eqs. (1.13) and (1.17) can be used to define the threshold in terms of the velocity at particle level, the so-called bottom velocity. The threshold is often conveniently defined using the critical value of the shear velocity, with the help of the definition appearing under eq. (1.13). However, the specification using shear stress undoubtedly has the most immediate physical meaning.

Being a boundary, the motion-threshold can be apprehended only in terms of the observations "I see no particle motion" and "I see particle motion" (Yalin, 1972). But what constitutes particle motion? A measure of agreement is at last emerging after many years during which investigators using different materials and apparatus disagreed widely between themselves. Neill and Yalin (1969) gave a practical definition of the threshold of motion which, although depending on the observation of moving grains, is quantitative and capable of consistent use. Paintal (1971) and Tsuchiya and Kawata (1970) have perhaps solved the problem altogether by defining the motion-threshold as that flow condition obtained when measurements of the sediment transport rate over a range of conditions are extrapolated to zero rate.

Laboratory and other measurements of the motion-threshold have led to many empirical descriptions of this condition. Some workers state the condition in terms of a critical velocity, either the mean velocity (Hjulström, 1935; C.A. Wright, 1936; Marsal, 1950; Menard, 1950a; Neill, 1967, 1968a, 1968b; Baker, 1973; Yang, 1973), the bottom velocity (Rubey, 1938; Mavis and Laushey, 1948; Helley, 1969), or the velocity at a fixed distance above the bed (Sundborg, 1956). J. Allen (1942) and Futterer (1977, 1978) express the initiation of motion of single particles, respectively cubes, and bivalve and gastropod shells, using the mean flow velocity. A shear-velocity criterion has appealed to other workers (Chepil, 1945a, 1945b; Inman, 1949; Bagnold, 1954b; Horikawa and Shen, 1960; Tsuchiya and Kawata, 1970; Greely et al., 1974; Iversen et al., 1976; Miller and Komar, 1977; Miller et al., 1977), particularly for sand in air. The most widely used criterion, however, is that of Shields (1936), who expressed the threshold of motion using the group on the left in eq. (2.5), now called the Shields-Bagnold non-dimensional boundary shear stress θ_{cr}, and the grain Reynolds number $Re = U_{*(cr)}D_1/\nu$, where $U_{*(cr)}$ is the shear velocity at the motion-threshold. Figure 2-4 gives curves applicable to mineral-density sand in water and to sand in air based on the best available information, largely summarized by Miller et al. (1977). Many other workers report or summarize results applicable to sand–water (Shields, 1936; C.A. Wright, 1936; Vanoni, 1964; Bagnold, 1966; Birkeland, 1968; Grass, 1970; S.J. White, 1970; Inokuchi and Takayama, 1973; Baker, 1973; Mantz, 1973, 1977b; Baker and Ritter, 1975). Sternberg (1971, 1972) finds a Shields-type criterion to be helpful in marine environments where currents can be regarded as unidirectional. Miller and Komar (1977) present criteria which they claim should apply under extreme conditions, as on Mars, where the atmosphere is tenuous, and in the deep oceans, where low-density foraminifera are widely transported. Ward (1969) gives a single criterion valid for large density differences.

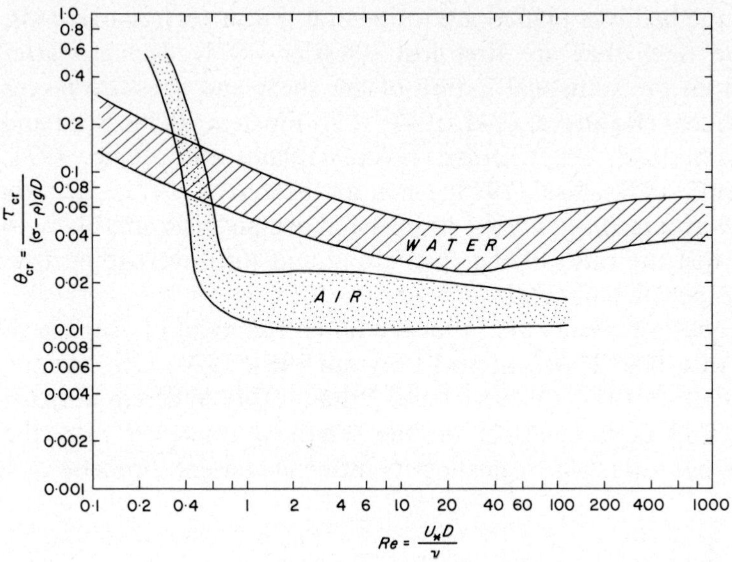

Fig. 2-4. Summary diagram of the experimental relationship between non-dimensional critical boundary shear stress for particle entrainment in water and air and particle size expressed as the particle Reynolds number. The shaded zones indicate the limits of the bulk of the experimental scatter for each medium. After Miller et al. (1977), who give full details of the extensive data base.

Figure 2-4 displays three regions. At $Re \lesssim 2$, when the grains lie deep within the viscous sublayer, the threshold stress decreases with Reynolds number. At sufficiently large Reynolds numbers, say $Re \gtrsim 100$, when the flow is fully rough, the stress appears to become constant. Between is a transition region in which the threshold stress falls to a minimum value of approximately 0.03, the flow changing from smooth to fully rough. An explanation for the shape of the curve may lie in the changing influence of lift. That this is a significant force in entrainment, under some conditions similar to or greater in magnitude than the drag force, has been amply demonstrated (Einstein and El-Samni, 1949; Chepil, 1958, 1961; Coleman, 1967; Benedict and Christensen, 1972; Cheng and Clyde, 1972; Lyles and Woodruff, 1972; Aksoy, 1973; Willetts and Drossos, 1975; Davies and Samad, 1978). Under conditions of smooth flow, the lift acts downward, augmenting the particle weight, with the result that the non-dimensional stress is relatively large. When the flow is rough the lift force seems to act upward.

The scatter portrayed in Fig. 2-4 is contributed by many factors, not least of which are the difficulties encountered in apprehending the threshold of motion. Sand and comparable grains of average or larger size that stand proud of the bed tend to be moved more readily than others (Fenton and Abbott, 1977), suggesting that the sorting of the bed material and the manner in which the bed was planed are influential. From very coarse beds, however, it is the fines that are first lost (Gessler, 1971). Much scatter probably arises from the statistical nature of the shear and pressure forces that cause entrainment (Kalinske, 1942, 1947; H.A. Einstein, 1950; Bisal and Nielsen, 1962; Sutherland, 1967; Grass, 1970; Williams and Kemp, 1971; Lyles and Woodruff, 1972; Bisal, 1973; Chen and Carstens, 1973). Criteria framed using time-averaged quantities, as in the above analysis, are likely to be unique only when the ratio of the fluctuating and time-average parts is uninfluenced by experimental conditions.

There is some question about the influence on entrainment of seepage of fluid into or out of a bed (Harrison and Clayton, 1968, 1970; C.S. Martin, 1970; Watters and Rao, 1971; Oldenziel and Brink, 1974). Whereas seepage has a significant and definite effect on the sediment transport rate, the results to date are contradictory regarding its influence on entrainment.

Unsteady flows

Particle entrainment within the oscillatory boundary layers due to wind-waves and the tide involves explicitly unsteady forces, irrespective of whether the flow is also turbulent. As well as lift, drag, and body forces, there are others related to horizontal gradients of fluid pressure, and to the acceleration of the fluid past the bed. Although of theoretical importance, analytical criteria for the plane-bed threshold of sediment motion in oscillatory boundary layers incorporating these forces have not surprisingly met with but little success (Taylor, 1946; Manohar, 1955; Eagleson et al., 1957; Eagleson and Dean, 1961; Kamphuis, 1967; Carstens et al., 1969; Silvester, 1970; Cacchione and Southard, 1974). An empirical approach, or one based on dimensional considerations, has appealed more.

Several non-dimensional groups can be formed from the variables governing particle entrainment in oscillatory flows, for example:

$$\frac{\sigma}{\rho}, \frac{d}{D}, \frac{U_{max}D}{\nu}, \frac{\rho U_{max}^2}{(\sigma-\rho)gD}, \frac{\rho U_{max}}{(\sigma-\rho)gT}$$

in which σ and ρ are the particle and fluid densities, respectively, d the orbital diameter of a near-bed water particle, D the particle diameter, U_{max} the maximum orbital velocity of the near-bed fluid, ν the fluid kinematic viscosity, g the acceleration due to gravity, and T is the wave period. The second group is the relative orbital diameter, the third a Reynolds number, the fourth a term proportional to the non-dimensional boundary shear stress and corresponding to the left-hand group in eq. (2.5), and the fifth the ratio of acceleration to gravity forces acting in the boundary layer. Bagnold (1946), apparently the first to investigate the matter, correlated laboratory measurements on particle entrainment by waves using a dimensional criterion practically identical with the non-dimensional stress. Others have also used the non-dimensional stress (Komar and Miller, 1973; Sternberg and Larsen, 1975; Dingler, 1979). Komar and Miller (1974, 1975a), with Madsen and Grant (1975), attempted plots in the same general form as Shields (1936) (Fig. 2-4). Rance and Warren (1969) introduced the use of the final group listed above as a means of correlating laboratory data on the motion-threshold beneath waves, and the same, or a closely similar, group was employed by Carstens et al. (1969), Chan et al. (1972), and Komar and Miller (1973). A variety of other correlations have been suggested (Manohar, 1955; Vincent and Ruellan, 1957; Vincent, 1958; Goddet, 1960; Horikawa and Watanabe, 1967; Lofquist, 1978; Dingler, 1979). Hammond and Collins (1979) consider entrainment in combined oscillatory and steady flow.

It is as difficult experimentally to define the motion-threshold beneath oscillatory boundary layers as beneath steady ones. Manohar (1955) distinguished between "initial motion", when just a few projecting grains are disturbed, from "general motion", at values of U_{max} 10–15% greater, when

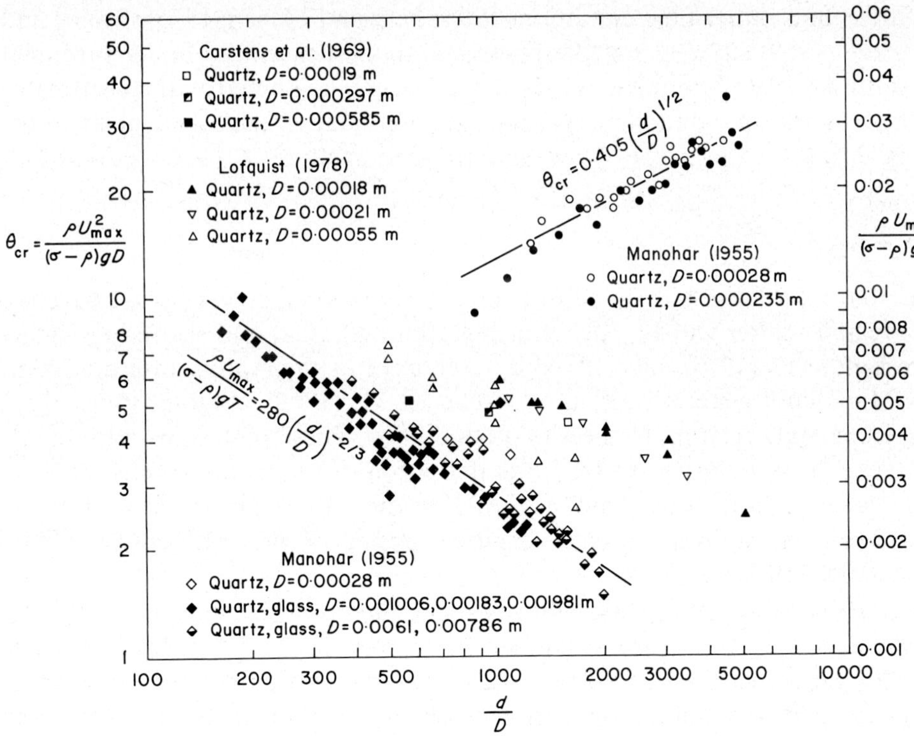

Fig. 2-5. Experimental data on the threshold conditions for particle entrainment under the action of oscillatory flows. The curve to the upper right defines the criterion for entrainment under laminar conditions, and is based on the experimental data summarized on the right. The criterion under turbulent conditions is given by the graph toward the lower left, based on the data summarized to the left (those of Carstens et al. and Lofquist appear for comparison only).

the entire top layer of grains is set moving. Rance and Warren's (1969) conception of initial motion is comparable with that of Manohar. Carstens et al. (1969) approached the problem similarly to Neill and Yalin (1969), identifying the motion-threshold with the movement of 10% of the surface grains. Lofquist's (1978) criterion—the motion of 10–20 grains per square centimetre—is sensitive to grain size and therefore unsatisfactory.

Experiment shows that mineral-density solids can be entrained in laminar as well as turbulent aqueous oscillatory boundary layers, contrary to experience with unidirectional flows. Figure 2-5 shows Manohar's (1955) data for the initial motion of two sizes of quartz sand and glass beads on a planed bed in a laminar boundary layer in water. As Komar and Miller (1973) point out, these data are satisfactorily correlated using the non-dimensional stress and the relative orbital diameter. The representative empirical formula is:

$$\frac{\rho U_{max}^2}{(\sigma-\rho)gD} = 0.405\left(\frac{d}{D}\right)^{1/2} \tag{2.6}$$

for $900 \leqslant d/D \leqslant 5000$. A similar correlation has been found applicable to quartz sands under field conditions (Sternberg and Larsen, 1975). Rance and Warren (1969) showed that the plane-bed motion-threshold under turbulent conditions is well-defined by $\rho U_{max}/(\sigma-\rho)gT$ combined with the relative orbital diameter. Manohar's (1955) data for quartz sands and glass beads (Fig. 2-5) afford practically the same trend, namely:

$$\frac{\rho U_{max}}{(\sigma-\rho)gT} = 280\left(\frac{d}{D}\right)^{-2/3} \tag{2.7}$$

for $160 \leqslant d/D \leqslant 2000$. His results for general motion, however, lie above the curve and scatter widely. Also plotting above eq. (2.7) and scattering widely are Lofquist's (1978) data, based on a concept of the motion-threshold akin to Manohar's general motion. Carstens et al. (1969) obtained only three measurements relating to threshold conditions. These plot above Manohar's curve, as may be expected from the definition of the motion-threshold used by these authors, but yield a flatter graph. A rough criterion for the turbulent entrainment of near-spherical grains of mineral-density is that $D \geqslant 0.0005$ m.

Little is known of the behaviour of markedly non-spherical particles under wave-action, such as are abundant in carbonate environments. Bosence (1976), however, has studied in a wave tank the entrainment of the branching, free-living coralline alga *Lithothamnium corallioides,* finding the more densely branched forms to be the most easily entrained.

EROSION OF COHESIVE BEDS

Fluid-stressing

The three mechanisms (Allen, 1971c) by which a bed of cohesive mud or rock may lose particulate matter are: (1) *fluid-stressing,* (2) *corrasion,* and (3) *cavitation erosion.*

Fluid-stressing involves in currents of normal speed the direct action chiefly of tangential fluid stresses upon the bed, which responds primarily according to its strength. Much as a cohesionless sediment, soupy mud is eroded flake-by-flake, or at least in only tiny aggregates that may correspond in some degree to primary floccules (Allen, 1969a). Muds that are neither soupy nor strong enough to be moulded freely in the hand respond plastically to sufficiently powerful currents (Migniot, 1968; Allen, 1969a; Einsele et al., 1974), the bed becoming shaped into transverse and/or streamwise ridges, from the crests of which mud in shreds and larger masses is repeatedly torn. Strong beds respond by brittle fracture, and require power-

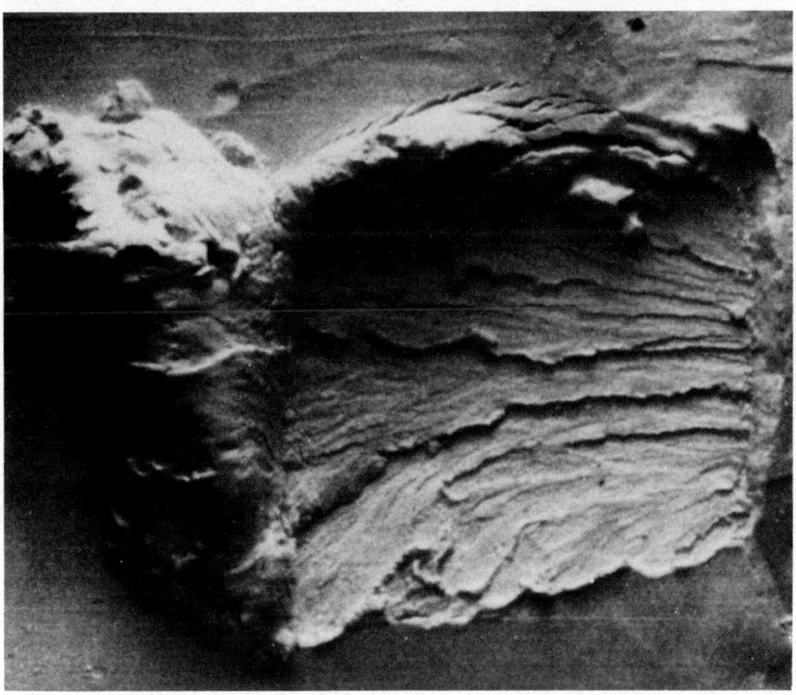

Fig. 2-6. Fluid stressing illustrated by the effect of an aqueous current of velocity=0.50 m s^{-1} and depth=0.065 m acting from right to left on a bed of moderately stiff kaolinite mud. The surface of parting carries a plumose mark. Length of bed shown approximately 0.085 m. Photograph courtesy of G. Einsele (see Einsele et al., 1974); reproduced by permission of the International Association of Sedimentologists.

ful currents to cause erosion. Muds of this type, strong enough to be freely moulded in the hand, fail in the form of thin, narrow sheets elongated parallel with flow that spring from the bed at some point of weakness, and are thereafter rolled up downcurrent like a carpet (Fig. 2-6). The failure-surface beneath the sheet, formed parallel with the lamination or grain-fabric in the sediment, characteristically bears plumose markings (N.J. Price, 1966; Cegla and Dzulynski, 1967; Hobbs et al., 1976), which give it a frondescent appearance and often a deeply crinkled margin. There is a strong resemblance (Einsele et al., 1974), in both appearance and orientation relative to flow, between this embayed surface and the so-called cabbage-leaf marks found as projecting moulds beneath turbidites (Kuenen, 1957; Dzulynski and Slaczka, 1958; Ten Haaf, 1959; Dzulynski and Sanders, 1962a, 1962b; Dzulynski, 1963; Dzulynski and Walton, 1963, 1965; Dzulynski, 1965). These sole markings probably record the tearing of strips of mud from a relatively strong bed, rather than the forceful injection of sediment from the base of a turbidity current, as favoured by Dzulynski and Walton.

So many ill-understood factors appear to control the response of a cohesive mud bed to an aqueous current that a satisfactory unique criterion for the threshold of motion remains to be devised, though Lambermont and LeBon (1978) make a brave attempt. Laboratory experiments suggest, however, that the most important factor is the yield stress of the bed, commonly estimated as its approximate surrogate, the inverse of the water content (Dunn, 1959; Smerdon and Beasley, 1959; Moore and Masch, 1962; Flaxman, 1963; Masch et al., 1965; Postma, 1967; Migniot, 1968; Partheniades and Paaswell, 1970; Peirce et al., 1970; Partheniades, 1971; Southard et al.,

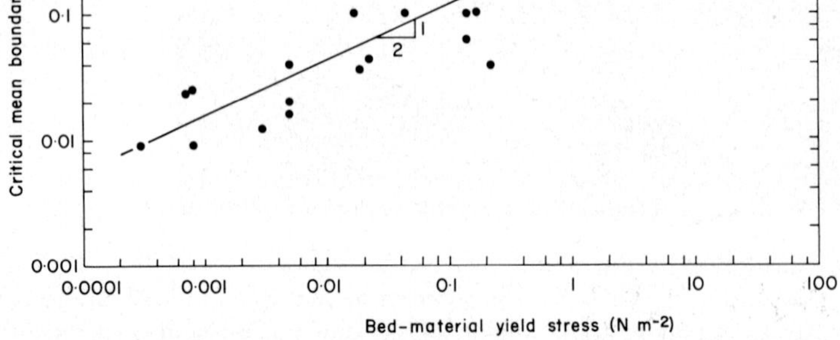

Fig. 2-7. Experimental erosion threshold for muddy sediments as a function of yield strength. Data of Migniot (1968).

1971; Lonsdale and Southard, 1974; Owen, 1977; Ariathurai and Arulanandan, 1978; Young and Southard, 1978). In Migniot's experiments (Fig. 2-7), the critical mean boundary shear stress for the erosion of a mud bed in water increases approximately as the square root of the yield stress, up to a yield stress of about 3 N m^{-2}, and approximately as the first power at larger stresses. The scatter is considerable, particularly at the lower yield stresses, and the threshold stress is fairly strongly influenced by: (1) the sediment provenance (i.e. grain size and mineralogy) (Migniot, 1968; Peirce et al., 1970), (2) the electrolyte in which deposition occurred (Migniot, 1968), (3) the presence of organic carbon and the extent of reworking of the sediment by organisms (Young and Southard, 1978), (4) temperature, an increase of which lowers the threshold stress (Ariathurai and Arulanandan, 1978), and (5) the time elapsed since deposition (Southard et al., 1971).

Theory and experiment show that a mud bed experiences erosion at an increasing rate as the applied fluid stress is raised above the threshold (Partheniades, 1965, 1972; Grissinger, 1966; Kendrick, 1972; Raudkivi and Hutchinson, 1974; Ariathurai and Arulanandan, 1978; Lambermont and LeBon, 1978). The precise form of the relationship remains uncertain, and the meagre published experimental studies suggest either a simple linear trend, a non-linear variation, or two linear laws joined at a critical value of the applied stress. As with the threshold stress, sediment provenance, water temperature, and the nature of the electrolyte all influence the erosion rate. The effect of increasing temperature on a particular mud is not always to raise the erosion rate for a given applied stress (Raudkivi and Hutchinson, 1974; Ariathurai and Arulanandan, 1978).

Corrasion

A cohesive bed experiences erosion through the mechanism of corrasion when pieces are either cut or broken from it as the result of the impingement of fluid-driven sedimentary particles acting as tools. Geologists have made a modest contribution to knowledge of this process (Blake, 1855; Thoulet, 1887; Hume, 1925; Kuenen, 1928, 1960; Alexander, 1932; Schoewe, 1932; Sharp, 1964), with our understanding coming mainly from the work of mechanical engineers concerned with sand-blasting as a finishing or shaping process, or as a cause of wear in industrial plant (Soo, 1977).

Bitter (1963a) indicated two modes of response of cohesive beds to particle impingement. In the brittle mode, resulting in *deformation wear,* particle impact fractures the bed, with the result that a fragment is knocked from it. In the ductile mode, affording *cutting wear,* the impinging grain acts like a chisel and cuts a sliver from the surface. Under certain conditions, the two modes of wear occur simultaneously, not only on metals and plastics (e.g. Bitter, 1963b; Finnie et al., 1967; Neilson and Gilchrist, 1968a; Tabakoff et al., 1979), but also on rock-like beds (Vickers et al., 1968). Sheldon

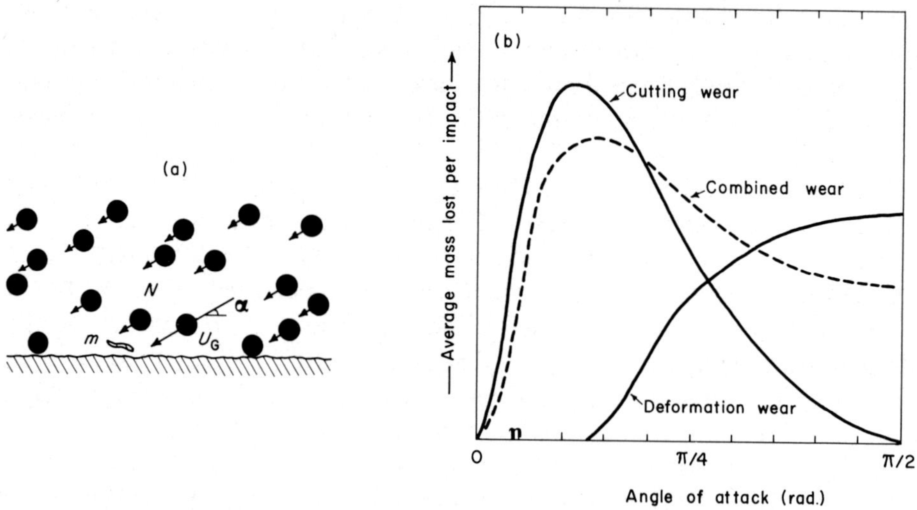

Fig. 2-8. Entrainment of material from cohesive beds (corrasion). a. Definition diagram for corrasion. b. Schematic variation of corrasion rate with angle of particle attack and mode of wear.

and Finnie (1966a), reviewing work by Klemm and Smekal (1941), Smekal and Klemm (1951), Puchegger (1952), and M.C. Shaw (1954), emphasised that the mode of erosion is strongly influenced by the size and velocity of the impinging grains. When particle velocity and size are sufficiently small, nominally brittle materials erode in the ductile rather than the brittle mode. The critical conditions increase, however, with increasing material hardness.

Consider in Fig. 2-8a the effect on a cohesive bed of grains of a uniform number concentration N per unit volume approaching at a uniform velocity U_G on paths inclined at a uniform angle of attack α. In the general case, when both deformation and cutting wear occur, the surface is eroded at a rate given by (Allen, 1971c):

$$\frac{dM}{dt} = -NU_G \sin \alpha (m_D + m_C) \tag{2.8}$$

in which M is the mass lost per unit surface area, t is time, m_D is the average mass lost per particle impact in deformation wear, and m_C the average loss per impact in cutting wear. Assuming for the moment that the losses per impact are independent of particle properties, the erosion rate should increase with $NU_G \sin \alpha$, which amounts to the frequency of impacts. Finnie (1962) and Bitter (1963a, 1963b) showed theoretically, however, that m_D and m_C are complex functions of both the properties of the bed and the particle immersed weight, approach velocity, and angle of attack (Fig. 2-8b; see also Laitone, 1979). In pure cutting wear, the average loss per impact is a maximum at an angle of attack close to 20–25°, the maximum loss increasing with particle immersed weight and velocity, but is zero at $\alpha = 0°, 90°$.

The maximum occurs at $\alpha = 90°$ in pure deformation wear, and decreases with smaller angles, to reach zero at a critical angle of attack controlled by particle and bed properties. In contrast, Head and Harr (1970) and Gibbings (1971) offer models of corrasion based on empirical or dimensional considerations.

The theoretical conclusions of Finnie (1962) and Bitter (1963a, 1963b) are well supported in many experimental studies (Stoker, 1949; Gladfelter et al., 1953; Finnie, 1960a, 1960b, 1972; Bitter, 1963a, 1963b; Sheldon and Finnie, 1966a, 1966b; Finnie et al., 1967; Neilson and Gilchrist, 1968a, 1968b; Goodwin et al., 1969; Raask, 1969; Sage and Tilly, 1969; Tilly, 1969, 1973; Sheldon, 1970; Smeltzer et al., 1970; Tilly and Sage, 1970; Mason and Smith, 1972; Sheldon and Kanhere, 1972; Hutchings et al., 1976; Mills and Mason, 1977a, 1977b; Finnie and McFadden, 1978; Christman and Shewman, 1979; Tabakoff et al., 1979). Many materials responded to impacting particles by a combination of deformation with cutting wear, affording loss-attack curves of intermediate form (Fig. 2-8b).

Combining all these results, it can be seen that the corrasion of hard, brittle natural materials, for example, crystalline or well-cemented rocks, should proceed most rapidly when the impinging grains have a large effective weight and approach on steep paths at large velocities, whereas mud beds, which may be classed as ductile, should respond most rapidly at small to moderate angles of attack. Whatever the mode of erosion, however, the tools, or at least their edges or corners, must approach the surface at a non-zero angle. Such motions can arise while the particles are caught up in fluid turbulence, when grains are dispersed in a separated flow reattaching to a surface (Fig. 1-9), and when comparatively large fluid-driven particles roll, tumble, or saltate.

Cavitation erosion

Cavitation is the appearance of bubbles of vapour within a liquid in non-uniform motion relative to a solid surface, as a consequence of the dynamical action of the flow (Eisenberg, 1961; Batchelor, 1967; Knapp et al., 1970). They emerge where the pressure in the fluid falls below the vapour pressure of the liquid at the prevailing temperature, but collapse and disappear on being carried into regions of higher pressure. By developing and expanding, the streams of vapour-filled cavities relieve the negative pressure in the liquid, which is generally unable to withstand tension. However, in order for cavitation to occur, the liquid must be in sufficiently violent relative motion with either a projecting obstruction on the flow boundary or with some immersed body. It can be shown from Bernoulli's theorem for the pressure in an inviscid fluid (Prandtl, 1952; Batchelor, 1967) that cavitation results wherever:

$$\frac{p - p_v}{\frac{1}{2}\rho U^2} < (k^2 - 1) \tag{2.9}$$

in which p is the absolute pressure in the fluid, p_v is the vapour pressure of the liquid, ρ the fluid density, U the flow velocity far upstream, and $k > 1$, equals the velocity of the accelerated fluid near the obstruction or immersed body relative to U. The development of cavitation is therefore favoured by a large fluid velocity and by severely restrictive obstructions or immersed bodies. It can also be shown that, in the case of free-surface flows, cavitation is favoured by shallow depths.

Cavitation will not occur in water unless the flow velocity measures many metres per second. It is therefore absent from rivers, except perhaps locally under extreme flood conditions, and also from tidal and wave-related currents. The occurrence of cavitation under natural conditions almost certainly is predominantly in englacial and subglacial drainage tunnels, where meltwater is driven hydrostatically rather than gravitationally, and therefore can reach exceptionally large velocities.

Cavitation erosion ensues where the vapour bubbles collapse and disappear in close proximity to a solid surface. Two phenomena attend the collapse: (1) the spread of a damaging shock wave and, if the cavity lies in a region of normal pressure gradient, as near a flow boundary or in a vortex, (2) the creation of a small but exceptionally violent jet of water which shoots from the high-pressure to the low-pressure side of the cavity. Theoretical and experimental work show that the jets resulting from cavity-collapse can exert on the flow boundary impulses sufficiently large as to break from it solid fragments (Shutler and Mesler, 1965; Benjamin and Ellis, 1966; Brunton, 1966; Leach and Walker, 1966; Joliffe, 1968). The cavitation erosion of metal objects, such as ship's propellers and pipe bends, and of hydraulic structures made of concrete or stone, is expressed by deep and irregular pits, commonly densely arrayed and large in size (Schröter, 1933; W.H. Price, 1947; Rao and Thiruvengadam, 1961; Leith and McIlquham, 1962; F.R. Brown, 1963; V.E. Johnson, 1963; Kenn, 1966, 1968; Kenn and Minton, 1968; Knapp et al., 1970). Under field conditions, however, corrasion often accompanies cavitation erosion, when the resulting forms tend to be relatively smooth. The factors that control the rate of cavitation erosion are little understood, but several investigators conclude that the rate is a very steeply increasing power of the flow velocity (J.M. Hobbs, 1966; Shal'nev et al., 1966; Kenn, 1968).

Cavitation erosion may explain certain naturally occurring erosion forms (Hjulström, 1935; Barnes, 1956; Dahl, 1965). Its occurrence is certainly possible, under the restrictions noted above, though it is doubtful if erosion by this mechanism in natural environments is ever unaccompanied by corrasion.

PARTICLE SETTLING

General

Bassett (1888, 1910), Odar and Hamilton (1964), and Hjelmfelt and Mockros (1967) find that a rigid particle released into a fluid is acted on by five forces: (1) that which accelerates the grain, (2) the resultant pressure, (3) the fluid drag on the particle, (4) the inertia of the virtual mass of fluid travelling with the particle (added mass), (4) the force due to the history of motion of the grain (Bassett term), and (5) the particle immersed weight (body force). A grain so released is observed first to accelerate, but ultimately to assume a uniform motion which, if the fluid is at rest, is measured by the grain *terminal falling velocity.* The magnitude of this velocity strongly influences particle behaviour during transport and deposition, and is determined by grain size, density, shape and surface texture, and by the viscosity, density, and intensity of turbulence in the fluid. Reviews and discussions of this important topic are given by Happel and Brenner (1965), from a primarily theoretical standpoint, and by Torobin and Gauvin (1959a, 1959b, 1959c, 1960a, 1960b), Soo (1967), Graf (1971), Yalin (1972), Raudkivi (1976), and Clift et al. (1978) from a more practical position.

Spherical particles

The simplest case of all is that of the steady, uniform motion of a smooth, rigid, spherical particle in an unbounded, stationary, Newtonian fluid, for the forces acting reduce to the particle immersed weight and the fluid drag. The first is:

$$F_G = \frac{4\pi}{3} a^3(\sigma - \rho)g \tag{2.10}$$

where σ and ρ are the particle and fluid densities respectively, g is the acceleration due to gravity, and a is the particle radius. The fluid drag acting on the projection area of the particle normal to the line of motion is:

$$F_D = C_D \pi a^2 \frac{\rho W^2}{2} \tag{2.11}$$

in which C_D is a non-dimensional drag coefficient, to be either calculated theoretically or determined empirically, and W is the falling velocity. Note the correspondence between eq. (2.11) for an immersed body and eq. (1.33) for flow in an open channel or pipe. Equating the body and drag forces, we derive:

$$W^2 = \frac{8}{3} \frac{a(\sigma - \rho)g}{C_D \rho} \tag{2.12}$$

whence falling velocity increases with grain size and relative density, but declines with increasing drag coefficient and fluid density. The remaining problem is to find the drag coefficient.

Neglecting fluid inertia, Stokes (1851) deduced theoretically that:

$$F_{\mathrm{D}} = 6\pi\eta a W \tag{2.13}$$

in which η is the fluid dynamic viscosity, whence from eq. (2.11):

$$C_{\mathrm{D}} = \frac{24}{Re} \tag{2.14}$$

in which $Re = 2aW\rho/\eta$ is the grain Reynolds number written in terms of diameter. Substituting into eq. (2.12), we obtain:

$$W = \frac{2}{9}\,\frac{a^2(\sigma - \rho)g}{\eta} \tag{2.15}$$

known as Stokes' law, by which the falling velocity is proportional to the square of the radius. A similar relationship applies to sufficiently small permeable grains, a model appropriate for clay-mineral floccules and some biogenic debris (Joseph and Tao, 1964; Ooms et al., 1970; Sutherland and Tan, 1970; Singh and Gupta, 1971; Yamamoto, 1971; I.P. Jones, 1973; Neale et al., 1973; Verma and Bhatt, 1974, 1976; Nir, 1976). The drag force on a permeable sphere corresponds to that on an impermeable sphere of reduced radius, where the actual and reduced radii are linked through the permeability.

Because inertia is neglected, Stoke's law is strictly valid only at very small Reynolds numbers, though experience shows it to be a good approximation up to $Re = 1$. Laboratory experiments reveal that the drag coefficient ceases to be inversely proportional to the Reynolds number as inertial forces increase in relative importance, but at sufficiently large Reynolds numbers, become essentially constant (Fig. 2-9) (Schlichting, 1961; Maxworthy, 1965; Pruppacher and Steinberger, 1968; Bailey and Hiatt, 1972; A.B. Bailey, 1974; Miller and Bailey, 1979). Hence from eqs. (2.12) and (2.14) the dependence of the falling velocity on particle size must change from a square law at small Reynolds numbers (Stokes range) to a square-root law at large Reynolds numbers, with a broad transition region between.

Attempts by analytical or numerical means to calculate the drag coefficient of spheres at Reynolds numbers above the Stokes range have met with some success up to $Re = 400$ (Lamb, 1911; Oseen, 1927; Goldstein, 1929; Proudman and Pearson, 1957; Chester et al., 1969; Le Clair et al., 1970). Drag coefficients at higher Reynolds numbers have been obtained empirically. Schiller and Naumann (1933) proposed using experimental data a falling velocity law for $Re \leqslant 1000$ that amounts to eq. (2.12) with:

$$C_{\mathrm{D}} = \frac{24}{Re}(1 + 0.150Re^{0.687}) \tag{2.16}$$

and Gibbs et al. (1971) give another applicable over a similar range. Rubey (1933) suggested a formula valid for all Reynolds numbers in which:

$$C_{\mathrm{D}} = \frac{24}{Re} + 2 \tag{2.17}$$

However, agreement with experiment is poor (Graf and Acaroglu, 1966), though R.L. Watson (1969) claims some improvement. Schiller and Naumann (1933) multiplied the expression equating the forces that led to eq. (2.12) by $(2a\rho/\eta)^2$, to obtain:

$$C_{\mathrm{D}} Re^2 = \frac{4}{3}\,\frac{(\sigma-\rho)g}{\rho}\,\frac{D^3}{\nu^2} \tag{2.18}$$

in which $D = 2a$ is the particle diameter, $\nu = \eta/\rho$ is the fluid kinematic viscosity, and the right-hand group is also non-dimensional. Yalin (1972) and Raudkivi (1976) use these forms in empirically based graphs from which, given the other quantities, either the particle diameter or falling velocity can be obtained. Figure 2-10 is Yalin's graph for falling velocity.

The variation in drag coefficient with Reynolds number shown in Fig. 2-9 reflects changes in flow pattern around a sphere and in the character of the boundary layer developed on it. Experiments made with spheres and cylinders (Roshko, 1954a, 1954b, 1955, 1961; Pruppacher et al., 1970) point to the gradual development with increasing Reynolds number of a stable region of recirculating separated flow, then a regime in which vortices are

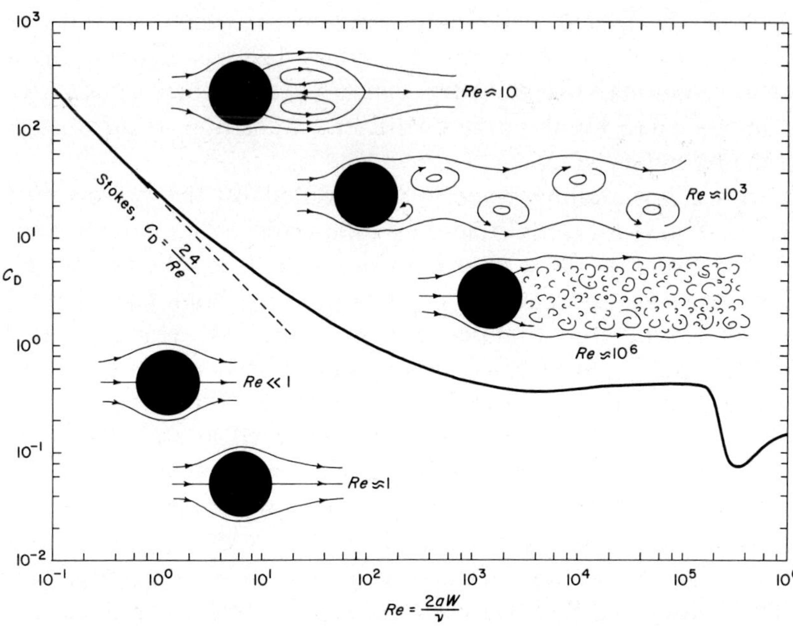

Fig. 2-9. Flow regimes for a spherical particle in relative motion with a fluid, superimposed on a graph for the drag coefficient as a function of particle Reynolds number.

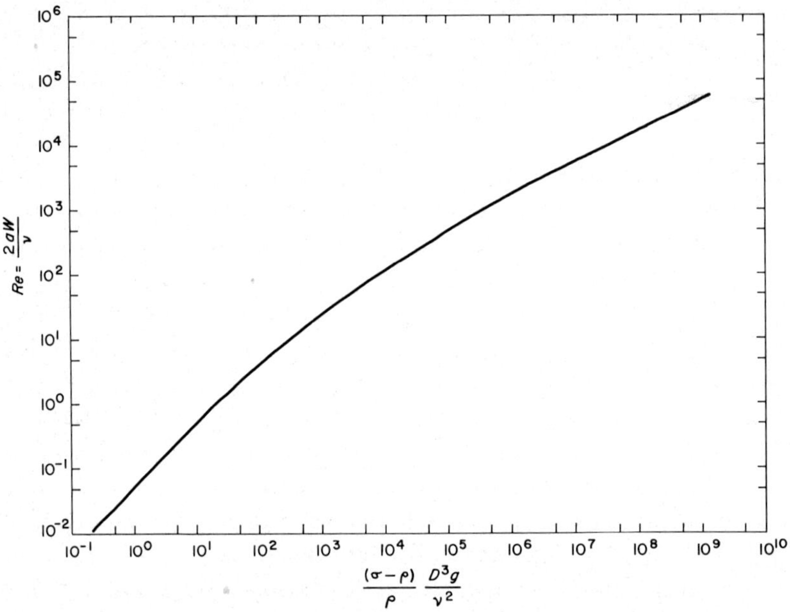

Fig. 2-10. The non-dimensional falling velocity of spherical particles as a function of particle and fluid properties (after Yalin, 1972).

regularly shed from the sphere, and finally a regime in which the wake is fully turbulent. Up to $Re = 3 \times 10^5$ the boundary layer is laminar on a smooth sphere in a stationary fluid. At this value, however, corresponding to a sudden dip in the curve for the drag coefficient, transition to turbulence occurs within the boundary layer.

It remains to see how falling velocity is modified by the presence of boundaries to the fluid, for example, settling-tube walls, a bed of sediment, or a fluid interface. Brenner (1961) showed theoretically that spheres obeying Stokes' law slow down on approaching a horizontal solid boundary or fluid interface (see also Happel and Brenner, 1965). The falling velocity is also reduced when grains settle adjacent to a single wall or between parallel walls in a vessel closed at the bottom (Fidleris and Whitmore, 1961; Happel and Brenner, 1965). The effect becomes increasingly marked as the particle is placed nearer to the wall or approaches the vessel in size.

Particles settle in non-Newtonian fluids at a slower rate than in the corresponding Newtonian ones, because of the influence of yield stress and other non-Newtonian properties (Slattery and Bird, 1961; Valentik and Whitmore, 1965; Ansley and Smith, 1967; Brookes and Whitmore, 1968; Ito and Kajiuchi, 1969; Whitmore, 1969; Lai, 1974; Pazwash and Robertson, 1975).

Non-spherical particles

Departure from spherical form can significantly affect the falling velocity of a particle and, as will be seen in a later chapter, its mode of settling. From this general standpoint, Happel and Brenner (1965) distinguish three classes of regular particle of interest in connection with sedimentation under natural conditions. *Spherically isotropic* particles have a form that is similarly related to three mutually perpendicular coordinate axes, for example, the sphere already considered, together with tetrahedra, octahedra, cubes, and cube-octahedra. Within the Stokes range such bodies fall vertically and stably in whatever orientation they initially possessed. A body which has three mutually perpendicular symmetry planes is called *orthotropic,* examples being ellipsoids, elliptical and hexagonal cylinders, and rectangular parallelepipeds. A third class comprises *bodies of revolution,* for example, right-circular cylinders, discs, double cones, and spherical caps and lenses. Many bodies of revolution and all orthotropic ones are said to be anisotropic, falling vertically only when oriented so that a principal axis of translation is parallel with the gravity field. The drag coefficient of non-spherical bodies is greater than that of the corresponding sphere, because of their increased ratio of surface area to mass and, in many cases, the presence of sharp edges and corners.

The drag coefficient of a non-spherical particle can be related to that of a sphere by introducing a settling coefficient K, defined by:

$$K = \frac{W_s}{W} \tag{2.19}$$

where W is the particle falling velocity and W_s is the falling velocity of a sphere of radius a_s with the same volume as the particle. With this definition, eq. (2.13) for the Stokes range becomes:

$$F_D = K(6\pi\eta a_s W) \tag{2.20}$$

while eq. (2.14) reads:

$$C_D = K\frac{24}{Re} \tag{2.21}$$

It will be seen that K has been defined so that $K = 1$ for a spherical particle. It will further be noticed that eqs. (2.16) and (2.17) are also of the form of eq. (2.21), which is a useful general equation when effects due to inertia, boundaries on the fluid, and particle shape and surface texture are to be accommodated. Other settling factors have been suggested, for example, by Pettyjohn and Christiansen (1948) on the basis of Wadell's (1934) particle sphericity, and by Heywood (1938, 1962) using particle projection areas.

Drag coefficients for a wide variety of spherically isotropic bodies are known experimentally over a large Reynolds number range as the result of work by Krumbein (1942a), Pettyjohn and Christiansen (1948), McNown and Malaika (1950), Heiss and Coull (1952), Chowdhury and Fritz (1959), and Graf and Mansour (1975). Tetrahedra deviate most from the sphere in behaviour, and cube-octahedra least.

Settling coefficients of many orthotropic particles and some bodies of revolution have been calculated theoretically at small Reynolds numbers, as well as determined experimentally over a wider range of conditions. Ellipsoids early attracted attention (Oberbeck, 1876; McNown, 1951; Aoi, 1955; Breach, 1961; Bowen and Masliyah, 1973; Chwang and Wu, 1976), with Masliyah and Epstein (1970) providing drag coefficients by a numerical procedure up to $Re = 100$. McNown and Malaika (1950) and Alger and Simons (1968) measured the behaviour of ellipsoids. Many workers have attacked theoretically the settling of circular cylinders and discs (Oberbeck, 1876; Bairstow et al., 1923; McNown, 1951; Aoi, 1955; Gupta, 1957; Marchildon et al., 1964b; Michael, 1966; De Mestre, 1973), with Dennis and Chang (1970) giving a numerical study of circular cylinders up to $Re = 100$. Experimental results for these shapes cover a broad range of Reynolds number (Krumbein, 1942a; McNown and Malaika, 1950; Heiss and Coull, 1952; Jones and Knudsen, 1961; Marchildon et al., 1964a; Christiansen and Barker, 1965; Alger and Simons, 1968; Gluckman et al., 1972). There are less data for elliptical cylinders (Tomotika and Aoi, 1953), prisms (Christiansen and Barker, 1965), and rectangular parallelepipeds (Heiss and Coull, 1952; Christiansen and Barker, 1965; Graf and Mansour, 1975). Fuller details appear in Happel and Brenner (1965).

Little is known of the settling of such bodies of revolution as circular pyramids (Gluckman et al., 1972; Futterer, 1977, 1978), double cones (McNown and Malaika, 1950), cones combined with spherical caps (Bowen and Masliyah, 1973), and spherical caps and concave-convex lenses (Payne and Pell, 1960; Dorrepaal, 1976; Dorrepaal et al., 1976). The latter are important models for many bivalve and brachiopod shells.

The settling of mostly irregular natural particles has been extensively studied experimentally. J.S. Owens (1911, 1912), Durand and Cohen de Lara (1953), and Graf and Acaroglu (1966) examined quartz sand, and Briggs et al. (1962) a variety of more dense naturally occurring minerals. Gravel particles have been investigated over a wide range of Reynolds numbers (Miller and M'Inally, 1936; Albertson, 1953; Alger and Simons, 1968; Stringham et al., 1969; Komar and Reimers, 1978). Fisher (1965) and Walker et al. (1971) measured the falling velocity of pyroclasts, and Berthois (1962) studied the settling of mica flakes. Biogenic debris has attracted less attention. Berthois and Calvez (1966), Maiklem (1968), Berger and Piper (1972), and Braithwaite (1973) studied the settling of a range of foraminifera. Futterer (1977, 1978) measured the falling velocity of bivalve and gastropod shells. Mehta et al. (1980) have also studied the settling of bivalve shells, quoting drag coefficients and shape factors. The settling of coral, bryozoan, and algal fragments, ranging from platy to cylindrical, was described by

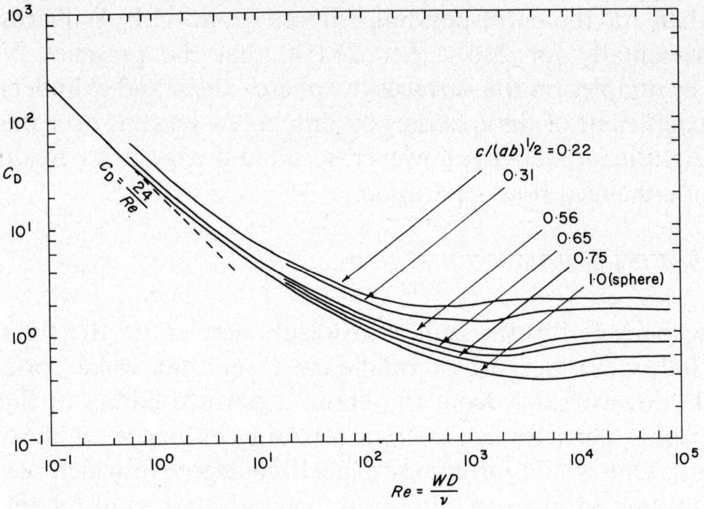

Fig. 2-11. Drag coefficient as a function of particle Reynolds number for spheres and for non-spherical particles characterized by the Corey shape factor (after Komar and Reimers, 1978).

Maiklem (1968) and Braithwaite (1973). However, data on biogenic particles are rarely given in terms of a drag coefficient, no doubt partly because of the difficulty of describing in a consistent manner the size of these often complex bodies.

In illustration of the results obtained with natural debris, Fig. 2-11 shows a graph for approximately ellipsoidal gravel particles prepared by Komar and Reimers (1978), using their own data for settling in glycerol ($Re < 1.5$) and that of Alger and Simons (1968) covering a range of much larger Reynolds numbers. The parameter is the Corey shape factor, a kind of settling coefficient and equal to $c/(ab)^{1/2}$, where a, b and c are respectively the long, intermediate and short particle axes. The drag coefficient and Reynolds number appear in terms of the nominal particle diameter $D = (abc)^{1/3}$. Departure from spherical form evidently increases the drag coefficient much more at Reynolds numbers greater than 10^3 than at intermediate values or in the Stokes range.

Surface roughness

The surfaces of most naturally occurring sedimentary particles possess a small-scale roughness, resulting from either breakage around crystal or grain boundaries, or from various mechanical and chemical effects associated with transport (e.g. Klein, 1963a; Krinsley and Doornkamp, 1973). One consequence of such roughness is slightly to increase the drag coefficient and, as Graf (1971) points out, to promote transition to turbulence at a lower

Reynolds number than for the corresponding smooth grain. G.P. Williams (1966) found experimentally for $230 < Re < 26{,}000$ that the presence of substantial grooves or dimples on the surfaces of spheres, discs and cylinders increased the drag coefficient of these bodies by only a few percent over the smooth particle. A significant increase, however, resulted if edges were made sharp, on account of enhanced flow separation.

Effects due to neighbouring particles

Because a sinking particle thrusts aside and locally accelerates the fluid through which it falls, and because a boundary layer and wake form respectively on and "downstream" from the grain, a particle falling in the presence of neighbouring particles can exert a retarding influence on them and they in turn on it. One would intuitively expect the degree to which one particle influences another to increase with grain concentration, and for the final effect to be a decrease in the falling velocity of the individual particle but an increase in the "viscosity" of the mixture. These intuitions are correct, but the component effects are complicated, and can seldom be calculated (Happel and Brenner, 1965). Their gross consequences under practical conditions are best treated empirically.

An empirical correlation of major importance is the Richardson-Zaki equation (Richardson and Zaki, 1954; Maude and Whitmore, 1958) for the falling velocity of monodisperse spherical particles, written as:

$$W = W_{\infty}(1 - C)^{n} \tag{2.22}$$

Here W is the falling velocity of a particle in the dispersion, W_{∞} is the falling velocity of one of the particles settling alone in the unbounded, stationary fluid, and C is the fractional particle volume concentration. The exponent n varies with the Reynolds number based on W_{∞}, being 4.65 for $Re \leqslant 1$, approximately 2.3 for $Re \geqslant 103$, and taking intermediate values at Reynolds numbers between. Equation (2.22) shows that the particle falling velocity decreases rapidly with increasing density of neighbours, the effect being strongest in and near the Stokes range.

SOME GENERAL CONCEPTS OF SEDIMENT TRANSPORT

Sediment transport is the general process whereby sedimentary particles are conveyed essentially horizontally from one place to another, either as a mixture of grains with an independent transporting fluid (e.g. as in a river, wind) or, when sufficiently densely aggregated, and usually with a fluid in addition, so as to form the transporting agent itself (e.g. mass-flows). The sediment transport rate is measured as the quantity of grains passed along a lane of unit width in unit time, where "quantity" can be defined in several ways. The transport is either *steady* or *unsteady* accordingly as the rate measured at a point remains constant or changes with time. Similarly it is either *uniform* or *non-uniform,* depending on changes with distance at a fixed time.

Without prejudice as to whether the transport is determinate in terms of grain, fluid, and flow properties, the instantaneous transport rate is simply the product of the average instantaneous grain transport velocity $\overline{U}_G$, and the total quantity of grains above unit bed-area parallel to the direction of the gravity field. For example, if the total dry-mass of grains above unit bed area is:

$$m = \sigma \int_0^h C(y) \cdot \mathrm{d}y \tag{2.23}$$

where σ is the solids density, h is the thickness of the transported layer, y is measured vertically upward from the base of the layer, and $C(y)$ is the fractional volume concentration of grains, then:

$$J = m\overline{U}_G \tag{2.24}$$

is the total dry-mass transport rate.

The particles in transport, normally combined with an intergranular fluid, constitute a *dispersion,* while the quantity m above constitutes a *load* (Bagnold, 1956, 1966), which it may or may not be necessary to support by the action of intergranular forces or forces arising within the fluid as the result of its motion. If a force or forces is necessary to sustain the load, then the transport rate is in principle determined by the grain, fluid, and flow properties. We must examine the stability of the grain dispersion, however, in order to establish to what extent the transport rate may be determinate in these terms.

Consider a dispersion of thickness h composed of spherical particles of a uniform density σ in a homogeneous fluid of density ρ. Two forces will invariably operate on the mass of grains present above unit bed-area, namely:

$$F_G = \sigma g \int_0^h C(y) \cdot \mathrm{d}y \tag{2.25}$$

the downward-acting particle weight, and:

$$F_{GD} = -\rho g \int_0^h C(y) \cdot \mathrm{d}y \tag{2.26}$$

the upward-acting buoyancy force due to the weight of displaced fluid. Now:

$$F_G + F_{GD} = \frac{(\sigma - \rho)}{\sigma} mg \tag{2.27}$$

is the immersed weight of grains above unit bed area, and is the *effective load* which must be balanced by any force or forces, F_M, arising between grains and/or within the fluid as the result of its motion. The stability of the dispersion is therefore determined by the magnitude and composition of:

$$F_{\mathrm{T}} = F_{\mathrm{G}} + F_{\mathrm{GD}} + F_{M} \tag{2.28}$$

in which F_{T} is the net force acting, and F_{M} may act either upward or downward as circumstances dictate.

Three classes of dispersion may be distinguished (Table 2-II). Class I is represented by neutrally buoyant particles either dispersed in a beaker on a laboratory bench (stationary case) or flowing with a liquid under gravity in a channel (translatory case). In neither case is an intergranular or intrafluid force needed to sustain a state of steady transport, that is, to maintain the centre of gravity of the grain-mass at a constant distance above the base of the dispersion. Sediment transport takes place in accordance with eq. (2.24) in the translatory case, but the rate is indeterminate in terms of flow properties, since the load is unconnected to the flow rate. Class IIIa corresponds to a dispersion of fluid and grains of contrasted density that has suddenly been "melted" in a stationary beaker or while sliding in a channel. There is again no intergranular or intrafluid force (at least not at the start in the translatory case), and the centre of gravity of the grain-mass must fall or rise accordingly as $\sigma > \rho$ or $\sigma < \rho$. Class II dispersions are of great theoreti-

TABLE 2-II

Dynamical classification of sediment dispersions

Relative density	Total force: $F_{\mathrm{T}}=0$	$F_{\mathrm{T}} \gtrless 0$
$\sigma=\rho$	*Class I* statically stable, stationary or translatory dispersions $F_{\mathrm{M}}=0$	not represented
$\sigma \gtrless \rho$	*Class II* dynamically stable, stationary or translatory dispersions $F_{\mathrm{M}} \neq 0$	*Class IIIa* statically unstable stationary dispersions, translatory under restricted conditions $F_{\mathrm{M}}=0$
		Class IIIb dynamically unstable, stationary or translatory dispersions $F_{\mathrm{M}} \neq 0$

cal and practical interest. They are exemplified by a beaker on a laboratory bench in which a statistically constant mixture of grains and less dense fluid is maintained by random agitation (stationary case), and by the steady, uniform flow of a fluid bearing grains of greater density over a bed of the same grains (translatory case). In each instance, the centre of gravity of the moving grain-mass is maintained at a constant level above the bed by the action of intergranular and/or intrafluid forces equal and opposite to the effective load. Dispersions of Class IIIb are also of great interest. If we change the intensity with which a stationary dispersion is agitated, its centre of gravity must inevitably change in position over a period, as a consequence of the change in the balance of forces. A similar effect will arise if, in the case of stream-borne grains, F_M changes with time and/or distance.

Dynamically unstable translatory dispersions of Class IIIb represent sediment transport under general unsteady and non-uniform conditions, that is, under the conditions that prevail in all natural environments. At any instant during such transport, the total effective load can always be divided between two parts: (1) that which exist as a dynamically *stable* dispersion under the conditions prevailing at that instant, and (2) that in deficit or excess of the dynamically stable component. Any excess corresponds to a statically unstable dispersion, and represents a quantity of grains available for deposition. A deficit indicates that the flow has a potential for net erosion.

That dynamically stable grain dispersions can exist at once implies that there are momentum transfer-mechanisms at work permitting the necessary supporting stresses to be transmitted ultimately from the solid boundary to the grains distributed in the fluid. Bagnold (1956, 1966) has pointed out that there are only two possible mechanisms by which this can be achieved: (1) by the transfer of momentum from grain to grain by intermittent near approaches or actual contact, and (2) by the transfer of momentum from one mass of fluid to another and thence to an otherwise unsupported solid. The first mechanism can operate only where grains are densely arrayed, as in an avalanche or at a stream bed, whereas the second, suggesting the action of fluid turbulence, should be most effective where grains are highly dispersed. Both mechanisms can be sustained by fluid shear.

FORCES ACTING ON TRANSPORTED PARTICLES

With drag between moving fluid and dispersed grains providing the ultimate motive force for sediment transport, and the immersed particle weight the tendency for grains to return to the bed, what other forces can influence the motion of transported solids?

Something has already been said in the preceding chapter and on the basis of Bagnold's (1954a, 1956, 1966) seminal work about the forces acting between densely arrayed transported grains. As they are sheared along in the fluid, particles of sufficiently large size repeatedly impact with each other and with the stationary grain bed, creating an intergranular force resolvable between tangential and normal dispersive components (eq. 1.29). Fluid viscosity dominates, however, when the grains become sufficiently small or of sufficiently little excess of density (eq. 1.28). Particles then no longer make direct contact, but seem to influence each other remotely, squeezing the fluid from between them as they make near approaches. At present neither mode of influence is particularly well understood. The nature of particle impacts, however, must depend on grain size, shape, and surface texture. Remote influences between particles have been extensively studied for still fluids (e.g. Happel and Brenner, 1965), but hardly at all in turbulent fluids under shear.

Comparatively large grains, especially when travelling close to the static bed, can be influenced by hydrodynamic lift forces, due to: (1) the influence of the nearby boundary, (2) fluid shear, and (3) particle spin (Magnus effect). There has for some time been an awareness of the nature and role of these forces individually (e.g. G.I. Taylor, 1917; Jeffreys, 1929; Lamb, 1932; E.F. Ford, 1957; Chepil, 1961; Happel and Brenner, 1965; Willetts, 1970; Vasseur and Cox, 1976, 1977; White and Schulz, 1977; Bearman and Zdravkovich, 1978), but steps to clarify their collective role are relatively recent (Lawler and Lu, 1971; Bagnold, 1974).

Jeffreys (1929) showed that a particle close to a stationary boundary and in relative motion with a fluid stream is acted on by a normal force due to the restriction placed on the motion of the fluid by the proximity of the bed. This lift force is effective so long as the free separation distance between bed and particle is less than the order of one particle diameter. The lift acts away from the bed when the particle travels more slowly than the surrounding fluid, for the flow is slowed between the particle and bed, with the result that the fluid pressure on the particle is the greater on the side facing the bed (Fig. 2-12a). But should the particle travel the faster, the adjacent fluid is speeded up, affording the larger fluid pressure now on the distant particle face and a lift toward the bed (a phenomenon recently exploited in motor racing) (Fig. 2-12b).

A particle in relative motion with a sheared fluid moves up the gradient of relative velocity and across the line of flow (G.I. Taylor, 1917; Saffman, 1965; Lawler and Lu, 1971). In Fig. 2-12c the grain travels more slowly than the fluid surrounding it, whence the larger fluid pressure occurs on its underside, giving an upward lift. But a particle moving faster than the stream experiences a negative lift, because the fluid pressure on its upper surface is the larger (Fig. 2-12d). Shear-lift increases with increasing particle size and fluid shear rate.

Lift also results when a grain spins in a fluid stream, the well-known Magnus effect (Batchelor, 1967; White and Schulz, 1977), but only if there is an overall non-zero relative velocity between particle and flow. As Fig. 2-12e, f shows, the lift force is always directed toward that side of the

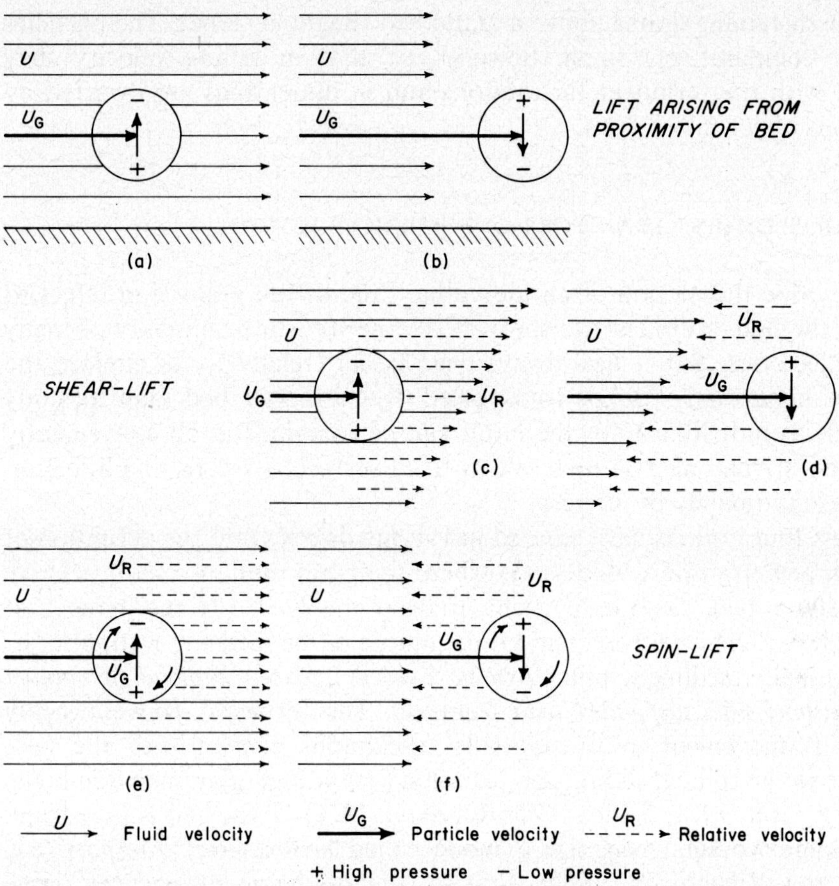

Fig. 2-12. Diagrammatic summary of the three sources of positive or negative lift on sediment particles during transport.

spinning particle moving in the same direction as the relative velocity. The lift increases with particle size and relative velocity.

An interesting case concerns a particle drifting and rotating freely in a sheared fluid under conditions of zero overall relative velocity. Under these conditions, which Bagnold (1974) calls autorotation, there can be neither a net shear-lift nor a net spin-lift. Because of their excess of density, however, real sedimentary particles can only rarely travel thus.

In a turbulent fluid, there is a fourth source of lift, seemingly vital to grain transport in suspension. The preceding chapter showed that the instantaneous velocity at a point in a turbulent fluid can be resolved into orthogonal components, whence the instantaneous drag force on a more dense particle immersed there can also be resolved in three mutually perpendicular directions. If the vertically up and vertically down parts of the force are on the average unequal, and the upward-acting part dominates, we have another means of sustaining grains above a static bed (Bagnold, 1966). The particles must be comparatively small, however, so that in falling velocity they compare with the turbulent fluctuations and in dimensions are dwarfed by the parcels of turbulent fluid.

MODES OF SEDIMENT TRANSPORT AND PARTICLE MOTION

To describe the motion of an individual fluid-driven grain is in effect to describe the bulk sediment transport mode, the statistical summary of many grain trajectories. But it has always been easier (relatively) to explore the collective behaviour of solids transported over a mobile bed than to study the grains as individuals, for the latter can be examined at all conveniently only when driven one at a time over a fixed surface, a severe simplification of the case ultimately of interest.

At least four notions have entered in varying degrees into the definition of bulk sediment transport modes: (1) where the grain motion occurs, relative to the static bed, (2) the particle grade(s) involved, (3) the type(s) of individual particle trajectory, and (4) the nature of the load-supporting force. Conventionally, sediment bulk flow is divided between *bedload* or *contact load transport* and *suspended load transport.* The former is conventionally defined as movement in substantially continuous contact with the bed, reference to the rolling, sliding and saltation (bounding) of grains often being made (e.g. Graf, 1971; Yalin, 1972; Raudkivi, 1976; Garde and Ranga Raju, 1978). Some workers recognize a mode called *bed-material transport* (e.g. Simons and Richardson, 1966), that is, the discharge of particles large enough to be found in appreciable quantities in the static bed. The suspended load, conversely, is usually recognized to comprise particles so small in size that they are distributed in comparable amounts throughout the whole flow. Such particles are what Simons and Richardson (1966) refer to as *fine sediment,* to be found in negligible amounts in the static bed. Usually the notion of turbulent suspension of the grains is associated implicitly or explicitly with the definition of suspended load transport (e.g. Graf, 1971; Yalin, 1972; Raudkivi, 1976; Garde and Ranga Raju, 1978). Evidently, grains that are bed-material or transportable as bedload under one flow condition may become "fine" sediment and move in suspension under a more severe one.

Some workers seek to remove these ambiguities using purely dynamical conceptions of bulk transport-mode. H.A. Einstein (1950) defined bedload as those particles which, while in motion, are supported by forces arising by contacts with the static bed and not immediately from the fluid. The weight of the suspended load, however, is directly borne by the surrounding fluid, through the action of turbulence. Bagnold (1956, 1966, 1973, 1977), while acknowledging no debt to Einstein, has proceeded similarly. According to

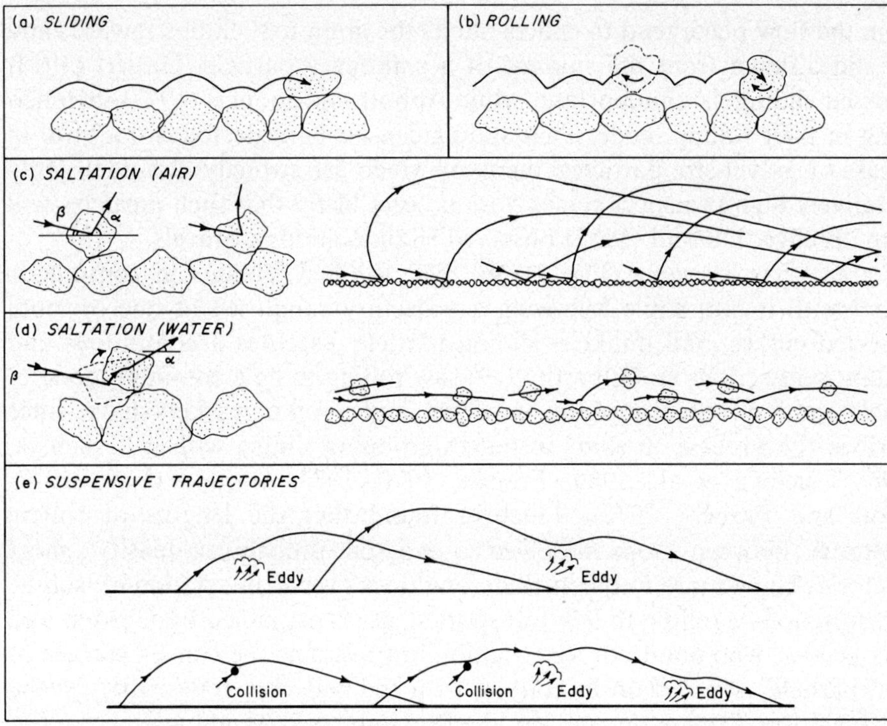

Fig. 2-13. Schematic representation of modes of particle motion during fluid-induced sediment transport.

him (Bagnold, 1977), bedload is that "solid material which is transported in a statistically dispersed state above the bed but which is not, however, suspended, i.e. the immersed weight is supported, on average, not by upward currents of fluid turbulence but by a combination of fluid and solid reactive forces exerted at intermittent contacts with the bed solids." Suspended load transport, however, is that mode "in which the excess weight of the solids is supported wholly by a random succession of upward impulses imparted by eddy currents of fluid turbulence moving upward relative to the bed" (Bagnold, 1973). Whereas in these terms bedload transport can occur in laminar as well as turbulent flows, as the experiments of Bagnold (1955), Parsons (1972) and Francis (1973) go to show, suspended load transport demands turbulent currents.

Individual grains disturbed by a current may travel according to one or more of four modes: (1) sliding, (2) rolling, (3) saltation, and (4) suspension. Their trajectories are determined completely by the three kinds of force previously described.

A particle *sliding* (Fig. 2-13a) over the static bed retains continuous contact with it but executes with negligible net rotation a trajectory consisting of one or more shallow connected curves. The short rotations of its long axis in the flow plane tend to cancel out as the grain first climbs towards and then slides down from the summit of a stationary particle. Gilbert (1914) dismissed sliding as unimportant, while Abbott and Francis (1977) included sliding in their rolling mode. That sliding can be so neglected is doubtful in the case of gravel-size particles, many of which are strongly discoidal. Only at relatively high transport stages does it seem likely that such particles will tip up on edge. Durand (1951) observed sliding amongst gravels.

A grain that is *rolling* (Fig. 2-13b) over the bed remains in continuous contact with it but, while following a trajectory composed of one or more connected curves, and unlike a sliding particle, executes a continuous and constant-sense rotation. Gilbert (1914) saw rolling to be a common mode of particle motion beneath gentle currents, and several workers have since described the process, in some instances grouping sliding with it (Tsuchiya, 1969b; Tsuchiya et al., 1969; Francis, 1970, 1973; Gordon et al., 1972; Abbott and Francis, 1977). Tsuchiya found that the lengths of rolling trajectories between stops followed a Poisson probability-density, short trajectories being more frequent than long ones over many stationary solids. The restriction of rolling to low transport stages is explained by Gordon and his associates, who point out that a grain, in traversing the convex surface of a bed particle, will remain in contact with the bed only for so long as the centrifugal force acting on it is less than its immersed weight.

Saltation is that mode of movement in which the solid takes relatively long leaps or bounds over the bed, touching the bed grains only at the start and finish of each trajectory. It is undoubtedly one of the two most important modes of grain motion. In air, where a thousand-fold density difference between solid and fluid prevails, the process of saltation is significantly different than in water, where the densities are of the same order. A large density difference means that the internal forces on a grain at each impact greatly exceed the fluid forces. Hence the grain rebounds whenever it collides more or less directly with a bed particle. A mineral-density grain in water, however, impinges so lightly on the stationary solids that rebound is practically impossible. So long as the contact is other than glancing, the grain is arrested and set in rolling motion for a short distance.

Saltation in air was first clearly described by J.S. Owens (1927) during an examination of sand movement on an East Anglian beach. It has since been much studied, both experimentally and using the equations of translational and, in at least one case, rotational motion (Bagnold, 1935, 1936, 1954b; Chepil, 1945a; Zingg, 1953; E.F. Ford, 1957; Horikawa and Shen, 1960; Bisal and Nielsen, 1962; Kawamura, 1964; P.R. Owen, 1964; Sharp, 1964; G.P. Williams, 1964; Chiu, 1967; Tsuchiya, 1971; Tsuchiya and Kawata, 1971, 1973; Ellwood et al., 1975; White and Schulz, 1977). Saltating particles either bounce off grains in the bed, which becomes disturbed in the process, or actually crater it, splashing up like a stone thrown into a pond a substantial number of bed solids. Grain trajectories are convex up, consisting of a short steep rise followed by a long, flat return to the bed (Fig. 2-13c). The rebound angle, denoted by α, is moderate to steep, Schulz and White measuring an average of 49.9° and a range from between about 20° and 100°. The impact angle β is much shallower, the same workers measuring a range of 4°–28° and an average of 13.9°, in agreement with Chepil, Bagnold, Bisal and Nielsen, and Tsuchiya. Trajectory lengths measure about ten times the heights, but both height and length exceed the grain diameter by one to many orders of magnitude (Fig. 2-14a), increasing with the fluid shear stress. There is evidence that the average trajectory height increases

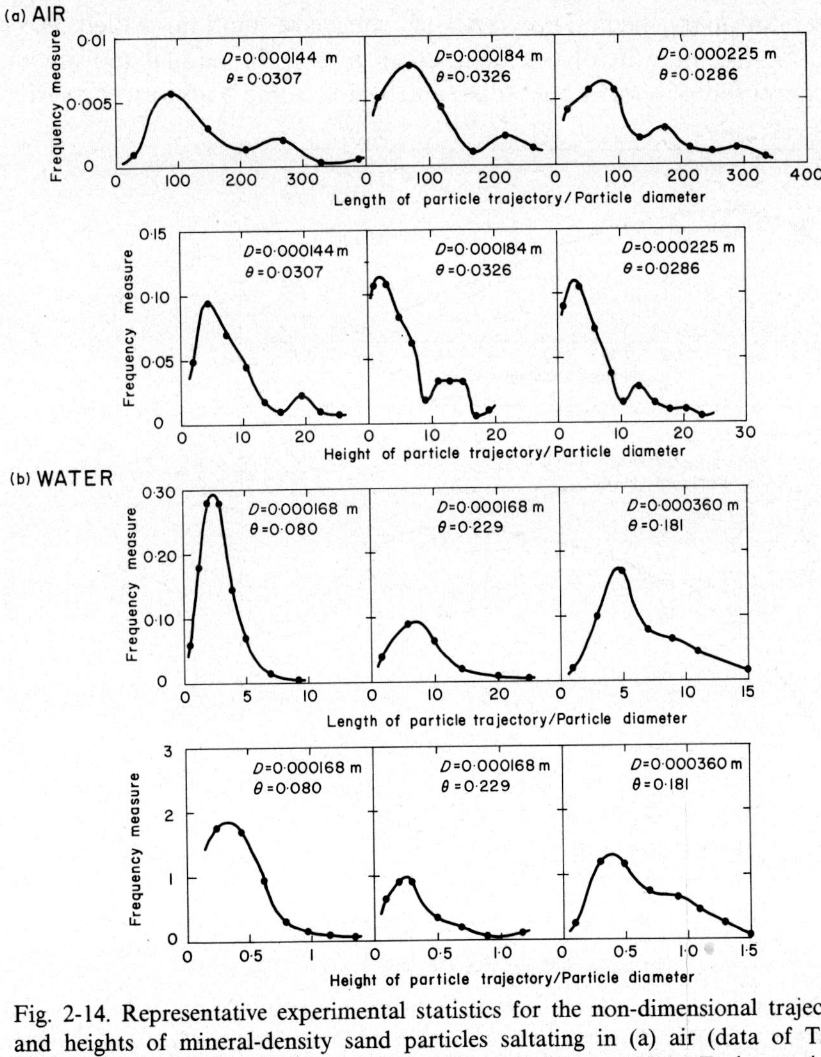

Fig. 2-14. Representative experimental statistics for the non-dimensional trajectory lengths and heights of mineral-density sand particles saltating in (a) air (data of Tsuchiya and Kawata, 1971), and (b) water (data of Tsuchiya and Aoyama, 1970). Each experimental curve is distinguished by the particle diameter (D) and the Shields-Bagnold non-dimensional boundary shear stress (θ).

with grain size (Bagnold, 1935; Zingg, 1953; P.R. Owen, 1964; Sharp, 1964; G.P. Williams, 1964; Chiu, 1967; Winkelmolen, 1969; White and Schulz, 1977) but declines with increasing particle flatness and angularity (G.P. Williams, 1964; Winkelmolen, 1969). A substantial proportion of grains during flight are seen rapidly to spin or oscillate (Chepil, 1945a; Bisal and Nielsen, 1962; White and Schulz, 1977), and it appears impossible to ignore the effects of lift on trajectories (White and Schulz, 1977). Bagnold's (1954b) assertion that rotation is rare and unimportant cannot be supported.

Most of the kinetic energy of grains saltant beneath the wind is passed on to bed solids which, under the continual bombardment, experience a slow net forward movement, largely within a few particle diameters of the ultimately stationary bed. This type of collective motion, called *creep* (Bagnold, 1954b), may involve a distinctive type of individual movement, perhaps composed of short hops, rolls, and slides, some backward or across

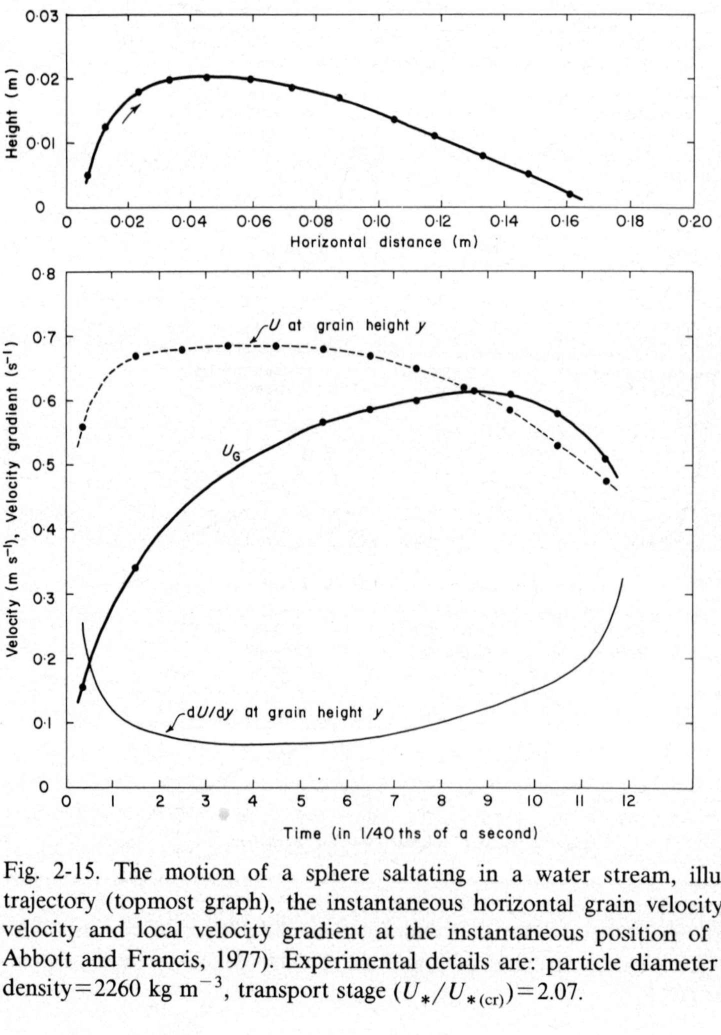

Fig. 2-15. The motion of a sphere saltating in a water stream, illustrated by a typical trajectory (topmost graph), the instantaneous horizontal grain velocity (U_G), and the flow velocity and local velocity gradient at the instantaneous position of the particle (data of Abbott and Francis, 1977). Experimental details are: particle diameter=0.00642 m, particle density=2260 kg m^{-3}, transport stage ($U_*/U_{*(cr)}$)=2.07.

the general line of transport but the majority forward.

Saltation in a water stream is different in several respects. The fluid forces acting on particles are considerably more important and it becomes necessary to distinguish saltation as clearly as possible from suspension. Abbott and Francis (1977) define saltation as a particle motion which, after the initial jump from the bed, is determined by a vertical acceleration that is always downward. Gilbert (1914), who early made a detailed experimental study of the process, rejected turbulence as the cause of saltation, proposing instead that a grain about to leap upward at first rolled a short distance over an already stationary bed solid until effectively its speed becomes so great that it is flung up into fast-moving parts of the flow. It was not until much later that brief rolling trajectories (Fig. 2-13d), during which the centrifugal force on the moving particle increased to balance the immersed weight, were observed to separate saltations (Gordon et al., 1972; Abbott and Francis, 1977). As observed or calculated from the equations of motion, the trajectories of grains saltant in water are relatively flat and short, with lower take-off angles than in air (Francis and Vickers, 1968; Tsuchiya, 1969a; Tsuchiya et al., 1969; Francis, 1970, 1973; Rossinskiy and Lyubomirova, 1970; Tsuchiya and Aoyama, 1970; Willetts, 1970; Gordon et al., 1972; Luque and van Beek, 1976; Abbott and Francis, 1977). Trajectory length is about 10 times the height, which increases with the applied fluid shear stress but rarely exceeds the order of one particle diameter (Fig. 2-14b). A typical trajectory appears in Fig. 2-15. The grain is accelerated over the first two-thirds of its flight, with any constriction and/or shear lift acting upward (Fig. 2-12), but in the final stage travels faster than the surrounding fluid, when its descent must have been steepened by negative lift.

The second major mode of individual particle motion is *suspension.* Abbott and Francis (1977) defined a trajectory as suspensive "when the vertical acceleration experienced by a grain, at any time between the upward impulses from the bed, is directed upward", but without restricting the cause of the upward acceleration. The resulting grain paths (Fig. 2-13e) denote conventional suspension only when, as Danel et al. (1953) and Rossinskiy and Lyubomirova (1970) also appreciated, fluid turbulence provides the necessary upward momentum flux between the inevitable (though possibly very infrequent) encounters with the bed. At large enough grain concentrations, suspensive trajectories can result from interactions between essentially saltant particles (Leeder, 1979a, b).

How relatively important are these different kinds of trajectory under given bed and flow conditions? This question cannot yet be precisely answered for bulk transport, but the experiments of Abbott and Francis (1977) on solitary grains repeatedly driven over a fixed bed of like particles gives us some clue. The relative proportion of the different modes was found to vary with the transport stage, defined as the ratio of the shear velocity U_* to the shear velocity $U_{*(cr)}$ at the motion-threshold of a fully mobile bed

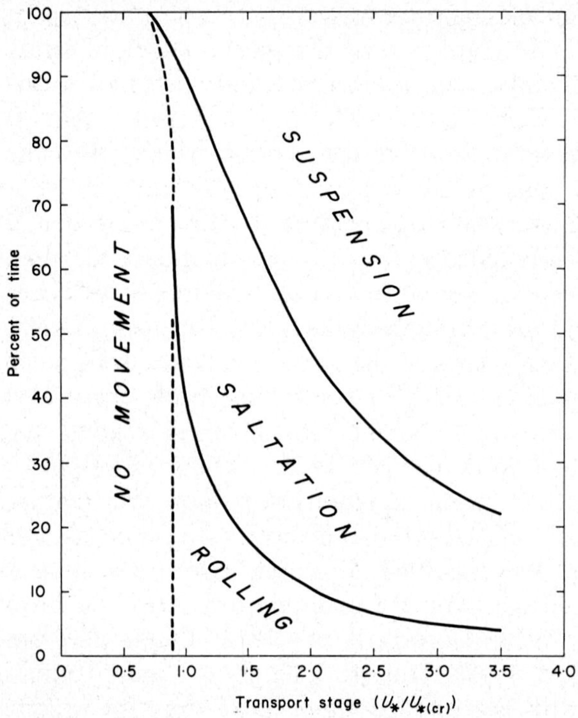

Fig. 2-16. The transport modes of large, solitary grains in a water stream, as a function of transport stage. Based on the analysis by Abbott and Francis (1977) of 716 separate trajectories using particles of four different densities within the mineral-density range. The percent of time in the rolling mode is read directly off the ordinate. The percent of time in suspension is read between the upper abscissa and the bound on suspension, counting from the top.

(Fig. 2-16). At stages close to unity, grains moved largely by sliding or rolling. With increase of stage, saltation and suspension become increasingly important until, at stages in excess of 2, suspension dominated. A theoretical criterion for suspension was suggested by Lane and Kalinske (1939), but on slender evidence. Bagnold (1966), with rather more justification, reasoned that suspension predominated when:

$$(W/U_*) \leqslant 1.25 \qquad (2.29a)$$

or, in terms of the Shields-Bagnold non-dimensional boundary shear stress:

$$\theta \geqslant 0.64 \frac{\rho W^2}{(\sigma - \rho) g D} \qquad (2.29b)$$

where W is the particle falling velocity. A closely similar criterion was deduced by Middleton (1977), and both his and Bagnold's have experimental support (Francis, 1973; Abbott and Francis, 1977). Physically, the criterion means that suspension cannot predominate until the vertical turbulent fluctuations, of the same order as the shear velocity, become comparable with the particle falling velocity. Furthermore, since W is proportional to $D^{1/2}$ for sufficiently large values of D, the non-dimensional stress becomes constant for these large sizes. In water streams, clay, silt, and the finer grades of sand are readily suspended, but in the wind only clay and the finer silt particles go easily into suspension.

The fact of such diversity in trajectory means that each grade of a particle mixture moves at its own distinctive velocity in a fluid stream. The differential transport (Bagnold, 1973; Chepil, 1957a, 1957b; Joliffe, 1964; W.H. Wood, 1970; Rana et al., 1973; Komar, 1977d; Deigaard and Fredsøe, 1978) thus implied provides a more acceptable explanation than differential particle abrasion for the downstream decline in average grain size found beneath the wind (Bagnold, 1954b), in rivers (USWES, 1935; NEDECO, 1959), and on beaches (Carr, 1969, 1975; Carr et al., 1970). Sediment transport is partly a stochastic process (Hubbell and Sayre, 1964; Grigg, 1970; Hung and Shen, 1976, 1979; Lee and Jobson, 1977), but both the mean transport velocity of particles while in motion and the transport velocity averaged over many trajectories and stops increase as the velocity of the fluid stream, and with the change from rolling and sliding, through saltation, to suspension (Ippen and Verma, 1955; Meland and Norrman, 1966, 1969; Francis and Vickers, 1968; Francis, 1970; Parsons, 1972; Luque and van Beek, 1976; Abbott and Francis, 1977). Complications are introduced by particle angularity (Winkelmolen, 1969) and flatness (N.C. Flemming, 1964), and by size differentials with respect to the stationary solids (Everts, 1973).

EQUILIBRIUM SEDIMENT TRANSPORT

Bedload transport rate

A legion of workers has studied bedload movement and devised equations for bedload transport rate under equilibrium conditions in water streams, as the reviews of Henderson (1966), Herbertson (1969), Graf (1971), Yalin (1972), Bogardi (1974), White et al. (1975), Raudkivi (1976) and Garde and Ranga Raju (1978) make clear. These formulae fall between five main groups, accordingly as the transport is explained in terms of either: (1) bed shear stress, (2) fluid velocity, (3) a probabilistic consideration of particle movement, (4) bedform celerity, and (5) energetics.

Formulae based on bed shear stress take the general forms:

$$J_{\mathrm{B}}\frac{(\sigma-\rho)g}{\sigma} \propto (\tau_0 - \tau_{0(\mathrm{cr})})\tau_0 \tag{2.30a}$$

$$J_{\mathrm{B}}\frac{(\sigma-\rho)g}{\sigma} \propto \tau_0^m \tag{2.30b}$$

or:

$$J_B \frac{(\sigma - \rho)g}{\sigma} \propto (\tau_0 - \tau_{0(cr)})^m \qquad (2.30c)$$

in which J_B is the dry-mass transport rate in kilograms per metre of width per second, σ and ρ are respectively the solid and fluid densities, g is the acceleration due to gravity, τ_0 is the mean bed shear stress, and $\tau_{0(cr)}$ is the stress at the threshold of sediment movement. The exponent m commonly is 1.5. A formula of this type was proposed by Du Boys in the last century and subsequently by many others (Schoklitsch, 1914; O'Brien and Rindlaub, 1934; Straub, 1935; USWES, 1935; Shields, 1936; Y. Chang, 1939; Zeller, 1963; Fleming and Hunt, 1976; Luque and van Beek, 1976). Relationships of the form of eq. (2.30) cannot be made dimensionally correct without the introduction of a dimensional coefficient and are therefore ultimately unsatisfactory.

The concept of velocity as the determinant of transport rate attracted several workers (e.g. Gilbert, 1914; Donat, 1929; Schoklitsch, 1930; Straub, 1939; Meyer-Peter and Müller, 1948), who developed formulae of the general type:

$$J_B \frac{(\sigma - \rho)g}{\sigma} \propto \bar{U}^2(\bar{U}^2 - \bar{U}_{cr}^2) \qquad (2.31a)$$

$$J_B \frac{(\sigma - \rho)g}{\sigma} \propto \bar{U}^n \qquad (2.31b)$$

$$J_B \frac{(\sigma - \rho)g}{\sigma} \propto (\bar{U} - \bar{U}_{(cr)})^n \qquad (2.31c)$$

and:

$$J_B \frac{(\sigma - \rho)g}{\sigma} \propto (q - q_{(cr)}) \qquad (2.31d)$$

in which $\bar{U}$ is the mean flow velocity, $\bar{U}_{(cr)}$ is the mean velocity at the sediment movement-threshold, $n \geqslant 3$ is an exponent (the lower limit generally applies at high transport stages), q is the unit fluid discharge, and $q_{(cr)}$ the unit fluid discharge at the motion-threshold. When $n = 3$ eqs. (2.31b) and (2.31c) corresponds to some forms of eq. (2.30), since the bed shear stress is given by the square of the mean flow velocity (eq. 1.19). Formulae of the form of eq. (2.31) have the same limitations as those based on excess stress. They are further unsatisfactory in that they involve no explicit notion of force. Partly theoretical bedload formulae based on aspects of the stochastic nature of sediment transport are given by H.A. Einstein (1942, 1950), Kalinske (1947), and by Engelund and Fredsøe (1977). Einstein's formula has the most elaborate derivation, which rests on the fact that the particles advance in discrete steps, the statistics of which determine the transport rate.

Bedload transport rate is often in practice measured as proportional to the product of the mean height of bedforms, such as ripples or dunes, with their mean celerity (Bagnold, 1954b; Znamenskaya, 1962; Simons et al., 1965a; Crickmore, 1970; Korchokha, 1972; Willis and Kennedy, 1977). Flow-separation effects, however, can in water introduce significant errors, by creating an actual bedform height substantially more than the effective height (Allen, 1969f; Crickmore, 1970).

Undoubtedly the theoretically most satisfactory bedload formulae are those founded in energetics, namely, the concept that sediment movement represents work done by the fluid stream acting as a transporting machine with a certain supply of available power. According to this idea, developed largely by Bagnold (1956, 1966, 1968, 1973, 1977, 1980), but also exploited by Rubey (1933), Knapp (1938), Yalin (1963, 1972), and Engelund and Hansen (1967) and Holtorff (1972), the relation can be generalized as either:

$$J_B \frac{(\sigma-\rho)g}{\sigma} = m_B \bar{U}_B \frac{(\sigma-\rho)g}{\sigma} \propto \tau_0 \bar{U} \tag{2.32a}$$

or:

$$J_B \frac{(\sigma-\rho)g}{\sigma} = m_B \bar{U}_B \frac{(\sigma-\rho)g}{\sigma} \propto \tau_0 \bar{U}_1 - \tau_{0(cr)} \bar{U}_2 \tag{2.32b}$$

in which m_B is the dry bedload mass above unit bed area, $\bar{U}_B$ the mean transport velocity of the grains, $\bar{U}$ the mean fluid flow velocity, $\bar{U}_1$ a fluid flow velocity appropriate to the stage represented by τ_0, and $\bar{U}_2$ a fluid flow velocity appropriate to the motion-threshold. The quantity $m_B \bar{U}_B (\sigma-\rho)g/\sigma$ has the quality and dimensions of work per unit bed area and time, while $\tau_0 \bar{U}$ clearly has the same dimensions and quality and is the available energy or power supply. Therefore, Bagnold (1966, 1973) reasons:

Rate of doing work = efficiency × available power

arriving at:

$$J_B = \frac{e_B}{\tan\alpha} \frac{\sigma}{(\sigma-\rho)g} \tau_0 \bar{U} \tag{2.33}$$

for the bedload transport rate, in which e_B is a theoretical bedload efficiency factor and $\tan\alpha$ is his coefficient of dynamic solid friction (Ch. 1). Setting aside the density term necessary to convert dry mass into immersed weight, the quotient $e_B/\tan\alpha$ is dimensionless, pointing again to the fundamental dimensional homogeneity of eq. (2.32). Equation (2.33) added to a similar expression for suspended load transport rate is in good agreement with laboratory data (Bagnold, 1966).

Recently, Bagnold (1977) advanced a partly empirical non-dimensional relationship that seems applicable to bedload transport on both laboratory and natural scales, reading:

$$\frac{J_B(\sigma-\rho)g}{\sigma(\omega-\omega_{(cr)})} = 1.6 \left(\frac{\omega-\omega_{(cr)}}{\omega_{(cr)}} \right)^{1/2} \left(\frac{h}{D} \right)^{-2/3} \tag{2.34}$$

in which $\omega = \tau_0 \overline{U}$ is the flow power at the stage of the transport, $\omega_{(cr)}$ is the power at the motion-threshold, h is flow depth, and D is particle diameter. The numerical factor appearing in eq. (2.34) before the power term to one-half is the practical value of $1/\tan \alpha$. This equation has a more acceptable form than eq. (2.33), and also eqs. (2.30b) and (2.31b), for bed-material movement must always cease at a finite fluid discharge. Moreover, it incorporates the important and familiar reduction of the bedload with increasing relative depth. A further empirical correlation has recently been suggested (Bagnold, 1980).

There are several formulae for bedload transport beneath the wind. An early version by O'Brien and Rindlaub (1936) for mineral-density solids has the form of eq. (2.31b), but all later ones treat the transport rate as a function of wind stress. Bagnold (1937a, 1954b) assumed that the quantity of saltant sand caused a loss of momentum by the air equal to the drag on the air due to the sand flow whence, treating the creep as a constant fraction of the total transport, he deduced a formula equivalent to eq. (2.30b) with an exponent of 1.5. Kawamura's formula of 1961 (Kawamura, 1964) has a similar basis, with the advantage that the threshold stress is included. A formula resembling these was developed by Hsu (1971, 1974), the accompanying coefficient being an empirical function of grain size. Zingg (1953) and Chiu (1967) also suggested transport formulae close to those of Bagnold and Kawamura. Field and laboratory tests show that Kawamura's and Bagnold's formulae describe sand transport rates well, particularly at high wind speeds, provided that appropriate coefficients are chosen (Chepil, 1945c; Horikawa and Shen, 1960; Belly, 1964; G.P. Williams, 1964; Svasek and Terwindt, 1974; C. Harris, 1975). Snow transport is commonly described by formulae of the form of eq. (2.31) (Mellor, 1965; Radok, 1977; Kobayashi, 1978).

Snow and sand particles during transport in the wind saltate much higher relative to their diameters than sand in water, making it easier to measure the vertical particle concentration and mass flux variations. Both the concentration and flux are found rapidly to decline upward from the bed (Chepil and Milne, 1939; Horikawa and Shen, 1960; Kawamura, 1964; Sharp, 1964; G.P. Williams, 1964; Mellor, 1965; Sommerfield and Businger, 1965; Chiu, 1967; Gillette and Goodwin, 1974; Radok, 1977; Kobayashi, 1978). There is little agreement on the laws that describe these changes, for both power and exponential functions have been suggested either theoretically or to fit observed distributions.

Suspended load transport rate

The transport of the suspended load—the mass of solids supported directly by fluid turbulence—has been attacked mainly by: (1) a consideration of the mass balance in a uniform turbulent stream bearing a fully developed suspended load (the classical approach), and (2) an application of energetics. The first approach seeks to specify the vertical distribution of suspended grains and then to calculate the total transport rate as the depth-integral of the product of grain concentration and transport velocity (assumed equal to the local flow velocity). As the approach is kinematic, however, only the relative concentration can be calculated, whence the total transport rate remains unknowable in the absence of empirical data. In the approach from energetics, we seek to relate the suspended load to the power supplied by the stream and to the efficiency of the turbulence. The approach therefore is dynamical, making the transport rate in principle calculable. Henderson (1966), Graf (1971), Ippen (1971), Yalin (1972), Bogardi (1974), Raudkivi (1976) and Garde and Ranga Raju (1978) provide helpful reviews.

The calculation of the suspended load transport rate from a consideration of mass balance begins with:

$$WC+\epsilon_S\frac{dC}{dy}=0 \tag{2.35}$$

where W is the particle free falling velocity, C is their time-averaged fractional volumetric concentration at a height y above the bed, and ϵ_S is the sediment diffusion coefficient. This is the simplest possible formulation of the diffusion equation (O'Brien, 1933), the space occupied by particles being neglected. Equation (2.35) has the following physical meaning. If we consider a unit horizontal area in the flow, then for a uniform streamwise transport of solids to persist, the rate WC at which grains are falling under gravity through that area must equal the rate $\epsilon_S dC/dy$ at which they are being discharged down the concentration gradient and away from the bed. Since suspended grains are small in size relative to the turbulence, and compare in falling velocity with the turbulent fluctuations, we can assume as a first approximation that $\epsilon_S = k\epsilon_M$, where ϵ_M is the coefficient for the diffusion of fluid momentum and k compares with unity. Equation (2.35) can therefore be integrated by prescribing (1) ϵ_M in relation to the shear stress and velocity gradient over the depth of the fluid, and (2) the shear stress and velocity as functions of height above the bed. The forms chosen will depend on whether we are modelling suspension transport in the air or a water stream with a free surface.

Ippen in 1934 and Rouse (1937) integrated eq. (2.35) using a logarithmic form of velocity distribution, to obtain for transport by a free-surface aqueous flow:

$$\frac{C}{C_{ref}}=\left[\frac{(h-y)}{(h-y_{ref})}\,\frac{y_{ref}}{y}\right]^n \tag{2.36}$$

where h is flow depth, C the solids concentration at a height y above the bed, and C_{ref} the concentration at a small reference height y_{ref}. Further:

$$n=\frac{W}{k\kappa(\tau_0/\rho)^{1/2}} \tag{2.37}$$

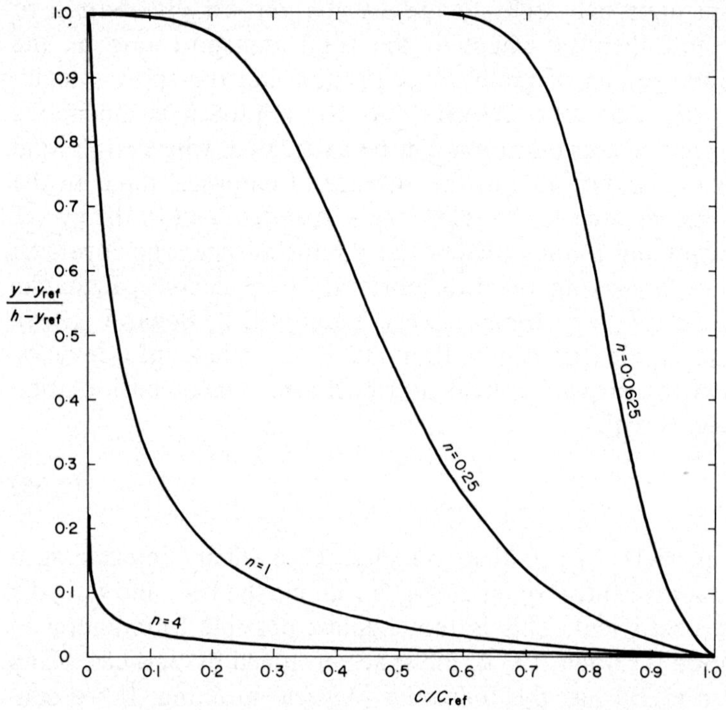

Fig. 2-17. Theoretical variation of the relative concentration of suspended particles in an equilibrium flow with relative height above the bed, eq. (2.36).

where κ, nominally equal to 0.4, is the Von Kármán coefficient in the velocity distribution. For a turbulent flow with a logarithmic velocity profile, the shear stress varies linearly with depth, while the distribution of ϵ_S is parabolic, having a maximum at $h/2$ but declining to zero at the bed and free surface.

Inspection of eq. (2.36) shows that the relative concentration increases from zero at the free surface to unity at the reference height, in practice the height at which the logarithmic distribution fails, and that grains of the smallest falling velocity achieve the most uniform distribution within the flow (Fig. 2-17). Hence when an ill-sorted mixture of particles of the same density is transported in suspension, we should find the largest grains mainly close to the bed but the smallest ones distributed much more evenly over the depth. Equation (2.36) agrees well with suspended sediment concentrations measured in laboratory channels and rivers, particularly for the finer solids (E.G. Richardson, 1934; Christiansen, 1935; Straub, 1936; A.G. Anderson, 1942; J.W. Johnson, 1943; Vanoni, 1946, 1953; Ismail, 1952; Colby and Hembree, 1955; Einstein and Chien, 1955; Nordin and Dempster, 1963; Nordin and Beverage, 1965; Toffaleti, 1965; N.L. Coleman, 1969, 1970; Zeller, 1963; Nordin, 1971a). In atmospheric dust storms a similar upward decline in the concentration of suspended particles is observable (Chepil, 1957b; Chepil and Woodruff, 1957).

Agreement with eq. (2.36) is less satisfactory for relatively large grains, and observed values for the exponent can differ substantially from the theoretical, eq. (2.37), for several reasons. The particle free falling velocity is used to calculate the exponent, whereas during turbulent transport the grains are accompanied by others and are moving relative to a violently agitated fluid, the effective falling velocity being reduced, often substantially, from the value for a solitary particle in a still unbounded fluid. The presence of solids also modifies the velocity profile and substantially reduces the Von Kármán coefficient (Vanoni, 1953; Einstein and Chien, 1955; Hino, 1963). Suspended clay minerals seem to have a complex influence on turbulence and bed shear stress (Gust, 1976; Gust and Walger, 1976). Furthermore, ϵ_S is not distributed precisely as ϵ_M (N.L. Coleman, 1969, 1970).

Several workers have tried to improve on the simple analysis represented by eq. (2.36). Hunt (1954, 1969) developed a mass-balance equation that took into account the volume of sediment, but the resulting formulae, although representing concentration distributions better than eq. (2.36), are complicated to use in practice. Zagustin (1968) took into account the fact that ϵ_M commonly is not zero at the free surface. Willis (1969, 1979) claims that the distribution of suspended sediment is best fitted by an error function. Bogardi and Szucs (1970), Drew (1975), and Drew and Kogelman (1975) explored energy and momentum balance equations, but these are far from being readily applied in practice. A similarity approach is described by Navntoft (1970).

The calculation of the suspended sediment transport rate from energetics is due to Bagnold (1966), who proposed the relationship (a bedload is also present):

$$J_S\frac{(\sigma-\rho)g}{\sigma}=m_S\bar{U}_S\frac{(\sigma-\rho)g}{\sigma}=\frac{e_S\bar{U}}{W}(1-e_B)\tau_0\bar{U} \tag{2.38}$$

in which m_S is the suspended sediment load, $\bar{U}_S$ is the mean transport velocity of the suspended grains, and e_S is the suspended sediment efficiency factor, it being assumed in the final expression that the suspended solids travel at the same speed as the fluid, on account of their relatively very small falling velocities. The term $\bar{U}/W$, analogous to $1/\tan\alpha$ in eq. (2.33), expresses the fact that the turbulence is pushing the suspended load up a notional frictionless incline of slope $W/\bar{U}$. Turbulence supports the load because, according to Bagnold, the upward vertical turbulent fluctuations are more vigorous than the downward-acting ones, whence there is a residual upward momentum flux, a conclusion supported by Irmay's (1960) rigorous treatment. Bagnold estimated the flux from experimental data and went on to derive $e_S = 0.015$ as a universally constant efficiency factor. Little attempt has been made to test eq. (2.38) as such, though in combination with eq. (2.33) it is consistent with laboratory data (Bagnold, 1966).

Total and bed-material load transport rate

It is often of more practical concern to know the total or bed-material load transport rate than the rates for bedload and suspended load separately. According to Einstein's (1950) and Bagnold's (1966) partly analytical models, the total load transport rate is obtained simply by adding the separately calculated bedload and suspended load transports. Thus Bagnold (1966) gave as his practical total-load transport-rate formula:

$$J = \frac{\sigma}{(\sigma - \rho)g}\left(\frac{e_B}{\tan\alpha} + 0.01\frac{\bar{U}}{W}\right)\tau_0 \bar{U}_0 \qquad (2.39)$$

in which only the bedload efficiency and $\tan\alpha$ appear as empirical requirements. All other relationships for total load (Lane and Kalinske, 1941; Laursen, 1958; Bogardi, 1965; Colby, 1964; Bishop et al., 1965; Barr and Herbertson, 1968; Graf and Acaroglu, 1968; Herbertson, 1968; Maddock, 1969, 1976; Toffaleti, 1969; Willis and Coleman, 1969) are more or less empirical, though commonly guided by dimensional considerations or similitude principles.

Two of the most interesting recent developments are the models of C.T. Yang (1972), and Yang and Stall (1974), and of Ackers (1972), Ackers and White (1973), and T.D. Yang (1979). In each model, underpinned by Bagnoldian considerations, much empirical data is analysed to provide coefficients for a general transport relationship. That of Ackers gives the non-dimensional transport in terms of non-dimensional measures of sediment mobility and grain size. The presence of bedforms is automatically accounted for in empirical data giving the coefficients. Generally good predictions are obtained for rivers using this model (Ackers and White, 1973; White et al., 1973a, 1973b).

SEDIMENT TRANSPORT AND DEPOSITION IN VARYING FLOWS

General

In the preceding sections, we considered sediment transport effectively under steady, uniform equilibrium conditions. Now the sediment continuity equation reads (Allen, 1977a):

$$\frac{\mathrm{d}m}{\mathrm{d}t} = -\left(\frac{\partial J}{\partial x} + \frac{J}{w}\frac{\partial w}{\partial x} + \frac{1}{\bar{U}_G}\frac{\partial J}{\partial t}\right) \qquad (2.40)$$

in which m is mass per unit area transferred between bed and flow, J the total-load transport rate, w the flow width (or distance between two pathlines between which the transport is considered confined), $\bar{U}_G$ the mean grain transport velocity, and x and t are respectively streamwise distance and time.

Under equilibrium conditions, eq. (2.40) sums to zero for steady flow in a channel uniform in every respect. The bed experiences neither deposition (dm/dt + ve) nor erosion (dm/dt − ve), whence its vertical position cannot be altered, nor can its relief undergo permanent change. Neither condition favours the preservation of sedimentary structures.

Sedimentary structures have the best chance of formation and preservation when the sediment transport rate varies in space and time, because the flow conditions similarly vary. Erosion and deposition can then act to alter the vertical position and relief of the sedimentary boundary, making and preserving structures indicative of the flow regime and the sediments available in flow and bed. Equation (2.40) does not then sum to zero, and the sediment load may not be dynamically stable (Table 2-II). Bagnold (1968) very clearly made the point that equilibrium sediment transport is of interest only because we ultimately wish to know what happens when, as in all natural environments, conditions continually change with space and/or time.

Unidirectional flows

Rivers are the most important of the natural agencies which transport sediment at a varying rate. Are these rates then calculable, and do the bedload and suspended loads differ significantly in their time relationship to the unsteady and non-uniform flow?

Field observations by NEDECO (1959, 1961) on the Niger system, and laboratory experiments by Griffiths and Sutherland (1977), strongly suggest that in rivers the bedload transport rate responds almost instantaneously to changes of flow. This is understandable because the bedload grains seldom travel higher than a few diameters above the bed or more than a few tens of diameters downstream between contacts with the stationary boundary. Even though the mean velocity and boundary shear stress due to the river are changing with space and time, the bedload can be interpreted as a dynamically stable dispersion (Table 2-II), therefore becoming calculable in terms of the steady-state formulae discussed above and the instantaneous flow conditions (but see De Vries, 1965, and Vreugdenhil and De Vries, 1967).

But suspended load is distributed over the whole flow depth, with its centre of mass at a significant height above the bed. Since the load comprises grains of a relatively small falling velocity, it should respond with considerable time and spatial lags to changes of flow. For example, a mineral-density silt particle of diameter 4×10^{-5} m has a falling velocity of about 1×10^{-3} m s^{-1}. If such a grain, originally at the free surface of a river 10 m deep flowing at 2 m s^{-1}, were no longer to be supported by available fluid forces, it would take 1×10^{4} s to reach the bed, during which time it would have been carried 20 km downstream. Sumer (1977) presents a sophisticated treatment of this type of problem. Similarly, material supplied to the suspended load from the bed can take a substantial time to become distributed throughout the flow depth (Hjelmfelt and Lenau, 1970). There is evidence from many environments of such lags in the suspended sediment.

In rivers, the suspended load transport and the depth-averaged suspended sediment concentration show a variable lag with respect to the passage of individual flood peaks, as well as with the annual or semi-annual flood wave. Typically, in the upper and middle reaches of a river system, the largest transport rates and concentrations are measured when stage is rising and before the peak discharge is reached (Hjulström, 1935; Einstein et al., 1940; J.W. Johnson, 1943; Sundborg, 1956; NEDECO, 1959, 1961; Allen and Welch, 1967; Douglas, 1967; Walling and Gregory, 1970; Church, 1972; Culbertson et al., 1972; Lee Wilson, 1972; Temple and Sundborg, 1973; Walling, 1974; Østrem, 1975; P.A. Wood, 1977a, 1977b). Consequently, plots of load or concentration against discharge reveal an often huge scatter (e.g. Nordin, 1963; P.R. Jordan, 1965; Axelsson, 1967; Müller and Förstner, 1968), and are of little value as rating curves. In the lower reaches, however, it is common to find that the suspended-sediment concentration or load peaks after the fluid discharge (A.D. Lewis, 1921; USWES, 1939; Heidel, 1956; Leopold and Wolman, 1956). The change of phase with downstream distance is largely explained by the fact that the flood wave moves at 1.50–2.0 times the speed of the sediment wave (A.D. Lewis, 1921; Heidel, 1956), but a contributary factor undoubtedly is that the sediment travels slightly more slowly than the fluid. The upstream appearance of the sediment peak before the maximum of fluid discharge may record during the unfolding of each flood event the declining availability of suspendable sediment on the stream bed and/or interfluves. This decline could be related to rates of weathering, as a means of providing transportable material, or to the development of bed armour. Whatever the explanation for these lag effects, it seems clear that the suspended loads of rivers, including much sand at times of flood, must be regarded as dynamically unstable dispersions (Table 2-II).

Reversing (oscillatory boundary-layer) flows

Sediment transport by reversing flows is altogether more complex and less satisfactorily understood than in one-way currents. Two cases are particularly important to natural environments: (1) bedload and suspended load transport by tidal currents, and (2) bed-material transport by wind-wave action, both outside and within the breaker zone.

Practically nothing is known, directly or otherwise, of bedload transport by tidal currents, though field measurements show that the mean boundary shear stress lags the current speed (Bowden et al., 1959; McCave, 1973a; Gordon, 1975a; Bohlen, 1976; Dyer, 1976), with the consequence that bedload transport is enhanced during the later part of the ebb above what the speed itself would suggest (Kachel and Sternberg, 1971). Gordon explains the effect as possibly related to the influence on turbulence of the adverse pressure-gradient generated during flow deceleration.

There is much more empirical data on suspended sediment, predominantly silt and clay, in tidal currents. The mud content of tidal waters can increase several-fold between the calms of summer and the winter, when frequent storms enhance wave action and create surges (W.H. Jackson, 1964; Halliwell and O'Connor, 1965, 1966; Terwindt, 1971a). On the scale of the spring-neap cycle, the greatest mud contents tend to be observed over the time of the strongest tidal currents (spring tides) (e.g., Inglis and Allen, 1957; W.H. Jackson, 1964; Halliwell and O'Connor, 1965, 1966), apparently with a small time-lag (Delft Hydraulics Laboratory, 1962). The most striking patterns, however, accompany the semi-diurnal or diurnal tide.

Beginning with Gry (1942), who plotted mud content in the form of phase diagrams, it has become clear that the sediment suspended in the tidal current at a site varies in quantity in a complex pattern both with height above the bed and with time (Postma, 1954, 1961; Inglis and Allen, 1957; Delft Hydraulics Laboratory, 1962; Halliwell and O'Connor, 1965, 1966; Pestrong, 1965, 1972a, 1972b; Schubel, 1968, 1969; R.W. Sheldon, 1968; Buller et al., 1971; F.E. Anderson, 1973; D'Anglejan and Smith, 1973; Gallenne and Salomon, 1975; Gordon, 1975a; Thorn, 1975a, 1975b; D'Anglejan and Ingram, 1976). At a fixed height, the concentration varies with time on the same period as the tidal velocity but with a degree of delay. The lag is generally small close to the bed but in mid-flow and higher up can be significant. At a fixed time, the concentration normally declines upward, though locally a maximum at mid-depth may be found. Even at slack water, much silt and clay remains dispersed.

The lag can be explained in various ways (Postma, 1954; Van Straaten and Kuenen, 1957, 1958; Groen, 1967; Gordon, 1975a; A.G. Davies, 1977), of which the most appealing is the influence on tidal turbulence of pressure gradients resulting from the repeated acceleration and deceleration of the flow. In all events, formulae for steady-state suspension transport seem valueless in the tidal case, if only because of the substantial amounts of sediment remaining in the water column each time the flow velocity passes through zero.

McCave (1969, 1970, 1971b, 1973b, 1975) and Odd and Owen (1972), working from the ideas and experimental results of H.A. Einstein (1968) and Einstein and Krone (1962), developed a helpful model for mud deposition from tidal and other marine currents, for example, the oceanic western boundary undercurrents (e.g. Biscaye and Eittreim, 1977). Underlying McCave's analysis is the idea that there is a plane parallel with and close to the bed below which particles can only continue to settle, perfect transfer being impaired by the random upward bursts of fluid from within the viscous sublayer and buffer layer. An empirical formulae for the deposition rate

(Odd and Owen, 1972; McCave and Swift, 1976), equivalent to McCave's analytical result, reads:

$$\frac{\mathrm{d}m}{\mathrm{d}t} = CW\left(1 - \frac{\tau_0}{\tau_{0(\mathrm{cr})}}\right) \tag{2.41}$$

in which C is the near-bed dry-mass particle concentration and $\tau_{0(\mathrm{cr})}$ is a threshold mean boundary shear stress above which deposition is impossible. Experimentally, the critical deposition stress is very small, in the order of 0.05 N m^{-2} (McCave and Swift, 1976). Equation (2.41) requires more testing in natural environments.

Bed-material transport caused by the action of wind-waves outside the breaker zone can be considered at two levels, assuming that the transported solids come from the wave-affected bed. It was explained in Chapter 1 that wave-generated currents combine a periodic (oscillatory) part with a generally much weaker but unidirectional mass-transport component, which in natural environments may be reinforced, balanced, or reversed by steady currents due to tides, wind stress, and thermohaline circulation. Once sediment threshold conditions are exceeded, however, the oscillatory part will entrain grains, disperse them some distance upward into the water column, and give rise to a non-zero transport, on scales of measurement that are less than the near-bed water particle orbital diameter and the wave period. Such transport may be characterized as "instantaneous". Of much greater interest, however, are the net transport rates due to waves, that is, the transport in the long term and on the scale of many orbital diameters. The rate at this scale is determined largely by the mass-transport component (and any accompanying unidirectional currents), since all that the oscillatory flow can accomplish is the movement of the same grains to-and-fro over an identical path. The net transport rate due to waves is therefore:

$$J = \sigma \int_0^h C(y) \cdot U_{\mathrm{G}}(y) \cdot \mathrm{d}y = m\overline{U}_{\mathrm{G}} \tag{2.42}$$

in which $C(y)$ and $U_{\mathrm{G}}(y)$ are respectively the fractional volumetric grain concentration and net grain transport velocity as a function of distance y above the bed, h is flow depth, m is the dry mass of grains dispersed above unit bed area, and $\overline{U}_{\mathrm{G}}$ is the overall average net grain transport velocity. Physically, C and m depend primarily on the oscillatory part of the motion, whereas U_{G} is determined by the mass-transport alone, under conditions of pure wave motion.

Instantaneous transport rates were measured experimentally on plane cohesionless beds by Kalkanis (1964), Abou-Seida (1965), and Sleath (1978). Using all available data, Sleath showed that the rate, averaged over one-half of the oscillation period, increased with the boundary shear stress to the power 1.5, an exponent commonly applied under steady conditions, eq. (2.30). The data scatter considerably, however, and his relationship is of little value for the calculation of net transport, since the grain concentrations and transport velocities are not separately distinguished.

An important initial problem in the calculation of net transport under wave action concerns the form of $C(y)$. Analytical studies suggest that the concentration declines steeply upward, both in progressive (Hom-ma and Horikawa, 1963a; Hom-ma et al., 1965; Liang and Wang, 1973) and in standing waves (Hattori, 1969). This result is confirmed experimentally above rippled beds (Fairchild, 1959; Hom-ma and Horikawa, 1963a, 1963b; Hom-ma et al., 1965; Horikawa and Watanabe, 1970; Bhattacharya and Kennedy, 1971; M.M. Das, 1971; Kennedy and Locher, 1972; T.C. MacDonald, 1977; Nakato et al., 1977), both power laws (M.M. Das, 1971) and exponential functions (T.C. MacDonald, 1977) appearing to fit the data according to circumstances. The local concentration above rippled beds varies in a complicated way with time and space, but by integrating profiles of suitably averaged concentration it becomes possible to relate the sediment load to wave conditions. Calculating the load non-dimensionally as $mg/(\sigma - \rho)gD$, where D is the particle diameter, the data of Fairchild (1959), Hom-ma and Horikawa (1963a), Hom-ma et al. (1965), Horikawa and Watanabe (1970), Bhattacharya and Kennedy (1971), and Nakato et al. (1977) are fairly well correlated by further groups (Fig. 2-18), using the formula:

$$\frac{mg}{(\sigma-\rho)gD} = 50{,}000\left(\frac{\rho U_{max}^2}{(\sigma-\rho)gD}\,\frac{\nu^{1/2}}{U_{max}T^{1/2}}\right)^4 \tag{2.43}$$

in which U_{max} is the maximum near-bed water particle orbital velocity, ν is the fluid kinematic viscosity, and T is the wave period. The first group on the right is a Shields-Bagnold criterion, and the second a measure of the influence of acceleration forces due to the waves.

Figure 2-18 shows substantial scatter, but not much more so than Sleath's (1978) graph for the plane-bed instantaneous rate. It is probably attributable to differences of bed roughness between experiments and the difficulty in measuring transient concentrations. Equation (2.43) has a substantially larger exponent than Sleath's comparable formula (allowance being made for the difference of a grain transport velocity), suggesting that the vortices generated at rippled beds (Chapter 11) are powerful sediment dispersing agents.

Since normally the mass-transport is small compared with the unsteady velocities, the net sediment transport rate under wave action is found theoretically (Liang and Wang, 1973; Wang and Liang, 1975), experimentally (Scott, 1954; Vincent, 1958; Abou-Seida, 1965), and by field measurements (Cook and Gorsline, 1972) also to be small. When mass-transport alone contributes the unidirectional flow component, the net sediment transport is normally in the direction of wave propagation.

Longshore sediment transport resulting from wave-action at the coast is

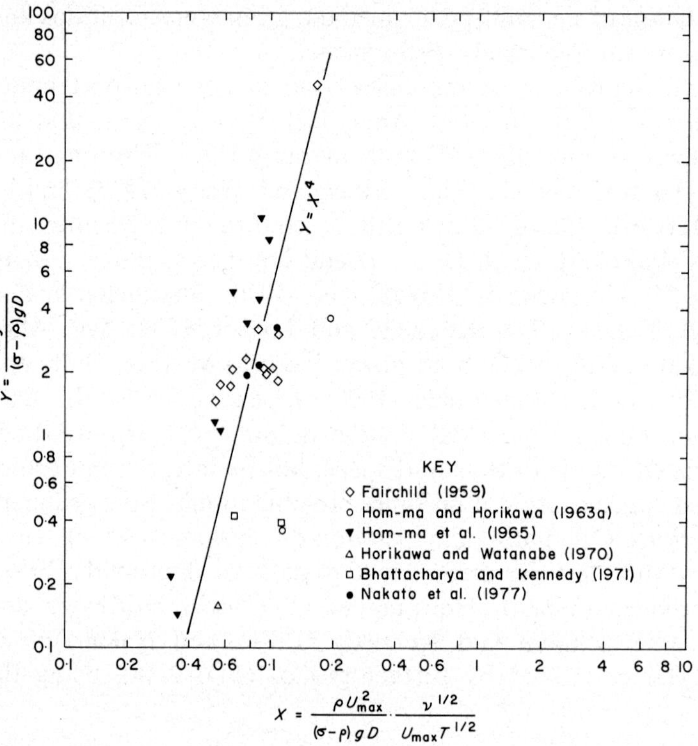

Fig. 2-18. Experimental data illustrating the relationship between the steady non-dimensional sediment load (mineral-density solids) capable of being maintained by wave action above a remote bed, as a function of wave conditions, and bed and fluid properties.

important to some of the larger scale littoral sedimentary structures. Progressive surface waves advancing on the coast carry energy with them which, if the waves are obliquely incident, has a longshore component, expressed by a longshore flow of water and entrained sediment between the breaker zone and the beach (Longuet-Higgins, 1970). Because of its practical importance, longshore sediment transport has been widely studied from many aspects, as Niemeyer (1974) and Komar (1976) show in their reviews. The most successful attack, however, which rests on energetics, was begun by Bagnold (1963) and by Inman and Bagnold (1963), and greatly extended by Komar and Inman (1970) and Komar (1971, 1977b, 1977c, 1977d, 1978).

This work shows that the total-load dry-mass transport rate, averaged across the nearshore zone, can be described by:

$$J = 0.28\frac{\sigma}{(\sigma-\rho)g}(Ec_{p}k)_{B}(\bar{U}_{L}/U_{max}) \tag{2.44}$$

in which $(Ec_pk)_B$ is the wave energy flux evaluated at the breakers, $\bar{U}_L$ is the longshore current velocity averaged over the nearshore zone, and U_{max} is the maximum water particle orbital velocity due to the breaking waves. In the bracketed group (Ec_pk), E is the energy density of the waves, and k the fraction of that energy travelling at the wave phase velocity c_p. We further have:

$$\bar{U}_L = 2.7U_{max} \sin \alpha_B \cos \alpha_B \qquad (2.45)$$

in which α_B is the angle between the wave crests and breaker zone evaluated at the breakers, and:

$$U_{max} = (2E/\rho h)_B^{1/2} \qquad (2.46)$$

where h is the breaker-zone water depth. Equation (2.45), which is partly empirical, is consistent with a theoretical relationship derived by Longuet-Higgins (1970).

Equation (2.44) shows that the total-load transport rate increases with the energy flux, proportional to the square of wave height, being for given waves greatest at intermediate angles of incidence. The maximum local transport rate occurs just on the beach side of the breaker zone (Bijker, 1971; Thornton, 1973; Komar, 1977c) and bedload seems greatly to predominate over suspended load (Komar, 1978). There are two general modes of longshore sediment transport, either conventionally as bedload and suspended load over most of the nearshore, or as a zig-zag movement of grains within the zone of wave swash and backwash. The selective transport of particle sizes is complicated in the nearshore; on certain beaches the coarser grains appear to travel faster than the finer (Komar, 1977d).

Internal gravity waves seem capable of setting sediment in motion where they break on submarine slopes and lake beds (Southard and Cacchione, 1972; Emery and Gunnerson, 1973; Cacchione and Southard, 1974) and, if obliquely incident, could drive that sediment parallel with bottom contours in a similar manner to surface waves.

UNUSUAL MODES OF SEDIMENT TRANSPORT

Sedimentary particles can be transported in other ways than the above, and these have occasional relevance to sedimentary structures. Thus dry or moist sand grains are sometimes carried on the surface tension film of water (O.F. Evans, 1938a, 1938b; Parea, 1970), and detritus can be moved over long distances amongst the roots of land plants, as the holdfasts of marine algae, or in the stomachs of swimming animals (Emery, 1963).

SUMMARY

The shape, size, and composition of sedimentary particles profoundly influences their response to and behaviour in moving fluids. Beds of sediment can be divided between cohesive and cohesionless. Erosion of the former involves the removal of multi-granular masses either by the direct action of fluid stresses, corrasion, or cavitation-erosion. Cohesionless beds when eroded lose grains one at a time. Grains dispersed into a fluid settle at a rate determined by their size and excess of density and, under certain conditions, by the fluid viscosity. However, sedimentary particles in bulk settle more slowly than when alone, as a consequence of particle interactions in the fluid. The coarser grades of particle are transported mainly as bedload, in which the grains slide, roll or saltate close to the bed, making frequent contact with the bed, whereas the finer debris travels in suspension, buoyed up by turbulence and dispersed throughout the whole flow. The rate of transport in each mode can be described in terms of the power supplied by the transporting agent, that is, the product of a flow stress and velocity. In order for sedimentary structures to be preserved, and in order for many of them to form, the transport rate must vary with time and/or space, so that there can be deposition and erosion.

Chapter 3

PARTICLE MOTIONS AT LOW CONCENTRATIONS: GRADING IN PYROCLASTIC-FALL DEPOSITS

INTRODUCTION

Explosive volcanic eruptions provide the commonest and most spectacular circumstances under which extensive layers of detritus are accumulated as the result of the fall or settling of debris through a fluid which only accidentally serves as a transporting agent. Such eruptions are consequent on the violent exsolution of dissolved gases following the release of pressure on ascending magma, or result from the combination of the magma with water existing at the eruption site.

Huge amounts of energy are released during explosive eruptions. Some is used to propel volcanic bombs and blocks of country rock from the crater on to the surrounding cone. These large pyroclastic fragments travel ballistically, for their path, like that of an artillery shell, is determined primarily by the angle of ejection and initial velocity, which may amount to several hundred metres per second. More of the energy is used in forcing upward, commonly for many kilometres, a column of gases and magma fragmented to grades smaller than bombs or blocks, and typically down to sizes as small as dust. The particles transported upward in an eruption column are for the most part insufficiently weighty to behave ballistically, but instead settle along paths controlled by falling velocity and wind speed. Their concentration in the column is very low, whence settlement is substantially unaffected by the presence of neighbours. Wind strength has but little influence on the paths of the ballistic particles, but strongly affects the spread of the finer debris. Only when the wind is gentle will this accumulate fairly evenly around the crater. A strong wind acting on a tall eruption column will cause debris to be deposited over a plume considerably elongated downwind from the vent. Some particles may be transported for hundreds of kilometres before reaching the ground, and the very finest, spewed into the upper atmosphere, may remain suspended for months or years.

A marked vertical and lateral grading should normally arise in the deposits making up a plume, for the coarser debris will tend to settle out earliest and closest to the volcano. An axial symmetry, increasingly distorted as the ambient wind grows stronger, should mark the lateral grading. The acquisition of these patterns by *pyroclastic-fall deposits* (Sparks and Walker, 1973) forms the subject of the present chapter.

PYROCLASTIC DEBRIS

The solid matter ejected by volcanoes, known as *tephra,* is extremely varied in both size and composition, Ollier (1969), MacDonald (1972), and P.J. Francis (1976) providing useful introductory descriptions. Texturally, *ashes* are here taken to be particles less than 0.004 m across, *lapilli* are fragments between 0.004 and 0.032 m in diameter, and *bombs* or *blocks* are pieces of debris larger than 0.032 m. Compositionally, the fragments range from acidic to basic glass, with or without phenocrysts, to whole or broken mineral crystals, to chunks torn from the walls of the volcanic vent or crater. Debris of the latter origin ranges widely in size and is often described as accidental or lithic. Much ash consists of angular glass shards (vitric fragments) displaying characteristically curved re-entrant margins that once bounded gas bubbles. Whole or broken crystals and pieces of vesicular pumice are also commonly of ash grade. Lapilli are of many kinds, ranging from relatively dense single crystals, lithic fragments, or blobs of lava (Pelée's tears), through moderately dense clinker-like debris known as scoria, to highly vesicular pumice. Accretionary lapilli are smooth, roller-shaped to spheroidal masses which result from the layer-by-layer accretion of ash around a wet nucleus, such as a rain-drop or a wetted crystal. Bombs are of many kinds. Some are simply large, irregular pieces of scoria, whereas others are aerodynamically moulded masses that approach a regular form. Blocks are angular fragments, either torn from the vent, or formed by the disruption of lava already partly cooled.

Because of the compositional variability of tephra, particle size alone is an uncertain guide to its behaviour either in the atmosphere or when sinking through water. Although most of the crystals encountered in tephra are of quartz and feldspar, ferromagnesian minerals of a significantly higher density are commonly found. Pumice varies greatly in shape and much has so small a bulk density that the pieces at first float when showered on to the sea or a lake.

CLASSIFICATION OF EXPLOSIVE VOLCANIC ERUPTIONS

Volcanologists recognize that volcanoes differ widely as to the scale and violence of the explosive eruptive behaviour displayed. Attempts to distinguish between modes of behaviour have been made using such terms as "Plinian", "Hawaiian", and the like, derived from particular volcanoes or historic eruptions, but these have been difficult to define precisely, and their application has depended greatly on the chance observation of an actual eruption.

G.P.L. Walker (1973) has pointed out that an explosive eruption is characterized not only by its scale, as measured by the volume of ejecta produced, but also by its explosive violence, which involves the rate of energy expenditure and the way in which the energy is used. Ejecta produced by a violent eruption is spread as a sheet over a large area around the volcano, whereas the debris from a weak explosion accumulates as a cone localized on the vent. The degree of fragmentation of the ejecta, and the particle shapes that are represented, should also be a function of the violence of eruption. Finely comminuted debris should result from violent eruptions, whereas relatively gentle ones should yield much coarse scoriaceous material and Pelée's tears.

These ideas become the basis for a practical classification of eruptive violence when measures of dispersal and fragmentation are defined. G.P.L. Walker (1973) measured dispersal as the area of a pyroclastic-fall deposit enclosed by the isopach at which the thickness is 10^{-2} times the maximum thickness. His measure of fragmentation is the percentage of the deposit that is less than 0.001 m in size at the place where the axis of the plume is crossed by the isopach at which the thickness is 10^{-1} times the maximum. Figure 3-1 summarizes Walker's classification (see also Booth et al., 1978). Strombolian eruptions are only mildly explosive and create eruption columns less than 1 km high above steep-sided cones of coarse scoriaceous to semi-plastic debris. At the other end of the scale, Plinian eruptions are extremely violent, leading to eruption columns often 20–30 km tall, and to considerable magma fragmentation. The debris is angular and predominantly pumice, glass shards, and crystals. Intermediate in character is the sub-Plinian (Volcanian) eruption, of moderate violence. Surtseyan eruptions occur in aqueous environments and are associated with extremely fragmented tephra, but as are general class are unrestricted as to extent of dispersal.

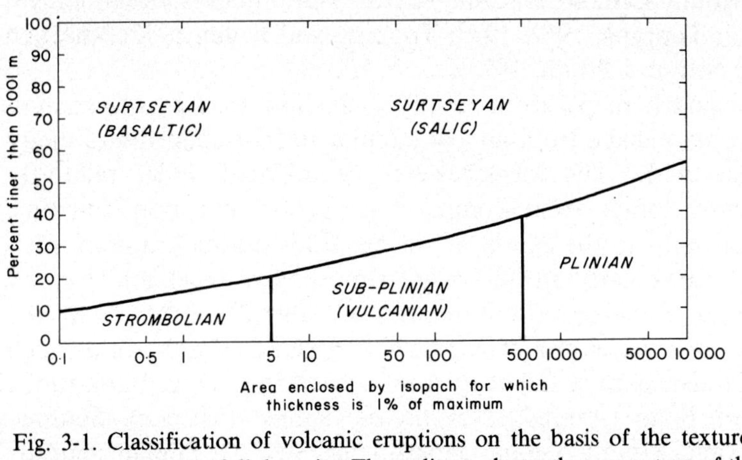

Fig. 3-1. Classification of volcanic eruptions on the basis of the texture and spread of the resulting pyroclastic-fall deposits. The ordinate shows the percentage of the fall deposit that is finer than 0.001 m at the place where the axis of the plume crosses the isopach at which the fall has one-tenth of the maximum thickness. The abscissa shows the area in km^2 enclosed by the isopach for which the thickness is 1% of the maximum. Adapted from G.P.L. Walker (1973) and Booth et al. (1978).

THICKNESS CHANGES IN PYROCLASTIC-FALL DEPOSITS

The successful study of thickness and other changes in pyroclastic-fall deposits depends crucially on an ability unequivocally to recognize the deposit of a particular eruption wherever it is exposed. This is rarely possible for more than short distances in the older stratigraphic record (e.g. Slaughter and Early, 1965), partly because tephra is prone to alteration. However, exceptions are provided by the widespread pyroclastic-fall deposits in the Precambrian of Australia (Trendall, 1965) and South Africa (Lowe and Knauth, 1978), the Ordovician of the Appalachian Mountains (Eaton, 1964), and the Eocene of Denmark (S.A. Andersen, 1937; Norin, 1940), which as individual beds can be traced in some instances for more than 100 km.

Many more recently erupted tephra plumes can be traced over most of their original extent, allowing thickness changes to be portrayed in considerable detail. Numerous falls related to volcanoes in the Mediterranean (Lirer and Pescatore, 1968; Booth and Walker, 1973; Lirer et al., 1973; Bond and Sparks, 1976; Watkins et al., 1978) and Atlantic areas (Eaton, 1963, 1964; Walker and Croasdale, 1971; Booth, 1973; Self et al., 1974; Self, 1976; Booth et al., 1978; Sparks et al., 1981) have been treated in this way. There are several studies of the same kind from the Americas (Larsson, 1937; Segerstrom, 1950; Eaton, 1963, 1964; Fisher, 1964; Kittleman, 1973; Koch and McLean, 1975; Bloomfield and Valastro, 1977; Bloomfield et al., 1977; Rowley, 1978; Crandell, 1980) and the Pacific region (Minakami, 1942b; Wilcox, 1959; Eaton, 1963, 1964; Lerbekmo and Campbell, 1969; Lerbekmo et al., 1975; Pain and Blong, 1976). Many detailed studies have been undertaken of New Zealand tephras, particularly those associated with the Tongariro and Mount Egmont volcanic centres (Eaton, 1963, 1964; Lloyd, 1972; Neall, 1972; Topping, 1972, 1973; Topping and Kohn, 1973; Vucetich and Pular, 1973; Self and Booth, 1975).

The selected isopach maps given in Fig. 3-2 show that pyroclastic-fall deposits range in plan shape from almost circular to lobe-shaped and many times longer than wide. The former were accumulated under relatively windless conditions, often from comparatively short eruption columns, whereas the latter point to the action of strong winds during eruption, such winds occurring particularly in the high atmosphere, reachable by tall columns. The falls cited differ enormously in lateral extent, from the vent-localized Strombolian scoria of Heimaey in Iceland (Fig. 3-2a), through the mainly sub-Plinian pumice fall of Furnas in the Azores (Fig. 3-2b), to the Plinian tephra of Fogo (Azores), Nevado de Toluco (Mexico), Somma-Vesuvius (Italy), and Mount Asama (Japan) (Fig. 3-2c–f). Plinian deposits even more extensive than these have come from Quizapu Volcano (Larsson, 1937), Mount Bona (White River Ash) (Lerbekmo et al., 1975), Katmai (Wilcox, 1959), and Santorini (Bond and Sparks, 1976; Watkins et al., 1978).

Does the decline in thickness of pyroclastic-fall deposits with increasing

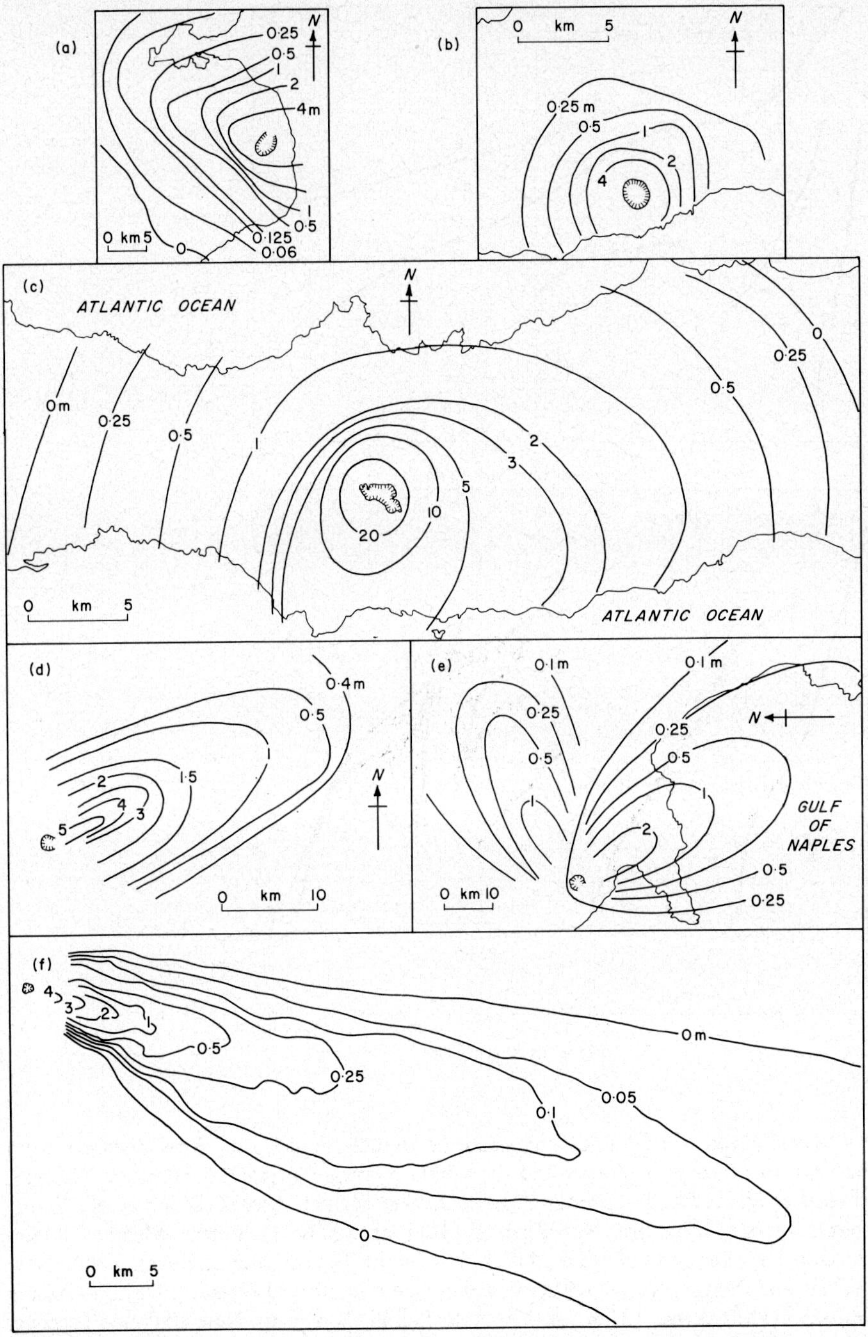

Fig. 3-2. Isopachs in metres illustrating the spatial variation of thickness of pyroclastic-fall deposits. a. Heimaey Scoria Fall, Iceland, from 19 January to 1 February, 1973 (after Self et al., 1974). b. Pyroclastic-fall of 1640 A.D. from Furnas, Azores (after Booth et al., 1978). c. Pumice A erupted from Fogo Volcano, Azores (after Walker and Croasdale, 1971). d. Upper Toluca Pumice erupted from Toluca Volcano, Mexico (after Bloomfield et al., 1977). e. Pompei and Avellino Pumice, Somma-Vesuvius, Italy (after Lirer et al., 1973). f. Temmei Pumice erupted from Asama Volcano, Japan (after Minakami, 1942b).

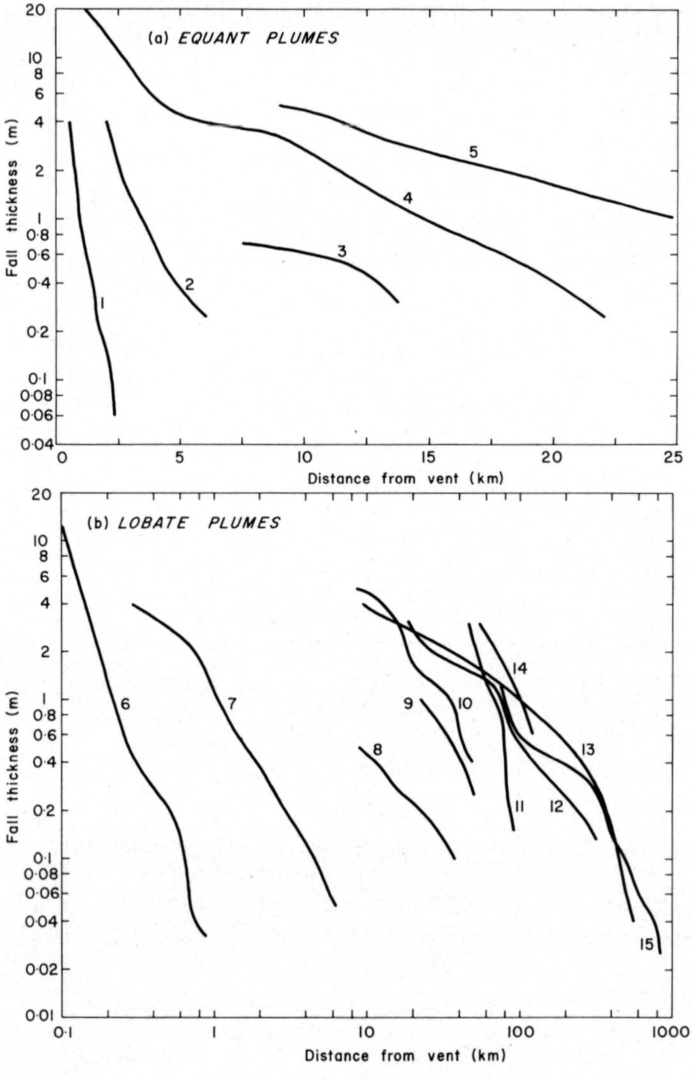

Fig. 3-3. Generalized variation in thickness of pyroclastic-deposits with distance from volcanic vent, in (a) equant plumes and (b) lobate plumes. Curves: 1—Heimaey Volcano, Iceland (Self et al., 1974); 2—Furnas 1740 A.D., Sao Miguel, Azores (Booth et al., 1978); 3—Papakai Tephra, Tongariro, New Zealand (Topping, 1973); 4—Furnas Member A, Sao Miguel, Azores Walker and Croasdale, 1971); 5—Tomba Tephra, Mount Hagen, Papua New Guinea (Pain and Blong, 1976); 6—Etna Volcano, Sicily (Booth and Walker, 1973); 7—Asama Volcano, Japan (Minakami, 1942b); 8—Te Rato Lapilli, Tongariro, New Zealand (Topping, 1973); 9—Avellino Pumice, Somma-Vesuvius, Italy (Lirer et al., 1973); 10—Upper Toluco Pumice, Nevado de Toluca Volcano, Mexico (Bloomfield et al., 1977); 11—Mazama Ash (northeast lobe), U.S.A. (Fisher, 1964); 12—Katmai, Alaska (Wilcox, 1959); 13—Minoan Tuff, Santorini, eastern Mediterranean (Bond and Sparks, 1976; Watkins et al., 1978); 14—Mazama Ash (northwest lobe), U.S.A. (Fisher, 1964); 15—White River Ash, Alaska (Lerbekmo et al., 1975).

distance from source (Fig. 3-2) follow any definite rule? The representative examples sketched in Fig. 3-3 suggest that a distinction can be made between comparatively equant plumes, say, less than 1.5 times longer than wide, and lobe-shaped ones. Thickness measured radially along the axis (if one is recognizable) of an equant plume decreases exponentially with distance from the vent, the smaller plumes tending to give the steeper slopes. Lobe-shaped plumes afford concave-upward curves which, on logarithmic scales, plot out as power functions except close to source. The effect of wind therefore is to distort the normal tendency for tephra to accumulate exponentially in amount around a volcano. The slope of the thickness–distance graph, measured either overall or near the intercept with the ordinate, could be a more reliable measure of dispersal than G.P.L. Walker's (1973), for his is sensitive to the accidental effects of wind in altering the shape of the plume.

Booth (1973) has discussed another way of analyzing the thickness distribution of pyroclastic-fall deposits. He gives a plot showing that in the majority of tephra plumes the area enclosed by a given isopach is a negative power function of the isopach thickness. What is interesting is that the graphs have similar slopes, approximately -1.2, though differing widely in position. This points to a marked similarity between pyroclastic-fall deposits of different scales and origins.

VERTICAL GRADING

Tephra showered on to water before settling seems invariably to form normally graded (coarse → fine) beds (Fig. 3-4). Norin (1940) and Pedersen and Surlyk (1977) found such grading to be well developed in the Danish Eocene pyroclastic-fall deposits, at sites far down the strongly lobe-shaped depositional plumes (S.A. Andersen, 1937). Normal grading was also reported by Slaughter and Early (1965) from bentonites in the Cretaceous Mowry Formation of Wyoming. Many individual falls in the Ordovician Cwm Clwyd Tuff of Wales are normally graded, though some are non-graded (Brenchley, 1972). A marked normal grading, of accretionary lapilli passing up into coarse to medium ash, characterizes the Precambrian pyroclastic-fall deposits described by Lowe and Knauth (1978).

Most pyroclastic-fall deposits that were accumulated on land, while mantling older topography and showing moderate to good stratification (e.g. Sparks and Walker, 1973), show little consistency as regards their vertical textural grading, to judge from available descriptions and records (Walker and Croasdale, 1971; Booth, 1973; Lirer et al., 1973; Topping and Kohn, 1973; Vucetich and Pullar, 1973; Koch and McLean, 1975; Self, 1976; Bloomfield et al., 1977; Booth et al., 1978). Assessment is complicated by the fact that some falls are composite, consisting of sub-units that represent pulses within a single eruption. The sub-units may individually show one type of grading while the fall as a whole displays another. Many comparatively thin unitary falls and sub-units are normally graded, whereas the thicker ones, particularly at sites close to source, are either reverse-graded (fine → coarse) (Fig. 3-5) or non-graded. Some thick beds show normal followed by reverse grading, the coarsest pumice and lithic fragments occurring in the middle levels. However, as Walker and Croasdale's grain-size analyses clearly demonstrate, the pumice and lithic fragments are invariably graded similarly in each unit or sub-unit.

Since pumice, widely variable shape and density, abounds in many pyroclastic-fall deposits, it might be asked whether the observed patterns of textural grading reliably indicate the aerodynamic grading of the sediment. A reverse textural grading need not mean a parallel increase in particle falling velocity. This aspect of grading has rarely been explored, but Lirer et al. (1973) found the Pompeii and Avellino falls ejected from Somma–Vesuvius to be reverse-graded both texturally and aerodynamically.

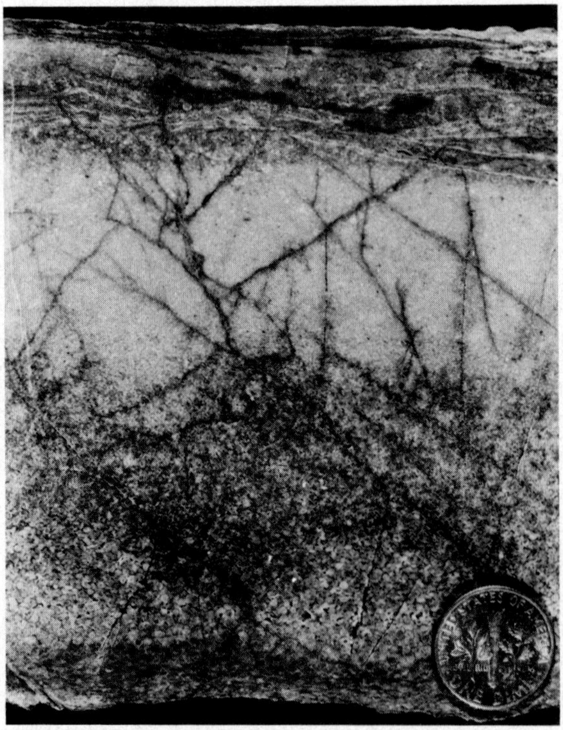

Fig. 3-4. An entire (thickness approximately 0.07 m) normally graded pyroclastic-fall into water, Middle Marker, Hooggenoeg Formation, Onverwacht Group, Barberton Mountain Land, South Africa. Photograph courtesy of D.R. Lowe (see Lowe and Knauth, 1978).

Fig. 3-5. A reverse-graded pyroclastic-fall deposit, Lower Member of Upper Toluca Pumice formation, Nevado de Toluca Volcano, central Mexico. Bed approximately 0.58 m thick. Photograph courtesy of K. Bloomfield (see Bloomfield et al., 1977; Bloomfield and Valastro, 1977).

LATERAL GRADING

Pyroclastic-fall deposits rapidly become finer grained with increasing distance from source. Basically, two methods have been used to document these changes: (1) measurement of the average diameter of a fixed number of the largest fragments exposed at a site, and (2) a full size-frequency analysis of the coarsest sediment, or of a channel sample. The first has the advantage of speed and applicability to both loose and lithified tephra, but is somewhat rough and ready, and subject to errors related to exposure size. The second, readily applicable only to loose material, is time-consuming but allows characterization in terms of both central tendency and spread of size. Moreover, the data allow pyroclastic-fall deposits to be compared texturally with tephra accumulated in other ways.

Minakami (1942b) characterized the tephra produced in the 1783 eruption of Mount Asama, Japan, by measuring the mean diameter of the five largest pumice clasts at a series of sites along the axis of the markedly elongate

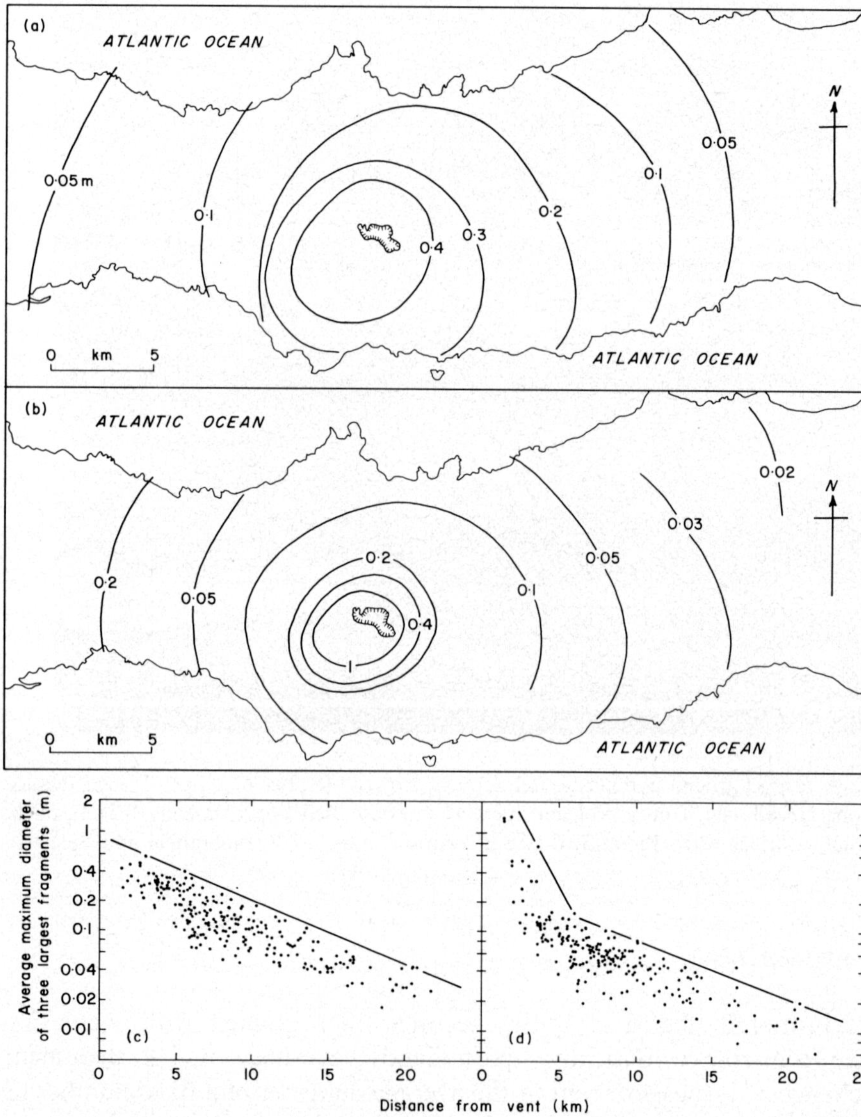

Fig. 3-6. Spatial variation in textural characteristics of Pumice A erupted from Fogo Volcano, Azores (after Walker and Croasdale, 1971). a. Isograde map for pumice, showing the average maximum diameter in metres of the three largest pumice fragments at a site. b. Isograde map for lithic fragments, giving the average maximum diameter in metres of the three largest fragments at a site. c. Variation with distance from the vent of the average maximum diameter of the three largest pumice fragments at a site. d. Variation with distance from the vent of the average maximum diameter of the three largest lithic fragments at a site. Sample stations omitted from (a) and (b) but equal the numbers of points in (c) and (d), and may be seen in the original papers.

tephra plume (Fig. 3-2f). The average maximum size decreased steeply near to the crater but more gradually as the distance increased. This technique has since been widely applied to a range of pyroclastic-fall deposits, with similar results to Minakami's (Fisher, 1964; Walker and Croasdale, 1971; Topping, 1972; Booth, 1973; Booth and Walker, 1973; Lirer et al., 1973; Self et al., 1974; Bond and Sparks, 1976; Self, 1976; Bloomfield et al., 1977; Booth et al., 1978; Sparks et al., 1981). Representative of these data are the observations of Walker and Croasdale (1971) on the distribution of the average maximum size of pumice and lithics in Pumice A erupted from Fogo Volcano in the Azores (Fig. 3-6). Pumice size follows a simple exponential trend, but two graphs are necessary to describe the change in the lithic fragments. The scatter in Figs. 3-6c and d is considerable, and much greater than in Minakami's plot, primarily because no regard is paid to the position of sample points relative to the axis of the plume.

The kind of size–distance pattern established by Walker and Croasdale (1971) for the lithic fragments of the Fogo A deposits is representative of many other falls (e.g., Booth and Walker, 1973; Self, 1976; Booth et al., 1978), in some of which it is also repeated by the pumice. Significantly, the change of slope in all such cases occurs at an average maximum size of about 0.1 m or a little less, putting the debris represented by the steeper curve into the grade of bombs and blocks. The explanation of the change in slope in graphs such as Fig. 3.6d is thought by Walker and Croasdale (1971) to lie in a size-dependent shift in the aerodynamic behaviour of particles. Bombs and blocks, represented by the steep branch of the curve, are suggested to have behaved ballistically, in contrast to the smaller particles which are believed to have followed paths determined by their terminal falling velocities. Few studies specifically of the distribution of bombs and blocks have been made. On Mount Asama, Minakami (1942a) found them to occur within a few kilometres of the vent. Explosions directed vertically upward yielded bombs of a wide range of size distributed in a narrow annular zone. Asymmetrical distributions, in which the largest bombs lay furthest out, resulted from explosions that were directed away from the vertical. During an eruption of Arenal Volcano, Costa Rica, large bombs were projected more than 5 km from the vent (Fudali and Melson, 1972).

The median size of pyroclastic-fall deposits declines with increasing distance from the vent, while the degree of sorting of the tephra generally improves, as shown by the reduction in a sorting coefficient. Fisher (1964) and Walker (1971) first pointed to these trends, each analyzing existing information while adding new data of his own. Further confirmation comes from more recent studies (Walker and Croasdale, 1971; Topping, 1972; Booth, 1973; Booth and Walker, 1973; Kittleman, 1973; Lirer et al., 1973; Self et al., 1974; Bloomfield et al., 1977; Sparks et al., 1981). As an illustration, Fig. 3-7 shows how the textural composition and median size (phi units) vary in the Fogo A pumice with distance from the crater along

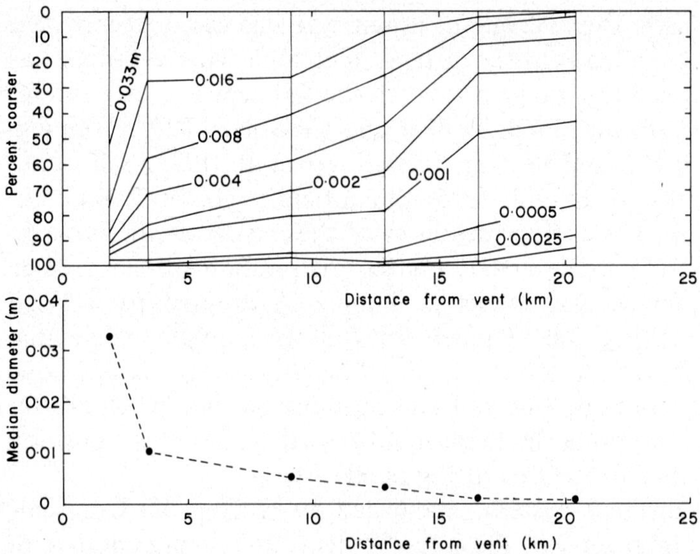

Fig. 3-7. Variation in textural character with distance from vent in Fogo A pyroclastic-fall deposit resulting from a Plinian eruption of Agua de Pau, Sao Miguel, Azores. The upper diagram shows the variation of textural composition of the coarsest sample collected from each of six localities on the axis of the plume (except for the locality nearest the vent, which lies off the axis). The lower diagram shows the variation in median diameter of these samples with distance from the vent. Adapted from Walker and Croasdale (1971, fig. 27) with additions.

the axis of dispersal of the plume. Juvigné (1977) established the trend in a fall by measuring the change in size of the crystal fraction.

Recognizing that tephra consists of variable proportions of constituents of a wide range in density, and that various mechanisms of fractionation operate during the accumulation of pyroclastic falls, it is logical to describe the lateral grading of these deposits in terms of the median terminal falling velocity. For the particles settling together at any point should be those which tend to be aerodynamically equivalent rather than similar in physical size. Kittleman (1973) and Lirer et al. (1973) attempted this task using grain size-frequency distributions as a basis. In the Crater Lake (Mazama) pumice, Kittleman found that the median terminal velocity declined as a power of the distance from the vent. With more data at their disposal, Lirer and his associates detected an exponential decline in the Pompeii and Avellino pumice-falls of Somma–Vesuvius. Self et al. (1974) and Self (1976) obtained exponential relationships for other falls. In analyses of this sort, however, reliable transformations of particle size into falling velocity are hard to achieve, because of variable particle shape and roughness.

Air-resistance neglected

The simplest analysis of the distribution of coarse volcanic ejecta around a vent rests on the equations of motion of a particle neglecting air resistance.

The equations of motion of a particle projected from the origin (crater mouth) of the coordinate system in Fig. 3-8 with an initial velocity V at an angle α may be written:

$$x = Vt \cos \alpha \tag{3.1}$$

$$y = Vt \sin \alpha - \tfrac{1}{2}gt^2 \tag{3.2}$$

where g is the acceleration due to gravity and t is the elapsed time. Substituting for t from eq. (3.1) into eq. (3.2) yields:

$$y = x \tan \alpha - \frac{g}{2V^2}x^2(1+\tan^2\alpha) \tag{3.3}$$

Let the particle travel above an inclined plane (sides of volcano):

$$y = x \tan \beta \tag{3.4}$$

where β is the slope of the trace of the plane in the vertical plane of the particle trajectory. Putting eq. (3.4) into eq. (3.3) leads to:

$$x = \frac{2V^2}{g}\,\frac{\tan \alpha - \tan \beta}{(1+\tan^2\alpha)} \tag{3.5}$$

whence the range r of the particle measured down the plane is:

$$r = \frac{2V^2}{g}\,\frac{\tan \alpha - \tan \beta}{(1+\tan^2\alpha)}(1+\tan^2\beta)^{1/2} \tag{3.6}$$

Equation (3.6) shows that particle range is a steeply increasing function of initial velocity. On differentiating this equation, and putting $\mathrm{d}r/\mathrm{d}\alpha = 0$, we

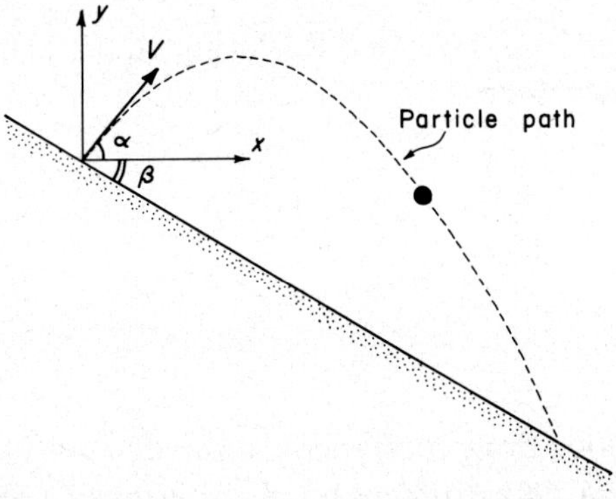

Fig. 3-8. Definition diagram for the motion of a projectile above an inclined plane.

find that the range is a maximum when $\alpha = \frac{1}{2}(90° - \beta)$, both larger and smaller projection angles resulting in a shorter range for the same velocity at the start. Equation (3.5) can be differentiated with respect to x to yield an expression for the angle at which the particle strikes the plane. Because of the neglect of air-resistance, eqs. (3.1) and (3.2) are independent of particle mass.

The above equations are clearly inappropriate to volcanic ejecta of small size, where the particle surface-area is large compared to the mass, but should represent satisfactorily the motion of sufficiently large bombs and blocks. Minakami (1950), Gorshkov (1959) and Herdervari (1968) have used them to calculate the initial velocities of volcanic bombs, with the ultimate aim of estimating the kinetic energy of eruption. But the equations seriously underestimate the initial velocity even of what might be thought to be large bombs. Fudali and Melson (1972) showed that for a bomb of diameter 0.8 m travelling over a range of 5 km, the simple equations above gave an initial velocity only about one-third as large as the initial value with air-resistance taken into account. In fact, the diameter would have to be 5–10 m before air-resistance became negligibly small, so that the velocity over the trajectory remained effectively constant.

Air-resistance included

Consider the motion of a solid particle travelling freely in the atmosphere with respect to a coordinate system in which the x-axis is horizontal and the y-axis vertically upward. A wind with velocity $U(x, y)$ is blowing. At some instant during flight, the particle has a velocity component u measured horizontally and v vertically. The drag force on the particle, however, is exerted in the direction of relative motion. Assuming that the vertical component if the wind velocity is small compared to the horizontal component, the magnitude of the relative particle velocity V is:

$$V = \left[(U-u)^2 + v^2\right]^{1/2} \tag{3.7}$$

and the inclination of the trajectory is $\tan^{-1} v/(U-u)$. Restating eq. (2.11), the components of the fluid force are therefore:

$$\frac{C_D}{2}\rho A V^2 \frac{(U-u)}{V}$$

horizontally and:

$$\frac{C_D}{2}\rho A V^2 \frac{v}{V}$$

vertically, in which C_D is the instantaneous drag coefficient, ρ the density of the air, and A the particle projection area. Substituting for A in terms of the particle diameter D and density σ, the general equations of motion of the particle in this case become:

$$\frac{d^2x}{dt^2}=\frac{3}{4}\frac{C_D(V,x,y)\rho(x,y)}{D}V[U(x,y)-u] \quad (3.8)$$

$$\frac{d^2y}{dt^2}=\frac{3}{4}\frac{C_D(V,x,y)\rho(x,y)}{D}Vv-\frac{\sigma-\rho(x,y)}{\sigma}g(x,y) \quad (3.9)$$

Here g and ρ are shown as functions of position because they display noticeable variation over the maximum vertical and horizontal ranges of volcanic ejecta (say, 50 km vertically by 100 km horizontally). The drag coefficient, being a function of Reynolds number, is dependent on both position and the value of V itself. The positional dependence arises because the molecular viscosity and density of air vary vertically and horizontally.

These equations are hard to solve for several reasons. Each acceleration component affects the other and a complete knowledge of $\rho(x,y)$ and $g(x,y)$ is generally not available. Finally, the drag coefficient is known only empirically and for simple shapes and steady conditions, the effects due to high speed and large accelerations being uncertain.

A numerical solution to the equations is described by Lionel Wilson (1972), on the assumption that the wind can be neglected, that g and ρ vary only vertically, and that the kinematic viscosity of air depends on the absolute temperature alone. The particles are assumed to be cylindrical in calculating the drag coefficient. The solution is given as tables of particle range as a function of particle density, initial velocity, and projection angle, and as graphs of fall-time as a function of density, size and height of release. Application of the solution to the bombs of the 1968 eruption of Arenal Volcano, Costa Rica, gave initial velocities agreeing with values obtained by Fudali and Melson (1972). Minakami (1942a) included air-resistance in calculating the flight of bombs produced during the eruptions of Mount Asama, but in a less satisfactory manner than Wilson. Further discussions on the behaviour and trajectories of bodies accelerating in viscous fluids are given by Torobin and Gauvin (1959a, 1959b, 1959c), Heywood (1962), Brush et al. (1964), Coulson and Richardson (1965), Odar and Hamilton (1964), Odar (1966), Hjelmfelt and Mockros (1967), Ockendon (1968), and Holland-Batt (1972a, 1972b).

Air-resistance predominant

Just as ballistic behaviour is an appropriate model for sufficiently large pyroclastic debris, so we can assume that air-resistance is overwhelmingly predominant in controlling the motion of particles of sufficiently small size. Such particles having been projected upward in an eruption column then sink to the ground at their terminal falling velocities while travelling horizontally effectively at the wind speed. We can thereby obtain useful insights into

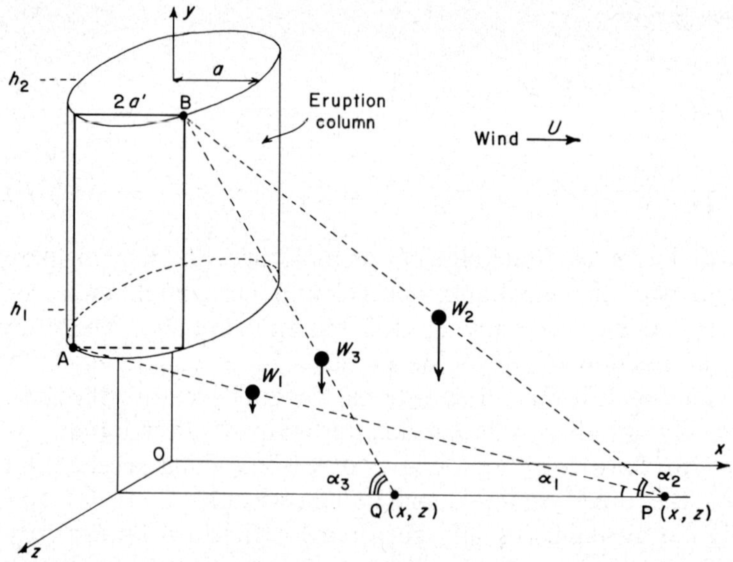

Fig. 3-9. Definition diagram for calculation of extreme textural composition and sorting of tephra, on the assumption that particles descend from a cylindrical eruption cloud at a horizontal velocity equal to that of a uniform wind.

the character of pyroclastic-fall deposits at distances from source further than would be reached by bombs and blocks.

Imagine that the eruption column produced in a volcanic explosion can be represented in shape by a right-circular cylinder with a vertical axis coincident with the y-axis of the coordinate system shown in Fig. 3-9. The column is of radius a with a base of height h_1 (crater rim) and top of height h_2 above ordinary ground level. The column was thrust rapidly upward into a uniform wind of velocity U blowing in the positive x-direction. Consider a point P with horizontal coordinates (x, z), $z \leqslant a$, lying downwind from the volcano. Neglecting the slight tendency of the column to spread through diffusion (Pasquill, 1962; Csanady, 1973), the tephra falling at P must follow paths in a plane parallel with the xy-plane but at a distance z from it, and therefore originate within the section of the cylinder bounded by the points $(-a', h_2)$, (a', h_2), (a', h_1) and $(-a', h_1)$.

If particles of any desired characteristic falling velocity are present everywhere in the column, and at concentrations that are too small for particle interactions to be significant, what falling velocities can be represented at P and in which order will they have appeared?

Inspection of Fig. 3-9 quickly answers the first part of this question. The particle of smallest terminal falling velocity that can reach P originates at the point A$(-a', h_1)$ in the section through the cylindrical column and takes the shallowest average path AP to the ground. Similarly, the particle of largest falling velocity to reach P must come from B(a', h_2) and take the steepest possible average path, shown as BP. If α_1 and α_2 are the average inclinations from the horizontal respectively of AP and BP, then:

$$\tan \alpha_1 = \frac{W_1}{U} \qquad x \geqslant a' \tag{3.10}$$

$$\tan \alpha_2 = \frac{W_2}{U} \qquad x \geqslant a' \tag{3.11}$$

in which W_1 and W_2 are the respective characteristic terminal falling velocities. But from the geometry of the problem:

$$\tan \alpha_1 = \frac{h_1}{(x+a')} \qquad x \geqslant a' \tag{3.12}$$

$$\tan \alpha_2 = \frac{h_2}{(x-a')} \qquad x \geqslant a' \tag{3.13}$$

whence, eliminating the tangents, we obtain:

$$W_1 = \frac{Uh_1}{(x+a')} \tag{3.14}$$

$$W_2 = \frac{Uh_2}{(x-a')} \tag{3.15}$$

for the minimum and maximum falling velocities as functions of wind speed, eruption-column geometry, and particle range. Subtracting eq. (3.14) from eq. (3.15), we derive:

$$\frac{\Delta W}{U} = \frac{h_2(x+a') - h_1(x-a')}{(x^2 - a'^2)} \tag{3.16}$$

for the relative velocity difference, where $\Delta W = (W_2 - W_1)$. Equation (3.16) becomes:

$$\frac{\Delta W}{U} = \frac{h_2}{x} \tag{3.17}$$

when a' is small compared to the range and h_1 is small compared to h_2. This is tantamount to writing eq. (3.15), with W_2 standing for the velocity range rather than maximum.

Consistent with what has been described (e.g. Fig. 3-6), eqs. (3.14) and (3.15) show that in a pyroclastic-fall deposit composed of particles of reasonably uniform density, the average maximum size of Minakami (1942b), Fisher (1964), and Walker and Croasdale (1971), to which W_2 corresponds, will decline with increasing distance along the plume. Similarly, the median size should decrease downwind, since both W_2 and W_1 simultaneously decline in this direction. The quantity ΔW measures the overall sorting of the deposit present at each station. Equation (3.16) shows that ΔW decreases

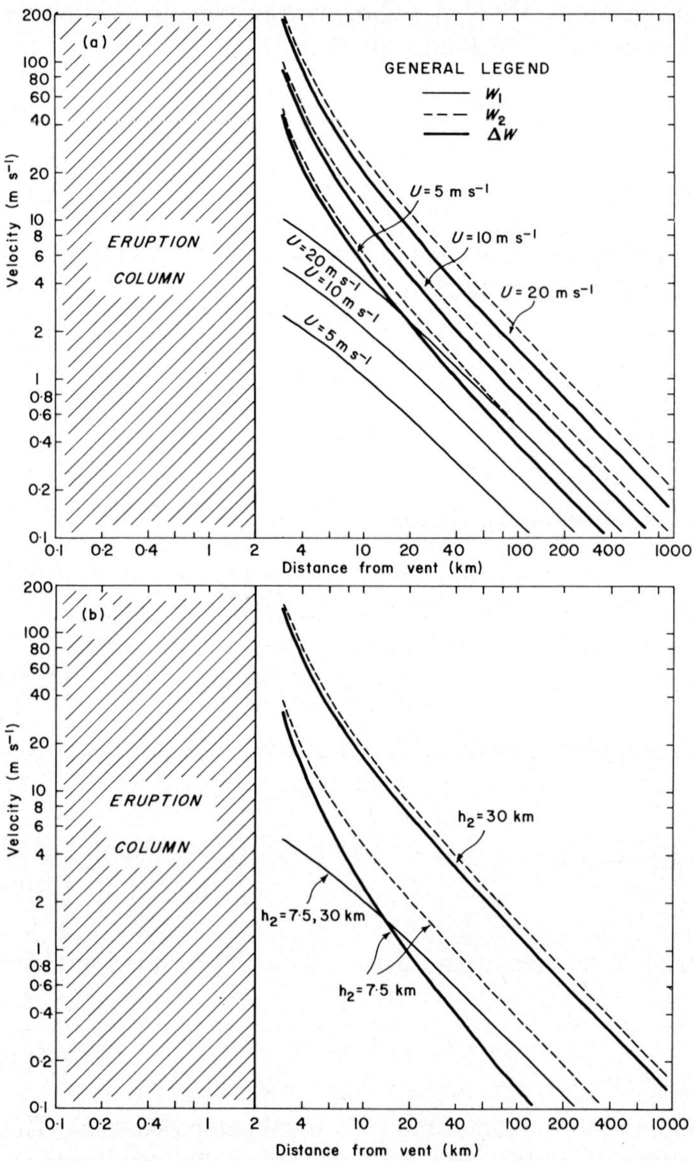

Fig. 3-10. Variation in coarseness (denoted by characteristic particle falling velocity) with distance from vent for pyroclastic fall deposits calculated from eqs. (3.14) and (3.15). (a) Effect of varying the wind-speed U for $h_1=2.5$ km, $h_2=10$ km, and $a=2.5$ km. (b) Effect of varying the height of the eruption column for $a=2.5$ km and a constant wind-speed of 5 m s^{-1}.

downwind, whence the fall improves in sorting distally. It is evident from eq. (3.17) that the decline in each case is approximately hyperbolic, the deposit extending to an infinite distance downwind, on the supposition that there is no limit on the available falling velocities. Figure 3-10a shows the effect of changing the wind speed on the downwind spread and variation of size and sorting in pyroclastic falls. The main effect of increasing the height and height-difference in the eruption column (Fig. 3-10b) is respectively to widen the spread of the deposit and worsen the sorting.

The order of arrival of particles at P is also easily deduced, assuming that fall commences at a single time. A particle starting at A requires a time $t_1 = h_1/W_1$ to arrive at P, whereas one starting at B takes $t_2 = h_2/W_2$. But as $h_1 = (x + a') \tan \alpha_1$ and $h_2 = (x - a') \tan \alpha_2$, with the tangents being given by eqs. (3.10) and (3.11), we find that $t_1 = (x + a')/U$ and $t_2 = (x - a')/U$. Evidently $t_2 < t_1$, whence the deposit at each station is normally graded, the particles of the largest characteristic terminal falling velocity having arrived first.

Equation (3.16) is unsuitable for the calculation of the distance of an unknown volcano from a site on one of its pyroclastic-fall deposits. Referring to Fig. 3-9, let W_3 be the largest characteristic terminal falling velocity measurable at a point Q situated at a distance x_3 from the source of tephra. At the previous site P, situated $x_3 + \Delta x$ from the source, the largest characteristic terminal falling velocity is W_2 as before. The particles giving these velocities originate at point $B(a', h_2)$ within the slice, whence:

$$\tan \alpha_3 = \frac{h_2}{x_3 - a'} \tag{3.18}$$

$$\tan \alpha_2 = \frac{h_2}{x_3 - a' + \Delta x} \tag{3.19}$$

Substituting from:

$$\tan \alpha_3 = \frac{W_3}{U} \tag{3.20}$$

$$\tan \alpha_2 = \frac{W_2}{U} \tag{3.21}$$

and eliminating h_2 between the resulting equations, we finally obtain:

$$x_3 = \frac{W_2 \Delta x + a' \Delta W}{\Delta W} \tag{3.22}$$

where $\Delta W = (W_3 - W_1)$, which contains only a' as an unknown, since W_2, W_3, and Δx can be measured at sites P and Q.

The distribution of tephra thickness at sites within a plume deposited from an eruption column can be calculated if we know the initial distribution of the particles of each characteristic falling velocity. Consider in Fig. 3-11 the particles settling on paths contained in a plane parallel with the (x, y) plane but separated from it by a constant distance z. The particles of smallest and largest falling velocity that can reach point P at (x, z) on the ground are, as before, typified by respectively $W_1 = Uh_1/(x + a')$ and

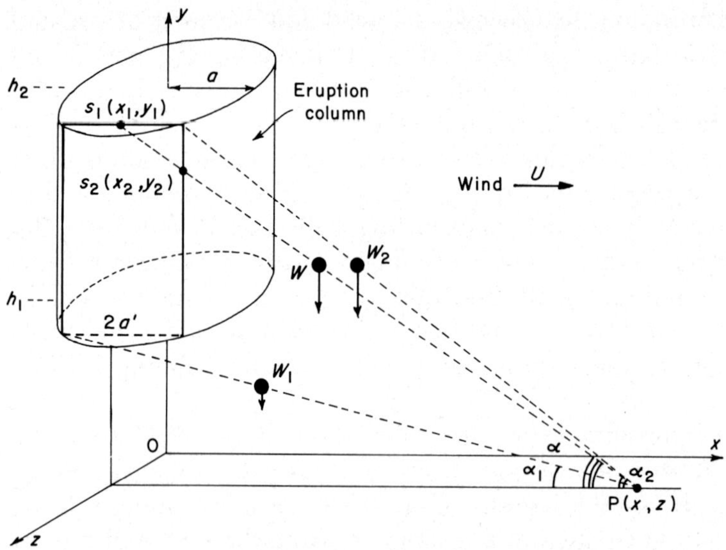

Fig. 3-11. Definition diagram for calculation of tephra thickness and textural composition, on the assumption that the particles descend from a cylindrical eruption cloud with a horizontal velocity equal to that of a uniform wind.

$W_2 = Uh_2/(x-a')$. Intermediate grains $W_1 < W < W_2$, follow a path that cuts the edges of the column at points $s_1(x_1, y_1)$ and $s_2(x_2, y_2)$. Therefore the total mass of tephra of characteristic velocity W falling over unit area at P is:

$$m = \sigma \frac{\tan \alpha}{(1 + \tan \alpha)} \int_{s_1}^{s_2} C(s) \cdot ds \tag{3.23}$$

where $C(s)$ is the fractional volumetric concentration of tephra in the eruption column as a function of distance along the particle path s_1P. On dividing m by the bulk density of deposited tephra, we obtain the thickness at P of the accumulate typified by the value W.

Examination of Fig. 3-11 and eq. (3.23) shows that the overall tephra thickness will decline downwind, at first rapidly and then more gradually, provided that initially particles of a wide range in W are reasonably uniformly distributed within the eruption column, or provided that those of larger W are concentrated toward its base. Figure 3-12 shows on the basis of eq. (3.23) the thickness and textural composition (in terms of characteristic terminal falling velocity) of the tephra plume deposited from an eruption column containing tephra at a fractional volume concentration of 0.025 instantaneously emplaced into a uniform, steady wind of 10 m s^{-1}. It is assumed that the tephra is divided equally between seven characteristic terminal falling velocities and that it was initially distributed evenly within the eruption column. The grains are assumed to have a uniform density of 1250 kg m^{-3} and to accumulate with a bulk density of 60% of this value. It

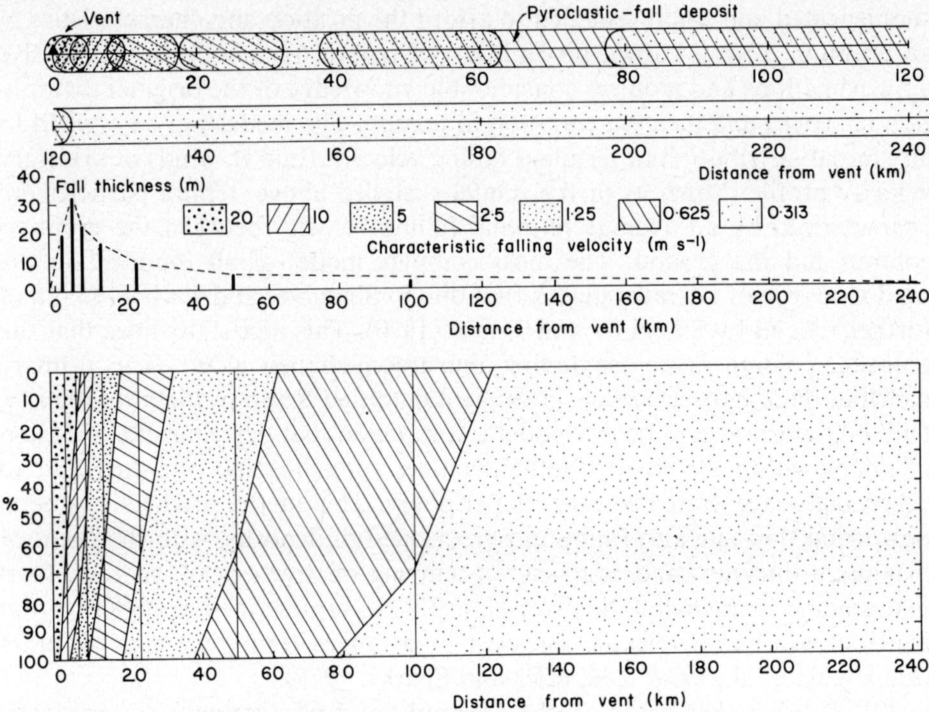

Fig. 3-12. Variation in coarseness (denoted by characteristic particle falling velocity) with distance from vent for pyroclastic-fall deposit calculated from eqs. (3.14), (3.15) and (3.23) for $U=10\ \mathrm{m\ s^{-1}}$, $h_1=2.5$ km, $h_2=7.5$ km, $a=2.5$ km, a uniform tephra fractional volume concentration of 0.025 in the eruption column, and for characteristic particle falling velocities of $W=20$, 10, 5, 2.5, 1.25, 0.625, and 0.313 m $\mathrm{s^{-1}}$. The upper diagram shows the deposit in plan (note the drastic effect of neglecting diffusion), the middle graph the variation in total thickness of the fall at seven intermediate localities, and the lower diagram the variation in the falling-velocity composition of these samples.

will be seen that the tephra ranges far downwind, and becomes finer textured and better sorted in this direction, being normally graded at each station. The plume should be compared with the lobe-shaped spreads of Figs. 3-2 and 3-3 and with Fig. 3-7. There is a qualitative similarity between the theoretical and actual examples, even though lateral diffusion is ignored, making the calculated plume of a uniform width. However, it is clearly inadequate to ignore diffusion, particularly in the case of fine particles which travel far downwind before settling.

Several models of pyroclastic-fall deposits related to the above have been described. Norin (1940) developed a method similar to that outlined for estimating the position of a volcano from the characteristics of the tephra at two stations a known distance apart, but assumed the eruption column to be negligibly wide. He also used his model to obtain wind-speed on the assumption of column-height. The model of Knox and Short (1964) is more sophisticated and again intended to afford the position and characteristics of an eruption from a known distribution of tephra. It operates by successive approximations and requires considerable knowledge of the original distribution of tephra and its areal variation in texture. The model can be applied to any initial distribution of terminal falling velocities and to winds of arbitrary velocity profile. Much as in the model sketched above, tephra particles are characterized by an average terminal falling velocity between the eruption column and the ground. The most complex model of all founded on the predominance of air-resistance is that due to Slaughter and Early (1965) and further refined by Slaughter and Hamil (1970). This model assumes that the eruption column is an expanding mushroom-shaped cloud. The primary distribution of particle sizes is given by a modified Rosin probability-density, and an important part of the model is a function that describes the diffusion of grains within the cloud, the lighter ones spreading more rapidly and so becoming concentrated toward its periphery. This model predicts a general downwind thinning and fining in pyroclastic-fall deposits, with the point of maximum thickness and coarseness occurring relatively near the vent. Normal grading is predicted. Shaw et al. (1974) have described a model for the distribution of fine volcanic ash over long ranges from very high altitudes (see Huang et al., 1979; Ledbetter and Sparks, 1979).

All of the models based on the assumption of predominant air-resistance carry two main weaknesses: (1) the assumption is strictly allowable only for particles of very small falling velocity compared to the wind, and (2) eruption columns develop over time (albeit short in many cases compared to particle descent-times) and not instantaneously as these models assume (even the Slaughter-Early-Hamil model demands an initial condition for the cloud). Nonetheless, the models give a qualitatively accurate picture of the areal distribution of thickness and texture in pyroclastic-fall deposits, and provide useful insights into the controls on these attributes. But they do not explain the frequent occurrence in tephra sheets of reverse textural grading, commonly linked to a compositional grading. Lirer et al. (1973) and Koch and McLean (1975) associate reverse grading with a gradual increase in energy over the time-span of an eruption and with the tapping of progressively deeper layers in a stratified magma chamber.

A model of pyroclastic-fall deposits that only involves peripherally the prime assumption of the others described was presented by Scheidegger and Potter (1965, 1968). According to this model, tephra deposition is governed by the decay of turbulence within the erupted cloud as it drifts with the wind. It is true that the turbulence due to eruption would influence the subsequent behaviour of the tephra dispersed upward, but it is doubtful if the effect is of over-riding importance, and questionable that any debris could be sustained in a state of transport by the turbulence. For transport to occur the turbulence would have to be anisotropic, as Bagnold (1966) has explained, but this state is unlikely in the absence of shear against rigid boundaries.

BOMB SAGS

When a freely moving body strikes a surface underlain by loose grains much smaller than itself, a crater is formed as the body penetrates to a distance dependent on its mass, shape and velocity at the moment of impact, and on the properties of the granular layer. Such impact craters, or bomb sags, are commonly formed around volcanoes when blocks and bombs, travelling at velocities in the order of 100 m s^{-1} and more, strike surfaces underlain by lapilli or ash. The diameter of the crater is a useful guide to bomb character (Fudali and Melson, 1972), but cannot generally be ascertained from buried tephra. In the latter case, buried craters are seen in the form of a bomb sag (Fig. 3-13), a buried depression, associated with depressed strata, that contains a block or bomb. Crater depth is in these circumstances the obvious measure of projectile properties, for it is comparable with the distance of penetration for all but gently to moderately inclined trajectories. A similar cratering, by blocks fallen from the roof or sides, is reported from layered basic igneous rocks (Irvine, 1965; Thompson and Patrick, 1968). Most work on cratering has been done in the fields of weaponry and astrophysics, and is either inaccessible or inappropriate to the problem of volcanic bomb sags. Allen et al. (1957) and Fuchs (1963), however, present general surveys of the penetration of granular materials by projectiles of low to moderate speed. The deceleration of a projectile

Fig. 3-13. Sag due to boulder-size bomb which penetrated the Minoan Tuff, Santorini. Geological hammer below bomb for scale. Photograph courtesy of R.S.J. Sparks.

entering a granular medium can be described using:

$$-\frac{dV}{dt} = aV^2 + bV + c \tag{3.24}$$

in which V is the projectile velocity at time t, and a, b and c are coefficients influenced by the mass and shape of the projectile and the nature of the medium. Deceleration occurs because the kinetic energy of the projectile is used in: (1) pushing aside the granular material, (2) breaking down the fabric of the granular material, and (3) distorting and fracturing the grains. At low speeds energy expenditure related to the first two items is predominant. Losses due to distortion and breakage become significant as well only at sufficiently high speeds. Equation (3.24) therefore exists in various simplified forms, according to the range of conditions of interest. In the Robins-Euler form, the deceleration is regarded as constant ($b=0$, $c=0$). Chisholm sets $a=0$ in his equation, appropriate to quasi-elastic impacts, whereas Poncelet puts $b=0$ in his version, a form with some appropriateness to high-speed cases. The form of the equation due to Résal has $c=0$. Wang (1971) derived an equation for the penetration of a low-speed projectile, on the supposition that the energy is used only in imparting motion to the receiving material and in disrupting its fabric.

By the time they strike the ground, volcanic bombs and blocks are unlikely to have velocities less than several tens of metres per second and are probably travelling at rates of several hundred metres per second (Walker et al., 1971). During the early stages of penetration into tephra, when bomb speeds are still high, fracturing of the disturbed pumice and other debris contributes substantially to the deceleration. In the final stages, however, the bomb loses energy primarily through disturbing and thrusting aside the tephra. Wang (1971), for example, quotes a projectile speed of 100 m s^{-1} as sufficient to cause grain fracture during the cratering of quartz sand, and the value for pumice probably is substantially lower. The experiments made by Allen et al. (1957) cover these two ranges of behaviour and may prove useful in the eventual interpretation of bomb sags. The penetration of the projectiles they used could be represented by two simple equations with empirical coefficients, one equation covering moderate speeds in granular material and the other low speeds.

SUMMARY

During explosive volcanic eruptions, large amounts of fragmented magma and debris originating in the country rocks of the vent are thrust violently upward into the atmosphere. Fragments on the scale of bombs and blocks behave essentially ballistically, following projectile-like trajectories and accumulating close to the vent. Their fall commonly results in the formation of impact craters and bomb sags. Lapilli and ash, presenting a much larger ratio of surface area to mass than bombs and blocks, drift downwind from the eruption column essentially under the influence of air-resistance, and may be deflected several hundred kilometres laterally by the wind before reaching the ground. The pyroclastic-fall deposits formed by explosive eruptions become thinner, finer grained, and better sorted with increasing distance downwind from source. Models that attribute the dispersal of tephra from an eruption column to control by air-resistance and the wind afford results in excellent qualitative agreement with field observations. These models further predict that pyroclastic-fall deposits are everywhere normally graded, as well as fining downwind. However, many deposits show no systematic local grading or are reverse-graded, particularly at relatively proximal sites, apparently because of an increase in the energy of the eruption during its course. The models thus far deviate from reality by assuming the instantaneous emplacement of the eruption column.

Chapter 4

PACKING OF SEDIMENTARY PARTICLES

INTRODUCTION

The particles forming detrital sediments assume at deposition a certain mutual relationship, the geometry of which is their primary packing. A packing may be described either by reference to the relative amount of the particles and by its relative emptiness, or in terms of local variations in the amount of particles, or again by a statement of the average number of contacts between a particle and its neighbours. The character of natural primary packings depends on the distributions of grain size and shape in the sediment, and on the forces acting on particles during deposition. These forces invariably include gravity, which pulls the particles towards the bed, and generally also a tangential fluid force associated with the transporting medium. The fluid force may act either directly or, at large enough concentrations of moving grains, indirectly through particle collisions. However, interparticle collisional forces may also arise in the absence of significant horizontal fluid movement, provided there is a sufficiently concentrated downward rain of grains. For completeness, we should add friction between particles and, in the case of sufficiently small grains (especially clay minerals), attractive forces.

Packing is a sedimentary structure of wide significance. It influences the amount of pore fluid which can be held in a sediment, the ease of movement of fluids through the deposit, the extent to which dissolved or dispersed materials can be introduced, and the strength of the aggregate under shearing or vertical load. Great attention to packings has therefore been paid in hydrogeology and petroleum engineering (Slichter, 1899; Graton and Fraser, 1935; Scheidegger, 1960; Von Engelhardt, 1960), soil mechanics (Terzaghi, 1925; Deresiewicz, 1958), and earthquake engineering (Prakash and Gupta, 1971; Seed and Silver, 1972). Packings are also of interest in fields of science and technology more distant from the earth sciences (B.S. Neumann, 1953; Crosby, 1960; Ridgway and Tarbuck, 1967; Gray, 1968; Morrow and Graves, 1969). They pose exciting mathematical problems (e.g. Minkowski, 1904; Broch, 1932; Fejes Tóth, 1953; Coxeter, 1961), and are used as models of the states of matter (e.g., Wollastan, 1813; Barlow, 1883; Rice, 1944; Metropolis et al., 1953; Bernal, 1959, 1960a). Reynolds (1885, 1886) studied sphere packings in conjunction with his theory of gravitation. Marvin (1939), Matzke (1939, 1946) and R.E. Williams (1968) see their relevance to an understanding of the geometry and arrangement of living cells. Particle packing is studied in chemical engineering, because packed beds provide a favourable environment for many reactions (e.g., Furnas, 1929; Haughey and

Beveridge, 1969), and it is relevant in branches of powder technology (e.g., Heywood, 1946; Dallavalle, 1948), ranging from pharmaceuticals, to ceramics, to metallurgy. Meldau and Stach (1934) and Rosin (1937) early stressed the importance of an understanding of packing in fuel technology. Most of us have some personal experience of the fact that the engineering quality of concrete aggregates and mortars depends critically on the correct choice for the blend of particle sizes, that is, on the final packing of the mixture (e.g., Furnas, 1931; Anderegg, 1931; D.A. Stewart, 1951). The packaging and storage of materials or goods may also demand a knowledge of packing.

A substantial though far from complete insight into particle packing and its controls now exists. Much early work gave attention to the ordered packing of uniform-sized bodies of simple form, notably spheres. More recent developments have stressed the empirical and theoretical study of disordered packings of monodisperse systems and of simple polydisperse aggregates. Although of limited significance, as several workers have rightly stressed (Carman, 1938; Fatt, 1956; Scheidegger, 1960; Smalley, 1970), this work provides an essential basis for understanding the complex packings of natural sediments, typified by variable particle size and shape.

SOME DEFINITIONS

Extending the rigorous definition of Rogers (1958), a packing is a system of closed solid particles arranged in space in such a way that no two particles have any inner point in common and each particle is in contact with at least one other. This is a more general definition than that of Graton and Fraser (1935), who required in addition the action of gravity as a stabilizing force. Packings are thus of three kinds (Gray, 1968), *ordered* (or systematic), *haphazard,* and *random*.

Ordered packings include the "piles" of Bernal (1964, 1965), the "regular" packings of Smalley (1970), and the regular and irregular assemblages of Manegold et al. (1931) and Manegold and Von Engelhardt (1933). Made only from bodies of the same size and shape, for example, spheres of equal radius, or spheroids of the same size and axial ratio, an ordered packing has a periodic character, in that a definite structural unit is regularly repeated within it. This unit may be represented by reference to the *point lattice* formed by the particle centres. The simplest arrangement of the centre-points which fully defines the repeating unit is the *unit lattice*. It in turn defines the *unit cell,* the simplest regular geometrical figure by which to characterize the assemblage, formed by joining some or all of the unit-lattice centre-points. The unit cell may be described by its face angles or, more simply, by the shapes of its faces. Ordered packings may be distinguished as either *simple* or *non-simple* (Smalley, 1970). The unit cell of a simple packing has centre-points only at the corners. That of a non-simple packing has centre-points on its edges and/or within its interior. Strictly, ordered packings exist only theoretically, since even ball bearings, the commonest modelling material, are made within definite tolerances.

Another theoretical packing is the random packing in which centre-points are randomly distributed. A random packing cannot be attained, though it may be reached, by real particles, because always during accumulation there are forces acting to guide particles to preferred sites. Real particles therefore assume haphazard packings, the graphically described "heaps" of Bernal (1964, 1965), in which centre-points are arranged in a mainly disorderly manner but where small ordered particle-groupings may exist locally. There are two reproducible limiting types of haphazard packing for any chosen material, *loose haphazard packing,* in which the centre-points are very nearly at their maximum possible mean spacing, and *dense haphazard* packing, in which the spacing is practically the minimum possible. Random and haphazard packings are *monodisperse* when composed of particles of a single size and shape, but *polydisperse* when formed from a mixture of sizes of the same or different shapes.

Any packing may be described by its *concentration, porosity,* and *coordination.* The concentration, C, is the fraction of the total space occupied by the assemblage which is actually taken up by particles. The porosity, P, is the fraction occupied by voids (i.e., $P = 1 - C$). Concentration and porosity are here defined as overall values, but it is also possible to define values of more local significance. For some tasks it is preferable to define the concentration or porosity linearly (Bagnold, 1954a; Sherman, 1970). The coordination, N, of the packing is the mean number of particle-contacts per particle. It is a true mean in all haphazard and random packings and in some ordered packings. In most ordered ones, it is the same for all particles that lie sufficiently distant from the assemblage limits.

All haphazard and all realizable random packings occur as finite assemblages, defined either physically, e.g. container walls, or by the memory of a computer. The values of C, P and N measured from such finite assemblages are found to differ from the values, C_∞, P_∞, N_∞, that can be obtained for infinitely large but otherwise identical packings. This is because of a boundary or wall effect, the particles packing less densely against the limits of the assemblage than in the interior. We find that $C < C_\infty$, $P > P_\infty$, and $N < N_\infty$, the extent of the difference increasing as the particle size approaches the assemblage size. Only with ordered packings do we find that the concentration, porosity and coordination for the unit cell are numerically equal to the respective quantities obtainable from the same packings of infinitely large size.

Simple packings

The simplest unit lattices comprise eight centre-points, and the simplest of these has the points arranged at the corners of a cube-shaped unit cell. This rudimentary packing is made by arranging eight equal spheres in contact at the corners of a cube. By suitable translations (rotations not permitted) of one of the layers of four lattice-points, the cube-shaped cell can be continuously transformed into a range of other parallelepipeds, terminating in a rhombohedron of six equal rhombus faces with face-angles of 60° and 120°. The translations create new packings, of increasing concentration and coordination. Smalley (1970) recently has elegantly demonstrated the character of this transformation, though it is implicit in the earlier work of Slichter (1899), Hrubisek (1941), and Kezdi (1966).

The resulting packings may be described in various ways of which those based on crystal symmetry have found wide popularity. However, perhaps the most versatile is Smalley's (1970), based on the shape of the faces of the unit cell. The untransformed cube-shaped unit cell has six equal square faces, whereas the ultimate rhombohedron comprises six equal rhombus faces. Intermediate transformations may be represented by combinations of

TABLE 4-I

Simple sphere packings based on either eight or twelve uniform particles

Number of centre-points defining unit cell	Symbol	N	C	Face angles	Symmetry
8	600	6	0.5236	90°, 90°, 90°	cubic
	420	6		90°, 90°, α_1	orthorhombic
	240	6		90°, α_1, α_2	triclinic
	060	6		α_1, α_2, α_3	triclinic
	402	8	0.6046	90°, 90°, 60°	hexagonal
	204	12	0.7405	90°, 60°, 60°	triclinic
	006	12	0.7405	60°, 60°, 60°	rhombohedral
	042	8		60°, α_1, α_2	triclinic
	024	10	0.6981	60°, 60°, 75°31′	tetragonal
	222	8		90°, 60°, α_1	triclinic
12	2600	5	0.4031	120°, 90°, 90°, 90°	hexagonal
	2060	5		120°, α_1, α_2, α_3	monoclinic
	2006	6	0.4654	120°, 75°31′, 60°, 75°31′	monoclinic
	2240	5		120°, 90°, α_1, α_2	monoclinic
	2204	9	0.6134	120°, 90°, 60°, 60°	monoclinic

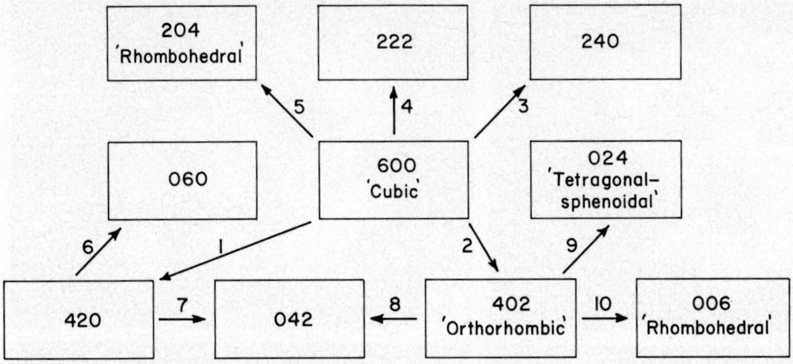

Fig. 4-1. The set of simple sphere packings derivable from a unit lattice of six centre-points arranged at the corners of a cube. The symmetry terms are those used by Graton and Fraser (1935).

square faces, the ultimate rhombus faces as specified, and intermediate rhombus faces. Following Smalley, let A denote a square face, B an intermediate rhombus face, and C a terminal rhombus face. The initial cube-shaped unit cell may therefore be represented by the symbol 6A0B0C, simplifying to 600. Visually the packings may be represented by face diagrams, in which the faces are depicted laid out flat with their correct shapes.

Table 4-I lists the packings Smalley (1970) distinguished in the transformation of a cube-shaped into a rhombohedral unit cell, and Fig. 4-1 gives the network of successive layer-movements connecting them. The symbol α denotes undetermined face angles, developed when intermediate rhombus faces are present. Because of an insufficient number of crystal classes, the nomenclature based on symmetry is ambiguous, in contrast to Smalley's unambiguous code. In this system of packings the coordination number and concentration increase broadly with growth in the number of ultimate rhombus faces. Whereas the coordination increases in steps, the concentration varies continuously within several sub-ranges spread over the total system range (e.g. Hrubisek, 1941). Taking the transformation of the cube-shaped cell as involving a rhombohedron with a uniform acute face-angle α, the concentration may be written:

$$C = \frac{\pi}{6(1 - \cos\alpha)(1 + 2\cos\alpha)^{1/2}} \tag{4.1}$$

as established by Slichter (1899).

The packings distinguished in Table 4-I and Fig. 4-1 are more numerous than have commonly been recognized. The 600-packing (Fig. 4-2a), early figured by Wollaston (1813), is the *reguläre 6er Packung* of Manegold and von Engelhardt (1933), Graton and Fraser's (1935) "cubic" packing, and Hrubisek's (1941) system B. It is incorrectly thought by Graton and Fraser

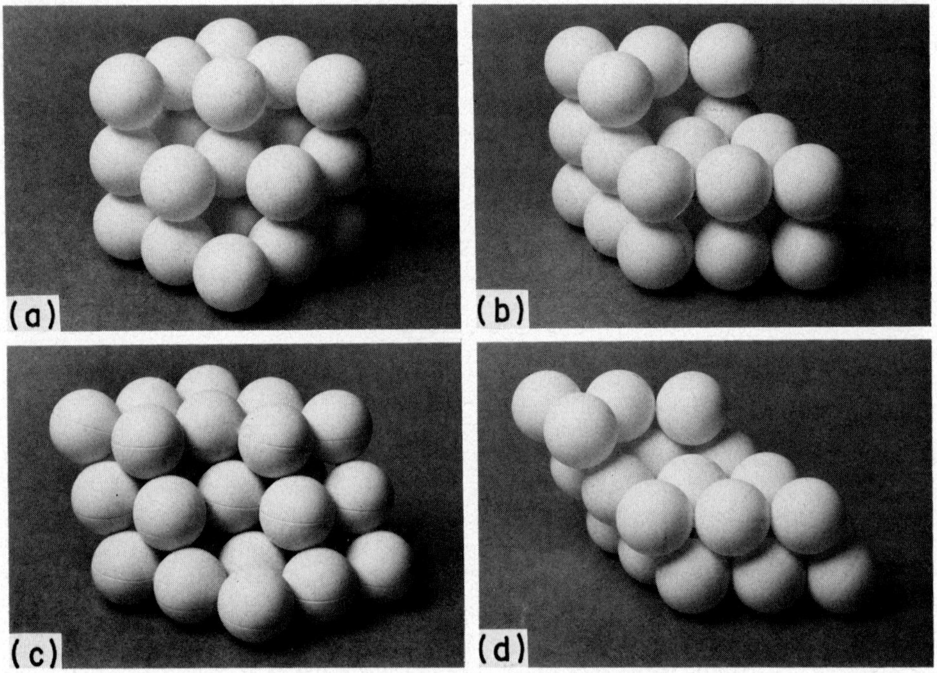

Fig. 4-2. Simple sphere packings. a. 600 (cubic) packing. b. 402 (orthorhombic) packing. c. 024 (tetragonal–sphenoidal) packing. d. 006 or 204 (rhombohedral) packing.

to be the loosest stable ordered packing. The 402-packing (Fig. 4-2b), Hrubisek's system D, is the "orthorhombic" packing of Graton and Fraser, and Manegold and von Engelhardt's *irreguläre 8er Packung.* Graton and Fraser call the 024-packing "tetragonal-sphenoidal" (Fig. 4-2c). It is Hrubisek's system F and the *irreguläre 10er Packung* of Manegold and Von Engelhardt. Twelve-coordinated packings (Fig. 4-2d) are illustrated by Wollaston (1813) and are discussed by Barlow (1883), Barlow and Pope (1907), Broch (1932), Ackerman (1945), Foord (1945), and Melmore (1947, 1949). There are two such arrangements, differing in symmetry, represented by Hrubisek's system A and Smalley's 006 and 204 packings. Manegold and Von Engelhardt noted a *reguläre 12er Packung,* and Graton and Fraser a twelve-coordinated "rhombohedral" packing. Empirically, twelve-coordinated packings are the densest that can be achieved with equal spheres.

Another set of simple packings arises from a unit lattice of twelve centre-points, arranged at the corners of a right-hexagonal prism with the angles of the hexagon faces uniformly 120°. This packing comprises twelve equal spheres arranged in contact in two hexagonal layers. In transforming this cell, the hexagonal sphere-layers are translated either parallel with a diagonal joining opposite corners of the hexagon, or along a line connecting

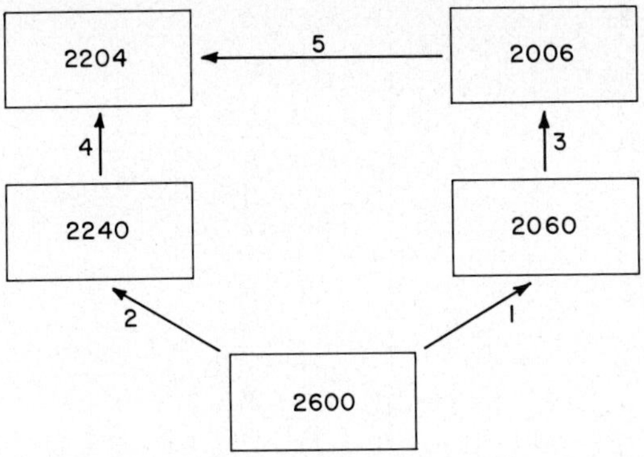

Fig. 4-3. The set of simple sphere packings derivable from a unit lattice of twelve centre-points arranged at the corners of a right-hexagonal prism of uniform face angle.

the mid-points of opposite sides. The four types of face involved are the hexagon faces (A_h), the square faces (A_s), the intermediate rhombus faces (B_{srh}), and the ultimate rhombus faces (C_{srh}) with face angles of 60° and 120°. The untransformed cell therefore has the symbol $2A_h6A_s0B_{srh}0C_{srh}$, simplifying to 2600. The notation therefore employs A, B and C to denote respectively the initial, intermediate, and ultimate faces. The first subscript on the intermediate and ultimate faces denotes the shape of the initial face, and the second the new shape (s = square face; rh = rhombus face).

Table 4-I lists the packing obtained involved in the transformation, Fig. 4-3 gives the connecting network, and Fig. 4-4 shows the two limiting assemblages. The use of a more open sphere arrangement means that a lower range of concentrations and coordinations is achieved than with the more versatile cube-shaped lattice. Barlow and Pope (1907) appear first to have distinguished the hexagonal arrangement of equal spheres, and it was later

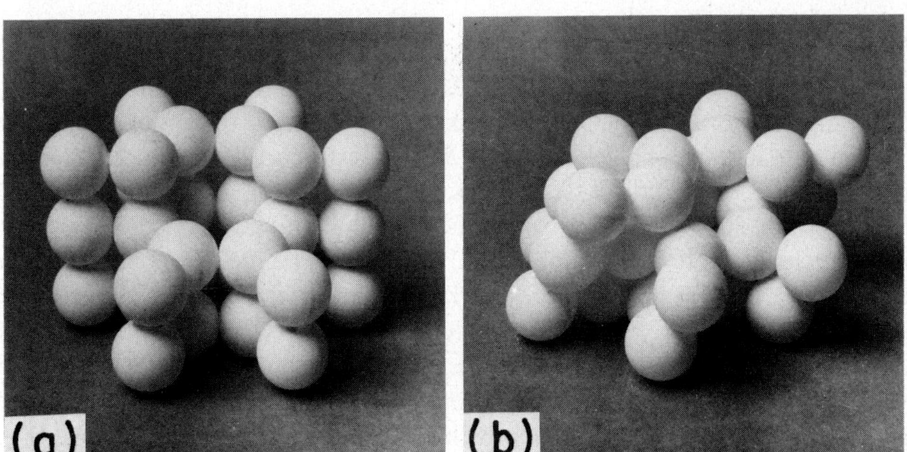

Fig. 4-4. Simple sphere packings. a. 2600 packing. b. 2204 packing.

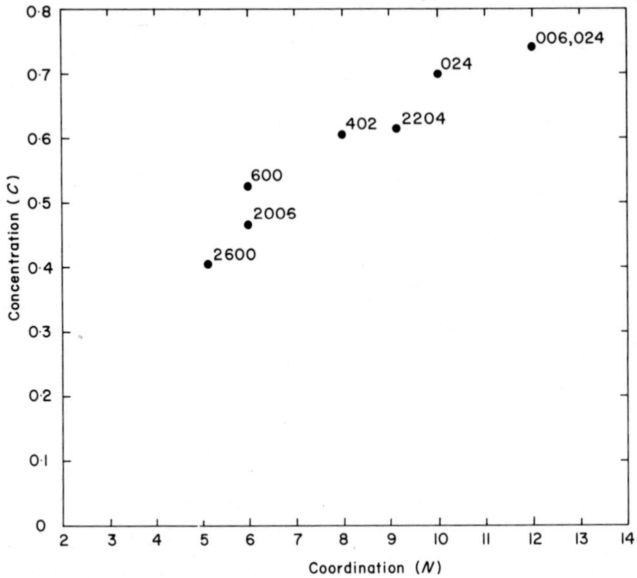

Fig. 4-5. Concentration as a function of coordination in some simple sphere packings. The number beside each point identifies the packing in terms of Smalley's notation.

proposed by Manegold and Von Engelhardt (1933, *irreguläre 5er Packung*), Salsburg and Wood (1962), and Coutenceau Clarke (1972, packing 8). The 2204-packing has also been identified (Manegold and Von Engelhardt, 1933, *irreguläre 9er Packung*; Hrubisek, 1941, system G; Coutenceau Clarke, 1972, packing 14). The 2006-packing has not previously been distinguished.

Figure 4-5 compares these simple packings in terms of concentration as a function of coordination. As may be inferred from the nature of the generating layer-movements, there is no single continuous function linking these two properties. Further simple packings are possible.

Non-simple packings

Graton and Fraser (1935) and many later workers did not distinguish non-simple packings because of an unduly narrow concern with the gravitational stability of particle assemblages. Numerous non-simple packings are gravitationally stable, however, and others of this kind achieve stability when interparticle attractions predominate. Many of the packings are known (Table 4-II), chiefly through the work of Manegold and Von Engelhardt (1933), Heesch and Laves (1933), Coutenceau Clarke (1972), and A.J. Smith (1973), and more are thought to exist. A wide range of concentration and coordination is covered and some packings are of great openness and complexity (e.g. Melmore, 1942a, 1942b). As with simple packings (Fig. 4-5), the concentration increases with the coordination (Fig. 4-6).

Some examples will illustrate non-simple packings. Figure 4-7a shows

TABLE 4-II

Non-simple packings of equal spheres

Packing	N	C	Authority
1	11	0.7182	Manegold and Von Engelhardt (1933), fig. 17k
2	10	0.6982	Coutenceau Clarke, pers. comm. (1973)
3	10	0.6982	Coutenceau Clarke (1972), packing 17
4	10	0.6658	Coutenceau Clarke (1972), packing 15
5	10	0.6658	Coutenceau Clarke (1972), packing 16
6	9	0.6481	Coutenceau Clarke (1972), packing 18
7	8	0.6802	Manegold and Von Engelhardt (1933), fig. 17g
8	8	0.6173	This book, Fig. 4-7a
9	8	0.6046	Coutenceau Clarke (1972), packing 12
10	8	0.6046	Coutenceau Clarke (1972), packing 13
11	8	0.5553	A.J. Smith (1973), packing 25
12	7	0.5612	Manegold and Von Engelhardt (1933), fig. 17e
13	7	0.5182	A.J. Smith (1973), packing 21
14	6.75	0.5084	This book, Fig. 4-7d
15	6	0.4861	Coutenceau Clarke (1972), packing 10
16	6	0.4467	This book
17	6	0.4535	Coutenceau Clarke (1972), packing 9
18	6	0.3702	A.J. Smith (1973), packing 23
19	6	0.5204	Meissner et al. (1964), open-bar cubic
20	6	0.5204	Meissner et al. (1964), open-bar hexagonal
21	5.75	0.4212	This book
22	5.5	0.4155	This book, Fig. 4-7c
23	5	0.4031	Coutenceau Clarke (1972), packing 8
24	5	0.3593	A.J. Smith (1972), packing 19
25	5	0.3593	This book, Fig. 4-7b
26	5	0.3240	Coutenceau Clarke (1972), packing 7
27	5	0.2604	A.J. Smith (1973), packing 20
28	5	0.2029	A.J. Smith (1973), packing 26
29	4	0.6582	A.J. Smith (1973), packing 23
30	4	0.3401	Manegold et al. (1931), fig. 6
31	4	0.1235	Heesch and Laves (1933), fig. 12
32	3	0.2234	Manegold and Von Engelhardt (1933), fig. 17a
33	3	0.1720	Hrubisek (1941), System N
34	3	0.0560	Heesch and Laves (1933), fig. 10
35	3	0.0450	Melmore (1942b)
36	3	0.0420	Melmore (1942a)

packing 8 of Table 4-II, which comprises two hexagonal layers of spheres arranged above and below a layer of three spheres at the corners of an equilateral triangle. Hence the unit cell is a right-hexagonal prism, but taller than in the simple 2600-packing, and the unit lattice possesses three interior centre-points. Several non-simple packings involve layers of eight spheres whose centre-points are divided between the sides of a square of length $(2+2\sqrt{2})$ radius units. In packing 25 (Table 4-II, Fig. 4-7b), of tetragonal

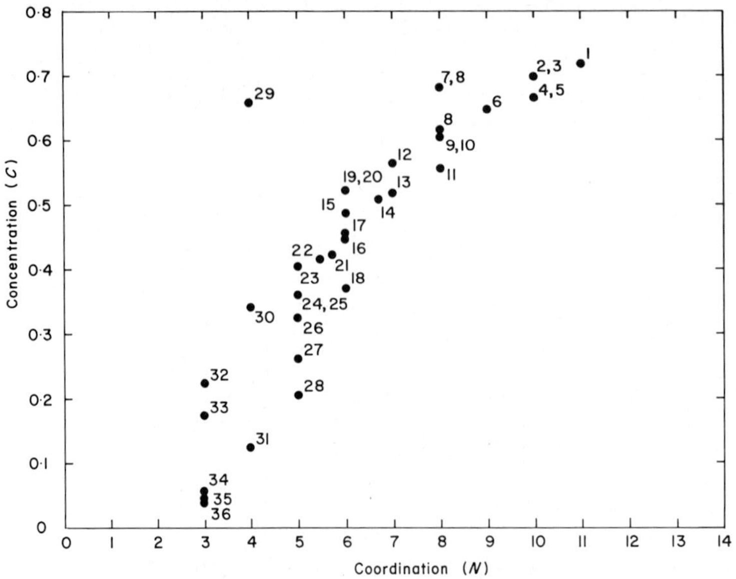

Fig. 4-6. Concentration as a function of coordination in the non-simple sphere packings numbered in Table 4-II.

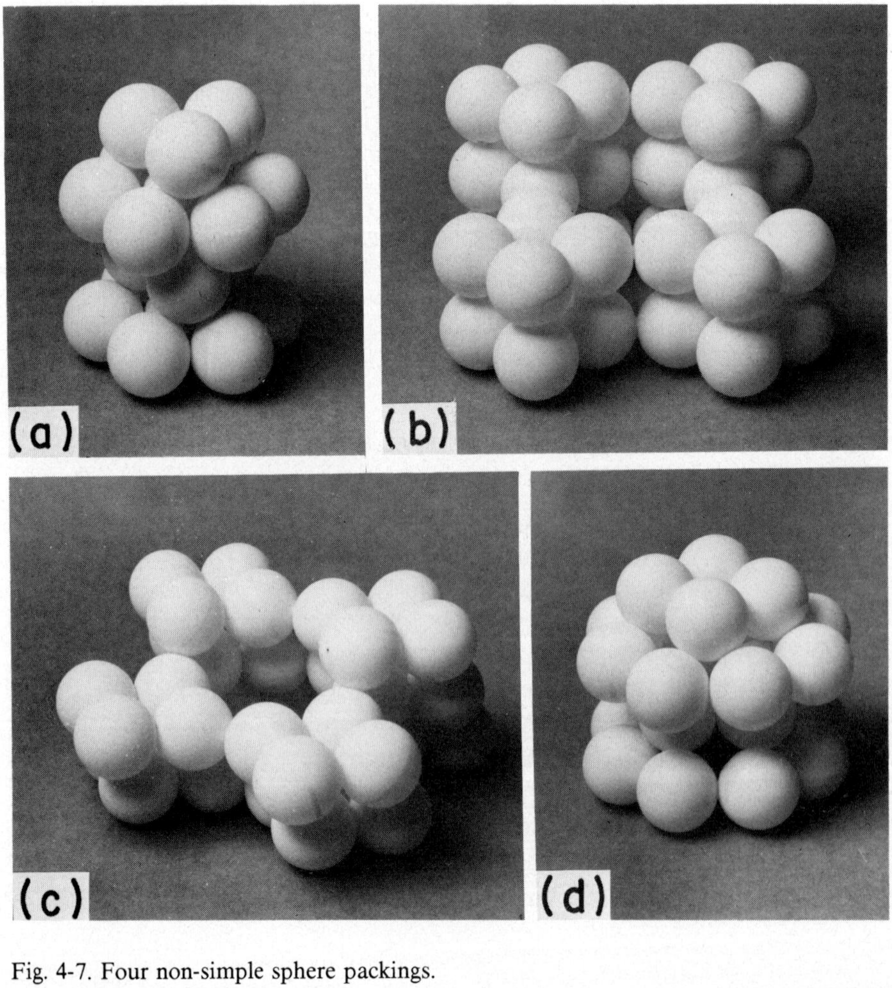

Fig. 4-7. Four non-simple sphere packings.

Fig. 4-8. A non-simple sphere packing composed of open tetrahedra.

symmetry, one such layer directly overlies another. Packing 22 (Table 4-II, Fig. 4-7c), having monoclinic symmetry, arises from packing 8 by a layer translation parallel with the diagonal of the square face of the unit cell. Packing 14 (Table 4-II, Fig. 4-7d) differs from packing 8 in having a square layer of spheres sandwiched between two layers of eight, the unit lattice therefore having four interior centre-points. Finally, Fig. 4-8 shows the open packing of tetrahedra recognized by Manegold et al. (1931).

Voids and their infilling

Slichter (1899) described general features of the complex void shapes encountered in simple sphere-packings based on the cube-shaped cell. Meldau and Stach (1934) and, independently, Graton and Fraser (1935), gave fully detailed accounts of these shapes. In the 600-packing, for example, the void contained in the unit cell—what may be called the unit void—is a concave octahedron, a body bounded by eight spherical triangles and six concave

TABLE 4-III

Effect of adding to packings based on the cube-shaped cell a secondary sphere just able to fill the largest available unit void

After Manegold et al. (1931), Horsfield (1934), and White and Walton (1937)

Primary packing		Packing modified by secondary spheres		
symbol	C	a_2/a_1	C	improvement in concentration (%)
600	0.5236	0.7320	0.7290	39.2
402	0.6046	0.5276	0.6933	14.7
204, 006	0.7405	0.4142	0.7931	7.1

squares. The twelve-coordinated packings contain two kinds of unit void, the larger, single void being a concave-cube and the two equal smaller ones forming concave-tetrahedrons.

Much attention has been given to the question of the size of largest sphere that can be fitted into the unit voids of packings based on the cube-shaped unit cell. Manegold et al. (1931) made calculations for the 600, 402 and twelve-coordinated packings, the densest of these also being investigated by Horsfield (1934) and by White and Walton (1937). The results appear in Table 4-III, where a_1 is the radius of the primary spheres and a_2 the radius of the largest secondary sphere fitting the largest unit void. The most marked improvement of concentration occurs for the 600-packing, with its relatively large unit void. The insertion of secondary spheres into the primary unit voids of course defines new voids, which may in their turn be occupied by further spheres, and so on. This sequence has been investigated for the dense packings (Horsfield, 1934; White and Walton, 1937; Deresiewicz, 1958) and for the 024-packing (White and Walton, 1937), the results of Deresiewicz appearing in Table 4-IV. Broadly, the concentration increases at first rapidly

TABLE 4-IV

Effect on concentration of adding the largest secondary, tertiary, etc. spheres to an 006 primary sphere-packing

After Deresiewicz (1958)

Type of sphere:	Primary	Secondary	Tertiary	Quaternary	Quinternary
Total number of spheres in packing:	8	16	32	96	160
a_n/a_1, (n=1, 2, 3, 4, 5)	1	0.4142	0.2247	0.1766	0.1163
a_n/a_{n-1}	–	0.4142	0.5425	0.7859	0.6586
C	0.7405	0.7931	0.8099	0.8426	0.8519

and then more slowly as further spheres are added, while the sphere radii halve and the grand total of spheres doubles. By introducing very small spheres into the remaining voids, concentrations comparable with 0.96 are theoretically attainable. These results suggest the qualitative influence of particle size-distribution on the packing of natural materials, but are of doubtful quantitative significance, because they rest on the assumption that the packing remains undisturbed as each type of sphere is inserted.

ORDERED SPHEROID PACKINGS

General comments

We may fairly argue that the isometric sphere is a poor model of natural detrital sedimentary particles, which are anisometric, and that equal sphere packings are poor models for natural detrital aggregates, which consist of polydisperse particles with generally a degree of preferred spatial orientation. The sphere, as a model particle, has been imposed on sedimentologists by workers (e.g. Wadell, 1935) narrowly concerned with such properties as sphericity and roundness. Impressed by the morphometric studies of Sneed and Folk (1958), Moss (1962, 1963, 1966), and N.C. Flemming (1965), I proposed the spheroid as the ideal particle shape for the study of the packing of natural detrital sediments (Allen, 1969d, 1969e, 1970a). Spheroid packings are not new, however, for Wollaston (1813) illustrated some in connection with his work on crystal structure, and W. Thomson (Lord Kelvin) (1889) alluded to them. White and Walton (1937) calculated the concentration of a "cubical" packing of prolate spheroids, and Peleg (1972) and Tsutsumi (1973) studied spheroid packings from the same set. Deelman (1974a, 1974b) has experimented with ellipses in two-dimensional studies of loaded packings.

The application of terms like "cubical", "orthorhombic", and "rhombohedral" by White and Walton (1937) and Allen (1969d, 1970a) to ordered spheroid packings is significant. It indicates a certain correspondence, realized intuitively rather than explicitly, between ordered spheroid and ordered sphere packings. W. Thomson (Lord Kelvin) (1889), who earlier saw this correspondence, considered that spheroid packings were strained arrangements of spheres. This notion is helpful, but it does not cover all of the possibilities where ordered spheroid packings are concerned. The spheroid is anisometric and may therefore be assigned an orientation in space, unlike the isometric sphere. Hence, when constructing ordered spheroid packings, it is necessary to specify the orientation of the particles relative to the edges of the unit cell. Consequently, a wider range of orderd packings is possible than for spheres; in some the spheroids have a perfect preferred dimensional orientation, whereas in others the particle orientation is less well ordered. In many spheroid packings, the concentration becomes a function of particle shape, as expressed by the ratio of the major and minor axes. For each type of packing, however, the coordination is unchanged by varying the axial ratio. Open packings with large voids are readily obtained.

Simple packings

The building of simple ordered spheroid packings under the usual condition that the constituent particles are touching, requires the following sequence of choices to be made: (1) unit cell, (2) spheroid shape (either oblate or prolate), and (3) particle orientation(s) relative to cell faces. The simplest and most general choice for the unit cell, for example, is an orthogonal parallelepiped of unequal sides. Since the possibilities implicit in this choice are enormous, demanding a monograph for their complete development, we shall here limit discussion to representative cases using equal prolate spheroids of major semi-axis a and minor semi-axis b.

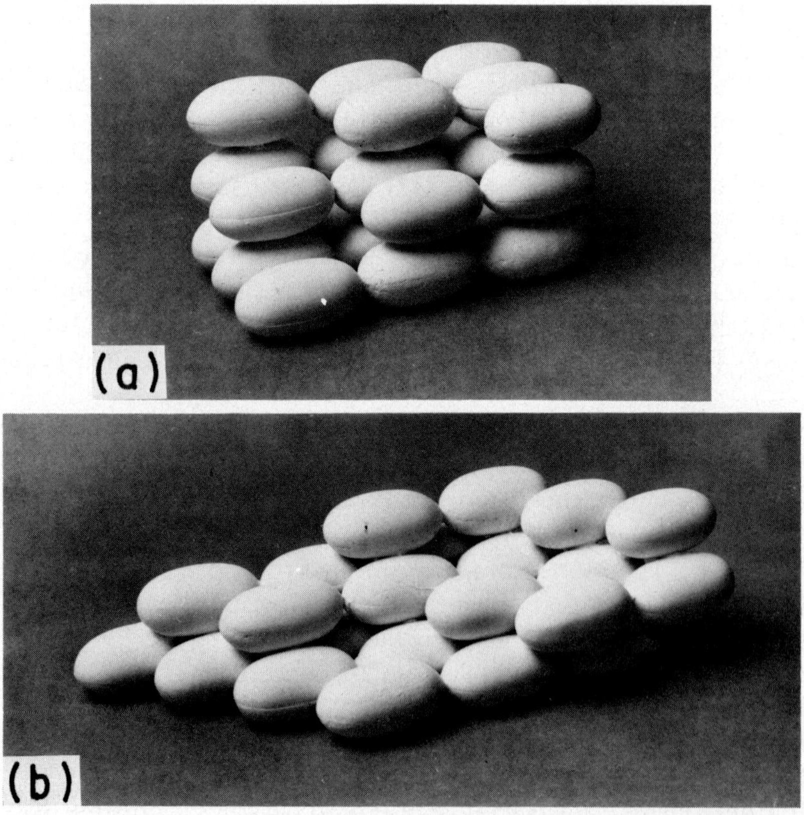

Fig. 4-9. Simple packings of prolate spheroids. a. 24000000 or "cubical" packing. b. 00000024 or "rhombohedral" packing.

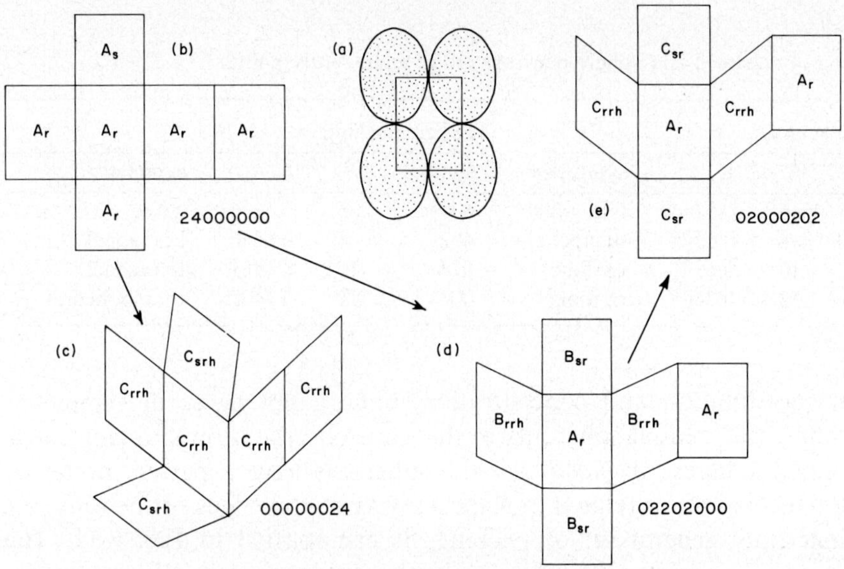

Fig. 4-10. Some of the set of packings of prolate spheroids derivable from a unit lattice of six centre-points arranged at the corners of a right-square prism.

We first use a rudimentary unit cell which is a right-square prism. The packing (Fig. 4-9a) consists of two layers each of four prolate spheroids, the major axes of which lie in the plane of the rectangular prism faces and parallel with the longer edges of these faces. Figure 4-10a shows the arrangement in a layer and Fig. 4-10b is the face diagram for the rudimentary unit cell. Keeping Smalley's (1970) notation, the rectangular faces are denoted by A_r and the square faces by A_s. By Smalley-type transformations, the rudimentary packing may be converted into a set exactly corresponding to the set based on the cube-shaped unit cell (Table 4-I). The ultimate packing of this new set (Fig. 4-9b), my "rhombohedral" packing (Allen, 1969d, 1970a), has a face diagram composed only of rhombus shapes (Fig. 4-10c). One sort of rhombus is derived from the initial square faces, and the second from the initial rectangular surfaces. They may be denoted as before by C_{srh} and C_{rrh}, and there are, of course, intermediate rhombus faces B_{srh} and B_{rrh}. A first layer-movement parallel with a longer face-edge alters each square face into an intermediate rectangle B_{sr} and eventually into the ultimate rectangle C_{sr} (Fig. 4-10d, e). Hence these packings demand for representation an eight-unit symbol, in the case of the rudimentary unit cell, for instance:

$$2A_s4A_r0(B_{sr}B_{srh}B_{rrh}C_{sr}C_{srh}C_{rrh})$$

simplifying to 24000000 (Fig. 4-10b). Table 4-V compares in detail the sets of sphere and spheroid packings. As was demonstrated (White and Walton, 1937; Allen, 1969d, 1970a), corresponding members in the two sets have an

TABLE 4-V

Comparison of spheroid and sphere packings, all of eight centre-points

Spheroid packing				Sphere packing			
symbol	*N*	*C*	symmetry	symbol	*N*	*C*	symmetry
24000000	6	0.5236	tetragonal	600	6	0.5236	cubic
04000200	8	0.6046	monoclinic	402	8	0.6046	hexagonal
00020004	10	0.6981	triclinic	024	10	0.6981	tetragonal
00000024	12	0.7405	triclinic	006	12	0.7405	rhombohedral

identical coordination and concentration, but do not agree in symmetry. Thus we may pack equal spheroids to the same coordination and concentration as equal spheres, *provided* that the spheroids have a perfect preferred dimensional orientation related in a special way to the edges of the unit cell.

A remarkably general set of packings is exemplified in Fig. 4-11. The essential packing consists of eight equal spheroids with parallel major axes, arranged in two rectangular layers. One sub-set is generated as the longer sides of the upper layer are moved parallel with the longer sides of the lower. In the rudimentary packing, corresponding to the 24000000-packing, each of the remaining larger faces of the unit cell is a rectangle; in the packings derived as above it is a rhombus of acute face-angle α. The spheroid major axes lie in planes parallel with the planes of the rhombus faces and make a uniform angle β with the plane containing each layer of spheroid centres. From the geometry of the configuration, we have $C = \pi/6$ when $(\alpha - \beta) =$

Fig. 4-11. Simple packing of imbricate prolate spheroids with parallel major axes.

90°, and otherwise obtain:

$$C=\frac{\pi ab}{6\sin\alpha\left\{\dfrac{\left(\dfrac{a^2b^2}{b^2+a^2\tan^2\beta}\right)^{1/2}\left(\dfrac{a^2b^2}{b^2+a^2\tan^2(\alpha-\beta)}\right)^{1/2}}{\cos\beta\cos(\alpha-\beta)}\right\}} \tag{4.2}$$

which is a maximum for each axial ratio when $\tan\alpha/2=\tan\beta=b/a\sqrt{3}$. The maximum improvement in concentration is a few percent for an axial ratio of 2, a fairly common value for natural sedimentary grains. The coordination in the ultimate packing is 8 and in the rudimentary and intermediate packings is 6. This sub-set, and therefore eq. (4.2), is subject to Rogers' (1958) condition that the spheroids share no interior points. We may in addition translate the rectangular layers parallel with their shorter sides, so completing the full set of packings, of which the set described in Fig. 4-10 and Table 4-V is a limiting sub-set. The ultimate coordination is 12.

A final set of simple packings arises by placing spheroids in directly opposite layers at the corners of a rhombus, such that the spheroid major axes on opposite sides of the rhombus are parallel with those sides. The set is derived by varying the angles of the rhombus faces. When the concentration is a minimum for each axial ratio, the packing is my "cardhouse" arrange-

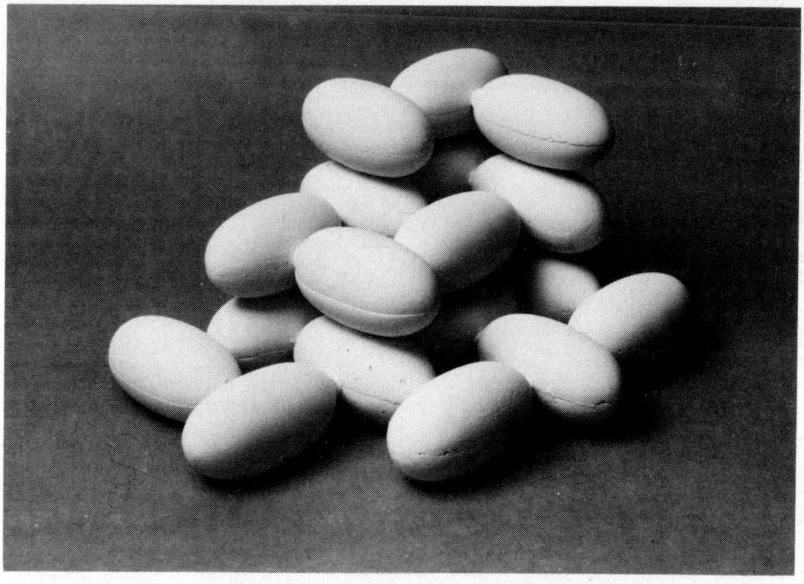

Fig. 4-12. Simple "cardhouse" packing of prolate spheroids.

ment (Fig. 4-12) (Allen, 1969d, 1970a), for which:

$$C = \frac{2\pi ab}{3(a+b)^2} \tag{4.3}$$

the unit cell then being orthogonal. The coordination is 6 in the rudimentary (cardhouse) and intermediate packings, but is 8 in the ultimate arrangement, through which the set is linked to the assemblages of Table 4-V and the general set above. The packings of the cardhouse set are relatively open, primarily because the spheroids are permitted two possible orientations.

Non-simple packings

Being anisometric, the spheroid offers fewer possibilities for the construction of non-simple packings than the sphere. Nevertheless, a few arrange-

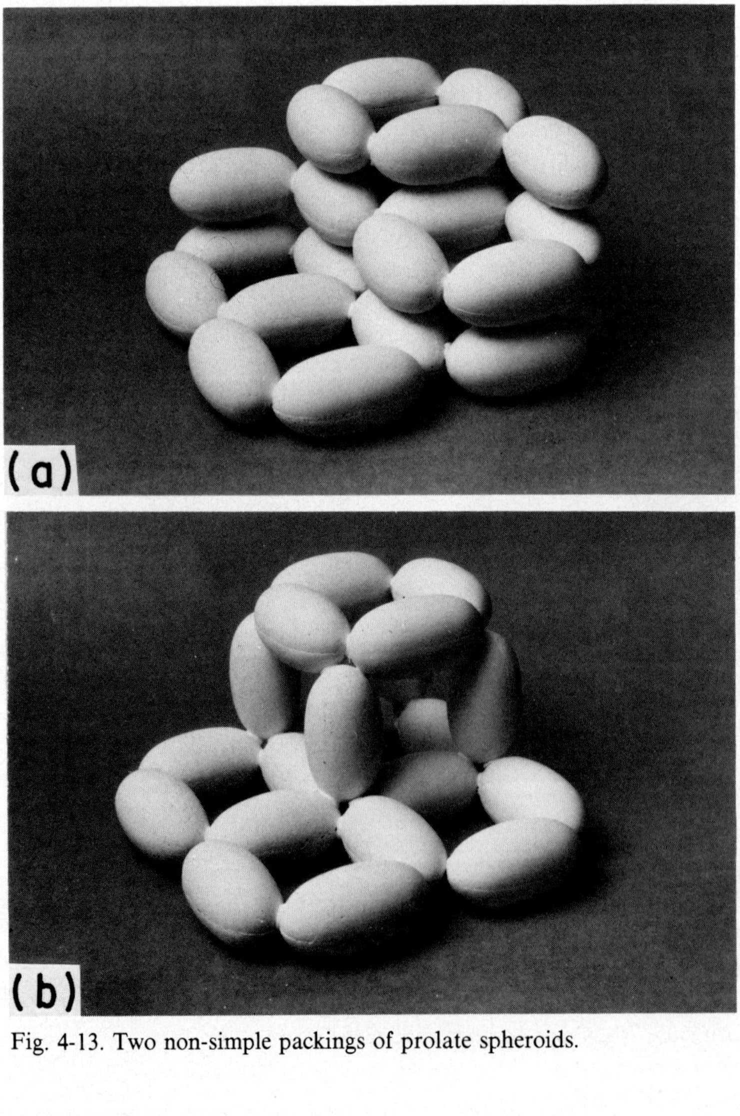

Fig. 4-13. Two non-simple packings of prolate spheroids.

ments have been proposed (Allen, 1969d, 1970a), all of a relatively open character and intermediate coordination.

An interesting set may be made by arranging spheroids in directly opposite layers on the mid-points of the equal sides of a rhombus, so that the major axes are coincident with those sides. Fig. 4-13a shows the rudimentary packing, when each rhombus face is a square, for which:

$$C = \frac{\pi ab}{3(a^2 + b^2)} \tag{4.4}$$

the coordination being 6. On putting $a = b$ this packing becomes the 600-packing. A related 8-coordinated packing arises when twelve touching spheroids are placed with their centres at the mid-points of the edges of a cube and their major axes concident with those edges (Fig. 4-13b). We then have:

$$C = \frac{\pi ab^2}{2(a^2 + b^2)^{3/2}} \tag{4.5}$$

and find that the packing is related to A.J. Smith's (1973) packing 25. In these packings the spheroid orientation is less than perfect, either two or three attitudes being assumed.

Two final examples rest on a unit lattice of twelve centre-points defining a unit cell in the shape of a right-hexagonal prism. In each packing the spheroids lie in directly opposite layers, with their centres at the mid-points of the sides of an equal-sided hexagon and their major axes in the plane of the hexagonal layers. The coordination is 6 and the packings differ only in the orientation of the spheroid major axes relative to the sides of the hexagonal faces. In one arrangement, the axes are coincident with the sides, giving:

$$C = \frac{\pi ab\sqrt{3}}{(9a^2 + 3b^2)} \tag{4.6}$$

In the second, the major axes lie at right-angles to the sides, whence the concentration is:

$$C = \frac{\pi ab\sqrt{3}}{(3a^2 + 9b^2)} \tag{4.7}$$

Within the plane of the hexagonal layers, the spheroid orientation approaches a uniform distribution in both packings. These examples have affinities with Coutenceau Clarke's (1972) packing 17 (Table 4-II).

ORDERED PACKINGS OF OTHER REGULAR SHAPES

Few workers have studied theoretically the packing of other non-spherical shapes, particularly such as could serve as models for biogenic particles, for example, cylinders (coral and bryozoan sticks) or spherical shells (bivalve shells). W. Thomson (Lord Kelvin) (1889) and Minkowski (1904) investigated dense packings of general convex bodies, and White and Walton (1937) the packings of cylinders, which in the closest arrangement assumed a significantly higher concentration than equal spheres. Glastonbury and Bratel (1966) examined disc packings, and Allen (1974c) made calculations for conical, cylindrical and spherical shells. Prompted by observations similar to those of Greensmith and Tucker (1968), Sanderson and Donovan (1974), and D.F. Ball (1976), nested and opposed arrangements of these shapes in various stackings were described. As with spheroids, open arrangements of low concentration prove possible, particularly when the particle thickness is small compared with the longest dimension, as is true of most bivalve shells. Details may be found in the paper cited. Recent experimental work by Vinopal and Coogan (1978) confirms the low concentrations predicted for shell packings.

RANDOM SPHERE PACKINGS

Completely random sphere packings exist only in theory and so are as ideal as the ordered packings thus far discussed. They are nonetheless much closer in their properties than are ordered arrangements to the haphazard packings of the real world. Numerous models have been proposed for random packings, chiefly in connection with theories on states of matter.

The simplest model of a random packing is due to Smith, Foote and Busang (1929). It was independently given by Euler (1957), and Gotoh (1971), and was used successfully by Kunii and Smith (1960) and Allen (1969e, 1970b, 1970c). The model rests on the postulate that the magnitude of some property of the packing is the same as if the assemblage consisted of a random mixture of small elements divided between at least two ordered packings in the proportion required to yield the specified magnitude. A successful theoretical justification of this postulate is so far lacking (see Gotoh, 1971), and the model clearly is limited to overall properties.

We now generalize the Smith-Foote-Busang model for a two-component system in which the particles are of the same size and shape. Let y_∞ be the required property considered at infinitely large assembly size, and let y_1 and y_2 be the magnitudes of the same property for, respectively, the chosen ordered packings 1 and 2. Packing 1 exists in the mixture in a volume fraction k_1 and packing 2 in a volume fraction $(1 - k_1)$. If n_1 and n_2 are the numbers of particles in unit volume of the respective packings, then:

$$y_\infty = \frac{n_1 y_1 k_1 + n_2 y_2 (1-k_1)}{n_1 k_1 + n_2(1-k_1)} \tag{4.8}$$

It is often convenient to relate y_∞ to the concentration whence, putting C_∞ as the overall concentration, and C_1 and C_2 as the concentrations in the respective ordered packings:

$$C_\infty = k_1 C_1 + (1-k_1)C_2 \tag{4.9}$$

$$k_1 = \frac{C_\infty - C_2}{C_1 - C_2} \tag{4.10}$$

allowing k_1 to be substituted for in eq. (4.8). Smith et al. (1929) used these equations, with the 600 and 006-packings as the ordered arrays, to calculate the coordination number of random packings. Fair agreement was noted.

Higuti (1961) gave a statistical model for a random packing of mixed sphere sizes. He supposed that the spatial coordinates and sizes of the particles were drawn as independent random samples from, respectively, a uniform and a lognormal distribution. By considering the joint probability density of the sphere positions and sizes, Higuti developed expressions for the areal concentration, for the radial distribution of spheres about a fixed sphere (radial distribution function) and, tentatively, for the coordination. This model suffers from a single crippling weakness, namely, that Rogers' (1958) definition of a packing is violated.

An interesting pair of theories for equal spheres consistent with Rogers' definition may be grouped as the local sphere-shell model (Haughey and Beveridge, 1966, 1967; Beresford, 1969). Essentially, spheres are grouped around and in contact with a base sphere, until either a predetermined number of spheres is reached or until there are no more holes large enough to receive spheres able to touch the base sphere. These spheres form the first shell. In the Haughey-Beveridge model, the n spheres of this shell are arranged in two alternative modes. Either the first three lie on the base sphere as a close-packed triangular cluster, subsequent spheres being placed in contact with at least two other first-shell spheres, or the n spheres are arranged equidistant over the base sphere. In Beresford's version the n first-shell spheres are distributed on a chance basis, the operation of a random process allowing a subsequent readjustment of the position of each. Haughey and Beveridge place the p second-shell spheres so as to fit in the triangular holes formed by the first-shell particles whence, if $n < 6$, they also touch the base sphere. The near neighbours of Beresford, equivalent to a second shell, are also distributed on a chance basis.

Haughey and Beveridge calculated the concentration and coordination of random packings, and the radial distribution function. The model is unrealistic, to the extent that a substantial order is imposed on the arrangement of spheres. Beresford's model is more appealing in that the spheres are arranged by chance and the introduction of a random process permits the action of gravitational or cohesive forces on the growing assemblage to be explored. He developed expressions for several important properties of a random packing, notably the distribution of coordination amongst the particles.

Adams and Matheson (1972), in an independent development, similar to that of Haughey and Beveridge and of Beresford, calculated the dense packing of equal spheres permitted to touch but not interpenetrate. Distribution functions were obtained and the particle concentrations proved relatively high.

A Monte Carlo approach to the random packing of equal spheres is widely favoured (for significant early review, see Fluendy and Smith, 1962). Studies have been made of one, two and three-dimensional systems. Streett and Tildesley (1976) studied dumbell-shaped particles.

Smalley (1962) analysed the random packing of equal spheres along a line, obtaining linear concentrations between 0.75 and 0.78, in fair agreement with Griffith's (1962) theoretical value of 0.7476. Later Round and Newton (1963) studied the random packing without overlap of spheres on a plane, the computational scheme allowing for some readjustment in the position of existing spheres as new ones were added. The areal concentrations were as large as 0.785. However, Kausch et al. (1971), and Visscher and Bolsterli (1972a, 1972b), packing circles or discs, produced concentrations in the order of 0.82–0.83 for particles of a single size and somewhat large concentrations for many binary mixtures. Metropolis et al. (1953) used a modified Monte Carlo approach to the packing of equal spheres in a plane; an initially ordered lattice is repeatedly disturbed until complete disorder prevails. In this way the radial distribution function for a prescribed concentration of spheres could be obtained. Herczynski (1975) gives a theory for this function in two and three dimensions.

Metropolis's perturbation technique has been successfully extended to spheres in three dimensions (Rosenbluth and Rosenbluth, 1954; Alder, 1955; Alder et al., 1955; Alder and Wainwright, 1957; Wood and Jacobson, 1957; Wood and Parker, 1957; E.B. Smith and Lea, 1960; Rotenburg, 1965). Bernal (1960b, 1964, 1965) constructed by Monte Carlo methods a three-dimensional random packing of equal spheres, on the basis that no two spheres approached closer than a certain minimum distance. As Herczynski (1975) later found, the concentration obtained is low, on account of the prohibitive number of trials required for the fitting of late-entering spheres. Bernal reproduced the calculated packing in the form of a physical "ball-and-spoke" model, which he then condensed into an assemblage with a concentration of 0.61. Radial distribution functions were obtained for the two arrangements.

These Monte Carlo techniques are sedimentologically objectionable, on the grounds that the assemblies are not constructed in ways that parallel reality. Most real packings form by a process of external accretion, in which new particles are added along one boundary of the growing assemblage.

Several workers used computers to simulate haphazard packings formed by the settling of particles, either equal or polydisperse spheres or rod-like or other shapes of particle fashioned using appropriate strings of smaller spheres. A base particle is positioned in a defined space and further particles selected and added according to prescribed rules allowing the influence of gravitational and/or cohesive forces to be expressed. The sedimentation of floccules was studied by Vold (1959a, 1959b, 1963), Hutchison and Sutherland (1965), Sutherland and Goodarz-Nia (1971), and Goodarz-Nia and Sutherland (1975) using single spheres and sphere clusters. Each fresh particle, upon touching a particle already in the assemblage, was halted in the attitude of movement. Loose packings were consequently obtained, having concentrations comparable with 0.2 or less, and mean coordinations of not more than about 2.4. These models remind one of the more open non-simple ordered packings (Table 4-II) and would seem relevant to real systems where cohesive forces act between particles, for example, in powders (Pilpel, 1969) and dispersions of clay minerals. Several models stress the effects of gravity in the packing of sedimented spheres, each particle being allowed to roll into an ultimate position more stable than that of initial contact (Tory et al., 1968, 1973; Visscher and Bolsterli, 1972a, 1972b; Gotoh et al., 1978a), as seems likely in many real cases of sedimentation. Packings with concentrations comparable with 0.58–0.60 and mean coordinations of about 6 are obtained. These models clearly have the same advantages as Beresford's (1969) and are more appealing than attacks of the Rosenbluth type.

In a final group of models, stress is placed on the geometrical aspects of particle packings, notably the way particle centres in various ways define shapes in space. For example, by joining centres using spokes, polyhedra are defined which may be sorted as to size and shape. Alternatively, we may divide the space into its component Voronoi polyhedra, whose faces are planes drawn normal to the spokes at their mid-points, and whose edges are the intersections of these planes. The Voronoi polyhedra may likewise be evaluated to obtain the properties of the packing; the average polyhedron should have 13.56 faces (Coxeter, 1961). Considerations of these kinds appear in the partly statistical work of Boerdijk (1952), Wise (1952, 1960), Hogendijk (1963), Norman et al. (1971a, 1971b), Dodds (1975) and Watson (1975). They dominate the largely empirical studies of Bernal (1960b, 1964, 1965), and appear later in the work of Bernal and Finney (1967), J.L. Finney (1970a, 1970b), and Gotoh and Finney (1974). Levine and Chernik (1965) produced Voronoi polyhedra using a computer technique. They obtained an ensemble of about 3000 of the polyhedra, representing a sphere packing with a coordination of 8.52 and a mean concentration of 0.609. G. Mason (1971) calculated an ensemble of tetrahedra on the basis of an observed radial distribution function. However, Gotoh and Finney's particular statistical-geometrical model is an especially important and powerful development. Arguing that the average coordination in a random packing of equal spheres should be precisely 6, they calculate the most probable Voronoi polyhedras of the packing, and thence obtain concentrations falling between the bounds of 0.6099 and 0.6472. These are in remarkably good agreement with observation, as will be seen.

WALL AND RELATED EFFECTS

Every real packing is finite and bounded, either by containing walls or by surfaces across which grain size changes appreciably, as between sedimentary laminae. Such boundaries disturb the packing as exemplified by the mode in the centre of the mass, and may impose special modes of packing adjacent to them. Boundary effects adjacent to smooth walls are graphically shown in drawings and photographs of equal sphere packings given by Graton and Fraser (1935), Brown and Hawksley (1947), Susskind and Becker (1966), and Gray (1968). The spheres adjacent to the wall are commonly packed in an orderly or "crystalline" manner, as noted by Wadsworth (1960, 1963), and a careful attention to boundary shape can lead to extensive crystalline arrangements (Rocke, 1971). Even so the sphere concentration adjacent to the wall is visibly less than in the interior of the packing. Locally, the spheres near the wall form arches enclosing comparatively large voids, the concentration in the wall layer being even smaller than when the particles are ordered. Finally, as the spheres approach the packing in size, the effects penetrate to the very centre of the arrangement. The effects influence other properties of packings besides concentration (e.g. Beavers et al., 1973), and their analysis has taken several courses in the effort to obtain an understanding.

A mathematical approach is widely favoured. Carman (1937) analyzed the concentration of equal spheres packed in a cylinder, but under severe restrictions as to relative size. Later workers have been less restrictive (Dunagan, 1940; Heywood, 1946; McGeary, 1961; Lees, 1969; Pillai, 1977), finding that container size had no significant effect in reducing concentration, provided that loosely packed spheres were at least one order of magnitude smaller than the container, and densely packed ones were smaller by at least two orders. Rose (1945), Ayer and Soppett (1965, 1966), and McGeary (1961) confirmed these results experimentally. Gotoh et al. (1978b) describe an approach based on computer modelling.

A decisive advance in the understanding of boundary influences was made by Verman and Banerjee (1946) in their critique of Brown and Hawksley's (1945) study of the packing of broken coal. Verman and Banerjee recognized two effects, one due to the bounded nature of real packings, noted earlier by Dunagan (1940), and the other that of the boundary itself. The first effect may be illustrated by supposing that an imaginary plane is inserted into the central regions of an extensive packing. If the plane were a boundary to the packing, all the particles cut by the plane would have to be removed in calculating the overall concentration. Hence if we have a packing in a cubical container whose dimensions are n times particle diameter, then:

$$C = \left(\frac{n-1}{n}\right)^3 C_\infty \tag{4.11}$$

Now the container walls not merely interrupt the packing but actually order the distribution of particles close to them. Assuming that this second effect is confined to a single layer of particles immediately adjacent to the walls, Verman and Banerjee modified eq. (4.11) to read:

$$C = \left(\frac{n-1}{n}\right)^3 C_\infty + \left[1 - \left(\frac{n-2}{n}\right)^3\right] \Delta C \tag{4.12}$$

where ΔC (to be obtained empirically) is the deviation of the concentration in the wall layer from C_∞. Fig. 4-14 gives eqs. (4.11, 4.12) for the plausible combination $C_\infty = 0.60$ and $\Delta C = 0.20$, from which it will be seen that Verman and Banerjee's two effects are comparable in magnitude and significant for all but very large values of n. Brown and Hawksley (1946) modified eq. (4.12) to a linear form, in which C_∞ appeared as the intercept constant, replacing $(n-2)$ by $(n-3)$, in recognition of the persistence of the ordering

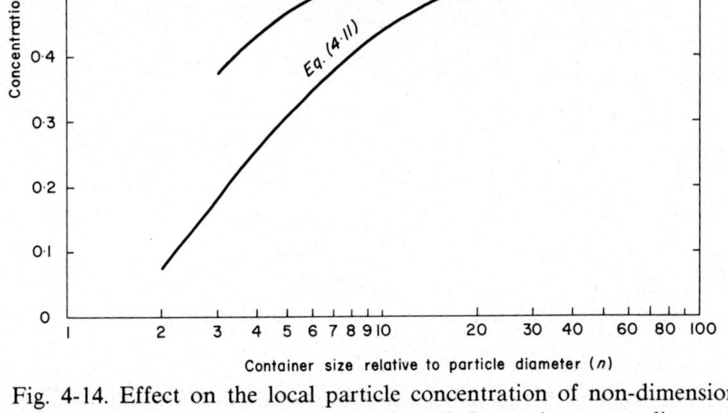

Fig. 4-14. Effect on the local particle concentration of non-dimensional distance from the edge of a packing formed in a smooth-walled container, according to Verman and Banerjee (1946). The graphs plotted are for $C_\infty = 0.60$ and $\Delta C = 0.20$.

effect deeper into the packing than one particle diameter. Coulson (1949) and Pillai (1977) have also calculated the wall effect.

The concentration (or porosity) of a packing independently of the effects of container size and the wall may be obtained by plotting measured concentrations for the packing (of constant mode) against a measure of packing size. Extrapolation to infinitely large size yields C_∞ or P_∞ in the form of the intercept constant. Dunagan (1940), and later Denton (1957), used as the measure of size the ratio of the external surface area to the volume of the packing. G.D. Scott (1960), and Scott and Kilgour (1969), packed equal spheres into cylinders of various sizes and shapes and employed either the reciprocal of the height of the packing or the reciprocal of the cylinder radius as the measure of size. Allen (1974c) used the reciprocal of the sample mass in a study of shell packings. Non-dimensional measures of packing size have been widely adopted (Leva and Grummer, 1947; Ayer and Soppet, 1965, 1966; B.P. Hughes, 1960, 1962; Sonntag, 1960; Sommer and Soeder, 1963; Jeschar, 1964; Lees, 1969), covering polydisperse as well as monodisperse systems and a broad range of particle shapes. A linear relationship was in most cases observed between the concentration or porosity and the measure of packing size.

Ingenious techniques exist for the practical investigation of the local effects of boundaries on particle packing (Roblee et al., 1958; Sonntag, 1960; Benenati and Brosilow, 1962; Ridgway and Tarbuck, 1966; Thadani and Peebles, 1966; Kondelik et al., 1968). Benenati and Brosilow, for instance, filled up the voids between lead shot haphazardly packed in a cylinder with a

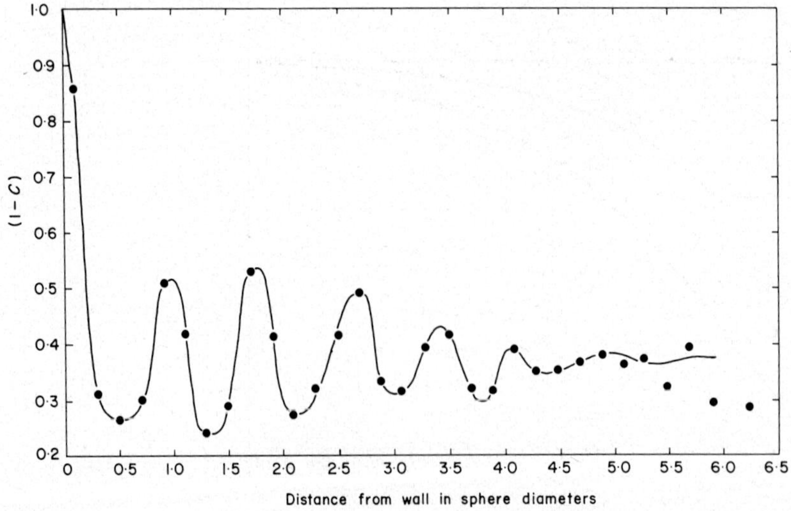

Fig. 4-15. Experimentally determined inward radial distribution of particle concentration in a haphazard assemblage of equal spheres packed in a cylinder (data of Ridgway and Tarbuck, 1968a).

cureable plastic. By machining down the cylinder in stages on a lathe, they could measure the concentration of particles in annular shells, each a known distance from the original boundary of the packing. The results of experiments of these kinds, in good agreement with the semi-empirical theory of Ridgway and Tarbuck (1968a), may be plotted as a distribution function related to that for a single particle. Along a normal to the boundary of an equal sphere packing, Ridgway and Tarbuck (1968a) found that there was a damped cyclical variation in the local mean particle concentration, which persisted over a distance of 4–5 particle diameters into the assemblage, the oscillations pointing to a strong ordering or "crystallinity" amongst the spheres (Fig. 4-15). The greater part of the wall effect derives, however, from the outermost one or two layers of particles, as surmized by Verman and Banerjee (1946), and by Brown and Hawksley (1946). Packings of non-spherical or polydisperse particles appear to be random at distances from the container wall less than the 4–5 particle diameters observed for equal spheres. No data are available for natural sediments.

HAPHAZARD PACKINGS

Concentration in packings of equal spheres

Table 4-VI compares real with theoretical assemblages of equal spheres, the emphasis being placed on the dense and loose haphazard packings (G.D. Scott, 1960).

A *dense haphazard packing* may be achieved by tapping, jolting or vibrating a particle assemblage, a technique used by every housewife when packing sugar or rice into a barely adequate jar. Accurately measured concentrations for this packing at infinite sample size are 0.6366 ± 0.0005 (Scott and Kilgour, 1969) and 0.6366 ± 0.0004 (J.L. Finney, 1970a), in fair agreement with the theoretical value of 0.6472 (Gotoh and Finney, 1974). G.D. Scott (1960) obtained a concentration of 0.637, and Finney reported a value of 0.6342 ± 0.0012 after a re-analysis of Scott's data. Other values at infinite sample size are 0.684 (Leva and Grummer, 1947), 0.641 (Sonntag, 1960), 0.641 (Rutgers, 1962), and 0.635 (Ayer and Soppett, 1965). Comparable but uncorrected results were obtained by numerous workers (Westman and Hugill, 1930; Oman and Watson, 1944; Kohn and Gonell, 1950; Denton, 1957; McGeary, 1961; Yerazunis et al., 1965; Debbas and Rumpf, 1966; Parsick et al., 1966; Susskind and Becker, 1966; Sempere, 1969; Allen, 1970c). Berg et al. (1969) obtained the unusually low value of 0.615. Sempere found that the concentration was influenced by the fluid medium in which the packing was formed, the largest values arising in air.

The other packing widely regarded as reproducible—*loose haphazard packing*—can be obtained either by allowing spheres to avalanche during

TABLE 4-VI

Concentration and coordination in some equal sphere packings

Author	Assemblage	C	N
Dense haphazard packing			
Bernal and Mason (1960), G.D. Scott (1960)	ball bearings	0.637	8.5 [a]
G.D. Scott (1960, 1962)	ball bearings	0.637	9.3 ± 0.8 [b]
G.D. Scott (1960), G. Mason (1968)	ball bearings	0.637	9.39 [b]
Gotoh and Finley (1974)	theoretical	0.6472	6.0 [c]
Loose haphazard packing			
Bernal and Mason (1960), G.D. Scott (1960)	ball bearings	0.601	7.1 [a]
G.D. Scott (1960), Scott et al. (1964)	ball bearings	0.601	8.0 [b]
Gotoh and Finley (1974)	theoretical	0.6099	6.0 [c]
Other packings			
Smith et al. (1929) [d]	lead shot	0.641	9.14
	lead shot	0.628	9.51
	lead shot	0.574	8.06
	lead shot	0.553	6.92
Levine and Chernick (1965)	theoretical	0.609	8.52
Beresford (1969)	theoretical	0.62	5.452
Tory et al. (1968)	theoretical	0.59	6.114
Tory et al. (1973)	theoretical	0.581	6.01

[a] Based on contacts and nearest neighbours to 1.05 diameters between centres.
[b] Based on contacts and nearest neighbours to 1.1 diameters between centres.
[c] Assumed value.
[d] Some packing size, near neighbour, and boundary effects likely.

emplacement, or by dumping them into a container. Measured concentrations at infinite size are 0.665 (Leva and Grummer, 1947), 0.601 (G.D. Scott, 1960), 0.608 (Sonntag, 1960), 0.596 (Rutgers, 1962), and 0.575–0.611 (Scott and Kilgour, 1969), Gotoh and Finney's (1974) theoretical value being 0.6099. Even discounting Leva and Grummer's result as too high, this packing is evidently less reproducible experimentally than the dense one. Are we correct to recognize loose haphazard packing as a definite state?

Concentrations intermediate between the values for loose and dense packings have been measured after pouring equal spheres into containers (O.K. Rice, 1944; Denton, 1957; Wadsworth, 1960).

Packings looser even than loose haphazard packings are obtainable using other methods of emplacement than avalanching or dumping. Steinour (1944) obtained concentrations of 0.513–0.546 by very slowly sedimenting virtually equal and spherical tapioca grains in lubricating oil. By inverting the packing container, Oman and Watson (1944) found values of 0.531–0.536, and Happel (1949) a concentration of 0.56. Similar results were given by Eastwood et al. (1969). Parsick et al. (1966) found that a bed of equal spheres settled after fluidization to a concentration between 0.59 and 0.61, though Ergun and Ornung (1949) obtained somewhat lower values.

Particle mass may influence the concentration obtainable by each method of emplacement. Working with narrowly graded glass spheres less than 0.001 m in diameter which were dumped or rapidly sedimented in air or liquids, Allen (1970c) obtained concentrations between 0.560 and 0.602, the value increasing with particle size, in agreement with Sempere (1969).

We may compare with these results the concentrations measured from the loosest and densest packings of other equal and regular but non-spherical particles. Oman and Watson (1944) found that short cylinders packed to concentrations between 0.540 and 0.635, and Coulson (1949) gave similar results. Allen (1970c) packed a variety of shapes. Long cylinders gave concentrations between 0.446 and 0.599, the relatively low limiting values perhaps being due to bridging between particles. Lentils, resembling spherical caps, gave concentrations between 0.599 and 0.659, and rice grains of subspheroidal form afforded values between 0.586 and 0.653, both shapes packing like spheres. Coulson (1949) found that equal cubes and hexagonal prisms also packed similarly to spheres, though thin plates gave much lower concentrations. Hagemeyer (1960) found that plates packed loosely compared with needles and rhombs. Bivalve shells pack at especially low concentrations, in the order of 0.1–0.2 (Allen, 1974c).

Well graded angular particles produced by crushing give concentrations between 0.33 and 0.64 (F.H. King, 1899; Furnas, 1929; Fraser, 1935; Brown and Hawksley, 1945; Shergold, 1953; Huang et al., 1963; Evans and Millman, 1964; Ayer and Soppett, 1966; Koerner, 1969; Lees, 1969). However, angularity alone (Fraser, 1935; Shergold, 1953) cannot explain these unusually loose arrangements, since the upper limiting concentration is similar to that for densely packed spheres. Probably the ability of the particles to nestle closely together is inhibited by their high surface roughness (Macrae and Gray, 1961; Ayer and Soppett, 1966; Scott and Kilgour, 1969), which would arise naturally from the crushing. The grains forming natural sands are generally smoother and better rounded than freshly crushed materials, but not as "slippery" as new ball bearings. Cuts from sands pack to concentrations between 0.53 and 0.70, a range comparable with that for equal spheres (Kolbuszewski, 1948a, 1948b, 1950a; Gaither, 1953; Walker and Whitaker, 1967; Sohn and Moreland, 1968). Powders pack much more loosely (Heywood, 1946; Neumann, 1953).

Coordination in packings of equal spheres

The mean coordination of packed particles may be found after great labour either directly by counting contacts or, indirectly, from the measured radial distribution function. Smith et al. (1929) and Bennett and Brown (1940) used chemical methods to mark contacts in packed beds. Bernal and

Mason (1960) and Wadsworth (1960) flooded packed beds with paint which, on being drained off and allowed to dry, left a solid meniscus at each contact between spheres. G.D. Scott (1962) worked from the radial distribution function.

Measurements on equal-sphere packings give the results in Table 4-VI, in which some theoretical values also appear. The coordinations measured experimentally are based on the numbers of true contacts and of "near neighbours", where a near neighbour lies closer to a particle than 0.05–0.10 diameters. This is simply due to the practical difficulty of satisfactorily distinguishing close approaches from physical contacts. However, the coordinations obtained theoretically by Beresford (1969) and Tory et al. (1968, 1973) are in good agreement with the mean number of "close contacts" (6.4 per sphere) measured by Bernal and Mason (1960). It would seem that haphazardly packed spheres are in contact with approximately six others and have either two or three particles as near neighbours, conclusions which are supported by the work of Marvin (1939) and Wadsworth (1960, 1963). Hence there is a measure of justification for Gotoh and Finney's (1974) assumption from stability considerations that the mean coordination in such a packing is 6. We may again note how well the concentrations calculated on this assumption agree with the measured values.

It is here worth noting some of the properties of the polyhedra which can be calculated from sphere packings or formed from them by compression. Marvin (1937, 1939) and Matzke (1939, 1946) found that the polyhedra formed by compressing lead shot had 14.17 faces on the average, whereas bubbles in foam had 13.70. Bernal (1959) later squeezed "Plasticine" balls to form polyhedra with on average 13.3 faces. Two measured sphere packings analysed as Voronoi polyhedra gave averages of 14.28 ± 0.05 and 14.251 ± 0.015 faces respectively (Bernal and Finney, 1967; J.L. Finney, 1970a), a little in excess of Coxeter's (1961) theoretical 13.56 faces.

Radial distribution function in packings of equal spheres

The sedimentological importance of the radial distribution function is not yet clear, though Smalley (1964) argues that the packing of natural sands, commonly assessed in terms of grain contacts or porosity (e.g. Coogan and Manus, 1975), is better represented by this function than in any other way. A decision on whether or not he is correct must await future theoretical developments and actual trials. Because of its relevance to the structure of liquids, however, the function for haphazardly packed equal spheres is relatively well known from measurements of the Cartesian coordinates of particles. Bernal et al. (1970) discuss techniques.

The measured functions give evidence of some degree of local ordering within sphere assemblages. Morell and Hildebrand (1936), who used glass or gelatine spheres, showed a damped cyclical variation of the density of

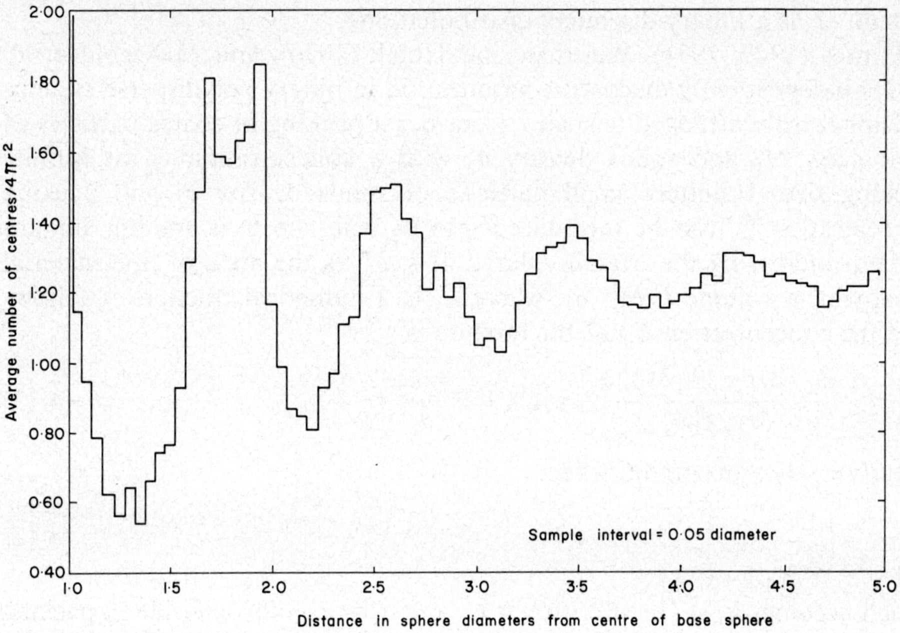

Fig. 4-16. Radial distribution function for an experimental dense packing of 7994 equal spheres. Data of Finney (1970a).

particle centres relative to a base sphere, similar to that in Fig. 4-15. Improved results were obtained by G.D. Scott (1962), Scott et al. (1964), Mason and Clark (1965, 1966), Iczkowski (1966), and G. Mason (1968). J.L. Finney's (1970a) function, based on a dense packing of 7994 spheres, is probably the most accurate yet obtained (Fig. 4-16). The distribution becomes random at about five sphere diameters away from the base sphere, but at nearer distances a degree of ordering appears. The peak at unit diameter represents the nearest neighbour sphere, and is not of further interest, but that at approximately 1.73 diameters hints at the presence of tetrahedral arrangements, while the peak at nearly 2 diameters suggests three-member collineations (Bernal, 1964). Angular distribution functions have also been measured (Scott and Mader, 1964).

Concentration in polydisperse systems (discrete size-distributions)

Natural sediments and many manufactured packings are polydisperse. Typically, the manufactured ones are blends of two or more narrowly graded sizes of particle, their size-distributions being discrete, whereas in natural deposits particle size generally is continuously distributed. But no sharp distinctions separate the two kinds of system. A gravel with a sand matrix, for example, may be treated as either a bimodal continuously distributed system, or as a binary discretely distributed one.

Furnas (1929, 1931), Westman and Hugill (1930), and Manegold et al. (1931) independently analysed concentration in binary polydisperse systems of haphazardly arranged particles. Consider a packing of coarse particles of total mass M_2 and solids density σ_2 with a concentration C_2 at infinite packing size. Infinitely small particles of solids density σ_1 and packing concentration C_1 can be introduced into its voids up to a limiting amount without increasing the overall volume. If $k_1 M_2$ is the mass of fine material occupying a volume $k_1 M_2/\sigma_1$, where k_1 is a numerical fraction, it follows that the concentration C_m of the mixture is:

$$C_m = \frac{(M_2/\sigma_2)+(k_1 M_2/\sigma_1)}{M_2/C_2\sigma_2} = C_2\left(1+\frac{k_1\sigma_2}{\sigma_1}\right) \tag{4.13}$$

Clearly C_m is a maximum when:

$$k_1 = \frac{\sigma_1}{\sigma_2}\left(\frac{C_1}{C_2} - C_1\right) \tag{4.14}$$

which becomes $k_1 = (1-C)$ if $C_1 = C_2$ and $\sigma_1 = \sigma_2$. Now consider a packing of infinitely small particles of total mass M_1 to which coarse ones are added at random without introducing voids. Putting k_2 as another numerical fraction, such that $k_2 M_1$ is the mass of coarse particles added, it follows that:

$$C_m = \frac{(M_1/\sigma_1)+(k_2 M_1/\sigma_2)}{(M_1/C_1\sigma_1)+(k_2 M_1/\sigma_2)} = \frac{C_1(\sigma_2+k_1\sigma_1)}{(\sigma_2+k_1 C_1\sigma_1)} \tag{4.15}$$

If it is now supposed that the two kinds of particle differ in all respects except size, we may write:

$$C_m = \frac{C_1 C_2[k_1\sigma_2+\sigma_1(1-k_1)]}{[k_1 C_2\sigma_2+(1-k_1)C_1\sigma_1]} \tag{4.16}$$

in terms of the factor k_1. Note that C_m = const. when $C_1 = C_2$ and also $\sigma_1 = \sigma_2$.

Equations (4.13) and (4.15), intersecting at a single point, completely define with eq. (4.16) the existence field for a binary mixture (Fig. 4-17a). The field is a triangle with apex upward, an important theoretical justification of this feature being given by Dodds (1975).

Westman and Hugill (1930) studied the concentration of a general polydisperse system of n components. Their analysis, based on the same assumptions about mixing as above, may be illustrated by the ternary system, its components 1, 2, and 3 being regarded as coarse, medium, and fine respectively, with an infinite diameter ratio. We now have three pairs of binary mixtures whose graphs may be combined to give a solid figure (Fig. 4-18a); the ordinates show the reciprocal of the concentration, and the abscissae the

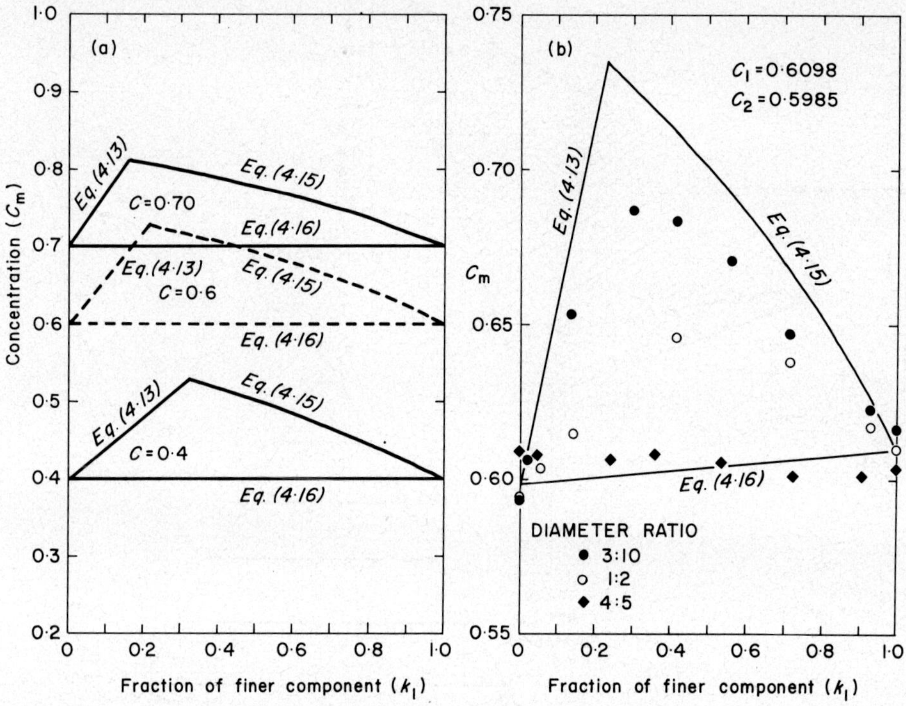

Fig. 4-17. Concentration of a binary mixture of particles of unequal size as a function of the fraction of the finer component. a. Theoretical curves for equal-density particles of three uniform concentrations in the pure state. b. Concentrations observed by Ridgway and Tarbuck (1968b) in sphere packings at three diameter ratios, the calculated curves being given on the basis of the average of the concentrations of the pure components.

volume fraction of each component. The three intersecting surfaces formed from the binary graphs have the equations:

$$\frac{1}{C_{m,1}} = \frac{k_1}{C_1} \tag{4.17}$$

$$\frac{1}{C_{m,2}} = k_1 + \frac{k_2}{C_2} \tag{4.18}$$

$$\frac{1}{C_{m,3}} = k_1 + k_2 + \frac{k_3}{C_3} \tag{4.19}$$

where the additional subscripts on C_m refer to the apex through which the surface passes, and:

$$k_1 + k_2 + k_3 = 1 \tag{4.20}$$

The mixture has a maximum concentration given by:

$$\frac{1}{C_m} = \frac{k_1 k_2 k_3}{k_1 k_2 + k_2 k_3 + k_1 k_3 - (k_1 + k_2 + k_3) + 1} \tag{4.21}$$

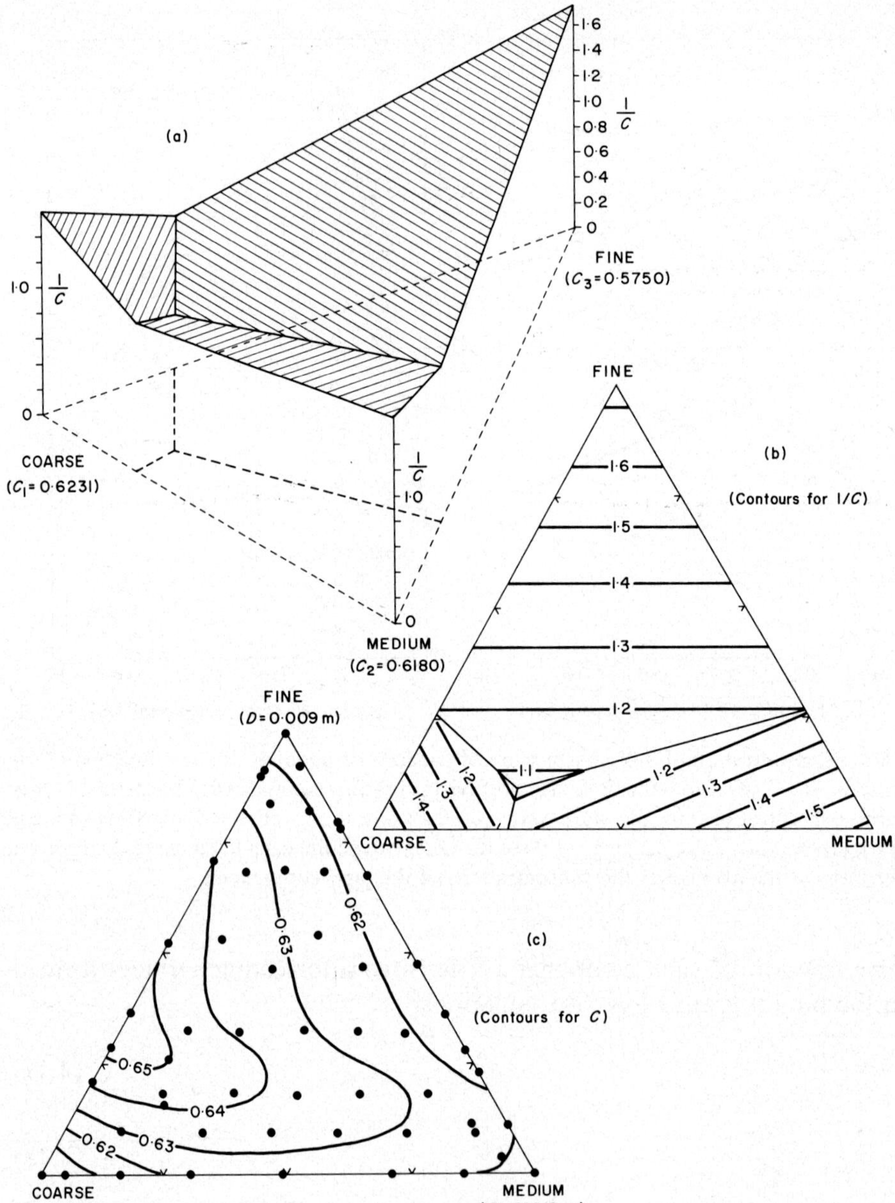

Fig. 4-18. Concentration of a ternary mixture of particles of unequal size as a function of composition. a. Three-dimensional composition diagram defined by graphs for the three binary mixtures, given in terms of the reciprocal of the concentration, calculated for the same combination of concentrations in the pure state as used by Westman and Hugill (1930). b. Compositional triangle calculated for the system shown in (a). c. Experimental concentrations in a ternary system of unequal spheres (after Ridgway and Tarbuck, 1968b).

and, when $C_1 = C_2 = C_3$:

$$\frac{1}{C_m} = \frac{k_1^3}{(1-k_1)^3 + k_1^3} \tag{4.22}$$

in terms of k_1. Furnas (1931) derived an equation containing a geometric series for the maximum concentration of polydisperse systems. Figure (4.18b) is a representative calculated ternary diagram for C_m. It will be seen from Table 4-VII how rapidly the maximum concentration of a polydisperse sphere packing increases with the number of sizes present (see also Tables 4-III and 4-IV). The concentration may be reduced, however, by mixing particles of different shapes as well as size (Ben Aim and Le Goff, 1968a).

Binary systems have been repeatedly experimented on, the results broadly confirming the analysis of Furnas (F.H. King, 1899; Furnas, 1929; Andreasen and Andersen, 1930; Westman and Hugill, 1930; Anderegg, 1931; Manegold et al., 1931; Tickell et al., 1933; Fraser, 1935; D.A. Stewart, 1951; Manglesdorf and Washington, 1960; McGeary, 1961; Hausner, 1962; Yerazunis et al., 1962, 1965; Naar et al., 1963; Sommer and Soeder, 1963; Jeschar, 1964; Ayer and Soppett, 1965, 1966; Ben Aim and LeGoff, 1967; Ridgway and Tarbuck, 1968b; Eastwood et al., 1969). Figure 4-17b gives the data of Ridgway and Tarbuck who, like other workers, found that the measured concentrations departed increasingly from eqs. (4.13) and (4.15) as the particle size ratio became smaller, approaching eq. (4.16) instead.

Ternary systems have likewise been widely studied (Andreasen and Andersen, 1930; Westman and Hugill, 1930; Manegold et al., 1931; Fraser, 1935; White and Walton, 1937; Busby, 1950; McGeary, 1961; Epstein and Young, 1962; Ridgway and Tarbuck, 1968b; Dexter and Tanner, 1971; Standish and Burger, 1979). Figure 4-18c gives Ridgway and Tarbuck's experimental data. Ternary systems also depart increasingly from the above equations based on the assumption of an infinite size ratio, as the actual

TABLE 4-VII

The effect of the number of particle sizes present on the concentration of a polydisperse sphere packing

After Westman and Hugill (1930)

Number of sizes	Calculated maximum concentration
1	0.627
2	0.858
3	0.946
4	0.980
5	0.992

ratio approaches 1:1:1. McGeary (1961) experimented on complex mixtures, producing a quaternary system which, remarkably enough in view of its exceptional concentration of 0.949, could be poured from a container.

Clearly, a completely successful analysis for the concentration of discretely varying polydisperse systems has yet to appear. It would seem that a wholly satisfactory theory must cover an effect arising from the relative size of the domains occupied by the different sizes of particles, another akin to the influence of container walls, and a third related to the relative curvature of the particles. Although empirical or semi-empirical attempts on these effects exist (Furnas, 1931; McGeary, 1961; Epstein and Young, 1962; Yerazunis et al., 1962, 1965; Sommer and Soeder, 1963; Ayer and Soppett, 1965, 1966; Ridgway and Tarbuck, 1967), only the approach of Ben Aim and LeGoff (1967) at present looks at all promising.

Concentration in polydisperse systems (continuous size-distributions)

Surprisingly, natural sand-size and coarser sediments with continuous size-distributions have in situ and "remoulded" concentrations similar to those achieved by haphazardly packed equal spheres (e.g., F.H. King, 1899; Athy, 1930; Fraser, 1935; Carman, 1938; Von Engelhardt and Pitter, 1951; Von Engelhardt, 1960; Allen, 1970c; Pryor, 1971a).

As may be expected from our knowledge of ordered sphere packings (e.g. Horsfield, 1934) and discrete systems (e.g. Westman and Hugill, 1930), the concentration of continuously varying polydisperse systems is significantly increased as the particle sorting grows poorer (Fig. 4-17b). There exists limited field evidence for this conclusion (e.g. Pryor, 1971a), but our understanding rests chiefly on experiments with synthetic mixtures of sand or glass beads, made up in conformity with prescribed density functions. Sohn and Moreland (1968) studied Gaussian distributions and, with Rogers and Head (1961) and Wakeman (1975), worked on log-normal mixtures. Bo et al. (1965) and Koerner (1969) examined sands made up in other ways. There is only qualitative agreement between the theory of Bierwagen and Saunders (1974) and the available data.

Although the measurement of a radial distribution function from continuously distributed packings has not so far been attempted, for all Smalley's (1964) advocacy, there exist some results on coördination in two-dimensions. J.M. Taylor (1950), working on synthetic sands, measured a mean coordination of 1.6. Later, Gaither (1953) obtained mean coordinations of 1.00–1.52 from polished surfaces of artificially cemented sands, and of 0.21–0.51 from thin-sections. Kahn (1956a, 1956b) has also discussed the two-dimensional coordination of natural sands.

We may surmise from the exploratory work of Ben Aim and LeGoff (1968b) that both the mean and range of coordination in continuously varying systems is a function of the characteristics of the size-distribution. A large grain, for example, may easily be in contact with many more than the six others required for stability, whereas a small particle could be stable directly on top of a large one, so achieving a coordination of unity.

A survey of continuously distributed packings would not be complete without brief reference to muddy sediments, in which interparticle cohesive forces are vital in controlling packing. Particle concentration in these sediments at deposition is relatively low compared with sands and gravels, a fact early recognized (e.g. Athy, 1930; Boswell, 1961). But reliable and comprehensive data are of recent origin. G.H. Keller (1969) and Keller and Bennett (1970), reviewing an extensive literature, report concentrations between 0.09 and 0.55 from the Pacific Ocean, and between 0.15 and 0.34 from the Atlantic Ocean. Continental shelf muds typically have concentrations close to 0.4, but may range as low as 0.1 and as high as 0.7 (Almagor, 1967; Bryant et al., 1967; Chassefière and Leenhardt, 1967; Einsele, 1967; Gouleau, 1968). These higher concentrations largely reflect the smaller clay-mineral content of shelf muds as compared with their deep-sea equivalents.

EFFECTS OF MODE OF DEPOSITION AND MATERIAL PROPERTIES ON THE PACKING OF COHESIONLESS PARTICLES

General

We have seen that particle shape, angularity, surface roughness, and size-distribution each influence packing. It is now appropriate to consider the effect of material properties, and to explore further the controls exerted by the medium and the mode of particle emplacement.

It has been claimed that, under laboratory conditions, reproducible loosest and densest particle packings can be formed by an appropriate choice of mode of emplacement. Under natural conditions also, emplacement conditions appear to influence concentration. Fraser (1935) early noted this, though he produced no decisive evidence in its favour. However, Bagnold (1954b) has graphically described the "pools" of fluid quicksand formed where sand had avalanched down the sides of desert dunes, and has compared these deposits with the firm, closely packed sands accreted on near-horizontal beds. Simons et al. (1961) also described the loose nature of avalanched sands, but in the context of subaqueous deposition, and parallel observations come from sea beaches (Allen, 1972a). Typically, the wave-beaten parts of a beach are underlain by parallel-laminated closely packed sand, so firm and strong as scarcely to take the imprint of the feet. The rippled areas, in contrast, commonly take a deep impression, and clearly consist of more loosely packed and in places also cavernous sand.

Experimental evidence

Packing concentration increases, though not indefinitely, with the distance through which the particles must fall to reach the accreting surface. This was shown using quartz sands (Kolbuszewski, 1948a, 1948b, 1950a; Walker and Whitaker, 1967; Butterfield and Andrawes, 1970) and relatively large spheres and other shapes of metal (Coulson, 1949; Macrae and Gray, 1961). Other things being equal, there is an upper limit of distance of fall beyond which no further change of concentration can be produced. In effect, an increase in the distance of fall means an increase in the velocity at which the particles strike the bed, up to the limit in each case by the resistance of the fluid medium in which the packing is being formed. Kolbuszewski (1948b) gives a graph of packing concentration as a function of particle velocity for quartz sand packed in air, water, and ether. He later found that the concentration of aeolian sands increased with wind speed (Kolbuszewski, 1950b, 1953), apparently an expression of the same effect, since the velocity of saltating grains measured normal to the bed increases with the wind.

A strong relationship exists between packing concentration and intensity of particle supply (Fig. 4-19), where the intensity is measured as the dry mass of grains falling towards the bed in unit time through unit area. Kolbuszewski's (1950a) extensive experiments with quartz sand show that, at least when air is the medium, the concentration varies between an upper (densest

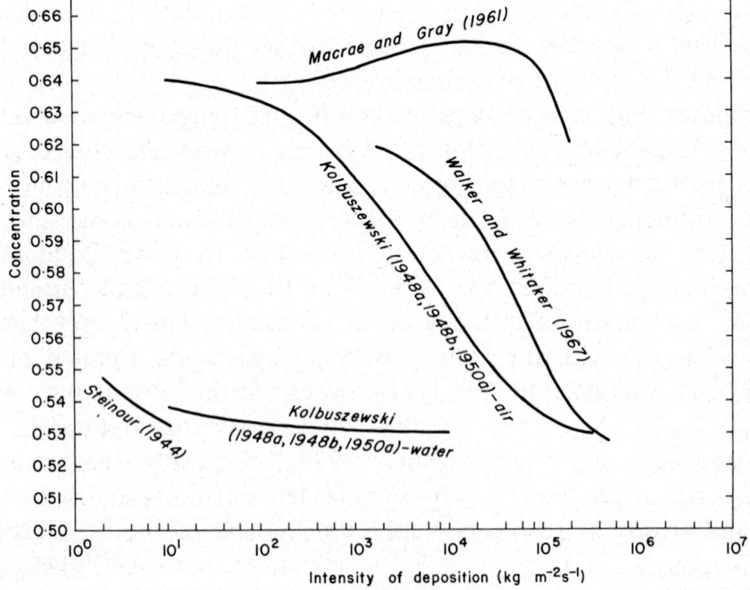

Fig. 4-19. Summary of experimental results on the variation in the concentration of an assemblage of particles formed by sedimentation from above, as a function of intensity of supply.

haphazard packing) and a lower (loosest haphazard packing) limit with increase in the intensity of particle supply. Steinour's (1944) experiments with tapioca in lubricating oil extend Kolbuszewski's curve for water considerably to the left. A graph similar in trend to Kolbuszewski's may be plotted from the calibration of a sand spreader by Walker and Whitaker (1967), the intensities having been recalculated to a state of continuous deposition. Macrae and Gray (1961) experimented on the packing of lead, steel, phosphor-bronze, glass and polystyrene spheres in air. Their data for glass, the material most resembling the quartz sands of other investigators, plot close to Walker and Whitaker's curve. This seems consistent with the fact that both groups of investigators supplied the grains first to a relatively small area on the bed, from which particles subsequently spread away in various ways.

Macrae and Gray (1961) investigated the effect of particle resilience on packing concentration. Generally, the more resilient materials like glass and steel gave the largest concentrations for a given intensity of supply and distance of fall. Careful observation of the growing beds showed that particles were in motion for some distance below the nominal instantaneous surface of accretion. This "active" layer was five or six sphere diameters thick in the case of steel and phosphor-bronze particles dropped through a distance of 0.45 m or more. The steel spheres jostled each other much more than the phosphor-bronze balls, however, to a degree sufficient to induce "crystalline" arrangements in all but a central zone in the packing. This effect could be attributed to the greater resilience of the steel.

Interpretation of experimental data

Kolbuszewski (1950a) tentatively interpreted these effects when he wrote: "With high particle velocities there is sufficient energy available for a dense packing to be achieved but with high intensities of deposition there is insufficient time for this close packing to be achieved owing to the "locking" action of the newly arrived grains". Macrae and Gray (1961) and Allen (1972a) further developed this suggestion, and additional notes are presented here. Note that by "intensity of deposition", Kolbuszewski means the intensity of supply as defined above; his argument, however, is not invalidated.

The effect of intensity of supply on packing concentration must demand a statistical analysis. We are concerned with the likelihood that a particle newly arrived on the accreting bed will roll or otherwise move into a gravitationally stable position in the hopper formed by a group of adjacent particles already incorporated into the bed, before that hopper is blocked up or bridged over through the arrival of other grains. If the particle reaches the hopper, it is likely to achieve a higher coordination, and enter into a more dense arrangement with its neighbours, than if it alights alongside other grains close enough to furnish premature lateral support. In each packing situation, then, there is a characteristic mean time t which is required by the particles if they are to move into the most gravitationally stable positions in hoppers from their initial points of contact. We must compare t with the other relevant time in the problem, namely, the mean period, T, between particle arrivals at a fixed point on the bed. We should expect the bed concentration to diminish below the value for densest haphazard packing as T becomes smaller relative to t.

Following Eagleson et al. (1957, 1958), the value of t should grow larger with increasing particle radius and increasing viscosity of the packing medium. The time will decrease, however, as the density difference between particle and medium grows larger. An increase in the surface roughness should augment t by increasing rolling friction and fluid resistance, and there should be effects arising from any particle anisometry.

The value of T is given by the product of the linear concentration of the particles approaching the accreting bed with their relative velocity of approach. Assuming a steady supply of grains:

$$T = \left(\frac{4\pi a^3}{3C_\mathrm{f}} \right)^{1/3} \frac{1}{(U_\mathrm{f} - U_\mathrm{b})} \tag{4.23}$$

where C_f is the fractional volume concentration of the particles approaching the bed, a is the particle radius, U_f is the particle falling velocity relative to the ground (measured positive downward), and U_b is the velocity relative to the ground of bed accretion (measured positive downward). Since the intensity of supply to the bed $R = \sigma C_\mathrm{f} U_\mathrm{f}$, where σ is the solids density, we may from continuity also write:

$$(U_\mathrm{f} - U_\mathrm{b}) = \frac{C_\mathrm{b} R}{\sigma C_\mathrm{f}(C_\mathrm{b} - C_\mathrm{f})} \tag{4.24}$$

where C_b is the concentration in the packed bed at infinite size. The fact that U_b and C_b must be included in the expression for T means that the function connecting the concentration in the bed with T/t will be complex. Nevertheless, the effects of increasing R indicated by eq. (4.23) seems generally consistent with observation. For example, if the medium and particles are kept the same and R is small, then T will vary in inverse proportion to R, and T/t will decrease, as should the concentration.

The second influence suggested by Kolbuszewski—the energy supply to the bed representing by the particles accreting upon it— is analysed in detail by Macrae and Gray (1961). The supply, $\sigma C_\mathrm{f} U_\mathrm{f}^3 = R U_\mathrm{f}^2$, is dissipated in collisions following small relative movements of the particles in the "active" layer, as a consequence of which a denser packing is achieved, presumably because particle bridges are destroyed and interparticle friction is increasingly overcome. Hence particle concentration should *increase* with increasing intensity of supply and also with an increase in the approach velocity of the particles. The effect of material properties is that, at a constant energy supply, the packing concentration decreases as particle resilience declines.

These considerations suggest that the energy supply is a dominant control in a different regime than the intensity of particle supply. When small particles settle in a viscous liquid medium to form a packing, the energy supply is very small and there can be no active layer; particle hindering should then predominate. This seems to be borne out by Kolbuszewski's (1948a, 1948b, 1950a) results for quartz sands in water and other liquids, and by Steinour's (1944) work on tapioca. In the other regime, energy supply is dominant, as exemplified by the work of Macrae and Gray (1961). However, much more work needs to be done on this problem, in view of its importance to an understanding of a number of deformational sedimentary structures, as we shall see in Volume II of this work.

SUMMARY

Particle packings may be ordered, random or haphazard. Regular particles, whether isometric or anisometric, may form packings of all three kinds, whereas natural particles, which are irregular, can only form packings of the random and haphazard kinds. Regular packings of spheroids are of more sedimentological interest than regular packings of spheres, because in such packings particle concentration is a function of particle shape and orientation.

Packing concentration is increased by a widening of the range of particle sizes present in a mixture, but is decreased by an increase of particle angularity and surface roughness, and by the inclusion of strongly anisometric particles. The concentration is decreased by an increase in the rate at which material is supplied to build up a packing, but increased by increasing particle energy and resilience.

Chapter 5

ORIENTATION OF PARTICLES DURING SEDIMENTATION: SHAPE-FABRICS

INTRODUCTION

If a sediment's packing is the density and mutual arrangement in space of constituent particles, then its fabric is the attitude in space and degree of preferred orientation displayed by those grains. Such a pattern is a *shape-fabric,* because it involves elements each of which has an orientation by virtue of an inequality of dimensions. A mass of spheres therefore cannot have shape-fabric, though it may be packed according to several different schemes, because a sphere is isotropic. However, an assemblage of prolate spheroids must have a shape-fabric, for each such particle has one dimensional axis longer than any other. This is why the spheroid is more acceptable than the sphere as an ideal sedimentary particle.

The existence of shape-fabrics in sedimentary deposits was recognized at an early date. Jamieson (1860) noticed the shingling or imbrication of flat stones on river beds, and H. Miller (1884) observed the preferred orientation of large clasts in glacial till. Rowland (1946) and Griffiths and Rosenfeld (1953) measured shape-fabrics from sandstones. The shape-fabrics of muddy sediments have for long been studied in both engineering and geological circles (e.g. Lambe, 1960a, 1960b), but can only be investigated with difficulty. The earlier workers used thin sections of conventional thickness (Urbain, 1937; Mitchell, 1956). More recently, freeze-dried clays have been studied in ultra-thin section, or fractured surfaces scanned, using the electron microscope (e.g. Bowles, 1968; Gillott, 1969). X-ray techniques are useful for consolidated materials (e.g. Odom, 1967). Organic remains, such as shells, bones, and driftwood, are also commonly found to have a preferred orientation in sediments. This fact was early recorded by Hall (1843) and later exploited by Ruedemann (1897), and subsequently many others. Many organisms living together in large numbers assume during life a preferred orientation which may in time become fossilized, as in the case of certain bivalves (e.g. Seilacher, 1953, 1961; S.M. Stanley, 1970; Eager, 1971). These preferred orientations also may be described as shape-fabrics.

Genetically, shape-fabrics are of three kinds: (1) appositional, (2) rheotactic, and (3) deformational. An *appositional shape-fabric* is the response of the sediment particles to the system of forces in operation at the time of transport or deposition. This system usually comprises a number of rheological forces together with forces associated with the Earth's gravity and magnetism. Appositional fabrics are therefore primary fabrics. *Rheotactic shape-fabrics* may also be regarded as primary, for they express the fact that many organisms, usually burrowing or sessile, take up a preferred orientation relative to the prevailing currents, because that orientation favours either (1) ease of growth, (2) ease of feeding and waste disposal, or (3) retention of a hold on the substrate. *Deformational shape-fabrics* are always secondary. They form after sediment deposition, by the action of external forces, as during the flow of an unconsolidated sediment. They almost always form during consolidation, as a more or less substantial modification of a primary or early deformational fabric, a fact to be remembered in the interpretation of fabrics from the older stratigraphic record.

Shape-fabrics are sedimentologically interesting for two reasons, both to do with the dependence of fabrics on systems of forces. Firstly, fabrics can provide information about patterns of currents, both local and regional. Secondly, they can in principle yield insights into the processes and mechanisms of transport and emplacement of different kinds of sediment, and may be particularly instructive when direct observation is difficult or impossible (e.g. emplacement by glacier ice, turbidity currents). Shape-fabrics as indicators of mode of emplacement have been studied for a much shorter period than as current indicators. In this chapter we shall emphasize the physical causes of preferred orientation.

MEASUREMENT AND REPRESENTATION OF SHAPE-FABRICS

Two general classes of method, *particulate* and *bulk*, can be used to measure shape-fabrics. The former depend on the measurement, one particle at a time, of the orientation of a sufficiently large sample of grains. Bulk methods rely on the measurement of the orientation of the anisotropy of some mass physical property of the deposit, for example, magnetic susceptibility, permeability, or dielectric strength. Of these the study of magnetic susceptibility offers the greatest prospect of sedimentological usefulness (see Rees, 1961, 1964, 1971, 1979; Ab Iowerth et al., 1967; Rees et al., 1968; von Rad, 1970; N.L. Hamilton and Rees, 1971; Aziz-ur-Rahman et al., 1975; Rees and Woodall, 1975; Gough et al., 1977). Samples of very large mass compared with the individual particle must be used and, generally speaking, the results initially require interpretation in terms of shape-fabrics determined particulately on the same deposits. Although attractive on several practical grounds, bulk methods shed little light on the forces responsible for preferred particle orientation, and will not be discussed further.

The study of shape fabrics by particulate methods begins with the measurement, in the field or laboratory, of the orientation of a series of particles relative to suitable orthogonal coordinates, either based on magnetic north or an independently established current direction. Many ways of measuring orientation have been devised, as Potter and Pettijohn (1963) show in their thorough review, but these can nevertheless be grouped between two general procedures.

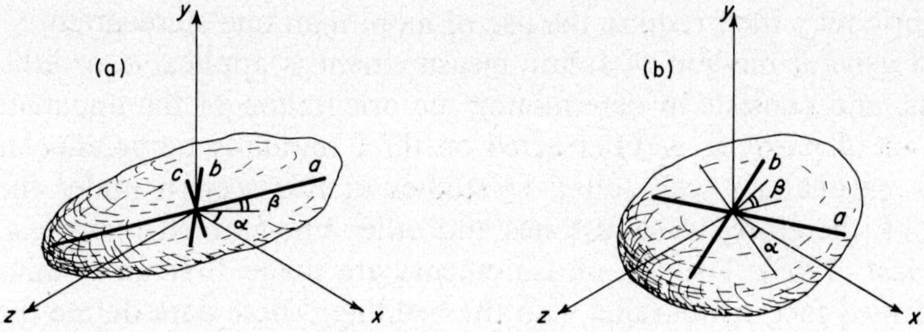

Fig. 5-1. Definition diagrams for the orientation of (a) elongate and (b) discoidal or blade-shaped sedimentary particles.

If the sediment is an unconsolidated sand or gravel, the orientation of each particle can be specified completely, once the particle has been assigned a long axis a, an intermediate axis b, and a short axis c. When the particle resembles a prolate spheroid, it is customary to measure the bearing α and plunge β of the a-axis relative to the chosen coordinates (Fig. 5-1a). The orientation of a particle resembling an oblate spheroid is usually specified by measuring the bearing of the strike α and the maximum dip β of the ab-plane (Fig. 5-1b). Usually in field investigations the coordinate system chosen for the *measurement* of the orientation takes the positive x-direction as magnetic north and the xz-plane, to which the y-direction is normal, as lying in the horizontal. In a laboratory study arbitrary coordinates may be chosen or, if the work is experimental, coordinates related to flow direction may be selected.

Shape-fabrics so measured are generally shown as stereograms. The orientation of a particle resembling a prolate spheroid is represented on a stereographic net (usually the Schmidt net) by the projection of the intersection of its a-axis with the lower hemisphere. In the case of a disc-like particle, the orientation is expressed by the projection of the intersection of the pole from the ab-plane with the lower hemisphere. Customarily, in these modes of representation, the coordinate system used for the original measurements is rotated so as to show in the simplest way the particle orientations relative to the original bedding and to the current direction. Normally, the positive x-direction becomes the current direction and the xz-plane the bedding.

Slight difficulties of representation occur with axisymmetric particles having no plane of symmetry normal to the symmetry axis, for example, bivalve shells, high-spired gastropod shells, egg-shaped pebbles. In the case of a concavo--convex shell, two poles can be drawn from the ab-plane containing the rim of the valve: one passes through the material of the valve (intersecting pole), whereas the other does not (free pole). A similar distinction is possible between the two portions of the a-axis of a cone-like high-spired gastropod shell or pebble with unequal ends. A full representation of the fabric may then require the use of more than one stereogram.

The second general method of fabric measurement is applicable to lithified sediments, and consists in determining the orientation of the apparent long-dimensions of particles as intersected on three mutually perpendicular surfaces. It is particularly well suited to studies in thin-section under the microscope and, therefore, to sandstones and other fine-grained rocks (e.g. Potter and Mast, 1963). Usually measurements are made first on a thin-section or polished face cut parallel with the bedding. These data define the x-direction of the fabric. Measurements are then made in two further thin-sections or polished faces, one parallel with the xz-plane and the other parallel with the xy-plane. A fabric obtained in this way may be represented by three "wind-rose" diagrams, one for each of the planes of measurement (Fig. 5-2).

The interpretation of sedimentary shape-fabrics in terms of particle transport and emplacement owes much to the analysis of deformational fabrics by structural geologists (Sander, 1930; Turner and Weiss, 1963; Ramsay, 1967). Two kinds of symmetry, orthorhombic and monoclinic, are mainly encountered amongst shape-fabrics. Figure 5-3a illustrates the orthorhombic symmetry of the poles to the *ab*-plane of an assemblage of flat-lying oblate spheroids. This kind of symmetry may be described by three mutually perpendicular symmetry planes and three two-fold axes of symmetry. The girdle fabric of Fig. 5-3b also has orthorhombic symmetry, but represents an assemblage of prolate spheroids randomly oriented close to a single plane. If the spheroids of Fig. 5-3a were to become shingled, or those in Fig. 5-3b assumed a subparallel imbricated arrangement, a monoclinic symmetry

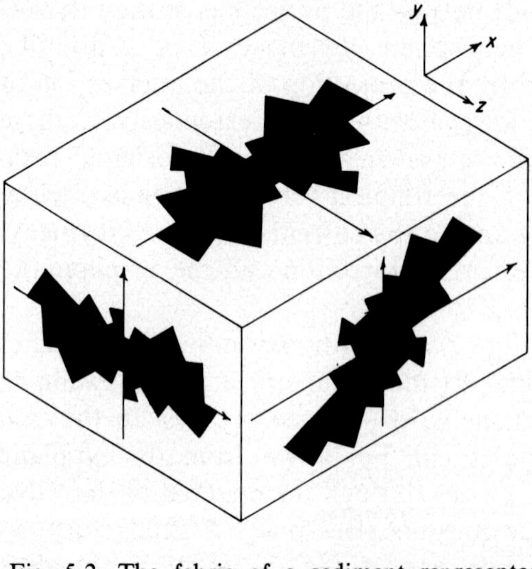

Fig. 5-2. The fabric of a sediment represented by the orientation of apparent particle long-axes on three mutually perpendicular planes.

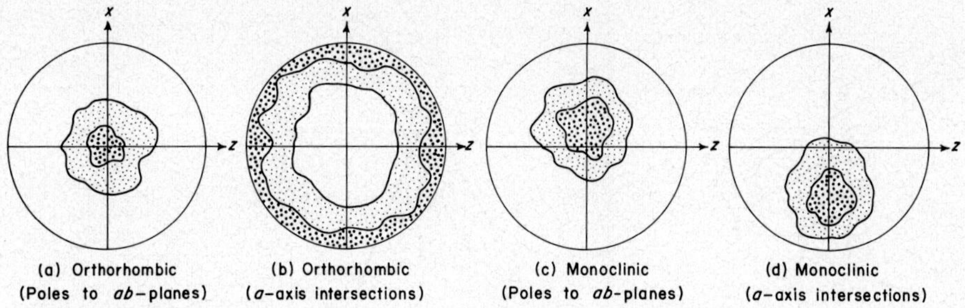

Fig. 5-3. Schematic representation of the common shape-fabrics of sediments as represented in stereographic projections on the lower hemisphere of (a, c) poles to *ab*-planes, and (b, d) *a*-axis intersections.

would exist (Fig. 5-3c, d). There is now only one symmetry plane and a single two-fold symmetry axis. Triclinic symmetry, with neither planes nor axes of symmetry, is rare amongst shape-fabrics.

SHAPE-FABRICS DUE TO SETTLING IN THE FIELD OF GRAVITY

Theory

The simplest situation in which a sediment may acquire a shape-fabric is when widely dispersed particles settle under gravity in a still fluid of large extent compared with the particle size. Two problems demand consideration: (1) the particle orientation while settling, given some initial orientation, and (2) the final stable orientation of the particle on the bed, assuming that the particle first touches the bed with the orientation it has in the fluid.

The orientation of homogeneous rigid particles settling in a fluid may be obtained theoretically at very small Reynolds numbers and for a limited number of isotropic, orthotropic and axisymmetric particle shapes. In Chapter 1 we briefly considered the resistance and settling velocities of such bodies, for the restricted case of motion parallel with an axis or plane of symmetry. On relaxing this restriction, it can be shown that orthotropic and axisymmetric particles experience lateral forces causing translation during fall, for only isotropic bodies can settle vertically regardless of their initial orientation. If the particle shape is made arbitrary, or a shape is chosen with no well-defined symmetry, a hydrodynamic couple may exist causing the particle to spin and perhaps also wobble as it descends. The combined forces may induce a downward-spiralling path.

Happel and Brenner (1965) showed that at very small Reynolds numbers, homogeneous ellipsoids, discs and cylinders retained their initial orientation during fall, no matter what that orientation. However, the horizontal components of the resisting force cause a sideways drift, of a magnitude directly

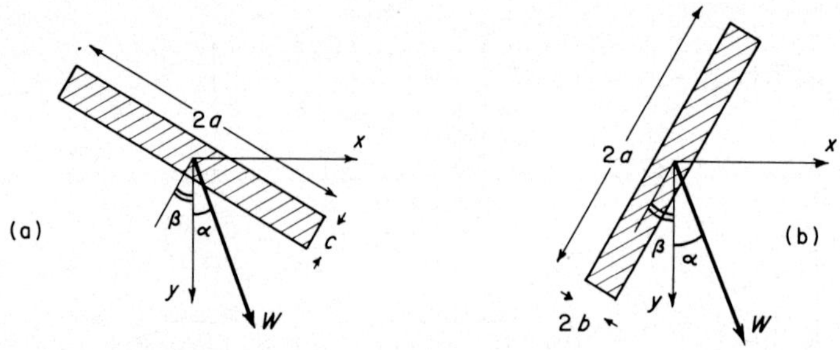

Fig. 5-4. Definition diagrams for the orientation while settling under gravity of (a) discs, and (b) cylinders.

proportional to the immersed particle weight, inversely proportional to the fluid dynamic viscosity, and affected by the particle orientation. In the case of a flat circular disc of radius a and thickness c falling symmetrically relative to the space-coordinates (Fig. 5-4a), the horizontal component W_x of the settling velocity is:

$$W_x = \frac{\pi ac(\sigma - \rho)g}{64\eta} \sin 2\beta \tag{5.1}$$

where σ and ρ are the particle and fluid densities respectively, g is the acceleration due to gravity, η is the fluid viscosity, and β is the angle between the vertical and the normal to the plane of the disc. The vertical downward component W_y is:

$$W_y = \frac{\pi ac(\sigma - \rho)g}{64\eta}(5 - \cos 2\beta) \tag{5.2}$$

giving:

$$\alpha = \tan^{-1}\left(\frac{\mathrm{d}x}{\mathrm{d}y}\right) = \tan^{-1}\frac{W_x}{W_y} = \tan^{-1}\left(\frac{\sin 2\beta}{5 - \cos 2\beta}\right) \tag{5.3}$$

where α is the angle between downward vertical and the particle path. The corresponding expression for an obliquely falling cylinder of length $2a$ and radius b ($a \gg b$) (Fig. 5-4b) is given approximately by:

$$\alpha = \tan^{-1}\left(\frac{-\sin 2\beta}{3 + \cos 2\beta}\right) \tag{5.4}$$

for a symmetrical descent. For the disc α reaches a maximum of 11.5° when $\beta = 39.2°$, the corresponding maximum for the cylinder being $\alpha = 19.5°$ when $\beta = 54.8°$. Note that in the coordinate system of Fig. 5-4b, the value of α turns out to be negative, meaning that the actual path of the cylinder is to the left.

According to these results, particles settling obliquely at very small Reynolds numbers may retain their original orientations indefinitely, and

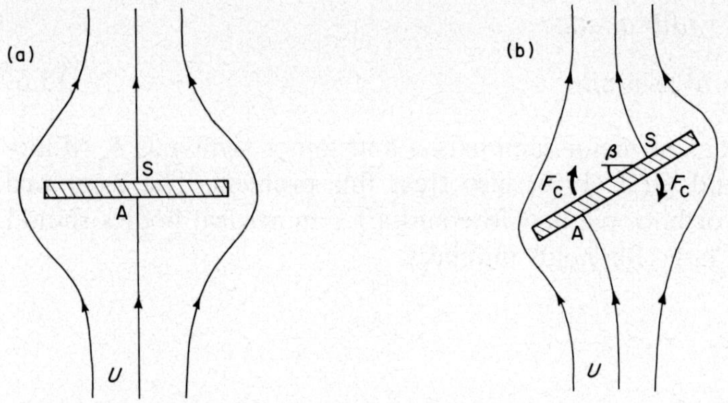

Fig. 5-5. Development of turning moments (couple) on a lamina set transversely to a uniform inviscid fluid stream.

also become dispersed because of drift to the side.

Happel and Brenner (1965) also discuss for very small Reynolds numbers the stability of settling bodies. The relevant principle is analogous with that in considering the static stability of immersed bodies, where stability exists only if the centre of gravity lies vertically above the centre of mass. A particle will settle stably in any position (e.g. the homogeneous disc above) if the centres of mass, buoyancy and hydrodynamic reaction coincide. However, a unique terminal orientation is possible for homogeneous bodies that are axisymmetric but lack fore-and-aft symmetry. Although the centres of mass and buoyancy coincide, the centre of hydrodynamic reaction lies at some other point within the body. A stable orientation is possible during settling if the reaction is directed vertically upwards; any slight change of orientation calls into play a couple tending to restore the original position. The stable orientation can be calculated for some simple shapes, but generally must be found experimentally.

The stability of bodies at large Reynolds numbers is examined by Lamb (1932), who considered the motion of an incompressible inviscid fluid past a fixed infinitely long cylinder of elliptic cross-section, and its degenerate case, the lamina (Fig. 5-5). As was verified by Hele-Shaw (1898), the streamlines are symmetrical in the plane of flow only when the current is directed normal to the lamina or parallel with the minor axis of the elliptic cross-section of the cylinder. On tilting the body relative to the stream, the stagnation points move towards opposite edges of the body. A turning force also appears, tending to restore the body to broadside-on, since the separation and attachment points, S and A, are points of maximum pressure. Writing U as the flow velocity and β as the angle between the stream and the body, the magnitude of the turning force F_C is:

$$F_C = -\frac{\pi}{2}\rho U^2 a^2 \sin 2\beta \tag{5.5}$$

for a lamina of half-width a, and:

$$F_C = -\frac{\pi}{2}\rho U^2(a^2 - b^2)\sin 2\beta \tag{5.6}$$

for an elliptic cylinder of major semi-axis a and minor semi-axis b. Milne-Thompson (1955) and Graf (1965) also treat this problem. The combined results suggest that orthotropic and fore-and-aft symmetrical bodies should fall broadside-on at large Reynolds numbers.

Experimental results

The effects of shape on particles settling singly in still fluids have been widely studied experimentally, partly to test but largely to extend available theoretical treatments. This work shows that the attitude and behaviour of particles while settling may be related to two parameters, the particle Reynolds number based on the falling velocity, and the non-dimensional moment of inertia. The latter serves as a stability parameter. Willmarth et al. (1964) introduced and defined it as:

$$I = \frac{i}{32\rho a^5} \tag{5.7}$$

where i is the particle moment of inertia about a suitable axis (e.g. diameter in case of disc), and the quantity $32\rho a^5$ is proportional to the moment of inertia of a rigid sphere of fluid of density ρ and radius a equal to the particle radius.

Table 5-I summarizes the available results. The smallest Reynolds numbers refer to the Stokes range, where inertia can be neglected. Reference to Fig. 2-9 will show that the intermediate range covers particles with a steady separation bubble and those shedding vortices regularly. At large Reynolds numbers the particles have disturbed wakes and may carry turbulent boundary layers.

At small Reynolds numbers, as theory indicates, homogeneous particles of the listed shapes fall steadily with their initial orientation. This was established for a range of isotropic shapes by Pettyjohn and Christiansen (1948), for oblate and prolate spheroids by McNown and Malaika (1950), for circular discs by Schmiedel (1928) and McNown and Malaika (1950), and for cylinders by McNown and Malaika (1950) and Jayaweera and Mason (1965). I have experimented on spherical shells at very small Reynolds numbers; these shapes, too, fall with their initial orientation.

At intermediate Reynolds numbers, orthotropic and axisymmetric particles fall broadside-on, presenting the greatest surface area normal to the motion (Schmiedel, 1928; McNown and Malaika, 1950; Marchildon et al., 1964a, 1964b; Willmarth et al., 1964; Jayaweera and Mason, 1965; Stringham et al., 1969; Futterer, 1977, 1978). Their descent is steady and any tendency

TABLE 5-I

Summary of the experimental attitudes and behaviour of a range of regular homogeneous solids during fall through still fluids, as a function of particle Reynolds number

PARTICLE SHAPE	REYNOLDS NUMBER Small, $Re \lesssim 0{\cdot}1$	Intermediate, $0{\cdot}1 \lesssim Re \lesssim 250$	Large, $Re \gtrsim 250$
Cube-octhedron Cube Tetrahedron	Any orientation stable	Stable with one face normal to motion	Spins during descent
Oblate spheroid	Any orientation stable	Stable when broadside-on	Stable when broadside-on, or pitches and tumbles
Prolate spheroid	Any orientation stable	Stable when broadside-on	Pitches broadside-on
Circular disc	Any orientation stable	Stable when broadside-on	Pitches broadside-on → Swings and pitches → Pitches and tumbles → Tumbles
Circular cylinder	Any orientation stable	Stable when broadside-on	Stable when broadside-on. May pitch or swing
Spherical shell	Any orientation stable	Stable-when broadside-on and concave-up	Stable broadside-on and concave-up. May pitch or tumble
Right-circular cone	—	Stable apex-up if $\alpha < \pi/4$; Stable apex-down if $\alpha > \pi/4$	Tumbles
Double cone (fore-and-aft symmetrical)	Any orientation stable	Stable broadside-on if $\alpha < 64°$; Stable vertical if $\alpha > 64°$	—
Double cone (fore-and-aft asymmetrical)	—	Stable if $\alpha_1 < 3\pi/4 - \alpha_2/2$; Stable if $\alpha_1 > 3\pi/4 - \alpha_2/2$	—

to rock about a horizontal axis is heavily damped. The attitude of single right-circular cones, however, depends on the apical angle. The flat base faces downward when the angle is less than $\pi/4$, but upward for larger angles. Double cones with fore-and-aft symmetry fall with apices pointing up and down only if the apical angle is greater than approximately 64°. For smaller angles the axis of symmetry becomes perpendicular to the motion. Jayaweera and Mason found that the attitude of unsymmetrical double cones depended on the relative magnitude of the apical angles. Cube-octahedra, cubes and tetrahedra were found by Pettyjohn and Christiansen (1948) to fall stably with one face normal to the motion. I observed that spherical shells fell concave-up and broadside-on. If released concave-downward such particles rapidly turned over, after following a path curved similarly to particle shape itself.

At large Reynolds numbers the pattern of fall depends strongly on particle shape, the Reynolds number, and the stability parameter. Isotropic particles and single cones tumble as they fall, sometimes also spiralling (Pettyjohn and Christiansen, 1948; Christiansen and Barker, 1965; Jayaweera and Mason, 1965). I found that spherical shells behaved much more stably, descending concave-up and broadside-on with little or no pitching. A shell with a central angle of 180°, for example, fell in this way at a Reynolds number of 10^5. As the central angle was made smaller, however, and the shell became more disc-like, a pitching and tumbling pattern appeared at progressively smaller Reynolds numbers. Middleton (1967b) observed that shell-like particles fell concave-up at $Re \approx 400$.

Oblate spheroids fall steadily broadside-on until quite large Reynolds numbers are reached (Stringham et al., 1969), but discs behave unsteadily upward from Reynolds numbers close to 10^2 (Schmiedel, 1928; Krumbein, 1942a; Becker, 1959; Willmarth et al., 1964; Christiansen and Barker, 1965; Stringham et al., 1969). At the lower Reynolds numbers, the disc oscillates regularly about a diameter perpendicular to the path of fall, with little if any horizontal translation. At larger numbers the disc swings from side to side in a vertical plane as it pitches, shedding a vortex at the end of each swing (Willmarth et al., 1964). For a constant Reynolds number the character of the oscillation depends on the magnitude of the stability parameter (Willmarth, 1964; Willmarth et al., 1964; Stringham et al., 1969). At large I the swing is of small amplitude but large frequency. Conversely, for small I the frequency becomes small and the amplitude very large, comparable with many disc diameters. Discs tumble at very large Reynolds numbers.

Cylinders fall unsteadily broadside-on (Krumbein, 1942a; Becker, 1959; Marchildon et al., 1964a, 1964b; Christiansen and Barker, 1965; Jayaweera and Mason, 1965; Stringham et al., 1969; Kajikawa, 1976). At the smaller Reynolds numbers they pitch about a horizontal axis perpendicular to the long-dimension, as vortices are shed alternately from opposite ends. Cylinders as they descend may also oscillate in a horizontal plane about a vertical axis and swing from side to side in a vertical plane.

Jayaweera and Mason (1966) experimented on the complex behaviour of cylinders and discs asymmetrically loaded with a small particle.

Final attitude on the bed

The above theoretical and experimental results suggest that particles may reach a depositional surface with any orientation when: (1) the Reynolds number is so small that the Stokes equations are valid, and (2) when the Reynolds number is so large that they swing markedly or tumble during descent. A large number of such particles released from above would effectively have a random orientation, if each particle was considered at the moment it touched the bed. If the bed and particles are of a suitable character, the particles could be held in substantially the attitudes in which they struck the bed, and a triclinic fabric should be observed. Otherwise gravity would rotate the particles into statically more stable positions. The resulting fabric, if *a*-axes were measured, should be an orthorhombic girdle.

The results also indicate that there is an intermediate range of conditions when particles fall broadside-on with or without pitching; for some shapes these conditions persist for $Re \leqslant 10^5$. The broadside-on orientation is identical with or similar to the statically stable one for the same particles, which should therefore experience little or no gravitational rotation after striking the bed. An orthorhombic fabric should be observed, a girdle in the *xz*-plane in the case of elongated particles, and a concentration of poles around the *y*-direction for discoidal shapes. It may be noticed that the stable orientation of spherical shells and similar particles when static or when settling is the opposite of that when acted on by tractional currents.

Applications

Little attention is given to the shape-fabrics of sediments that may have accumulated by settling in a substantially still environment.

Carey and Ahmed (1961), Spjeldnaes (1965), and Bjørlykke (1968) suggested that a random shape-fabric should typify debris dropped from floating ice on to a muddy sea-floor or lake-bed. There exist many deposits apparently accumulated in this way (D.J. Miller, 1953; Armstrong and Brown, 1954; Hardy and Leggett, 1960; Hübner, 1965; Whetten, 1965; Rattigan, 1967; F.E. Anderson, 1968; Frakes and Crowell, 1969; Brückner and Anderson, 1971; Crowell and Frakes, 1971; Frakes et al., 1971; Howarth, 1971; Lindsey, 1971; Spencer, 1971; Nystuen, 1976; Ojakangas and Matsch, 1980). Usually the sediments are muds or mudstones, commonly laminated or containing thin graded beds, in which sand, granule and gravel-size stones lie scattered indiscriminately. The larger fragments— the so-called drop-stones— evidently sank some way into the mud after reaching the bed, since they disturb the lamination. From the symmetry of the disturbances, it would appear that drop-stones are more often buried in the attitude at which they strike the bed than otherwise. The oblate stones, for example, are generally flat-lying, in the position of settling of oblate spheroids at comparably large Reynolds numbers. If such stones had sunk edge-on and then toppled over on reaching the bed, they would not be associated with symmetrical disturbances of the laminae. Hunkins et al. (1970) photographed flat-lying drop-stones on the bed of the Arctic Ocean.

There is but one published study of drop-stone fabric. Spencer (1971) measured the apparent long-dimensions of drop-stones at five localities in a Precambrian deposit, with results suggesting that the observed fabrics are at least partly deformational. However, the drop-stones were markedly less well-oriented than stones in adjacent current-laid conglomerates, and therefore could have had an initially random fabric.

The orientation of the separated valves of bivalve molluscs preserved in turbidites may also be due primarily to settling. Compton (1962), Crowell et al. (1966), and Hoskins (1967) all observed that concave-up orientations were common, and Natland (1957) found that strong imbrication was frequent. Hoskins thought that the concave-up shells had been rotated into this attitude as the sediment packed down without significant internal shearing. Middleton's (1967b) experiments, however, point to an orientation imposed by settling.

He released into a tank 5 m long turbidity currents made of suspensions of plastic beads to which were added pieces of thin-walled plastic tubing cut parallel with the long axis. These pieces, simulating concavo-convex bivalve shells, settled in plain water at $Re \approx 10^3$, though in the experimental currents their Reynolds numbers would have been rather smaller. From the summary of results given in Table 5-II, we see that between 16 and 50% of the pieces were concave-up, the stable orientation of concavo-convex particles at intermediate and large Reynolds numbers. A substantial number were oriented vertically, an attitude that is statically unstable on a horizontal bed. The results suggest that increase of concentration favours the dynamically stable concave-up position. Thus a significant part of the total fabric can be explained by settling, the convex-up shells suggesting that traction currents had a limited success in turning over the clasts.

A predominantly concave-up orientation of bivalve shells due to settling may also arise under shallow-water conditions. Clifton and Boggs (1970) found that typically more than half the valves of the small bivalve *Psephidia* preserved in a shallow-marine sand were concave-up. Experimenting with the tractional transport of the same shells across rippled beds, they noticed that concave-down shells were "flipped over and partly buried" as they

TABLE 5-II

Attitude of concavo-convex particles deposited from experimental turbidity currents. Data of Middleton (1967b)

Fractional sediment concentration in turbidity current	Number of valves in deposit		
	concave-up (stable during settling)	vertical (statically unstable)	concave-down (statically stable)
0.223	9	1	15
0.228	5	2	24
0.430	20	4	24
0.437	24	9	15

glided over bed ripples, a mode of behaviour earlier alluded to by Menard and Boucot (1951). It would appear that the shells are swept by the current sufficiently far downstream from the ripple crests that they can turn over and settle stably in the sluggish flow to lee. My experiments showed that concavo-convex shapes turned over from concave-down to concave-up in falling a vertical distance comparable with their radius of curvature.

Rees (1966a) made several experiments in which quartz sand was allowed to settle through still water on to a sloping bed. The grains deposited on a slope of 20° from the horizontal lay with their long-axes aligned in the direction of the slope. Presumably this resulted from the slight rotation of each grain as it touched and settled on the bed, for the end of the grain that first made contact would serve as a pivot. The slope is much too gentle for avalanching to have created the fabric.

SHAPE-FABRICS DUE TO TRANSLATION IN SHEAR FLOWS

Theory

A rigid homogeneous particle of unequal dimensions placed in a simple shear-flow (Couette flow) of viscous fluid may: (1) spin about an axis, (2) rotate about an axis, (3) move relative to the fluid by virtue of an external force, (4) move perpendicular to fluid streamlines on account of a shear or spin lift (Chapter 2), and (5) interact directly or at a distance with any particles already in the flow in such a manner as to move in a modified way and to change their motion. If the rate of rotation of the axis of a particle contained in a fluid is dependent on the orientation of the particle, we have a second situation in which a preferred orientation may arise in a dispersion of particles.

Bretherton (1962a) calculated for small Reynolds numbers, and Tsien (1943) for an inviscid fluid, the forces acting on a circular cylinder placed with its axis normal to the plane of motion of a flow experiencing simple shear. The shear gives rise to a transverse lift-force on the cylinder, tending to move it towards the region of higher velocities. The magnitude of the lift force is little affected if the cylinder is permitted to rotate about its axis at a rate comparable with the rate of shear.

The behaviour of a sphere moving in Couette flow was calculated for small Reynolds numbers by Wakiya (1956, 1957), Rubinow and Keller (1961), and by Saffman (1965). The sphere is acted on by a couple causing it to spin, and by a lift force perpendicular to the line of movement of the sphere and to the spin-axis. According to Saffman, when the sphere moves parallel with the fluid (e.g. a particle sedimenting in a flow through a vertical pipe), the magnitude of the lift force is:

$$F_{\mathrm{L}} \approx 6.46\eta V a^2 \left(\frac{\mathrm{d}U}{\mathrm{d}y}\right)^{1/2} / \nu^{1/2} \qquad (5.8)$$

where V is the *relative* velocity between sphere and fluid at the streamline intersecting the centre of the sphere, a is the sphere radius, $\mathrm{d}U/\mathrm{d}y$ is the velocity gradient, and η and ν are respectively the fluid dynamic and kinematic viscosities. Inspection of eq. (5.8) shows that the lift force becomes zero when the particle travels with the local fluid velocity. We also notice that if the particle lags behind the fluid, the lift acts towards the region of higher velocities. The sphere moves towards the regions of lower velocity, however, when travelling faster than the fluid. These and related conclusions, sketched in Chapter 2, help to explain experimentally observed particle migrations in shear flows (e.g. Oliver, 1962; Segré and Silberberg, 1962; Shizgal et al., 1965; Lawler and Lu, 1971).

The unrestricted motion of a homogeneous particle of unequal dimensions in Couette flow was first calculated by Jeffery (1922), and later studied in relation to a wider range of flows by Saffman (1956), Bretherton (1962b), Giesekus (1962a, 1962b), Brenner (1964), Batchelor (1970, 1971), Cox (1970, 1971), D.G. Willis (1977), Gierszewski and Chaffey (1978), Harris et al. (1979) and Hinch and Leal (1979).

Jeffery calculated the hydrodynamic turning forces acting on a rigid ellipsoidal particle in an unbounded steady flow with a uniform velocity gradient $\mathrm{d}U/\mathrm{d}y$, on the supposition that the particle had the same density as the surrounding fluid, and that the motion was such that inertial effects could be neglected. His approach is therefore valid either for sufficiently slow motions or for sufficiently small particles. The positions of the particle axes proved to be functions of time, but the equations were solved completely only for ellipsoids of revolution (oblate and prolate spheroids). We may specify the position of the axis of symmetry of these particles by reference to the coordinate system of Fig. 5-6, where the *xy*-plane is parallel

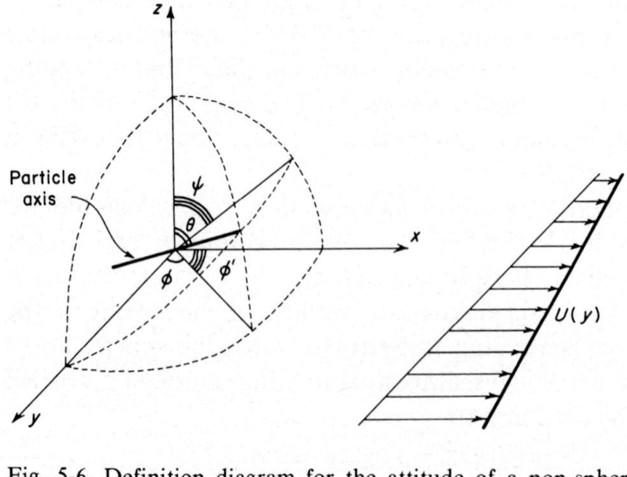

Fig. 5-6. Definition diagram for the attitude of a non-spherical particle in a fluid stream undergoing simple shear (Couette flow) in the *xz*-plane.

with the plane of the fluid motion and the positive x-direction is in the direction of flow, the origin of the coordinates moving with the particle centre. The angle ϕ lies between the y-direction and the intersection by the xy-plane of the plane containing the particle axis and the z-direction. The angle θ lies between the particle axis and the z-direction. Note that the axis of symmetry of a prolate spheroid is the major axis; that of the oblate spheroid is the minor axis. The particles may be described by an axial ratio $R(=a/b)>1$ for a prolate spheroid and $R<1$ for an oblate spheroid. Putting t as time, Jeffery obtained for the angular velocities the equations:

$$\frac{\mathrm{d}\phi}{\mathrm{d}t}=\frac{1}{(R^2+1)}(R^2\cos^2\phi+\sin^2\phi)\frac{\mathrm{d}U}{\mathrm{d}y} \tag{5.9}$$

and:

$$\frac{\mathrm{d}\theta}{\mathrm{d}t}=\frac{(R^2-1)}{4(R^2+1)}\sin 2\phi\sin 2\theta\left(\frac{\mathrm{d}U}{\mathrm{d}y}\right) \tag{5.10}$$

showing that the axis of symmetry of the particle rotates on an orbit about the particle centre. At the same time the particle spins about its axis of symmetry with an angular velocity:

$$\Omega=\frac{1}{2}\frac{\mathrm{d}U}{\mathrm{d}y}\cos\theta \tag{5.11}$$

Integration of eq. (5.9) shows that the particle rotates about its axis of symmetry with a period:

$$T=\frac{2\pi}{\left(R+\frac{1}{R}\right)}\left(\frac{\mathrm{d}U}{\mathrm{d}y}\right)^{-1} \tag{5.12}$$

whence:

$$\tan\phi=R\tan\left(\frac{2\pi t}{T}\right) \tag{5.13}$$

Dividing eqs. (5.9) and (5.10) and integrating yields:

$$\tan\theta=\frac{KR}{(R^2\cos^2\phi+\sin^2\phi)^{1/2}} \tag{5.14}$$

where K is a constant of integration defined as the orbital constant.

These equations show that the ends of the axis of symmetry of the particle describe a pair of symmetrical spherical ellipses, rather like a spinning top. The eccentricity of these ellipses is defined by the value of the orbital constant; their principal axes are $\tan^{-1}K$ at $\phi=0$, π and $\tan^{-1}KR$ at $\phi=\pi/2$, $3\pi/2$. Moreover, the orbit is invariant with time, since the orbital constant expresses only the initial attitude of the spheroid, and may take any value between zero and plus infinity. When $K=0$ ($\theta=0$ at all ϕ) the axis of symmetry of the spheroid lies normal to the plane of the flow, and the particle merely spins about this axis with an angular velocity $\Omega=\frac{1}{2}(\mathrm{d}U/\mathrm{d}y)$. If $K=\infty$ ($\theta=\pi/2$ at all ϕ) the axis of the spheroid rotates in the xy-plane without spinning. It will be noticed for the simple case of the sphere that $\mathrm{d}\theta/\mathrm{d}t=0$ and $T=4\pi\,(\mathrm{d}U/\mathrm{d}y)^{-1}$.

Bretherton (1962b), Brenner (1964), and Cox (1970, 1971) generalized Jeffery's problem to arbitrary bodies and to sharp-ended axisymmetric shapes like cylinders and discs. They found that each such body, so far as its rotation is concerned, is equivalent to a particular ellipsoid of revolution. Hence Jeffery's equations may be applied, with an equivalent axial ratio substituting for the axial ratio defined above. Cox (1971) calculated the relationship of the equivalent and Jeffery axial ratios for double cones and cylinders. Experimental relationships for cylinders and discs are given by Goldsmith and Mason (1967), and by Anczurowski and Mason (1968). Gierszewski and Chaffey (1978), and Hinch and Leal (1979), give numerical and analytical solutions for non-axisymmetric ellipsoids, the rotation of these bodies proving to have a doubly periodic structure. D.G. Willis (1977) found that the attitude of a non-spheroidal body of unequal dimensions was determined both by strain history and particle shape.

Jeffery (1922) was concerned that, according to his investigations, the

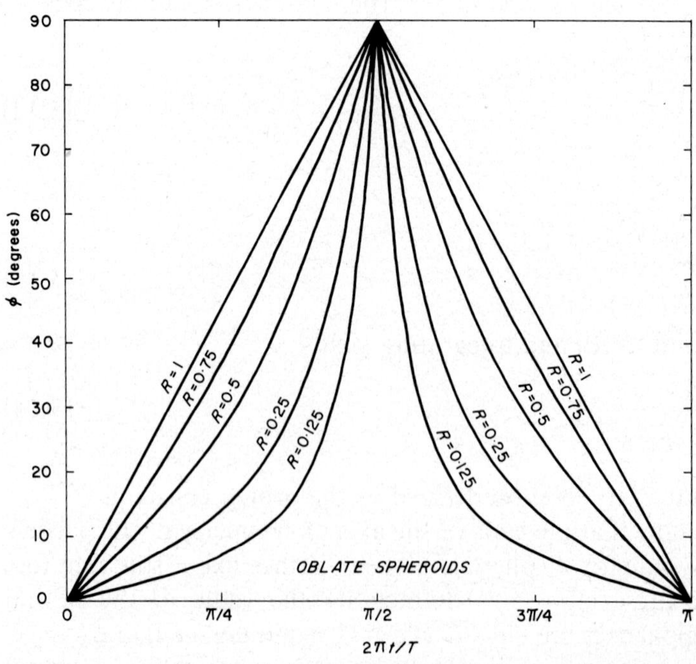

Fig. 5-7. The simple-shear deviation (ϕ) of the axis of symmetry of an oblate spheroid from the y-direction as a function of time (t), with the axial ratio (R) as a parameter.

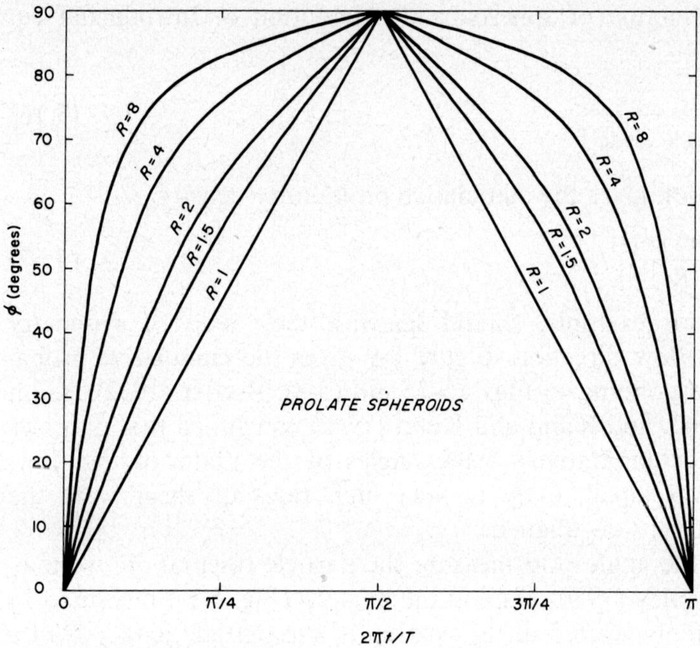

Fig. 5-8. The simple-shear deviation (ϕ) of the axis of symmetry of a prolate spheroid from the y-direction as a function of time (t), with the axial ratio (R) as a parameter.

orbital constant could take any value. He considered this indeterminacy to arise from his neglect of inertia, and proposed that particles would in reality "tend to adopt that motion which, of all the motions possible under the approximated equations, corresponds to the least dissipation of energy". Minimum overall energy dissipation corresponds to the axis of symmetry parallel with the z-direction for prolate spheroids and lying in the xy-plane for oblate spheroids. Saffman (1956) and Bretherton (1962b) also speculated on the factors that might lead to a preferred orbit.

The angle ϕ is the complement of the apparent angle of imbrication of a particle. Figures 5-7 and 5-8 show orbits calculated from eq. (5.13) for realistic axial ratios. These graphs, as also of course eq. (5.9), show that $d\phi/dt$ is a minimum at $\phi = \pi/2$, $3\pi/2$, etc. and at $0, \pi$, etc. for an oblate sphere. Hence a single prolate spheroid will spend most of its time in an orientation such that its long-dimension lies close to the flow direction. On the other hand, an oblate spheroid will spend most time nearly edge-on to the flow. The quantity $(d\phi/dt)^{-1}$ is clearly proportional to the likelihood of finding spheroids of orientation ϕ in a dispersion of a large number of non-interacting spheroids. If such a dispersion has reached a steady state, we may write:

$$p(\phi) = \left(T \cdot \frac{d\phi}{dt}\right)^{-1} \tag{5.15}$$

where $p(\phi)$ is the fraction of spheroids of orientation ϕ. Introducing eq. (5.12):

$$p(\phi)=\frac{R}{2\pi(R^2\cos^2\phi+\sin^2\phi)} \tag{5.16}$$

which upon integration gives the cumulative probability density:

$$P(\phi)=\frac{1}{2\pi}\tan^{-1}\left(\frac{\tan\phi}{R}\right) \tag{5.17}$$

Prolate spheroids, for example, should lie with their axes of symmetry mainly close to the flow direction. Figure 5-9 gives the cumulative probability densities corresponding to Figs. (5-7) and (5-8). Boeder (1932), Kuhn (1933), Peterlin (1938), and Kuhn and Kuhn (1945) concluded that particles became oriented at comparatively steep angles in the plane of the flow, provided dU/dy was small. Only at very high rates of shear were the particles expected to be flow-aligned.

Instead of using the angle ϕ to measure the particle orientation, we may with Mason and Manley (1956) employ the angle ψ (Fig. 5-6), measured in the xz-plane and simply related to the bearing of the particle long-axis in a

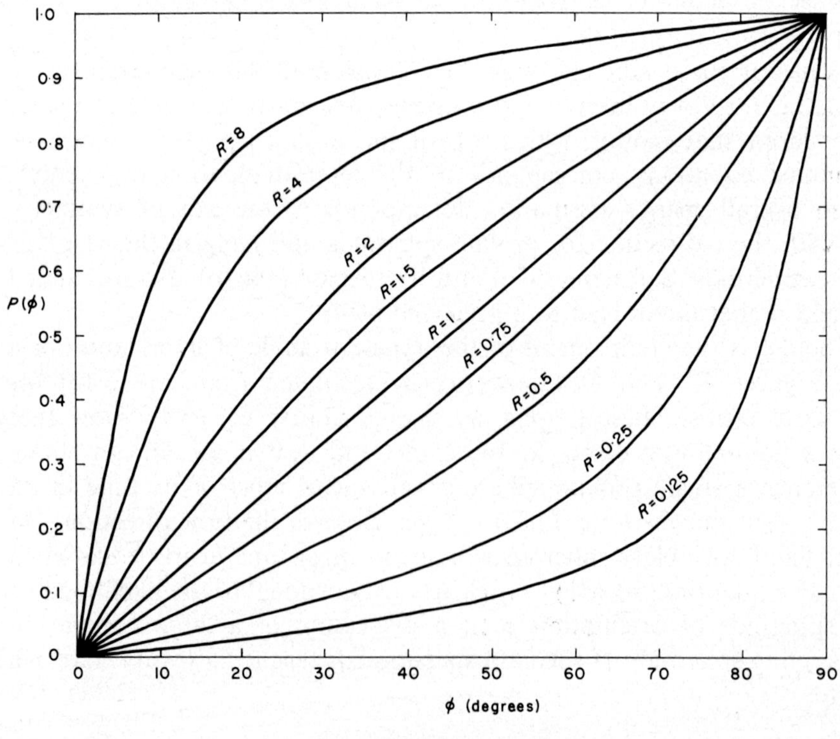

Fig. 5-9. The fractional cumulative probability (P) that the axis of symmetry of spheroids in simple shear deviates by the angle ϕ from the y-direction.

real shape-fabric. Jeffery's equations yield:

$$\tan \psi = KR \sin\left(\frac{2\pi t}{T}\right) \tag{5.18}$$

whence the corresponding expression to eq. (5.16) is:

$$p(\psi) = \frac{\sec^2 \psi}{\pi\left(K^2R^2 - \tan^2\psi\right)^{1/2}} \tag{5.19}$$

from which it follows that for each value of K and R, $p(\psi)$ is a minimum when $\psi = 0$. As viewed in the xz-plane, therefore, most prolate particles lie close to the flow direction. However, in order to calculate $p(\psi)$ explicitly, we must know the value of K.

For each dispersion there is obviously a critical fractional volume concentration of particles above which particle interactions must be significant. Jeffery's equations will not then apply and orientations will be assumed that depend to an extent on the behaviour of particles as they collide with each other. S.G. Mason (1954) and Mason and Manley (1956) showed that the critical concentration C_{crit} is reached when:

$$\frac{Q_{\text{E}}(K)C_{\text{crit}}}{Q_{\text{P}}} = 1 \tag{5.20}$$

where Q_{P} is the volume of the particles and $Q_{\text{E}}(K)$ is the volume of fluid swept out by the particles in the time for one orbit. Assuming dispersions for which $K = \text{const.}$, we have for cylinders:

$$Q_{\text{E}}(K) = \frac{2\pi(b + Ka)(a + Kb)(b + KR_{\text{E}}a)}{(1 + K^2)(1 + K^2R_{\text{E}}^2)^{1/2}} \tag{5.21}$$

where $2a$ is the length of the symmetry axis, b is the particle radius, and R_{E} is the equivalent axial ratio. For a disc:

$$Q_{\text{E}}(K) = \frac{2\pi(a + KR_{\text{E}}b)}{(1 + K^2R_{\text{E}})^{1/2}}\left(b^2 - \frac{(a + KR_{\text{E}}b)^2}{3(1 + K^2R_{\text{E}}^2)}\right) \tag{5.22}$$

where $2a$ is the thickness and b the radius. The value of $Q_{\text{E}}(K)$ for cylinders is a maximum for $0 < K < \infty$, but for the disc is a maximum at $K = \infty$.

Goldsmith and Mason (1967) summarize these results as applied to non-rigid bodies in shear flows, for example, liquid drops, flexible rods, and chains of loosely connected particles. Gay (1968a, 1968b) developed a theory for the motion of deformable particles in simple shear, and also studied motions in finite pure shear (Gay, 1966, 1968c), as will be seen below. Owens (1974) applied the Jeffery model to the magnetic fabric of deformed rocks.

Experimental justification

Jeffery's theory has been extensively tested experimentally, on account of its relevance to several rheological problems.

Observed rates of rotation of rigid spheres in a Newtonian fluid in Couette flow are in excellent agreement with theory (Trevelyan and Mason, 1951; Manley and Mason, 1952; Bartok and Mason, 1957; Mason and Bartok, 1959). Goldsmith and Mason (1962a) found good agreement with theory in the case of spheres in Poiseuille flow, provided the spheres were small in size compared with the pipe. G.I. Taylor (1923) found qualitative agreement between the observed motion of aluminium spheroids (prolate and oblate) in water-glass and that indicated by Jeffery's theory. Data for prolate spheroids consistent with theory were obtained by Goldsmith and Mason (1962a), and by Anczurowski and Mason (1968).

Many experiments have been made using cylinders and discs, as these shapes are easy to procure and occur in many applications. Forgacs and Mason (1959), and Goldsmith and Mason (1962b), observed that the spin of these particles in Couette flow was consistent with theory. There are numerous studies in which ϕ and T were measured, showing that the rotation of cylinders and discs in shear flows is in accordance with Jeffery's theory, provided that an equivalent axial ratio is used rather than the true axial ratio (Trevelyan and Mason, 1951; Manley and Mason, 1952; Bartok and Mason, 1957; Forgacs and Mason, 1959; Bretherton and Lord Rothschild, 1961; Goldsmith and Mason, 1962b; Karnis et al., 1966; Darabaner et al., 1967; Anczurowski and Mason, 1968). Quantitative agreement has also been obtained for other kinds of shear flows (Chaffey et al., 1965). Qualitatively consistent behaviour was observed by Eirich et al. (1936) for long cylinders in both Couette and Poiseuille flows.

Theory suggests that two equal spheres in contact will rotate together as if they were a prolate spheroid of $R_E = 2$, and this is confirmed by Manley and Mason (1952) and by Bartok and Mason (1957). The behaviour of longer chains of spheres and discs, investigated by Goldsmith (1966) and by Zia et al. (1966, 1967), is also consistent with theory. Long flexible cylinders deform under simple shear into snake-like or helical forms (Forgacs and Mason, 1959).

Particularly interesting investigations have been made of the steady-state distribution functions of ϕ and ψ in dispersions containing large numbers of particles, and of the distribution of values of the orbital constant. Mason and Manley (1956), Goldsmith and Mason (1962b), and Anczurowski and Mason (1967) obtained data in good agreement with theory for discs and cylinders in Couette flow and for cylinders in Poiseuille flow. Most cylinders, for example, lay with long-axes close to the flow direction and had values of K less than unity, the distribution of values lying between $K = 0$ and $K = \infty$ for all particles, required under Jeffery's (1922) hypothesis of minimum energy dissipation, and by the results of calculations based on the suggestion of Eisenschitz (1932) that the particles are uniformly distributed at the onset of motion.

Experiments shed some light on the indeterminacy in Jeffery's theory. At small Reynolds numbers, and provided great care is taken to obtain neutral buoyancy, there is no tendency for a single particle sheared in a Newtonian liquid to change its orbit, even after hundreds of rotations (Manley et al., 1955; Anczurowski et al., 1967). These are, of course, the conditions for which Jeffery's theory is valid. At larger Reynolds numbers, inertial effects become important, and neutrally buoyant particles in Newtonian liquids drift into orbits of maximum energy dissipation, though the rotation of the particles is unaffected (Binder, 1939; Karnis et al., 1963, 1966). Particles sheared in viscoelastic fluids, however, drift into terminal orbits apparently of minimum energy dissipation (Saffman, 1956; Karnis et al., 1963; Karnis and Mason, 1966). These results seem to be in conflict, and it would appear that the influence of inertial and non-Newtonian effects on particle orbits remain very largely to be clarified, though there are recent signs of progress (C.C. Ferguson, 1979).

Application to the shape-fabrics of mass-flow deposits

The preceding theory, with its extensive empirical backing, has intuitively been thought applicable to the shape-fabrics of debris-flow deposits (Lindsay, 1968), and it may also be relevant to other sediments of mass-flow origin, amongst them flow tills, solifluction deposits, and creeping soils.

Debris flows, or mud flows, are characterized by a content of clay, silt and water sufficiently large as to create a rapid, surging flow within which the larger stones appear uniformly dispersed and suspended (Sharpe, 1938). Debris flows are important agents of transport and deposition in semi-arid regions, where they appear chiefly on alluvial fans after floods (Blackwelder, 1928; Sharp and Nobles, 1953; Blissenbach, 1954; Beaty, 1963; W.B. Bull, 1964; Lustig, 1965; Hooke, 1967). They are known from mountainous areas generally, often in association with other kinds of mass flow (e.g. Tricart, 1957; Clapperton and Hamilton, 1971), and in volcanic districts occur as the awesome *lahars* (Scrivenor, 1929; Iida, 1938; Van Bemmelen, 1949; Schmincke, 1967; Crandell, 1971). Mud flows can also be important in the denudation of clay cliffs (Htuchinson, 1967). Related to debris flows are flow tills, formed as the result of the collapse and subsequent downslope movement of somewhat liquid material released from bands of englacial debris melted out on the surface of a glacier. Boulton (1967, 1968) describes several examples. Solifluction—a much slower process than mud-flow— is defined by Andersson (1906) as "the slow flowing from higher to lower ground of masses of waste saturated by water". Deposits of solifluction are common in semi-arid and temperature regions, but are best developed under cold conditions (Sharpe, 1938; Kirkby, 1967; Washburn, 1967; Benedict, 1970). Creeping soils are ubiquitous.

In order for Jeffery's complete theory to apply strictly to shape-fabrics developed during mass flow, it is necessary that: (1) the immersing fluid is Newtonian, (2) the velocity gradient be uniform, (3) the particles are ellipsoids of revolution and not too concentrated, (4) the particles are neutrally buoyant, and (5) an appropriately defined Reynolds number is sufficiently small.

How far are these conditions met in reality, and to what extent may any of them be safely violated? The first condition certainly is not satisfied, for A.M. Johnson (1970) has convincingly shown that debris flows behave in a non-Newtonian manner. Probably their behaviour is most closely described by the rheological model for a Bingham plastic, in which a finite yield strength is associated with a constant apparent viscosity, eq. (1.2). Although natural sedimentary particles often closely approximate to ellipsoids, the third condition cannot be regarded as exactly satisfied. That part of it concerned with particle concentration can be tested by eq. (5.18) above. Likewise the condition of neutral buoyancy is practically never satisfied. Violation of the second condition, concerned with the uniformity of the velocity gradient, is perhaps not so serious, since there can generally be found some particle size which is sufficiently small compared with the flow thickness. It is particularly important for the fifth condition to be satisfied, however, for otherwise the debris should be found to have drifted into limiting orbits. An appropriate Reynolds number is one formed using the velocity gradient:

$$Re = \frac{|\mathrm{d}U/\mathrm{d}y|a^2\gamma}{\eta_a} \tag{5.23}$$

in which a is the particle long dimension, γ is the debris-flow bulk density, and η_a its viscosity. In the case of a non-uniform velocity gradient, the Reynolds number will change in value from level to level in the flow,

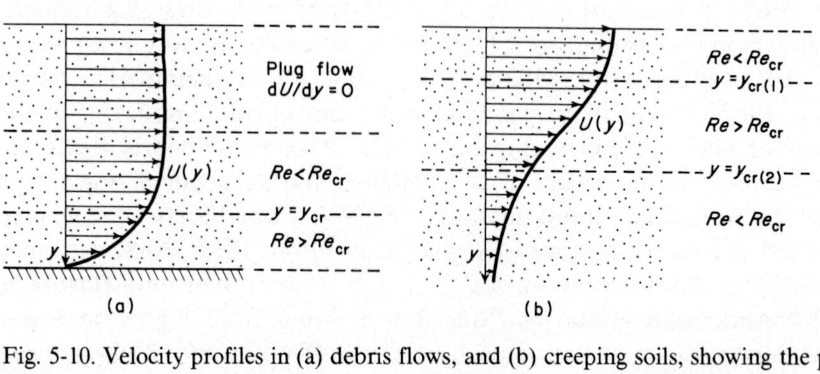

Fig. 5-10. Velocity profiles in (a) debris flows, and (b) creeping soils, showing the permissible depth ranges of Jeffery's mechanism of preferred particle orientation.

irrespective of whether there is also a viscosity change with depth. Hence there may exist levels in the flow at which the Reynolds number is too high for Jeffery's theory to be valid.

We suggested in Chapter 1 that the velocity profile of a mud flow is of parabolic form and that in a solifluction deposit is essentially exponential (Fig. 5-10). If U is the local velocity, measured parallel with the x-direction, and y is distance measured down from the surface, we have for a debris flow on land:

$$-\frac{\mathrm{d}U}{\mathrm{d}y}=\frac{1}{\eta_{\mathrm{a}}}(\gamma g \sin \beta y-\tau_{\mathrm{cr}}) \tag{5.24}$$

and for a solifluction deposit or creeping soil:

$$-\frac{\mathrm{d}U}{\mathrm{d}y}=\frac{\gamma g \sin \beta}{\eta_{\mathrm{a}(0)}} \exp(-ky)y \tag{5.25}$$

where β is the angle of the slope. In eq. (5.24), η_{a} is the apparent viscosity and τ_{cr} the yield strength. The quantity $\eta_{\mathrm{a}(0)}$ in eq. (5.23) is the viscosity measured at the surface, and k is a coefficient describing the rate of downward increase.

If Re_{cr} is the critical value of the Reynolds number given by eq. (5.23), we can further define a level y_{cr} above (or below) which Jeffery's theory should apply. Substituting for $\mathrm{d}U/\mathrm{d}y$ from eq. (5.23), the theory should be valid in a debris flow for:

$$\frac{\tau_{\mathrm{cr}}}{\gamma g \sin \beta}<y<\frac{\eta_{\mathrm{a}}^2}{a^2\gamma^2 g \sin \beta}\left(Re_{\mathrm{cr}}+\frac{a^2\gamma\tau_{\mathrm{cr}}}{\eta_{\mathrm{a}}^2}\right) \tag{5.26}$$

where the smaller limit will be recognized as defining the base of plug flow. In solifluction deposits and creeping soils, Jeffery's theory will apply where $y<y_{\mathrm{cr}(1)}$ and $y>y_{\mathrm{cr}(2)}$, with:

$$Re_{\mathrm{cr}}=\frac{a^2\gamma^2 g \sin \beta \exp(-ky_{\mathrm{cr}})y_{\mathrm{cr}}}{\eta_{\mathrm{a}(0)}^2} \tag{5.27}$$

As sketched in Fig. 5-10, these inequalities suggest that for particles of a given long dimension, there can in debris flows be but one range of depths within which Jeffery's is valid but, in solifluction deposits and creeping soils, two possible depth ranges. Presumably a random fabric accompanies plug flow in a debris flow. The fabric is unlikely to be random, however, where in a creeping soil, solifluction deposit, or debris flow the velocity gradient exceeds the critical.

Since the critical depths are functions also of particle size, we should find that different sizes of particle depart from Jeffery's conditions at different depths. Hence shape-fabrics based on particles of a *range* of sizes may change qualitatively and quantitatively from level to level, as well as the

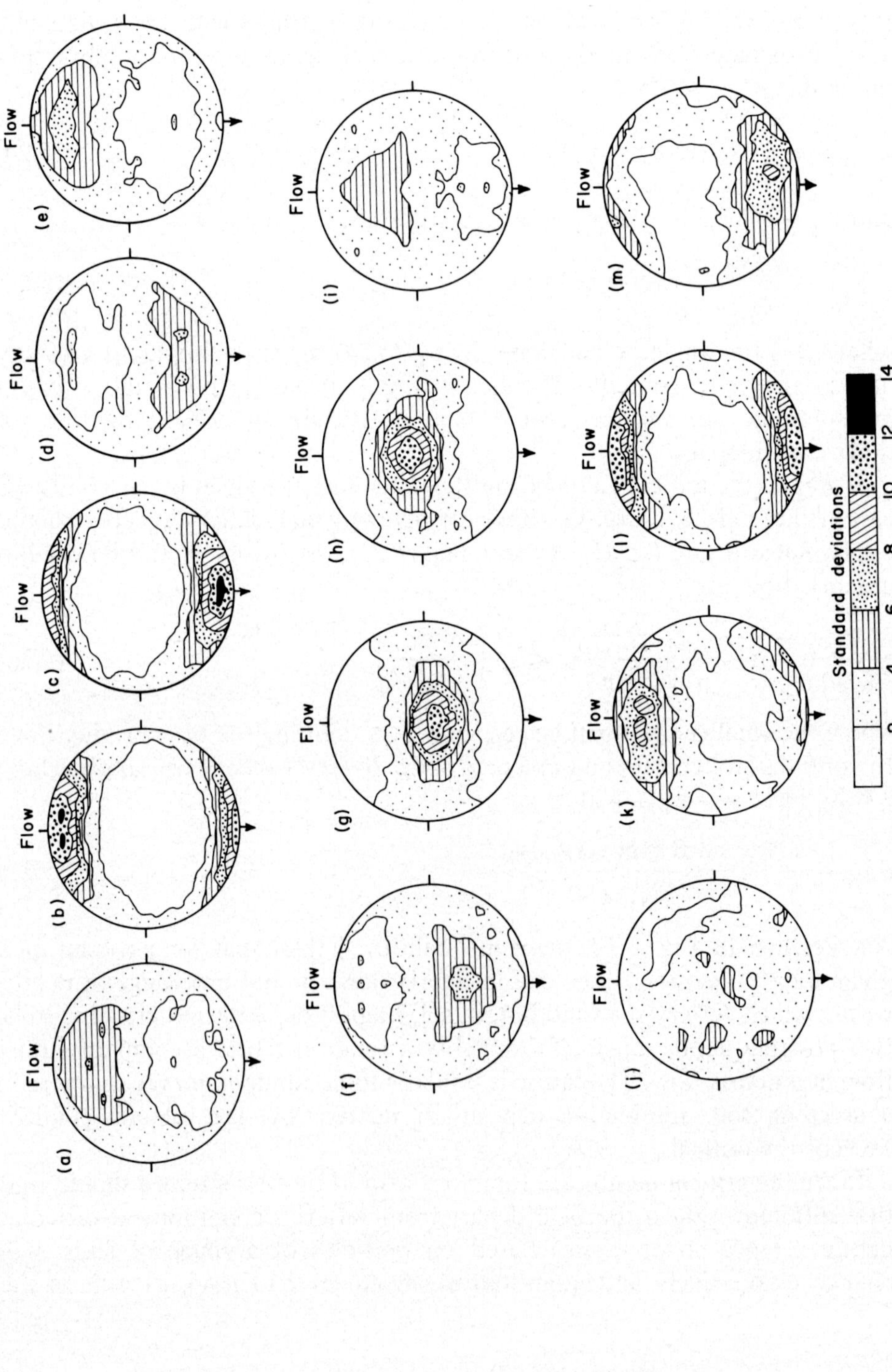

Flow
(a)
(b)
(c)
(d)
(e)
(f)
(g)
(h)
(i)
(j)
(k)
(l)
(m)
Standard deviations
2
4
6
8
10
12
14

fabrics measured on a *single* size of debris.

In practice, the apparent viscosity of solifluction deposits and creeping soils is so large that the Reynolds number is everywhere below the critical. Lundqvist (1949, 1951) measured the azimuths of the long-axes of stones in solifluction deposits and found a strong orientation parallel with the direction of the slope and the inferred movement. Similar observations were made by E. Watson and S. Watson (1967), by E. Watson (1969), and by Mottershead (1971) and C.A. Baker (1976), some of whom noticed also a weak upcurrent imbrication. Benedict (1969, 1970) measured the long-axis orientation of sand grains from a solifluction lobe in the Niwot Range, Colorado, and found strongly developed fabrics with a weak to marked imbrication. These data are very limited, though the fabrics are broadly what may be expected.

Although debris flows on land reach maximum speeds of many metres per second and maximum thicknesses of several metres, the available descriptions (Iida, 1938; Sharp and Nobles, 1953; Curry, 1966; Waldron, 1967; A.M. Johnson, 1970) suggest that the typical flow has a thickness of order 1 m, a mean velocity of order 1 m s^{-1}, an apparent viscosity of order 10^2 N s m^{-2}, a yield strength of order 500 N m^{-2}, a bulk density of approximately 2300 kg m^{-3}, and flows on a slope comparable with 10°. Boulton's (1968) observations of active flow tills are consistent with this categorization of the typical debris flow.

Now suppose we measure the shape-fabric of a debris flow as specified above. If stones for which $a = 0.05$ m are chosen, and $Re_{cr} = 0.1$, we can expect inertial and non-Newtonian influences to have had no effect on the fabric only between approximately 0.13 and 0.17 m downwards from the surface, a very small percentage of the active flow. But if we choose sand grains for which $a = 0.001$ m, the Reynolds number is less than the critical throughout the whole flow below the plug. Jeffery's complete theory may therefore have little relevance to debris flows, at least so far as stone-fabrics are concerned.

Lindsay (1968) applied Jeffery's equations to debris-flow fabrics, assum-

---

Fig. 5-11. Computer simulations of the development in time of clast fabrics according to the Jeffery (1922) equations. After Lindsay (1968). a–e. Development with increasing time from an initially uniform distribution of the long-axis fabric (intersections with lower hemisphere) of prolate spheroids of $a/b=2$. f–i. Development with increasing time from an initially uniform distribution of the short-axis fabric (intersections with lower hemisphere) of oblate spheroids of $a/b=0.5$. j–m. Development with increasing time from an initially random distribution of the long-axis fabric (intersections with lower hemisphere) of prolate clasts of mean $a/b=1.77$. The diagrams are projected on a horizontal plane and are contoured using the method of Kamb, which uses a counting circle such that, if the particle population lacks preferred orientation, the number of points expected to fall within the area is three times the standard deviation of the number of points that will actually fall within the area under random sampling of the population.

ing that the above conditions were all satisfactorily met. He prepared a computer simulation in which the shape-fabric of a group of initially uniformly or randomly distributed spheroids was followed as a function of time, the strength of their preferred orientation waxing and waning on a period of $T/2$. In our typical debris flow, for example, this cycle would be completed in a matter of a fraction of a second or a few seconds. Figure 5-11 illustrates these interesting fabrics. That for prolate spheroids of $R = 2$ begun from a uniform distribution combines at its strongest development monoclinic symmetry with a bold girdle (Figs. 5-11a–e). The particles are aligned mainly in the flow direction, a weak upcurrent imbrication appearing as the fabric strengthens, and an opposite dip as it decays. Lindsay also calculated the corresponding fabrics of oblate spheroids of $R = 0.5$, observing the periodic growth and decay of an orthorhombic symmetry associated with a weak girdle in the *yz*-plane (Figs. 5-11f–i). His final calculations were for the fabric of an initially randomly distributed population of prolate spheroids of $R = 1.77$, regarded as close to real sedimentary particles. An essentially monoclinic fabric with a girdle in the *xz*-plane and a reversing imbrication emerged (Figs. 5-11j–m). Reed and Tryggvason (1974) attempted some limited calculations on fabrics produced by simple shear, their conclusions agreeing with Lindsay's.

Little is known of shape-fabrics from debris flows. From modern flows Rapp (1960a) observed flow-parallel as well as transverse stone orientations, and Bull (1964) noticed that shale fragments, presumably platy, were generally flat-lying. Piper (1972) reported flow-parallel and flow-transverse orientations of stones from Pleistocene deposits interpreted as debris flows. Lindsay (1966, 1968) also measured the three-dimensional fabric of stones in several fossil flows. Long-axis orientations that formed strong girdles slightly tilted from the *xz*-plane were observed from the Pagoda Formation, Antarctica. These fabrics do not resemble the computer simulations. A better agreement with simulated fabrics was found in the case of debris-flow deposits from the Mackellar Formation, Antarctica, and the Kullatine Formation, New South Wales. Their fabrics, however, all showed a consistently upcurrent imbrication, whereas the calculated ones reversed in imbrication periodically and were unimbricated at their maximum strength. Harrison (1957) studied the orientation of the *ab*-planes of blade and disc-shaped stones at several stations in a single debris flow. An upstream imbrication of these planes was observed. Boulton (1968) found that elongated stones in flow tills lay parallel with the direction of movement.

Because the stones carried in debris flows are not neutrally buoyant, Lindsay (1968) briefly discussed the effects of particle settling on shape-fabrics. As may be inferred from Table 5-I, settling should have no effect where the Reynolds number is small and the particles do not interact. Fall broadside-on at higher Reynolds numbers should strengthen the orthorhombic features of a fabric.

Bhattacharyya (1966) explained by Jeffery's theory the sub-parallel orientation of minerals in igneous and metamorphic rocks showing evidence of flowage. Gay (1966, 1968c), however, considers that these fabrics in metamorphic rocks are better explained by a model assuming pure and not simple shear.

SHAPE-FABRICS DUE TO TRANSLATION IN PURE SHEAR

Theory

Many materials can deform very gradually by creep, for example, ductile metals, some rocks, and glacier ice. Often this mode of deformation involves a predominant, or at least substantial, element of pure shear, that is, a deformation without rotation (stretching).

Gay (1966, 1968c) extended Jeffery's (1922) general equations to the motion of a rigid ellipsoidal particle in a viscous fluid deformed by pure shear. The particle may again be considered in the coordinate system of Fig. 5-6, where the x-direction is now parallel with the direction of maximum elongation, and the y-direction is parallel with the direction of maximum contraction. The intermediate axis of the strain ellipsoid, parallel with the z-direction, remains constant. Therefore we have plain strain, with the xy-plane as the deformation plane. Gay solved the equations for an ellipsoid of revolution, specifying the instantaneous position of the symmetry axis by the angle θ in Fig. 5-6 and the angle ϕ', the complement of the angle ϕ previously used. The equations of motion for the ellipsoid of revolution are:

$$\frac{\mathrm{d}\phi'}{\mathrm{d}t}\cos\theta = 0 \tag{5.28}$$

$$\frac{\mathrm{d}\phi'}{\mathrm{d}t} = -\left(\frac{R^2-1}{R^2+1}\right)\sin 2\phi'\frac{\mathrm{d}\epsilon}{\mathrm{d}t} \tag{5.29}$$

and:

$$\frac{\mathrm{d}\theta}{\mathrm{d}t} = \frac{1}{2}\left(\frac{R^2-1}{R^2+1}\right)\sin 2\theta\cos 2\phi'\frac{\mathrm{d}\epsilon}{\mathrm{d}t} \tag{5.30}$$

where ϵ is the natural strain.

Rearranging eq. (5.29) to read:

$$\int_{\phi_1'}^{\phi_2'}\frac{\mathrm{d}\phi'}{\sin 2\phi'} = -\left(\frac{R^2-1}{R^2+1}\right)\int_0^{\epsilon}\mathrm{d}\epsilon \tag{5.31}$$

and then integrating gives:

$$\ln\tan\phi_2' = \ln\tan\phi_1' - 2\left(\frac{R^2-1}{R^2+1}\right)\epsilon \tag{5.32}$$

for the position of the symmetry axis after a strain ϵ, where the subscripts 1 and 2 refer to the initial and final positions. Similarly, from eq. (5.30) and in terms of $\phi = (90° - \phi')$:

$$\frac{\tan\theta_2}{\tan\theta_1} = \left(\frac{\sin 2\phi_1}{\sin 2\phi_2}\right)^{1/2} \tag{5.33}$$

which gives the variation of θ with ϕ. Equation (5.30) affords θ when ϕ is either 0° or 90°.

These equations are clearly non-periodic, unlike the corresponding equations for a particle in simple shear. In pure shear, an ellipsoidal particle rotates towards a limiting position, parallel with the direction of maximum elongation, but attains that position only after an infinitely large time and strain. Equation (5.28) implies that in approaching the limit the particle experiences no axial spin; in simple shear there is no spin only when the particle symmetry axis lies in the plane of shear. Equation (5.30) shows that with increasing strain the particle symmetry axis approaches progressively closer to the x-axis (Fig. 5-12), the direction of maximum elongation. Thus a prolate spheroid starting from an arbitrary position will tend to line up parallel with flow, as Reed and Tryggvason (1974) confirmed for a particle assemblage, whereas an oblate spheroid will approach broadside-on. A

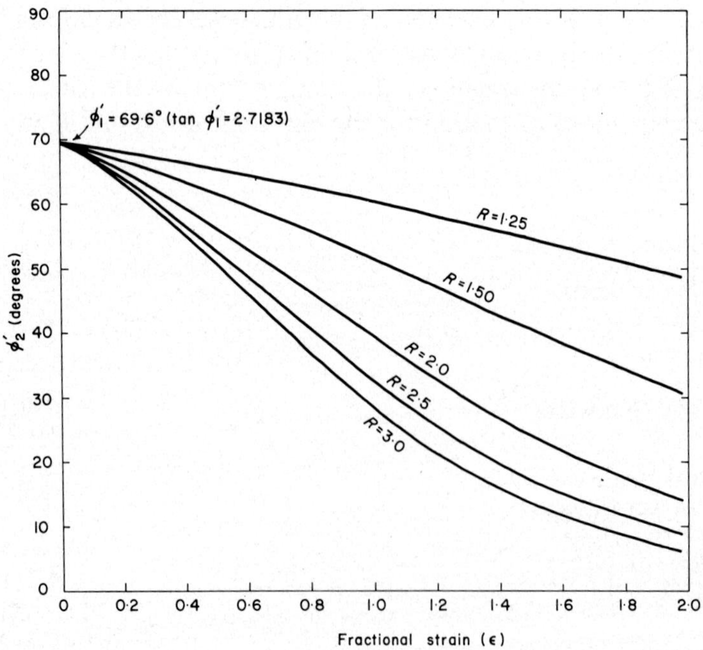

Fig. 5-12. The pure-shear deviation (ϕ_2') of the axis of symmetry of a prolate spheroid from the flow or x-direction, as a function of the total strain (ϵ). The axial ratio (R) appears as a parameter, and the particles all possess the same initial orientation (ϕ_1').

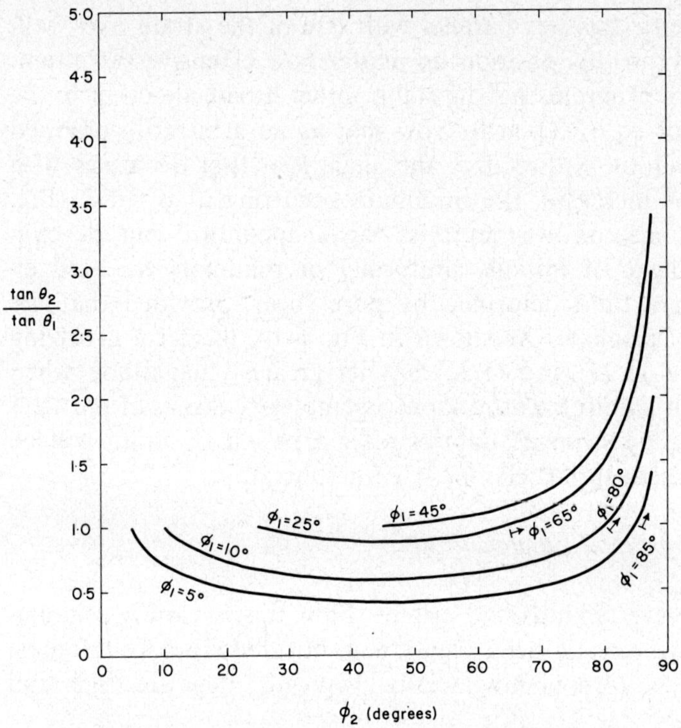

Fig. 5-13. The pure-shear deviation (θ) of the axis of symmetry of a prolate spheroid from the z-direction as a function of the final deviation (ϕ_2) from the y-direction, with the initial deviation from the y-direction (ϕ_1) as a parameter.

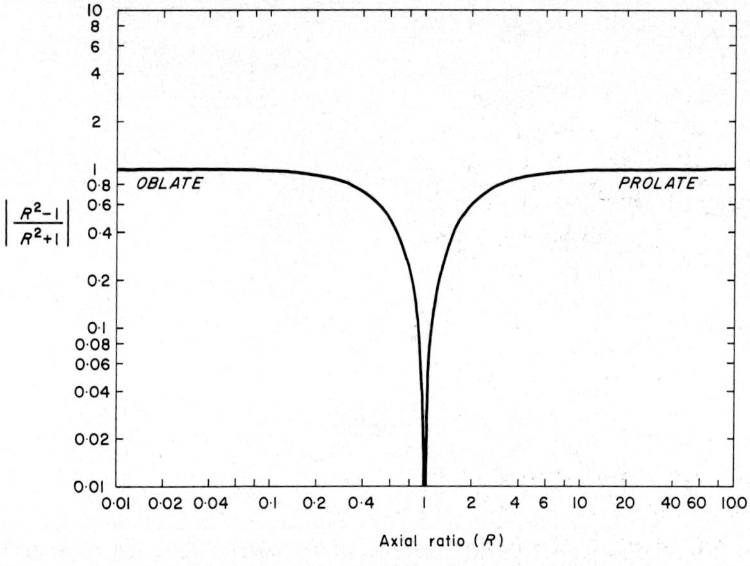

Fig. 5-14. Variation of the magnitude of the axial ratio term, $(R^2-1)/(R^2+1)$, as a function of the axial ratio.

particle whose symmetry axis lies parallel with one of the strain axes will, however, remain stably in this position no matter how extensive the strain. An oblate spheroid, for example, may lie stably either broadside-on or in the xz-plane. Inspection of eq. (5.33) will show that as an arbitrarily oriented particle rotates towards the x-direction, the angle θ at first decreases with increasing ϕ and then increases, the minimum occurring at $\phi = 45°$ (Fig. 5-13). The equations also show that a strong shape-fabric can develop rapidly in an assemblage of initially uniformly or randomly oriented ellipsoids embedded in a fluid deformed by pure shear, provided that the particles are relatively elongate. As shown in Fig. 5-14, the term involving the axial ratio in eqs. (5.29) and (5.32) has its greatest magnitude when $R = 0, \infty$. Hence when a uniform or random assemblage consists of particles of a range of shapes, the strongest fabrics after a period of strain will be shown by the least spherical particles ($R = 1$ for sphere).

Application to shape-fabrics of subglacial tills

Glacial tills are of several kinds and origins. Flow tills, a variety originating supraglacially, have already been mentioned, but undoubtedly the most important tills are those formed subglacially. Typically, they are thick and

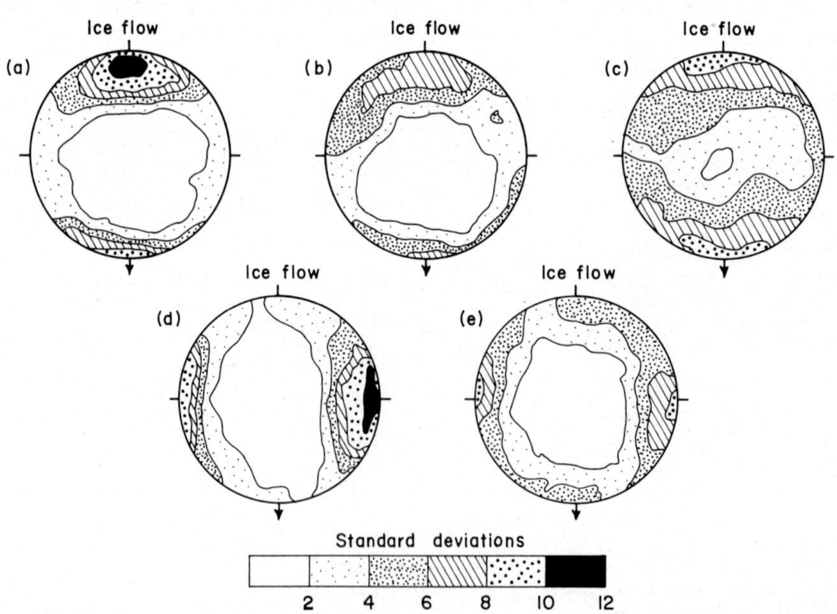

Fig. 5-15. Examples of clast fabrics (long-axis intersections with lower hemisphere projected horizontally) from glacial tills in the Palaeozoic Pagoda Formation, Transantarctic Mountains, Antarctica. After Lindsay (1970a). The majority of the fabrics recorded from this formation are about equally divided between patterns similar to (b) and (c) and patterns resembling (d) or (e). Diagrams contoured as in Fig. 5-11.

massive, and comprise stones scattered through an abundant, poorly sorted matrix of clay, silt and sand. The fabric of these elements, now to be described, may be partly explicable in terms of Gay's theory for the motion of an ellipsoid in pure shear, though other models have been proposed.

K. Richter (1932, 1933, 1936) was first to undertake the systematic regional study of the shape-fabrics of stones in subglacial tills. He found that the more elongate particles were generally aligned close to the direction of ice movement as indicated by independent evidence. The same general conclusion has been reached by many other workers (Krumbein, 1939; Holmes, 1941; Hyyppä, 1948; Molder, 1948; Virkkala, 1951, 1960, 1961; Hoppe, 1952, 1957; Galloway, 1956; West and Donner, 1956; Donner and West, 1957; Harrison, 1957; MacClintock and Terasome, 1960; Flint, 1961; Gillberg, 1961; Groth, 1961; Schulz, 1961; Moss and Ritter, 1962; Pettijohn, 1962; Wright, 1962; Schytt, 1962; Hälbich, 1964; Andrews, 1965; Bjorlykke, 1968; Harris, 1967, 1968; Lindsey, 1966, 1969; Kirby, 1968, 1969a, 1969b; Penny and Catt, 1967; H.G. Johansson, 1968; Saunders, 1968; Casshyap, 1968; Burke, 1969; Boulton, 1970b, 1971; Lindsay, 1970a, 1970b; McKenzie, 1970; Price, 1970; Beaumont, 1971; Hill, 1971; Mickelson, 1971, 1973; Casshyap and Qidway, 1974; Drake, 1974, 1977; Rose, 1974; Olszewski and Supryczynski, 1975; H.H. Mills, 1977a, 1977b). Figure 5-15 illustrates some till fabrics. Although these workers gave their results differently—some by stereographic plots and others by wind-rose diagrams—the recorded shape-fabrics are strikingly similar. Most reveal a weak upcurrent imbrication of the primary long-axis mode and a secondary long-axis mode at right angles to the direction of ice-flow and in the *xz*-plane. Occasionally, however, the transverse mode is equally as strong as or stronger than the parallel mode (e.g. Donner and West, 1957; Harris, 1967; Boulton, 1970b; McKenzie, 1970). The most detailed investigations point to a further weak long-axis mode parallel with the *y*-direction (Holmes, 1941) and show that the short-axes of the flatter stones tend to lie parallel with *y* (Krumbein, 1939; Harrison, 1957).

Little is known of the fabrics of the finer-grained constituents of such tills. The sand-sized particles seem generally to lie parallel with the direction of ice-flow, yielding fabrics similar to those for stones (Von Seifert, 1954; Harrison, 1957; Gravenor and Meneley, 1958; Ostry and Deane, 1963; Penny and Catt, 1967). Sitler and Chapman (1955) and Sitler (1968) noticed that silt-sized flaky minerals in the till matrix lay nearly parallel with the *xz*-plane. This was attributed to plastic deformation of the till.

Attempts have been made to relate the strength and character of till fabrics to stone shape, roundness and size. The earliest study of this type is by Holmes (1941), who concluded, after an elaborate but essentially qualitative analysis, that these factors indeed predisposed stones to a particular orientation. Andrews and King (1968), and Krüger (1970), partly supported by Holmes, noticed that the more elongate stones gave the strongest fabrics.

Lindsay (1970c), studying a fossil till, found evidence that fabric strength decreased with increasing stone size, but increased with the proportion of fine-grained till matrix. Harris (1967) observed no relation between fabric strength and the shape, size, and degree of rounding of the stones. Drake (1974, 1977) finds that the transverse mode is given not by markedly elongate stones but by clasts of little to moderate elongation. Several workers directed attention to the local variability in strength and orientation of till-fabrics, partly with the problems of sampling in mind (Harrison, 1957; Gravenor and Meneley, 1958; Kauranne, 1960; J.A.T. Young, 1969; Andrews and Smith, 1970; Rose, 1974). Till-fabrics appear to be moderately to highly variable, particularly in the vertical at a site.

A few data on the orientation of englacial stones are available which may help to explain till-fabrics. Donner and West (1957) found that in thin till bands enclosed between ice layers, the stone long-axes lay as often transverse as parallel with the direction of ice movement. In the thicker bands, however, the parallel orientation of long-axes was typical. Harrison (1956, 1957) also observed strongly developed transverse and parallel orientation; one of the samples with transverse orientation came from a shear zone in the ice. Schytt (1962) and Lindsay (1970a) likewise noticed parallel orientations. Boulton (1970a, 1970b, 1971) recently gave some valuable observations from Svalbard glaciers. In stagnant ice at Erikabreen he measured long axes that were parallel with the direction of ice-flow; statistically indistinguishable fabrics were observed from the associated tills. Long axes oriented mainly parallel with flow were also observed from Makarovbreen, where the occurrence of open crevasses showed the ice to be in tension. Transverse fabrics were found in crevasse-free zones in the same glacier, and also from a till with included ice layers deposited upstream of a *roche moutonée*. Boulton found that long-axes lay parallel with the direction of movement in a till that had been strongly sheared and foliated. He observed from Makarovbreen one fabric based on the short-axes of flat stones. These axes were clustered around the *y*-direction. Some elongated stones trapped in glaciers have been observed to rotate, rather in the manner of Jeffery's ellipsoids, as the ice bore them along.

The fabrics just summarized may have been influenced by some or all of the following: (1) the orientation assumed by debris when emplaced in the ice, (2) the changes of clast orientation during ice transport, (3) the mode of deposition of the enveloping till, and (4) deformation of the till after deposition.

The debris found in glaciers becomes emplaced in several ways. It may tumble on to the glacier surface, or be washed into crevasses by meltwater. Debris can also be entrained as the result of thrusting at the glacier base (Goldthwait, 1951), or by basal cavitation and regelation (Weertman, 1961; Lliboutry, 1965; Boulton, 1970a). The extent to which these modes of emplacement induce preferred orientations is unknown, and it seems best to assume, with Glen et al. (1957), that the debris is initially randomly oriented.

The changes in orientation experienced by debris during glacier-transport depend on particle shape and concentration, and on the mode of ice flow. Laboratory tests show that ice deforms in a non-Newtonian manner (Glen, 1958) and field observations, summarized by Paterson (1969), indicate that glaciers flow partly by simple shear and partly in pure shear. Moreover, if meltwater occurs throughout the whole thickness of the ice, slip may occur between the glacier and its bed, leading to a reduced velocity gradient in the ice. Nye (1952, 1957) has analyzed some of these cases.

The similarity between till-fabrics and the orientations of englacial debris suggests that the fabrics of subglacial tills are largely determined by the behaviour of the debris while in transport. Several models have been proposed for the orientations assumed by transported debris.

Glen et al. (1957) considered that Jeffery's (1922) model for the motion of an ellipsoid in Couette flow explained the flow-parallel stone elongations so frequently observed from subglacial tills. They advanced in support of this view a long-axis fabric calculated on the assumption of a steady state, but their fabric shows neither the subordinate transverse mode nor the upcurrent imbrication commonly observed. Predominantly transverse orientations were attributed to prolonged transport in the ice, on the supposition that the particles would ultimately drift into orbits of minimum energy dissipation. However, parallel and transverse englacial fabrics can be found closely adjacent. Transverse orientations were also thought to be due to stone collisions.

Lindsay (1970a) thinks that glacier ice deforms like a polycrystalline metamorphic rock, and therefore he relates englacial fabrics to the system of "shear domains" that he expects to be present in such a tectonite. The primary long-axis mode parallel with the direction of ice-flow, together with the preferred orientation of short-axes nearly parallel with y, is attributed to a flat-lying shear domain tilted gently upglacier. The generally weaker transverse orientation of long-axes is believed to reflect a nearly vertical shear domain.

The model perhaps most closely consistent with knowledge of the behaviour of real glaciers is a combination of Jeffery's (1922) simple-shear model with Gay's (1966, 1968c) scheme for an ellipsoid in pure shear. Such a combination explains the parallelism of long-axes with the direction of ice-flow, for this orientation occurs in the steady state for simple shear and is approached asymptotically in pure shear. The weak, and sometimes predominant, transverse orientation of long axes is also explained by the combined models, because in pure shear a particle parallel with a strain-axis will remain parallel with that axis. The stronger transverse orientations may be expected where the ice is in compression, the principal strain-axis lying perpendicular to the general direction of ice-flow. This appears to be consistent with, say, Boulton's (1970a) observations. The model also seems able to explain the other weak modes that are sometimes observed in till-fabrics, because of the component of pure shear. The mathematical aspects of the combined model remain to be worked out.

Andrews and Smith (1970) suggest that the fabrics of subglacial tills depend primarily on till flowage beneath the weight of ice. This proposal is not supported by the evidence (e.g. Boulton, 1970b; Lindsay, 1970a), though some fabrics are certainly explicable as deformational modifications of earlier patterns (Andrews, 1963; MacClintock and Dreimanis, 1964; Andrews and Smithson, 1966; Banham, 1966; Penny and Catt, 1967; Cowan, 1968; Price, 1969, 1970). There is often in these cases independent morphological or structural evidence of deformation.

SHAPE-FABRICS OF FLOWS OF DENSELY ARRAYED PARTICLES

Theory

Strong shape-fabrics are commonly developed during the flow of densely arrayed particles, for example, in sand avalanches down the lee slopes of dunes, and in talus slides on mountain screes. Typically, rod-shaped particles lie with their long-axes parallel with the flow direction but imbricated upslope. Blades and discs, on the other hand, lie with their *ab*-planes nearly parallel with the slope, with a gentle upslope dip. Since these fabrics can also be made in a vacuum, and the grains are always seen to be densely arrayed, it follows that the fabrics may depend only on the interparticle impact and frictional forces called into play during granular shear. In these cases gravity merely provides a driving force.

The impelling force may otherwise depend on a fluid flow. Thus the bedload of a vigorous stream lies in a thin layer in contact with the static bed beneath. The grains of the bedload layer are comparatively densely arrayed, and those in the lowermost levels travel much more slowly than those higher up. At these concentrations, the collisional forces between grains in relative motion may be expected to induce a preferred particle orientation, provided that the viscosity of the intergranular fluid is sufficiently small. Even when the viscosity is not small, so that mainly hydrodynamic forces are called into play, a preferred orientation is to be expected, for there will be only one position relative to these forces in which the particle is stable.

Rees (1968) sketched a theory of preferred orientation in gravity-driven avalanches of loose grain and related flows. His detailed argument is confined to the two-dimensional inertial case, and he touches only lightly on the three-dimensional problem, and on the effects of intergranular friction. We repeat some features of his argument in what follows, but attack the problem in a substantially different and more general way.

Bagnold (1954a, 1956, 1966) showed that in a gravity or fluid-impelled layer of concentrated elastic particles sheared steadily over a bed of similar particles, there exists a time-average resisting force R which can be resolved into a normal component P, opposing the downward-acting particle weight, and a tangential component T, opposing the motion (Chapter 1). In the inertial case these forces derive solely from particle impacts. When the effects of viscosity predominate, the resisting force represents a more remote interaction, unlikely to involve actual physical contacts. In all cases P is dispersive, the grains being less closely packed than in a static heap.

Figure 5-16 shows these forces; their ratio, T/P, defines Bagnold's dynamic friction coefficient $\tan\alpha$. The particles, assumed to be smooth, perfectly elastic, and elliptical in cross-section, are flowing parallel with the x-direction at a local mean velocity $U(y)$. Individuals follow a zig-zag path, on average parallel with the x-direction, as they approach, collide with, and rebound from grains in adjacent layers. Now particles may take up a statistically preferred orientation as the result of these collisions only if they can achieve some stable orientation relative to the direction of action of the average resultant R. This implies that collisions between particles which are unstably oriented must result in turning couples tending to move those grains into stable attitudes.

To see how this might arise, we suppose as an analytical device that the particles at level I in Fig. 5-16 define in Fig. 5-17 a single rigid plane, inclined at an angle α to the y-direction, and upon which particles from layer II will impinge. Since R acts on average normal to this plane, and the grains are assumed smooth and perfectly elastic, the impact force F_{IP} has the same orientation as R. Particles such as B and C in Fig. 5-17 therefore experience when they strike the plane a clockwise moment tending to bring their long-axes parallel with XY. Grains such as E and F are turned anticlockwise

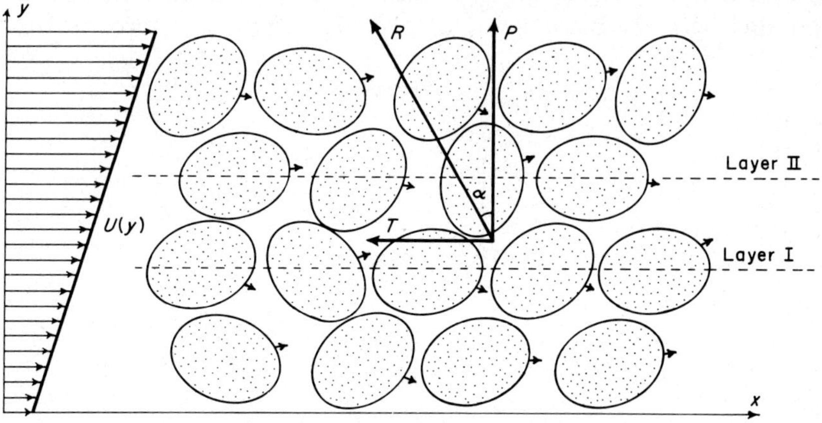

Fig. 5-16. Schematic representation in the flow plane of the shearing of non-spherical particles. The resisting force R, averaged over time and a large number of particles, is resolved into tangential (T) and normal (P) components.

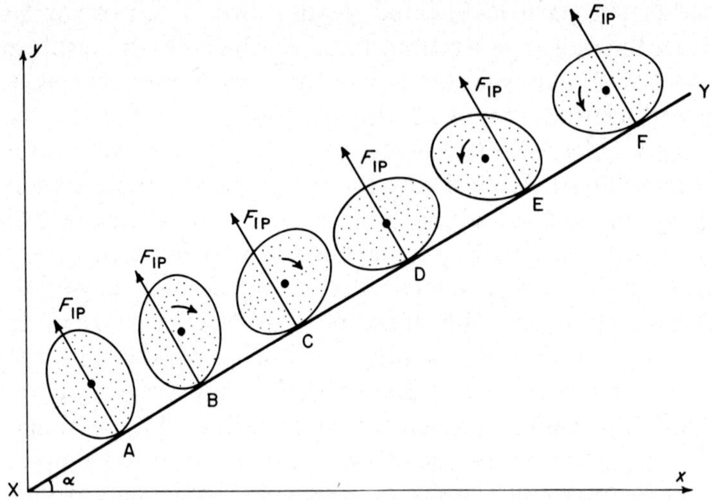

Fig. 5-17. The stability and sense of rotation of non-spherical particles striking a notional rigid plane inclined from the direction of grain motion at the dynamic friction angle (α).

on impact, so that their long-axes also tend to lie parallel with XY. Evidently the stable particle attitude is the orientation of particle D, whose long-axis lies parallel with XY. A grain such as A experiences like D no turning moment, since the impact force acts through the centre of mass. But A is unstable, since any slight deviation results at impact in the turning force discussed. The orientation of D should therefore dominate, with but few grains adopting the attitude of A. It follows that prolate spheroids will tend to lie with their long-axes parallel with a plane tilted upcurrent at the angle α.

Whether or not the long-axes are parallel with this plane will depend on the interparticle forces acting in the third-dimension, normal to the plane of flow and parallel with the plane containing R. The dispersive force in this

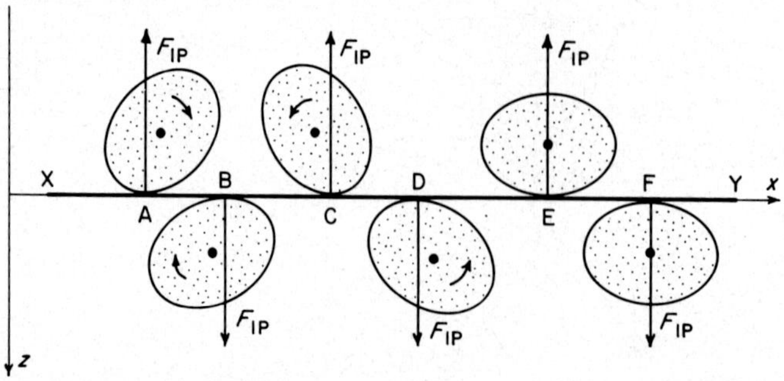

Fig. 5-18. The stability and sense of rotation of non-spherical particles striking a notional rigid plane parallel with the plane of grain flow. Particles on both sides of the plane are shown because the process is symmetrical (compare Fig. 5-17).

dimension must be symmetrical if the flow is uniform, and is likely to be much smaller than either P or T. Let XY in Fig. 5-18 be a section parallel with the x-direction through an imaginary rigid plane parallel with the flow plane. The grains will again zig-zag with respect to their mean path parallel with x. If they are prolate spheroids, the turning moment on impact with XY is zero only when the long-axis is parallel with the x-direction. Combining this and the previous result, the long-axes of an array of prolate spheroids should be oriented statistically parallel with the flow direction and dip upcurrent at the angle α.

This reasoning would appear to suggest that oblate spheroids will lie with their maximum-projection planes parallel with the xy-plane. This orientation can be achieved, however, only if the sideways dispersive force is larger than either P or T, and so produces the dominant moments. Since the sideways force appears to be much the smaller, oblate spheroids should be oriented with the maximum-projection planes parallel with the plane normal to R. These preferred orientations must be modified in aggregates of real particles by the effects of surface friction, if not by additional factors, as follows.

Consider in Fig. 5-19a a perfectly elastic but frictional particle impinging on a rigid plane. The impact force F_{IP} can now be resolved into a normal component F_{IN} and a tangential component F_{IT}, their ratio, F_{IT}/F_{IN}, defining the friction angle δ between particle and plane. The impact force fails to act through the centre of mass of the grain, and so a clockwise moment is set up turning the particle to a more stable attitude. But provided the friction angle is not too large, there are clearly impact positions on the grain circumference where the impact force would act through the centre of

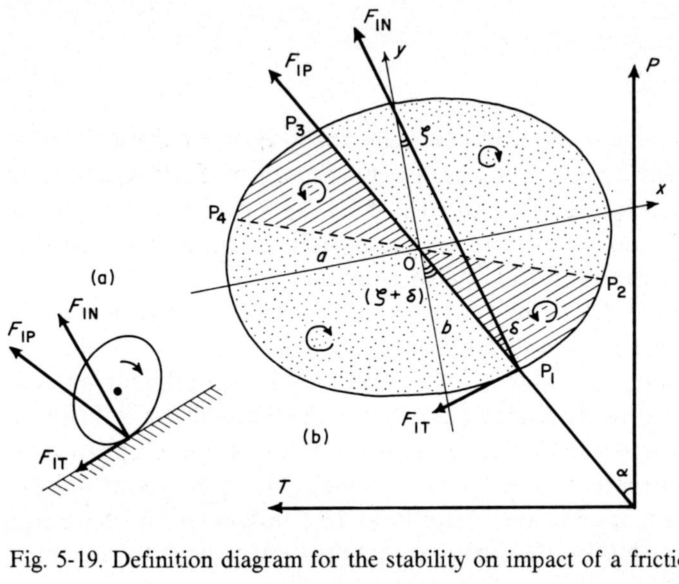

Fig. 5-19. Definition diagram for the stability on impact of a frictional elliptical particle.

mass. Impact at one of these points is a stable impact in the previous sense, but would not occur where a principal axis emerged from the particle. Impacts at other positions must create a turning moment, rotating the particle either clockwise or anticlockwise, accordingly as their position lies relative to a point of stable impact. It follows that, since the position for stable impact does not coincide with the point of emergence of a principal axis, the preferred orientation of frictional particles must differ from α by an amount dependent on the friction angle and axial ratio.

Referring to Fig. 5-19b, in which F_{IP} acts as required through the particle centre, the new angle of imbrication is equal to $\alpha - (\delta + \zeta)$, where ζ is the angle between the normal component F_{IN} and the minor axis of the ellipse. Putting a and b as the lengths respectively of the major and minor semi-axes:

$$\tan \zeta - \frac{b^2}{a^2} \tan(\delta + \zeta) = 0 \tag{5.34}$$

whence:

$$\frac{a^2}{b^2} \tan \delta \tan^2\zeta - \left(\frac{a^2}{b^2} - 1\right) \tan \zeta + \tan \delta = 0 \tag{5.35}$$

Each of the roots of eq. (5.35) corresponds to a position of stable impact, for example P_1 and P_2 in Fig. 5-19b, when the force F_{IP} acts through the particle centre. These positions correspond in the frictionless case to the points of emergence of principal axes.

Equations (5.34) and (5.35) show that in the frictional case the positions of stable impact depend only on the friction angle and grain shape, but that a stable orientation is attainable only if anticlockwise as well as clockwise moments can arise during collisions. From eq. (5.35), both kinds of moment will arise provided that:

$$\left(\frac{a}{b} - \frac{b}{a}\right) > 2 \tan \delta \tag{5.36}$$

when eq. (5.35) has two real roots. When such roots exist, an anticlockwise moment is created by an impact between P_1 and P_2, and a clockwise turning force if the impact lies between P_1 and P_4. As δ approaches closer to the critical value, P_1 and P_2 move nearer along the particle circumference, further limiting the possibility of an anticlockwise moment. When δ is equal to the critical value, eq. (5.35) has equal roots and P_1 and P_2 coincide; only clockwise turning is then possible, and the particle cannot become stable. At values of δ greater than the critical, the grain should spin about an axis normal to the plane of flow. It is seen from eq. (5.36) that spheres will spin if there is the slightest friction between them. As the grain elongation is increased, however, an increasing degree of friction can be tolerated without a statistically preferred orientation being lost. The introduction of friction has no sensible effect either on the flow-parallel attitude of prolate ellipsoids, as projected on to the xz-plane, or on the orientation of oblate spheroids such that their maximum-projection planes are normal to the plane of flow.

These conclusions may be compared with real shape-fabrics if values are assigned to α and δ. We saw in Chapter 1 that tan α is related experimentally to the balance between the inertial and viscous forces acting during the flow of particles, as expressed by the Bagnold number (Bagnold, 1954a, 1956, 1966). When inertia predominates, $\alpha = \tan^{-1} 0.37 \approx 20°$. The value rises to approximately 37° when viscosity predominates. Since δ has not yet been experimentally determined, but must take some non-zero value, these magnitudes afford an upper bound on the imbrication angles to be expected under natural conditions.

Application to gravity-controlled deposits

The above arguments are most directly applicable to the shape-fabrics of deposits formed by avalanching of debris under gravity.

Several workers, including Hamelin (1958) and Andrews (1961), noticed that debris is often arranged on screes so that particle long-axes lie parallel with slope and dip slightly upslope with respect to the surface of slope. Rapp (1959, 1960a, 1960b) reports many examples of slope-parallel long-axis orientation, though Gardner (1971) found some particles that had a strike-parallel attitude. Caine (1967, 1969) noticed little or no preferred orientation of the debris on screes she studied, but was commonly obliged to work with particles of no marked elongation. Scree deposits are emplaced by the avalanching beneath the atmosphere of thick layers of coarse debris. Very probably their shape-fabrics form under wholly inertial conditions of particle shearing. Predominantly if not wholly inertial conditions are also likely to have controlled the shape-fabrics of wind-laid cross-bedded sands, and of water-laid cross-bedded gravels and gravelly sands. Water-laid cross-bedded sands, however, are emplaced substantially if not wholly under the influence of viscosity.

Wadell (1936) made an early quantitative study of the shape-fabrics of water-laid cross-bedded gravels. He observed long-axes generally parallel with the dip-direction of the depositional slope, and imbrications of up to 30° upcurrent relative to the slope. A similar long-axis orientation was reported by Kalterherberg (1956), Wright (1957), Sengupta (1966), and Bandyopadhyay (1971), though Kalterherberg and Wright also observed strike-parallel pebble alignments. Kalterherberg reported a negligible angle of imbrication from the cross-bedded gravels he studied. C.E. Johansson (1960, 1963, 1965, 1976) comprehensively examined the fabrics of cross-bedded gravels and gravelly sands, by means of laboratory experiments as well as in the field. He confirmed that long-axes were statistically parallel with the cross-bed dip (Fig. 5-20), provided that the data were not drawn from near the disturbed tops and bottoms of the foresets, and measured

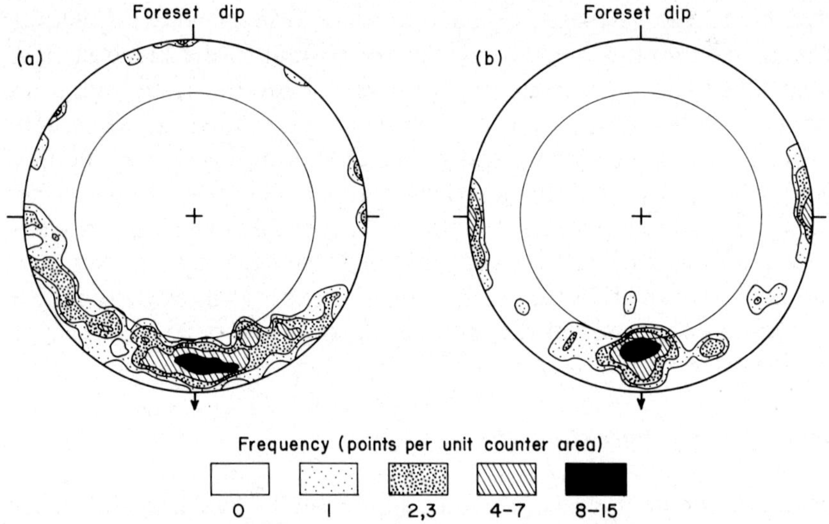

Fig. 5-20. Clast long-axis fabrics (intersection with lower hemisphere projected horizontally) in sandy gravels deposited as large avalanched foreset beds in two different runs in a laboratory flume. The inner circle in each diagram represents the magnitude of the foreset dip, whence the long-axes dip less steeply than the foreset bedding. After Johansson (1963).

upcurrent imbrications as great as 15°. Doeglas (1962) reported shape-fabrics from gravels deposited on steep slopes, but avalanching was not for certain involved.

Potter and Mast (1963) assembled a comprehensive set of data on the shape-fabrics of cross-bedded sandstones, showing that long-axes are statistically parallel with foreset dip, and that the long-axis imbrication is on average upcurrent at 15° relative to the foreset slope. Shelton and Mack (1970), using a dielectric method of fabric measurement, came to broadly similar conclusions about the fabric of cross-bedded sandstones. Rees (1968) reported long-axes statistically parallel with the depositional slope from the avalanche-face of a modern wind-blown dune.

Laboratory experiments have so far contributed little to our knowledge of the shape-fabrics of avalanched sediments. Schwarzacher (1951) reported a faint upcurrent imbrication, but found that the orientations of the long-axes and the slope did not agree. An earlier avalanching experiment, by Dapples and Rominger (1945), failed to establish a preferred particle orientation. Rees (1979), however, has measured flow-parallel particle elongations and imbrications of up to 27.5° from experimental sand avalanches.

There is a general consistency between the expected and observed fabrics, and to this extent the theoretical model described above is acceptable. The typical observed imbrications are upcurrent and of the magnitude expected. They do not violate in value the upper bound placed on imbrication by the model for smooth particles, but it is so far impossible to distinguish, on the basis of fabric, inertial from viscous shearing.

Other deposits from high-concentration flows

The above model should also apply to deposits formed from thick bedload layers with high particle concentrations, such as are formed at large transport stages. It may therefore be relevant to the shape-fabrics of flood gravels, parallel-laminated sands, the graded and parallel-laminated lower parts of turbidity current sandstones, and to mass-flow deposits with large concentrations of coarse debris.

Krumbein (1940, 1942b) measured flood-gravel shape-fabrics and found that long-axes were generally flow-aligned and imbricated slightly upcurrent. Members of the Sedimentary Petrology Seminar (1965) plotted the shape-fabric of gravel on the bars of a flashy stream. The maximum-projection planes of the mainly slabby and platy particles were observed to dip upcurrent on the average at approximately 23°. The predominantly transverse alignment of the long-axes of these particles is perhaps not to be compared with the parallel orientation that would be expected of prolate forms.

Numerous investigations combine to show that parallel-laminated sand-grade deposits have typically a monoclinic shape-fabric characterized by long-axes parallel with the current and imbricated at a low upstream angle (Curray, 1956; Nachtigall, 1962; McBride and Yeakel, 1963; Potter and Mast, 1963; Allen, 1964a; Picard and Hulen, 1969; Shelton and Mack, 1970). Imbrications are very variable. Potter and Mast measured a mean angle of 12° upcurrent, but observed a range between 24° upcurrent and 26° downcurrent, whereas Allen found only upcurrent imbrications of between 8° and 10°.

Turbidite shape-fabrics have attracted considerable attention. Kopstein (1954), the first to make a comprehensive study in this facies, concluded that grain long-axes in Welsh Cambrian turbidites were aligned parallel with the current direction. According to Bassett and Walton (1960), however, Kopstein's fabrics are deformational, a view consistent with the strong folds and cleavage in the rocks. Fabrics unequivocally of appositional origin have nevertheless been measured from many turbidite formations (Hand, 1961; Bouma, 1962; McBride, 1962, 1966; McBride and Kimberley, 1963; Spotts, 1964; Spotts and Weser, 1964; Sestini and Pranzini, 1965; K.M. Scott, 1967a; Colburn, 1968; Henningsen, 1968; Onions and Middleton, 1968; Parkash and Middleton, 1970; Von Rad, 1970; Hiscott and Middleton, 1980). Most of the fabrics are monoclinic and, where independent evidence of current direction is available at the horizon of the sample, show broadly flow-parallel long axes. Typically, the imbrication is upcurrent between 5° and 20°. Maximum upcurrent imbrications between 30° and 40° are reported by McBride, Onions and Middleton, and Parkash and Middleton. In approximately half the samples measured by Sestini and Pranzini, the imbrication was downcurrent, to a maximum of approximately 25°. Rees and Woodall (1975) measured flow-parallel orientation and small angles of imbrication from experimental turbidity current deposits. Slurries and slumps also gave flow-parallel alignments, but with often steeper angles of imbrication (Wing-Fatt and Stacey, 1966; Rees and Woodall, 1975). In a submarine mass-flow deposit, Hendry (1976) found the *ab*-planes of discoidal clasts to dip gently upcurrent.

Some of the atypical turbidite fabrics deserve mention. Bouma (1962) found that long-axes, in general, were at right angles to flow but parallel with the plane of the bedding. He also recorded bimodal fabrics and some that appeared random. Ballance (1964a) measured fabrics from turbidites with deformation structures on their soles. The long-axes were unimbricated and parallel or subparallel with the elongation of the deformations, which lay transversely to the inferred current path. Parkash and Middleton (1970) found that, of the 175 samples they studied, 57 showed no significant preferred grain orientation. Several samples had bimodal fabrics, the two modes lying approximately 90° apart. Some of the uniform fabrics came from the massive, unlaminated upper parts of the turbidites, presumably emplaced by settling.

Again a general consistency is apparent between the observed shape-fabrics and the model, though the variability of the imbrication angle invalidates a more searching comparison. It seems unnecessary to postulate with Parkash and Middleton (1970) that the typical shape-fabric of turbidites is due to the shearing of a "quick" bed, since it would appear that a fabric similar to that observed may have existed in the layer of densely arrayed grains driven over the bed and from which the deposit was formed.

Application to creeping flows of liquidized sand

In Vol. II, Chapter 8, it will be shown that waterlogged unconsolidated sand can be transformed under suitable conditions into a very viscous liquid, which may then flow, either under gravity, creating a fast but essentially laminar current, or under the action of a small current-applied force, giving rise to a creeping motion. After the latter, original lamination remains preserved, though of course in a new geometrical configuration on account of the flow. Particle concentrations are large, however, preventing free grain movement, and the total strain is generally comparatively small. Bagnoldian intergranular forces therefore have little opportunity to act, and it is perhaps not surprising that the available evidence suggests that the fabrics of sands deformed after liquidization seem mainly to record the rotation of primary fabrics toward the plane of shear (Yagashita, 1973; Yagashita and Morris, 1979).

PREFERRED ORIENTATIONS OF PARTICLES LODGING ON A HORIZONTAL BED

Theory

Current-driven grains may become preferentially oriented as they undergo deposition, because of their interaction during lodgement with particles already stationary on the bed. Preferred orientations so caused are associated with low concentrations of transported debris, for interactions between moving particles are then unimportant. Such preferred orientations arise chiefly when: (1) moving and stationary grains are comparable in size, and (2) moving particles are much larger than the majority already in the bed. Certain gravel and sand fabrics are covered by the first case. The second is related mainly to the orientations assumed on sand beds by shells and other large biogenic fragments.

Even in the second case, which is the simplest, a complete theoretical treatment is not yet possible, though some progress can be made by referring organic remains to simple shapes, for example, cylinders (groups of crinoid colunals), cones (gastropods), shells or concavo-convex lenses (separated bivalve and brachiopod shells), or discs (ammonites).

An isolated cylinder during transport over a plane bed of smaller particles should roll transversely to flow. The resistance offered by the bed to the rolling object is theoretically a minimum in this orientation, as also is the fluid drag on the cylinder. Schwarzacher (1963) showed, again theoretically, that this orientation of a cylinder also corresponds to the minimum force necessary for its entrainment.

A conical particle of small apical angle transported in contact with a planed granular bed cannot maintain a stable transverse alignment. While rolling it must also rotate about the point of intersection between its axis of symmetry and the surface of the bed, until the moment of the fluid drag equals that of bed friction. The particle in its final orientation should lie with its symmetry axis subparallel with flow and the apex pointing into the current.

Isolated shells and lenses are less straightforwardly treated. At conditions not too far removed from entrainment, such particles should glide concave-down over a smooth bed, to which they remain close and occasionally touch. Two main reasons suggest this conclusion: (1) the drag is a minimum for the concave-down orientation and, (2) the lift force is negative, making the particle difficult to overturn. The orientation of the long-dimension of an elliptical shell or lens will depend on several factors. In the case of a homogeneous shell of uniform thickness, the long-axis should be flow-transverse, for the same reason as an elliptical cylinder falls broadside-on (Lamb, 1932). But if the mass of the particle lies more towards one end than the other, a flow-parallel orientation may be expected, with the greater mass upstream. For if the object is broadside-on in a uniform flow field, the weightier end makes more frequent contact with the bed than the lighter one, there appearing a turning force tending to rotate it parallel with flow about the more massive part. Hence only the flow-parallel orientation is stable, for then the drag and frictional forces act on the same line.

Some bivalves, for example, the burrowing clam *Mya*, have an essentially elliptical shell with, centrally placed along the hinge-line of one valve, a large outward-projecting process about which the shell can pivot when transported convex-up. These shells should assume a stable transverse orientation on the bed, because otherwise there exists a turning force due to the fact that the parts of the valve on either side of the process present unequal projection areas. A similar effect is known to occur with trees which retain their roots after being torn up by floods (Axelsson, 1967).

Two stable orientations seem possible for discs. At low transport velocities, a disc may glide over the bed, the *ab*-plane lying nearly parallel with the bed. At higher velocities, when the disc can be lifted onto its edge, the only possible stable orientation occurs when the plane of the disc lies parallel with the plane of flow, the disc then bowling along like a wheel (J.S. Owens, 1908). In this orientation the turning forces acting on the disc are balanced and the gyroscopic effect further promotes stability.

Particles like non-uniform shells and lenses that assume a flow-parallel stable orientation during transport are unlikely to have that orientation significantly changed when they for some reason lodge on the bed. In contrast, cylinders and uniform shells with a flow-transverse stable orientation may rotate during lodgement. Consider in Fig. 5-21 the homogeneous cylindrical particle A of length b and radius a that has rolled up to the fixed particle B. The two touch at 0, taken as the origin of the coordinate system. The x-axis is parallel with the current, the y-axis perpendicular to the bed and the plane of the diagram, and the z-axis parallel with the bed but perpendicular to the xy-plane. Cylinder A approaches B such that its axis of

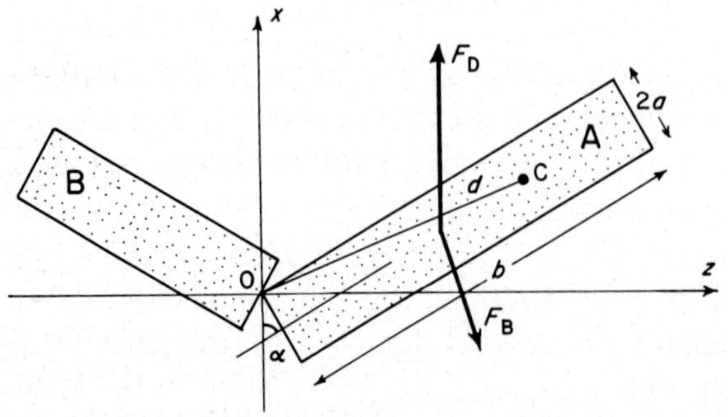

Fig. 5-21. Definition diagram for the motion of a particle A, as it rotates over the bed under the action of a current around the already stationary grain B. The bed is considered to be formed of particles much smaller in size than A or B.

symmetry make an angle α_0 with the x-direction. It is assumed that the flow field is uniform, the bed uniformly frictional, no hydrodynamic interaction occurs between A and B, and the drag coefficient C_D for particle A is constant and uniform.

The cylinder A is at each instant acted on by a fluid drag force F_D and a bed friction force F_B. The drag may be considered to operate through the centre of the particle projection-area, and it tends to turn the cylinder anticlockwise about 0. The turning action of the drag is opposed by the bed force, acting through the central point on the line of contact between cylinder and bed. No other forces need be considered, if it is further supposed that A is free only to turn about Oy. Putting M as the effective mass of A, d as the distance between O and the centre of gyration C, and Ω as the angular velocity of the axis of symmetry measured in the xz-plane, the equation of motion is:

$$-Md^2\frac{\mathrm{d}\Omega}{\mathrm{d}t}=\sin\alpha\left(a\cot\alpha+\frac{b}{2}\right)F_D-\left(a^2+\frac{b^2}{4}\right)^{1/2}F_B \qquad (5.37)$$

where t is time, α is the instantaneous angle between the symmetry axis and the x-direction, and the minus sign indicates that the cylinder is being arrested.

When A is brought to rest, the equation of motion reduces to:

$$\sin\alpha\left(a\cot\alpha+\frac{b}{2}\right)F_D=\left(a^2+\frac{b^2}{4}\right)^{1/2}F_B \qquad (5.38)$$

showing that the cylinder assumes a final orientation such that the turning moments of the drag and bed forces are equal. But the final orientation differs from the initial bearing α_0 only if, at the instant of contact with B:

$$\sin\alpha_0>\frac{\left(a^2+\frac{b^2}{4}\right)^{1/2}F_B}{\left(a\cot\alpha_0+\frac{b}{2}\right)F_D}. \qquad (5.39)$$

When the bed is frictionless, the final orientation of the cylinder is given by:

$$\tan\alpha=-\frac{2a}{b} \qquad (5.40)$$

the minus sign indicating measurement to the left of the positive x-direction. Therefore eqs. (5.39) and (5.40) define the range of possible orientations for A. According to circumstances, the particle may assume any orientation between flow-transverse and nearly flow-parallel. According to the analysis, the flow-parallel orientation is progressively approached as the fluid drag increases relative to the bed friction force.

When the cylinder radius is small compared to length, the equation of motion simplifies to:

$$-Mb\frac{d\Omega}{dt}=\tfrac{3}{2}(F_D \sin\alpha - F_B) \tag{5.41}$$

since a can be neglected compared with b and for these conditions $d=b/\sqrt{3}$. At equilibrium the drag force on A is:

$$F_D=\frac{C_D\rho U^2}{2}(2ab\sin\alpha) \tag{5.42}$$

where U is the flow velocity, ρ the fluid density, and the term in brackets is the particle projection area. Equation (5.41) then simplifies at equilibrium to:

$$C_D\rho U^2(ab\sin^2\alpha)=F_B \tag{5.43}$$

from which the azimuth of the cylinder may readily be calculated. By rearranging eq. (5.43) so that the two forces are compared, it will be noticed that $\sin^2\alpha$ is proportional to the first power of a linear dimension of the particle, since the bed friction force is dependent on the particle mass. To the generalization following eqs. (5.38) and (5.39) may be added the further conclusion that, for a given solids density and flow velocity, the largest cylinders are likely to be deflected least.

If the complete motion of the cylinder is required, F_D must be calculated recognizing that the slip velocity between fluid and moving cylinder varies along its length. If h is the axial distance measured from the end of the cylinder nearest to O we have, under the preceding simplification:

$$F_D=C_D\rho a\int_0^b (U-\Omega h\cos\alpha)^2\cdot dh \tag{5.44}$$

giving the value of the force to be used in eq. (5.41).

The first problem mentioned, that of the orientations assumed when the moving and stationary particles are comparable in size, can at present be treated only in outline. Instead of lodging on a large number of bed grains, the moving particle now comes to rest on just a few, perhaps very often the three particles that are the minimum necessary for a stable support. The forces acting are numerous and can be predicted only if, in addition to the flow, the bed shape is completely specified. However, Rusnak (1957a) has sketched the qualitative arguments for the orientation assumed by prolate ellipsoids under these conditions.

He concluded that, because of interactions with stationary grains, prolate ellipsoids would lodge with their long-axes parallel with flow. The moving particles are regarded as pivoting about the stationary ones, and his argument invokes the same turning forces as are suggested above to explain the orientations of cylindrical and related particles. In order to explain the shingling or imbrication of ellipsoidal particles, Rusnak (1957a) considered the forces acting in the plane of flow on a single grain. He concluded that the particle was most difficult to dislodge from the bed, that is, it was most stable, when the long-axis was tilted upcurrent, for then the fluid force pressed the particle on to the bed. Otherwise the particle tended to be overturned. However, the upstream imbrication of a particle just lodged on the bed can be maintained when the current ceases only if the particles are frictional. Taking ellipsoids of axial ratio and coefficient of friction typical of natural grains, it is easily shown that imbrications of 10°–30° are possible.

Shell orientations

Numerous experimenters have studied the behaviour during transport, and the orientation after deposition, of shells and other biogenic particles, or regular bodies intended to represent them. The results are sometimes conflicting.

Schwarzacher (1963) worked with cylindrical particles substituting for groups of crinoid ossicles. Entrainment was easiest when the cylinders lay nearly transverse to flow. The effect of bottom friction on their final orientation, consistent with the theoretical model proposed above, is well illustrated by his experiments in which the same particle was repeatedly carried down the flume into a region of reduced velocity, where it lodged on the bed. Three different bed-materials were used, fine sand, mud, and smooth metal, in that order of decreasing frictional resistance. Most cylinders on the sand bed lodged with their long-axes almost at right angles to flow. On the mud bed, presenting less resistance to particle rotation, the final value of α was about 55° either side of the flow direction. An even smaller final value for α, about 35°, was obtained on the smooth metal. Schwarzacher also released cylinders one after the other into the flume, measuring their orientations after they came to rest. The cylinders often interfered in lodging on the bed, to give T-shaped and what he described as upstream-pointing arrow-head configurations (see also Futterer, 1977). In these experiments flow-transverse and flow-parallel cylinder orientations were about equally well developed. Cain (1968), experimenting with natural crinoid ossicles, found that these were readily entrained, but made no observations on orientations.

Experiments show that the orientations assumed by conical gastropod shells beneath unidirectional currents depends strongly on shell shape. Trusheim (1931) found that cones with small apical angles turned apex downstream, whereas stubby cones rotated so that the apex pointed upstream. Nagle (1967) repeated these observations in a laboratory one-way flume, and again in tidal channels and a stream. He used *Turritella mortoni* and *T. plebia,* both with a small apical angle, and *Busycon auranum,* a stout-shelled, tubercular, stubby form. The turritellids arranged themselves mainly apex-downstream (Fig. 5-22a), as noticed experimentally by Brench-

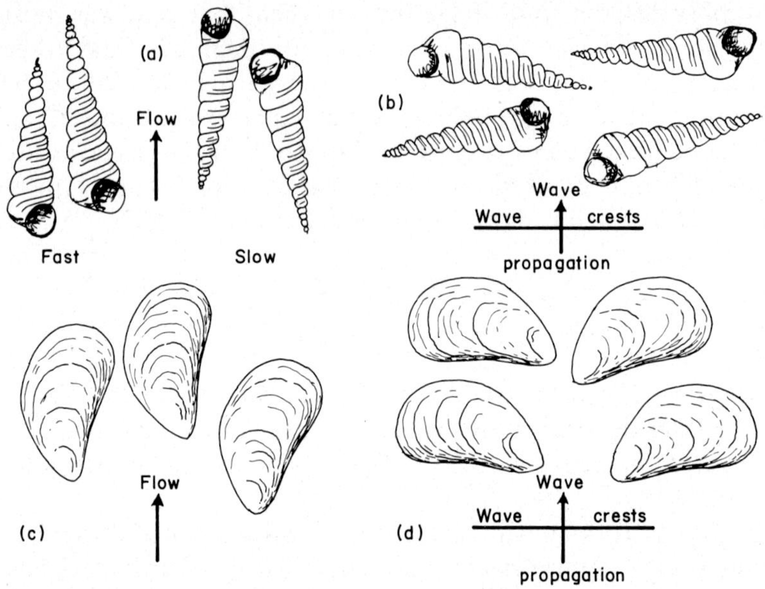

Fig. 5-22. Schematic representation of the attitudes assumed by high-spired gastropods (*Turritella*) and by elongated bivalves (*Mytilus*) under the action of (a, c) unidirectional currents, and (b, d) oscillatory wave-generated currents.

ley and Newall (1970). *Busycon* in Nagle's work developed no consistent preferred orientation, apparently because the tubercles hindered the turning of the shell. Independently, Kelling and Williams (1967) experimented with the dog whelk *Nucella lapillus,* a smooth, strong-shelled, and rather globose form. These shells, arranged on sand or silt beds, were acted on either by a slowly increasing current or by a surge. Rather variable orientations were observed, including the effects of interference as in Brenchley and Newall's study (cf. Nagle), but most shells ended up apex-downstream, the proportion increasing with flow velocity. Erickson (1971) found that the stubby land snail *Helminthoglypta arrosa* assumed under wind-action an orientation such that the aperture opened downwind.

Nagle (1967) also examined the response of gastropod shells to wave-generated oscillatory currents. In the laboratory, he found that the turritellids aligned themselves transversely to the direction of wave propagation, the apices pointing in roughly equal numbers to each side of this direction (Fig. 5-22b). Similar orientation patterns were observed for the shells under the action of natural waves. *Busycon* again showed little tendency to assume a preferred orientation.

The behaviour of separated bivalve and brachiopod valves under experimental conditions presents a less clear picture. There are, for instance, three different published accounts of the response to unidirectional currents of the separated valves of the moderately elongated bivalve *Mytilus.* Nagle (1967) found that the valves tended to line up with the long-axis parallel with flow and the beak pointing upcurrent (Fig. 5-22c). The same result was obtained at sufficiently large flow velocities by Kelling and Williams (1967), on silt as well as sand beds, regardless of whether the current was a surge or one developed gradually. Brenchley and Newall (1970), however, found the valves to be mainly flow-transverse on sand and mud beds. Two factors may explain these differences. Firstly, Brenchley and Newall placed the shells concave-up at the start of the runs, whereas Kelling and Williams put them concave-down; Nagle's practice is not stated. Secondly, Brenchley and Newall worked at flow velocities significantly smaller than Nagle, and Kelling and Williams. At these relatively low speeds, the shells could generally be rotated and often dragged a short way, but not for the most part become overturned and afterwards transported. Branchley and Newall found several other transverse orientations, notably from the elliptical bivalves *Gari* and *Venerupis,* which seem explicable in the same way. The bivalve *Donax variabilis* is also an elongated form. Wobber (1967) observed that on beaches it became aligned by wave backwash so that the blunt anterior end pointed towards the sea. Seilacher (1961) and Folk and Robles (1964) also report shell alignments parallel with beach slope. Reineck (1960d) described the parallel alignment of the tubes of the marine worm *Pectinaria.* The attitude of the separated valves of relatively equant bivalves, such as species of cockle (*Cardium*), is sensitive to the relative position of the centre of gravity (Futterer, 1974). The centre of gravity in *C. echinatum* lies comparatively close to the beaks, which therefore point upcurrent, whereas that of *C. edule*, assuming the opposite orientation, is more central in position.

The orientation of *Mytilus* and other shells by waves was described by Nagle (1967). The valves become aligned transversely to the direction of wave propagation, the beaks pointing each side of this direction and at a small angle towards the incoming crests (Fig. 5-22d). The elliptical valves of the brachiopod *Spirifer cyclopterus* lined up with the hinges parallel with the waves.

Theoretically, concavo-convex shells exposed on a bed to sufficiently strong currents are stable only when convex-up. Moreover all the evidence available from the laboratory and modern environments where these conditions exist suggests that such shells will be preserved more often convex-up than concave-up (Sorby, 1908; R.G. Johnson, 1957; Kornicker et al., 1963; J.B. Wilson, 1967; Wobber, 1967; Lever and Thijssen, 1968; Clifton, 1971; Reyment, 1971). There are nevertheless processes acting during the accumulation of shells that can give substantial numbers a concave-up orientation (Menard and Boucot, 1951; Kornicker and Armstrong, 1959; Clifton and Boggs, 1970). Emery (1968) and H.E. Clifton (1971) report that in shallow waters unaffected by strong currents, shells can be oriented dominantly concave-up, apparently through the disturbing action of scavengers and bottom burrowers.

The orientations of fossils, for example, graptolites, conical nautiloids, ammonites, gastropods, tentaculitids, and pieces of driftwood, have often been measured from the stratigraphic record, primarily in order to obtain information on current directions (Ruedemann, 1897; Kindle, 1938; R.L. Clifton, 1944; Müller, 1951; Crowell, 1955; Pelletier, 1958; Colton and de Witt, 1959; Seilacher, 1959, 1960; Krinsley, 1960; Sullwold, 1960; Schwarzacher, 1963; B.S. Clarke, 1964; Wobber, 1966; Nagle, 1967; Reyment, 1968; Schleiger, 1968; Enos, 1969a; Moors, 1969; Barrett, 1970; O.A. Dixon, 1970; Jones and Dennison, 1970; Jawarowski, 1971; Gauss and House, 1972; Brenner, 1976; Kelling and Moshrif, 1977; Loubere, 1977). A few of these orientations were interpreted in the light of independent but associated evidence of current direction. The orientations of crinoid stems measured by Schwarzacher, for instance, show very clearly and in accordance with the theoretical model the influence of friction between particle and bed. A few others, for example, those obtained by Nagle, can be interpreted with some confidence in terms of observations from the laboratory and from modern environments. Many are equivocal, primarily because not enough is understood, theoretically as well as experimentally, about the behaviour during lodgement of these complex particles.

Land and Hoyt (1966) measured the orientation of grass stalks in a modern estuary. On rippled sand-flats the stalks lay parallel with the ripple crests and, therefore, at right angles to the current. Similar orientations were measured from the mud flats.

Sand and gravel shape-fabrics

Limited and somewhat conflicting experimental results exist on the fabrics of sand and fine gravel deposited from flows with comparatively low concentrations of moving debris. Dapples and Rominger (1945) found that sand grains took up positions mainly with the long-axis parallel with flow and the larger end of the particle facing the current. Schwarzacher (1951) found a similar orientation and noted a slight upcurrent imbrication. Vollbrecht (1953) and C.E. Johansson (1963) noted that the larger particles in moving over the bed tended to roll with the long-axis transverse to flow. On meeting an obstacle, the particle was often rotated more nearly parallel with flow, as is expected on theoretical grounds, and thereafter tilted slightly upcurrent, often as it tipped back into an upstream scoured hollow (C.E. Johansson, 1960; Fahnestock and Haushild, 1962). Monoclinic fabrics with weak girdles and a low upcurrent imbrication were measured by Johansson. Nachtigall (1962), experimenting with sand, also emphasized the rolling of particles and their rotation about obstacles during lodgement, but found that flow-transverse and flow-parallel long-axis orientations were about equally common over a comparatively wide range of flow velocities. It is not clear why Nachtigall's results conflict with those of other experimenters.

Some parallel results are available from modern streams. K. Richter

(1936) noticed that in slow currents stones were rolled along and deposited in a transverse orientation. In faster currents a flow-parallel orientation was assumed on lodgement, presumably a reflection of the decreased importance of bed friction relative to fluid drag. Cailleux (1938) also reported rolling. Usually stream gravels show an upstream imbrication, with most stones lying parallel with flow and a number transverse (Schlee, 1957; Unrug, 1957; Doeglas, 1962; Potter and Pettijohn, 1963; C.E. Johansson, 1965; Dal Cin and Sperandio, 1966; Church, 1972; Klimek, 1972; Teisseyre, 1978). The flow-transverse orientation is in some cases pronounced and in others dominant, including cases where stones lie scattered over a relatively smooth bed (Kürsten, 1960; Schiemenz, 1960; Rust, 1972b, 1975).

Comparatively little is known, and less understood, about the orientations assumed by sand and gravel where waves are active. Nachtigall (1962) found that sand grains on wave-related ripple marks lay with their long-axes parallel with the ripple crest and, therefore, with the wave-crests. On a sandy shore with bars and troughs he found that, except in the exposed swash-backwash zone, grains lay as commonly transverse as parallel with the coast. Wave, tidal, and other currents produced the complex orientation patterns found by Maxwell et al. (1961) in the shallow-water sediments of Heron Island Reef. Fraser (1935) observed that gravel particles tended to remain parallel with wave-crests during movement on beaches, though they could be rotated on meeting obstacles. Cailleux (1938), and Williams and Gulbrandsen (1977), reported pebbles aligned parallel with the beach. Other workers have reported a confusion of fabrics. In places the flatter stones and sometimes the shells are found standing on edge, while in others they are inclined gently seaward, and in others again dip landward (Krumbein, 1939; Cailleux, 1938, 1945; Bluck, 1967; Dionne, 1971; Grinnell, 1974; D.F. Ball, 1976; Ricketts and Donaldson, 1979). However, flow-parallel but vertically oriented stones are also known from mass-flow deposits (Patton, 1910). These fabrics are to some extent associated with the different facies that can be recognized in gravel beaches, and it is clear from the work of Cailleux (1938, 1945), Bluck (1967) and N.C. Flemming (1964, 1965) that they are a further expression of the processes, so far little known because of their complexity, that lead to shape and size-sorting under wave action.

Sand and gravel fabrics are frequently reported from the geological record, largely for the light they shed on transport directions (e.g. Griffiths, 1950; W.S. White, 1952; Rusnak, 1957b; Nairn, 1958; Schiemenz, 1960; McCann, 1961; Chen and Goodell, 1964; Laming, 1966; Lützner, 1966; Nilsen, 1968, 1969; Lené and Owed, 1969; Matalucci et al., 1969; Dziedzic, 1971; Martini, 1971; Clague, 1974; Liboriussen, 1975; Teisseyre, 1975). Normally, the *a*-axis orientation is taken to be the current direction.

SHAPE-FABRICS OF MUDDY SEDIMENTS

General

The muddy sediments (Boswell, 1961) comprise a mixture of fine-grained solids and water. The solids, present in substantial amounts, are dominated by clay-mineral particles of microscopic to near-colloidal dimensions, but usually include grains of detrital mica, feldspar, and quartz generally larger in size than the clay minerals. Recently deposited muddy sediments have porosities comparable with 50–75 percent and often contain representatives of all the important groups of clay minerals—the kaolinites, montmorillonites, illites and chlorites—which are complex layered silicates (Grim, 1962, 1968). The clay-mineral composition of the older and more consolidated muddy sediments is usually much simpler, restricted to illite and chlorite, on account of the effects of heat and pressure during burial. These sediments have porosities comparable with 5–30%.

The clay minerals generally occur as plate-like crystals, though some species are fibrous (Beutelspacher and Van der Marel, 1968). Because of the small size and the shape of these crystals, muddy sediments possess huge surface areas per unit of mass (Searle and Grimshaw, 1959). It is therefore reasonable to suppose that the fabrics of muddy sediments, whether at the time of deposition or after a degree of consolidation, depend more on the forces relating to the clay-particle surfaces than on the particle weight. The nature of the surface forces affecting clay minerals is discussed from various standpoints by Van Olphen (1963), Meade (1964), Gillott (1968), and Ingles (1968).

Clay-mineral particles suspended in aqueous media are in theory acted upon simultaneously both by repulsive and attractive forces. In the clay minerals, the crystal faces parallel with the silicate layer-structure are negatively charged. The exchangeable cations present in the interstitial water are attracted to such a face and become concentrated near it, forming a diffuse "atmosphere", or electrical double-layer, the outer part of which is positively charged. Hence two particles face to face will repel each other because of the like charges carried by the outer parts of their double-layers. But the edge-surfaces of clay-mineral crystals differ in atomic structure from the layer-parallel surfaces, and are positively charged. The double-layer associated with the edge will therefore be negatively charged in its outermost parts. Hence two particles placed edge to face are mutually attracted; the particles may be either perpendicular to each other or lie face-to-face but overlapping. The second attractive force between colloidal particles is the van der Waals force, varying inversely as a large power of the particle spacing. The effectiveness of this force is little influenced by the dissolved substances associated with the clay.

The repulsive forces, conversely, are for each clay-mineral species strongly dependent on the nature and concentration of the cations in the water. At low cation-concentrations the double-layer is relatively thick, so that the combination of attractive and repulsive forces is expressed as a net repulsive force. The particles do not aggregate in any way but remain dispersed, or unflocculated, throughout the medium. The double-layer grows thinner as the cation concentration is raised, leading to a progressive reduction in the effective range of the repulsive forces. At sufficiently high cation-concentrations the net force between particles becomes attractive, and the clay crystals flocculate into loose open clusters.

Fabrics of freshly deposited clays

Consideration of the forces between clay-mineral particles and the ionic composition of natural waters has led to many workers to suggest idealized fabrics for freshly deposited clays (see review by Moon, 1972). Van Olphen's (1963) scheme is by far the most comprehensive so far advanced. He concluded that clay particles might be associated together in two basic ways. The fabric elements could be either *single* flakes, or groups of flakes, which he called *aggregates.* Further, the fabric elements might be either unflocculated (Van Olphen's "deflocculated") or flocculated. If unflocculated, the elements could be arranged either edge-to-face or edge-to-edge. The flocculated elements might be arranged either edge-to-face, edge-to-edge, or partly edge-to-face and partly edge-to-edge. Codes identifying these combinations appear in Fig. 5-23a–c.

The SU and AU types involve elements that are by and large mutually repulsive, and are perhaps to be found when the pore waters are exceptionally poor in salts. The elements may show some degree of preferred orientation, subparallel with the surface of accumulation. A combination of these types is implied by Von Engelhardt and Gaida (1963) for clays poor in salt.

The SFEF fabric, or *cardhouse arrangement,* was proposed by U. Hofmann (1942, 1952), and later accepted by Schofield and Samson (1954), Tan (1959) and Lambe (1960b) as a model of flocculated clays commonly formed in fresh water. The AFEF and AFM types of fabric are closely related to the conception of the salt-flocculated fabrics of Lambe (1953, 1960a) and of Schofield and Samson (1953, 1954), and also to Sloane and Kell's (1966) bookhouse fabric. In all these fabrics there is little or no preferred orientation of the elements.

Several other models have been proposed. They include Terzaghi's (1925) *honeycombe fabric,* later developed by Casagrande (1932) and by Mitchell and Houston (1969), in which the particles lie in long linked chains on a polygonal pattern (Fig. 5-23h), and the *tactoid* model of Emerson (1959, 1962, 1963), essentially an AFEE type of association.

All of the models in Fig. 5-23 involving aggregates are *turbostratic* (Biscoe and Warren, 1942), in that they comprise elements which as individuals are

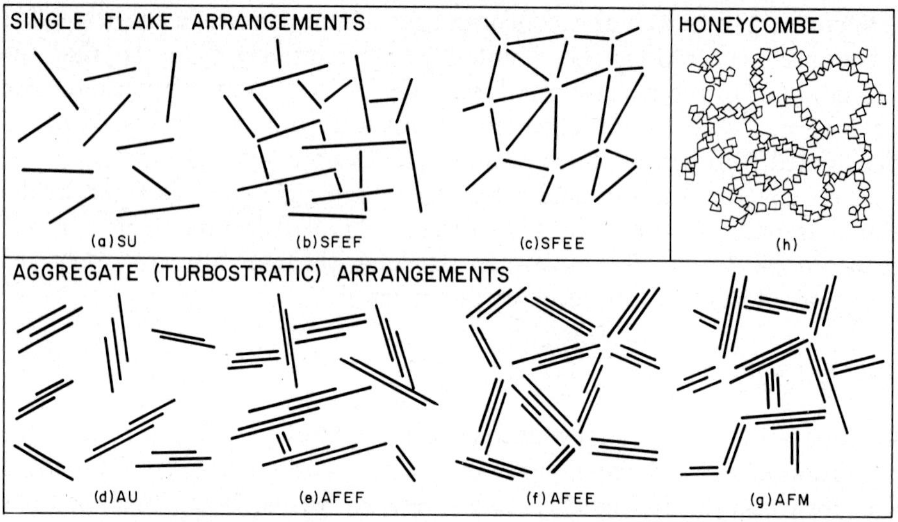

Fig. 5-23. Schematic representation of models which have been proposed for the fabrics of fine sediments dominated by clay minerals. Code: A—aggregates; E—edge; F—-face; mixed edge-to-edge and edge-to-face; S—single particles.

highly ordered internally but between which there is no agreement as to orientation (Aylmore and Quirk, 1959, 1960, 1962). It should be carefully noted that the term turbostratic gives recognition to different degrees of order; it is applicable to fabrics of low as well as high porosity.

Confirmatory evidence from freshly deposited clays is confined to the laboratory. O'Brien and Suito (1969) found that flocculated kaolinite had a highly porous random fabric with affinities with the AFEF, AFEE and AFM types of Fig. 5-23. Some resemblance to Terzaghi's honeycombe model may also be noted, but little of this fabric remained after drying a flocculated kaolinite (O'Brien, 1970a). Flocculated illite also gave a highly porous and random turbostratic fabric (O'Brien, 1970b), the flakes in some of the aggregates having the stepped face-to-face arrangement (type AU) noted by Smalley and Cabrera (1969). Pusch and Arnold (1969) also observed aggregates in salt-flocculated illite. Sides and Barden (1971) studied unflocculated as well as flocculated kaolinite and illite; the fabrics developed in montmorillonite were undecipherable. Turbostratic arrangements were observed under all conditions, though in the unflocculated samples there was a strongly preferred particle orientation parallel with the depositional surface. The flocculated clays had a more open and less ordered fabric, the illite more so than the kaolinite.

Fabrics of lightly consolidated natural muddy sediments

Extensive investigations have for practical reasons been made in Scandinavia and Canada into the fabrics of muddy sediments compacted under overburdens of the order of a few metres to a few tens of metres thick. The sediments in question are mainly Pleistocene in age and those of fresh-water origin differ in fabric from those accumulated in salt water. The fabrics as preserved are apparently slight modifications of the original appositional fabrics.

Lightly consolidated marine clays generally have a random or poorly oriented turbostratic fabric and a moderate to high porosity. These features were first noticed by Mitchell (1956), studying thin sections of conventional thickness optically, but confirmed by numerous workers using less crude techniques (Rosenqvist, 1958, 1962; Pusch, 1966, 1970; Crawford, 1968; Gillott, 1968, 1969, 1970; O'Brien and Harrison, 1969; O'Brien and Suito, 1969; Burnham, 1970; Silva and Hollister, 1973). The fabric reported by Pusch was interpreted by him as composed of aggregate elements connected by linkages of single flakes. It therefore combines features of the AFEF, AFEE and AFM models of Fig. 5-23 with Terzaghi's (1925) honeycombe structure. The same model is helpful in interpreting many of the fabrics that Bowles (1968) figures from ocean-floor muds of recent date. Some of the more deeply buried muds he studied showed a moderate degree of preferred particle orientation but still had large (though somewhat flattened) open pores.

Less is known of fresh-water muds. According to Mitchell (1956), Penner (1963), and Burnham (1970), they possess an indistinct turbostratic structure and a moderate to well developed preferred particle orientation of the fabric elements parallel with bedding. The higher degree of order than in the marine deposits is consistent with the lower concentrations of dissolved salts in the depositional waters.

Soil clay-minerals show a range of special fabrics (Brewer, 1964), among them the arrangement of clay flakes parallel with the surfaces of the larger detrital grains (Brewer and Haldane, 1957; Lafeber, 1964). The same fabric, however, can also arise in a suspension of clay with silt particles (Sides and Barden, 1971), and is not uncommon in a variety of deposits (Barden and Sides, 1970b; Barden, 1972a, 1972b).

Fabrics of strongly consolidated natural muddy sediments

Strongly consolidated muddy sediments have been covered by an overburden of the order of hundreds or thousands of metres thick. Naturally, they are generally much older than the lightly consolidated deposits described above. These strongly consolidated materials have very low to low porosities, and their fabrics, as now seen, may be quite unrelated to the modes of particles association at the time of deposition. This is to be attributed to the considerable deformation suffered by these deposits and, in the view of several workers (Keller, 1946; Siever, 1966; G. Müller, 1967; Heling, 1970), also to the modification and growth of the fabric-elements as the result of recrystallization of the clays.

Rocks of this type vary greatly in fissility, from the highly fissile paper shales to the blocky-fracturing mudstones. Those with marked fissility invariably show a strongly developed bedding-parallel preferred orientation of the clay crystals (Kaarsberg, 1959; White, 1961; Gipson, 1965, 1966; Odom, 1967; O'Brien, 1968a, 1968b, 1970a; Gillott, 1969). As the fissility grows poorer, however, the degree of preferred orientation of the clay minerals weakens until, in blocky-fracturing mudstones, random fabrics prevail (Gipson, 1965, 1966; Odom, 1967; Burnham, 1970; O'Brien, 1970a). Burnham has observed turbostratic structure both in moderately fissile marine clays and in blocky siltstones of fresh or brackish water origin.

Several workers report that highly consolidated muddy sediments with high fissility and strongly preferred particle orientation can be found in close association with poorly fissile deposits having disordered fabrics (Rubey, 1930; Grim et al., 1957; White, 1961; Gipson, 1966; Odom, 1967). This relationship may depend on grain size, since the poorly structured sediments generally have the largest content of silt grains. These during consolidation seem to act as rigid elements which confer some strength on the deposit and around whose surfaces the clay flakes become moulded. Meade (1968), however, tentatively identified this relationship between grain size and preferred orientation only in the marine clays he studied. Although in the other muddy sediments the degree of preferred orientation was closely linked to depositional environment, the specific factors responsible for the association could not be conclusively identified.

Progress of consolidation

It has long been understood from measurements on field samples that the porosity of muddy sediments decreases with increasing overburden thickness (Athy, 1930; Hedberg, 1936; Weller, 1959). Soil scientists, civil engineers, and geologists have studied aspects of this process in the laboratory, using either artificially prepared or natural unconsolidated sediments. The samples, usually laterally confined, were compressed under loads equivalent in some studies to a few metres of overburden and in others to hundreds or thousands of metres of superincumbent material. Meade (1964, 1966) fully reviewed earlier work in this field. Significant works are by Bolt (1956), Mitchell (1956, 1960), Olson (1962), Von Engelhardt and Gaida (1963), West (1964), Martin (1965), Quigley and Thompson (1966), Sloane and Kell (1966), Morgenstern and Tchalenko (1967a, 1967b), Smart (1967), Bowles (1968), Bowles et al. (1969), Barden and Sides (1970a), Clark (1970), Greene-Kelly and Mackney (1970), Lee and Morrison (1970), and Olson and Mesri (1970).

So far as fabrics are concerned, this work shows that with increasing overburden there is a simultaneous decrease in the porosity and, in most of the experimental materials, an improvement in the degree of preferred orientation of the clay particles or aggregates. A strong, but local, preferred parallelism of fabric elements can also be induced during the shearing of clay (Seed and Chan, 1959; Morgenstern and Tchalenko, 1967c, 1967d; Smart, 1967).

Meade wrote in his survey of 1966 that: "Although preferred orientation has been produced under a wide range of pressures in laboratory experiments on clays, no unequivocal evidence shows an increase in preferred orientation with increasing depth of burial in nature." The experiments, however, successfully reproduce porosity reduction observed during natural consolidation, and there seem to be no reasons (particularly when a natural clay is the experimental material) to regard them as giving results irrelevant to the formation of natural fabrics. The detailed observations of workers such as Odom (1967), outlined in the preceding section, together with the experimental results, suggest that Meade's conclusion may have to be revised to exclude the muddy sediments that fall below that grain size when silt particles can no longer confer strength on the deposit and impose special fabrics on the clay flakes.

SUMMARY

The shape-fabrics of particulate sediments are either appositional, rheotactic, or deformational, the second being of minor importance compared to the first and last. Appositional fabrics in sands and gravels record the influence on sedimentary particles of the fluid and body forces acting during particle transport and/or emplacement in the static bed. Preferred particle orientations in these sediments can be attained as the result of: (1) the settling of solitary grains in a fluid, (2) particle shearing in a viscous medium, (3) the shearing of a dense array of similar grains, and (4) interference with already stationary particles while undergoing emplacement. Least is understood about appositional fabrics originating during emplacement, though perhaps these fabrics have been studied more frequently than any other. In contrast, the appositional fabrics of the muddy sediments, rich in clay minerals, are complicated and depend primarily on the character of particle surface forces, as determined by particle mineralogy and the chemistry of the fluid medium. These fabrics are rapidly and substantially modified during consolidation, acquiring features due to deformation. Deformational fabrics, such as arise during the flow of liquid-like sands and during the consolidation of muddy sediments, seem to be explicable in terms of models involving simple shear and/or pure shear.

Chapter 6

TRANSITION TO TURBULENCE AND THE FINE STRUCTURE OF STEADY TURBULENT BOUNDARY LAYERS: PARTING LINEATION AND RELATED STRUCTURES

INTRODUCTION

When any fluid streams pass a solid boundary, the fluid layers nearest the surface are retarded by frictional forces arising from fluid viscosity. The retarded zone is called the boundary layer, either laminar or turbulent, as we have seen. Osborne Reynolds (1883), in experiments now classic, first illustrated the difference between these two kinds. In a laminar boundary layer, streamline motion prevails and the fluid in adjacent layers mixes only on a molecular scale. By contrast, the turbulent layer represents an unstable form of motion. It reveals large-scale random movements of fluid elements which are superimposed on the general motion. Reynolds found that turbulent replaced laminar motion when the inertial flow forces exceeded a certain ratio with the viscous ones.

The problem presented by the laminar-turbulent transition is to find, experimentally or mathematically, the conditions when the small disturbances inevitably present in real flows can extract energy from the mean flow, and so become amplified with the eventual production of a fully developed shear turbulence. Because of the huge difficulties, mathematical studies have uncovered little beyond the conditions for instability together with the general significance of non-linear and three-dimensional effects. However, experiments reveal that transition to turbulence is accompanied by an orderly sequence of well-organized phenomena in which small disturbances are amplified. They further show that some of the flow configurations and processes accompanying transition closely resemble others associated with fully developed turbulent boundary layers.

The role of turbulence in sediment entrainment, transport and deposition has already been discussed. In a new sedimentological connection, Karcz (1970) suggests that there is a possibly significant similarity in morphology and sequence of development between some common bed configurations, at least in embryonic form, and the flow configurations in transitional and turbulent flows. Less controversially, there is growing evidence that several common sedimentary structures directly express the generation of turbulence in natural flows. These structures, the chief of which is parting lineation, will be explored in this chapter against the background of a sedimentologically orientated approach to transitional and fully developed turbulent boundary layers. Additional information and background may be obtained from the many reviews in which the physical, mathematical or engineering aspects of transition are emphasized (Lin, 1955, 1958; Dryden, 1959; Tollmien and Grohne, 1961; Stuart, 1963, 1965; Lin and Benney, 1964; Drazin and Howard, 1966; Betchov and Criminale, 1967; Tani, 1967, 1969; Reshotko, 1976). There exist several major surveys of turbulent boundary layers in general (Clauser, 1956; Kovasznay, 1967; Schlichting, 1960; Townsend, 1956, 1958, 1970, 1976).

OUTLINE OF TECHNIQUES

Three main techniques dominate the experimental study of transitional and turbulent layers: flow visualization, hot-wire anemometry, and wall-pressure measurement (P. Bradshaw, 1964, 1971). Anemometry using water as the experimental fluid was impracticable until recently, but all three techniques are successful with gases.

In flow visualization the fluid is marked in some way so that the flow configurations are directly revealed, the patterns being recorded usually with the help of high-speed photography. Small particles, ideally neutrally buoyant, may be introduced into the flow and tracked using dark-ground illumination. Often the fluid motion is visualized with dye or smoke, introduced into the body of the flow from a syringe, or from a hole or transverse slot in the wall. Dyed elements can be introduced into water as the result of an electrolytic reaction at a wire or other electrode suitably arranged in the flow. The data are suitable for quantitative treatment, since regular pulses of current can be passed through the electrode. An elegant related technique lending itself to quantitative analysis is hydrogen bubble visualization in water (Schraub et al., 1964; Merzkirch, 1974). This method places sequentially and geometrically ordered markers into the flow, namely, lines of hydrogen bubbles produced electrolytically at suitably placed wires by controlled pulses of current. Usually in boundary layer studies one of the wires lies transversely very close to the wall, while the other is placed a little downstream normal to the wall. The markers swept downstream are photographed with a high-speed ciné-camera. An idea of flow structure, together with estimates of instantaneous, fluctuating and time-average flow velocities, can all be obtained from the film provided that suitable precautions are taken (Hama, 1962).

Hot-wire anemometry is well-established technique based on the fact that the electrical resistance of a metallic wire or ribbon carrying a heating current depends on the temperature of the wire, in turn controlled by the speed of the experimental fluid sweeping the wire (P. Bradshaw, 1971). In practice, the hot-wire anemometer comprises one or more pieces of thin tungsten or platinum wire a few millimetres long, mounted at one end of a more robust support. The probe may consist of a single straight wire, or two identical wires arranged in a cross, depending on the velocity components to be measured. It forms one arm of a Wheatstone bridge yielding an amplified signal whose fluctuations correspond to the varying flow past the wire. The signal can be processed electronically to yield the root-mean-square value, frequency spectrum and probability distribution of the fluid velocity. Using probes at two or more points in the flow, the extent in time and space of individual turbulence elements or other structural features can be expressed by means of an appropriate correlation. Even more elaborate descriptions are possible with computerized signal-processing (e.g. Kovasnay et al., 1970).

Less extensive use has been made of measurements of wall-pressure fluctuations (Willmarth, 1975), perhaps because the technique is less versatile than hot-wire anemometry. These fluctuations may be measured using miniature condenser or piezoelectric microphones mounted in the wall of the wind-tunnel or flume. They yield electrical signals that can be processed electronically to yield data similar to but less extensive than hot wires. They respond of course to fluctuations or pressure originating throughout the whole thickness of the boundary layer. Such devices are sometimes used in combination with hot-wires (e.g., Willmarth and Tu, 1967).

TRANSITION TO TURBULENCE

Mathematical solutions

Successful analyses are limited to boundary layers assumed to be parallel, incompressible, one-phase, two-dimensional flows upon which small two-dimensional disturbances are superimposed. The basic flow is assumed steady and a function of y only, the normal distance from the wall. Then:

$$\left.\begin{aligned} U &= \phi_1(y) \\ V &= 0 \end{aligned}\right| \quad (6.1)$$

where U is the local velocity of the basic flow in the boundary layer measured parallel with the streamwise or x-direction, and V is the local velocity of the basic flow measured parallel with y. With t as time, the disturbances are described by:

$$\left.\begin{aligned} u' &= \phi_2(x, y, t) \\ v' &= \phi_3(x, y, t) \end{aligned}\right| \quad (6.2)$$

in which u' and v' are, respectively, the instantaneous components of fluctuating velocity in the x and y directions. These components are assumed small, allowing the differential equations of flow to be linearized. Hence the total velocities measured at a point in the flow are:

$$\left.\begin{aligned} u &= U + u' \\ v &= v' \end{aligned}\right| \quad (6.3)$$

measured in the x and y directions, respectively.

With ρ as the fluid density, ν as the kinematic viscosity, and p as the pressure due to the disturbances, the equation of continuity is:

$$\frac{\partial u'}{\partial x}+\frac{\partial v'}{\partial y}=0 \tag{6.4}$$

and the Navier-Stokes equations of fluid motion in terms of the disturbances are:

$$\frac{\partial u'}{\partial t}+U\frac{\partial u'}{\partial x}+v'\frac{\partial U}{\partial y}=\nu\left(\frac{\partial^2 u'}{\partial x^2}+\frac{\partial^2 u'}{\partial y^2}\right)-\frac{1}{\rho}\frac{\partial p}{\partial x} \tag{6.5}$$

$$\frac{\partial v'}{\partial t}+U\frac{\partial v'}{\partial x}=\nu\left(\frac{\partial^2 v'}{\partial x^2}+\frac{\partial^2 v'}{\partial y^2}\right)-\frac{1}{\rho}\frac{\partial p}{\partial y} \tag{6.6}$$

after linearization by neglecting the sums and products of the disturbance velocities and of their differential coefficients. These are the basic equations that must be solved to discover when a boundary layer as prescribed above becomes unstable.

Tietjens (1925) first solved the equations for velocity profiles consisting of short rectilinear segments, but did not establish a critical Reynolds number for instability. Later, Tollmien (1929), Schlichting (1933, 1935), Lin (1945) and many others (e.g. Fasel, 1976) gave solutions for curved velocity profiles similar to those found in real boundary layers. Critical Reynolds numbers were calculated and curves of neutral stability established.

Figure 6-1 is Lin's (1945) neutral stability curve, in the form of the non-dimensional frequency parameter, $2\pi f\nu/U_\infty^2$, as a function of Reynolds number Re_s, where f is the disturbance frequency, U_∞ is the velocity outside the boundary layer, and the Reynolds number is defined in terms of this velocity and the boundary-layer displacement thickness δ_s. The curve divides the graph into two fields, the unstable field enclosed by the neutral curve where disturbances are amplified, and the stable field without, where disturbances are damped. There is one Reynolds number, the critical value Re_{cr}, below which the flow is stable to two-dimensional disturbances. Tollmien (1929) calculated $Re_{cr}=420$ and Lin (1945) and Shen (1954) obtained very similar values. Jordinson (1970) recently found the critical Reynolds number slightly to exceed 500. For Reynolds numbers greater than critical, Fig. 6-1 shows two branches, defining a band of frequencies and wavelengths to which the flow is unstable. Notice that as the Reynolds number increases, the flow is unstable to disturbances of progressively longer wavelength.

Experiments show that non-linear and three-dimensional effects develop rapidly during transition, but modelling of these difficult problems has so far met with limited success (Görtler and Witting, 1958; Benney and Lin, 1960; Benney, 1961, 1964; Criminale and Kovasnay, 1962; Stuart, 1962; Craik,

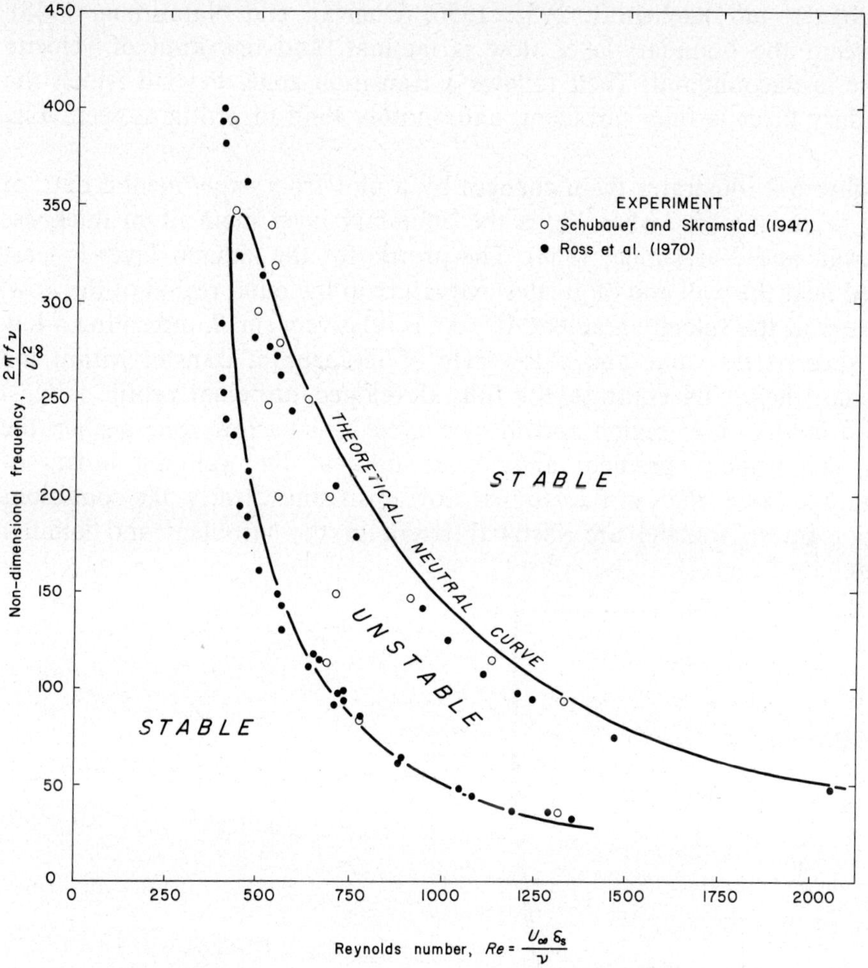

Fig. 6-1. The stability of laminar boundary layers to small disturbances, as illustrated by Lin's (1945) calculated neutral stability curve in an alternative form, and the experimental data of Schubauer and Skramstad (1947) and of Ross et al. (1970). After Ross et al. (1970).

1971). We are nevertheless beginning to obtain insights into the mechanisms favouring the particularly rapid growth of three-dimensional disturbances. For example, small two-dimensional roughness elements, by altering the velocity profile close to the boundary, destabilize the flow often so quickly that transition occurs close to the element (Klebanoff and Tidstrom, 1972).

CHANGES IN VELOCITY PROFILE

The profile of the local time-average velocity U changes markedly as a laminar boundary layer undergoes transition on a pipe wall or flat plate

(Schubauer and Klebanoff, 1955, 1956; Dhawan and Narasimha, 1958). Upstream the boundary-layer flow is laminar, and one kind of velocity profile is encountered. Then follows a transition zone, beyond which the boundary layer is fully turbulent, and another kind of profile is seen (Fig. 1-6).

Figure 6-2 illustrates these changes by a plot from experimental data of U/U_∞ against y/δ_m, where δ_m is the boundary layer momentum thickness (Dhawan and Narasimha, 1958). The profile for the laminar layer is least curved near the wall and of greatest curvature in the outer region of the flow. At the wall the velocity gradient dU/dy is relatively small, indicating a low wall shear-stress value and a low rate of momentum transfer within the boundary layer. By contrast, the fully developed turbulent profile is least curved in the outer region and most curved in a narrow zone against the wall. The velocity gradient and shear stress at the wall are orders of magnitude larger than in the laminar flow upstream. Clearly, the conditions of momentum transfer are vastly different in the turbulent and laminar layers.

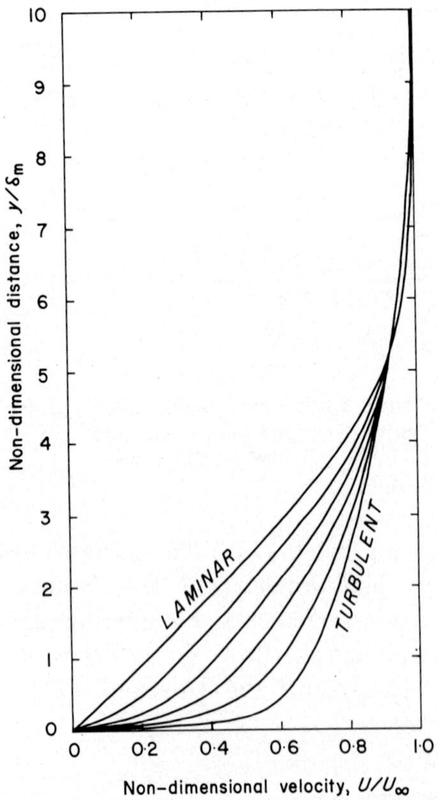

Fig. 6-2. Transition to turbulence in a laminar boundary layer, as illustrated by changes in the non-dimensional velocity with non-dimensional distance from the boundary. The labelled profiles represent the end-member boundary-layer states. After Dhawan and Narasimha (1958).

When the velocity structure of the turbulent boundary layer is analyzed, it becomes apparent that the flow may be conveniently divided from the wall outwards into a number of regions, each partly characterized by a unique velocity profile (Clauser, 1956). These regions may be defined by choosing an appropriate ratio between the mean velocity U at the boundary in question and the shear velocity $U_* = (\tau_0/\rho)^{1/2}$, where τ_0 is the wall shear-stress. Thus U^+, the non-dimensional velocity ratio, is given by U/U_*. But because U is an experimental function of y, the regions may also be distinguished by the corresponding non-dimensional distance $y^+ = yU_*/\nu$ from the wall.

The innermost region, called the laminar or viscous sublayer, is so close to the wall that viscous effects overwhelm inertial ones and the profile of mean velocity is linear. Its extent is commonly taken to be $0 < y^+ \leqslant 10$, though some authors prefer a smaller or slightly larger range.

The next region is the buffer layer ($10 < y^+ \lesssim 50$), where the velocity profile is logarithmic. A well-defined deterministic flow configuration capable of being impressed on a deformable bed is associated with the buffer layer and the flow immediately outside it. Turbulence production is a maximum within the layer, in which the configuration referred to plays a decisive part.

Beyond the buffer layer is a large region, usually known as the logarithmic region, in which the velocity profile follows a second logarithmic law. The outermost and rather indeterminate portion of the boundary layer is sometimes distinguished as the wake region, where another kind of velocity profile is found.

Visualization of transition

Experimental observations coalesce to reveal several stages in the transition of a boundary layer on a wall (Stuart, 1965). Proceeding downstream, they are: (1) laminar flow with all disturbances damped, (2) laminar flow unstable to small wavy disturbances (Tollmien-Schlichting waves), (3) laminar flow with amplifying three-dimensional waves, (4) laminar flow with amplifying three-dimensional waves and streamwise vortices, (5) laminar flow with regions of vorticity-concentration and shear-layer development, (6) breakdown of shear layers and production of embryo turbulent spots, (7) growth and agglomeration of turbulent spots, and (8) fully developed turbulent flow.

Prandtl (1933) seems to have been the first to visualize transition, and he has been followed by many others (Emmons, 1951; Tani and Hama, 1953; Wortmann, 1953; Mitchner, 1954; Fales, 1955; Hama et al., 1957; Bergh, 1958; Bergh and Van den Bergh, 1958; F.N.M. Brown, 1959, 1965; Elder, 1960; Thwaites, 1960; Kline, 1967; Fischer, 1972; Wygnanski and Champagne, 1973; Rao, 1974; Wygnanski et al., 1976; Cantwell et al., 1978). The work of Brown and Hama is most complete and is the basis for what follows.

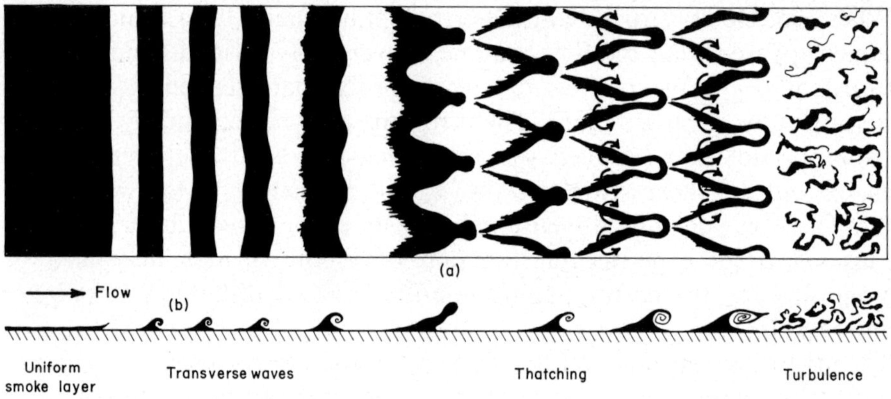

Fig. 6-3. Schematic representation of transition to turbulence in a laminar boundary layer, shown visually by the break-up and further modification of a continuous smoke sheath released into the boundary layer established on an axisymmetric body. After F.N.M. Brown (1959, 1965).

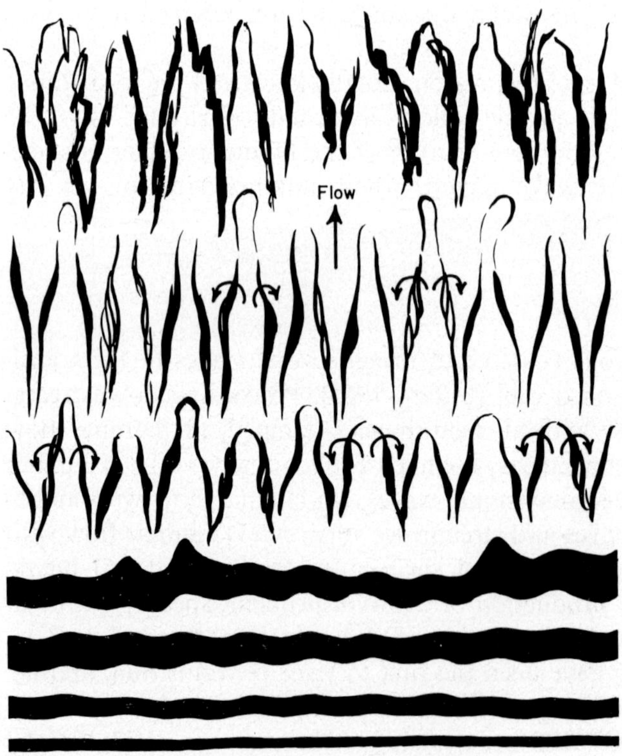

Fig. 6-4. Schematic representation of transition to turbulence in a laminar boundary layer, shown visually by the downstream modification of transverse cylinders of coloured fluid introduced into the boundary layer by an electrochemical reaction at a fixed transverse wire. After Hama et al. (1957).

Brown (1959, 1965), in a visual study of natural transition, used a continuous sheath of smoke on an axisymmetric body. Hama et al. (1957) worked with a "controlled" transition set off by a trip-wire that shed dye-marked transverse line-vortices into the boundary layer. The observations are summarized in Figs. 6-3 and 6-4. R.F. Blackwelder (1979) also presents a physical picture of the development of boundary-layer transition.

In Brown's experiments (Fig. 6-3), stage 1 is marked by a smoke-layer of essentially uniform thickness, and stage 2 by a spatially periodic thickening and thinning of the layer to form transverse ring-like waves ("tiger stripes"). The stripes, which express Tollmien-Schlichting waves, have a wavelength of several times the boundary-layer thickness and their celerity is roughly one-third U_∞. Stage 3 is denoted by the growth on the waves of spanwise sinuosities, or thinnings and thickenings, indicating that parts of each wave are moving away from the wall and therefore travelling faster than other parts. In Hama's experiments (Fig. 6-4), the spanwise waviness had a definite periodicity of several times the boundary-layer thickness.

Stages (4) to (6) are difficult to separate but collectively are distinctive (Figs. 6-3 and 6-4). Together they constitute Brown's "thatching". The sinuosities on the waves developed first into triangular patches of smoke, as the head of each sinuosity was dragged up into the outer, faster-moving parts of the boundary layer, and the base was pushed closer to the wall. Hama's line-vortices correspondingly became zig-zag in plan. In later stages, each smoke patch developed well-marked limbs, coming to resemble a wishbone or hairpin. Hama's zig-zags grew into similar shapes, which he likened to milk bottles. However, whereas Hama's configurations usually were in phase between the line-vortices, Brown's hairpins were generally out of phase by one-half of a wavelength. The two series of experiments again differed slightly as regards the eventual fate of the hairpin configurations, though in both the configuration began to break down as a localized burst of turbulence appeared in the flow at about the level of the shoulder of the configuration. Brown recorded that a complete ring was swept away from the head of the configuration, whereas Hama found that the ring remained incomplete and attached to the limbs by faint threads of marked fluid. Each configuration in decaying may be regarded as an embryo turbulent spot.

These experiments show clearly how the vorticity of the flow becomes concentrated first of all into periodically arranged transverse zones, as Kline (1967) later found. These begin to interact with three-dimensional motions in the flow, resulting in the stretching of the vortex lines and the amplification of vorticity. Prior to the appearance of bursts of turbulence, the flow is organized into three-dimensional vortex loops in which the vorticity is largely concentrated.

Unfortunately, this work tells little about the seventh stage of transition, on account of the confused appearance of the marked fluid. Emmons (1951) from his experiments on this stage concluded that, though the embryo

Fig. 6-5. View through the transparent side of a water channel showing a large turbulent spot (width approximately 0.2 m) visualized by means of dispersed aluminium powder. Flow velocity outside boundary layer $=0.12 \text{ m s}^{-1}$, with current from left to right. Photograph courtesy of B. Cantwell (see Cantwell et al., 1978), reproduced by permission of Cambridge University Press.

turbulent spots formed spontaneously and randomly in time and space on a smooth boundary, they could also be induced by introducing disturbances (e.g. dipping in a wire). Subsequently, it was proposed, each spot grew in size and spread laterally as it was convected downstream. The semi-angle representing the spread sideways was measured to be 9.6°, similar values being later obtained by Mitchner (1954), Schubauer and Klebanoff (1956), Wygnanski et al. (1976), and Cantwell et al. (1978). The growth of the spots and their progressive agglomeration during convection downstream caused an increasing proportion of the flow to be turbulent. Sufficiently far downstream, the whole flow was turbulent as the result of the enlargement and agglomeration of spots. Elder (1960) has photographed dyed turbulent spots and their interactions. Kline (1967) and Cantwell et al. (1978) also visualized the spots (Fig. 6-5), and noted that a streaky structure, found in fully developed turbulent flows, occurred beneath them. Zilberman et al. (1977) successfully traced spots generated in the laminar part of a boundary layer downstream into the turbulent part, where they seemed to be the major structures present.

Hot-wire anemometry

Schubauer and Skramstad (1947) investigated the stability of a laminar boundary layer to natural disturbances, loudspeaker-induced perturbations, and oscillations made by vibrating a ribbon. Bennett and Lee (1955) investigated natural transition. These studies yielded data confirming the upper branch of the neutral curve for $Re > 1300$ and, in the case of the ribbon-induced perturbations, the lower branch and the value of Re_{cr} (Fig. 6-1). A later study of controlled transition by Ross et al. (1970) gave even better agreement between theory and observation (Fig. 6-1).

Schubauer (1958) and Klebanoff et al. (1962) traced in detail the fate of

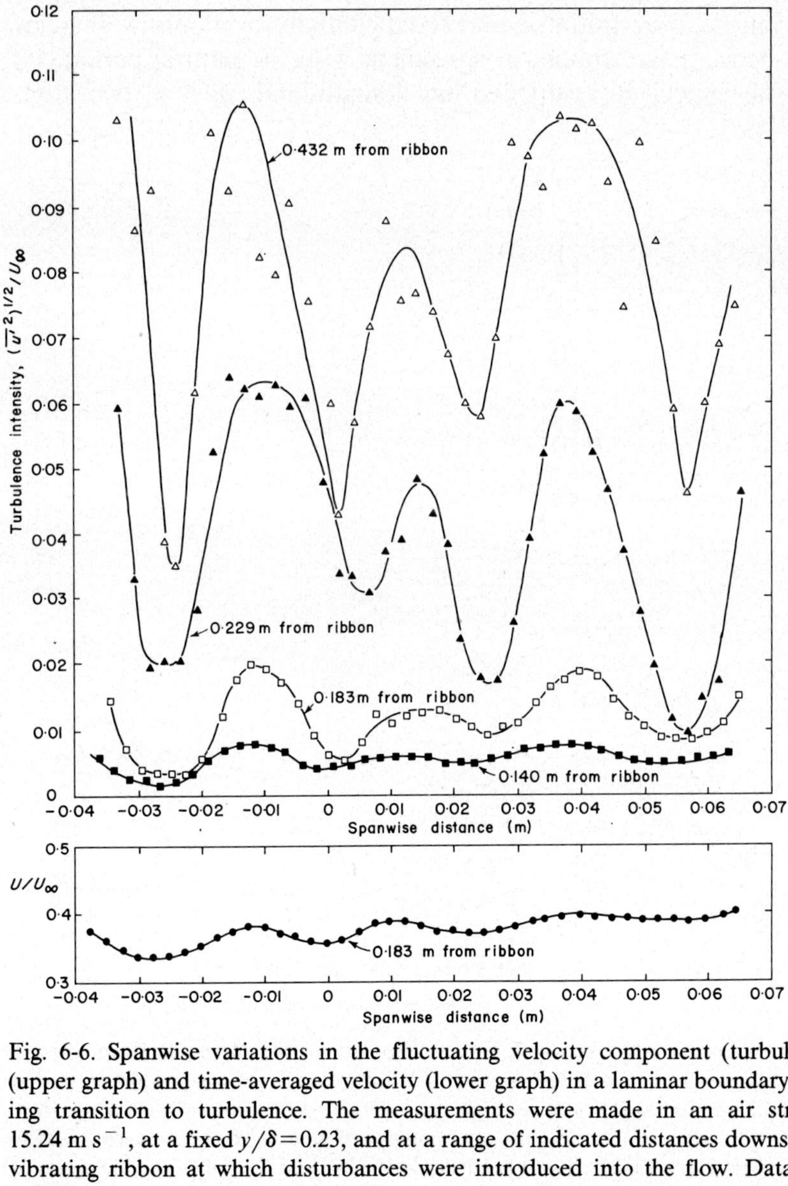

Fig. 6-6. Spanwise variations in the fluctuating velocity component (turbulence intensity) (upper graph) and time-averaged velocity (lower graph) in a laminar boundary layer undergoing transition to turbulence. The measurements were made in an air stream of $U_\infty = 15.24\ \mathrm{m\,s^{-1}}$, at a fixed $y/\delta = 0.23$, and at a range of indicated distances downstream from the vibrating ribbon at which disturbances were introduced into the flow. Data of Schubauer (1958).

ribbon-induced oscillations as they were convected downstream in the transitional portion of a flat-plate boundary layer. To do this the root-mean-square value $(\overline{u'^2})^{1/2}$ of the component of fluctuating velocity parallel with the streamwise direction was measured at different points within the layer. The velocity ratio $(\overline{u'^2})^{1/2}/U_\infty$ increased downstream from the ribbon and, at each downstream position, showed a markedly periodic spanwise variation (Fig. 6-6). The periodicity took a scale of several boundary-layer thicknesses and depended on the slight spanwise irregularities existing naturally even in the carefully formed experimental flow. Later (Klebanoff et al., 1962), the spanwise variations were initiated and fixed spatially by thinning the flow with spacers beneath the ribbon, in agreement with the natural periodicity.

The flow was conceived as divided into longitudinal zones of two kinds,

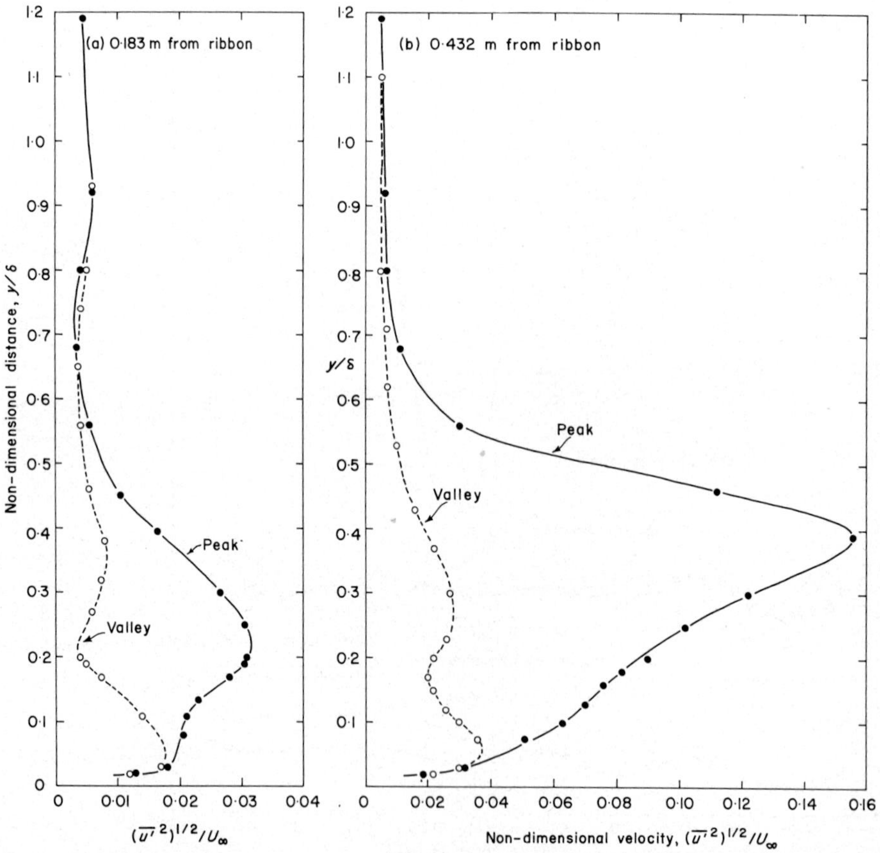

Fig. 6-7. Variation between "peaks" and "valleys" in the profile of the non-dimensional fluctuating velocity-component as a function of non-dimensional position within a laminar boundary layer undergoing transition to turbulence under the influence of disturbances from a vibrating ribbon. Experimental conditions the same as in Fig. 6-6. The peaks correspond in Fig. 6-6 to the streamwise zones of large turbulence intensity, and the valleys to the zones of low intensity. After Klebanoff et al. (1962).

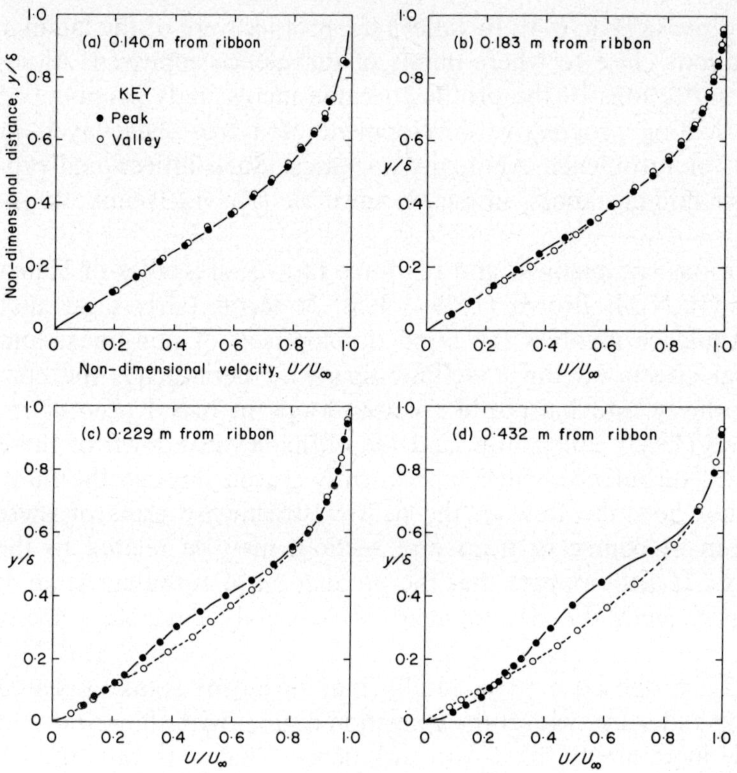

Fig. 6-8. Variation between "peaks" and "valleys" in the non-dimensional time-averaged velocity as a function of non-dimensional position within a laminar boundary layer undergoing transition to turbulence under the influence of disturbances from a vibrating ribbon. Experimental conditions as in Fig. 6-6. After Klebanoff et al. (1962).

alternating in the spanwise direction (Figs. 6-7 and 6-8). Those called "valleys" were marked at all levels in the boundary layer by relatively low values of $(\overline{u'^2})^{1/2}/U_\infty$. They also showed the lowest values of U/U_∞ in the runs without spacers beneath the ribbon but, when the spacers were present, gave relatively large values of this ratio. In the zones described as "peaks", lying between valleys, $(\overline{u'^2})^{1/2}/U_\infty$ was relatively large, particularly at low to intermediate positions. The disturbances whose amplification in the boundary layer is here recorded were at first laminar, that is, $(\overline{u'^2})^{1/2}$ described only low-frequency sinusoidal fluctuations. When each wave shed from the ribbon had travelled a sufficient distance downstream, however, a burst of high-frequency turbulence (an embryo turbulent spot) appeared in each peak at about one-third of the boundary-layer thickness from the wall. The relative amount of fluid affected by turbulence increased downstream, and fully developed turbulence was encountered not far beyond the first appearance of the bursts.

Parallel changes occurred in the profiles of mean velocity $U(y)$ measured in the valleys and peaks (Fig. 6-8). In valleys the profiles were of the laminar type, even at stations close to where bursts of turbulence appeared. Along peaks, however, inflections in the profiles became increasingly pronounced downstream, indicating progressive development of a free shear-layer, in which the bursts of turbulence eventually appeared. Such inflectional profiles are, in other circumstances, inviscidly unstable (Kelvin-Helmholtz instability).

These observations complement and reinforce the visual studies of Hama et al. (1957) and F.N.M. Brown (1959, 1965). It seems fairly clear that transition to turbulence involves the rapid development of non-linear and three-dimensional effects having a definite spanwise periodicity, the concentration of vorticity into hairpin-like vortex loops in broad accordance with Theodorsen's (1955) conception, and the ultimate breakdown of these loops into spots of turbulence which subsequently spread through the fluid. Figure 6-6 implies how the flow in the nearby streamwise arms of these vortex loops, seen of course in transverse section, may be related to the peaks and valleys. It also appears that the production of turbulent spots is associated in space with the development of inviscidly unstable velocity profiles.

Emmons (1951) proposed experimentally that turbulent spots appeared randomly in time and space within the transition region, with the result that the fluid became increasingly filled with turbulence. That is to say, at each point in the boundary layer the flow is turbulent for a fraction of the time that is determined by its position relative to the start of spot-production. Narasimha (1957), and Dhawan and Narasimha (1958), extended Emmons' analysis to show that this fraction depended exponentially on the length of the transition region, a relationship that experimental data fit quite well (Schubauer and Klebanoff, 1956; Dhawan and Narasimha, 1958). Further details of the shape and growth of turbulent spots are available (Mitchner, 1954; Schubauer and Klebanoff, 1956; Elder, 1960; McCormick, 1968; Wygnanski et al., 1976; Cantwell et al., 1978).

SEDIMENTARY STRUCTURES AND TRANSITION CONFIGURATIONS

Karcz (1970) has raised the possibility "that transition configurations are instrumental in triggering the deformation of the streambed" during the course of sedimentation in natural environments. He noted that transition configurations arise in a clear and reproducible sequence within the developing steady boundary layer, and that bed configurations occurred in a definite order in relation to monotonically changing flow conditions, as will be discussed in Chapters 7 and 8. He also recognized certain broad geometrical similarities between, on the one hand, transition flow structures and bedforms and, on the other, between transition flow structures and the flow patterns associated with bedforms.

Three groupings of bed configurations were considered. Current ripples (Chapters 7 and 8) were divided between relatively two-dimensional "low-energy" types and comparatively three-dimensional "high-energy" forms, all associated with separated flows. These were confronted with the Tollmien-Schlichting waves of early transition. Parting lineations and sand ribbons are each features elongated parallel with flow and of a regular transverse spacing. These Karcz matched against the flow-elongated vortices implied by the smoke patterns and peak-and-valley structure observed at an intermediate stage of transition. Finally, he linked initiation of erosional flute marks on mud beds, considered in Vol. II, Chapter 7, with the hairpin vortices and turbulent spots of late transition.

Two problems dog these ideas. The first is practical. Will it be possible to identify and characterize the particular flow configuration which, while being convected with the flow, gave rise at a fixed station on the bed to the embryonic form of some bed feature? The second is perhaps weightier. Which if any of the configurations of transition are also to be found in fully developed turbulent boundary layers, given that the configurations are defined dynamically as well as kinematically? This is an important question, because, while there are many different causes of instability in one-phase flows, and even more when flows are two-phase, there are few flow structures by which those instabilities may be expressed. Although much remains unknown about transitional and fully turbulent flows, it already seems fairly clear that the Tollmien-Schlichting waves of early transition do not occur in the fully developed flows, and that the streamwise vortices of transition may differ from bursting streaks and the larger vortices of fully developed turbulent flows. Inasmuch as streaks are recorded beneath turbulent spots (Kline, 1967), and similar streaks are associated with the wall region of fully developed turbulent flows, parting lineations and some flute marks may be the only sedimentary structures to be triggered as Karcz suggests. The work of Williams and Kemp (1971) links the initiation of current ripples in fully developed turbulent boundary layers with the accumulated action of several streaks, rather than with two-dimensional waves.

FLOW CONFIGURATIONS OF TURBULENT BOUNDARY LAYERS

General conceptions

Observations of many kinds show that in turbulent boundary layers there exist "eddies" of a wide range of scales, and of remarkable degrees of coherence, which tend to increase in size outwards from the wall. For example, measured wall-pressure fluctuations point to the presence of a broad spectrum of eddies which, no matter what their scale, travel downstream a distance of several times this scale before decaying (Willmarth and

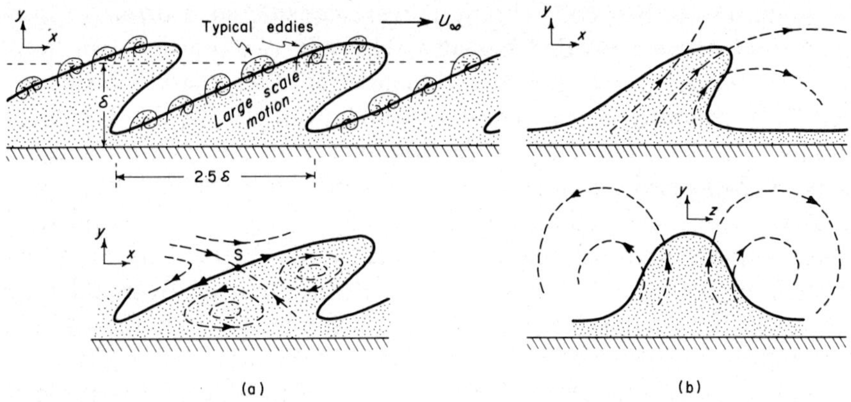

Fig. 6-9. The larger scale motion in the turbulent boundary layer. a. As envisaged by Falco (1977) and by Brown and Thomas (1977). The lower diagram shows the motion with respect to an observer travelling at the speed of the saddle point S. b. As envisaged by Townsend (1958). The motion depicted is that relative to an observer travelling near the base of the layer.

Woolridge, 1962; Bull, 1967; Blake, 1970; Wills, 1971). The character of the flow configurations represented by these eddies is significant for several problems, but it is elusive in experiments and speculation has consequently been rife.

Townsend (1956) at first proposed a simple, roller-like configuration for flow within the eddies. Grant (1958) suggested that a jet-like motion occurred within the larger eddies found in the outer part of the boundary layer (see Fiedler and Head, 1966; Tritton, 1967; Kovasznay et al., 1970). These eddies (Fig. 6-9a) occur at a streamwise spacing of the order of the boundary-layer thickness and appear to be inclined upstream at about 20° from the horizontal (Brown and Thomas, 1977; Falco, 1977). Instead of being jet-like, however, they seem to be horseshoe-shaped, with the arms pointing upstream in three dimensions. For the smaller eddies associated with the viscous and buffer layers, closer to the wall, Townsend (1958) postulated a markedly elongated jet-like or double-roller configuration (Fig. 6-9b), structures much like the vortex loops of transition. Because it occurs close to the wall, the configuration could deform a suitable boundary into a streamwise feature. Later Townsend (1970) slightly modified the concept of the double-roller, primarily to cover eddies of a much wider range of scales. Perhaps the most complicated proposal is that of Busse (1970), who postulated a cascade of eddies.

Configurations in the wall-region

The viscous sublayer conceived as a region of uniform thickness and near-laminar steady flow is but a convenient fiction. Extensive work by

Laufer (1951, 1954) and Klebanoff (1955) showed that significant velocity fluctuations occurred in and near the sublayer, and later studies offer nothing contradictory to this result (Einstein and Li, 1958; Clyde and Einstein, 1966; Bakewell and Lumley, 1967; Clark, 1968; Morrison et al., 1971; Gupta et al., 1971). The sublayer seems to be subjected to such vigorous disturbance by the outer flow that it must be regarded as continuously in a state of critical stability (Clauser, 1956), which may be expressed by a possible cyclic growth and decay of the layer or by some other type of intermittent behaviour (Hanratty, 1956; Einstein and Li, 1958).

Additional evidence strongly favouring an unsteady sublayer comes from the observation that dye injected into the layer is speedily mixed into the outer flow by a kind of ejection process (Einstein and Li, 1958), after becoming concentrated into longitudinal streaks. Indeed, at a much earlier stage, Fage and Townend (1932) had demonstrated visually using tracer particles the unsteady nature of the sublayer, though the implications of their results were for long unappreciated.

We now know a great deal about the character and role of the streaky structure observed in the wall region of turbulent flows. Dye and hydrogen-bubble visualizations contribute most data (Mitchell and Hanratty, 1966; Kline, 1967; Kline et al., 1967; Clark and Markland, 1971; Grass, 1971; Kim et al., 1971; Nychas et al., 1973; Offen and Kline, 1974, 1975), which is supported by the complementary results obtained from measurements of pressure or velocity fluctuations at a fixed station (Bakewell and Lumley, 1967; Coantic, 1967; Willmarth and Tu, 1971; Clark, 1968; Gupta et al., 1971; Morrison et al., 1971; Rao et al., 1971; Wallace et al., 1972; Bremhurst and Walker, 1973; Py, 1973; Lee et al., 1974; Lu and Willmarth, 1973; Strickland and Simpson, 1975; Ueda and Hinze, 1975; Antonia et al., 1976; Blackwelder and Kaplan, 1976; Heidrick et al., 1977; Oldaker and Tiederman, 1977; Nakagawa and Nezu, 1977; R.L. Simpson et al., 1977). Corino and Brodkey (1969), Johnson et al. (1976), and Praturi and Brodkey (1978), revitalizing the older techniques of Fage and Townend (1932), contribute a third body of relevant data. Related observations have begun to be made from naturally occurring boundary-layer flows (Gordon, 1974, 1975b; Heathershaw, 1974; Antonia et al., 1976; Gordon and Witting, 1977).

The observed streaky structure (Fig. 6-10) reflects the organization of the flow near the wall into alternate streamwise zones of contrasted velocity, reminiscent of the peak-and-valley structure of transition (Schubauer, 1958; Klebanoff et al., 1962). The dye or other marker is concentrated into the zones of slow-moving or low-momentum fluid. The intervening zones are clear because they consist of faster-moving or high-momentum fluid that in some way has penetrated towards the wall from the outer flow. The streaks are shifting and wavy in appearance and form randomly in space and time. They have a characteristic transverse spacing, $\lambda_z^+ = \lambda_z U_*/\nu$, and a characteristic longitudinal wavelength, $\lambda_x^+ = \lambda_x U_*/\nu$, both dimensions scaling on

Fig. 6-10. Streaky structure in a turbulent boundary layer, as revealed by dispersed aluminium powder. The lower portion of the photograph shows the view from outside through the transparent side of a water channel. In the mirror above is shown the view vertically down on the free surface of the boundary layer developed against the wall. Flow from right to left. Velocity outside boundary layer $=0.15$ m s^{-1}. Boundary layer thickness approximately 0.06 m. Photograph courtesy of B. Cantwell (see Cantwell et al., 1978).

parameters describing the wall region of the flow (Table 6-I). These characteristic dimensions are comparatively insensitive to the pressure gradient and practically independent of the Reynolds number calculated in terms of the boundary-layer momentum thickness. Note that the characteristic transverse spacing is several times the characteristic thickness of inner parts of the flow, where the velocity gradient is steep, just as the peak-and-valley structure in transition has a spacing of several times the boundary-layer thickness.

Table 6-I refers to the flow of Newtonian fluids only over smooth boundaries, the effects of surface roughness on the scales of the streaks being unknown. Grass (1971) believes that the dimensional scales may grow rapidly with increasing roughness. But the evidence cited for this view—the streaks of sand or snow several decimetres transversely apart beneath the wind blowing over smooth beaches or asphalt roads (e.g. Fisk, 1959)—seems inconclusive, because such surfaces are virtually smooth hydrodynamically. The other evidence advanced is also difficult to accept. Elongated patterns of glassy and ruffled water on the surface of the sea or a lake often express a near-surface secondary circulation driven by the combined action of wind and waves (Vol. II, Chapter 1). However, the convected herring-bone patterns formed by the wind on fields of fully grown but not yet ripe wheat, as

TABLE 6-1

Experimental measurements of the non-dimensional transverse and streamwise wavelength of bursting streaks in turbulent boundary layers on smooth walls

Authority	Reynolds number (momentum thickness)	Pressure gradient, dp/dx	$\lambda_z^+ = \lambda_z U_*/\nu$	$\lambda_x^+ = \lambda_x U_*/\nu$	Experimental method
Mitchell and Hanratty (1966)	500–3500	≈ 0	50	–	water in pipe; visual electrochemically formed dye
Bakewell and Lumley (1967)	1900	≈ 0	80–100	–	glycerine in pipe; hot-wire correlations
Coantic (1967)	2500	–	110–130	–	air; hot-wire correlations
Kline (1967); Kline et al. (1967)	1200	$=0$	131, 136	–	
	885–1160	≈ 0	131–138	–	
	567, 615	$\ll 0$	223, 312	–	water channel; visual, dye and hydrogen bubbles
	556–1680	≈ 0	83–131	–	
	588–1560	≈ 0	117–177	–	
Willmarth and Tu (1967)	40,000	≈ 0	200	–	air; wall-pressure correlations
Clark (1968)	30,000	$=0$	100	–	water channel; hydrogen bubbles
Clark and Markland (1971)	270	$=0$	100	440	water channel; visual, hydrogen bubbles
Grass (1971)	700	$=0$	80	–	water channel; visual using quartz sand
Gupta et al. (1971)	2200–4700	–	89–110	–	air; hot-wire correlations
Kim et al. (1971)	2000	$=0$	100	–	water channel; visual, dye and hydrogen bubbles
Morrison et al. (1971)	1000–5000	≈ 0	135	630	air in pipe; hot-wire fluctuations
Lee et al. (1974)	34,700	–	106	–	water in pipe; correlation based on electrochemical reaction
	37,200	–	105	–	
	40,900	–	107	–	
R.L. Simpson et al. (1977)	6020, 10100	$\ll 0$	95, 155	–	air in duct; laser anemometry based on smoke marker
	13600, 18400	>0	65, 95	–	
	21400, 25700	$\gg 0$	12, 18	–	

the stalks are bent according to the local air flow, may be more acceptable as evidence in support of his view.

Increasing attention is being given to the effects of adding small quantities of long-chain polymers, such as may be present in natural waters flowing off richly vegetated areas, to turbulent flows of Newtonian fluids. One effect is to reduce the resistance of flow relative to the corresponding untreated current, often to a significant degree. A related effect, proportional to the amount of polymer in solution, sees an increase in the characteristic dimensions of the streaks as measured against the wall (Achia and Thompson, 1977; Oldaker and Tiederman, 1977; Willmarth and Bogar, 1977). How the presence of the polymer modifies the structure of the near-wall flow is uncertain, but it is possibly related to the ability of such long-chain molecules to resist stretching motions.

Richardson and Beatty (1959) explained the streaky structure by the action of stable contra-rotating streamwise vortices close to the wall, and related conceptions were later developed by Bakewell and Lumley (1967), Clark and Markland (1971), Lee et al. (1974), and Praturi and Brodkey (1978). These models do not account completely for all the features of the motion that can be observed. Morrison et al. (1971), following J. Sternberg (1962) and Schubert and Corcos (1967), modelled the streaky structure as a three-dimensional wave propagated through the wall region. This is an interesting suggestion but again difficult to reconcile completely with the observations subsequently made by Kim et al. (1971), Grass (1971), Offen and Kline (1974), and Johnson et al. (1976). Fage and Townend's (1932) observation of "corkscrew" or "wavy" motions in the wall region are critical to Morrison's model, but equally fit models of other types (e.g. Kim et al., 1971). Willmarth and Tu (1967), Willmarth and Lu (1972), and Wygnanski and Champagne (1973) all favour a hairpin vortex similar to that observed during transition to explain the streaky structure. An amalgamation of the models proposed by Grass (1971), Kim et al. (1971), Offen and Kline (1975), and Blackwelder and Eckelman (1979) seems to fit the observations best.

Figure 6-11 summarizes the behaviour and possible character of the streaks as they are transported by the flow: (1) the long periods of quiescence, (2) the periodic eruption (bursting or ejection) of the low-momentum fluid away from the wall and into the outer parts of the flow, and (3) the close association in time and space between a bursting streak and violent inrushes of fluid (sweeps of Corino and Brodkey, 1969) into the laterally adjacent zones of high-speed fluid. The model is presented largely as a combination of "frozen" kinematic structures, reflecting the importance of visualization techniques in the study of the inner zones of the turbulent boundary layer. These structures are, however, being convected downstream with the flow, and are changing with both distance and time. An understanding of the relationship between, on the one hand, the results of visualization studies (Fig. 6-11) and, on the other, the non-visual hot-wire measurements

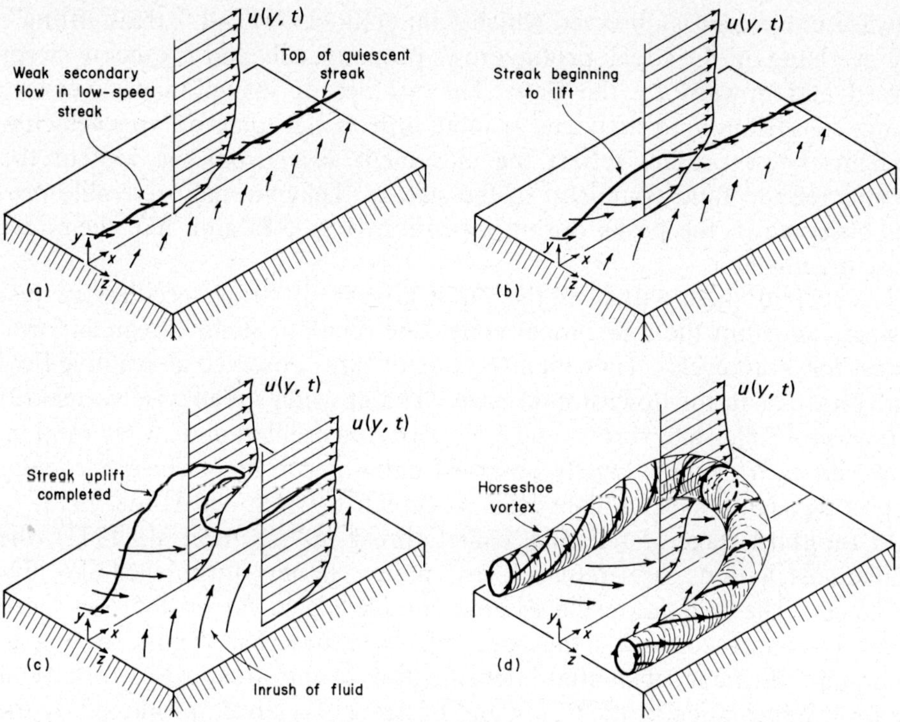

Fig. 6-11. Behaviour of sublayer streaks (a–c) up to the completion of lift-up, and (d) the horseshoe vortex present by this stage.

made at a fixed station, is greatly helped by the scheme for the classification of the fluctuating velocities advanced by Brodkey et al. (1974). They have introduced the terms "ejection" ($u' < 0$, $v' > 0$) (bursts or eruptions of other workers), "sweep" ($u' > 0$, $v' < 0$), "inward interaction" ($u' < 0$, $v' < 0$), and "outward interaction" ($u' > 0$, $v' > 0$), permitting the two kinds of observation to be linked. Streak-bursting coupled with inrush is a continuous and repeating cycle of events which appears to be controlled by the movement of the larger turbulent eddies (Fig. 6-9a) past the relatively more retarded fluid in the immediate vicinity of the wall (Brown and Thomas, 1977; Praturi and Brodkey, 1978). In order to see in more detail what happens during the cycle, let us ride with a streak as it is convected downstream (Fig. 6-11).

In the first stage of bursting, the streak gradually lifts away from the wall while being transported downstream, the lifting taking place over a relatively large downstream distance. In the plane of flow, the straight or weakly curved top of the streak increases in roundness, until a cusp appears near the crest, followed later by an overhang. The profile of instantaneous velocity $u(y, t)$ resembles the profile of mean velocity $U(y)$. However, when the streak attains a certain critical distance from the wall, a much more rapid outward movement is observed which Kim et al. (1971) call "streak-lifting". The overhang on the streak profile grows pronouncedly and is quickly swept foward and upward by the flow. The profiles of instantaneous velocity change significantly, to show one or more inflectional zones of large velocity-gradient. These profiles reflect the movement away from the wall of the low-momentum fluid contained in the streaks. They strongly resemble profiles observed in the peaks during transition (Fig. 6-8) and, like them, are inviscidly unstable.

The second stage witnesses the rapid growth of strong oscillations just downstream from the inflectional zone. The oscillations, in a regular form, persist for 3–10 cycles. They manifest one of three observed alternative flow configurations in the downstream zone: (1) a growing streamwise vortex, (2) a transverse roller-like vortex, and (3) a repeated oscillation or wavy motion. However, each of these locally observed patterns can represent a particular section through a hairpin-like vortex as depicted (Offen and Kline, 1975).

In the third stage of bursting, called "break-up" (Kim et al., 1971), the regular oscillations are replaced by more chaotic turbulence-like disturbances. The bursting cycle is now completed, so far as the low-speed streak is concerned, and a quiescent phase ensues. The events leading to break-up indicate a substantial intensification of the streamwise vorticity of the flow. Kinematically, the lifting and bursting of a streak, promoted by the passage of one or more large eddies overhead, may be regarded as leading to the development of temporary streamwise zones of attachment and separation which are convected by the flow.

Whereas the dimensions of boundary-layer streaks are set by parameters describing the wall-region, the frequency with which bursting occurs at a fixed point near the bed is found to depend on the properties of the outer flow. Rao et al. (1971) showed from their own data, and those of Willmarth and Tu (1967) and of Kim et al. (1971), that the non-dimensional period of bursting is $U_{\infty}T/\delta_s \approx 32$, independently of the Reynolds number, where T is the mean interval between bursts. It is difficult to measure this interval, but data supporting Rao's result have also been obtained by Laufer and Narayanan (1971), Lu and Willmarth (1973), Blinco and Simons (1975), Ueda and Hinze (1975), Antonia et al. (1976), and Heidrick et al. (1977). In naturally occurring boundary layers, the same non-dimensional periodicity broadly holds (Gordon, 1974, 1975b; Heathershaw, 1974, 1979; Antonia et al., 1976; R.G. Jackson, 1976c; Gordon and Witting, 1977). This is further evidence in support of the idea that lifting and bursting are controlled by events taking place in the outer flow. The frequency of bursts is reduced in drag-reducing polymer solutions (Gyr, 1976; Achia and Thompson, 1977). In hypersonic boundary-layers, however, the controls may be different (Owen and Horstman, 1972).

The effects of boundary roughness on the frequency and strength of bursts and inrushes are not yet well known. Sabot et al. (1977) find that violent bursts are less frequent near rough than smooth walls. Nakagawa and Nezu (1977), in a particularly detailed study, show that in rough flow the inrushes near the wall are generally more intense than the bursts. In smooth flow, by contrast, it is the bursts that are the more violent.

Corino and Brodkey (1969) and Kim et al. (1971) concluded that a major part of the turbulence production could be attributed to the bursting of low-speed streaks in the wall-region. From the work of Grass (1971), the remaining, comparably substantial part must be due to the violent inrushes to the high-momentum streaks. Earlier, hot-wire studies had shown that production occurs chiefly in the wall-region (e.g. Klebanoff, 1955).

GENERAL EFFECTS OF STREAKS ON DEFORMABLE BEDS

Natural boundaries composed of deformable sediment may be expected to respond in some semi-permanent or permanent manner under the action of the fluctuating forces associated with the sweeps and bursting streaks found in the inner parts of turbulent flows. The mode of deformation should depend on the bed material, whether cohesive (mud or rock) or cohesionless (sand), and it may also be influenced by the nature of the deforming forces, if the material is cohesive. Whatever the mode, however, the resulting sedimentary structures should be: (1) elongated parallel with flow, because of the greater streamwise than spanwise scale of the causative flow configuration, and (2) of positive relief where streamwise flow separation occurred. It is also reasonable to expect their spanwise scale to compare with that of the boundary-layer streaks, though the bed may have the ability to retain in a limited and imperfect way the history of streak-development (Mantz, 1978). For example, a sand bed is a boundary covered with readily moveable flow markers, able to respond fairly quickly to the local bottom flow and, in particular, to roll from the zones of temporary flow-attachment (inrushes or sweeps) into the zones of separation where streak-lifting occurs. The response of cohesive beds is more difficult to anticipate, because of their multiple modes of erosion but separation could cause puckering.

Several sedimentary structures seem explicable by streak-bursting and inrush. They are parting lineations on sand beds, and various longitudinal ridges and furrows shaped on cohesive surfaces. In Vol. II, Chapter 1, structures dependent on mass-transfer but probably related to the streaks will be described.

PARTING LINEATION

General character

Sorby (1859, 1908) observed that the action of a current drifting sand grains over a level bed of the same sand was to form on the surface a

Fig. 6-12. Parting lineations (and also parting-step lineations) on bedding surfaces in a fine-grained parallel laminated sandstone, Brownstones Group (Old Red Sandstone), Forest of Dean, England. The width of the surface shown is approximately 0.34 m.

streamwise graining or striping. He also noticed sandstones composed of extensive, flat-lying, parallel laminae arranged like the leaves of a book. The surfaces of these laminae bore the same type of graining (Fig. 6-12).

This type of sandstone is now called flat-bedded or parallel-laminated, and the graining goes under several names. Cloos (1938) next after Sorby described the facies, but his account is not detailed. Häntzschel (1939) noticed a "fine parallel graining" on the surfaces of tidal sand bodies, which Plessman (1961), referring to earlier descriptions (Seilacher, 1953; Rabien, 1956), divided between coarsely spaced (*Strömungsriefung*) and finely spaced types (*Strömungsstreifung*). Meanwhile, Stokes (1947) introduced the name "primary current lineation", though Crowell's (1955) term "parting lineation" is less inclusive and therefore preferable. Plessman's two categories represent in my opinion the same structure, on slightly different scales. Conybeare and Crook (1968) have unnecessarily introduced the term "streaming lineation" for Crowell's parting lineation. Picard and High (1973), who follow them, have confused a number of structures under this term. McBride and Yeakel (1963) introduced the name "parting-step lineation" to describe the secondary structure formed when the laminae preferentially break across, in a direction parallel with the surface striping and the average grain long-axis orientation. In a confused analysis, Picard and High (1973) include this structure in their parting lineation.

Parting lineation has been described from parallel-laminated sandstones of many different ages and from flat-bedded sands deposited on modern beaches and river bars. It is found chiefly in coarse silt-grade to medium sand-grade deposits (Picard and Hulen, 1969; Allen, 1974a), and only rarely in coarse-textured parallel-laminated sediments.

According to Allen (1964a), lineations similar in orientation can be found over the surfaces of laminae measuring a few to many square metres in area. The lineations are low, parallel ridges and hollows with a relief seldom greater than a few grain-diameters. The hollows between ridges are commonly flat-bottomed, whereas the ridges themselves are generally rounded in profile. Hollows and ridges are off-set, each ridge leading downstream into a hollow. In a sample of the lineations, Allen found the transverse ridge-spacing to vary from 8 to 17 per 0.10 m. In very fine sandstones, the ridges were 0.035–0.12 m long but reached 0.05–0.30 m in length on medium grained rocks. Since these orders of spacing and length appear to be representative, the longitudinal scale of the ridges may be said to be 5–20 times the spanwise scale. Often dark-coloured heavy minerals and mica flakes are preferentially distributed between the ridges and hollows, the coarser sand being heaped on to the ridges (Allen, 1964a; Schröder, 1965).

It should not be supposed that parting lineation is found only in association with parallel-laminated sand-grade sediments. It also occurs in the erosional environment on the backs of ripples and dunes (Allen, 1968c; Karcz, 1974).

Sand shape-fabric

Several workers measured the sand shape-fabric of deposits with parting or parting-step lineation and compared the preferred grain orientation with the trend of the surface markings (McBride and Yeakel, 1963; Potter and Mast, 1963; Allen, 1964a; McBride, 1966; Picard and Hulen, 1969; Lafeber and Willoughby, 1970; Shelton and Mack, 1970). A generally good agreement was observed between the preferred grain long-axis orientation and the trend of the lineation or parting-steps (Fig. 6-13). Allen (1964a) found in most of the cases he examined that these fabrics were symmetrically bimodal, the two modes lying usually 20–40° apart. Other workers have frequently recorded the same property (McBride and Yeakel, 1963; Picard and Hulen, 1969), though none of McBride's (1966) three samples showed bimodality. The grain imbrication is upcurrent at a low angle relative to the bedding. Potter and Mast (1963) measured imbrications averaging 12° and Allen (1964a) obtained angles between 8° and 12°. Figure 6-14 summarizes the character of the structures discussed.

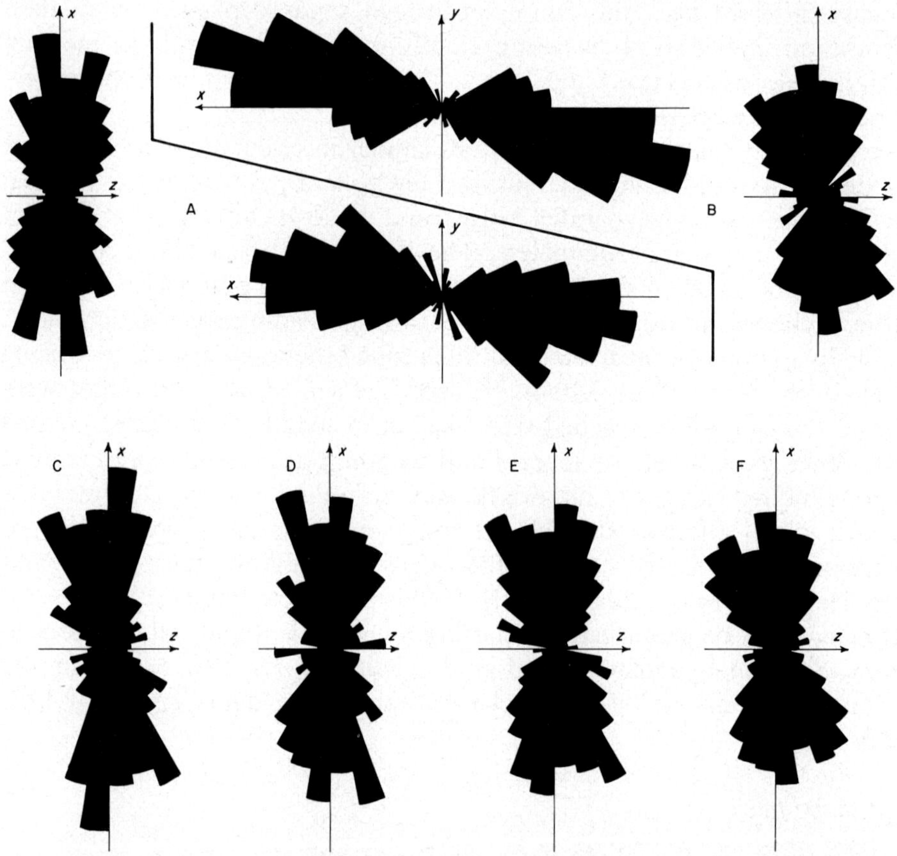

Fig. 6-13. Particle fabrics in representative sandstones combining parallel lamination with parting lineations. Rocks of six different ages and localities are represented, and in each case the grain long-axis azimuths in the plane of the bedding are plotted, except in A and B where the distribution of axes in the vertical plane parallel with the lineations also is shown. Data of Allen (1964a).

Interpretation

If parting lineation arises during streak-bursting and inrush under conditions of natural turbulent flow, its nature should be consistent with the characteristics of the streaks.

The agreement seems excellent in all salient respects. Parting lineations have the requisite streamwise elongation. Their length is 5–20 times the spanwise scale, comparing with the λ_x^+/λ_z^+ ratio for the streaks (Table 6-I). In some experiments, Grass (1971) visualized the streaks using sand grains thickly scattered over a smooth bed. He saw that during inrushes into high-speed streaks, grains were driven sideways across the bed to form streamers, elongated parallel with the general flow, beneath low-speed streaks.

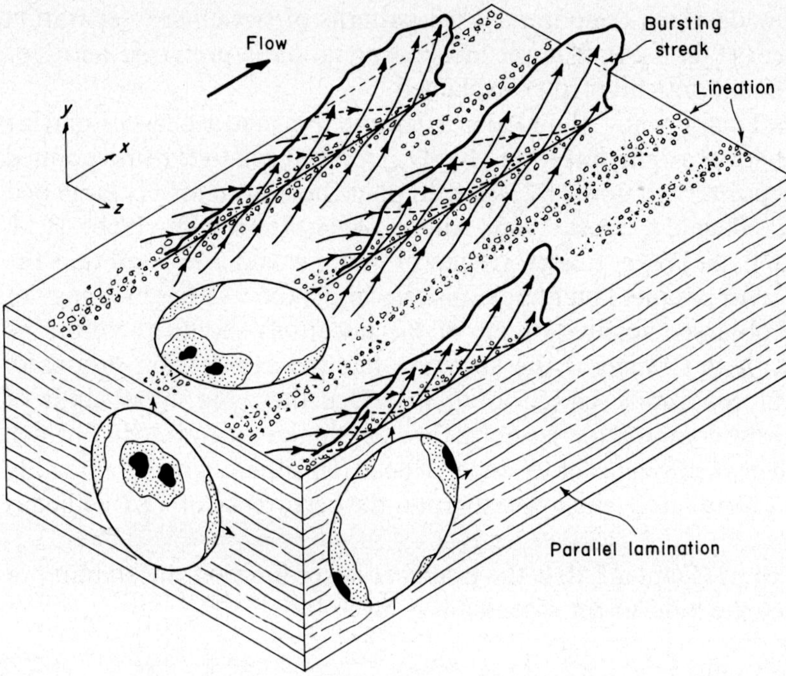

Fig. 6-14. A model for the origin of parting lineations by the action of boundary-layer streaks, in which the macroscopic structure and grain fabric (particle long-axis intersections with lower hemisphere in each plane as viewed) are integrated with the flow configuration (transient zones of flow separation and attachment associated with lifting and bursting streaks and associated inrushes).

Typically, a divergence of 10–20° in the plane of the bed was observed between the average flow direction and the paths of grains forced into the low-speed streaks where streak-lifting was occurring. Sand visualizations by Karcz (1974) revealed similar effects. These observations are convincing proof of Allen's (1968c) suggestion that the bimodal shape-fabrics associated with parting lineation reflect the alignment of sediment grains by near-bed currents with a significant spanwise component. The same observations also suggest that the ridges of the lineation formed beneath low-speed streaks, since grains should accumulate preferentially in zones of converging flow at the bed. The concentration of the coarser grains into the ridges also fits this model. In a zone of temporarily converging flow, where an upward motion must also occur, small particles could easily be lifted in suspension off the bed, while the less mobile larger grains lagged behind.

Allen (1964a, 1968c) linked parting lineation with systems of implicitly steady longitudinal vortices in the near-bed flow, and Schröder (1965) also connected the lineation with vortex action. Later, Allen (1968b, 1970g), Williams and Kemp (1971), Karcz (1974), and Mantz (1978) explained parting lineation in terms of boundary-layer streaks, in association with which the flow develops temporary, local patterns of streamwise separation and attachment (Fig. 6-14). The second interpretation is preferred here, for reasons which will now be further explained.

The required consistency should also extend to dynamic aspects of parting lineation and streaks. The lineation chiefly occurs in relatively fine grained sand-grade deposits, laid down by water on plane beds at relatively large bed shear stresses (Bagnold, 1956, 1966; Allen, 1964a; H.M. Hill, 1966; H.M. Hill et al., 1969). At lower stresses, ripples or dunes would arise, and thus the combination with parallel lamination appears to denote a definable range of stresses. The observed spanwise scale of the lineations should therefore be compatible with this range, if the structure has the explanation proposed. Under natural conditions, chiefly of free-surface flow, the upper limit of occurrence of the combination is set by flow depth and mean velocity, which determine when the flow becomes supercritical. This limit cannot be readily predicted but, fortunately, the lower limit on the occurrence of a plane bed is accessible.

We shall see in Chapter 7 that the practical lower limit on the stability of a plane bed of the kind under discussion is given by:

$$\frac{\tau_{cr}}{(\sigma-\rho)gD} = C_0 \tan\alpha \tag{6.7}$$

in which τ_{cr} is the critical bed shear stress, σ and ρ are the solids and fluid density respectively, g is the acceleration due to gravity, D is the mean sediment diameter, C_0 is the fractional solids volume concentration in the static bed, and $\tan\alpha$ is the dynamic friction coefficient of the sediment. Reminding oneself that $U_* = (\tau/\rho)^{1/2}$ and that $\lambda_z^+ = (\lambda_z U_*)/\nu$, eq. (6.7) can be rewritten to eliminate τ_{cr}. However, at the lower limit of stability of the plane bed, the mean transverse spacing of the boundary-layer streaks is a maximum, so that on the hypothesis proposed the observed mean spacing, L_z, of the lineations in parallel-laminated sands should not exceed this value. Equation (6.7) then affords the inequality:

$$L_z \leqslant \frac{\lambda_z^+ \nu}{\left(\dfrac{\sigma-\rho}{\rho} gDC_0 \tan\alpha\right)^{1/2}} \tag{6.8}$$

where $\lambda_z^+ \approx 100$ on the supposition of a zero pressure gradient (Table 6-I).

Consider a quartz sand of $D = 0.00015$ m and $\sigma = 2650$ kg m^{-3} in plain water at 15°C. Values of $C_0 = 0.65$ and $\tan\alpha = 0.75$ are plausible. The limiting value of L_z in this case is therefore 0.0033 m. Allen (1964a) measured a mean transverse ridge spacing of $0.0059 \leqslant L_z \leqslant 0.0125$ m from very fine to medium grained parallel-laminated sandstones. The observed spacings are of the correct order, but larger than expected, perhaps because not all streaks gave rise to discernible lineations.

It may be objected that when sand grains are transported over a plane sand bed, the conditions are quite different from those of one-phase flow over smooth beds that gave rise to the data summarized in Table 6-I. Although Grass (1971) could visualize streaks with typical dimensions using sand thickly sprinkled over a bed of varnished plywood, a continuous granular boundary is significantly much rougher, and he suggests that an increase in bed roughness may cause an increase in streak spacing. Another possible explanation for the greater observed spacings is that, on lamina after lamina, the lineations that Allen (1964a) measured were formed during the terminal phases of flows so unsteady that the plane-bed condition represented could not be replaced by the rippled or duned beds that those final stages would otherwise have warranted.

In summary, several lines of evidence converge to suggest that parting lineation is the response of a sand bed to flow configurations arising during turbulence production (Figs. 6-11 and 6-14). If this is correct, the lineation combined with parallel lamination should arise wherever natural turbulent currents attain or exceed a specific and definable strength. The transverse scale of the lineation may therefore be useful for estimating flow parameters. The lineation is a useful indicator of current path (e.g. Crowell, 1955); the shape-fabric associated with it gives the current direction.

Environmental distribution

The combination of lineation and lamination is widely distributed, in accordance with the ubiquitous natural occurrence of turbulent flows. Its environmental significance is therefore limited.

Allen (1964a) and Karcz (1974) produced the combination experimentally, and Jopling (1964a), Williams and Kemp (1971), McBride et al. (1975a) and Mantz (1978) reported the lineation from their flumes. From modern river sands, G.E. Williams (1971) illustrated combined lineation and lamination, Picard and High (1973) reported parting-step lineation, and Karcz (1972) recorded the lineations. Parallel laminations themselves are well known from river deposits (e.g. Harms et al., 1963; Harms and Fahnestock, 1965; McKee et al., 1967; G.E. Williams, 1970a). Parting lineation on plane beds abounds in the swash zone of modern beaches (Fig. 6-15) and can also be found in tidal run-off channels (Häntzschel, 1939; Seilacher, 1953; Stokes, 1953; Trefethen and Dow, 1960; Plessman, 1961; Bajard, 1966; Allen, 1968c; Conybeare and Crook, 1968; Lafeber and Willoughby, 1970; Hayes, 1972; Donovan and Archer, 1975; Knight and Dalrymple, 1975; P. Wright, 1976).

The combination of structures is also well known from ancient shallow-marine deposits (Von Bertsbergh, 1940; Schindewolf and Seilacher, 1953; Allen, 1964a; P.F. Williams, 1968; Picard and Hulen, 1969), and from rocks ascribed to a fluvial origin (Stokes, 1953, 1961; Potter and Glass, 1958; Fahrig, 1961; Potter, 1963; Allen, 1962a, 1964a, 1970d; Friend, 1965; Schröder, 1965; Allen and Friend, 1968; Stanley, 1968; Way, 1968; Brynhi, 1978).

Fig. 6-15. Parting lineations in fine grained sand beneath the swash zone of a modern beach, Mundesley, Norfolk, England. Hammer has handle approximately 0.27 m long and points upbeach.

Lineation and lamination and parting step lineation also occur in sandstones attributed to turbidity currents (Crowell, 1955; Rabien, 1956; Dzulynski, 1963; McBride, 1966; R.G. Walker, 1966; K.M. Scott, 1967a; Ricci-Lucchi, 1970; Schenk, 1970; Tanaka, 1970; Stanley, 1974), suggesting that the natural currents are fully turbulent during some part of their lives. Banerjee (1973) records possible parting-step lineation from glacial-lake varves.

LONGITUDINAL GROOVES IN MUD BEDS

General character

Under some flow conditions, weakly cohesive mud beds become sculptured into longitudinal grooves and ridges, of which two kinds, meandering and rectilinear, are known.

Dzulynski and Sanders (1962a, b) described from turbidites the moulds of what were called "meandering rill marks" (see also Dzulynski, 1963; Dzulynski and Walton, 1965), and Craig and Walton (1962) reported another example. Allen (1969a) produced the structure experimentally, naming it "longitudinal meandering grooves". A similar structure was produced by

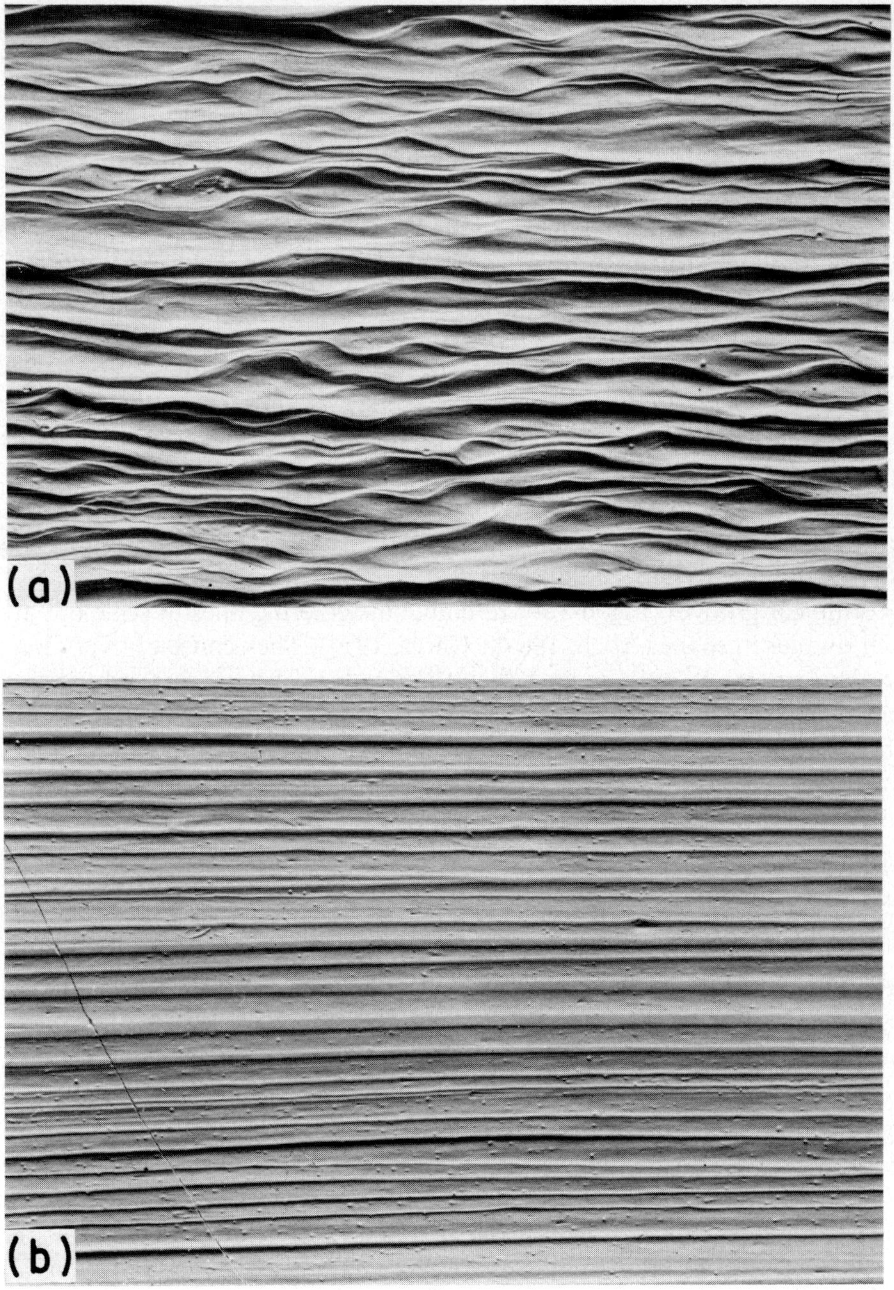

Fig. 6-16. Moulds in Plaster of Paris of experimental flow-parallel structures made by the action of a turbulent boundary layer on a weakly cohesive bed of kaolinite mud. a. Longitudinal meandering grooves. Flow from left to right at mean velocity=0.378 m s^{-1}. b. Longitudinal rectilinear grooves. Flow from left to right at mean velocity=0.298 m s^{-1}. The length of bed shown in each case is about 0.23 m.

Lonsdale and Southard (1974) during the experimental erosion of a Pacific red clay. The other kind of longitudinal groove is rectilinear and is only known experimentally (Allen, 1969a, 1971c; Karcz, 1974).

Experimental meandering grooves (Fig. 6-16a) consist of long, sinuous grooves with a fairly regular transverse spacing, which locally join at characteristic downstream-pointing Y-shaped junctions. Transversely, the grooves are broad and generally shallow, with a narrow, flat floor. The intervening ridges are, in contrast, gently convex-upward and sometimes flat. The spanwise distance between adjacent grooves varies over a ten or twenty-fold range, averaging 0.005–0.0075 m, and the streamwise sinuosities have a characteristic wavelength several times the transverse scale. In a representative run, the mean spanwise scale was 0.0067 m and the mean sinuosity wavelength was 0.034 m. The sinuosities on adjacent grooves seem to be roughly π rad out of phase. These experimental grooves closely resemble in size and shape the natural marks figured by Dzulynski and Sanders (1962b). Flute marks accompany these grooves, and they were also found with grooves in the experiments.

Rectilinear grooves (Fig. 6-16b) resemble meandering ones in scale but are much simpler in shape (Allen, 1969a; Karcz, 1974). They consist of very long and almost perfectly rectilinear hollows and ridges of low relief, which may branch and rejoin under the influence of large-scale secondary flows. In transverse profile the grooves show steep sides and flat floors; the ridges between are chiefly flat and less often convex-upward. The grooves vary in transverse spacing over an approximately ten-fold range, but in three of Allen's runs averaged 0.0056, 0.006 and 0.008 m, respectively.

Mode of origin

Dzulynski and Sanders (1962b) recognized the erosional origin of meandering grooves, and it was later suggested that these may have formed in response to longitudinal vortices created at down-cut flute marks (Dzulynski and Walton, 1965). The experimental evidence supports the erosional origin of both types of groove, and suggests that transverse components of flow, associated with boundary-layer streaks, are involved in the shaping of the bed (Allen, 1969a; Karcz, 1974). However, the existence of flute marks does not seem to be necessary for the appearance of grooves. Neither Karcz nor Allen observed flute marks with rectilinear grooves, and of two of Allen's runs with meandering grooves, only one yielded flutes in addition.

Allen (1969a) and Karcz (1974) experimented at fairly low Reynolds numbers within the turbulent range, using beds of weakly cohesive mud settled from suspensions of kaolinite or montmorillonite clay. These beds showed vertical gradients of mass physical properties, a thin soup-like upper layer grading down into stronger and thicker lower strata. Hence the mud at each level in the bed responded in its own particular way to the eroding current. In Allen's experiments the soupy layer was eroded chiefly flake by flake and locally as abruptly detached small aggregates that speedily were broken into their constituent particles. The suspension thus formed close to the bed became organized into wavering and periodically bursting streamwise streaks, indistinguishable in scale and behaviour from the dyed boundary-layer streaks studied by Kline (1967) at an early stage. The "gusts" of Southard et al. (1971) and the "wandering lineations" of Mantz (1978) seem to be similar. Progressively less clay suspension was produced as the current bit down into stronger parts of the bed. With the thinning of the suspension that obscured the bed, it was seen that grooves had begun to develop (see also Southard et al., 1971). Each groove contained at the bottom a narrow stream of clay suspension and, rolling along in it here and there, scattered aggregates of clay. Karcz (1974) produced the grooves similarly, by allowing quartz sand grains to roll along the bed. Erosion of the bed now seemed to be limited to the vicinity of these aggregates, and appeared to depend on a local heightening of the bed shear stress as the flow is accelerated round each grain. The meandering grooves formed at higher flow velocities and bed shear stresses than the rectilinear kind (Allen, 1969a, 1971c).

Rectilinear and meandering grooves seem to be the response of a weakly cohesive mud bed to boundary-layer streaks (Allen, 1969a). The coexistence in experiments of clay-marked streaks and juvenile grooves at once points to a connection. If the flow configuration can drive clay flakes transversely across the bed, resulting in visible streaks, clay aggregates or other grains introduced into the flow should also become concentrated beneath the low-speed streaks, just as Grass (1971) and Karcz (1974) observed was true of quartz sand on rigid surfaces. The low-speed streaks thus become the loci for erosion, because of the locally heightened stresses associated with the flow round each aggregate and, perhaps, because of the cutting action of the rolling grains. The grooves initiated in this way would tend to trap other grains or aggregates rolling over the bed, and so would be the cause of the intensification of their own relief. The transverse spacing of the grooves is consistent with the process here described, whereby the scale of the flow configuration becomes impressed on the bed. Allen (1969a) found good agreement between measured groove and calculated streak spacings.

It is worth drawing attention again to the fact that the streaks are generated randomly in time and space during one-phase turbulent flow over a rigid bed. The process just described from mud beds is one of several allowing a spatially and temporally random flow configuration to generate spatially *fixed* deformations. To what extent do grooves and ridges, when established on the bed, determine where in space the contemporaneous flow configurations arise?

Meandering grooves are rare structures in nature, and rectilinear ones are so far known only experimentally. They lie parallel with flow, and indicate soft mud bottoms, but are otherwise of uncertain geological significance.

SUMMARY

The transition from laminar to turbulent boundary-layer flow has been studied both mathematically and experimentally. Transition is marked by the development in space and time of an orderly sequence of deterministic flow configurations, which evolve from two-dimensional (Tollmien-Schlichting waves), to three-dimensional (hairpin vortices) to partly chaotic (random turbulence). The more advanced of these configurations may occur within fully developed turbulent boundary layers, and could be responsible for triggering the deformation of natural streambeds. What is now clear is that the wall region of fully turbulent boundary currents is typified by well-ordered streamwise flow configurations (periodically bursting streaks) which are capable of shaping deformable sedimentary surfaces. Parting lineation in sands and a variety of streamwise grooves in mud beds can all be convincingly attributed to streak action. These structures when fossilized are an indication of the path and turbulent nature of currents, but are not otherwise environmentally diagnostic.

Chapter 7

MODELS OF TRANSVERSE BEDFORMS IN UNIDIRECTIONAL FLOWS

INTRODUCTION

Once the threshold of motion is exceeded in the unidirectional flow of a viscous fluid over a loose granular boundary, particles become transported over the bed at a rate which is a steeply increasing function of the flow strength. The bed in the process remains plane under some conditions, but under others is shaped into transversely oriented wave-like features, for example, the current ripples studied by Blasius (1910) and Kindle (1917), the tide-shaped sand waves and the desert dunes which so attracted Cornish (1914), and the puzzling antidunes that Gilbert (1914) studied. These bedforms travel beneath the current, take part in the sediment transport and, as a result of their effectively episodic movement, give a characteristic imprint to the enclosed deposits. The features have been intensively studied for more than a century, because of their beauty of appearance and wide distribution, their relevance to human endeavours, and their utility in reconstructing past depositional environments and transport systems.

The purpose of this chapter and two subsequent ones is to give an account of these bedforms. Why should a granular boundary over which there is sediment transport be plane under some conditions but wavy under others? What determines the travel direction of bed features and what controls their ultimate shape and size? These are some of the questions that will be explored in the present chapter, from two main standpoints. There exist many physical models of bedforms. Most models are concerned with but one type of feature, or with a particular growth stage of a restricted group of forms. On the other hand, the mathematical approach initiated by Exner (1920, 1925) has led in the last two decades to theories of bedforms of considerable power and generality (A.J. Reynolds, 1976), though retaining important limitations. Whereas we can now often state what kind of bed feature will arise under given conditions, yet the ultimate attributes of the forms cannot generally be predicted, and reliance upon empiricism is still essential. Observations and correlations of transverse bedforms, however, are mainly relegated to Chapters 8 and 10, and no more than a short reminder is now needed.

CHIEF TRANSVERSE BEDFORMS

These are *current ripples* and *dunes,* formed by unidirectional water streams, *ballistic ripples, barkhan dunes* and *transverse dunes* fashioned by the wind, and *antidunes* generated by free-surface aqueous flows. Current ripples are restricted to the finer grades of quartz sand and the aeolian ripples to sediments of the granule and finer grades. Dunes shaped by flowing water occur more commonly in mineral sands than gravels.

Current ripples (Figs. 8-8, 8-9) are transverse ridges, much steeper on the downcurrent or lee side than on the upcurrent or stoss side, with a height of less than 0.04 m and a wavelength below 0.6 m (Allen, 1968c). Water-shaped dunes (Figs. 8-14, 8-15, 8-16) exceed these dimensions, and may reach many metres in height and several hundreds of metres in wavelength (Allen, 1968c). Ripples and dunes are sharp-crested and travel downstream, sediment being eroded from the stoss and redeposited in the lee. By contrast, antidunes migrate downstream only rarely and typically are either virtually stationary or upstream-travelling. They are bed waves (Figs. 10-16, 10-19) of low amplitude and sinusoidal streamwise profile, substantially in phase with somewhat steeper surface waves (Kennedy, 1961, 1963). Wavelengths in natural flows rarely exceed 10 m. Typically, antidunes travel by lee-side erosion and stoss-side deposition, the converse of what is true of ripples and dunes. Ballistic ripples (Figs. 8-4, 8-5) resemble current ripples, but commonly are somewhat flatter, reaching a wavelength of 10–20 m in the coarser textured sediments. Aeolian barkhan and transverse dunes (Figs. 8-10, 8-11) have gentle upwind but steep downwind slopes, attaining dimensions even greater than their water-shaped counterparts, which they resemble in mode of travel.

PHYSICAL MODELS OF TRANSVERSE BEDFORMS

Initiation of bed features

Concepts of bed-feature initiation on plane granular surfaces involve mainly the influence of boundary defects, either pre-existing ones or those directly induced by the flow.

Perhaps the oldest and most detailed observations relating to initiation pertain to current ripples. Bertololy (in Bucher, 1919) reported in 1900 that the first ripples formed about some of the larger grains scattered over the bed, and Sundborg (1956) took a similar view. USWES (1935), Rathburn and Guy (1967), and Southard and Dingler (1971) reported the growth of ripples from slight hollows or mounds on sand beds, either developed in some way by the flow or induced artificially. Inglis (1949) and Raudkivi (1963, 1966b) were also impressed by the role of local boundary defects, attributing these to intermittent and uneven grain transport by the turbulent current itself. This model was further elaborated by Grass (1970), Williams and Kemp (1971, 1972), and Costello (1974), who all linked the spatially and temporally varied transport to the action of boundary layer streaks

(Chapter 6). According to the Inglis-Raudkivi model, the initial local irregularity, whether pre-existing or current-induced, causes flow deceleration and a heightening of turbulence to lee and, if standing sufficiently proud of the surface, a more or less sustained flow separation. Hence the defect acts as a trap for near-bed grains swept over it. Once it exceeds a critical height, in the order of a few grain diameters according to Williams and Kemp, or the order of the thickness of the viscous sublayer (Etheridge and Kemp, 1979), scour is greatly enhanced where the separated flow reattaches to the bed downstream, with the result that the irregularity becomes even more effective a trap. However, because of a fall in turbulence intensity downstream from reattachment (Vol. II, Chapter 3), particles entrained at the scour are redeposited within but a short streamwise distance, and a second feature begins to grow downstream from the first. As USWES (1935), Antsyferov (1969), B.D. Taylor (1971), and Southard and Dingler (1971) show, a ripple-train of triangular ground plan (apex upstream) spreads laterally and downstream over the bed (Fig. 7-1). Cornish's (1897) and W.H.J. King's (1916) similar conception of ballistic ripple initiation cannot be supported observationally and is inconsistent with the inertia-dominated behaviour of wind-blown sand.

Although some aeolian dunes may arise at pre-existing irregularities (e.g. Cornish, 1897, 1914), Bagnold (1935, 1954b) has pointed out with experi-

Fig. 7-1. Banner-shaped area filled by current ripples formed in a flume downstream from an initial mound on planed sand bed. The viewer looks obliquely down on to the 0.17 m wide bed (ripples reflected in glass walls), with flow from the left. Photography by courtesy of J.B. Southard (see Southard and Dingler, 1971).

mental justification that a chance patch of sand on a hard or coarser textured surface can readily grow, because the grain transport rate is attenuated as the wind crosses the patch. W.H.J. King (1916), Gripp (1961b, 1961c), and Jäkel (1980) have vividly described the transformation of sand patches into barkhans apparently in this way. Thus an inherent tendency towards aggregation seems to exist wherever a little sand drifts over a contrasted surface. Cornish (1914) with great insight gave a different explanation of dune initiation, but similar to that now accepted for current ripples. Noting the wind's gustiness and turbulence, he proposed that "if we follow in imagination any particular particle or parcel in its onward course, its velocity will likewise vary greatly" with the result that "in one place the air will become suddenly surcharged with sand owing to loss of speed, and will drop the excess, forming a mound, and at the same moment in another place it will scour a hollow." These surface defects are thought gradually to grow into dunes by the action of the decelerated or actually separated flows to which they give rise. Some of Bagnold's chance sand patches might be formed as the result of the proposed action of the wind turbulence.

Movement of bed features

Ideally, transverse bedforms are features which stretch across the current without change of size or shape. The most fundamental way of viewing their movement is therefore in vertical streamwise profile.

Gilbert (1914), Kennedy (1961), Middleton (1965), and Hand (1974) show that antidunes typically travel upcurrent and are associated with nearly

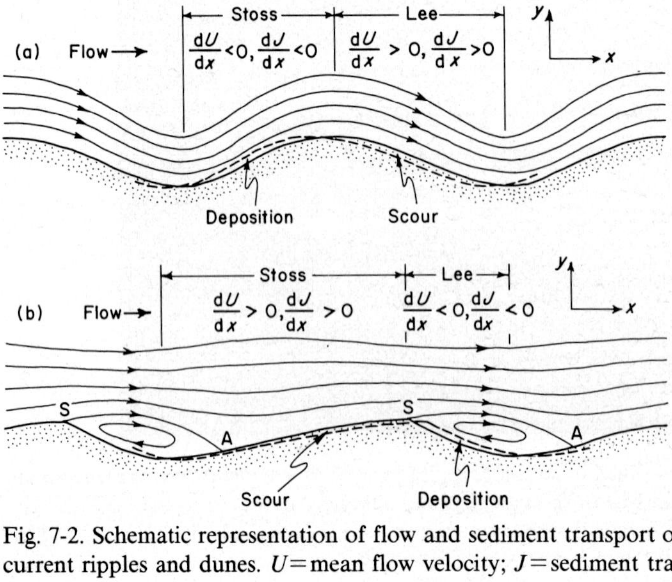

Fig. 7-2. Schematic representation of flow and sediment transport over (a) antidunes, and (b) current ripples and dunes. U=mean flow velocity; J=sediment transport rate.

in-phase surface waves of a greater height than the bed waves (Fig. 7-2a). Deposition on the stoss, from the decelerating flow pushing grains uphill, is matched by scour beneath the flow accelerating over the lee. Flow separation is generally no more than incipient or weak, occurring only when the waves steepen so as inevitably to break upstream. At most times streamlines can be traced smoothly over the bed waves.

A host of observers (e.g. Darwin, 1884; Mikhailova, 1952; Mikhailova and Naidenova, 1953; Bagnold, 1954b; Jopling, 1961, 1965, 1967; Allen, 1968c) have described the movement of current ripples and dunes in water and dunes beneath the wind (Fig. 7-2b). The recirculating quasi-steady lee eddy is generally several times longer than high, and the flow reattaches at a point somewhat higher up on the stoss than the base of the trough. There is sediment transport downstream from a reattachment point to the next crest, where the grains are projected forward into the eddy and, after sinking through it, settle on and ultimately avalanche down the lee. A lesser sediment transport occurs upstream from each reattachment point towards the lee.

Bed features and kinematic structures in natural currents

Because of a wide interest in undulatory phenomena generally during the late nineteenth and earlier twentieth centuries (see Cornish, 1914, 1934; Kaufmann, 1929), transverse bedforms came to be compared to water waves and to Kelvin-Helmholtz instability waves. Thence flowed the idea that transverse bed features (and also many of a streamwise orientation) somehow depend for their origin and scale upon the prior existence in natural currents of more or less regularly spaced kinematic structures, able to scour or build up a granular boundary, and so fashion bedforms. There are two chief developments of this idea, accordingly as attention is focussed upon quasi-steady vortices and/or waves in the current, or upon statistically ordered random turbulence.

The notion that transverse bedforms, particularly ballistic ripples and aeolian dunes, arise to match either waves or transverse, roller-like vortices in the current has been espoused by many workers (Baschin, 1899, Högbom, 1923; Dobrowolski, 1924; Bourcart, 1928; Matschinski, 1954, 1955; Clos-Arceduc, 1967a, 1969; I.G. Wilson, 1972b, 1972c). It is most elaborately and fancifully expressed by Folk (1976, 1977), in his general theory of bedforms. He believes that by increasing the rate of shear of a natural current, transverse rollers are transformed into hair-pin shaped eddies, which in turn change into streamwise vortices, and that this evolutionary sequence of kinematic structures is generated in three distinct ranges of shear, with a "chaos zone", represented by antidunes, between the ranges. Except in a limited way for certain streamwise bedforms (Vol. II, Chapter 1), these ideas have no empirical or theoretical support. For example, the antidunes of

Folk's chaos zone demand a free-surface flow of large Froude number falling in a rather narrow range, but the atmosphere has no free surface. Leeder (1977a) gives other criticisms of Folk's theory.

More persuasive as an explanation at least of subaqueous and aeolian dunes is the idea that the occurrence and equilibrium character of these forms depends on the action and scale of the larger eddies of turbulence in a current. This concept originated amongst Russian workers, notably M.A. Velikanov and N.A. Mikhailova (see Kondrat'yev, 1962; Yalin, 1972). A quantitative development was given by Yalin (1971, 1972), who argued that the larger eddies, having dimensions comparable with the flow or boundary layer thickness, were not only organized spatially along the flow at any instant, but also were correlated as to whether they scoured or built up the bed. He concluded that dune wavelength is approximately equal to $2\pi h$, where h is the boundary layer or flow thickness. The theory has some empirical support for dunes in water (Yalin, 1964; Hino, 1968; Pratt and Smith, 1972), though the numerical coefficient may be nearer 10 than 5 (Allen, 1977a) and it is not inconsistent with the scale of aeolian dunes. Unexplained, however, are: (1) how eddies convected at something like the flow speed become coupled to bed features of a celerity, even when of little height, several orders of magnitude smaller, and (2) why dunes have a scale much smaller than their equilibrium size when first formed on an artificially smoothed bed (Raichlen and Kennedy, 1965). Antsyferov (1969) found the Velikanov-Mikhailova theory inapplicable to current ripples, which apparently scale on grain size rather than flow thickness (Yalin, 1964, 1972). R.G. Jackson (1976c) recently revived the theory in a slightly different garb, claiming from the work reviewed in Chapter 6 that "bursting produces dunes". This cannot be true, for bursting accompanies turbulent flow over rippled and plane granular beds as well as over dune beds. Hence it is not the ultimate cause of dunes though, under the right set of conditions, bursting may be amongst the factors which determine their equilibrium scale.

Bed features and the lee-side eddy

Full separation of flow is rarely associated with antidunes, because of the relatively gentle bed-slopes involved and the slow streamwise slope changes, but may be expected at the sharp crests of ripples and dunes. Amongst the earliest to study bed features experimentally was Darwin (1884), who proposed that current ripples "are due to the vortex which exists in the lee of any inequality of the bottom; the dominant current carries sand up the weather slope and the vortex up the lee slope". Cornish (1897, 1900, 1901a, 1901b, 1902, 1908, 1914) seized on this discovery and vigorously advocated that all matured transverse ripples and dunes depended on the existence and role, as a combined sediment trap and erosional agent, of the separated flow to lee.

Cornish's theorizing prompted controversy, on the one hand as to whether or not the eddy existed and, on the other, as to its possible efficacy. W.H.J. King (1916) and Sharp (1963, 1966) rejected the postulated action of the eddy in the case of ballistic ripples, an opinion consistent with the behaviour of wind-driven sand (Bagnold, 1954b). Sharp (1963) and Khanna (1970) could find little evidence even for the existence of an eddy coupled to these forms. The eddy nonetheless vitally influences the shaping of current ripples (Allen, 1968c, 1969b, 1973a), for the grains and fluid are then of the same order of density. Many workers dismissed a lee eddy as irrelevant to the maintenance of wind-shaped dunes (Beadnell, 1910; W.H.J. King, 1916, 1918; Hume, 1925; Bagnold, 1937a, 1954b; Sharp, 1966; Cooper, 1958, 1967), and some in making their objection even denied that an eddy existed. The existence of the eddy as a quasi-steady structure seems unquestionable, however, in view of measurements of wind speed and direction and the observed movement of smoke and detritus over dunes (Sidwell and Tanner, 1938; F.A. Melton, 1940; Bagnold, 1954b; Volkov, 1957; Cooper, 1958; Verlaque, 1958; Coursin, 1964; Hoyt, 1966; Inman et al., 1966). Several investigators report evidence for slight sediment erosion and transport beneath the lee eddy of the aeolian forms (Sevenet, 1943; Hoyt, 1966; Sharp, 1966; Glennie, 1970). That much scour and transport is effected by the eddy coupled to dunes in water cannot be doubted (Jopling, 1961; Simons et al., 1961; Guy et al., 1966; Allen, 1968c), though R.G. Jackson (1976a) has questioned whether separation occurs.

The place of the lee-side eddy by itself in the maintenance of dunes and ripples has perhaps been falsely emphasized, for this vortex is but one element in the kinematic and dynamic situation (Fig. 7-2b) created when an internal flow separates at a boundary discontinuity (Vol. II, Chapter 3). The influence on sediment transport of the new boundary layer begun at each reattachment point could be equally important, and perhaps is the significant element in any coupling between flow and bed on the scale of dunes. J.D. Smith (1970) and Costello (1974) point out that, for a large enough region of separated flow, there exists in the reattached current downstream a point where the sediment transport is a maximum because the mean bed shear stress is a maximum. Hence there also exists a station where a new bed feature must inevitably lie.

Bed features and kinematic waves

Bagnold (1935) at first explained natural ballistic ripples in terms of the effect that saltating particles would have on the bigger grains creeping forward under the bombardment. He reasoned that these grains travelled at a speed proportional to their degree of exposure to the rain of sand. A grain just downwind from another received some shelter, and so travelled relatively slowly, trapping its neighbour following behind, whereas a grain which had no near neighbours was hurried along. He therefore argued that there were "among the bigger grains alternate traffic blocks and empty spaces", a state of instability that persisted until either the supply of large grains ran out or the ripple slopes became steep enough to prevent further creep. Although this explanation is not exhaustive, and was soon extended by Bagnold (1936, 1937a, 1954b) himself, such "blocks" and "empty spaces" do seem to occur amongst the bigger grains. Cornish (1897) observed during the early stages of sand rippling that the coarser fractions became grouped to give "the mottled appearance which precedes the formation of regular ridges".

What Bagnold here implies is that ballistic ripples can be treated in terms of kinematic waves, a class of waves whose main properties are described mathematically using a continuity equation and a velocity relationship of which only empirical knowledge is necessary (Exner, 1920, 1925; Polya, 1937; Lighthill and Whitham, 1955a, 1955b; Kluwick, 1977). Examples are the bunching of traffic along a road, flood waves in rivers, and clustered sand or gravel particles moving with currents. J. Müller (1969) explicitly treats ballistic ripples in terms of Lighthill and Whitham's theory. Langbein and Leopold (1968) in addition recognize current ripples, subaqueous dunes, and river gravel bars as kinematic waves, because these forms consist of moving debris which in time passes through the forms. Costello (1974) emphasizes that a kinematic wave need not have a physical wave form, for it is ideally a point moving with a certain velocity and carrying with it in space and time a constant value of some attribute. For example, consider a mound of grains driven by a current over an otherwise flat bed, and let the attribute in question be bed height above a parallel datum (Fig. 7-3a). Let x be distance in the flow direction and let t be time. The locus of any point y_1, y_2, y_3 etc. on the bed profile, starting at stations x_1, x_2, x_3 etc., can be

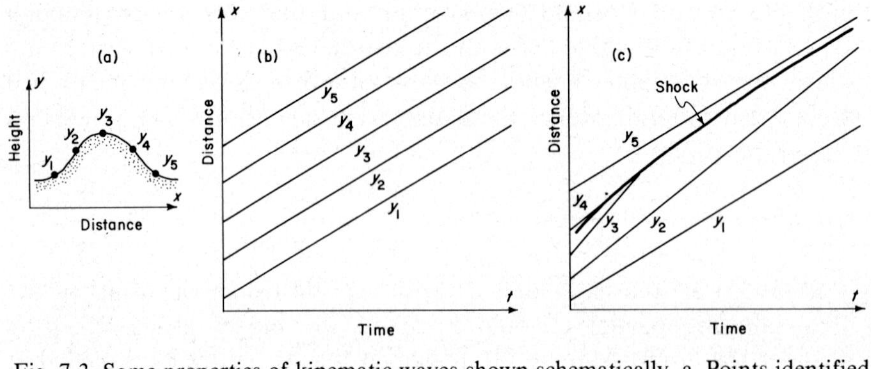

Fig. 7-3. Some properties of kinematic waves shown schematically. a. Points identified on the surface of a bed wave. b. Movement of the points when their velocity is uniform and independent of height. c. Differential movement of the points to form a shock when their velocity is an increasing function of height.

represented in the space–time plane by a characteristic curve, the slope of which is proportional to the velocity of the point. Suppose that the velocity of the points is independent of their height above the datum, i.e. is a constant. The characteristic curves are parallel, the kinematic waves remain unchanging in their properties, and the mound advances unaltered (Fig. 7-3b). Now let the velocity of points on the bed profile be an increasing function of height. The characteristic curves in this case diverge, so that a discontinuity in height arises after a certain stage in the progression of the mound (Fig. 7-3c). This discontinuity, expressed physically as a sharp crest towards the leeward side, is a kinematic shock wave with its own distinctive celerity. Costello (1974) regards the flatter dunes, which he calls bars, as examples of these waves, though he admits that current ripples and the steeper dunes may be similarly interpreted.

Kinematic shock waves tend to attenuate with increasing age. If current ripples or dunes are such shocks, individuals should ultimately die away, and their place be taken by similar but new forms. Birth and death processes seem to influence ripple and dune populations alike (Allen, 1976a), and the grounds on which Costello (1974) singles out his bars for special interpretation as shocks appear slender.

Bed features, boundary layers and instability

The possibility that transverse bed features are instability phenomena has existed as a general idea for many years and has been extended qualitatively along several lines. Mathematical developments of the notion are much more recent and require separate treatment.

Karcz (1970, 1974) pointed to the similarity between the kinematic structures which arise in sequence during the instability of laminar boundary layers (Chapter 6), and the coupled bed shapes and flow configurations associated with certain bed forms. The parallel is interesting but uncertain of meaning, as virtually all bedforms appear beneath fully turbulent and not transitional boundary layers. His work nonetheless points to the ultimate need to include three-dimensional effects in the mathematical stability analysis of bed features, and it may be directly relevant to liquid flow in films and thin sheets.

Kelvin-Helmholtz instability—that between superposed fluid layers in relative motion—has been intermittently invoked to explain transverse bed forms. Bucher (1919) thought that subaqueous dunes and antidunes expressed this instability, between the layer of sand-laden fluid moving over the bed and the clearer water above, and Von Kármán (1947, 1953) explained ballistic ripples similarly. Yalin (1972) postulated that current ripples recorded an unstable interaction between the flow and the granular bed beneath, which he regarded as a “plastic medium” behaving as a continuum.

Liu's (1957) analysis is not dissimilar. According to him, current ripples express the instability of the viscous sublayer, which is sandwiched between the granular bed, supposedly a fluid of infinitely large viscosity, and the outer, turbulent part of the boundary layer. Sundborg (1956) also assigned current ripples to an instability of the viscous sublayer, existing when the flow is hydraulically smooth, whereas dunes he considered to be influenced by the whole flow. Yalin (1964, 1972) and R.G. Jackson (1975, 1976c) have taken a similar view. These models present several difficulties. A granular bed is doubtfully a continuum at the low sediment transport rates which prevail at the threshold of motion when current ripples appear. Jackson's arguments link current ripples and dunes with processes in the wall and outer regions respectively of the turbulent boundary layer, but do not reveal how these processes determine the scale of the bed features. Although ripples are not found in rough flows and dunes are invariably lacking in hydraulically smooth ones (Yalin, 1972), there is a large transition region between these two regimes, in which the bedforms are either dunes or ripples, there being no corresponding transitional forms.

Bagnold (1936, 1937a, 1954b), extending his original interpretation of ballistic ripples, reasoned that the "traffic blocks" and "empty spaces" within the creeping load would interact with the saltating grains until the ripple wavelength and the length of the saltation path fell into a stable mutual adjustment. Although these two lengths are similar (Chepil, 1945a, 1945B; Bagnold, 1954b; Borszy, 1973), Sharp (1963) doubted that the salta-tion path controlled the wavelength, on the grounds that, when ripples first appeared on a smooth surface, their spacing increased with time (Cornish, 1897; W.H.J. King, 1916; Bagnold, 1936). He considered instead that the wind shear and the sediment coarseness mainly determined the wavelength. Folk (1976) roundly rejected Bagnold's interpretation, claiming, incorrectly, that if it were true "ripples of coarse sand should be closer spaced (because coarse grains have shorter saltation paths) than finer sand".

What Sharp ignored in discarding Bagnold's proposal was the influence of the imperfect sorting of natural sands. It is of course true that ripple wavelength for a given natural sand increases with the wind force (Chepil, 1945b; Bagnold, 1954b; Wilcoxson, 1962; Chiu, 1967; Borszy, 1973). It is also true that wavelength and grain size increase together under natural conditions (Cornish, 1914; Sharp, 1963; Stone and Summers, 1972; I.G. Wilson, 1972b, 1972c). The grain sizes reported, however, do not include the saltating particles, but refer only to the ripples. White and Schulz (1977) convincingly showed that the length of the characteristic saltation path is independent of particle size for each wind speed, provided that the saltating and bed grains are of the *same* size. Ellwood et al. (1975) demonstrated that for each value of the wind shear, the path length increases steeply as the bed grains—in practice the creeping load—become progressively coarser than the saltating particles. The observations that led Sharp (1963) to reject

Bagnold's concept therefore represent a non-linear effect, the influence on the saltation path during ripple growth of the progressive sorting consequent on the sand transport itself. As more and more sediment is worked over, an increasing textural differentiation should occur between the fine grained saltation load, consisting of the finer grains available, and the coarser creeping load. Referring to Ellwood et al. (1975), the length of the saltation path must inevitably increase for a given shear as the leaping and the creeping grains grow more disparate. But their differentiation is limited by the grading of the primary sand, and so a constant, equilibrium wavelength is ultimately reached.

Even more influential and widely quoted are Bagnold's (1956) criteria for the instability of a plane granular bed over which there is steady water-driven sediment transport. These derive from two contentions: (1) that the ultimate shear strength of a static granular bed is that of its topmost layer, (2) that there is a critical value of the fluid-applied stress above which the stress is wholly carried by encounters between bedload grains. At subcritical stresses, the fluid-applied stress is borne partly by the grain load and partly by stationary bed grains capable by themselves of withstanding the residual applied tangential stress not already opposed by the bedload resistance. This residual stress may be idealized as a thin statistical solid layer lying beneath the plane of the bed. It is defineable using the coefficient of static grain friction, whereas the coefficient of dynamic friction is involved in the prescription of the bedload resistance. But because the two coefficients generally differ, a deficit of shear resistance may arise, with the result that more grains are eroded than can be transported. But the deficit can be made up if these grains are redeposited as raised bed features able to offer an equivalent form drag.

It is convenient to state Bagnold's criteria in non-dimensional Shields-Bagnold form by introducing the unit stress $(\sigma-\rho)gD\cos\beta$, where σ and ρ are the sediment and fluid densities, respectively, g is the acceleration of gravity, D is the sediment particle diameter, and $\tan\beta$ is the bed slope (positive downward in the direction of flow). The bed is plane if:

$$\theta > C_0 \tan\phi \tag{7.1}$$

where θ is the non-dimensional mean boundary shear stress, C_0 is the static-bed fractional grain concentration, and $\tan\phi$ is the coefficient of static friction. The condition that a deficit of granular resistance just disappears is, from Bagnold's (1956) eq. 23:

$$\theta\left[1-\frac{\tan\phi}{\tan\alpha}-\frac{\theta_{cr}\tan\phi}{(C_0\tan\phi-\theta_{cr})\tan\alpha}-\frac{\theta_{cr}\tan\beta}{(C_0\tan\phi-\theta_{cr})\tan\alpha}+\frac{\theta_{cr}}{(C_0\tan\phi-\theta_{cr})}\right]$$

$$+\theta_{cr}\left[\frac{\tan\phi}{\tan\alpha}+\frac{\tan\beta}{\tan\phi}+\frac{\theta_{cr}\tan\phi}{(C_0\tan\phi-\theta_{cr})\tan\alpha}\right.$$

$$+\frac{\theta_{cr}\tan\beta}{(C_0\tan\phi-\theta_{cr})\tan\phi}-\frac{\theta_{cr}}{(C_0\tan\phi-\theta_{cr})}-1\Bigg]=0 \qquad (7.2)$$

in which $\tan\alpha$ is the coefficient of dynamic friction, and wherein it is assumed that the residual stress has an inverse linear relationship with θ between the threshold stress, θ_{cr}, when $\theta=\theta_{cr}$, and zero, when $\theta=C_0\tan\phi$, as dictated by Bagnold's second contention.

Figure 7-4 shows these criteria in the stress-grain size plane, and also an empirical curve for the threshold of motion of mineral-density solids in water. Equation (7.1) defines a field of large stresses and plane beds, and another, of smaller stresses, in which raised features—current ripples according to Bagnold (1956)—will exist given a deficit of shear resistance. The concentration C_0 is somewhat variable, depending on depositional conditions, but a value of 0.65 is fairly typical (Allen, 1970c). Allen and Leeder (1980) claim that ϕ is identifiable with the angle of initial yield (Bagnold, 1967; Allen, 1970c), giving a coefficient of static friction of approximately 0.84 based on laboratory measurements on natural sands (Allen, 1970c). Hence the critical value of θ afforded by eq. (7.1) is approximately 0.55.

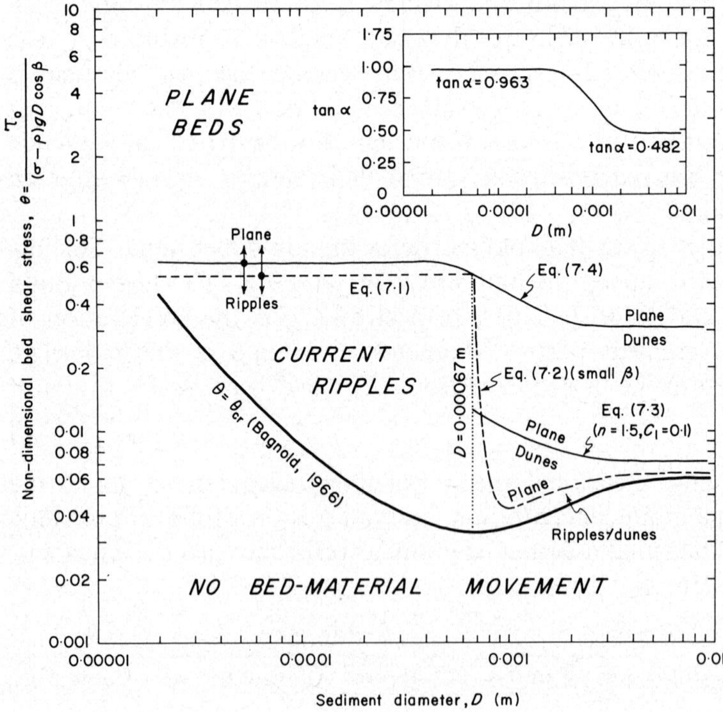

Fig. 7-4. Criteria for the instability of plane granular beds, according to Bagnold (1956, 1966) and H.M. Hill (1966). The inset graph is a suggested relationship between $\tan\alpha$ and particle diameter for naturally occurring grains, based on the proposals of Allen and Leeder (1980).

Bagnold (1956, 1966, 1973) consistently chose an implausibly low value for $\tan\phi$, obtaining a critical θ of approximately 0.4. Now there is a certain value of D below which $\tan\alpha > \tan\phi$, and above which $\tan\alpha < \tan\phi$ (Bagnold, 1954a, 1966). Since in uniform, steady free-surface flow, $\tan\beta$ is always positive, it follows from eq. (7.2) that a deficit of shear resistance always exists for D smaller than this value when eq. (7.1) is simultaneously not satisfied. Otherwise with eq. (7.1) not satisfied, the bedform is determined by $\tan\beta$. Hence for each slope eq. (7.2) defines a curve separating rippled from plane beds. The low slopes encountered in experimental and natural channels yield a curve which rapidly approaches extremely close to the threshold condition as D increases, so that its steep part defines a practical upper limit of D for ripple occurrence. Bagnold (1956) estimated this limit to be approximately 0.00067 m, for which there is considerable empirical support (Inglis, 1949; Chabert and Chauvin, 1963; Maggiolo and Borghi, 1965; Guy et al., 1966; G.P. Williams, 1967, 1970; Williams and Kemp, 1971; Southard and Boguchwal, 1973; Costello, 1974), though some dissent (Sahgal and Singh, 1974). The agreement may be fortuitous, however, as his estimate depends on values of $\tan\alpha$ measured from smooth waxy spheres (Bagnold, 1954a). Natural sands should give larger values of $\tan\alpha$ (Allen and Leeder, 1980), but as $\tan\phi$ is also greater for these sands than for the spheres, Bagnold's estimate may not be much in error. Leeder (1980) suggested that plane beds of coarse sand do not develop into ripple-like features because flow separation at small defects on such beds is inhibited by the strong vertical turbulent mixing promoted by the roughness due to the relative large grains. Another possibility is that, on account of the large particle sizes, flow *through* the bed prevents separation at all but very large (compared to the grains) defects.

One theoretical difficulty with Bagnold's criteria concerns his second contention, which Leeder (1977b) questioned from a somewhat indirect analysis of the experiments of G.P. Williams (1970) on plane-bed transport at $0.30 \leqslant \theta \leqslant 1.93$. Even at these high flow stages, Leeder calculated, the bedload resistance was only about one-half of the total resisting force, rather than the whole of it as Bagnold presumed. Luque and Van Beek (1976) drew the opposite conclusion, estimating experimentally that the fluid contribution to the total resistance disappeared when $\theta \approx 0.12$. They showed that grain wakes (added mass) contributed to the occlusion of the bed by the load in transport, and that complete occlusion could occur at quite low bedload concentrations, in the order of those estimated by Leeder (1977b).

A second difficulty is the meaning of eq. (7.1) at supercritical grain sizes, for eq. (7.2) also limits a field of plane beds. However, Hill (1966) has given a reason why a plane bed at high flow stages could become unstable as θ falls and $\tan\alpha < \tan\phi$. He recognized that the bedload transport system could fail if the bedload layer became sufficiently reduced that particle size began to limit its thickness. If a sufficient thinning occurred as the stress was lowered, the load to be carried would ultimately exceed the applied stress, but could be brought into balance by a local downward tilting of the bed. Raised features which he identified as dunes should appear on the bed when:

$$\theta > nC_1 \tan \alpha \tag{7.3}$$

is no longer satisfied, in which n is the number of layers of grains derived from the static bed into the bedload, and C_1 is the mean fractional bedload concentration. Leeder's (1977b) estimates of n and C_1 suggest that Hill's criterion would define a boundary (Fig. 7-4) well below that given by eq. (7.1). The condition separating Bagnold's plane bed from Hill's dunes, both at large grain sizes, is as yet unknown.

Lately Bagnold (1966, 1973) has advanced what may be called a universal criterion for the persistence of a plane bed:

$$\theta = C_0 \tan \alpha \tag{7.4}$$

The reasons for this revision are not made clear, but it appears that he feels that $\tan \alpha$ approximates to $\tan \phi$, which he regards as essentially constant for natural sediments. This condition is also plotted in Fig. 7-4 for $C_0 = 0.65$ and $\tan \alpha$ adjusted upward to account for the likely influence of the non-spherical and rough-textured character of natural sand particles (Allen and Leeder, 1980).

Lags between property variations

In the steady uniform transport of sediment over a plane grain bed, no downstream variation should occur in the local values of the flow depth and velocity, the turbulence characteristics, and the rate of transport of sediment, whether as bed or suspended load. But if regularly arranged bed features are present, these properties should each exhibit a regular downstream perturbation, on the same wavelength as the bed elevation but not necessarily on the same or even a common phase.

Perhaps the most exciting of recent developments in bedform theory is the recognition that these spatial lags or phase differences are critical to the stability of current-swept erodible beds. Bagnold (1937a, 1954b, 1956) was perhaps the first to give empirical evidence for a spatial lag between the sediment transport rate and a governing flow property, and to discuss its implications for bedform development. In mathematical studies of bed stability, spatial lags first appeared as arbitrary devices affording escape from constraints implicit in particular flow models, for example, the neutral stability of a bed perturbation under an inviscid treatment (Cartwright, 1959; Kennedy, 1963, 1964), or the damping ordained by a simple hydraulic approach (Exner, 1920, 1925; A.J. Reynolds, 1965). Lags as free parameters are rightly suspect, but ought not to be rejected out of hand (Costello, 1974), as they can be justified physically and experimentally (A.J. Reynolds, 1965,

1976; Engelund and Hansen, 1966; Kennedy, 1969). G. Parker (1975) has urged that lags should be implicit in the equations describing sediment transport mechanics, but this is a belated plea in view of the initiatives of Falcon (1969) and Engelund (1970). Most modern theories of bedforms implicitly include lag.

The total lag effect is built up from a possible maximum of three contributions: (1) in free-surface flow, a difference of phase between surface and bed waves, which controls the phases of flow depth and mean velocity, (2) a phase difference between the variation of mean bed shear stress and the bed wave, which may be influenced by the shift between the velocity and bed waves, and (3) differences of phase between the bedload and suspended-load transport rates and governing flow properties, themselves unlikely to be in phase with the bed wave. This system is further complicated by the fact that the two radically different transport modes generally occur together, at least in real environments.

Frictionless free-surface flows over a slightly wavy bed may be divided (Lamb, 1932) between three classes (Fig. 7-5), using the Froude number written in terms of the non-dimensional wave number, kh, where h is the mean flow depth and $k = 2\pi/L$ is the wave number, L being bedform wavelength. Flows of class I present surface waves of smaller amplitude than the bed waves and exactly π rad out of phase with the bed features. The velocity wave therefore exactly coincides with the bed wave, whereas the depth wave is exactly π rad out of phase. Low to moderate Froude numbers typify these flows. Class II flows, associated with Froude numbers in the order of unity, have surface and depth waves exactly in phase with the bed waves, but a velocity wave exactly π rad out of phase. The surface wave now exceeds the bed wave in amplitude. Class III flows, representative of large Froude numbers, present bed, surface, and velocity waves in phase, but a depth wave π rad out of phase with the bed. As in class I flows, the surface

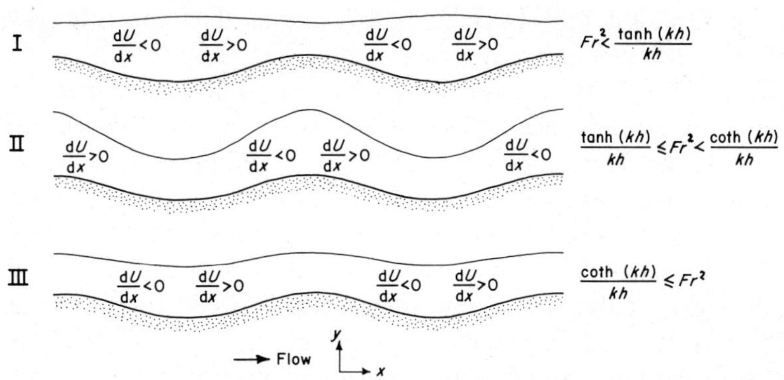

Fig. 7-5. Classification of inviscid free-surface flows over a wavy bed according to Lamb (1932).

wave has the lesser amplitude. Kennedy (1963) associated his dunes with class I flows and antidunes with classes II and III. Real fluids flowing over wavy beds, however, yield phase relationships commonly departing from these strict patterns (Kennedy, 1961; Simons et al., 1961; Raudkivi, 1963, 1966a; Engelund and Hansen, 1966; Robillard and Kennedy, 1967; Yuen and Kennedy, 1971). Theoretically, the introduction of friction in even a simple way (e.g. Henderson, 1964; A.J. Reynolds, 1965; Engelund and Hansen, 1966; Raudkivi, 1966a) permits substantial deviations from the inviscid phase relationships, thus offering some justification for the experimental results. Realistic theoretical treatments of real-fluid effects yield an extremely complex picture of possible phase relationships (Iwasa and Kennedy, 1968), which may never be fully exploited in bedform theory.

The mean boundary shear stress exerted by a deep turbulent current flowing over a wavy bed is theoretically out of phase with the bed wave by an amount only a little short of 2π rad. According to Benjamin (1959), the stress maximum occurs on the upstream side of the bed wave, at a distance of $L/12$–$L/6$ from the wave crest. Broadly similar conclusions were drawn by Engelund and Hansen (1966), Townsend (1972), Gent and Taylor (1976), Taylor et al. (1976), Taylor (1977), Taylor and Dyer (1977), Zilker et al. (1977), and Bordner (1978). Experiments by J.M. Kendall (1970) and by Hsu and Kennedy (1971) support these general results. The latter found that the maximum stress lay $0.050L$ upstream from the crest on a wavy bed of length/height ratio equal to 45, and $0.072L$ upstream when the bed was steepened to a ratio of 22.5. The shear-stress maximum was shifted but little upstream from the wave crests in the class I-type (Fig. 7-5) free-surface flows studied by Yuen and Kennedy (1971). Work with flows of a type parallel with class II gave a maximum in bed shear in the bed-wave troughs, that is, at a spatial lag of about $L/2$ with respect to the bed waves. In their class III-type flows, for which the depth and bed waves are broadly π rad out of phase, the maximum lay about $L/5$ upstream from the crests.

The streamwise intensity of turbulence should also vary out of phase with the bed wave, for Graham and Deissler (1967) argue that the intensity decreases with flow acceleration but increases with deceleration. In motions of classes I and III (Fig. 7-5), as represented by real fluids, the intensity should therefore decrease over the upstream slopes of the bed waves but increase over their downstream sides. Class II flows should show an increase over each stoss but a decrease over the lee. Limited experimental evidence confirms these suggestions for motions of class I, recognizing that these include flows in conduits with a wavy wall. Hsu and Kennedy (1971) and Khanna (1970) found that the streamwise turbulence intensity increased in the order of 50% from the bed-wave crest to its trough, the turbulence phase difference relative to the bed and velocity waves being approximately π rad. The other classes are unrepresented experimentally.

These effects are assigned to a single category because the lag between each flow property and the bed wave depends mainly on the scale of the bed wave itself. Another class is found on turning to lag associated with sediment transport, where the effect is controlled chiefly by the properties of the sediment and the basic flow, and not by the bed-wave scale.

Under uniform steady conditions, the bedload transport rate is a steeply increasing function of the stream power (Chapter 2) and, in the present non-uniform flows, may therefore be expected closely to follow the shear stress and flow velocity perturbations, the latter fixed by the bed waves in conduit flow and by the bed and surface wave phase-relationship in open-channels. But the transport rate should lag the changing stream power, since the grains of the load travel in discrete steps or saltations (Bagnold, 1954b; A.J. Reynolds, 1965; Engelund and Hansen, 1966). Hence the rate measured at a distance x along the stream will be that set by the stream power at a distance upstream in the order of $(x-d)$, where d is a characteristic grain path-length controlled by fluid and sediment conditions (Fig. 7-6a). The magnitude of d is therefore a measure of the spatial lag associated with the bedload transport. At flow stages but little above the threshold of motion in water, grains travel in a rolling mode (Francis, 1973; Abbott and Francis, 1977), in which true rolling (rotating motion involving continuous contact with granular bed) alternates with flat jumps no more than a few grain diameters long. For higher stages, Tsuchiya's (1969a, 1969b) calculations and experiments reveal that saltations of $10D-100D$ are likely before the onset of suspension, and the work of Francis (1973) and of Abott and Francis (1977) supports this general conclusion. Considerably longer saltations are possible in air. Leaps of 10^2-10^3D seem typical of equally coarse saltating and creeping grains. As the saltating load becomes finer relative to the creep, path lengths may rise to the general order of 10^5D based on the finer particles (Ellwood et al., 1975).

At sufficiently high flow stages, grains are transported in suspension, at a rate increasing steeply with flow velocity and turbulence intensity (Chapter 2). Unlike the bedload, however, the suspended load is spread throughout the whole flow, having a "centre of gravity" above the bed at a sizeable fraction

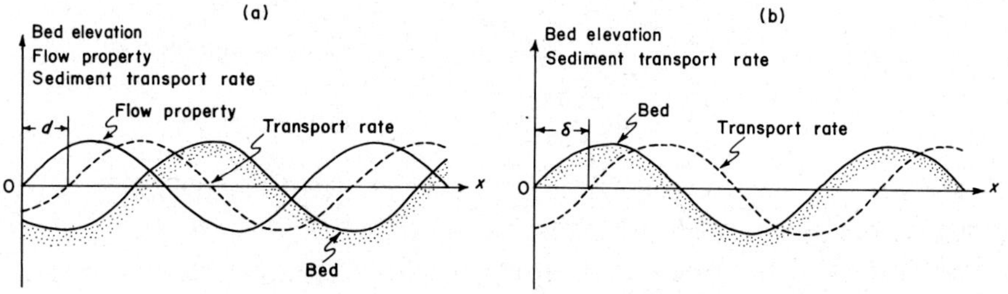

Fig. 7-6. Schematic representation of sediment transport lag over a wavy bed (a) with respect to a controlling flow property, and (b) with respect to the bed waviness itself.

of the flow depth. Therefore in a non-uniform flow, the magnitude of the suspended-load transport rate should lag the perturbations of flow properties by the streamwise distance necessary for the centre of gravity of the load to readjust to the changed flow. This length, akin to Kennedy's (1963, 1969) "transport relaxation distance" is plausibly in the general order of the flow depth itself, since the load lies so high above the bed and the mean flow velocity may be 1–3 orders of magnitude greater than the falling velocity of the suspended grains.

The bed and suspended loads travelling over a wavy bed therefore seem likely to be associated with lag distances of different but overlapping scales. The bedload spatial lag will probably influence most strongly the growth or damping of short-wavelength bed perturbations. Lag in the suspended-load transport is likely to be influential only when bed disturbances are of long wavelength.

Neglecting for a time the question of which transport mode may be operative, how will a general transport lag affect bed-wave stability? We may now conveniently measure the lag as a phase difference, δ, between the transport and bed waves (Fig. 7-6b). Table 7-I presents the four recognizable classes of behaviour, defined using the bed-wave celerity, c, and the time-rate of amplitude growth, where a is amplitude and t is time. A physical interpretation is given for free-surface aqueous flows, the case of greatest generality. If the transport maximum and wave crests exactly coincide

TABLE 7-I

Stability and response of a wavy granular bed to the sediment transport over it

Class	Phase difference between bed and sediment transport	Bed feature celerity	Rate of change of amplitude of bed feature	Physical interpretation
	$\delta=0$	$\frac{d^2c}{dt^2}=0$	$\frac{da}{dt}=0$	neutrally stable
A	$0<\delta<\pi/2$	>0	<0	damping ripples or dunes
	$\delta=\pi/2$	$c=0$		
B	$\pi/2<\delta<\pi$	<0	<0	damping antidunes
	$\delta=\pi$	$\frac{d^2c}{dt^2}=0$	$\frac{da}{dt}=0$	neutrally stable
C	$\pi<\delta<3\pi/2$	<0	>0	augmenting antidunes
	$\delta=3\pi/4$	$c=0$		
D	$3\pi/2<\delta<2\pi$	>0	>0	augmenting ripples or dunes
	$\delta=2\pi$	$\frac{d^2c}{dt^2}=0$	$\frac{da}{dt}=0$	neutrally stable

TABLE 7-II

Stability and response of a wavy granular bed to the sediment transport over it, where bedload and transport are separately considered

Phase of suspended load	Bedload phase			
	$0<\delta_b<\pi/2$	$\pi/2<\delta_b<\pi$	$\pi<\delta_b<3\pi/2$	$3\pi/2<\delta_b<2\pi$
$0<\delta_s<\pi/2$	I $\left(\frac{da}{dt}\right)_b<0$, $\left(\frac{da}{dt}\right)_s<0$ $c_b>0$, $c_s>0$ reinforcing; damping ripples or dunes	II $\left(\frac{da}{dt}\right)_b<0$, $\left(\frac{da}{dt}\right)_s<0$ $c_b<0$, $c_s>0$ reinforcing; damping stationary features	III $\left(\frac{da}{dt}\right)_b>0$, $\left(\frac{da}{dt}\right)_s<0$ $c_b<0$, $c_s>0$ neutralizing, stationary bed features	IV $\left(\frac{da}{dt}\right)_b>0$, $\left(\frac{da}{dt}\right)_s<0$ $c_b>0$, $c_s>0$ neutralizing; ripples or dunes
$\pi/2<\delta_s<\pi$	II $\left(\frac{da}{dt}\right)_b<0$, $\left(\frac{da}{dt}\right)_s<0$ $c_b>0$, $c_s<0$ reinforcing; damping stationary features	VI $\left(\frac{da}{dt}\right)_b<0$, $\left(\frac{da}{dt}\right)_s<0$ $c_b<0$, $c_s<0$ reinforcing; damping antidunes	VII $\left(\frac{da}{dt}\right)_b>0$, $\left(\frac{da}{dt}\right)_s<0$ $c_b<0$, $c_s<0$ neutralizing, antidunes	III $\left(\frac{da}{dt}\right)_b>0$, $\left(\frac{da}{dt}\right)_s<0$ $c_b>0$, $c_s<0$ neutralizing; stationary bed features
$\pi<\delta_s<3\pi/2$	III $\left(\frac{da}{dt}\right)_b<0$, $\left(\frac{da}{dt}\right)_s>0$ $c_b>0$, $c_s<0$ neutralizing; stationary bed features	VII $\left(\frac{da}{dt}\right)_b<0$, $\left(\frac{da}{dt}\right)_s>0$ $c_b<0$, $c_s<0$ neutralizing; antidunes	VIII $\left(\frac{da}{dt}\right)_b>0$, $\left(\frac{da}{dt}\right)_s>0$ $c_b<0$, $c_s<0$ reinforcing; augmenting antidunes	V $\left(\frac{da}{dt}\right)_b>0$, $\left(\frac{da}{dt}\right)_s>0$ $c_b>0$, $c_s<0$ reinforcing; augmenting stationary features
$3\pi/2<\delta_s<2\pi$	IV $\left(\frac{da}{dt}\right)_b<0$, $\left(\frac{da}{dt}\right)_s>0$ $c_b>0$, $c_s>0$ neutralizing; ripples or dunes	III $\left(\frac{da}{dt}\right)_b<0$, $\left(\frac{da}{dt}\right)_s>0$ $c_b<0$, $c_s>0$ neutralizing; stationary bed features	V $\left(\frac{da}{dt}\right)_b>0$, $\left(\frac{da}{dt}\right)_s>0$ $c_b<0$, $c_s<0$ reinforcing; augmenting stationary features	IX $\left(\frac{da}{dt}\right)_b>0$, $\left(\frac{da}{dt}\right)_s>0$ $c_b>0$, $c_s>0$ reinforcing; augmenting ripples or dunes

($\delta = 0, 2\pi$ rad), ripples or dunes advance downstream without change of form, a case of neutral stability. This is implicit in Bagnold's interpretation of ballistic ripples at equilibrium, the lag distance, or saltation path, being exactly one wavelength. Neutrally stable antidunes can exist when the phase lag is π rad, the transport maximum occurring in the wave trough. Other values of the phase lag cause either damping or augmentation of bed waves. Case D is of interest for ripple and dune growth. If the transport maximum occurs on the upstream side of a bed disturbance, and not too far from the crest, then the disturbance can grow.

Simultaneously occurring bed and suspended modes of transport probably do not afford identical phase lags. What may happen when the lags differ appears in Table 7-II, the subscripts b and s distinguishing respectively bedload and suspended-load quantities. Nine classes of behaviour are recognized, accordingly as the lags reinforce each other, either augmenting or damping an existing bed wave, or form a neutralizing combination. In the augmenting and damping categories, the disturbances must either travel downstream (ripples/dunes) or upstream (antidunes). A neutralizing combination can yield either (1) a plane bed when the one lag effect exactly offsets the other, or (2) either upstream or downstream-travelling waves if one transport mode dominates. Thus more effects seems possible when transport occurs on the two modes simultaneously than with either mode alone. A given transport mode can be stabilizing under one set of conditions but have a destabilizing influence under another. Engelund and Fredsøe (1974) are right to emphasize that the correct specification of the sediment transport laws is vital to the success of mathematical studies of bed stability.

MATHEMATICAL MODELS OF ERODIBLE BED STABILITY: THE TWO-DIMENSIONAL CASE

General requirements

It is clear from the above that a mathematical analysis of the stability of an erodible bed must seek to establish the fate of prescribed bed perturbations in the presence of a certain basic fluid flow, giving rise, with the bed sediment prescribed, to a certain basic sediment transport according to bed and/or suspended modes. The problem can be solved by the simultaneous satisfaction of universal balance relationships for fluid and sediment and appropriate boundary conditions, and by the specification of constitutive relations linking sediment transport to flow properties. Restriction to the two-dimensional case offers an immediate simplification. Maximum generality is obtained by considering bedforms in a free-surface flow.

The fluid flow may be treated as either ideal (inviscid) or real (viscous), but in every case the continuity equation connecting discharge, flow depth, and flow velocity is needed. In the simplest inviscid approach, the flow is treated as irrotational and is described using a velocity potential, on the general lines indicated by Milne-Thompson (1955). For example, the velocity potential, ϕ, representing a steady sinusoidal perturbation on a uniform stream of overall mean velocity U and satisfying dynamical conditions at the free surface is:

$$\phi = Ux + UA\left\{e^{ky} + \left(\frac{1-kU^2/g}{1+kU^2/g}\right)e^{-ky}\right\}\cos(kx) \tag{7.5}$$

$$u = \frac{\partial\phi}{\partial x}, \qquad v = \frac{\partial\phi}{\partial y} \tag{7.6}$$

in which x is distance in the streamwise direction, y is normal distance from the undisturbed free surface (positive upward), A is a real constant, "e" is the base of natural logarithms, g is the acceleration due to gravity, k is the wave number, and u and v are the local horizontal and vertical velocity components. A.J. Reynolds (1965) has preferred this equation to that earlier adopted by Kennedy (1963). The other inviscid approach treats the flow as rotational, and is widely used by Engelund's (1970) school. Starting from the vorticity transport equation, the flow pattern is described using a stream function, ψ, defined by:

$$u = -\frac{\partial\psi}{\partial y}, \qquad v = \frac{\partial\psi}{\partial x} \tag{7.7}$$

an approach immediately restricting the analysis to two dimensions. Exner (1920, 1925) initiated the third or hydraulic approach, which explicitly includes real-fluid effects. A simple example is A.J. Reynold's (1965) description of quasi-steady fluid motion over a wavy bed using the equation:

$$\frac{\partial u}{\partial x} = \frac{u}{h(1-Fr^2)}\frac{\partial y}{\partial x} + \frac{C_D}{g}\frac{u^3}{h^2(1-Fr^2)} \tag{7.8}$$

in which u is the local mean flow velocity, h is the local flow depth, y is the elevation of the bed above a datum, C_D is a bed friction coefficient, and $Fr = u/(gh)^{1/2}$ is the Froude number. The motion is called quasi-steady because the bed-wave celerity is ignored as very small compared with the flow speed.

The final balance relationship needed to solve the stability problem is a "continuity" equation relating the change of bed elevation, y, to the change along the stream of the sediment transport rate. Putting J as the total local dry-mass transport rate, the complete equation is:

$$\frac{\partial J}{\partial x} = -\sigma C_0\frac{\partial y}{\partial t} - \frac{\partial}{\partial t}\left(\sigma\int_y^{y+h} C_s \mathrm{d}y\right) \tag{7.9}$$

in which x is distance in the flow direction, σ is the sediment density, C_0 is the fractional volume concentration of sediment in the bed, C_s is the concentration in the suspended load, and h is the local flow depth. The second term on the right denotes the sediment stored in the suspension. It can generally be ignored, except when C_0 is very large, as the flow velocity always greatly exceeds the bed-wave celerity. Equation (7.9) less the storage term was developed by Exner (1920) and Polya (1937).

The constitutive relations seem more crucial than any in the analysis of bed stability but are the hardest to specify. Sediment transport-mechanics under steady uniform conditions are not yet well understood, let alone under the unsteady and non-uniform conditions associated with wavy beds. It is therefore generally considered for want of a better basis that transport over a bed perturbation can be described using the laws of uniform transport (Chapter 2). In potential flow analyses, it is usually assumed for consistency that the transport rate is a function of the flow velocity. Kennedy (1963) at first took the rate to be an arbitrary power of velocity, but Hayashi (1970) chose a specific power relationship, introducing a term to account for the influence of bed slope on the rate. Later Kennedy (1964, 1969) took the rate to be an arbitrary power of $(U - U_{cr})$, where U_{cr} is the velocity at the threshold of motion. In rotational and hydraulic models, extensive use has been made of the Meyer-Peter and Müller bedload equation, in which the rate is proportional to $(\theta - \theta_{cr})^{3/2}$ (e.g. Engelund, 1970; Furnes, 1976a, b). A gravity term allowing for variations in bed slope was recently introduced into this formula, with considerable effect on the stability analysis (Engelund and Fredsøe, 1974; Fredsøe, 1974a, 1974b). Suspension transport is dealt with in various ways. Falcon (1969) took the rate as proportional to the flow velocity, whereas Engelund and Fredsøe (1970) used a dependence on the square of the mean bed shear stress. Later, Engelund and Fredsøe (1974) adopted a semi-empirical method of describing the suspended-load transport. Use has been made of other transport laws (e.g. J.D. Smith, 1970; G. Parker, 1975), and sometimes of arbitrary functions (A.J. Reynolds, 1965; Gradowczyk, 1968; Callander, 1969).

Potential flow models

A.G. Anderson (1953) was probably the first to study bedforms in water using a potential flow model. He obtained an equation for their wavelength, corrected by Mercer (1971a), but did not define the existence of the different kinds. Kennedy (1963) explored the potential-flow model in detail, using a sediment transport law containing his arbitrary lag distance, appearing also in the form $j = d/h$, where d is the lag distance and h is the flow depth. He produced a set of bedform existence fields (Fig. 7-7) and obtained the upper limit for the occurrence of two-dimensional bed waves as:

$$Fr^2 = \frac{1}{(kh)} \tag{7.10}$$

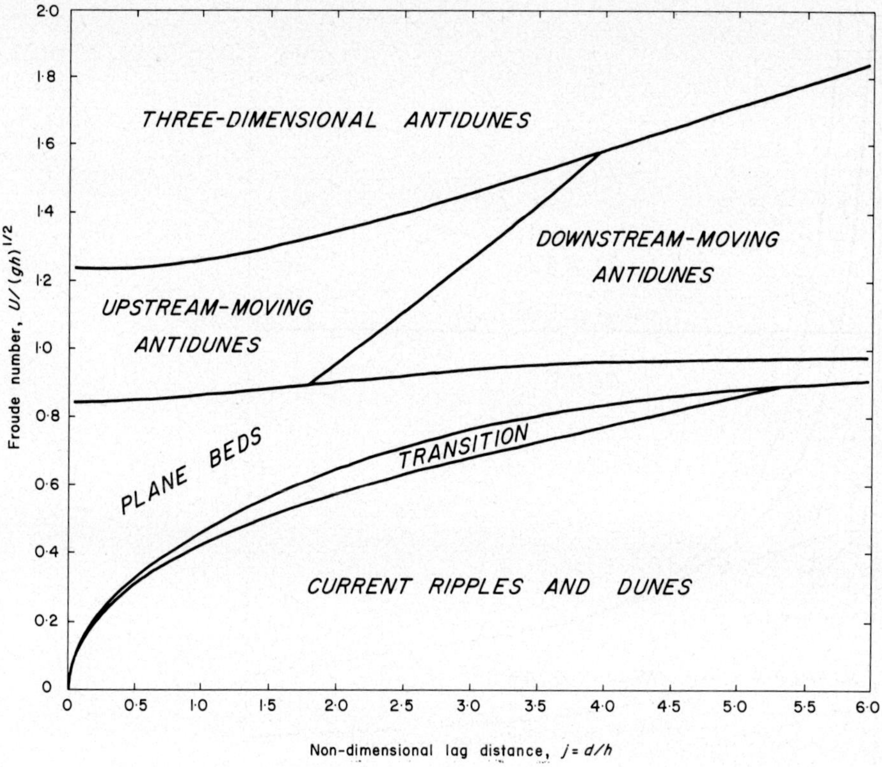

Fig. 7-7. Criteria for the existence of bedforms, in terms of Froude number and the non-dimensional sediment transport lag distance. After Kennedy (1963).

The condition separating ripples and dunes from antidunes was found to be:

$$Fr^2 = \frac{\tanh(kh)}{(kh)} \tag{7.11}$$

Kennedy (1964) extended the model to bedforms in closed conduits and in the atmospheric boundary layer, flows referable to class I of Fig. 7-5. A.J. Reynolds (1965) corrected weaknesses in Kennedy's analysis, deriving the relation:

$$Fr^2 = \frac{\coth(kh)}{(kh)} \tag{7.12}$$

for the upper limit of bed-wave occurrence. Tsuchiya and Ishizaki (1967) succeeded like Reynolds in obtaining eq. (7.11) as the condition separating dunes and ripples from antidunes, but deduced an equation for ripple-dune wavelength different from the Anderson-Mercer relation (Mercer, 1971a). Fig. 7-8 gives several of these equations. Kennedy (1969) subsequently elaborated on his earlier analysis, incorporating Reynold's suggestions. Hayashi (1970) extended Kennedy's model to include the influence of local

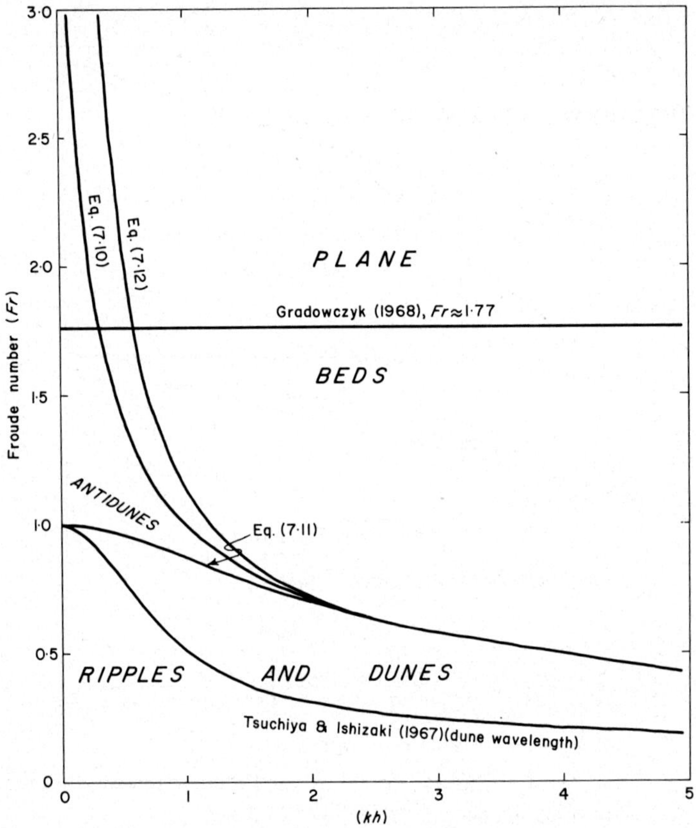

Fig. 7-8. Criteria for the stability of a granular bed according to Kennedy (1963), A.J. Reynolds (1965), and Gradowczyk (1968), in terms of Froude number and non-dimensional bedform wavelength. Also shown is Tsuchiya and Ishizaki's (1967) eq. (17) for dune wavelength.

bed slope on the sediment transport rate. He derived eq. (7.12) of Reynolds when the slope effect was set at zero, and also obtained eq. (7.11). When the bed slope was included, the upper limit of bed-wave occurrence was shifted to higher Froude numbers than given by eq. (7.12).

A major innovation was introduced by Falcon (1969), who obtained an implicit phase shift in the suspended-load transport by coupling a potential flow model with a diffusion equation for the sediment. The lag distance emerged as a function of the bedform wavelength, but no stability limits were graphed. A related but this time successful analysis restricted to suspended-load transport was made by Engelund and Fredsøe (1970). Equations (7.11) and (7.12) emerged as stability boundaries for two-dimensional flows, but on account of the stabilizing effect of the large lag distance associated with this mode of transport, the only bed waves found were antidunes. Unfortunately, antidunes are not restricted to flows with suspended load only, nor is suspension associated only with antidunes. Gradowczyk (1971) derived these equations as limits to ripples/dunes and to antidunes, as the result of a generalized potential-flow analysis. G. Parker (1975) also obtained an implicit lag effect by considering the inertia of the transported sediment, but found that antidunes were the only bed waves that could exist, the boundary being otherwise plane. Equation (7.11) yielded their lower limit of occurrence; the upper bound for instability induced by sediment inertia was obtained as:

$$Fr^2 = \frac{1}{(kh)\tanh(kh)} \tag{7.13}$$

somewhat reminiscent of Kennedy's (1963) eq. (7.10). Parker reconciled a dynamical description of the sediment transport with his choice of the potential flow model by treating frictional effects as limited to a thin region of the fluid adjacent to the bed. An attack from potential flow reminiscent of Liu's (1957) was made by Shirasuna (1973), who analysed bed forms as internal waves at the interface between the layer of transported sediment below and the clear fluid above. Existence fields similar to those in Fig. 7-8 were obtained, including at small *kh* a zone of plane beds dividing dunes and ripples from antidunes (see also Kennedy, 1963, 1969).

Rotational models

Engelund (1970) developed a powerful and flexible analytical framework for the study of bed stability, into which can be introduced most of the relevant physical processes without undue simplification. The flow is described using a stream function, derived from the vorticity transport equation, with the help of a uniform eddy viscosity, and a slip velocity at the bed. Real-fluid effects are thereby indirectly introduced, whence it becomes possible to describe the sediment transport realistically and to separate the influences of suspended and bedload transports. Limiting attention to suspended-load transport, Engelund found that only plane beds and antidunes resulted, pointing again to the stabilizing effect of this mode of transport. The actual stability field is defined in terms of the parameter U_*/W, in which U_* is the shear velocity of the flow and W the falling velocity of the transported sediment (Fig. 7-9). The absolute limiting stability boundaries are given by setting U_*/W infinitely large, when eqs. (7.11) and (7.12) from potential flow are regained. On admitting bedload transport, dunes in addition are obtained, showing that the bedload has a destabilizing role. Two of Engelund's stability fields appear in Fig. 7-10, for one value of the parameter U/U_* and two values of a sediment coarseness parameter $U_*/W \cdot Fr$. As may have been expected, the effect of reducing grain size is to broaden the field of plane beds at lower Froude numbers.

Fredsøe (1974b) significantly modified Engelund's basic analysis, by ad-

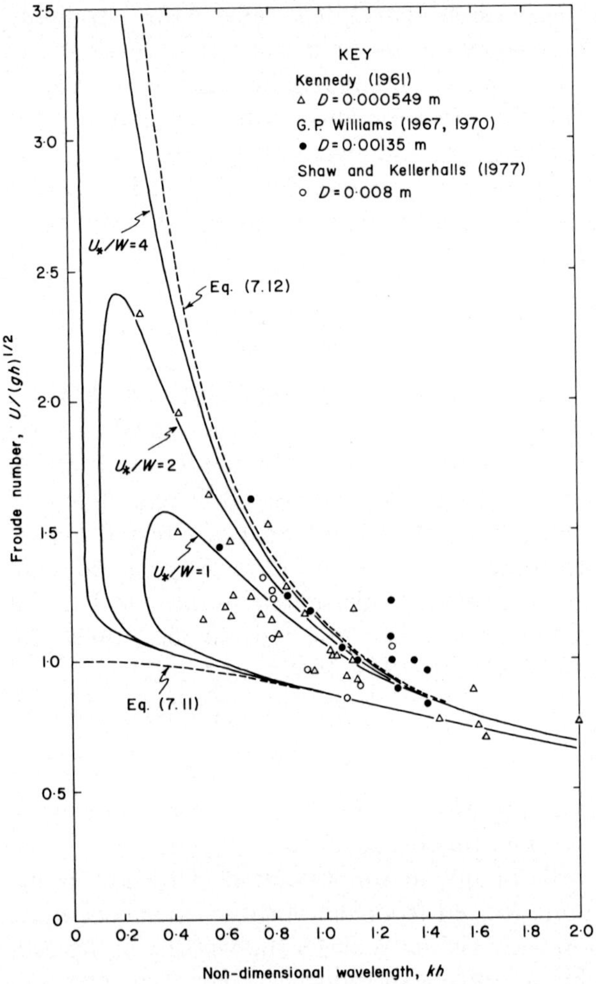

Fig. 7-9. Criteria for the instability of a granular bed in the presence of negligible bedload, calculated for the case $U/U_* = 17$, where U is the mean flow velocity, and expressed in terms of Froude number and non-dimensional bedform wavelength. The parameter is the ratio of the shear velocity to the sediment falling velocity. After Engelund (1970). The calculated curves are compared with experimental occurrences of antidunes in mineral-density sediments (Kennedy, 1969; G.P. Williams, 1967, 1970; Shaw and Kellerhals, 1977), the bedform consistent with the assumption of negligible bedload. Note how, in accordance with Engelund's theory, antidunes occur at larger Froude numbers with declining grain size.

ding to the bedload transport equation a term representing bed slope (see also Hayashi, 1970; Engelund and Fredsøe, 1974; Fredsøe and Engelund, 1975). The new term acts chiefly to limit the spread of the dune existence-field towards the larger values of kh. For example, Fig. 7-11 shows the calculated occurrence of dunes in moderately sorted quartz sands of median fall diameter 0.00028 m and 0.00093 m respectively in a flow 0.2 m deep. Com-

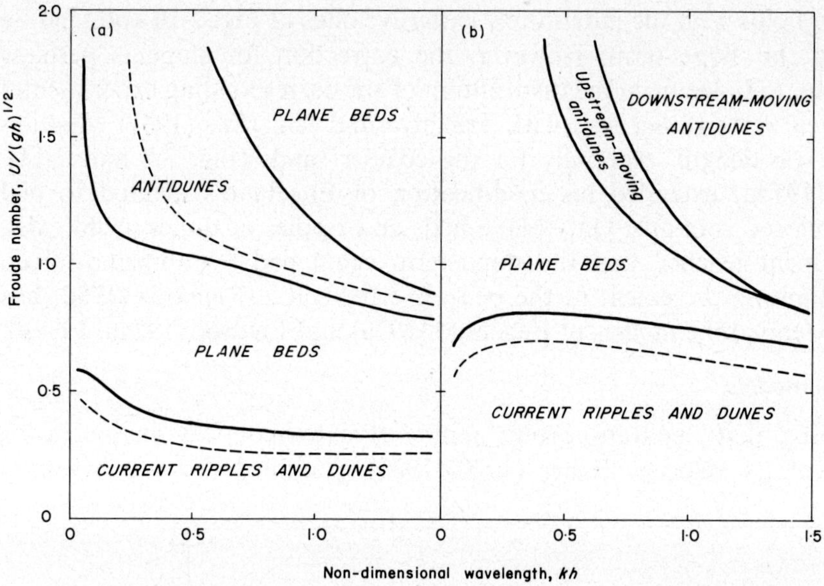

Fig. 7-10. Engelund's (1970) complete solution to the stability of a granular bed subject to bedload and suspended-load transport, in terms of Froude number and non-dimensional bedform wavelength. The curves represent the case $U/U_*=21$ for (a) $U_*/W \cdot Fr=2$, and (b) $U_*/W \cdot Fr=1$. The dashed lines represent the disturbances growing most rapidly in amplitude and, therefore, the wavelengths of the bedforms expected actually to appear. After Engelund (1970).

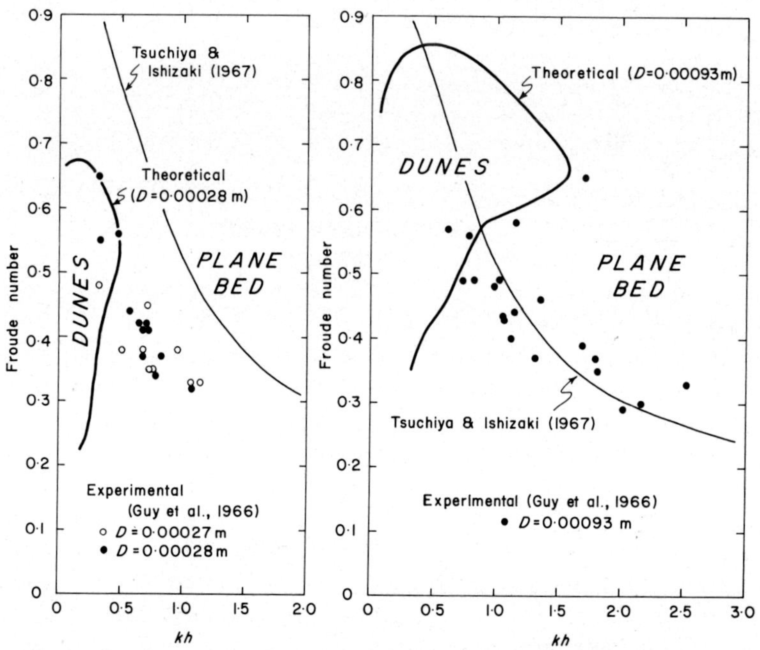

Fig. 7-11. Engelund and Fredsoe's (1974) theoretical limits for the occurrence of dunes for two grades of sand in water 0.2 m deep at 20°C, compared with the occurrence of this form under almost identical experimental conditions (Guy et al., 1966). The other curve plotted is Tsuchiya and Ishizaki's (1967) eq. (17) for dune wavelength.

pare these fields with the indefinitely extensive ones of Fig. 7-10 obtained by neglecting the slope term. However, the correction for slope is perhaps over-severe, to judge from the distribution of the corresponding experimental data, which agree rather well with Tsuchiya and Ishizaki's (1967) equation for dune wavelength, especially for the coarser sands (Fig. 7-8 and 7-11). Fredsøe (1974a) extended his modification of Engelund's method to bed forms in closed conduits. Only plane beds and ripples or dunes could exist, as experiment teaches, and the slope term again had a stabilizing effect, severely limiting the extent of the bed-wave field. K.J. Richards (1980) has further extended the models of Engelund (1970) and Fredsøe (1974a, 1974b).

Hydraulic models

Assuming that the transporting ability of a current was an increasing function of its velocity, Exner (1920, 1925) examined the stability of a

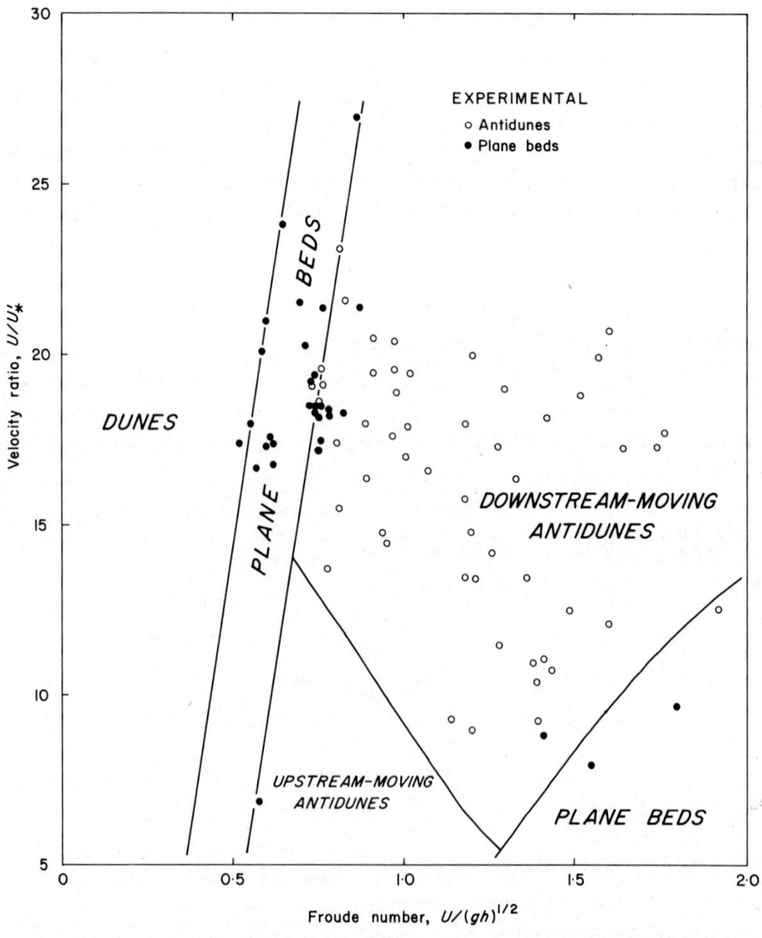

Fig. 7-12. Criteria for the existence of bedforms, in terms of U/U'_*, where U'_* is a restricted measure of the shear velocity, and the Froude number. After Engelund and Hansen (1966).

disturbance on an erodible bed to a flow described by a one-dimensional hydraulic equation involving bed friction. One effect disclosed was the progressive damping of the disturbance. A.J. Reynolds (1965) also found that friction was damping, unless Kennedy's arbitrary lag distance was introduced. Engelund and Hansen (1966) developed the hydraulic model in great detail, obtaining phase shifts between the bed shear stress and flow depth, and introducing a semi-empirical lag distance. Figure 7-12 shows their calculated existence-fields for two-dimensional bed forms; the parameter U/U'_* is a measure of the relative importance of friction, where U'_* is the shear velocity representing grain roughness. J.D. Smith's (1970) related work is restricted to low Froude numbers. Instability arose because of a combination of friction and local accelerations in the non-uniform flow, and either ripples or dunes could result. Gradowczyk (1968) studied a hydraulic model which accounted for the unsteady character of the flow over bed waves, and was then able to consider the interaction of surface waves with the bed. Of particular interest is his prediction of an absolute upper hydraulic limit to bed-wave occurrence ($Fr \approx 1.77$), at which the water surface itself becomes unstable and so prevents transport over other than a plane bed (Fig. 7-8).

Engelund (1971) and Fredsøe (1972) showed using a hydraulic model that a solitary travelling sand hummock could exist at sufficient small Froude numbers ($Fr < 0.6$) and within a certain narrow range of sediment transport conditions.

MATHEMATICAL MODELS OF ERODIBLE BED STABILITY: THE THREE-DIMENSIONAL CASE

Although the preceding investigations yield invaluable insights, we must recognize that naturally occurring bedforms are actually three-dimensional, and so cannot exactly satisfy a major restriction on the analyses. Study of the more general three-dimensional problem is, however, difficult and can only rest on either hydraulic or potential-flow models. Few workers have attempted it.

A.J. Reynolds (1965) examined the stability of an erodible bed to three-dimensional disturbances, using a potential-flow model, and choosing a perturbation of a three-dimensionality extreme even by comparison with natural bedforms. Fig. 7-13 shows his existence criteria (eqs. 7.11, 7.12) revised for this three-dimensional case. The consequence of giving the perturbation a finite crest length is to raise to higher Froude numbers the dune and antidune limits. Three-dimensional disturbances therefore seem to be destabilizing in the presence of bedload. According to Engelund and Hansen's (1966) hydraulic model, however, such disturbances have no consistent destabilizing influence in the presence of bedload, but nonetheless affect the bounds of several existence fields. For example, the field of plane

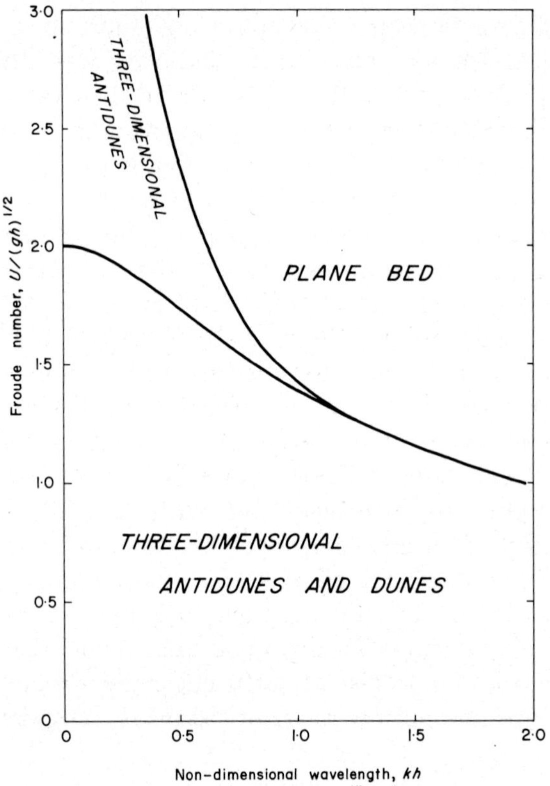

Fig. 7-13. Criteria for the existence of a class of (extremely) three-dimensional bed waves, in terms of Froude number and non-dimensional bedform wavelength. After A.J. Reynolds (1965).

beds at low Froude numbers in Fig. 7-12 is rotated clockwise and brought down toward moderate values of U/U'_*. Three-dimensional disturbances combined with suspended-load transport have a strong stabilizing influence (Engelund and Fredsøe, 1970). Callander (1969) predicted instability of an alluvial bed to all three-dimensional disturbances, but his analysis is restricted to very small values of kh, that is, to bedforms essentially at the scale of channel bars.

Engelund (1973) showed that many dunes in channels of transversely varying depth trend obliquely across the bed, each form lying furthest downcurrent where the flow is shallowest and slowest. It was later shown theoretically that such oblique dunes could exist as a stable bedform, and that their obliquity increased with decreasing kh for each Froude number (Engelund, 1974; Fredsøe, 1974c, 1974d; Furnes, 1976a, 1976b). These analyses seem in addition to explain the alternately left-handed and right-handed oblique portions of the lee sides of many individual ripples and dunes (e.g. Allen, 1968c).

It appears that periodic bedforms can occur in channel bends simply on account of the flow curvature (e.g. Zimmerman and Kennedy, 1973). Engelund (1975) explained these features by means of a stability analysis, although their sedimentological importance is at present uncertain.

BED-WAVE SHAPE AND SIZE

Stability analyses have occasionally led to proposals about the size and shape of bed waves, particularly ripples and dunes, and suggestions have also stemmed from other sources.

Can a bed wave retain the same shape and size during its history, or is change inevitable? Consider equation (7.9) above, restricting attention to bedload transport. On choosing a moving coordinate origin having a velocity identical with that of a bed wave, the equation can be rewritten as:

$$\frac{\partial J}{\partial x} = -\sigma C_0 \left(c\frac{\partial y}{\partial x} - \frac{\partial y}{\partial t} \right) \tag{7.14}$$

If the wave retains its shape and size as it travels, then the second bracketed term is zero, and we obtain upon integrating:

$$J = \sigma C_0 c y + b \tag{7.15}$$

where b is an integration constant. This equation becomes:

$$J = \sigma C_0 c(y - y_0) \tag{7.16}$$

since b can be interpreted in terms of an elevation, y_0, at which the bedload transport rate is zero. Because antidunes (Fig. 7-2a) have a negative celerity, y_0 in their case must lie wholly above the bed profile, in order to accommodate the downstream sediment transport with its maximum in the troughs. For ripples and dunes (Fig. 7-2b), y_0 must pass through the reattachment point on the stoss, in order to allow for the upstream transport beneath the separated flow (Allen, 1969f; Crickmore, 1970). Equations (7.15) and (7.16) state that the local sediment transport rate is linearly proportional to bed elevation, but can this be consistent with the local flow properties? With forms such as dunes and ripples in water, a linear dependence of J on y implies through continuity that J varies according to a linear function with the local mean flow velocity, since the surface waves have a small amplitude and the flow is generally several times deeper than the bed-wave height. In view of the strongly non-linear increase of J with flow velocity under uniform steady conditions, a similar local dependence seems plausible in a non-uniform flow, whence a single value of the celerity for the bed profile becomes most unlikely. Apparently each individual bed feature must change in shape and size during its history, as is indicated by considering bed forms as kinematic shock waves, and the involvement of ripples and dunes in creation and destruction processes seems inevitable.

Using a hydraulic approach, Exner (1925) found that a symmetrical bed wave became progressively steeper on the downstream side as it advanced, until a failure of granular cohesion generated a limiting slope to lee, and the form became a recognizable ripple or dune. A.J. Reynolds (1965), J.D. Smith (1970), and Gradowczyk (1971) confirmed this result, which depends on J being a steeply increasing function of y. Fredsøe (1974b) considered the growth of bed disturbances by accounting for flow properties neglected in Exner's model. The growing bed features progressively steepen to lee, but on account of the influence of the implicit lag in the sediment transport.

Attempts to model ripple and dune profiles have been few, and all implicitly assume that a stable profile is achievable. Ertel (1968) suggested an analytical model, but the profiles yielded are unsatisfactory as they lack sharp crests. The models developed by Mercer (1971b) and Mercer and Haque (1973) are more appealing, for calculated profiles possess a sharp crest and a convex-up stoss.

Noting that the preceding stability analyses all yield unstable bed features capable of unbounded amplitude growth, we now turn to the problem of predicting theoretically the scale of bed features in real environments. What wavelengths will first be expressed on the erodible bed, and what will be the ultimate wavelength and height of the forms, after non-linear effects and boundary conditions have created equilibrium?

The wavelength first to emerge follows directly from the stability analysis. Conventionally, it is that disturbance wavelength for which the time-rate of amplitude growth is a maximum, namely, the dominant wavelength. Kennedy (1963) gave an equation for this privileged wavelength, and Fredsøe (1974b) and Fredsøe and Engelund (1975) showed values graphically. Experimentally, the dominant wavelength is generally much less than the equilibrium value, for ballistic ripples (Cornish, 1897; W.H.J. King, 1916; Bagnold, 1936), as well as for current ripples and subaqueous dunes (Raichlen and Kennedy, 1965; Jain and Kennedy, 1971, 1974; Fredsøe and Engelund, 1975; Yalin, 1975). Two differently controlled "dominant" wavelengths emerge in the case of current ripples (Jain and Kennedy, 1971, 1974). The equilibrium wavelength may often be calculated by either assuming or seeking a neutrally stable sediment-transport phase shift (Table 7-I). Kennedy (1964, 1969) and Raichlen and Kennedy (1965) succeeded in developing relationships for ripple and dune wavelength, height, and height-wavelength ratio. For example, the height, H, in a free-surface flow creating bedload transport only is:

$$H = \frac{2}{nk} \frac{(U - U_{cr})}{U} \frac{\tanh(kh) - (kh)Fr^2}{1 - (kh)\tanh(kh)Fr^2} \tag{7.17}$$

in which n is the exponent in the velocity-based sediment-transport relation, U the mean flow velocity, U_{cr} the flow velocity at the threshold of motion, and kh the non-dimensional wave number. Tsuchiya and Ishizaki (1967) and Mercer (1971a) also calculated bed-wave scale (Fig. 7-8).

Yalin (1964, 1972) addressed the problem of the scale of two-dimensional current ripples and subaqueous dunes from a hydraulic standpoint. Arguing that the mean bed shear stress in the trough of the bed feature must be in the same order as the stress at the threshold of particle motion, τ_{cr}, he reasoned from dimensional and empirical considerations that:

$$\frac{H}{h}=\frac{1}{6}\left(1-\frac{\tau_{cr}}{\tau_0}\right) \tag{7.18}$$

in which h is the mean flow depth and τ_0 is the overall mean bed shear stress. Since τ_{cr} is never vanishingly small compared to τ_0, the ripple or dune height cannot exceed one-sixth of the mean flow depth. He further reasoned that current-ripple wavelength is in the order of 1000 times the bed-material diameter, and that dune wavelength is approximately five times the flow depth. These "linear" correlations were strongly criticized on publication, chiefly because of the large scatter in the empirical data from which Yalin drew numerical constants. Gill (1971) criticized eq. (7.18) more fundamentally, showing that the invariance of flow resistance with flow velocity is implicitly assumed. He was unable successfully to include a variable resistance, but produced from dynamical considerations the expression:

$$\frac{H}{h}=\frac{(1-Fr^2)}{2nb}\left(1-\frac{\tau_{cr}}{\tau_0}\right) \tag{7.19}$$

in which n is the exponent in a bed-load equation of Meyer-Peter and Müller type, and $1/2 \leqslant b \leqslant 2/3$ is a numerical coefficient related to the bedform cross-sectional shape. We see from eq. (7.19) that H/h can have a maximum in either Fr or in τ_0, since in free-surface flow the stress is proportional to Fr^2 for constant flow depth and flow resistance. According to Führböter (1967), the relative height is a decreasing function of the exponent in the sediment–transport relationship.

Using a physical argument, Yalin (1972) and Yalin and Karahan (1979) deduced that the height/wavelength ratio of current ripples and dunes is a bell-shaped function of the mean bed shear stress, each kind of feature reaching a maximum steepness at an intermediate stress within an appropriate stress range. Fredsøe (1975) used the principle of similarity to obtain a parallel relationship, restricted to dunes:

$$\frac{H}{L}=\frac{1}{8.4}\left(1-\frac{0.6}{\theta}-0.4\theta\right)^2 \tag{7.20}$$

where L is the wavelength, θ the Shields-Bagnold non-dimensional mean bed shear stress, and the numerical quantities are partly empirical. Subaqueous dunes therefore seem to be flattest under flow conditions near the lower and upper limits of the dune existence-field. If Yalin's (1964) proposal for dune wavelength is correct, eq. (7.20) becomes a similar statement to Gill's (1971) eq. (7.19), which can be rewritten in a form including a term in θ^{-1} and another in θ.

STATISTICAL ANALYSIS OF BEDFORMS

Much of the work surveyed in this chapter rests on the supposition that bedforms are deterministic, that each kind can exist only within a particular range of flow conditions, and that for each condition the size and shape of the forms are uniquely determined by that condition. But experience tells that features of a range of shapes and sizes—albeit of one kind—are shaped even by an equilibrium flow. The indeterminacy expressed by the variation between individuals is compounded by the effects of any unsteadiness of flow (Chapter 2), when it becomes even more difficult reliably to estimate bedform dimensions for the purpose of studying, say, sediment transport or bed roughness. Numerous workers, led by Nordin and Algert (1966), have therefore embraced the idea that bedforms are random phenomena best described in terms of stochastic processes. If there is any determinacy about bedforms, it must be restricted to population and not individual attributes.

Nordin (1971b) has usefully surveyed the statistical approach to bedforms. The bed elevation, y, measured with respect to the mean bed level, is a variable which depends on distance, x, measured along the flow, and time t. Then $y = y(x)$ and $y = y(t)$ is a random variable exhibiting a Gaussian distribution and $y = y(x, t)$ is a stochastic process which can be described using the autocovariance (autocorrelation) and spectral density functions. The spectral density function, which may be specified in either time or space domains, defines the contribution of variance that each frequency or wave number makes to the total process. The autocorrelation function describes the periodicity of the process.

Attention has so far been chiefly restricted to mainly experimental current ripples (Ashida and Tanaka, 1967; O'Loughlin and Squarer, 1967; Squarer, 1970; B.D. Taylor, 1971; Nordin, 1971b; Pratt and Smith, 1972; Jain and Kennedy, 1974) and subaqueous dunes (Nordin and Algert, 1966; Ashida and Tanaka, 1967; Fukuoka, 1968; Hino, 1968; Crickmore, 1970; B.D. Taylor, 1971; Nordin, 1971b; Annambhotla et al., 1972; Pratt and Smith, 1972). The most recent study is by Shen and Cheong (1977). Bed elevation is approximately Gaussian in distribution and both kinds of bed feature occur in populations that are ill-ordered even under equilibrium conditions. Current ripples are seemingly the less well-ordered features. Using spectral analysis, objective and reliable estimates of bedform population attributes are obtainable. There seem to be universal laws describing the spectral distributions of the bedforms, at least for the higher ranges of wave number and frequency (Hino, 1968; Engelund, 1969; Nordin, 1971b; Yalin, 1972;

Jain and Kennedy, 1974). Wave-number spectra follow a -3 power law and frequency spectra a power of -2. These distribution laws are consistent with the observation that ripples or dunes of small average height propagate faster than those of greater height under the same flow conditions.

An important theoretical development recently came from Jain and Kennedy (1974) who showed that, under equilibrium conditions, a stable spectrum of bedform wavelengths could be maintained by the operation of what was called a "variance cascade". They postulated that there is a continuous creation of features of small wavelength which in time evolve into longer-wavelength forms, only to disappear in various ways and so permit the creation of still further short-wavelength mounds. This reasoning points in essentially the same direction as the previous arguments about the influence of the strong non-linearity of the sediment transport rate on the stability of bed-wave profiles, and also parallels Costello's (1974) thesis about the attenuation of bed features treated as kinematic shock waves. Jain and Kennedy (1974) adduced no observational evidence for their proposed mechanism, but it is worth pointing out that "birth and death" processes have for some years been known or suspected to shape ripple and dune populations (Allen, 1968c, 1976a). A non-deterministic variance cascade operating around a deterministic central condition may therefore be sufficient to explain both the wide variation between individuals formed in the one flow, and the apparent consistency of the populations as a whole under sustained equilibrium conditions. Further study is warranted.

SUMMARY

Transverse bedforms such as ballistic ripples and dunes formed by the wind, current ripples and dunes shaped by flowing water, and antidunes in free-surface flows reflect a state of instability involving a sediment-transporting fluid and an erodible granular bed. In many cases, the forms arise at sufficiently large defects on the bed, and invariably grow up to a limit imposed by the general flow and sediment conditions. A continuous creation of short-wavelength features, their movement through a spectrum of increasing wavelengths, and their ultimate decease may well determine the character of many bedform populations. By applying the techniques of stability analysis to mathematical models of the sediment-flow system, it is possible to indicate the general conditions under which different kinds of bedform should exist. In free-surface flows, for example, we broadly expect the sequence ripples/dunes $\rightarrow$ antidunes $\rightarrow$ plane beds with increasingly severe flows. Bed stability is crucially influenced by the occurrence of property lags, between flow and bed, and between the sediment transport and flow. Lag can now be included implicitly in a stability analysis, and its physical reality should no longer be questioned. However, the mechanisms which promote bed instability are not necessarily those which determine the equilibrium characteristics of bedforms.

Chapter 8

EMPIRICAL CHARACTER OF RIPPLES AND DUNES FORMED BY UNIDIRECTIONAL FLOWS

INTRODUCTION

From the general nature and theoretical significance of transverse bedforms, we now direct attention to the detailed character and dynamical meaning of these features as observed in natural and laboratory environments. First to be discussed are ripples and dunes as developed in unidirectional currents, that is, in rivers, beneath the wind, and in marine environments on a scale such that the oscillatory motion of the tide can be ignored. What appearance and variability typify these configurations, and to what extent do their observed relationships to flow conditions lend further support to the theories, chiefly of bedform existence, previously sketched? Similar questions with regard to antidunes are deferred until Chapter 10. A cautionary word is appropriate. In comparing natural and experimental bedforms, it should always be remembered that natural currents are invariably unsteady, non-uniform, and multidirectional, not uncommonly in all these attributes simultaneously and to an extreme degree.

MORPHOLOGICAL ANALYSIS

Ripples and dunes occur for the most part on surfaces continuously underlain by erodible, granular material. In this context, an instantaneously perceived train or field of dunes or ripples constitutes a *statistical tesselation* (Allen, 1976a), for the available surface is fully covered, with neither gaps nor overlaps, by broadly similar elements. The investigator must solve the problem of unambiguously delineating these elements individually, in order to define and characterize the tesselation as a whole. A restricted one-dimensional solution to the problem of defining individuals is well known, but a general two-dimensional solution is now required, as an essential tool in obtaining a proper understanding of how individual features contribute, in the course of sediment transport, to bedform population attributes and to cross-stratification structures. Existing attempts at a solution—all tentative—rest on radically different interpretations of the individual feature (cf. Harms and Fahnestock, 1965; Allen, 1968c).

Figure 8-1a shows part of a simple ripple-train. As the earliest investigators understood, two-dimensional individuals may be defined one-dimensionally by marking off along a staff or taut line laid over the bed the normally projected positions of corresponding points on the vertical profile

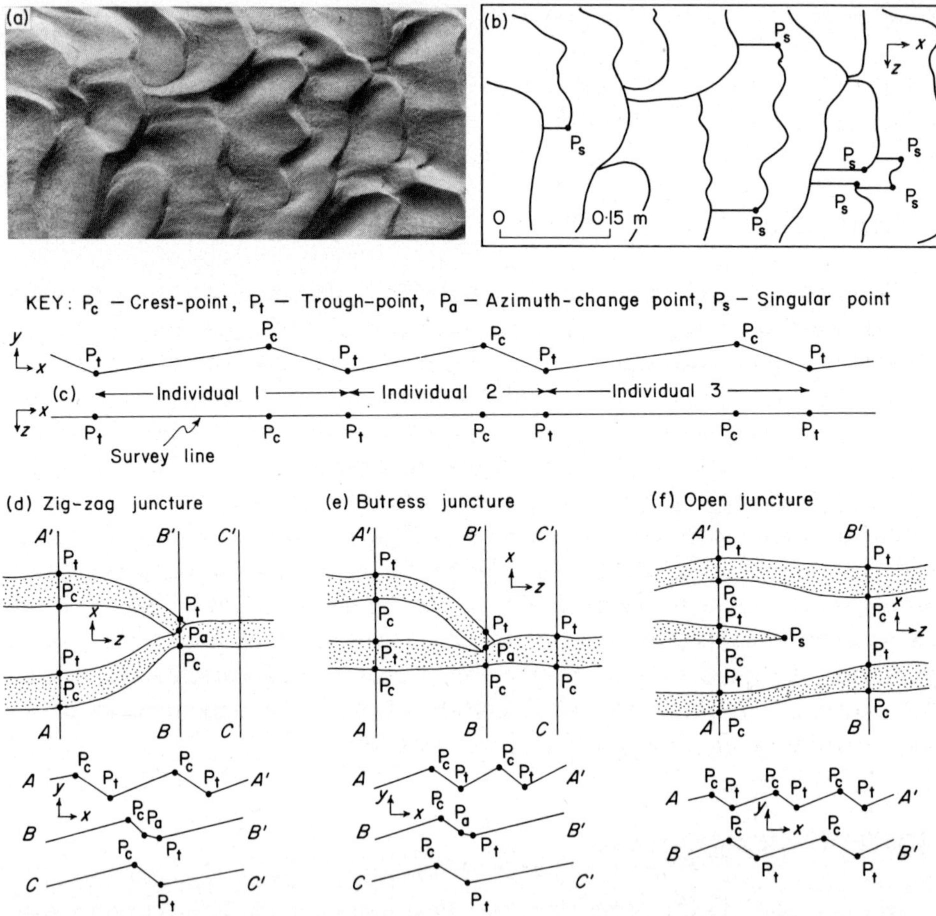

Fig. 8-1. Method of analysis of an assemblage of ripple marks into component forms (also applicable to wave-related ripple marks, dunes in aqueous and aeolian environments, and tidal sand waves). (d–f) also illustrate types of juncture.

of the surface (Fig. 8-1c). Conventionally, two particular kinds are emphasized (Allen, 1968c), points of maximum elevation (crest-points) and those of minimum elevation (trough-points) relative to a datum drawn parallel with the mean bed-level. This choice is implicitly based on the roles of the different parts of the surface when the forms are active. Ripples and dunes resemble water waves geometrically, but travel in a radically different manner, as the result of sediment transfers and transports. The interval on the bed from a trough-point downstream to a crest-point (stoss side) has a different role, which is to be eroded, than the interval from a crest-point downstream to a trough-point (lee side), which is to receive sediment derived largely if not wholly from the immediately upstream erosional interval (Fig. 7-2b). As a *functional* unit, therefore, the individual bed feature lies between two successive trough-points.

In order to delineate the tesselation two-dimensionally, it might be thought sufficient to connect in a plane the corresponding points identified as above on a sufficient number of parallel one-dimensional sampling lines. This method collapses in practice, for the zones thus mapped as lee side join from three directions in some places and in others die away (Fig. 8-1a), at Allen's (1968c) zig-zag, buttress and open junctures (Fig. 8-1d–f), with the result that some areas of lee and many of stoss are totally enclosed. At buttress and zig-zag junctures, however, lines of azimuth change—commonly actual discontinuities—can be mapped which separate portions of lee side of differing slope direction. As in profile B of Fig. 8-1d and e, these lines create in a one-dimensional mapping an azimuth-change point, lying downstream from a crest and upstream from a trough. In the conventional approach, recognizing only crest-points and trough-points, the dune or ripple lying between P_a and P_t would be counted with the individual immediately upstream. At each open juncture (Fig. 8-1f) lies a singular point at which crest and trough fuse. Here the arbitrary procedure of extending a bound from the singularity to the trough next upstream allows a functional individual to be mapped. In two dimensions, then, individual ripples or dunes can be mapped by identifying singular points and the lines connecting crest-points, trough-points, and azimuth-change points. The upstream and downstream boundaries of an individual must include a line of trough-points; the lateral limits must include either a singularity or a line of azimuth-change points. Figure 8-1a, b is a representative analysis made in these terms.

Figure 8-2 shows schematically the more important morphological features of ripples and dunes in unidirectional flows (Allen, 1968c, 1969b). The upstream or *stoss side* is comparatively long, gently sloping, and weakly convex-up. The downstream or *lee side* is relatively short and often com-

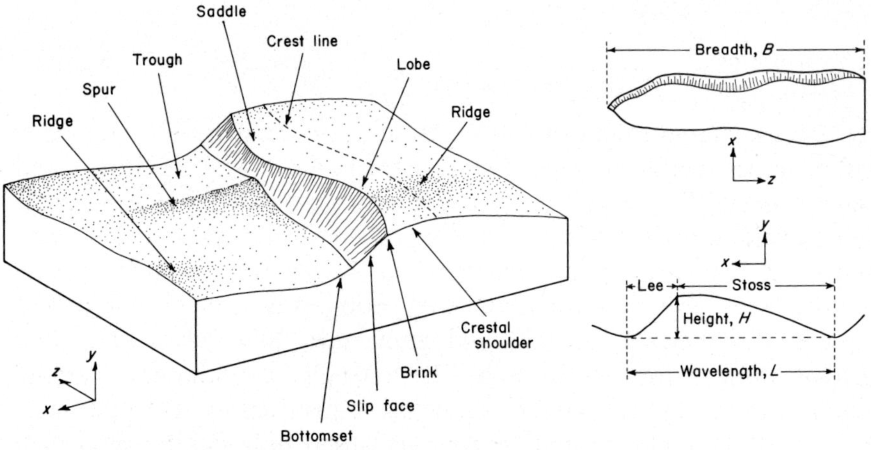

Fig. 8-2. Schematic representation of the chief morphological features of transverse bedforms (flow in positive x-direction).

posite, comprising as many as three parts, a gently sloping and weakly convex-up *crestal shoulder,* a *slip face* either plane or curved in one dimension only and ordinarily standing at the residual angle after shearing of the sediment (30°–35°), and a gently inclined weakly concave-up *bottomset.* Long-crested forms are often divisible transversely into sections on the basis of crestal curvature. Those portions of the lee side which project downcurrent afford *saddles,* whereas the re-entrant sections form *lobes.* Where an individual comprises several sections based on crestal curvature, the *crest line* may be either *sinuous* (lobes and saddles equally curved), *catenary* (saddles sharply pointed), or *cuspate* (lobes sharply pointed). Otherwise the crest line is *straight.* Lobes are commonly associated with low, round to sharp-crested, streamwise *ridges* which extend upstream some way over the stoss. Such a ridge which bisects an individual bed feature is a *median ridge.* Round to sharp-crested streamwise ridges which extend from the lee side of one individual partly over the stoss of the next downstream are called *spurs.* These usually lie downstream from saddles but can also be found at lobes.

The quantitative description of dunes and ripples is based on the use of such dimensions as the wavelength, height, length measured transversely to flow, length of lee side, and length of stoss side, usually combined into non-dimensional indices (Kindle, 1917; Bucher, 1919; Allen, 1963b, 1968c; Tanner, 1967; Reineck and Wunderlich, 1968b). The most useful are the *vertical form-index,* the ratio of wavelength, L, to height, H, and the *horizontal form-index,* obtained by dividing the wavelength into the breadth, B, of the feature measured transversely to flow.

Classification of transverse ripples and dunes are discussed by Jipa (1967, 1968), Allen (1968c), Glennie (1970), Clos-Arceduc (1972), I.G. Wilson (1972a), Cooke and Warren (1973), and Mainguet (1976). A broadly similar treatment to theirs is used below.

BALLISTIC RIPPLES

These are transverse ridges of sand, granules, or small pebbles generally less than 0.3 m in height (Glennie, 1970), shaped by the wind in deserts and along sandy coasts. For many years the ripples were divided into two classes considered distinct, the smaller forms being called sand ripples, *petites rides,* and impact or ballistic ripples, whereas the larger were named erosion ripples, grit waves, *grandes rides,* granule or pebble ridges, or residue ridges. Using plots of wavelength, height and grain size, Ellwood et al. (1975) proved that these forms all belonged to a single, continuously varying population, without natural breaks, a conclusion justified by the experimental results of Borszy (1973) and of Seppälä and Lindé (1978). Recorded wavelengths vary between 0.005 m (Bagnold, 1973a) and approximately 22 m (W.H.J. King, 1916), increasing almost linearly with the coarseness of the

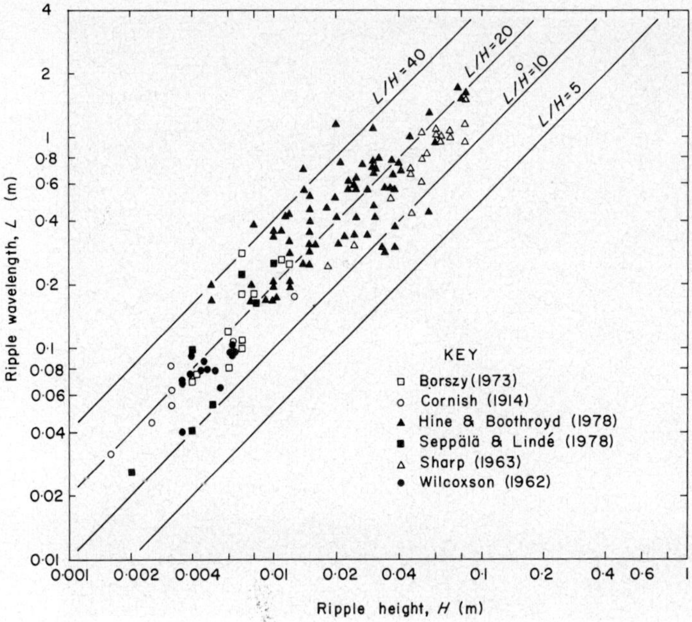

Fig. 8-3. Correlation between the wavelength and height of ballistic ripples, with the vertical form-index (L/H) as a parameter. Based on field observations by Cornish (1914), Hine and Boothroyd (1978), Sharp (1963), and Wilcoxson (1962), and on experiments by Borszy (1973) and Seppälä and Lindé (1978). The values plotted are group means only in the case of Borszy (1973), Seppälä and Lindé (1978), and Wilcoxson (1963).

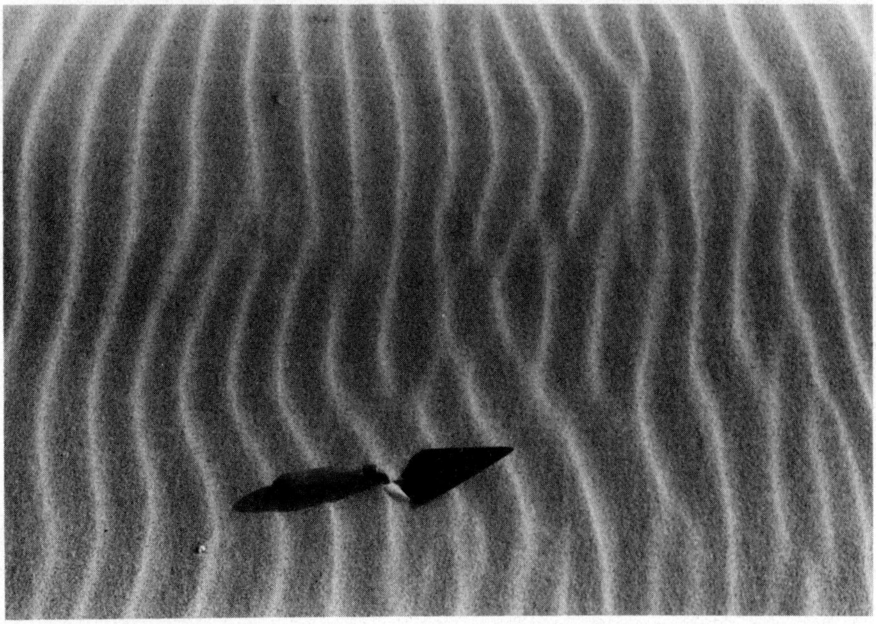

Fig. 8-4. Ballistic ripples in fine sand, coast near Burnham Overy Staithe, Norfolk, England. Trowel 0.28 m long points in wind direction.

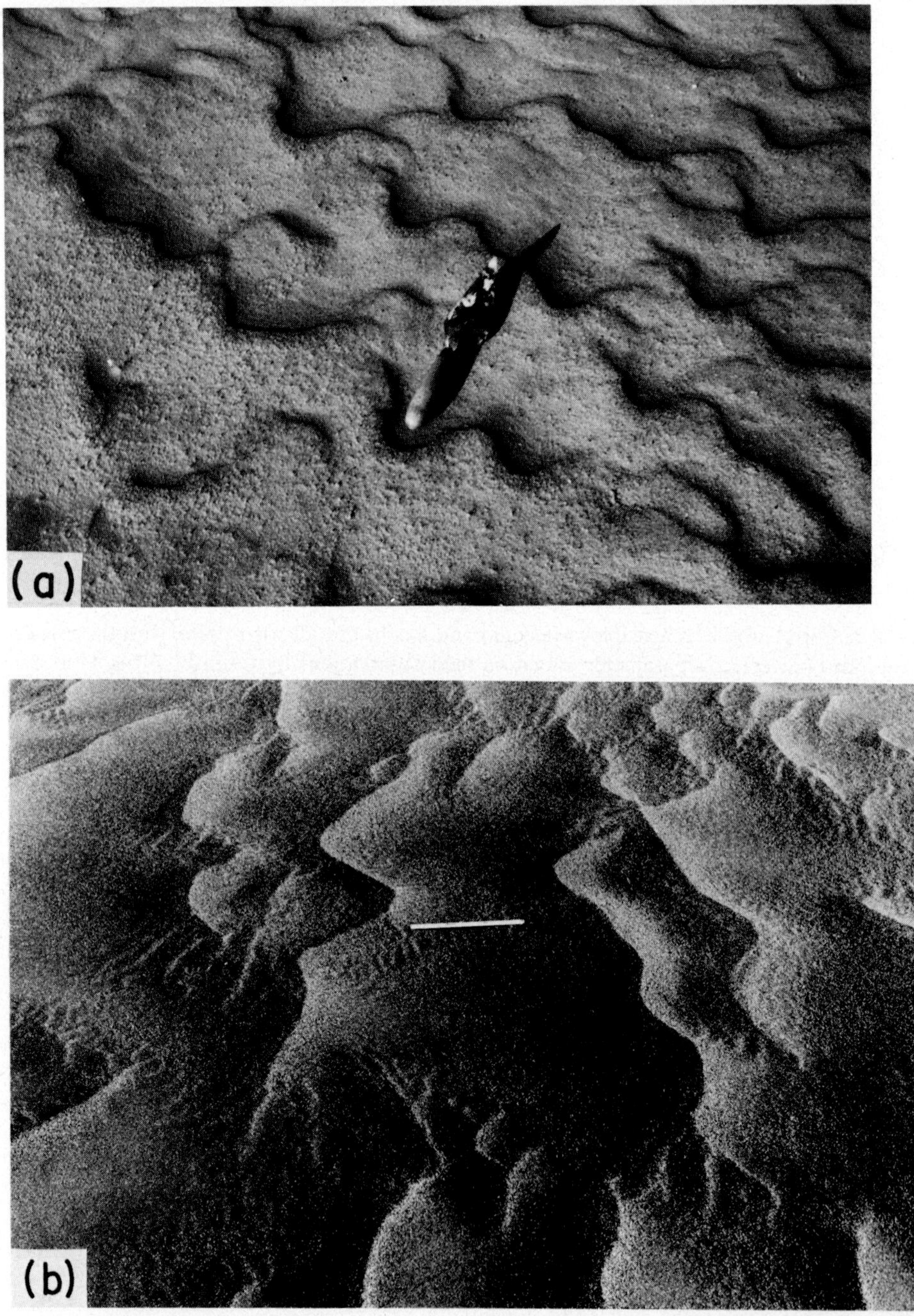

Fig. 8-5. Ballistic ripple marks. a. In illsorted medium to coarse sand, west of Holkham, Norfolk, England. Wind from bottom left to upper right. Trowel is 0.28 m long. b. In granule grade sediment, dry valley, Victoria Land, Antarctica. Scale 0.3 m long, wind from right to left. Photograph courtesy of M.J. Selby. Reproduced from Selby et al. (1974).

constituent sediment (Stone and Summers, 1972; I.G. Wilson, 1972a–c). The vertical form index is in the order of 20 (Fig. 8-3), but can range from 5–10 (e.g. Newell and Boyd, 1955; Selby et al., 1974) to at least 70 (Bagnold, 1937a).

The smaller ripples (Fig. 8-4), developed in very fine grained to medium grained sand, generally have long and weakly sinuous to remarkably straight crests (Cornish, 1901b, 1914; Kindle, 1917; Cressey, 1928; Kádár, 1934; McKee, 1945; Monod, 1958; Verlaque, 1958; Vaché-Grandet, 1959; Wilcoxson, 1962; Gripp and Martens, 1963; Martins, 1967; Folk, 1971; Bigarella, 1972; Stone and Summers, 1972; I.G. Wilson, 1972c; Ellwood et al., 1975). The horizontal form-index is of order 100. Stoss sides are plane to weakly convex-up and buttress junctures are typical (e.g. Cornish, 1897). Occasionally, these short-wavelength forms have strongly sinuous long crests (e.g. Ellwood et al., 1975). I.G. Wilson (1972c) recorded apparently rare linguoid forms arranged *en echelon.*

With increasing grain size, the horizontal form-index declines, stoss sides are more often weakly concave-up, and the vertical profile tends to greater symmetry. Ripples (Fig. 8-5) with long straight crests (Newell and Boyd, 1955; Sharp, 1963; I.G. Wilson, 1972c; Selby et al., 1974) no longer predominate over those with strongly curved and often short crests (Simons and Eriksen, 1953; Schiffers, 1957; Monod, 1958; Gripp, 1961a; Sharp, 1963; Schreiber et al., 1972; Borszy, 1973; Goldsmith, 1973; Selby et al., 1974; Hallier, 1976). Open terminations are as common as the buttres type, and some large ripples end in curious claw-like groupings of smaller ones (e.g. Selby et al., 1974). Hine and Boothroyd (1978) also record large ballistic ripples but give no details of their form.

Several factors control ripple wavelength. It increases with the duration of wind action on a flattened surface, up to a limit fixed by sediment sorting (Cornish, 1897; W.H.J. King, 1916; Bagnold, 1936; J. Müller, 1969; Seppälä and Lindé, 1978), the coarser grains accumulating on the ripple crests (cf. ripples and dunes shaped by water) (Cornish, 1897; Rim, 1953; Bagnold, 1954b; Norris and Norris, 1961; Sharp, 1963; Seppälä and Lindé, 1978). Wavelength also grows with general sediment coarseness and with an increasing textural disparity between the creeping and saltating loads (e.g. Simons and Eriksen, 1953; Sharp, 1963; Ellwood et al., 1975). The effect is well seen on wind-drifted sand patches, where average grain size and ripple wavelength simultaneously decline down-wind. Increase of wind speed causes wavelength growth, but also makes ripples flatter (Bagnold, 1954b; Chepil, 1945b; Wilcoxson, 1962; Chiu, 1967; Borszy, 1973; Seppälä and Lindé, 1978). Borowka (1980), however, found during his field studies no consistent influence on wavelength from wind strength. The temperature-controlled viscosity of the air may also influence wavelength. Ripples in cold climates (H.T.U. Smith, 1965a; Lindsay, 1973; Selby et al., 1974) seem to consist of coarser debris than forms of comparable wavelength shaped by hot winds (e.g. Bagnold, 1931, 1933, 1935; Newell and Boyd, 1955; J.E. Weir, 1962; Sharp, 1963).

It is worth remembering that ballistic ripples also arise in granular snow, though the wavelengths tend to be small (Cornish, 1902, 1914; Benson, 1962). Ballistic ripples are occasionally found in the stratigraphic record, McKee's (1945) account being the most convincing. Large-wavelength forms do not yet seem to have been distinguished, but may have been reported and ascribed to an aqueous origin because of their coarse sediment.

CURRENT RIPPLES

These are asymmetrical, transverse ridges of sand, typically of medium or finer grade, with a height of less than 0.04 m and a wavelength falling below 0.6 m, the steeper side facing downstream (Allen, 1968c). Other investigators have proposed broadly similar limits. Simons et al. (1965b) advocated a limiting height of 0.06 m and a bounding wavelength of 0.3 m. The same limiting height was used by Guy et al. (1966), but an upper wavelength limit of 0.6 m was suggested. These limits all derive from the comparatively recent quantitative demonstration that the asymmetrical transverse bedforms shaped by one-way water streams—now divided at least between ripples and dunes —do not form a continuous population (cf. development of concepts of ballistic ripples). Early investigators, such as Cornish (1901a, 1901b, 1914), Kindle (1917), Bucher (1919) and R. Richter (1926a), had recognized in a general way that these transverse forms took at least two distinct spatial scales. Inglis (1949) pertinently commented from experiments that "In the case of a ripple there are no secondary ripples on the upstream slope, whereas small secondary ripples appear at times on the upstream slopes of dunes". However, a number of geologists, and many engineers and fluid dynamicists, prominent amongst whom is Kennedy (1963), made no distinction between ripples and dunes. Van Straaten (1953a), for instance, felt that the asymmetrical structures varied continuously in their attributes, and so could not be subdivided except arbitrarily. The reality of a morphologically distinct class of ripples, and the validity of the proposed quantitative limits, is amply proved by the frequency distributions of wavelength and height prepared by Allen (1963b, 1968c) and G.E. Williams (1971) from the laboratory and field (Fig. 8-6).

Ripple wavelength and height are controlled by sediment coarseness and mean bed shear stress, the latter perhaps partly expressing relative roughness effects. Ripples arise in mineral-density sediments ranging from silt and very fine sand (Rees, 1966b; Vossmerbäumer, 1970; Reineck, 1974) up to medium-coarse sand (Inglis, 1949; Chabert and Chauvin, 1963; Maggiolo and Borghi, 1965; Guy et al., 1966; G.P. Williams, 1967, 1970; Williams and Kemp, 1971; Southard and Boguchwal, 1973; Costello, 1974). Yalin (1964,

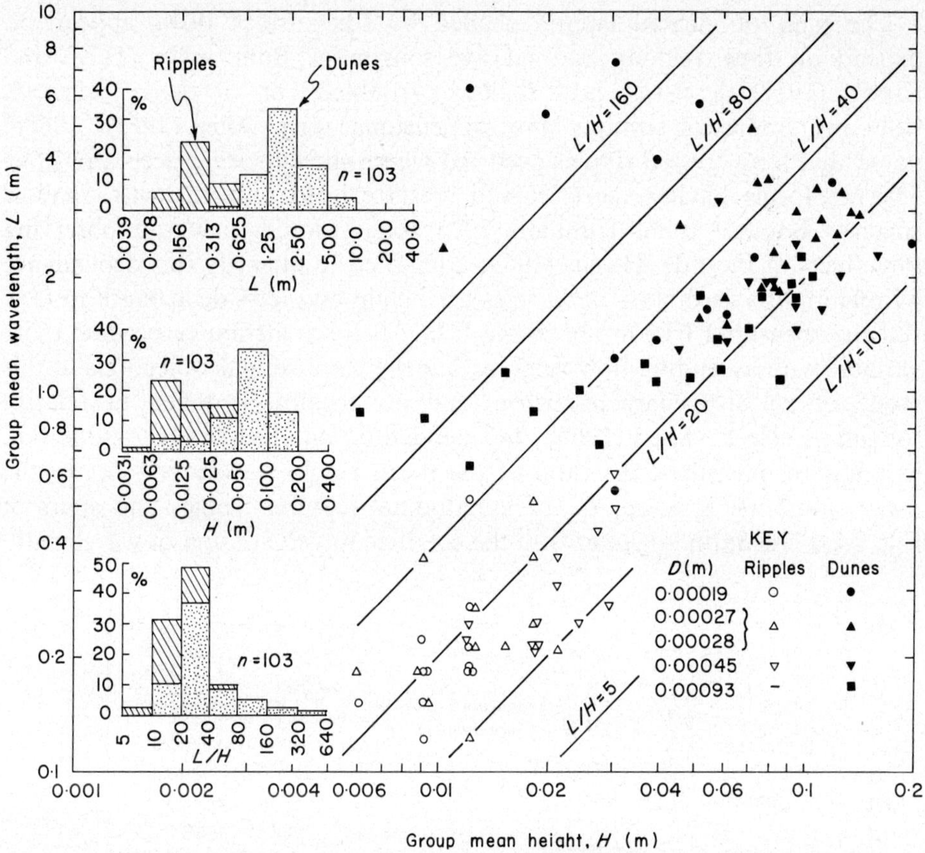

Fig. 8-6. Correlation between the group mean wavelength and group mean height of experimental aqueous current ripples and dunes under equilibrium conditions (data of Guy et al., 1966), together with number frequency distributions for group mean wavelength, group mean height, and the vertical form-index. Note the distinctness of current ripples and dunes in terms of wavelength, height, and vertical form-index (flatness).

1972) proposed that wavelength is approximately $1000D$, where D is the mean diameter of the bed material. This is broadly true, but the available data scatter widely, as is clear from his own graphs and from Allen's (1970g) plot of experimental results (Guy et al., 1966). Furthermore, at flow conditions near the upper limit of ripple existence, both wavelength and height increase rapidly for a small shear-stress increment (Allen, 1968c). Menard (1950b) found that a reduction in flow velocity could lead to the development of smaller forms. The vertical form-index of current ripples lies generally between 7 and 20 based on group mean wavelength and height, making them slightly steeper than ballistic ripples of a similar size, and their height appears to increase more rapidly than wavelength (Allen, 1963b, 1968c). According to Yalin (1972), the vertical form-index attains a maximum at an intermediate shear stress within the range appropriate to ripples.

The plan of current ripples shaped by free-surface flows appears to depend on flow velocity and relative roughness. Both Jukes (1872) and Kindle (1917) described how shallow "troubled" or "irregular" currents fashioned ripples of strongly three-dimensional form. Allen (1963b, 1968c) noted that long-crested ripples occurred where currents were deep and slow, whereas forms having short, curved crests existed only in swift, shallow currents. Boothroyd and Hubbard (1975) took a similar view after observing the effects of the tide. Harms (1969) attempted to quantify this dependence by relating the variability of wavelength, height and lee-side azimuth to flow velocity measured 0.03 m above the bed. All three attributes become more variable with ascending flow velocity, but there are serious objections to this mode of velocity characterization, and no account is taken of relative roughness effects. Allen (1969b) independently obtained similar results but, in addition, measured the ratio of the mean ripple wavelength, L_x, to the mean transverse spacing, L_z, of longitudinal features (ridges and spurs of Fig. 8-2). This ratio, together with the coefficients of variation of wavelength,

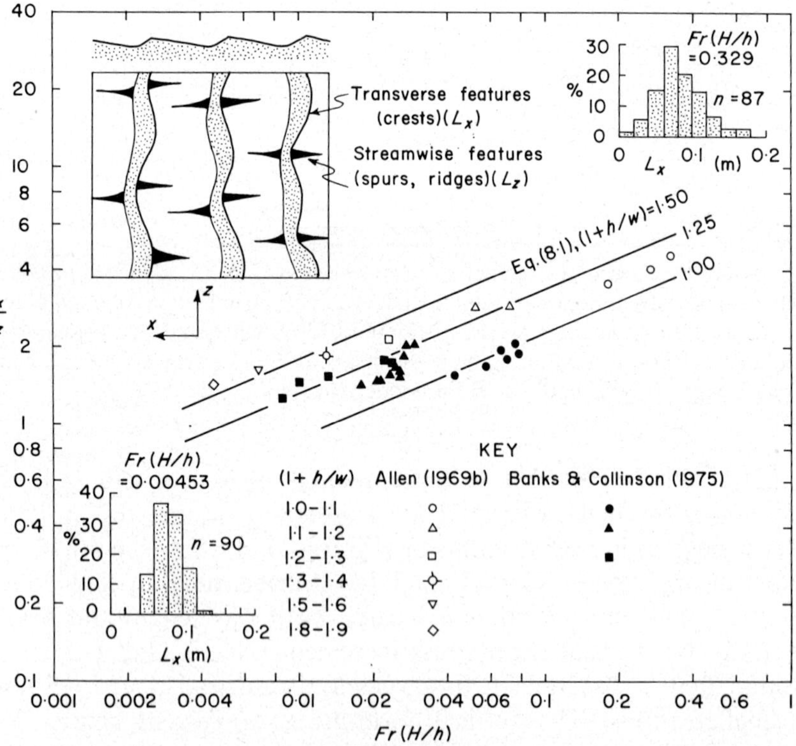

Fig. 8-7. The group mean relative spacing of transverse (crest lines) and longitudinal (spurs and ridges) features on current ripples, as a function of Froude number and group mean ripple height (H) relative to mean water depth (h), with the ratio of flow depth to width (h/w) as a parameter. Data of Allen (1969b) using a fine sand and of Banks and Collinson (1975) using a medium sand.

height and azimuth, was found to increase with ascending Froude number, Fr, and relative roughness (ratio of mean ripple height, H, to mean flow depth, h). Banks and Collinson (1975) repeated Allen's experiments with a much coarser sand and varying flow velocity rather than flow depth. They likewise found that L_x/L_z increased in value and the wavelengths spread more widely as the flow conditions became more severe, but did not obtain numerical agreement and were critical of the Froude number and relative roughness as parameters.

Allen (1977b) resolved these difficulties by showing that his and Banks and Collinson's data were in agreement when explicit account was taken of the width, w, of flow (Fig. 8-7). His empirical equations:

$$\frac{L_x}{L_z} = 5.85\left(Fr\cdot\frac{H}{h}\right)^{0.412}\left(1+\frac{h}{w}\right)^{1.71} \qquad (8.1)$$

$$Fr\cdot\frac{H}{h} = 0.026\left(\frac{L_x}{L_z}\right)^{1.73}\left(1+\frac{h}{w}\right)^{-5.20} \qquad (8.2)$$

relate ripple geometry to flow and incorporate both the flow strength and relative roughness effects. Some representative wavelength frequency distributions (Allen, 1969b) appear in Fig. 8-7 to illustrate the growth of disorder with these effects. We may note that the theory of bed-form crestal obliquity (Engelund, 1974; Fredsøe, 1974c, 1974d; Furnes, 1976a, 1976b) is supported

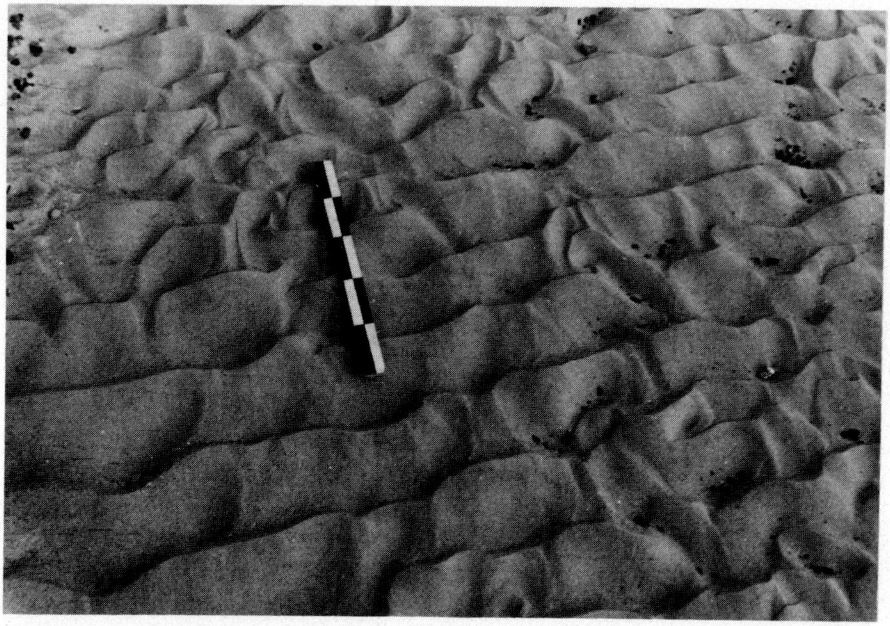

Fig. 8-8. Long-crested current ripples in fine grained sand, Wells-next-the-Sea, Norfolk, England. Scale 0.5 m long nearly parallel with current from bottom toward top.

qualitatively by these results (Allen, 1969b; Harms, 1969; Banks and Collinson, 1975), and that the main role of relative roughness seems to be to fix the scale of the secondary flows that control the spacing of the streamwise elements on ripples (Allen, 1969b).

Typically, ripples with low values of L_x/L_z have long and relative straight crests (Fig. 8-8), in detail either gently sinuous, or dominated by cusp-shaped or catena-like (lunate) elements in Allen's (1968c) terminology. Examples are widely known from modern stream and tidal channels and from flumes (Kindle, 1917; Inglis, 1949; Sundborg, 1956; Bajard, 1966; Guy et al., 1966; Rees, 1966b; Allen, 1968c, 1969b; Daboll, 1969; Harms, 1969; G.E. Williams, 1971; Picard and High, 1973; Banks and Collinson, 1975), and they are also shaped by deep-sea currents (Heezen and Hollister, 1971). These forms constitute what Harms called "low-energy current ripples", but this designation lacks precision and is also unsatisfactory because the proven effects of flow scale are excluded. In harmony with an increase of L_x/L_z go a shortening and curving of ripple crests (Fig. 8-9) (Kindle, 1917; McKee, 1939; Bajard, 1966; Guy et al., 1966; Allen, 1968c, 1969b; Harms, 1969; Heezen and Hollister, 1971; McGowen, 1971; G.E. Williams, 1971; Picard and High, 1973). Linguoid ripples (Blasius, 1910) are an especially common short-crested form in many environments. To categorize short-crested ripples as "high-energy" features (Harms, 1969) invites the same objections as before. Picard and High (1973) claim that sinuous, long-crested ripples arise

Fig. 8-9. Short-crested (linguoid) current ripples in fine sand, Wells-next-the-Sea, Norfolk, England. Trowel measures 0.28 m long and points in current direction from bottom to top.

in faster currents than short-crested forms, but offer no hydraulic data in support.

Field experience shows that there is a continuous morphological series between ripples formed by fairly steady one-way currents and the nearly symmetrical forms (wave ripples) shaped by the most balanced oscillatory flows. The intermediate forms, called "combined-flow ripples" (Harms, 1969) or wave-current ripples, can sometimes be distinguished using criteria developed by Reineck and Wunderlich (1968b) and Harms (1969), though supposedly diagnostic profiles sketched at flume walls should be treated cautiously. These ripples tend to be straighter crested and less strongly asymmetrical than current ripples (e.g. Allen, 1980b), but much remains to be explored in this complex area.

The rock record abounds in current-rippled surfaces preserved essentially because of a sudden reduction of flow strength. Most of the ripple types distinguished in modern environments are recognized (e.g. Kindle, 1917; McKee, 1954; Pepper et al., 1954; Hamblin, 1961a; Simon and Hopkins, 1966), though there is so far little attempt to exploit them palaeohydraulically. Nesbitt and Talbot (1966) and Mukherjee (1968) reported finding ripple-like forms in ultrabasic igneous rocks, the structures presumably having resulted by the deposition of crystals growing in and transported by the magma.

DUNES SHAPED BY WIND

Barkhans

These dunes are now recorded, often in great detail, from the Saharan region (Beadnell, 1910; W.H.J. King, 1918; Bourcart, 1928; Bagnold, 1931, 1933; Shaw, 1936; Capot-Rey, 1957; Schiffers, 1957; Monod, 1958; Verlaque, 1958; H.T.U. Smith, 1963, 1969; Coursin, 1964; Clos-Arceduc, 1967b, 1969; Grove and Warren, 1968; Mainguet, 1968; Warren, 1970; Mainguet and Callot, 1974; Sarnthein and Walger, 1974; Tsoar, 1974; Worrall, 1974; Jäkel, 1980), the Saudi Arabian and Iranian deserts (Bagnold, 1951; Kerr and Nigra, 1952; Holm, 1960; Glennie, 1970; Krinsley, 1970; Shinn, 1973; Dresch, 1975), the arid basins of central Asia (Doubiansky, 1928; Hörner, 1957; Petrov, 1962, 1976), the dry regions of eastern and southern Africa (Gevers, 1936; Grove, 1969; Kayser, 1973; Hay, 1976), the deserts of the southwestern U.S.A. (Rempel, 1936; Norris, 1956, 1966; Norris and Norris, 1961; Long and Sharp, 1964; R.B. Johnson, 1967; Stone and Summers, 1972; Ahlbrandt, 1975), and the Peruvian desert (Barclay, 1917; Simons and Eriksen, 1953; Simons, 1956; Finkel, 1959, 1961; Gay, 1962; Hastenrath, 1967; Lettau and Lettau, 1969). Curiously, only the Australian deserts seem to have prompted no accounts of barkhans. Cornish (1900, 1914), Kessler

(1971) and H.J. Walker (1973) have described small barkhans developed on river sand bars. Gripp (1961b, 1961c), Cowie (1963), Gripp and Martens (1963), Depuydt (1972), C.R. Harris (1974) and Borowka (1980) report the dunes from wind-swept coastal sands in the temperate zone.

A barkhan dune appears to form as the result of the growth in size of, and construction of a leeward slip-face upon, a small oval sand patch accumulating on a stony, rocky or cohesive surface in areas where loose sand is in meagre supply (W.H.J. King, 1918; Capot-Rey, 1957; Hörner, 1957; Gripp, 1961b, 1961c; Gripp and Martens, 1963; Hastenrath, 1967; Worrall, 1974). The matured dune (Fig. 8-10), standing in isolation, is crescent-shaped and bilaterally symmetrical in plan when sufficiently far removed from interfering neighbours. Its lateral "arms", "horns" or "wings" point down-wind and enclose a steep and slightly less extensive slip face shaped by sand avalanching, associated in some cases with a narrow crestal shoulder. Most barkhans have a height in the order of 2–20 m and are usually homogeneous in scale within any one area (Fig. 8-10b–d) (Finkel, 1959; Coursin, 1964; Long and Sharp, 1964; Norris, 1966; Hastenrath, 1967; Tsoar, 1974), but an extreme height of about 50 m is reached by some dunes, which then generally bear

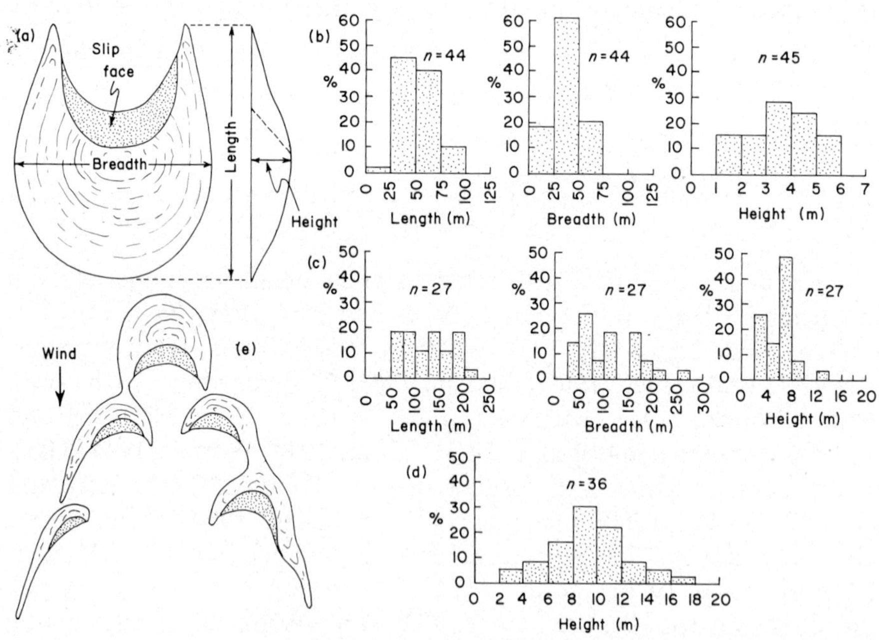

Fig. 8-10. Features of aeolian barkhan dunes. a. Schematic dune in plan and in vertical profile in the plane of flow. b. Frequency distributions for length, breadth and height measured from a sample of barkhans in Imperial Valley, California (Long and Sharp, 1964). c. Frequency distributions for length, breadth and height measured from a sample of barkhan dunes, Arequipa, Peru (Finkel, 1959). d. Frequency distribution of dune height in a group of Saharan barkhans (Coursin, 1964). e. Schematic representation of V-shaped formation of close-clustered barkhans.

smaller superimposed transverse forms (Beadnell, 1910; Simons and Eriksen, 1953; Simons, 1956; H.T.U. Smith, 1963; Shinn, 1973). Close packing causes barkhans to become distorted. Inflexions appear in the crest line, and one of the wings may become elongated relative to the other. A common mutual arrangement of close-packed barkhans is like a V-shaped formation of flying birds (Fig. 8-10e) (Kerr and Nigra, 1952; Clos-Arceduc, 1967b; Worrall, 1974), apparently for similar aerodynamic reasons. Sand movement off the barkhans is concentrated along a line down-wind from each wing, because there is a reduction of bed shear stress and a convergence of near-bed flow along and toward this trend. Birds in V-shaped formation conserve energy by flying in each others trailing vortices.

The movement rate of desert barkhans has been much studied (Beadnell, 1910; Kerr and Nigra, 1952; Hörner, 1957; Finkel, 1959; Gay, 1962; Coursin, 1964; Long and Sharp, 1964; Hastenrath, 1967; Lettau and Lettau, 1969) and has recently been successfully modelled (Howard et al., 1978). Speeds are generally in the order of 10–20 m annually, and inversely proportional to dune height. Much sediment sorting according to particle size, shape and density occurs during extensive barkhan movement. The coarsest grains tend to lodge on the skirts, wings and sometimes the summit of the individual dune (Bagnold, 1954b; Simons, 1956; Amstutz and Chico, 1958; Verlaque, 1958; Simonett, 1960; Tricart and Mainguet, 1965). Heavy minerals become concentrated in the dunes furthest upwind, whereas the finest grained quartz sands lie down-wind in each dune-train (Finkel, 1959; Hastenrath, 1967; Hay, 1976), as in rippled sand patches.

It has long been recognized that granular snow driven over ice can be shaped into barkhans (Cornish, 1902, 1914; Tschirwinsky, 1908; Oettli, 1917; Seligman, 1936; Benson, 1962). The dunes also form where small amounts of sand are driven across gravelly or rocky surfaces by river flows (McCulloch and Janda, 1964) or shallow tidal and other marine currents (Newell et al., 1951, 1959; Illing, 1954; Newell and Rigby, 1957; Potter and Pettijohn, 1963; Kenyon and Stride, 1968; Belderson and Kenyon, 1969; Drapeau, 1970; Auffret et al., 1972; Belderson et al., 1972; Werner and Newton, 1975b). Lonsdale and Malfait (1974) and Lonsdale and Spiess (1977) give excellent accounts of barkhans of foraminiferal sand which are moving over a scoured ocean bed in water several thousand metres deep. Current ripples also assume barkhan form when transportable sediment is meagrely supplied to a surface of glass, mud, or very fine silt (Ismail, 1952; Rees, 1966b; Allen, 1968c; Karcz, 1974; Mantz, 1978).

Transverse dunes (aklé and transverse draa)

As the transportable sand increases in quantity, isolated barkhans give place either upwind or downwind to short transverse ridges composed of a few catena-like sections fused laterally, and eventually to a more or less continuous sand cover decorated with asymmetrical, long-crested transverse dunes (F.A. Melton, 1940; Broggi, 1952; Hörner, 1957; H.T.U. Smith, 1963; Potter and Pettijohn, 1963; McKee, 1966b; Clos-Arceduc, 1967b). Similar morphological sequences, likewise related to a changing extent of sand cover, occur in aqueous environments (e.g. Newell and Rigby, 1957; Lonsdale and Malfait, 1974), their desert parallels being recognized by Allen (1963b, 1968c) and Kenyon and Stride (1968).

Transverse features with a wavelength in the general order of 10–100 m and a height comparable with 1–10 m are known as *aklé* or simply as transverse dunes (Fig. 8-11). Height and wavelength are fairly consistent within a single group of dunes (Fig. 8-12), and wavelength increases with the coarseness of the constituent sand (I.G. Wilson, 1972a, 1972b, 1972c). Typically, each dune has a weakly convex-up stoss side, a narrow gently shelving crestal shoulder where there is a saddle, and a steep lee side standing at 30–35° from the horizontal. The horizontal form-index is large and crests in plan are weakly sinuous to irregular; lobes occur generally where the lee side is tallest and saddles typically where it is lowest. Some

Fig. 8-11. *Aklé*, desert area in southwestern U.S.A., showing longitudinal elements (L) and slight crestal reversal (R). Wind from left to right. Photograph courtesy and copyright of Aerofilms Limited.

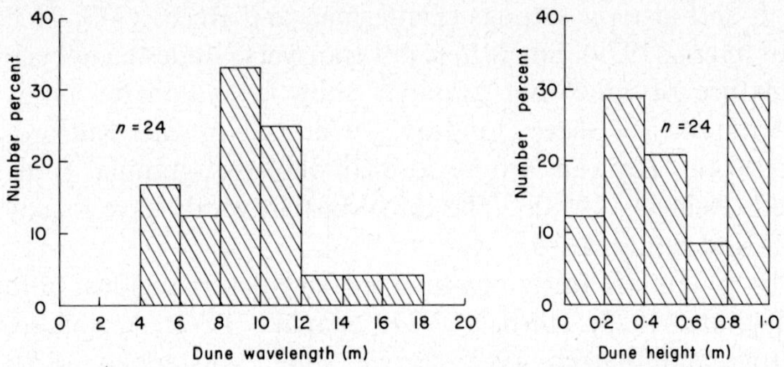

Fig. 8-12. Frequency distributions of the wavelength and height of transverse aeolian dunes, as measured on a single traverse across a sand bank in the Nile near Helwan, Egypt, during the dry season (Cornish, 1900).

dunes carry spurs, in places springing from saddles but usually from lobes (Cornish, 1900, 1914; Cooper, 1958; Monod, 1958). The spurs in some dune fields are asymmetrical in cross-section and do not lie exactly at right angles to the dune trend (Cooper, 1958; Monod, 1958). This asymmetry arises in some fields because the spurs are consistently eroded to a steeper slope on one side than the other. In other cases, it appears that accretion occurs preferentially on one particular side of the spurs. Their periodic transformation into features resembling full dunes may take place under the influence of seasonal cross-winds, and creates in a fairly extreme form net-like dunes with crests of L-shaped plan (e.g. F.A. Melton, 1940; Monod, 1958; McKee, 1966b; Solle, 1966).

Aklé have a wide distribution in hot deserts, like the Sahara (Cornish, 1900; Aufrère, 1935; Brosset, 1939; Capot-Rey, 1943, 1957; Sevenet, 1943; Matschinski, 1952, 1954, 1955; Monod, 1958; H.T.U. Smith, 1963; Solle, 1966; Clos-Arceduc, 1967b, 1969; I.G. Wilson, 1971b, 1972c), and those of Saudi Arabia (Holm, 1960; Glennie, 1972), southern Africa (A.D. Lewis, 1936a, 1936b), Asia (Doubiansky, 1928; Hörner, 1957; Higgins et al., 1974), and the Americas (MacDougal, 1912; F.A. Melton, 1940; Hack, 1941; Hefley and Sidwell, 1945; Broggi, 1952; Norris and Norris, 1961; McKee, 1966b; Sharp, 1966; R.B. Johnson, 1967; McKee and Douglass, 1971; McKee and Moiola, 1975). The dunes also occur on the drier and windier coasts (F.A. Melton, 1940; Cooper, 1958, 1967; W.A. Price, 1958; Cowie, 1963; H.T.U. Smith, 1963; Inman et al., 1966; Martins, 1967; Orme, 1973; Tsoar, 1974; Hine and Boothroyd, 1978), where their wavelength tends to increase away from the nourishing beaches (Cooper, 1958; H.T.U. Smith, 1963). Locally, these coastal *aklé* combine with barkhans and transitional forms into what W.A. Price (1958) called "banners", that is, triangle-shaped dune fields pointing upwind (see also H.T.U. Smith, 1963), which much resemble triangular trains of current ripples (e.g. USWES, 1935) initiated on a planed bed at some local irregularity (Fig. 7-1). Transverse dunes are occasionally to be found in river settings (Detterman and Reed, 1973; H.J. Walker, 1973; Koutaniemi, 1979). Small fields of transverse dunes have been shaped from a mixture of sand and granular snow by katabatic winds blowing off the Antarctic ice sheet (Lindsay, 1973; Calkin and Rutford, 1974; Selby et al., 1974). They can also be formed wholly of granular snow (e.g. Ljungner, 1930; Reineck, 1958b). The thin Martian winds have locally built up *aklé* (Cutts and Smith, 1973).

Aklé stabilized by plants are fairly common in once-desert regions, both periglacial (e.g. Högbom, 1923; Hörner, 1927; Seppälä, 1972; Aartolahti, 1973) and hot (Grove and Warren, 1968; Grove, 1969; Verstappen, 1970; Warren, 1970; White, 1971; Goudie et al., 1973; Bart, 1977), where they denote the former climate and wind direction.

I.G. Wilson (1971b, 1972a) gave the name transverse draa to transverse dunes having a wavelength in the order of 1–3 km and a height comparable with 100 m. Many dunes of this type are strongly asymmetrical, with enormously tall catena-like avalanche faces, and have *aklé* superimposed on their gentler upwind slopes (Hedin, 1904; Holm, 1960; Norris and Norris, 1961; Glennie, 1970). Subfossil draa fields composed of mixed transverse and barkhan-like forms are known (H.T.U. Smith, 1965b, 1968; Warren, 1976).

Zibar

Holm (1953) gave this name to flat nearly symmetrical transverse features devoid of slip faces which are widely developed from the ill-sorted and bimodal residual sands found in the corridors and basins between large dunes, or underlying extensive desert plains (Bagnold, 1931, 1933; Gabriel, 1938; Capot-Rey, 1947; Monod, 1958; Holm, 1960; H.T.U. Smith, 1963; McKee and Tibbitts, 1964; Capot-Rey and Grémion, 1965; Warren, 1971, 1972; I.G. Wilson, 1972c). Warren (1971, 1972) records wavelengths of 150–400 m and plan shapes from nearly straight to sinuous or irregular. The height of zibar seems to be a tiny fraction of the wavelength, putting the vertical form-index in the region of several hundred.

The association of zibar with deflated areas of poorly graded sands at first suggests their affinity with the larger ballistic ripples. The wavelength of zibar is, however, vastly longer than any conceivable saltation path, and another explanation must be sought for these obscure features.

Parabolic dunes and lunettes

The term parabolic, blow-out, or upsiloidal dune refers to a sand formation of U-shaped plan with generally parallel arms which point upwind, that is, opposite to a barkhan's horns (Hack, 1941; H.T.U. Smith, 1946). Parabolic dunes range widely in size. The length varies from comparable with to many times the width across the arms, and ranges from a few tens to many hundreds of metres. The scoop-like erosional hollow or blow-out lying between the arms is walled on all sides but the upwind by a round to sharp-crested ridge representing the culmination of an outward-dipping slip face. The down-wind portion of the lee side is almost invariably active, but usually the wings are fully stabilized by vegetation, which also occurs patchily in the blow-out itself. The crest can rise as much as 10 m above the floor of the hollow. Parabolic dunes are generally found together in small to large numbers. Typically they have an irregular, staggered arrangement, but may locally be nested.

These dunes early attracted attention because of their prevalence on temperate coasts (Steenstrup, 1894; Olsson-Seffer, 1910; Cressey, 1928; Enquist, 1932; Hack, 1941; Landsberg, 1956; Cooper, 1958, 1967; Van Straaten, 1961; Jennings, 1967; Gripp, 1968; Depuydt, 1972; Ritchie, 1972; Bigarella, 1975), though occurring also in warmer settings (Psuty, 1966; Thom, 1967; Orme, 1973), and occasionally in periglacial and hot deserts (e.g. McKee, 1966b; Verstappen, 1970; Bowler, 1971; Seppälä, 1971, 1972; Ahlbrandt and Andrews, 1978). Cooper (1958) admirably summarizes the conditions favouring their production, namely: (1) a generally stabilized surface, essential for wind attack at scattered places of weakness, (2) considerable initial thickness of sand, so that advance can be restricted to comparatively narrow fronts, and (3) a unidirectional effective wind. Once a sufficiently large blow-out is created, an active slip face of semi-circular plan can form and build forward, extending the arms of the dune progressively further down-wind (Hurault, 1966; Ritchie, 1972).

Akin to parabolic dunes in shape and orientation relative to the dominant wind are clay dunes (Coffey, 1909) or lunettes (Hills, 1940), composed of a variable mixture of primary particles, notably sand-sized aggregates of clay minerals, gypsum, and dolomite, with quartz sand (occasionally predominant) and the fragile shells of salt-tolerant molluscs (W.A. Price, 1963; Trichet, 1963; Bowler, 1973). Active lunettes are confined to areas of high temperature and seasonal dryness, notably the western Gulf of Mexico and the borders of the Sahara (Coffey, 1909; W.H.J. King, 1916; W.A. Price, 1933, 1958, 1963; Huffman and Price, 1949; Boulaine, 1954; Tricart, 1954a, 1954b; Jauzein, 1958; Fisk, 1959; Price and Kornicker, 1961; Perthuisot and Jauzein, 1975). Forms active in the comparatively recent past (Fig. 8-13) abound in southeastern Australia (Hills, 1939, 1940; Stephens and Crocker, 1946; Bettany, 1962; Campbell, 1968; Macumber, 1970; Bowler, 1971, 1973, 1976; Sprigg, 1978), and exist locally in southern Africa (Grove, 1969; Lancaster, 1978a, 1978b). Typically, the lunette is a solitary dune, invariably associated with some sort of depression. It varies in plan from crescentic when fringing the downwind margin of a pan, playa, or sabkha, to meandrine or irregularly linear when built on the leeward shore of a tidal channel or lagoon. Dune length depends on the size of the bordered depression. The

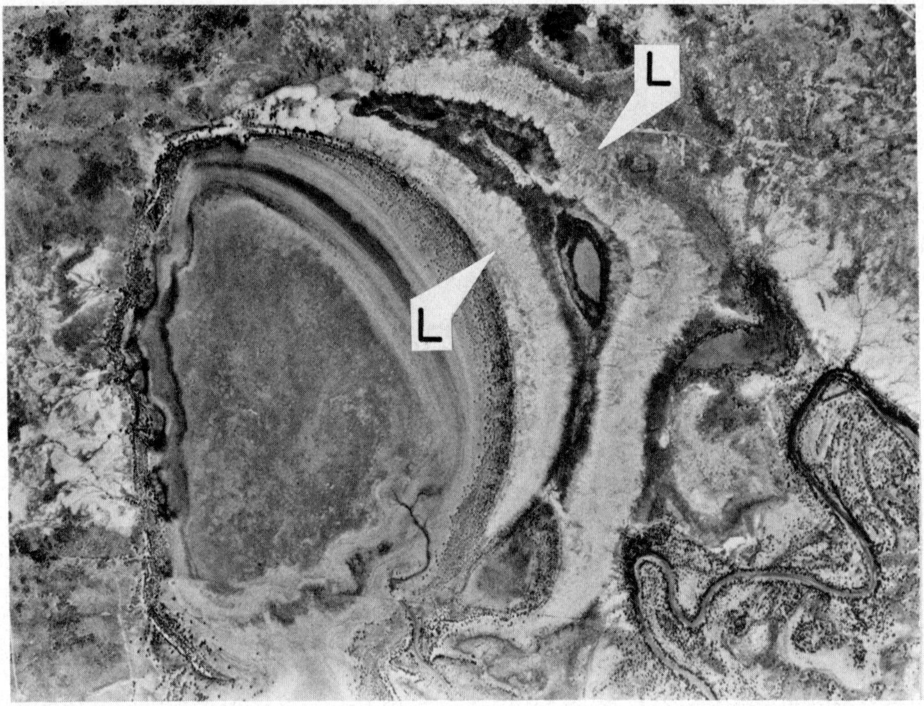

Fig. 8-13. Subfossil lunettes (L) on the eastern (downwind) margin of saline Lake Warrawenia (approximately 3 by 5 km), Darling River area, New South Wales, Australia. Photograph courtesy of J.M. Bowler and reproduced by permission of Crown Lands Office, New South Wales.

smallest lunettes have plan dimensions measured in tens of metres; many of the larger subfossil forms of Australia have crests several tens of kilometres long. Crest heights are uneven but seldom exceed 15 m. Cross-sections are inconsistent, ranging from asymmetrical, with either the upwind or the down-wind side the steeper, to symmetrical and rounded.

The origin of the Australian subfossil lunettes was for a long time debated (Bowler, 1973), but it is now clear from the work of Tricart (1954a, 1954b), Price and Kornicker (1961), and Price (1963) that clay dunes must be regarded as the result of the transport by a unidirectional effective wind of loose aggregates of clay, salts and shells from the bottom of some temporarily dried-up water body. Bowler (1973) discusses the extent to which palaeoclimatic inferences can be drawn from lunettes.

DUNES SHAPED BY FLOWING WATER

Since the U.S. Army Corps of Engineers (see Lane and Eden, 1940) detected them in the Mississippi, and Gilmore (1874), Cornish (1901a,

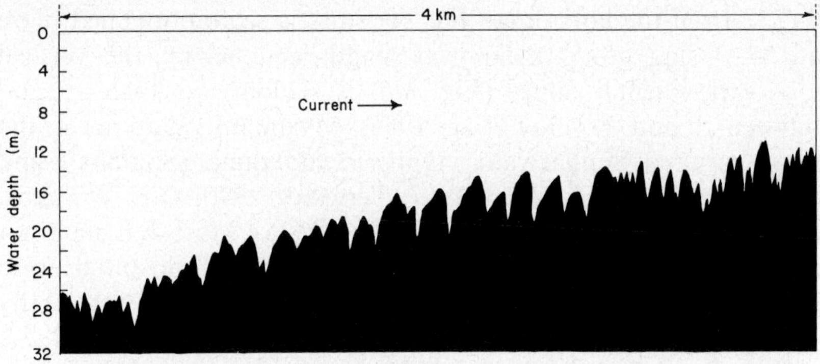

Fig. 8-14. Part of an echo-sounder profile from the Mississippi River, Baleshed Landing Dike Field, Choctaw Bend, Greenville, Mississippi. After Tietze (1975).

1901b, 1914), Sorby (1908), and Kindle (1917) reported tidal examples, dunes have been recognized as perhaps the commonest of all the bedforms shaped by flowing water.

Our knowledge of dune morphology is incomplete, particularly with respect to the larger forms, as the ridges can be adequately studied only on dried-out channel beds and sand banks, or using side-scan sonar. Echo-sounding affords excellent profiles across submerged dune fields, but gives no immediate information on three-dimensional bed shape. A strongly asymmetrical streamwise profile typifies dunes (Fig. 8-14). The lee slopes

Fig. 8-15. Long-crested dunes of 2–5 m wavelength in fine sand, east side of Scolt Head Island, Norfolk, England. Current from upper right to lower left.

down at 30–35° from the horizontal, but the stoss is seldom inclined more steeply than 5°. Using group mean wavelength and height, the vertical form-index of experimental dunes (Fig. 8-6) is seldom less than 15 and generally between 20 and 30 (Guy et al., 1966). Maximum values are in the order of several hundred. Similar values typify natural dunes (Shinohara and Tsubaki, 1959; Whetten and Fulham, 1967; Terwindt, 1970; G.E. Williams, 1971; Boothroyd and Hubbard, 1974, 1975; Hine, 1975). Most dunes are not strongly three-dimensional (Fig. 8-15). Crests are long (horizontal form index generally exceeds 10), in plan ranging from irregular but nearly straight to moderately sinuous (e.g. Cornish, 1901a; Newell and Rigby, 1957; Lundqvist, 1963; Pettijohn and Potter, 1964; Guy et al., 1966; Whetten et al., 1969; G.E. Williams, 1971; Boothroyd and Hubbard, 1974; Lonsdale and Malfait, 1974; Reinson, 1979). In the dune troughs downstream from weakly to moderately rounded lobes occur shallow hollows with streamwise spurs between. All three types of juncture are known, with perhaps the open type the most common. Strongly three-dimensional dunes take two main forms. The lunate type (Fig. 8-16), terminating a morphological series which begins with sinuous-crested forms (Allen, 1968c), has a similar breadth to wavelength, a strongly curved crest, and a deep trough (Reineck, 1960a, 1963; Allen, 1968c). Of a related character are the much-smoothed D-form of Häntzschel (1938) and Reineck (1963), and the "lunate megaripples" of Clifton et al. (1971). The other strongly three-dimensional kind of dune (Fig.

Fig. 8-16. Lunate dunes in fine sand, passing downstream into longer-crested forms, Barmouth Estuary, Wales. Scale 0.5 m long. Current from bottom to top.

Fig. 8-17. Oblique dunes in fine sand, Wells-next-the-Sea, Norfolk, England. Scale 0.5 m long and current from right to left.

8-17), Engelund's (1973) oblique form, is typified by series of bold streamwise spurs downstream from a relatively straight crest trending at 30–50° from the line of flow (e.g. Cornish, 1901b; Kindle, 1917; Harms et al., 1963; Harms and Fahnestock, 1965; Guy et al., 1966; Land and Hoyt, 1966; Allen, 1968c; Knight and Dalrymple, 1975).

Echo-sounder surveys reveal dunes on the sandy beds of most of the largest rivers (Lane and Eden, 1940; Carey and Keller, 1957; Sioli, 1965; Snischenko, 1968; J.M. Coleman, 1969; Stückrath, 1969; Anding, 1970; Peters, 1971, 1977; Tietze, 1975; Tricart, 1977). Many rivers of an intermediate scale also possess them in profusion (De Geer, 1911; Sundborg, 1956; NEDECO, 1959; Harms et al., 1963; Lundqvist, 1963; Brice, 1964; Harms and Fahnestock, 1965; Clos-Arceduc, 1967b; Galay, 1967; Whetten and Fullham, 1967; Neill, 1969; Whetten et al., 1969; Collinson, 1970b; Culbertson and Scott, 1970; N.D. Smith, 1970, 1971b; Culbertson et al., 1971; Nilsson and Martvall, 1972; Haushild et al., 1973, 1975; Singh and Kumar, 1974; R.G. Jackson, 1976a; Levashov, 1976; Cant, 1978; Gustavson, 1978). Small rivers are by no means sedimentologically negligible; many possess readily accessible dune covered beds (Shantzer, 1951; Lopatin, 1952; Shamov, 1959; Shinohara and Tsubaki, 1959; Martinec, 1967; Fahnestock et al., 1969; Korchokha, 1969; Bluck, 1971; G.E. Williams, 1971; Karcz, 1972; Erkek, 1973; Nordseth, 1973; Fahenstock and Bradley, 1974; Boothroyd and Ashley, 1975; Bridge and Jarvis, 1976; Levey, 1976; Koutaniemi, 1979).

Dunes abound in the tidally-influenced lower reaches of rivers, where they move chiefly during the flood season (Pretious and Blench, 1951; Ballade, 1953; Terwindt et al., 1963; Vollmers and Wolf, 1969; Terwindt, 1970; Nasner, 1973, 1974).

Most estuaries and tidal deltas, and some barrier coasts and shallow-marine platforms, present a complex of shoals and channels adjusted to horizontally segregated ebb and flood tidal currents. Consequently, the flow at many sites is effectively one-way and may give rise on the sand bed to trains of dunes. Cornish (1901a, 1901b, 1914) was the first to give a detailed account of estuarine dunes, and many further records have since become available (Kindle, 1917; Van Straaten, 1950; Allen, 1965a, 1968c; G. Evans, 1965; Land and Hoyt, 1966; Salsman et al., 1966; Völpel and Samu, 1966; Hoyt, 1967; Tricart, 1967; Hervieu, 1968; Swift and McMullen, 1968; G.P. Allen et al., 1969, 1971; Daboll, 1969; Haynes and Dobson, 1969; Farrell, 1970; Hartwell, 1970; Klein, 1970b; Göhren, 1971c; Mishra, 1971; M.M. Nichols, 1972; Wright et al., 1972, 1973, 1975; Luternauer and Murray, 1973; Boothroyd and Hubbard, 1974, 1975; Dalrymple et al., 1975, 1978; C.D. Green, 1975; Knight and Dalrymple, 1975; Allen and Friend, 1976b; Boggs and Jones, 1976; McCave and Geiser, 1979; Jago, 1980). The occurrence of dunes on tidal deltas, near-shore shoals, and beaches is described by several workers (R. Richter, 1926a, 1926b; Häntzschel, 1938, 1939; Hülsemann, 1955; Reineck, 1960a, 1963; Boersma, 1969; Newton and Werner, 1969, 1970; W.G.H. Maxwell, 1970; Clifton et al., 1971; W.R. Parker, 1973, 1975; Hine, 1975; Hubbard, 1975; Reinson, 1977, 1979). Swift et al. (1979) report forms from an open shelf. Dunes are also to be found in a similar range of carbonate environments (Newell et al., 1951, 1959; Illing, 1954; Newell and Rigby, 1957; McKee and Sterrett, 1961; M.M. Ball, 1967; Imbrie and Buchanan, 1965; Jindrich, 1969; Kendall and Skipwith, 1969; G.R. Davies, 1970; Farrow, 1971; Farrow and Brander, 1971; Hine, 1977). Evidence is accumulating that dunes can be formed in a wide variety of marine environments by other than tidal currents (Heezen and Johnson, 1969; Melières et al., 1970; Werner and Newton, 1970, 1975a, 1975b; Kelling and Stanley, 1972; Lonsdale et al., 1972; Kenyon and Belderson, 1973; Harms et al., 1974; Werner et al., 1974; B.W. Flemming, 1978, 1980).

The forms of water-shaped dunes are not readily preserved. Most remarkable of all are the dune fields shaped by Late Pleistocene break-out floods in the western U.S.A. (Thiel, 1932; Pardee, 1942; Bretz, 1959; Bretz et al., 1956; Whetten et al., 1969; V.R. Baker, 1973). These have modern counterparts in the gravel dunes shaped by recent break-out floods on the Knick River, Alaska (Fahnestock et al., 1969). Less impressive are trains of small dunes preserved in sandstones (Banks et al., 1971; Brasier et al., 1978; Hodgson, 1978), beneath shales (R. Richter, 1926a; McDowell, 1957; Swett and Smit, 1972), or below tuff (Cowperthwaite et al., 1972). Winn and Dott (1977) describe very large dune forms from a deep-water deposit. Dune forms are known from the fluviatile Old Red Sandstone, in the English West

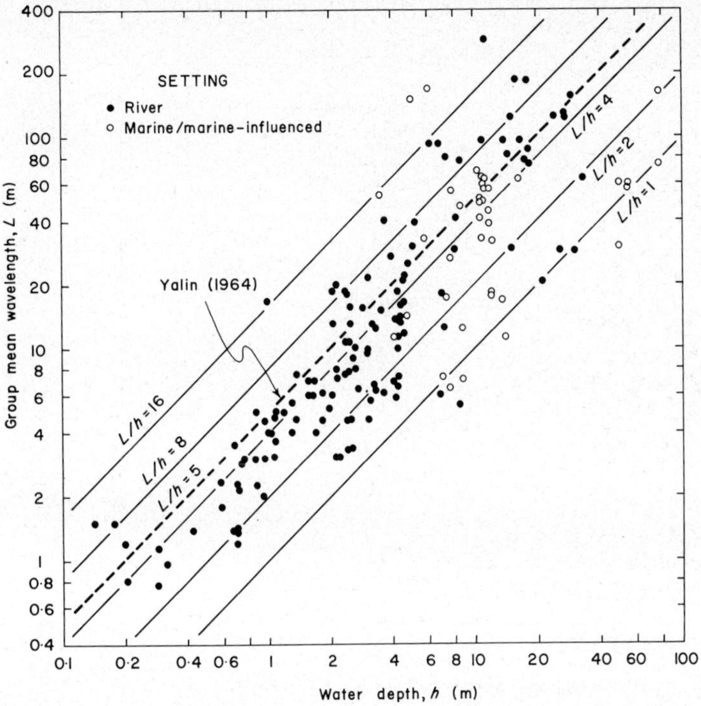

Fig. 8-18. Correlation between group mean wavelength and water depth for dunes in river and marine or marine-influenced setting. Data sources for river setting: Annambhotla et al. (1972), Carey and Keller (1957), Coleman (1969), Culbertson and Scott (1970), Galay (1967), Haushild et al. (1975), Jackson (1976a), Korchokha (1972), Lane and Eden (1940), Martinec (1967), Neill (1969), Nordin (1971), Shinohara and Tsubaki (1959), Snischenko (1968), Tietze (1975), Znamenskaya (1966). Data sources for marine or marine-influenced settings: Ballade (1953), B.W. Flemming (1979), Harms et al. (1974), Nasner (1974), Reinson (1979), Salsman et al, (1966), Werner et al. (1974), Wright et al. (1972). The scatter is large, particularly at the larger water depths, but the scale of dunes clearly increases with increasing scale of flow.

Midlands (Allen, 1974a), South Wales (Allen and Williams, 1978, 1979), and the Hook Head Peninsula, southeast Ireland. Fairchild (1980) found small dune forms preserved amongst late Precambrian carbonates.

The general scale of dunes in terms of wavelength is set by grain size, flow velocity, and boundary layer thickness, equivalent to mean flow depth for a river or unstratified tidal flow, but to some lesser and usually unknown depth in a stratified system (e.g. Kenyon and Belderson, 1973; Lonsdale and Malfait, 1974). Figure 8-18 is a plot of instantaneous group mean dune wavelength against water depth for a range of representative flow systems. The wavelength/depth ratio is in the general order of 5, broadly agreeing with Yalin's (1971, 1972) theoretical model, but the data are greatly scattered. Individual systems may reveal between wavelength and depth either a positive (e.g. G.E. Williams, 1971) or a negative (e.g. Werner et al., 1974; Hubbard, 1975) correlation, or virtually no correlation (R.G. Jackson, 1976a).

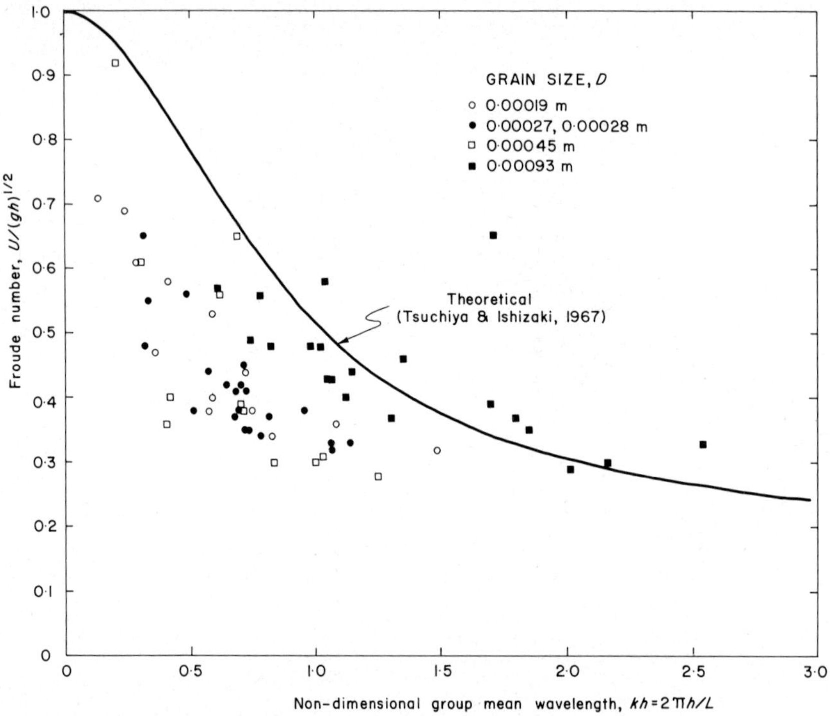

Fig. 8-19. Effect of bed-material calibre on the correlation between non-dimensional group mean dune wavelength and Froude number. The plotted curve, fitted well by dunes in the coarsest sand ($D=0.00093$ m), is that derived by Tsuchiya and Ishizaki (1967) for dune wavelength.

These differences at the system level seem partly attributable to non-uniformity and unsteadiness (Chapter 12), though the wavelength of steady-state equilibrium dunes is also not strongly correlated with flow depth (Guy et al., 1966; Allen, 1970g, 1977a). The influence of grain size (e.g. USWES, 1935; Simons and Richardson, 1965; Simons et al., 1965b; Guy et al., 1966), perhaps working through the ratio of bedload to suspended load, is well seen in Fig. 8-19, in which the non-dimensional group mean dune wavelength is plotted against Froude number. The actual wavelength declines with increasing grain size and, at the largest grain size plotted, when the fraction of the total load moving as bedload should be greatest, agrees well with the theoretical value of Tsuchiya and Ishizaki (1967).

The general height of dunes depends upon boundary layer thickness (Allen, 1963b, 1968c) and flow severity for each grain size. Figure 8-20 gives values for instantaneous group mean dune height and flow depth from a range of systems. The scatter is again considerable, possibly a reflection of individual unsteadiness and non-uniformity, and the values for relative height agree but poorly with Yalin's (1964, 1972) predictions. The detailed form of the height-depth correlation again varies widely between systems (cf.

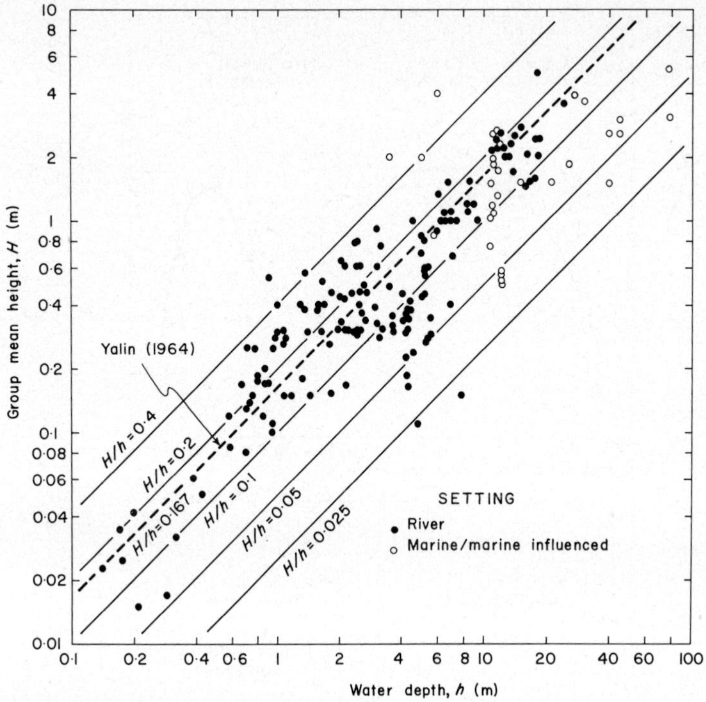

Fig. 8-20. Correlation between group mean height and mean water depth for dunes in river and marine or marine-influenced settings, with the relative dune height (H/h) as a parameter. Data sources for river setting: Annambhotla et al. (1972), Culbertson and Scott (1970), Galay (1967), Haushild et al. (1975), Jackson (1976a), Korchokha (1969, 1972), Lane and Eden (1940), Martinec (1967), Neill (1969), Nordin (1971b), Shinohara and Tsubaki (1959), Snishchenko (1968), Stückrath (1969), Tietze (1975), Znamenskaya (1966). Data sources for marine and marine-influenced setting: Ballade (1953), B.W. Flemming (1978), Harms et al. (1974), Nasner (1974), Reinson (1979), Salsman et al. (1966), Werner et al. (1974), Wright et al. (1972). Although the scatter is considerable, it is clear that dune height increases with the scale of the flow system.

Korchokha, 1969; G.E. Williams, 1971; Werner et al., 1974; Hubbard, 1975; R.G. Jackson, 1976a; B.W. Flemming, 1978). Experimentally, dune height increases with flow depth, but with much scatter (Guy et al., 1966; Allen, 1977a). Several workers showed in the laboratory that relative dune height tends to increase with flow-severity, in some cases to a definite maximum at an intermediate condition in the range of strengths appropriate to dunes (Shinohara and Tsubaki, 1959; Stein, 1965; G.P. Williams, 1967; Korchokha, 1969; Carstens and Altinbilek, 1972; Zanke, 1976a; Allen, 1977a). Representative data appear in Fig. 8-21 with the Shields-Bagnold θ as the measure of strength, the plots broadly confirming Gill's (1971) theory (eq. 7.19). Bell-shaped plots also emerge when the reciprocal of the vertical form-index is related to a measure of flow severity (Korchokha, 1972; McDonald and Vincent, 1972; Pratt and Smith, 1972; Yalin, 1972; Yalin and Karahan, 1979), in conformity with Fredsøe's (1975) similitude model (eq. 7.20).

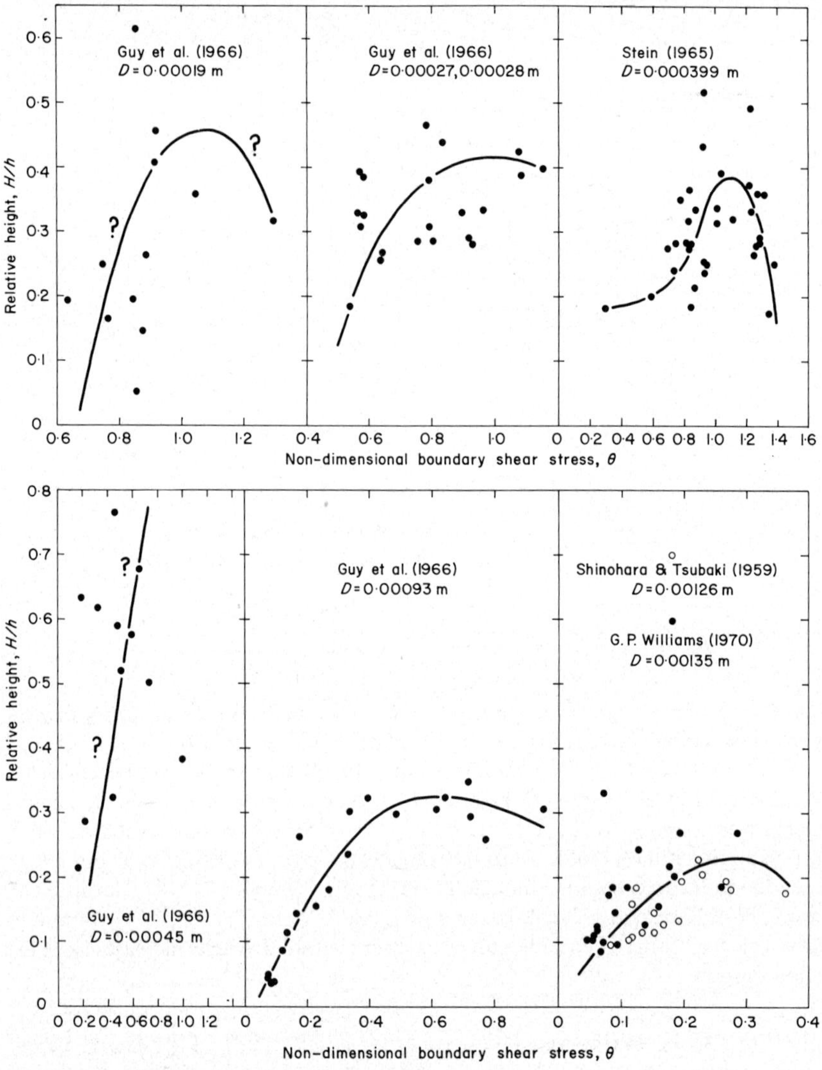

Fig. 8-21. Group mean dune height relative to mean water depth as a function under equilibrium conditions of the non-dimensional boundary shear stress and sediment calibre. Stress corrected for wall effects by procedure of G.P. Williams (1970).

INDIVISIBILITY OF DUNES

Instead of assigning to a single hydromorphological category (dunes as used above) the transverse bedforms equipped with an avalanche face that are intermediate in size between current ripples and the very largest forms in an aqueous system, some workers believe that they can recognize two morphologically distinct kinds of feature which, it is commonly implied or stated, are also distinct hydrodynamically. The one kind, generally the smaller, steeper, and more strongly three-dimensional, goes by such names as "megaripples" (Daboll, 1969; Farrell, 1970; Hartwell, 1970; Boothroyd and Hubbard, 1974, 1975; Dalrymple et al., 1975; Hine, 1975), "type 2 megaripples" (Dalrymple et al., 1978), and "simple dunes", "dunes" or "sand waves" (Klein, 1970b; Harms et al., 1974; R.G. Jackson, 1976a). To the other kind — flatter, straighter crested, and often of longer wavelength— is applied such terms as "sand wave" (Daboll, 1969; Farrell, 1970; Hartwell, 1970; Boothroyd and Hubbard, 1974, 1975; Harms et al., 1974; Dalrymple et al., 1975; Hine, 1975), "rippled sand wave" or "type 1 megaripple" (Dalrymple et al., 1978), "simple sand wave" (Klein, 1970b), "diminished dune" (N.D. Smith, 1971b), "scaloid sand wave" (Pryor et al., 1972), and "transverse bar" (R.G. Jackson, 1976a). Here also seem to belong many of the so-called transition bedforms of Guy et al. (1966) and Costello's (1974) "bars".

In river and tidal-shoal environments, the two supposedly distinct kinds of form often occur in separate areas with relatively narrow zones occupied by transitional forms between. These transitional kinds, for example, "linear megaripples" (Boothroyd and Hubbard, 1974), are intermediate in scale and/or dimensionality between the two supposedly different main forms. In some cases, however, the areas occupied the two forms overlap substantially. Harms et al. (1974) report dunes superimposed near the crests of some sand waves. R.G. Jackson (1976a, fig. 9) claims an invariable superimposition of dunes on active transverse bars in the River Wabash, but gives an echo-sounder trace, from a reach of active dunes, in which several bars lack superimposed smaller forms. The general spatial separateness of the two supposedly distinct kinds of form means that each kind is linked to currents that are distinctive as regards speed, depth, direction and, as well, duration and timing within the relevant hydraulic cycle(s). It appears that the more two-dimensional forms occur on the whole in gentler but only occasionally shallower currents than dunes, though the claimed overlap of conditions is substantial (N.D. Smith, 1971b; Klein and Whaley, 1972; Boothroyd and Hubbard, 1974; R.G. Jackson, 1976a; Dalrymple et al., 1978).

These observations do not by themselves prove that the two supposedly different kinds of form are hydrodynamically distinct in any fundamental sense, for the ridges were all observed under markedly unsteady and non-uniform conditions, and little if any information is available concerning their history. Hence the forms seen at each site and time could have been created elsewhere and at other times within the system, and so may have a shape and size unrelated to the flow conditions prevailing when the observations were made (Chapter 12). Theoretically, bimodal and even polymodal dune populations can be fashioned by the simplest unsteady flows (e.g. Allen, 1976b, 1976c, 1978a, 1978b). When as a consequence the equilibrium relationships between the attributes of dune-like bedforms and flow are examined, it becomes likely that the two kinds of form distinguished by the workers mentioned merely represent partly contrasted stages within a continuous, unclustered morphological series, and thus belong to but one hydrodynamic class (Allen, 1976e). Figure 8-21 suggests that relative dune height is an increasing and possibly bell-shaped function of flow strength, and we also know that the forms of smallest relative height are the flattest (Korchokha, 1972; Pratt and Smith, 1972; Yalin, 1972; Yalin and Karahan, 1979). Hence the two-dimensional "sand waves" and "megaripples" could simply be dunes which arose near to either the lower (probably) or upper (unlikely) limit of the dune existence field. The steep three-dimensional forms are likely to be dunes created under more "central" conditions. Costello's (1974) own experiments demand this interpretation, for his bars appeared in gentler currents than his steeper but comparably spaced dunes. Hence the variation in dimensionality amongst dunes—from two-dimensional forms in relatively deep, weak flows to three-dimensional in comparatively shallow and vigorous ones—seems largely to parallel the changes known to occur amongst current ripples as flow conditions are varied across their existence field (Fig. 8-7). There is perhaps a grain-size effect also to take into account, for Fig. 8-21 indicates that dunes in the coarser sands are relatively less tall than those in fine deposits, capable of forming a substantial suspended load, and so are most likely to be relatively two-dimensional in character. In summary, the evidence for the occurrence of two distinct kinds of dune-like bedform does not seem compelling and the matter merits further study.

BEDFORM EXISTENCE FIELDS FOR AQUEOUS ENVIRONMENTS

Mainly for practical reasons, engineers, sedimentologists and, latterly, geomorphologists have plotted the observed occurrence of one-way bedforms on graphs of selected flow variables, commonly arranged in non-dimensional groups, in order to define the existence of the features (see Chapter 7 for theoretical examples). The first such graphs, partly or wholly based on the results of Gilbert (1914) and USWES (1935), were given by Andersen (1931), Langbein (1942), and Menard (1950a). Most later graphs are also plots from equilibrium steady-state experiments, being intended for the prediction of either bed roughness (hydraulic engineering) or flow conditions/bedform (palaeohydraulics, sedimentological modelling). These plots are the best evidence to date on the meaning of bedforms, but are applied with difficulty to natural environments, where flow unsteadiness and non-uniformity create ill-understood effects (Chapter 12). Field data have at other times been plotted, generally for the purpose of comparing some natural with an experimental system. These graphs are useful descriptive devices and for the most part bolster faith in laboratory studies of bedforms, but may to a degree be challenged on the grounds that unsteady, non-uniform flows are unlike equilibrium ones. The recipe for successful prediction is perhaps not the combination of laboratory and field evidence sometimes favoured.

The variables relevant to bed state have been considered by many workers (Shields, 1936; Liu, 1957; Bogardi, 1961; Simons et al., 1961; Kennedy and Brooks, 1965; Barr and Herbertson, 1968; Blench, 1969b, 1970; Hill et al., 1969; Yen and Liou, 1969; Southard, 1971b; B.C. Yen, 1971; Vanoni, 1974). They are the fluid and sediment discharges per unit width, Q and J, respectively, the flow width w, the flow depth h, the mean flow velocity U, the bed or water-surface slope S, the bed friction coefficient f, the fluid density ρ, the fluid kinematic viscosity ν, the mean sediment diameter D, the sediment particle density σ, the sediment particle falling velocity W, and the acceleration of gravity g. Several of these variables are inter-related in various ways, either by balance or constitutive relationships, or by definition. Similarly, they can be variously arranged into non-dimensional groups, chosen either on practical grounds, or for some supposed physical significance. Many such groups also prove to be inter-related. One of the earliest sets suggested (Shields, 1936; Liu, 1957) is:

$$\frac{\rho g h S}{(\sigma-\rho) g D}, \quad \frac{U_* D}{\nu}, \quad \frac{U_*}{W} \tag{8.3}$$

where the first group is the Shields-Bagnold θ, the second a grain Reynolds number (the shear velocity $U_* = (ghS)^{1/2}$), and the third a grain Froude number. Another set, drawn upon by Hill et al. (1969) and Southard (1971b), comprises:

$$\frac{\sigma}{\rho}, \quad \frac{g h^3}{\nu^2}, \quad \frac{U^3}{\nu g}, \quad \frac{g D^3}{\nu^2} \tag{8.4}$$

in which the second and subsequent terms to the right are all Reynolds numbers, but can also be interpreted as non-dimensional forms respectively of depth, mean flow velocity, and sediment size. This set of groups brings out well the influence of changing viscosity in natural aqueous environments (a broadly two-fold variation occurs in the temperature range 0–30°C). A third widely used set (e.g. Vanoni, 1974) has for its major elements:

$$\frac{U}{(gh)^{1/2}}, \quad \frac{h}{D}, \quad \frac{U_* D}{\nu}, \quad \frac{W}{U} \tag{8.5}$$

where the first term is the flow Froude number and the last a grain Froude number or, alternatively, a mobility number.

Some additional inter-relationships may be noted. Bogardi (1965) showed that the Shields-Bagnold θ is also a density-adjusted grain Froude number, $(\rho/\sigma-\rho)(U_*^2/gD)$. We also find that, for a given sediment, fluid and temperature, the Shields-Bagnold group is proportional to the square of Shields' (1936) grain Reynolds number, $U_* D/\nu$. It is also proportional to the flow Froude number, for each sediment, fluid, flow depth and bed roughness. In addition, for grains sufficiently large that W is proportional to $D^{1/2}$, Liu's (1957) grain Froude number, U_*/W, increases as $\theta^{1/2}$. Bogardi (1965) pointed out that the flow Froude number and relative depth, favoured by Vanoni (1974) and many others, are related by definition through:

$$\frac{U}{(gh)^{1/2}} = \frac{U}{U_*}\left(\frac{U_*}{gD}\right)^{1/2}\left(\frac{D}{h}\right)^{1/2} \tag{8.6}$$

where U/U_* is a measure of flow resistance. A particularly interesting case is the grain Reynolds number, gD^3/ν^2, of Hill et al. (1969) and Southard (1971b). It corresponds to Bagnold's (1954a, 1956) grain Reynolds numbers, except that in the latter a "kinematic viscosity" may be formed from sediment and fluid properties. Bonnefille's (1965) non-dimensional grain size, $(\sigma - \rho)gD^3/\rho\nu^2$, is also closely related to Bagnold's Reynolds numbers, and is known to chemical engineers as the Galileo number.

Probably the most successful of the available existence diagrams is that proposed from experimental data by Simons and Richardson (1965, 1966, 1971) and Simons et al. (1965b), revised for coarse sediments and antidunes by Allen (1968c, 1969c, 1970g). The bedforms appear in the stream power-grain size plane, that is, the plane $U(\rho ghS)$–D, where the bracketed term will be recognized as the mean bed shear stress. Figure 8-22 is a new version of the plot, based on all available experimental data of acceptable quality. The power values are wall-corrected and temperature differences between experiments are compensated for by plotting the effective median physical sediment diameter at the standard temperature of 25°C. Hence the diagram can be applied at other temperatures simply by an appropriate shift of the existence-field boundaries, calculated from the settling velocity laws. The plot is attractive because it relates bedform and sediment calibre, each readily ascertainable in the stratigraphic record, to the stream power upon which the sediment transport rate depends. Like all practical existence diagrams, however, it does not separate dunes and ripples from plane beds particularly well. This is because the contribution to the total mean bed shear stress arising from the shapes of ripples and dunes has little effect on the sediment transport rate.

Several workers plotted field data in the stream power-grain size plane, but found only moderate agreement with the distribution of experimental bedforms (N.D. Smith, 1971b; Klein and Whaley, 1972; C.D. Green, 1975; Bridge and Jarvis, 1976). Harms (1969) used the plot for work on current ripples.

In the most widely used class of existence diagram, bedforms are shown in a graph in which D, usually in a non-dimensional form, appears on the abscissa with the Shields-Bagnold θ, or a related variable or non-dimensional group, on the ordinate. The θ–$U_* D/\nu$ plot of Shields (1936) is the oldest of these but does not separate the bedforms (particularly plane beds from

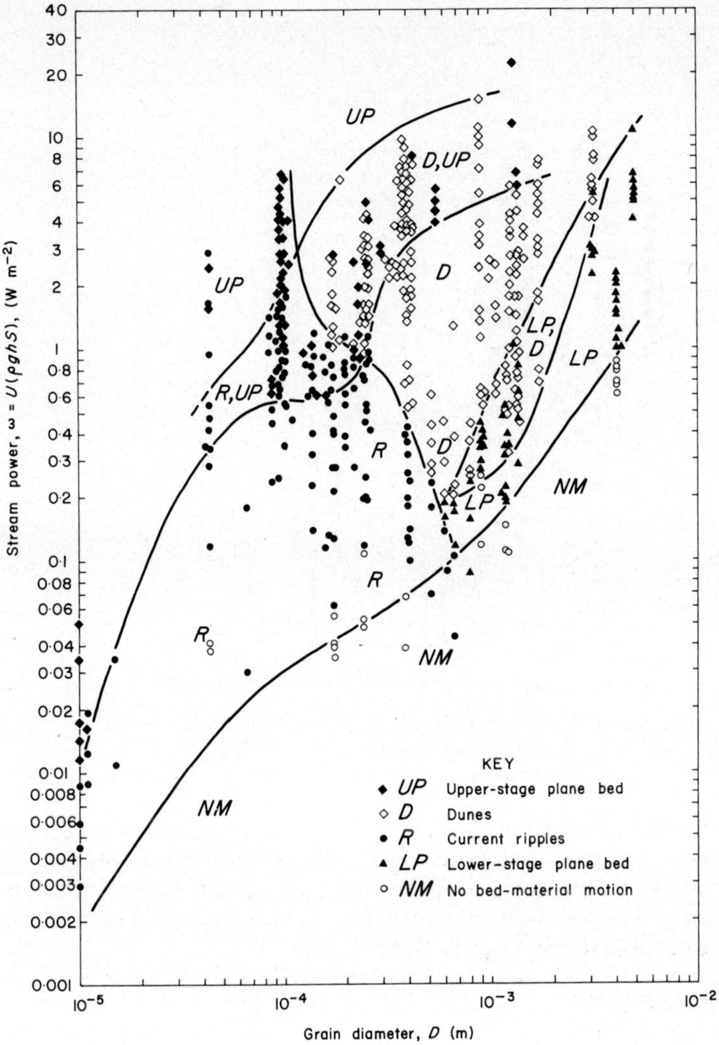

Fig. 8-22. Experimental existence fields for aqueous bedforms under equilibrium conditions, shown in the stream power (wall-corrected)-grain size plane at 25°C. A total of 566 observations is represented, but many data points (except critical or limiting ones) are omitted for clarity of presentation. Sources of data: Costello (1974, $D=5.1\times10^{-4}$, 6.0×10^{-4}, 7.9×10^{-4}, 1.15×10^{-3} m), Gee (1975, $D=3.1\times10^{-4}$, 1.05×10^{-3} m), Gilbert (1914, $D=1.71\times10^{-3}$, 3.17×10^{-3}, 4.94×10^{-3} m), Guy et al. (1966, $D=1.9\times10^{-4}$, 2.7×10^{-4}, 2.8×10^{-4}, 4.5×10^{-4}, 4.5×10^{-4}, 9.3×10^{-4} m), Hill et al. (1969, $D=8.8\times10^{-5}$, 1.5×10^{-4}, 3.1×10^{-4} m), Jopling and Forbes (1979, $D=4.5\times10^{-5}$ m), Kalinske and Hsia (1945, $D=1.1\times10^{-5}$), Kennedy (1961, $D=2.33\times10^{-4}$, 5.49×10^{-4} m), Mantz (1978, $D=1.5\times10^{-5}$, 6.6×10^{-5} m), Rees (1966b, $D=1.0\times10^{-5}$ m), Shinohara and Tsubaki (1959, $D=1.26\times10^{-3}$, 1.46×10^{-3} m), Stein (1965, $D=3.99\times10^{-4}$ m), B.D. Taylor (1971, $D=2.28\times10^{-4}$ m), Taylor and Vanoni (1972b, $D=1.38\times10^{-4}$, 2.28×10^{-4} m), USWES (1935, $D=4.08\times10^{-3}$ m), Vanoni and Brooks (1957, $D=8.8\times10^{-5}$, 1.37×10^{-4} m), Vanoni and Hwang (1967, $D=2.06\times10^{-4}$ m), G.P. Williams (1967, 1970, $D=1.35\times10^{-3}$ m), Willis et al. (1972, $D=1.0\times10^{-4}$ m). The stream power is based on wall-corrected (method of G.P. Williams, 1970) boundary shear stress values and the grain diameters listed are corrected to water at 25°C.

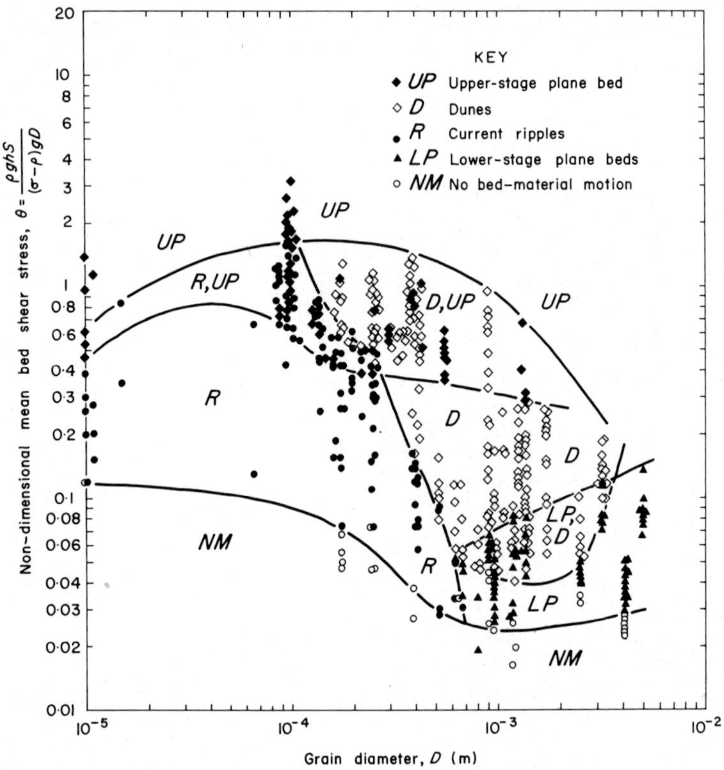

Fig. 8-23. Experimental existence fields for aqueous bedforms under equilibrium conditions, shown in the non-dimensional mean bed shear stress (wall-corrected)-grain size plane at 25°C. A total of 595 observations is represented, but many data points (except critical or limiting ones) have been omitted for clarity of presentation. Data as listed for Fig. 8-22 with the addition of the set of Chabert and Chauvin (1963, $D=9.6\times10^{-4}$, 2.48×10^{-4} m). The stress values are wall-corrected (method of G.P. Williams, 1970) and the grain diameters plotted are adjusted to water at 25°C.

ripples and dunes) any better than the stream power-grain size plot (Chabert and Chauvin, 1963; Acaroglu and Graf, 1968; R.G. Jackson, 1976c). It has the disadvantage that sediment transport is not as closely dependent on θ as on stream power. Closely related are the θ–D plots of Bagnold (1963), Harms (1969), and Srodon (1974). Figure 8-23 shows experimental bedforms in this plane using the same wall-corrected data set as before. Note the considerable extent to which the fields for plane beds overlap those for ripples and dunes. Figure 8-24 gives the field boundaries without the data points and, based on Fig. 8-21, shows tentative contours for H/h in the dune field. Other representations of a similar kind to Fig. 8-23 involve the use of the U–D plane (Menard, 1950a; Kádár, 1966; Southard and Boguchwal, 1973), the $gD/U^2{}_*$–D plane (Bogardi, 1961, 1965; C.D. Green, 1975), the $gD/U^2{}_*$–D plane (Bogardi, 1961, 1965; C.D. Green, 1975), the U_*–D plane

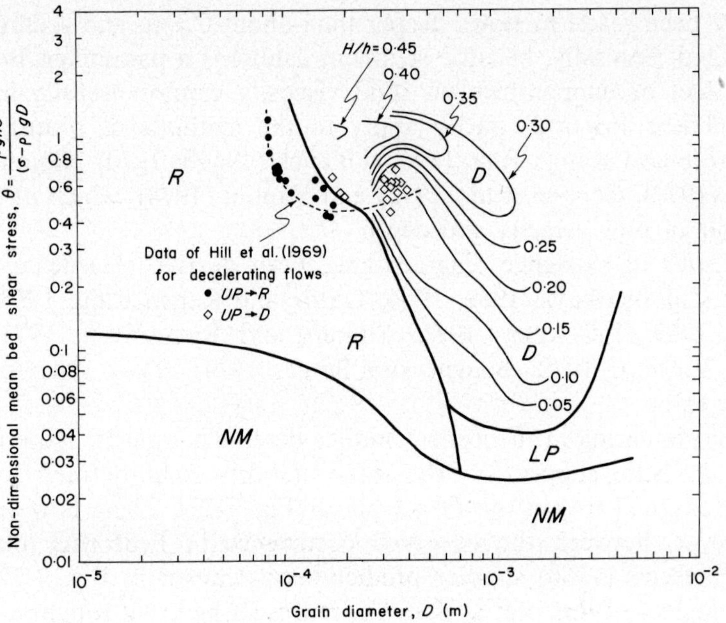

Fig. 8-24. Existence fields for aqueous current ripples and dunes in the non-dimensional mean bed shear stress (wall-corrected)-grain size plane at 25°C (see Fig. 8-23), to show the (tentative) variation of relative dune height (group mean height divided by mean water depth) as a function of flow and bed conditions (based on Fig. 8-21). Also plotted are the data of Hill et al. (1969), the only experimental results on the stability of upper-stage plane beds in effectively decelerating flows. At the temperature for which the graph is plotted, these data intersect the ripple-dune boundary at the triple point ($\theta = 0.46$, $D = 0.00025$ m).

(Bogardi, 1961), the $U/(gh)^{1/2}$–D plane (Engelund and Fredsøe, 1974), the U_*/W–U_*D/ν plane (Liu, 1957; Simons et al., 1961; Chabert and Chauvin, 1963; Culbertson and Dawdy, 1964; Sahgal and Singh, 1974), the gD^3/ν^2–U_*D/ν plane (Hill et al., 1969), a plane involving U_*D/ν and Bonnefille's $(\sigma-\rho)gD^3/\rho\nu^2$ (Bonnefille, 1965; Vollmers and Wolf, 1969), the U_*D/ν–WD/ν plane (Albertson et al., 1958; Liu and Hwang, 1961; Beckman and Furness, 1962; Mercer, 1971a), the $U/(gh)^{1/2}$–$gD^3/\rho\nu^2$ plane (Sahgal and Singh, 1974), a plane involving U/W, UD/ν and the flow Froude number (Simons and Richardson, 1961), and the U/W–$(\sigma-\rho)gD^3/\rho\nu^2$ plane (Zanke, 1976a, 1976b).

Simons et al. (1961) and Southard (1971b) advocated a diagram in which bedforms appear in the h–U plane for each value or range of D. It was later used in experimental work by Costello (1974) and in field studies by Klein and Whaley (1972), Boothroyd and Hubbard (1974, 1975), Harms et al. (1974) and R.G. Jackson (1976b). Such plots have the advantage that antidunes can validly be shown and are attractive as descriptive devices in that the only flow properties required are readily measurable ones. For prediction, however, two serious weaknesses exist. Firstly, few laboratory experiments have been made in flows deeper than about 0.3 m, and scaling factors are excluded. Secondly, because sediment calibre is a parameter, the non-negligible effect of temperature on fluid viscosity cannot usefully be accommodated. These strictures partly apply to the methods of plotting suggested by Garde and Ranga Raju (1963), Blench (1969b, 1970), Simons and Richardson (1971), Cooper et al. (1972) and Vanoni (1974), which also emphasize the role of flow velocity and depth.

Many other forms of existence diagram have been devised (Garde and Albertson, 1959; Znamenskaya, 1962, 1964; Garde and Ranga Raju, 1963; Yen and Liou, 1969; B.C. Yen, 1971; Simons and Richardson, 1971; McDonald and Vincent, 1972; Sahgal and Singh, 1974). They seem of doubtful or little value.

Two diagrams originating in theoretical studies deserve comment. Experimental work gives little support to the dune stability field defined by Engelund and Fredsøe (1974) in the *Fr–kh* plane (Fig. 7-11). There is good agreement, however, between the observed occurrence of bedforms and Engelund and Hansen's (1966) simpler predictions as shown in Fig. 7-12. The practical weakness of this plot is that U'_*, representing grain roughness only, is ordinarily unascertainable. The use of this shear velocity does, however, greatly reduce the overlap of current ripples and dunes on to the plane bed fields.

The virtual disappearance of overlap using the grain roughness shear velocity suggests that Figs. 8-22 and 8-23 can be employed in different ways for prediction, accordingly as a flow is increasing or decreasing in strength with time. With decreasing flows starting in the upper-stage plane bed field, this type of bed should persist to a limit defined by its lowest experimental occurrence. Bagnold's (1966) universal plane-bed criterion (eq. 7.4, Figs. 7-4 and 8-23) describes this limit fairly well, as assumed by Allen (1970f, 1971d) for decaying river and turbidity currents. It is also satisfactorily supported by the experimental results of Hill et al. (1969), which define effectively for decelerating flows the temperature-sensitive triple point between the upper-stage plane bed, ripple and dune fields (Allen, 1972c; Allen and Leeder, 1980). This point coincides with the viscous sublayer approximately equal in thickness to the diameter of the bed particles, a further confirmation of the views of such as Sundborg (1956) and Yalin (1972) that ripples and dunes reflect contrasted roughness regimes. In flows of increasing strength, the bed seems likely to remain rippled or dune-covered until the stream power or non-dimensional stress just exceeds the respective upper limits to these bedforms (Figs. 8-22 and 8.23).

Because of form roughness and flow-separation effects, the lower limit to the ripple field (Figs. 8-22 and 8-24) is not without ambiguity. Menard (1950a) and Rathburn and Guy (1967), for example, observed steady sediment transport on a rippled bed at lower mean flow velocities than entrained the same sediment on a flattened bed. In this region of the existence diagram, we can recognize at least four modes of response of an initially plane bed, that is, a bed devoid of features of relief taller than one grain diameter (Southard, 1971b; Southard and Dingler, 1971; Williams and Kemp, 1971, 1972; Southard and Boguchwal, 1973; Costello, 1974). Up to a relatively low mean flow velocity, no entrainment occurs, whether or not irregularities are imposed on the bed surface. At slightly larger velocities, no entrainment occurs on a plane bed, but the introduction of sufficiently large irregularities causes ripples to arise and propagate. In a very narrow range of still greater velocities, transport over a plane bed occurs and ripples propagate from sufficiently large defects. The fourth mode, observed in the highest range of velocities, is typified by the propagation of ripples over the bed whatever its initial state. The grading of the bed material seems to affect the response of the bed at the lower flow stages. After an initial period of weak sediment transport, commonly leading to local rippling, movement ultimately ceases as the finer grains roll into stable hoppers between the coarser and the bed develops a crust or "armour" (Williams and Kemp, 1971; Moss, 1975). Although these modes are of theoretical interest, it is difficult and of little importance under field conditions to distinguish between them. No attempt to separate them is made in Figs. 8-22 and 8-23.

SUMMARY

The common transverse bedforms fashioned in gravel and/or sand-sized sediments are ballistic ripples, barkhans, lunettes, parabolic dunes, and transverse dunes on two wavelength scales shaped by the wind, together with current ripples and dunes fashioned by flowing water. These features, with the exception of lunettes, are characterized by a gentle upcurrent side and a much steeper downcurrent slope. All of the forms listed have relatively sharp crests that lie across the line of the parent current. Barkhans, lunettes and parabolic dunes are strongly three-dimensional features. The shape of the other forms ranges in natural environments between strongly two-dimensional, when crests are long and comparatively straight and the lee sides are relatively uniform in height, to strongly three-dimensional, the leeward slopes then being very variable in height and crests short and strongly curved.

These bedforms participate in the sediment transport and express a state of instability in the sediment flow, in contrast to a plane bed with sediment movement, which is a stable bed configuration. The controls on the occurrence and attributes of transverse bedforms are best known for ripples, dunes and plane beds fashioned by flowing water. As the transport rate of fine sands is raised, the bedform changes from current ripples, to dunes, to a plane bed (upper stage). For relatively coarse sands under similar circumstances, the sequence observed with increasing severity of flow is plane beds (lower stage), to dunes, to plane beds (upper stage). Current-ripple wavelength is proportional to grain size, and there are indications that the vertical form-index is affected by flow strength, reaching a minimum at an intermediate flow condition in the range appropriate to ripples. The wavelength of ballistic ripples also increases with increasing coarseness of bed material, but for the reason that a textural differentiation occurs during wind transport, leading to coarse-debris concentration in the bed, a growth in the length of the saltation path of the finer grains, and a consequent increase of ripple scale. As a general rule, the wavelength of dunes in aqueous environments increases with the thickness of the effective flow. Dune height also scales with the flow and attains a maximum relative to flow depth at an intermediate condition in the range appropriate to dunes. Near the limits of this range, and in relatively coarse sediments, are developed rather flat dunes which have often been wrongly distinguished as a special class of bedform. Upper-stage plane beds appear only at high flow strengths, the lower limit of their occurrence being defined fairly satisfactorily by Bagnold's universal criterion.

Chapter 9

CLIMBING RIPPLES AND DUNES AND THEIR CROSS-STRATIFICATION PATTERNS

INTRODUCTION

Cross-strata are texturally and/or compositionally distinct layers of sediment that are more or less steeply inclined to the principal surfaces of accumulation of the formations in which they occur. In combination with ordinarily erosional bedding surfaces, often sub-parallel with these principal planes, they define a wide variety cross-stratified sedimentary units and cross-stratification patterns. Probably most cross-stratification is due to the movement of the ripples and dunes discussed in Chapter 8. Some units, however, seem to record the migration of isolated spits, solitary bars, and even scour holes (e.g. G.E. Williams, 1971; Asquith and Cramer, 1975). Cross-stratification involving erosional truncations and laminae steep enough to have been deposited from avalanches is occasionally reported from layered igneous intrusions (Wells, 1962; Dawson and Hawthorne, 1973; Mukherjee and Haldar, 1975; Umeji, 1975).

James Hall (1843) gave under the title of "diagonal bedding" perhaps the first detailed account of cross-stratification, in his case of thick units which would now be described as cross-bedded. A little later, Sorby (1859) summarized his field and experimental studies of sedimentary structures, and reported a similar structure to Hall's under the name "drift-bedding", showing that it could arise by delta-building (Gilbert type). He also described a smaller-scale cross-stratification, which he called "ripple-drift", and showed that it was formed when net sediment deposition occurred simultaneously with current-ripple migration. This structure is today known by many names, of which cross-lamination and climbing-ripple cross-lamination are perhaps the most popular. The latter term, although emphasizing a notable objective feature of cross-laminated deposits, is perhaps misleading, as every ripple and dune either "climbs" or "descends" when it travels and generates cross-stratification.

Interest in the hydraulic significance of cross-stratification languished for some decades after Sorby's work, but was rekindled for a time when attention was turned later in the century towards Quaternary stratified drifts (T.M. Reade, 1884; Gilbert, 1885; W.M. Davis, 1890; Spurr, 1894a, 1894b; Woodworth, 1901). Further stimulus arose from the controversy over certain structures in the Medina Formation of New York State (Gilbert, 1899; Fairchild, 1901), from which Hall (1843) himself had described cross-bedding. That cross-stratification was related to a wide range of bedforms was realized at about this time. Gilbert (1884, 1899) saw it beneath wave ripples, and Beadnell (1910), wetting the sand prior to excavation, found that desert dunes were internally cross-bedded. Kindle (1917) grasped the connection between dunes shaped by running water and cross-stratification, proposing that "cross-bedding in many instances represents one phase of a phenomenon called sand waves which are nothing more than a current made ripple-mark of mammoth proportions." Sorby's proof that the directional property of cross-stratification could be useful in palaeogeographical reconstruction (Sorby, 1859) lay unexploited for many decades. The revival of interest in his technique, promoted by Brinkmann (1933), Shotton (1937), Jüngst (1938), and Reiche (1938), has been followed by explosive activity (e.g. Pettijohn et al., 1972) and the procedure is now standard in sedimentological research, at both the regional and local levels.

NOMENCLATURE AND CLASSIFICATION OF CROSS-STRATIFICATION

There are no generally agreed procedures for naming and classifying cross-stratification. As with other sedimentary structures, however, some classificatory schemes emphasize dynamic and/or kinematic criteria (Zhemchuzhnikov, 1926; Illies, 1949; R.E. Elliott, 1965; Nagtegaal, 1965), whereas others stress geometrical and lithological attributes (Andrée, 1915; Andersen, 1931; McKee and Weir, 1953; Rukhin, 1958; Birkenmajer, 1959; Pettijohn, 1962; Allen, 1963b, 1963c, 1968c, 1970e, 1973b; Walker, 1963; R.E. Elliott, 1964; Imbrie and Buchanan, 1965; Jopling and Walker, 1968; Reineck and Wunderlich, 1968a; A.F. Jacobs, 1973a; Reineck and Singh, 1973). Some descriptive schemes rely only on what is evident in two dimensions (e.g. Botvinkina, 1959; A.F. Jacobs, 1973a), and so meet the immediate problems raised by cross-stratification structures as often found in the field. Other proposals, notably the important classification of McKee and Weir (1953), and Allen's (1963c, 1968c) scheme derived from it, ultimately stress the three-dimensional nature of cross-stratification, but are sometimes difficult to apply in the field. Accepting several of Allen's (1963c) categories, R.E. Elliott (1964) developed a nomenclature applicable to multidimensional exposures. It seems unquestionable that the ideal nomenclature and classification—still to be attained—is one which ultimately considers cross-stratification in three dimensions.

McKee and Weir (1953) recognized that cross-stratified units were made up of *cross-strata,* assembled in *sets* which in turn were often grouped into *cosets* (Fig. 9-1). Where the upper boundary of a set has the shape of a ripple or dune, and the internal cross-strata are consistent in attitude with this shape, the cross-stratified unit may be called a *form set* (Imbrie and Buchanan, 1965). Sets are otherwise *gradationally* or *sharply bounded* (Allen, 1968c). Additional descriptive terms are summarized in Fig. 9-2 (Allen, 1963c, 1968c; Crook, 1965; A.F. Jacobs, 1973a). It is useful here to introduce a

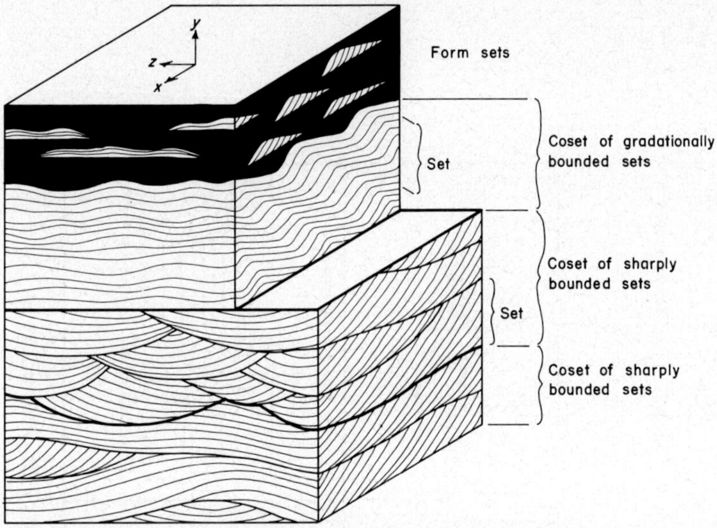

Fig. 9-1. Morphological features of cross-stratification sets.

coordinate system, in which the *x*-direction points with the current and the *y*-direction is normal to the generalized bed. The terms draw attention to: (1) the scale of the set as expressed by its thickness, (2) the degree of association of sets, (3) the character of the lower bounding surface of the set, (4) the shape of the cross-strata in a given plane, (5) the relationship of the cross-strata to the surface underlying the set, and (6) the degree of textural uniformity of the cross-strata composing each set.

Several terms listed in Fig. 9-2 need more explanation. So far as cross-stratified units shaped by aqueous ripples and dunes are concerned, a fundamental division of magnitude can be drawn at a set thickness of 0.04 m, corresponding to the low in the frequency distribution of ripple and dune height (see Fig. 8.6). The class of large-scale sets is a broad one, however, embracing set thicknesses from 0.04 m up to the order of 10 m, and may well merit arbitrary subdivision (e.g. A.F. Jacobs, 1973a). The grouping or association of sets seems particularly important. A *solitary* set is bounded vertically either by sedimentary units which are not cross-stratified or by cross-stratified sets of another attribute or type. Except for certain form sets, generated when incomplete ripples or dunes are driven over and then buried by a contrasted deposit such as mud (Chapter 12), we find that solitary sets do not depend on ripple or dune migration, but are related to bar, spit, or scour-hole movement. For sets to be described as *grouped* in a coset, they should occur in unbroken vertical sequence and be closely similar in size, shape, attitude and lithology. It is these sets which arise chiefly by ripple and dune migration. Cross-strata within sets are variable in their textural uniformity, a reflection of the steadiness of the depositional processes. Allen

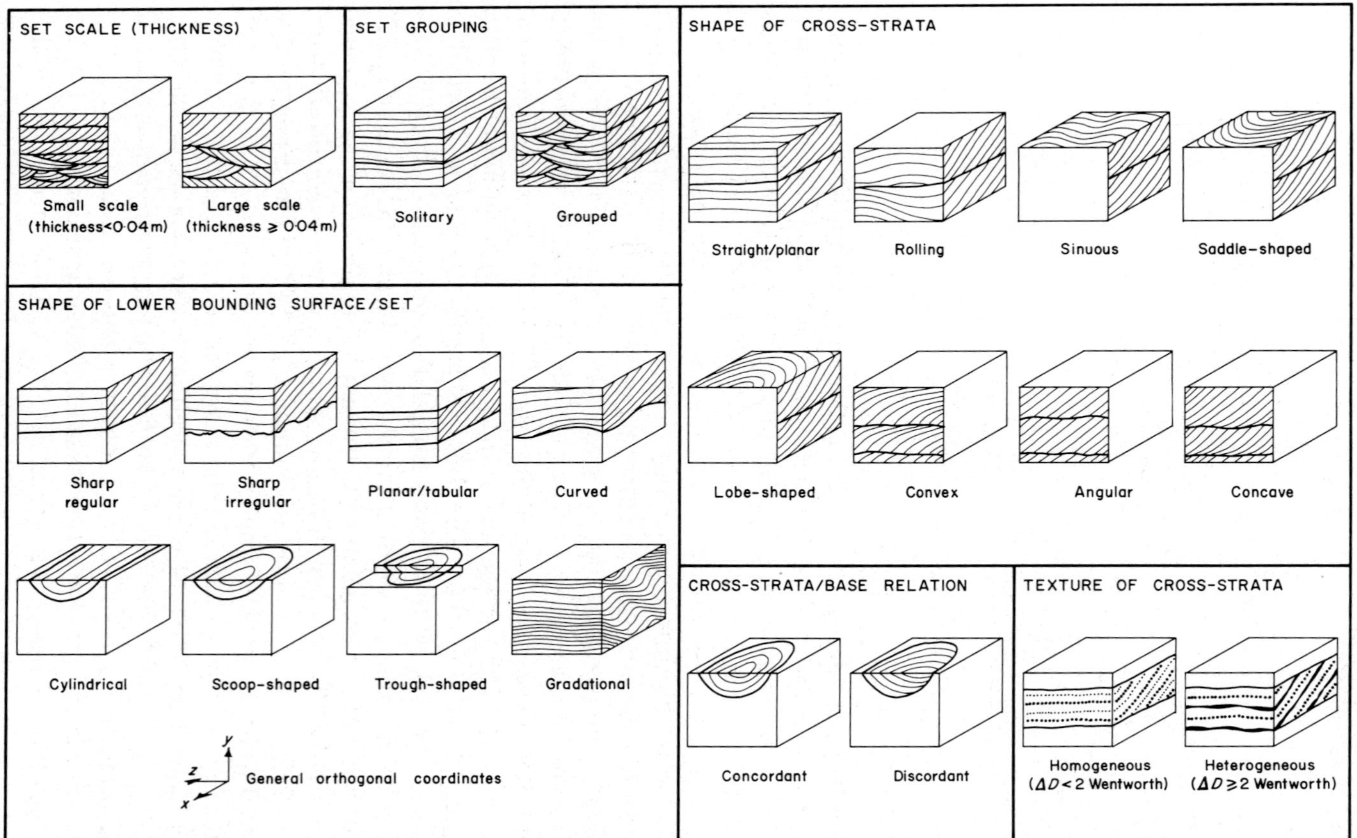

Fig. 9-2. Summary of morphological features of cross-stratified deposits, and terms descriptive of cross-stratification.

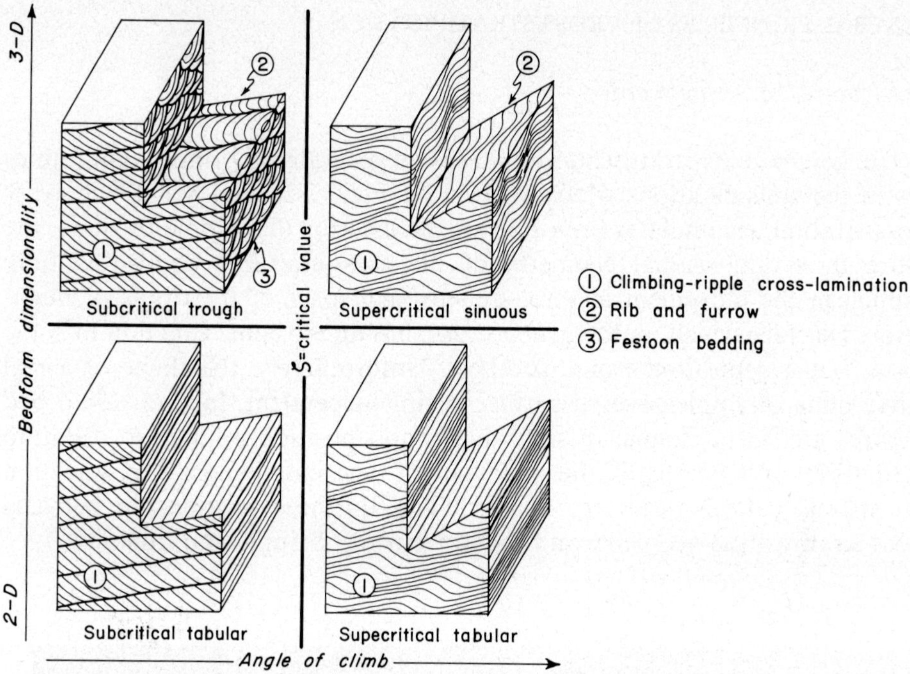

Fig. 9-3. End-member types of grouped cross-stratified sets, depicted in terms of bedform dimensionality and climb (relative deposition rate). The block diagrams depict the structure in vertical planes parallel and normal to flow, and in the plane of climb.

(1963c) called sets *homogeneous* if the cross-strata differed in grade by up to two Wentworth classes, and *heterogeneous* if the range was greater. This choice is not entirely arbitrary, as the processes acting under nominally steady conditions to the lee of continuously moving ripples and dunes and related forms, can produce textural differences in cross-strata amounting to one or two Wentworth grades.

Grouped sets of cross-strata are of many types, of which it is useful to distinguish four as practical end-members (Fig. 9-3). All four are represented amongst small-scale sets; the two labelled subcritical (term explained below) occur also at the large scale. We shall see below that the mutual relationship between the end-members is determined by ripple and dune dimensionality (see Chapter 8), and whether there is net sediment deposition or erosion overall. The latter influence may be summed up as the "climb" or "descent" of ripples and dunes. Several names are widely applied to the end-members in particular planes (Fig. 9-3), for example, climbing-ripple cross-lamination (e.g. Allen, 1970e), rib-and-furrow (Stokes, 1953, 1961), and festoon bedding (e.g. P. Allen et al., 1960).

GENERAL PRINCIPLES OF CROSS-STRATIFICATION

Bed features of a single order

The key to cross-stratification patterns as generated by ripples and dunes lies in the considerations which centred on Figs. 7-2b and 8-1. Since every cross-stratum accumulates on a part of the exterior of a ripple or dune, the entire three-dimensional geometry of the cross-stratified deposit is fully implicit in the movement in three-dimensional space of the ripple or dune-covered surface which, in the general case, has an erosional function in some places but a depositional one in others. Unfortunately, the shape of a real active dune or ripple-covered surface is time-dependent, for individual bed features are born, change in size, shape, and position relative to a datum during life, and eventually die in favour of new individuals. Simplification has therefore been necessary in modelling the inter-relationship between cross-stratification geometry and the movement of ripple and dunes.

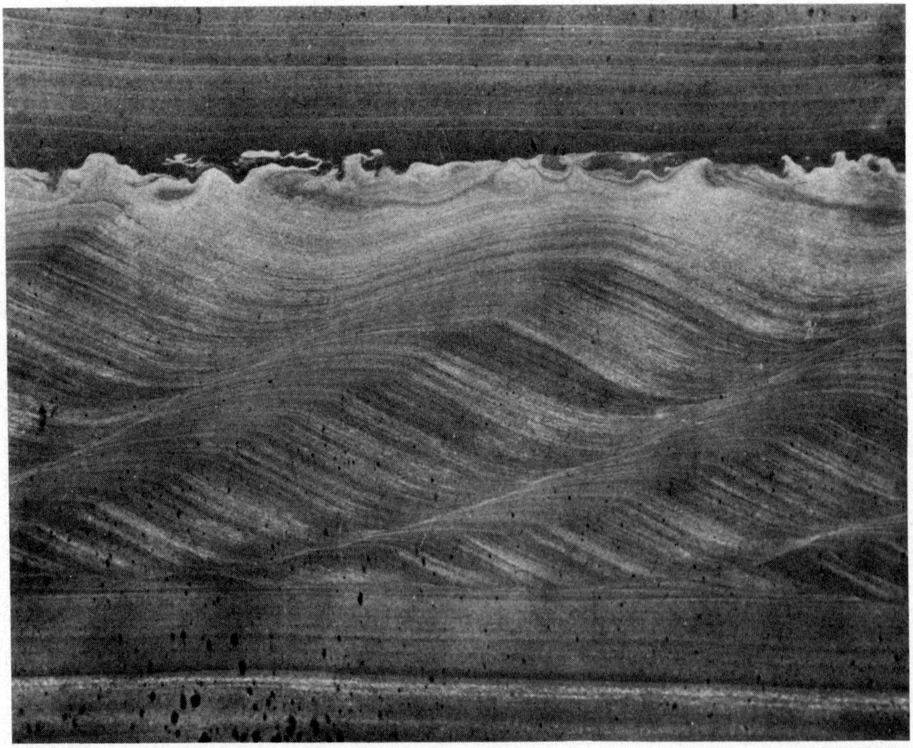

Fig. 9-4. Sorby's (1908) "ripple-drift" in the upper part of a normally graded volcaniclastic sediment unit, Langdale Slates (Ordovician), English Lake District. About one-half natural size. Photograph courtesy of D.W. Humphries, from material in the Department of Geology, University of Sheffield.

Sorby (1859, 1908) pioneered study of this problem. He analyzed the behaviour under the action of a steady current of two-dimensional current ripples of uniform shape and size, choosing for illustration "ripple-drift" from the Ordovician Langdale Slates of the English Lake District (Fig. 9-4). In 1859 he wrote: "If anyone reflect on the manner in which the 'ripple-drift bands' are formed, he will perceive at once that their thickness indicates the excess of material deposited on the sheltered side of the ripples over that washed up again from the exposed side, during the time required for each ripple to advance a distance equal to its own length, which time we may conveniently call its 'period'." This principle underpins all later analyses of ripple and dune cross-stratification, whether qualitative (e.g. Shantzer, 1951; E.D. McKee, 1939, 1966a; Jopling and Walker, 1968) or quantitative (e.g. Allen, 1963b, 1968c, 1970e, 1972b; Kuenen, 1967; J. Müller, 1969; R.G. Walker, 1969a; Banks, 1973a; Srodon, 1974).

Consider as in Fig. 9-5a the motion parallel with a steady current flowing in the positive x-direction of two-dimensional bed features of simple triangular streamwise profile (Allen, 1968c, 1970e). The features are uniform in size and shape, with a constant stoss slope ξ, height H, and wavelength L, their bases lying parallel with the principal surface of accumulation AB. Let the bedload transport rate be J_B units of dry mass per unit width and let there be a simultaneous net transfer of sediment between bed and flow of

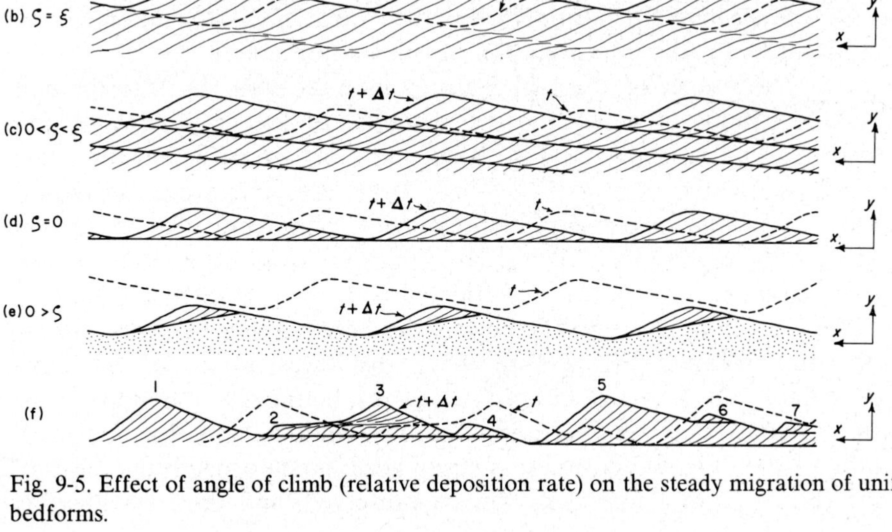

Fig. 9-5. Effect of angle of climb (relative deposition rate) on the steady migration of uniform bedforms.

R units of dry mass per unit area (parallel with AB) and time. The celerity, V_x, of the bed features parallel with the generalized bed is, therefore:

$$V_x = \frac{2J_B}{H\gamma} \tag{9.1}$$

where γ is the sediment dry bulk density. The celerity, V_y, in the direction normal to the generalized bed becomes:

$$V_y = \frac{R}{\gamma} \tag{9.2}$$

a positive value implying net transfer in the direction flow to bed. Hence the features are "climbing" relative to their bases at an angle ζ and a true celerity of V. But since $V_y = V_x \tan \zeta$, eqs. (9.1) and (9.2) yield:

$$\tan \zeta = \frac{RH}{2J_B} \tag{9.3}$$

whence the angle of climb is directly proportional to the rate of sediment transfer and the bedform height, and inversely related to the rate of bedload transport, the sediment bulk density being eliminated. The numerical factor of 2 in eq. (9.3) depends on the triangular form assigned to the features. Realistic profiles give slightly smaller form factors.

Many workers subsequent to Sorby (1859, 1908) effectively arrived at the relationship stated in eq. (9.3) (Reineck, 1961; E.D. McKee, 1966a; Kuenen, 1967; Kuenen and Humbert, 1969; J. Müller, 1969; Allen, 1970e). Nonetheless, some confusion remains. Jopling and Walker (1968) wrongly claim that the primary factor controlling the angle of climb is "the ratio of the suspended to traction load", but also refer to control by the ratio of "deposition from suspension" to "bedload movement". R.G. Walker (1969a) feels that "the angle of climb is dominantly controlled by the ratio of suspension/traction sedimentation", but this statement is too imprecise to be a useful basis for the interpretation of cross-stratification.

How does the form of the bed features interact with the transfer and transport conditions to determine the character of the cross-stratification seen in the special xy-plane? In Fig. 9-5a conditions are such that the angle of climb ζ exceeds the slope ξ of the stoss. Strata are therefore preserved on both stoss and lee slopes, and gradational sets accumulate. The case when $\zeta = \xi$ (Fig. 9-5b) represents a critical condition, for stoss slopes neither lose nor gain sediment. The normal set thickness $L \tan \zeta$ then just equals its critical value given by $L \tan \xi$. The boundaries between sets are sharp but are not erosional. When $0 < \zeta < \xi$ (Fig. 9-5c), the lowest point in each dune trough sweeps out a sharp but erosional set boundary, the normal set thickness $L \tan \zeta$ being smaller than its critical value $L \tan \xi$. In the special case of $0 = \zeta < \xi$ (Fig. 9-5d), no cross-strata accumulate beneath the bases of the bed features. Only form sets can be preserved, and their fossilization requires that bedload movement ultimately stops abruptly. In a final case (Fig. 9-5e), the bed features descend on a path below the generalized bed, to give form sets incompletely filled with cross-strata.

Allen (1968c) and J. Müller (1969) independently show that the relationship between ζ and ξ defines three important modes of cross-stratification (Figs. 9-1 and 9-3). When $\zeta \leqslant 0$ we have *form sets* and when $0 < \zeta < \xi$ sets that are *subcritical,* that is, separated by sharp, erosional boundaries inclined upwards relative to the bases of the parent bed features. With these sets, made by forms that behead their immediate predecessors, we find two orders of bedding surface, between cross-strata, and between sets. The case $\zeta = \xi$ represents a *critical* condition. The critical angle of climb of small-scale sets is generally between 10° and 20° (Allen, 1973b) but is usually much lower for large-scale ones, since dunes have gentler stoss slopes than ripples (Allen, 1963b). *Supercritical* sets related by gradational boundaries arise when $\zeta > \xi$. R.E. Hunter (1977a, 1977b) has independently described sets as either subcritical or supercritical, in the same senses as used above.

This analysis gives an essential insight into general principles but is unduly simplified, because of the restrictions applied. Let us now examine the effect of relaxing at least the condition of uniform bed features, though not the condition of constancy in time. Figure 9-5f shows the cross-stratification made by bed features of different sizes and attitudes at zero net sediment transfer. Feature 1 creates a set which is preserved in a deeply truncated form beneath the following features 2, 3 and 4. Similarly, feature 5 makes a set progressively thinned by the advance of features 6 and 7. Feature 2, however, because of its downward-tilted base, shapes a set with a sharp, erosional base but gradational top. The movement of a train of non-uniform bed features when there is a net deposition could, on account of the variable tilt of the bases, yield sets of a range of thickness compared to bedform height, some of the sets being erosional while others proved gradational.

Two orders of bed feature

The case summarized in Fig. 9-5 is unduly simplified for another reason. Because only one order of moving bed feature is considered, the generalized sedimentary surface is not subject to lateral movements. In reality, dunes and ripples travel over beds themselves shaped into larger features, commonly wave-like and migratory, for example, the undulations distinguished as pools and crossings in rivers, and systems of shoals and channels in tidal environments. After McKee's (1939) qualitative attack, Allen (1968c) and Banks (1973a) made the quantitative analyses which serve as the basis for what now follows.

Figure 9-6 represents a portion of a wavy bed, inclined downcurrent at an angle ξ_1 from the principal accumulation surface parallel with AB. Because

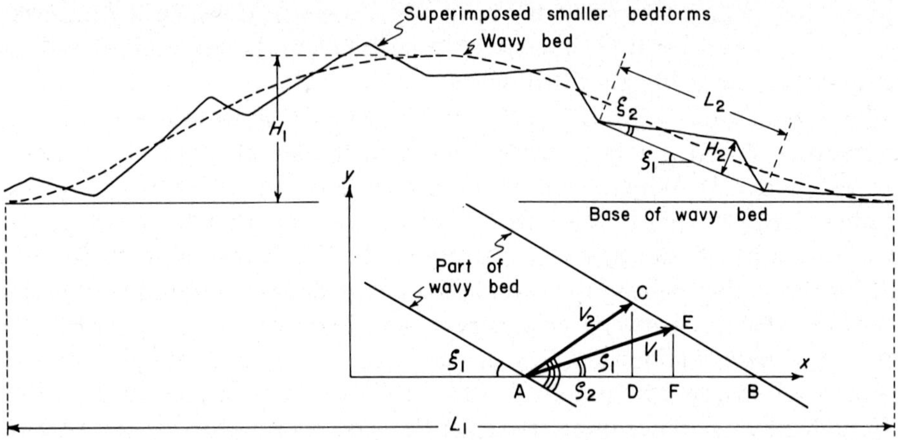

Fig. 9-6. Definition diagram for motion of transverse bedforms over a bed wavy on a larger scale.

of non-zero sediment transfer, the bed wave, of height H_1 normal to this surface, is travelling at a celerity V_1 along a path AE inclined at an angle ζ_1 upward from the generalized bed. The wave bears smaller features (ripples or dunes) of uniform wavelength L_2 and height H_2 (normal to base), travelling at a celerity V_2 along a path AC at an angle ζ_2 measured upward from the generalized surface of the wavy bed. Writing V_1 and V_2 in terms of their streamwise components, the two angles of climb are related by:

$$\tan(\zeta_2 - \xi_1) = \frac{(V_{1,x} - V_{2,x}) \tan \xi_1}{V_{2,x}} + \frac{V_{1,x} \tan \zeta_1}{V_{2,x}}. \tag{9.4a}$$

Banks (1973a) giving an explicit form of this equation. Retaining the notation, the relationship:

$$\tan(\xi_1 + \zeta_2) = \frac{V_{1,x} \tan \zeta_1}{V_{2,x}} - \frac{(V_{1,x} - V_{2,x}) \tan \xi_1}{V_{2,x}} \tag{9.4b}$$

applies to bed features on an upcurrent-dipping portion of a wavy bed.

When sets arise on a downcurrent-dipping portion of a wavy bed, the set boundaries dip upcurrent if:

$$0 < \frac{(V_{1,x} - V_{2,x})}{V_{2,x}} + \frac{V_{1,x} \tan \zeta_1}{V_{2,x} \tan \xi_1} \tag{9.5}$$

but downcurrent when:

$$0 > \frac{(V_{1,x} - V_{2,x})}{V_{2,x}} + \frac{V_{1,x} \tan \zeta_1}{V_{2,x} \tan \xi_1} \tag{9.6}$$

Real sets can form on an upcurrent-dipping portion of a wavy bed only when $(\zeta_2 + \xi_1) > \zeta_1 > \xi_1$, the set boundaries themselves then dipping upcurrent. In the special but probably predominant natural case when the net sediment transfer on the scale of the wavy bed is sufficiently small that it can be equated with zero, $\tan \zeta_1 = 0$ and eq. (9.4a) yields:

$$\tan(\zeta_2 - \xi_1) = \frac{(V_{1,x} - V_{2,x}) \tan \xi_1}{V_{2,x}} \tag{9.7}$$

whence:

$$0 < \frac{V_{1,x} - V_{2,x}}{V_{2,x}} \tag{9.8}$$

for upcurrent-dipping sets, and:

$$0 > \frac{V_{1,x} - V_{2,x}}{V_{2,x}} \tag{9.9}$$

for downcurrent-dipping ones. Equation (9.4b) of course gives only imaginary sets for this transfer condition. For all real sets, boundaries are subcritical when $0 < \zeta_2 < \xi_2$. Allen (1968c) tabulated the complete implications of eq. (9.7).

When bed-wave and bed-feature scale and shape are constant, the attitude of the cross-stratified sets in the case governed by eq. (9.4) is a function only of $V_{1,x}$, $V_{2,x}$, and ζ_1. As Allen (1968c) showed in the case of eq. (9.7), when $V_{2,x} > V_{1,x}$ the sets dip downcurrent, and are subcritical if also $0 < \zeta_2 < \xi_2$. In most real cases, however, $V_{1,x}$ and $V_{2,x}$ are not independent, for the sediment transport driving the smaller features also motivates the larger waves. Considered one type at a time, the bed waves and the superimposed smaller bedforms must therefore yield the same bedload transport rate. Hence, referring to eq. (9.1):

$$V_{1,x} = \frac{2J_B}{H_1} \tag{9.10a}$$

$$V_{2,x} = \frac{2J_B}{H_2} \tag{9.10b}$$

with the result that, upon eliminating J_B:

$$\frac{V_{1,x}}{V_{2,x}} = \frac{H_2}{H_1} \tag{9.11}$$

The preceding eqs. (9.4) and (9.7) can therefore be stated in angular and scalar terms alone.

In the simplified case described by eq. (9.7), upcurrent-dipping sets are unobtainable, and sets parallel with the generalized bed can exist only when $H_2 = H_1$, that is, when effectively there is only one class of bedform present, of base and motion parallel with the generalized bed. Figure 9-7 illustrates cross-stratification patterns calculated using eq. (9.7) for a range of values of H_2/H_1, and on the supposition that the wavy bed is of uniform slope. These

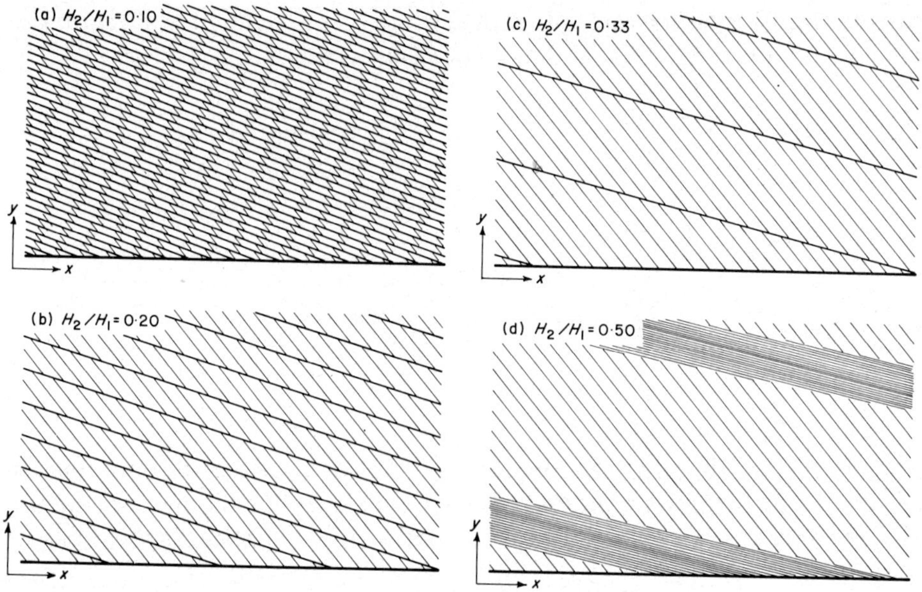

Fig. 9-7. Vertical sections parallel with flow illustrating patterns of downclimbing "compound" cross-stratification as a function of bedform height ratio, H_2/H_1, calculated on the supposition that the smaller forms have a vertical form-index of 10 and the larger a forward slope of 1:5.

patterns represent a kind of "compound" or multi-ordered cross-stratification. As Brookfield (1977, 1979), and Hubert and Mertz (1980), have recently re-emphasized, cross-stratified units of this type reveal a hierarchy of bedding surfaces. In addition to the surfaces between cross-strata and between sets, already distinguished as differing in order, we now have surfaces of a still higher order, at the base of each compound set. Because of accumulation on a wavy bed, the set boundaries are themselves inclined downcurrent, as well as the cross-strata within the sets. The surface beneath each compound set, however, is parallel with the generalized bed. Equation (9.4) is more general and yields a wider range of combination of dip and type of set relationship, for example, downcurrent dipping cross-strata, coupled with either erosional or gradational set-boundaries, and bases to compound sets that both dip upcurrent.

Minor features of cross-stratified sets

R.G. Walker (1969a) analyzed the connection in supercritical cross-stratification between the slopes, ξ' and ξ'', of the stoss and lee respectively, and the thicknesses, h' and h'', of the strata deposited on these surfaces. These quantities and the angle of climb are connected by:

$$\frac{h'}{h''} = \frac{\sin(\zeta - \xi')}{\sin(\zeta + \xi'')} \tag{9.12}$$

on the assumption, implicit in Walker's analysis, that strata formed with the same frequency on both sides of the bed feature, by the action of a single process which operated on an appropriately large spatial scale. Equation (9.12) shows that the steeper the angle of climb, the more nearly equal in thickness are the strata formed on the two surfaces. Walker concluded that "the angle of climb is a function of the angles of the lee and stoss slopes" but, referring to eq. (9.3), this can be true only if the bed-form wavelength is regarded as fixed, so that the height is defined by ξ' and ξ'' alone. Equation (9.12) is a necessary dependence but not a functional relationship.

It should be remembered that the climb of cross-stratified sets may have been lowered by compaction, particularly in the case of small-scale sets in fine-grained sediments holding substantial amounts of clay minerals, mica flakes, or macerated plant debris. If ζ_0 is the climb at deposition, and ζ_1 the angle after compaction, then:

$$\tan \zeta_1 = \tan \zeta_0 (1 - k) \tag{9.13}$$

where k is the fractional compaction experienced by the deposit (Borradaile, 1973). Compaction by half, for example, halves to a first approximation the angle itself.

TRANSCURRENT LAMINATION

What stratification might arise when bed waves migrate under conditions of very small net deposition rate, such that the diameter, D, typifying the transported sediment is not much less than $L \tan \zeta$? Two cases suggest themselves. In the first, the bed features are extremely flat, of marked uniformity and constancy, and climb at a near-critical angle, that is, $L \tan \xi$ also is not much greater than D. If the stoss and lee sediments differ texturally, as may be expected in real cases, a sequence of long, thin, substantially parallel, and sub-horizontal laminae should result. A similar lamination should result from bed features of a more normal steepness, provided that they also were markedly uniform and constant, and there was sufficient textural change across the stoss slope. All such laminae may be called *transcurrent,* because each involved deposition on a moving surface inclined to the principal plane of accumulation. Since the lamina thickness does not greatly exceed D, cross-stratification would not ordinarily be observable, though a tell-tale imbrication, indicating an apparent flow opposite to the true one, might be detectable.

N.D. Smith (1971a) found in the channel sands of the Platte River, Nebraska, a type of horizontal lamination (Fig. 9-8) which by direct observation of dune migration he could attribute to the general process sketched above. A similar fluviatile lamination was described by Picard and High (1973) as the supposedly distinct "horizontal parallel stratification" and

Fig. 9-8. Transcurrent lamination in fine- to coarse-grained glaciofluvial sands, Pleistocene, Banc-y-Warren, Cardigan, Wales. The coin is 0.025 m in diameter.

"horizontal discontinuous stratification". Their distinction seems unnecessary, however, as the accompanying photographs reveal but one set of characteristics, a critical feature being the frequently lens-like shape of the coarser laminae.

Two kinds of laminae differing in grade by one to two Wentworth classes occur in transcurrent-laminated sediments (N.D. Smith, 1971a; Picard and High, 1973). The coarser, up to 0.01 m in thickness but usually much thinner, may consist of sand up to coarse or very coarse grade. They are relatively persistent laterally, but gradually thicken and thin, often where thickest being cross-laminated internally. The finer laminae reach a maximum thickness of 0.018 m, but are typically much thinner and of greater lateral extent, up to several metres, than the coarser ones. Picard and High's transcurrent-laminated deposits appear to be finer grained on the whole than N.D. Smith's, and they are also more finely stratified. Transcurrent lamination occurs amongst the horizontally laminated flood deposits described by McKee et al.. (1967) and G.E. Williams (1971). Parting lineations cover the tops of the laminae observed by Williams, an association reported from "horizontal parallel stratification" by Picard and High (1973). The lamination may also be present in the tidally influenced deposits described by Singh (1969) and the outwash sands of Augustinus and Riezbos (1971). It is not objectively recognizable in the parallel laminated river sands described by Harms and Fahnestock (1965), J.M. Coleman (1969), and Kumar and

Singh (1978). Like most of the horizontally laminated Bijou Creek deposits (McKee et al., 1967), these sands are relatively fine grained and comprise thin, parallel, and laterally very persistent laminae. The coarser grained lenses typical of transcurrent lamination appear to be sparse or lacking. There nevertheless seems to be every gradation between transcurrent lamination and what has long been called parallel lamination or flat-bedding, with which parting lineations are also associated.

From his observations on the character, movement, and internal structure of flat, uniform dunes in the Platte River, N.D. Smith (1971a) concludes that transcurrent lamination forms when these bedforms advance with very little net deposition in shallow flows of moderate Froude number. The coarsest sand accumulates at the toe of each dune lee, as the result of down-slope avalanching, while the finer grains settle higher up. Laminae differing texturally therefore form together on the lee, but at different levels, and it is only at sufficiently steep climbs that the deposit is thick enough to show cross-stratification internally. Wunderlich (1967) proposed a somewhat similar but less detailed interpretation of horizontal laminations formed intertidally. Jopling (1964b, 1966, 1967) created experimentally a bedding similar to transcurrent lamination, as the result of the migration under steady or near-steady conditions of ripples or very flat dunes, which he called "rheologic micro-fronts". McBride et al. (1975a) made transcurrent lamination in a flume, but in a slightly different way from either Jopling (1964b) or that observed by Smith. Exceedingly flat current ripples formed of ill-sorted sand were allowed to climb at very nearly the critical angle. Because of their extreme flatness, uniformity, and constancy, the ripples generated alternately coarse and fine laminae of great regularity. The coarse ones accumulated in the lee, but on slopes too gentle for avalanching, while the fine grained laminae, bearing parting lineations, formed on the stoss sides.

In addition to the above mechanism, horizontal laminations in sand accumulated from unidirectional currents are attributed to: (1) velocity fluctuations (Pettijohn, 1957; Sanders, 1960, 1965; Lombard, 1963; Gagny, 1964), (2) repetition of currents (Kingma, 1958; Mangin, 1962), (3) laminar flow at the bed (Hsu, 1959), (4) the segregation of the coarser sediment into distinct clouds (Wood and Smith, 1959; Unrug, 1959), and (5) grain sorting in the bedload (Moss, 1962, 1963; Kuenen, 1965, 1966; Frostick and Reid, 1977). Kuenen (1966) satisfactorily disposed of the first four of these. Middleton (1970) in turn contends that the Moss-Kuenen hypothesis—that there is "a tendency amongst depositing grains for kind to seek kind" (Kuenen, 1966)—is itself not an explanation. Whatever the truth, their statement is unquestionably an accurate description of events during transport over a *nearly* plane (i.e. slightly wavy) bed, as the experiments of Jopling (1964b), McDonald and Vincent (1972), and McBride et al. (1975a) demonstrate. Bridge (1978) explains the laminations by bursting streaks.

There is much to favour the view that the forms of horizontal lamination present in sands deposited from unidirectional currents record the size and/or shape sorting of particles on an aggrading, slightly wavy bed. Many experimenters in addition to Jopling (1964b) found that nominally plane beds periodically bore extremely flat, decaying waves of symmetrical to strongly asymmetrical streamwise profile (Einstein and Chien, 1953; J.F. Kennedy, 1961; B.D. Taylor, 1971; Taylor and Vanoni, 1972b; McDonald and Vincent, 1972). These waves seem to arise in response to disturbances, for example, in the sediment transport rate or turbulence, and in free-surface flows are favoured by particular Froude numbers (Kennedy, 1961). Flow separation occurs only on the most asymmetrical waves, the symmetrical ones being sufficiently flat and rounded that parting lineations can appear on both lee and stoss slopes. Transcurrent lamination is created by the steeper waves, and ordinary parallel lamination by the flatter ones. This concept of the origin of horizontal lamination does no violence to the idea that a plane bed is a stable bed configuration, for the flow appears always to damp out the set of bed waves induced by each spontaneous fluctuation.

SUBCRITICAL CROSS-STRATIFICATION

Models, character and occurrence

Figure 9-9 shows three-dimensional models for the origin of subcritical tabular and trough cross-stratification from ripples and dunes. These are better than the over-simplified two-dimensional models in Fig. 9-5, not only

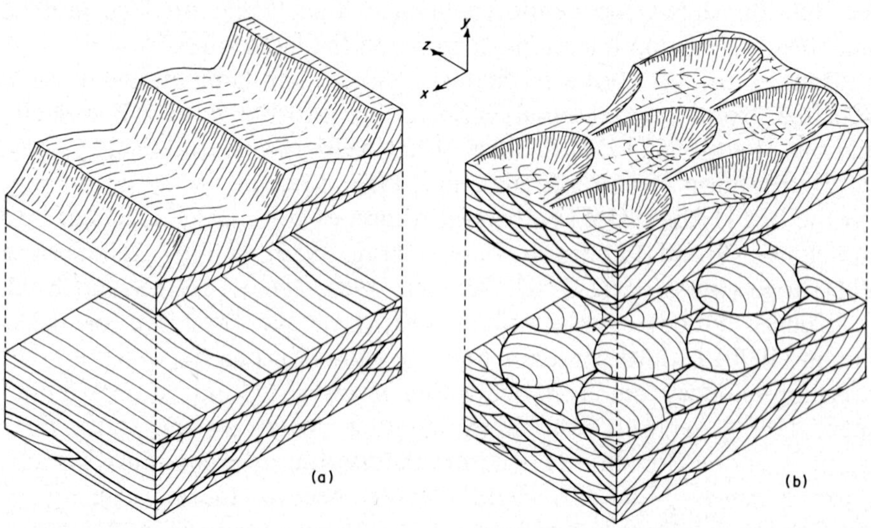

Fig. 9-9. Schematic forms of subcritical cross-stratification in relation to the shape of the parent bedforms.

in presenting the third dimension but also in showing features consistent with the behaviour of real bedforms, at least as exemplified by current ripples. Allen (1968c, 1973a) points out that, since current ripples are individually of finite life, the set each generates is necessarily restricted in the flow direction, beginning ordinarily at a concave-up scour. Moreover, since every feature varies in height and breadth during life, individual sets are of uneven thickness and width. Because of the non-uniformity of real bed features as perceived at an instant, and their changeability in time, even under equilibrium conditions there arise sets with a greatest thickness not much less than the height of the parent features (Jopling, 1967; Allen, 1973a; McDonald and Vincent, 1972; Jopling and Forbes, 1979). Downstream-dipping sets are then equally abundant with upstream-dipping ones.

The model of Fig. 9-9a applies to substantially two-dimensional bed features, regardless of scale or depositional environment. That of Fig. 9-9b is developed explicitly in terms of lunate dunes formed in water, but a similar pattern of cross-stratification is generated at low aggradation rates by linguoid to strongly sinuous or catenary current ripples (trough cross-lamination), or by strongly three-dimensional forms of aeolian or aqueous dune (through cross-bedding). As the bedforms migrate, the hollows in their troughs, or in the spaces between laterally adjacent features, sweep out concave-up erosion surfaces on which are stacked the lobe-shaped cross-strata deposited in harmony on the upstream sides of the hollows.

There are several models like these. Wurster's (1958a, 1958b, 1964) and Niehoff's (1958) are reconstructions of the bed form from the cross-stratification. Wurster thought subcritical trough cross-stratification was generated by closely spaced but nonetheless isolated barkhan-like features. As with Shotton's (1937, 1956) and Glennie's (1970) interpretations of some aeolian cross-bedding, this choice of bed feature is inconsistent with the cross-stratification pattern, for barkhan-like forms exist on flat, non-sandy surfaces. Depending on the scale of the cross-stratification, either linguoid ripples or long-crested, strongly lobed dunes make a better choice. In the same vein, Dzulynski and Slaczka (1958), Dzulynski and Zak (1960), and Dzulynski and Walton (1965) concluded that linguoid current ripples could not have made subcritical trough cross-lamination, even though the ripples immediately overlay the cosets. R.G. Walker (1965) also objected to the production of trough cross-lamination by linguoid ripples, on the grounds that an expected interfingering of cross-laminae was lacking. Examination of the troughs and interiors of the ripples shows that this expectation is false. Niehoff's reconstruction, for probably tidal trough cross-bedding, embodies a more realistic conception of dunes. J.H. Stewart (1961), like McDowell (1957, 1960), linked subcritical cross-bedding with dune migration, giving block diagrams which related tabular sets to relatively straight-crested forms and trough sets to strongly lobe-shaped features. His two-dimensional bed features look realistic, one of the dunes being terminated laterally, but the shape given to the three-dimensional features is that of Häntzschel's (1938) rare *D-form*. The models developed by Allen (1962b, 1963b, 1968c) and by Reineck and Singh (1973) use bed features of realistic shape, except that the "linguoid" ripples of the last-mentioned authors are atypical.

Subcritical cross-lamination mainly of the trough variety abounds in very fine to medium grained sands and sandstones, especially in those of fluviatile or related origin (Gürich, 1933; McKee, 1939; Shantzer, 1951; W.L. Stokes, 1953, 1961; Botvinkina et al., 1954; Jewtuchovicz, 1954; Niehoff, 1958; Dzulynski and Zak, 1960; Hamblin, 1961a; McBride, 1962; Allen, 1963b, 1972b; Harms et al., 1963; Potter, 1963; R.G. Walker, 1963; Coleman and Gagliano, 1964; Friend, 1965; Harms and Fahnestock, 1965; Jipa, 1965; D.K. Davies, 1966; Grumbt, 1966, 1974; Simon and Hopkins, 1966; Carrigy, 1967; Förstner et al., 1968; Jopling and Walker, 1968; Gradzinski, 1970; Gall, 1971; Aario, 1972; Costello and Walker, 1972; Van Beek and Koster, 1972; Singh, 1972a, 1972b; A.F. Jacobs, 1973b; Leflef, 1973; E. Lindström, 1973; Chanda and Bhattacharyya, 1974; R. Anderton, 1975; Banerjee and McDonald, 1975; Boothroyd and Ashley, 1975; Gustavson et al., 1975; Rust and Romanelli, 1975; Shaw, 1975a; Clague, 1976; P.K. Ray, 1976; Ruegg and Zandstra, 1977; Kumar and Singh, 1978). Unfortunately, even where unconsolidated deposits are available, few workers record the structure in more than one plane. Figure 9-10 therefore presents a composite view. Hamblin's (1961a) detailed description, under the name of "micro-cross-

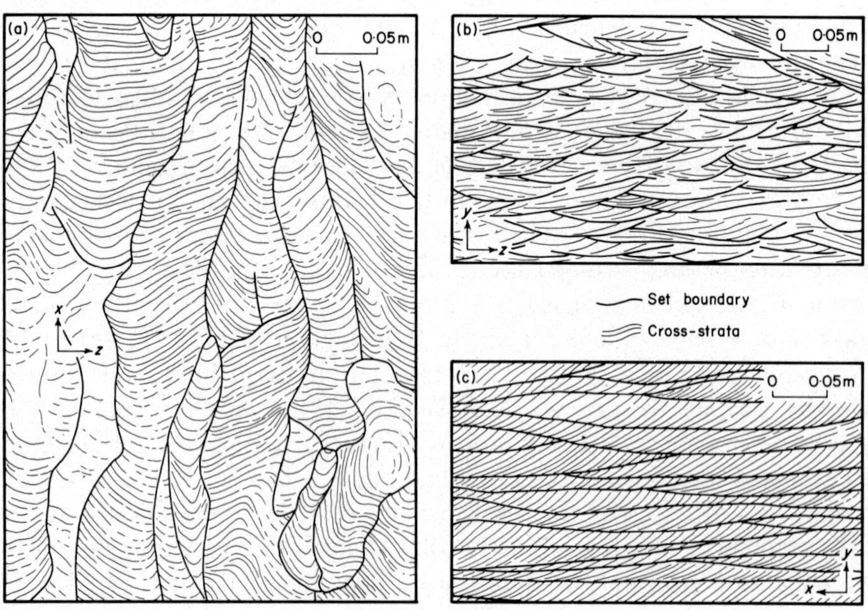

Fig. 9-10. Examples of trough cross-lamination in (a) plane of climb (after Wurster, 1964), (b) vertical plane normal to flow (after Harms et al., 1963), and (c) vertical plane parallel with flow (Mundesley Sands, Mundesley, Norfolk, England).

lamination", remains of outstanding value. Most subcritical cross-lamination is described from the *xy*-plane. It is then the "rolling incline-bedding" of Andersen (1931), R.G. Walker's (1963) "type 1 ripple drift", the "type A ripple-drift cross-lamination" of Jopling and Walker (1968), and the "type A climbing-ripple cross-lamination" of Allen (1973b). Trough cross-lamination

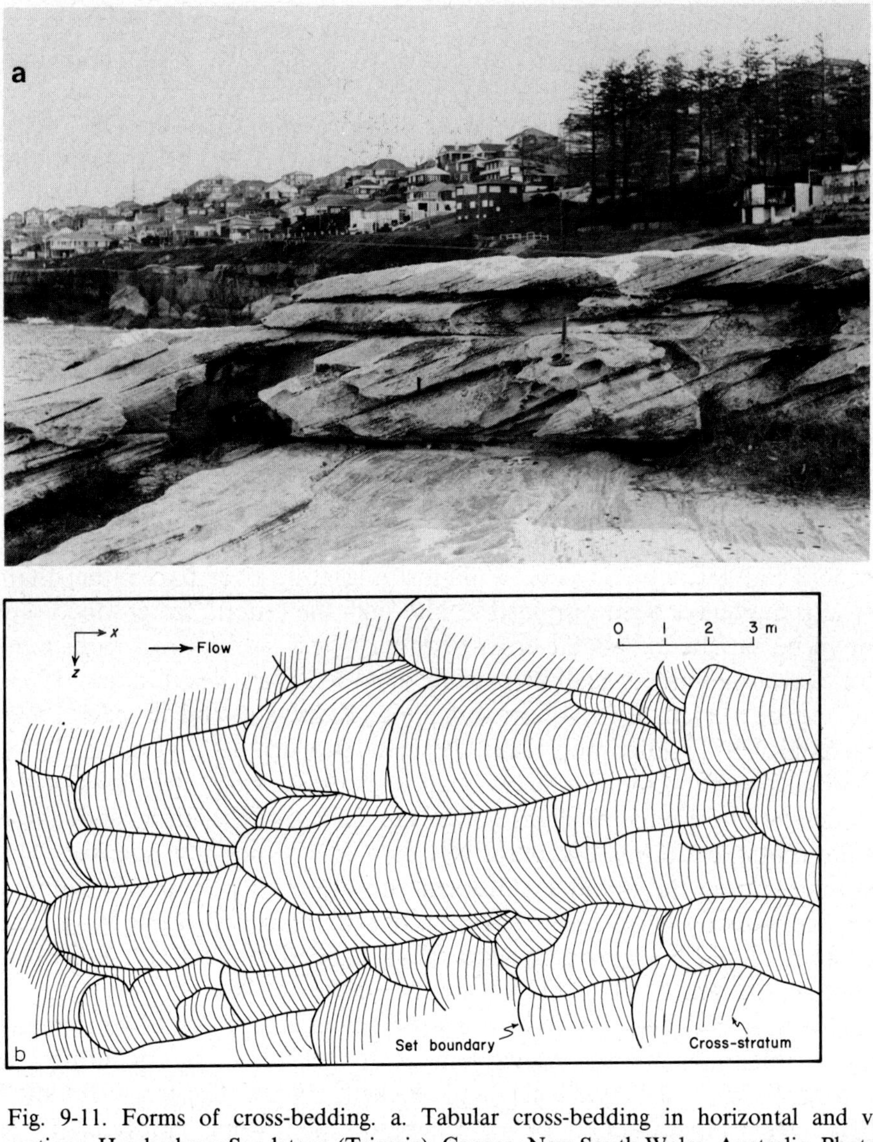

Fig. 9-11. Forms of cross-bedding. a. Tabular cross-bedding in horizontal and vertical sections, Hawkesbury Sandstone (Triassic), Coogee, New South Wales, Australia. Photograph courtesy of K.A.W. Crook, reproduced by permission of Bureau of Mineral Resources, Geology and Geophysics, Commonwealth of Australia. b. Trough cross-bedding in a section close to the plane of climb, Chinle Formation (Triassic), Canyon De Chelly, Arizona (after Stewart et al., 1972).

exposed on surfaces close to the *xz*-plane or the plane of climb is occasionally reported (e.g. Potter and Glass, 1958; Potter, 1963; Picard and High, 1964; Collinson, 1970b; Kumar and Singh, 1978), usually under the name "rib-and-furrow" proposed by W.L. Stokes (1953, 1961). His definition, however, also includes supercritical cross-lamination similarly exposed.

Subcritical cross-bedding is very widely developed in shallow-water and aeolian sandstones. The tabular variety is less common than the trough form, and gradational varieties exist.

Tabular cross-bedding (Fig. 9-11a) occurs in aeolian, fluviatile, and shallow-marine deposits. Records from aeolian deposits of sets often several or many metres thick include those by W.L. Stokes (1968) from the De Chelly Sandstone and the Navajo Sandstone, by Horne (1971, 1975) from the Siluro-Devonian Dingle Group of southwest Ireland, and by McKee (1966b) from modern dunes. Near-tabular units seem to be present in the Precambrian Waterberg Supergroup (Meinster and Tickell, 1975), the Casper Formation (Knight, 1929), the Gipsdalen Formation (Clemmensen, 1978), and the Arikaree Group (Bart, 1977). Stokes thinks that the parallel planar form of the bounding surfaces between sets may represent a downward scour limit fixed by the water table. Tabular cross-bedding in sets generally less than 1 m thick is reported from many fluviatile deposits (McDowell, 1957; G.P. Jones, 1961; J.H. Stewart, 1961; Shackleton, 1962; Potter, 1963; Friend, 1965; Malmsheimer, 1968; Beuf et al., 1971; G.E. Williams, 1971; Gustavson, 1978; Eriksson and Vos, 1979; Miall, 1979). In the cases reported by Harms and Fahnestock and G.E. Williams, migrating river bars rather than dune trains may have been responsible, although the criteria for grouped sets appear to be satisfied. Tabular cross-bedding is less well known from tide- and/or wave-dominated shallow-marine deposits (Van der Linden, 1963; Imbrie and Buchanan. 1965; Wermund, 1965; Knewtson and Hubert, 1969; Hrabar et al., 1971; Allen and Kaye, 1973; D.D. Carr, 1973; Button, 1976; Nio, 1976). Van der Linden noticed shallow depressions at the bases of the sets, which he interpreted as slight hollows in dune troughs. In most of these cases the cross-bedding is comparatively thin and bimodally oriented. The cosets representing the dominant dip azimuth are interspersed with solitary sets or, occasionally, with cosets having a broadly opposite orientation. The parent dunes could have been on sand shoals shaped by both ebb and flood tides (Allen and Kaye, 1973; D.D. Carr, 1973; Nio, 1976).

Trough cross-bedding (Fig. 9-11b) is profusely described. Sets up to many metres in thickness are recorded from aeolian sandstones, including the Waterberg Supergroup (Meinster and Tickell, 1975), Casper Formation (Knight, 1929; Steidtmann, 1974), Coconino Sandstone (McKee, 1933; Reiche, 1938), the Permo-Triassic dune sandstones of Great Britain (Shotton, 1937, 1956), and Navajo Sandstone (Kiersch, 1950; Sanderson, 1974). Contemporary coastal dunes in places are trough cross-bedded (e.g. Bigarella, 1965; Bigarella et al., 1969, 1971; Hine and Boothroyd, 1978), though on a more modest scale than in deserts, and may have ancient counterparts (Ghibaudo et al., 1974). Trough cross-bedding in sets generally less than 1 m thick occurs widely in contemporary fluviatile deposits, Pleistocene glaciofluvial and outwash sediments, and older river sediments (F.S. Mills, 1903; Hobbs, 1906; Shantzer, 1951; W.L. Stokes, 1953, 1961; Botvinkina et al., 1954; McKee, 1954; Crook, 1957; McDowell, 1957; Hamblin, 1958, 1961b; Potter and Glass, 1958; Wurster, 1958a; Fahrig, 1961; Frazier and Osanik, 1961; J.H. Stewart, 1961; Klein, 1962b; Harms et al., 1963; D.W. Lane, 1963; Ore, 1963; Potter, 1963; Allen, 1964b, 1974a; Conolly, 1964, 1965; Pick, 1964; Bigarella and Mousinho, 1965; Bigarella et al., 1965; Harms and Fahnestock, 1965; Meckel, 1967; Read and Johnson, 1967; Sarkar and Basumallick, 1968; G.E. Williams, 1968, 1971; McGowen and Garner, 1970; Picard and High, 1970b; Beuf et al., 1971; Mroczkowski, 1972; Stewart et al., 1972; Van Beek and Koster, 1972; McDonnell, 1974; Barrett and Kohn, 1975; Hume et al., 1975; Jackson, 1976a; Saunderson, 1976). Crook and Conolly describe types of cross-bedding transitional to the tabular form. Deposits formed under tidal and/or wave action occasionally reveal subcritical trough cross-bedding (Hülsemann, 1955; Hamblin, 1958, 1961b; Niehoff, 1958; Reineck, 1960a, 1963; Pryor, 1971b; Dott and Roshardt, 1972; Stone and Vondra, 1972; D.D. Carr, 1973; Stricklin and Smith, 1973; Howard et al., 1975b; H.D. Johnson, 1975; Davidson-Arnott and Greenwood, 1976). Combined ebb and flood influences are often apparent.

By no means all workers believe that grouped cross-bedded sets are explicable by dune migration. We have already seen that some tabular grouped sets may be formed by migrating bars and other features which are not strictly dunes but which are marked by dune-like properties and mechanisms (see also the delta experiments of A.L. Smith, 1909; Nevin and Trainer, 1927; McKee, 1957b; and Jopling, 1965, 1967). Hemingway and Clarke (1963a, 1963b) rejected dunes as the explanation of grouped cross-bedding, because set boundaries could generally not be observed to climb. Consideration of experimental results and dune shape show that this objection is invalid (Allen, 1963d).

Trough cross-bedding is particularly controversial. Knight (1929) proposes that each trough set in the Casper Formation represents the cutting by a current and then the filling of a scoop-shaped hollow, the two acts being separate in time and unrelated. Writing of trough cross-bedding in the fluviatile Salt Wash Member, W.L. Stokes (1953) suggests that vertically eddying masses of water, the "kolks" (vortices) of Matthes (1947), might simultaneously cut and fill hollows. This suggestion can be taken as implying an origin from three-dimensional dunes, to which powerful vortices are coupled as the result of flow separation, except that Matthes regards such kolks as not restricted to dune beds. Experimentally, McKee (1957b) assigned trough sets to the cutting and then filling of elongated channels, an explanation which is more appropriate to solitary than grouped sets. Sutton and Watson (1960) also explain subcritical trough cross-bedding by channel infilling. Harms et al. (1963) and Harms and Fahnestock (1965) broadly follow Knight. They suggest that each scoop-shaped hollow is first eroded and at some later time independently infilled by dune avalanche deposits. In every case it seems much more likely that the cross-bedding was due to the aggrading movement of trains of dunes, with their harmoniously migrating erosional stoss sides and depositional lee slopes.

Experimental studies

Subcritical cross-lamination, and on a small scale even subcritical cross-bedding, has been produced experimentally, though rarely under closely controlled conditions.

Subcritical cross-lamination arises experimentally under conditions of overall equilibrium (Jopling, 1967; Allen, 1973a; Jopling and Forbes, 1979), presumably because net erosion or net deposition can occur locally on a scale comparable to that of the bed features themselves, because of spatio-temporal variation in ripple shape and size. Under these conditions, roughly

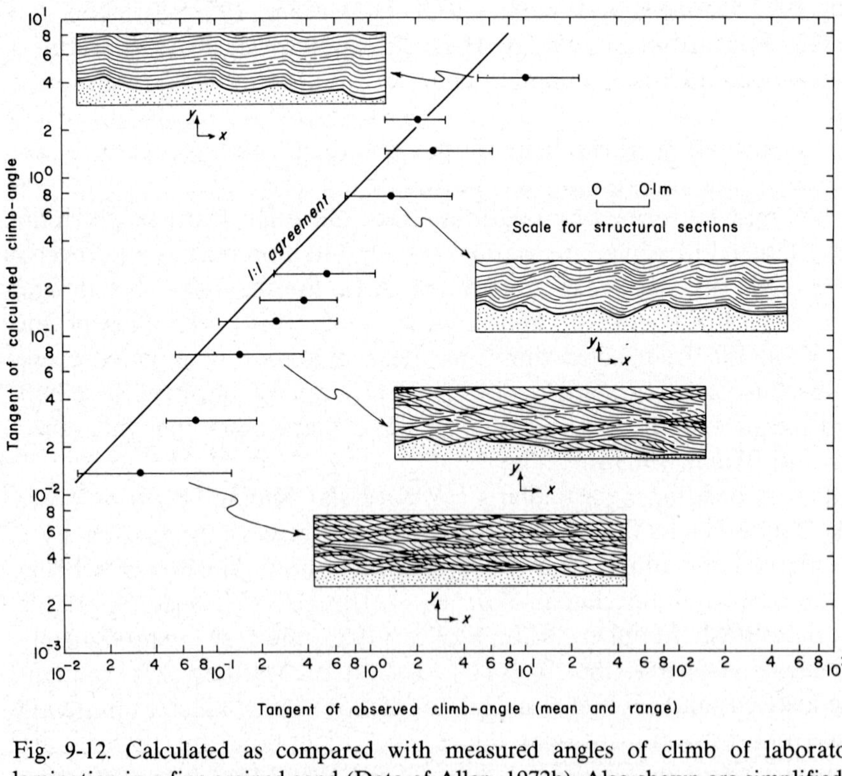

Fig. 9-12. Calculated as compared with measured angles of climb of laboratory cross-lamination in a fine grained sand (Data of Allen, 1972b). Also shown are simplified drawings of representative internal structures produced during the same experiments, as revealed by rubber latex relief casts taken from vertical sections in the deposits.

as many sets climb downcurrent as climb upcurrent, so that the average angle of climb is sensibly zero. McDonald and Vincent (1972) succeeded in making under equilibrium conditions in a large pipe a small scale example of subcritical cross-bedding. The erosional set boundaries on the average are parallel with the pipe base, but individually dip either upstream or downstream, though less steeply than smaller, ripple-generated structures.

We can form subcritical cross-lamination in at least four other ways, which ultimately have the same effect, by: (1) overloading a sand-bearing current (Reineck, 1961), (2) gradually increasing the depth of a sand-laden flow, so that net deposition is promoted (Jopling, 1963, 1966), (3) gradually decreasing the velocity of a sand-bearing stream, again causing net deposition (Kuenen, 1965, 1966, 1967; Kuenen and Humbert, 1969; Banerjee, 1977), and (4) feeding sand from above on to active ripples (McKee, 1957b, 1965; Rees, 1966b; Allen, 1971a, 1972b). In Kuenen's and Banerjee's studies, ripples were generated in a circular flume by a paddle-driven flow allowed gradually to decelerate. It appears that subcritical, gently climbing sets were produced at the lower experimental deceleration rates, corresponding to the lower rates of net deposition. Allen's controlled experiments were made in a straight flume, sand being fed uniformly on to active ripples of known celerity from a large box hung above the flow channel. Figure 9-12 shows that the observed mean tangent of the angle of climb compares well with the tangent calculated from the measured attributes of the ripples (see eq. 9.3). Some representative cross-lamination patterns in the *xy*-plane are also given. From peels made in the other two planes (Allen, 1972b), the experimental sets evidently belong to the trough type, and in no way deviate from Fig. 9-9b. Individual sets, of uneven in width and greatest thickness, begin at scoop-like scours and are seldom longer than a few ripple wavelengths.

SUPERCRITICAL CROSS-LAMINATION

Models, character and occurrence

As the net deposition rate increases relative to the tangential celerity of bed features, gradational contacts between successive cross-stratified sets begin to appear, at first occasionally and then more plentifully until, at sufficiently large rates, aggradation occurs on even the steepest stoss faces. The resulting structure, in which contacts are wholly or chiefly gradational, is supercritical cross-stratification, which occurs chiefly in water-laid deposits and may be restricted to small-scale sets. Figure 9-13 gives models for tabular and sinuous supercritical cross-lamination, such as could result from relatively two-dimensional current ripples on the one hand, and sinuous-crested ones on the other. Notice that laminae now generally extend unbroken over several consecutive ripples, becoming more even in thickness with steepening climb.

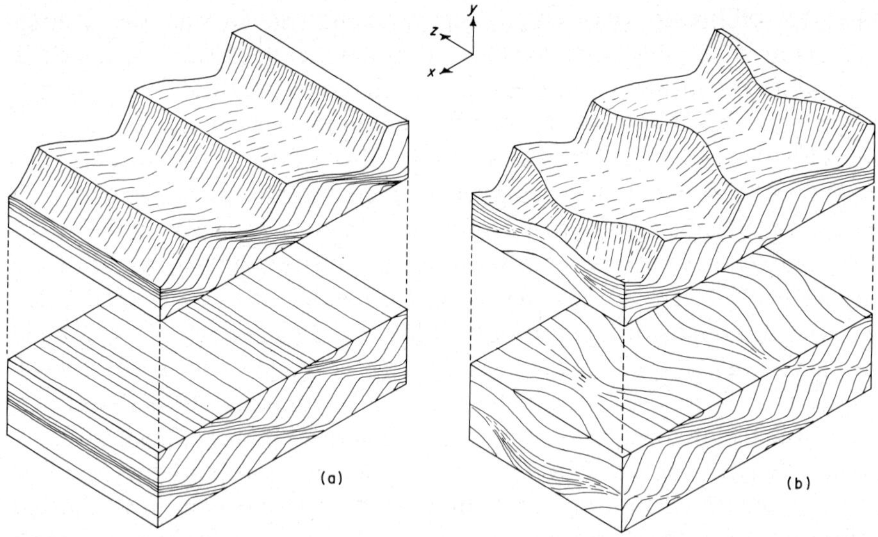

Fig. 9-13. Schematic forms of supercritical cross-stratification in relation to the shape of the parent bedforms.

Supercritical cross-lamination in the *xy*-plane is the "unilateral rolling" and "ordinary rolling" strata of Andersen (1931), the "cross-lamination developed from ripples in rhythm" of McKee (1939), R.G. Walker's (1963) "type 3 ripple-drift cross-lamination", the "types B and C ripple-drift cross-lamination" of Jopling and Walker (1968), and Allen's (1970e, 1973b) "type B" (subcritical $< \zeta \leqslant 60°$) and "type S climbing-ripple cross-lamination" ($\zeta > 60°$). Jopling and Walker (1968) had earlier distinguished as "sinusoidal ripple lamination" a cross-lamination typified by an angle of climb not much below the vertical and by rather rounded and often nearly symmetrical ripple shapes, a part of Allen's (1973b) type S. Lithologically similar laminae make up type S cross-lamination. Gustavson et al. (1975) gave the name "draped lamination" to a structure superficially resembling type S, but differing in that the laminae are texturally heterogeneous and the climb is not sensibly different from 90°.

The environmental distribution of supercritical cross-lamination is wide, with most examples coming from turbidites and from river or river-related deposits. Aeolian sands afford few descriptions, consistent with the migration of small ballistic ripples, and in one instance compatible with long and relatively straight-crested forms (Glennie, 1970; Hunter et al., 1972; Hunter, 1977a, 1977b). The cross-stratification of ballistic ripples is not especially clear and in the larger of these structures appears to climb little if at all (Sharp, 1966; Glennie, 1970; Piper, 1970a; Walker and Middleton, 1977; Fryberger et al., 1979). Supercritical cross-lamination apparently is also rare in tidally influenced shallow-marine and lacustrine deposits (R.G. Walker, 1963, 1969b; Singh, 1969; Wunderlich, 1970; Chanda and Bhattacharyya, 1974; Ruegg, 1975; L.S. Smith, 1976).

The structure often accompanies its subcritical relative as a major component of the well-known Bouma vertical sequence of turbidite textures and sedimentary structures (Bouma, 1962; R.G. Walker, 1965). A fine example in a (volcaniclastic) turbidite appears in Fig. 9-4, originally published by Sorby (1908). Others are described by Kuenen (1953), Wood and Smith (1959), McBride (1962), Bouma (1962), Sanders (1963, 1965), Jipa (1965, 1967), Crowell et al. (1966), Spalletti (1968), Ricci Lucchi (1969), and Jawarowski (1971). Banerjee (1973) reports the structure from glacial lake deposits possibly of turbidity-current origin.

The most astounding displays of supercritical cross-lamination are unquestionably in fluviatile and closely related sediments (Fig. 9-14). It is common in the channel and near-channel overbank facies of open streams and in stream-influenced deltaic environments (McKee, 1939, 1965, 1966a; W.L. Stokes, 1953; Botvinkina et al., 1954; Sundborg, 1956; Coleman et al., 1964; Coleman and Gagliano, 1965; D.K. Davies, 1966; Grumbt, 1966; McKee et al., 1967; Conybeare and Crook, 1968; Gradzinski, 1970; Wopfner

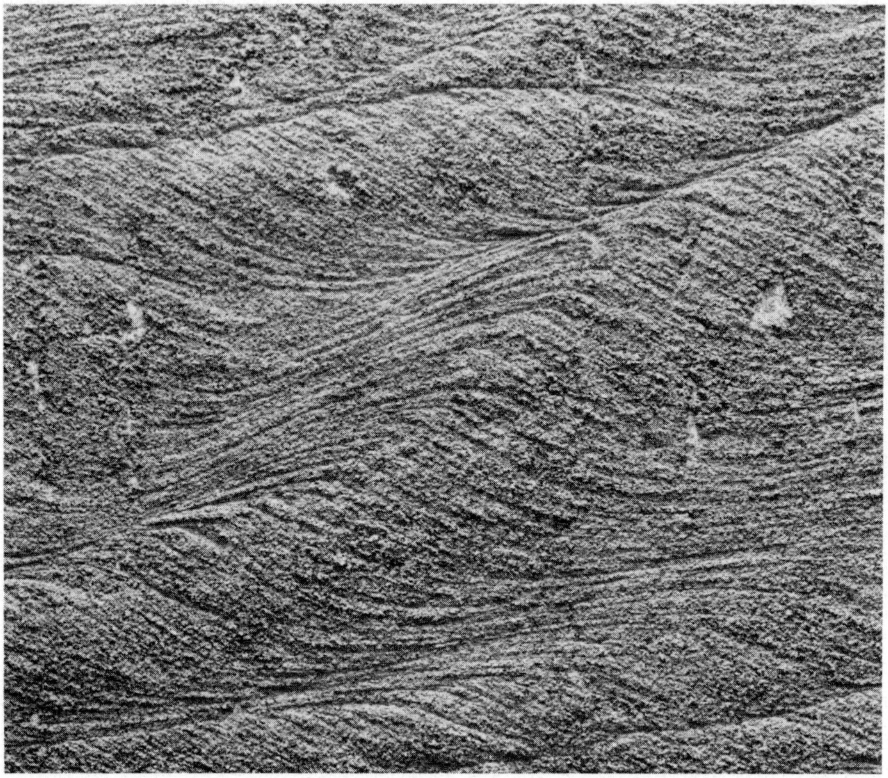

Fig. 9-14. Vertical rubber latex relief cast parallel with flow showing supercritical cross-lamination developed in fine to medium sands of the Uppsala Esker, Löfstalöt, near Uppsala, Sweden. Width of the cast 0.35 m. Current from left to right.

et al., 1970; G.E. Williams, 1971; Karcz, 1972; Singh, 1972a; Picard and High, 1973; McDonnell, 1974; P.K. Ray, 1976; Kumar and Singh, 1978), and occurs in great abundance and variety in ice-contact and pro-glacial deposits (T.M. Reade, 1884; Spurr, 1894a; Woodworth, 1901; Illies, 1949; Jewtuchowicz, 1954; Jopling and Walker, 1968; Aario, 1971, 1972; Allen, 1972b; Costello and Walker, 1972; Huddart, 1973; E. Lindström, 1973; Banerjee and McDonald, 1975; Gustavson et al., 1975; Helm and Roberts, 1975; Rust and Romanelli, 1975; J. Shaw, 1975, 1977). Most authors emphasize the structure as seen in the *xy*-plane, though McKee (1939) gives an important account in three-dimensions, and W.L. Stokes (1953) and Jipa (1965) figure it as a form of rib-and-furrow.

Experimental studies

There has been some success in making supercritical cross-lamination experimentally. McKee (1965) produced mildly supercritical sets by feeding sand on to active current ripples. Kuenen (1965, 1967) and Kuenen and Humbert (1969) were able under conditions apparently of relatively rapid flow deceleration to form supercritical sets with a moderate to very steep climb. The structure last to form in at least one run closely resembled the draped lamination of Gustavson et al. (1975). Jopling and Forbes (1979) produced supercritical sets locally under conditions of equilibrium overall.

In Allen's (1972b) controlled experiments, supercritical cross-lamination was produced at climbs from near the critical, when contacts between sets were locally erosional, up to nearly vertical, when bed traction was excluded by the choice of experimental conditions. Results in the form previously explained appear in Fig. 9-12, together with representative structures drawn from peels in the *xy*-plane. As in natural examples of supercritical cross-lamination, individual laminae can be traced over more than one ripple.

There remain difficulties over the meaning of supercritical cross-lamination of steep climb. The draped lamination of Gustavson et al. (1975) fairly clearly signifies a total absence of tractional sediment movement, because the angle of climb is indistinguishable from 90° and the laminae (some muddy) are lithologically heterogeneous. The significance of homogeneous cross-lamination at a near-vertical climb is less clear. Allen (1972b) found that a climb but little smaller than 90° arose when the flow velocity over an already rippled bed was less than the sediment entrainment threshold on such a bed. The slight forward shift of the ripple forms implied by the just non-vertical climb was attributed to the influence on the local aggradation rate of the local flow acceleration, deceleration and separation close to each ripple, the rate being greater on lee than stoss. An experiment at a flow velocity slightly above the entrainment threshold for the sediment used also gave a climb not far short of vertical, but a little steeper than in Jopling and Walker's (1968) examples of type S cross-lamination. These authors, however, conclude that traction is either absent or, at least, minimal when their sinusoidal ripple lamination is generated. Srodon (1974) also interprets this type as merely a gradually obscuring blanket deposit above "dead ripples". Clearly, more work is needed on the extent to which local flow properties can modify local deposition rates under conditions ranging about the sediment entrainment threshold on already rippled beds.

COMPOUND CROSS-STRATIFICATION

There abound examples of "compound" forms of cross-stratification which are consistent with the dependencies stated above in eqs. (9.4a) and (9.4b). Both small-scale and large-scale sets are represented.

Compound cross-stratification often involves small-scale sets. Units accumulated in some places on downcurrent-dipping surfaces and in others on upcurrent-dipping ones are reported by McKee (1939) from the Colorado River delta, California. Friend (1965), Wopfner (1970), and Grumbt (1974) described from fluviatile deposits upward-climbing gradational sets de-

Fig. 9-15. Downclimbing "compound" cross-bedding in the Ekkerøy Formation (Precambrian), Finnmark, Norway. Compass-clinometer gives scale. Photograph courtesy of H.D. Johnson.

posited on surfaces that dipped downcurrent. In Friend's example, sets climb at an apparent angle of approximately 11° relative to horizontal beds, whereas the surface across which the ripples moved has an apparent downcurrent slope of roughly 10°.

Figure 9-15 shows a typical example of compound cross-stratification formed from large-scale sets (see also Fig. 9-7). Structures of this kind were first described in detail by McKee (1962), from the mainly fluviatile Nubian sandstones of the Saharan region. The cosets, sandwiched between beds deposited apparently horizontally, consist each of several sharply-bounded sets of a similar thickness (e.g. 0.1–0.3 m) and downcurrent dip of the basal surface (e.g. 13–14°). Another account, which prompted an extension of Allen's (1968c) geometrical analysis, is given by Banks (1973a) from late Precambrian sandstones of northern Norway. Here the compound cross-stratification is represented by cosets up to 4 m thick, excellently exposed over streamwise distances of up to 100 m. Set boundaries are sharp, dipping downcurrent at angles of 8° or less relative to apparently horizontally deposited beds, and slope in virtually the same direction as the cross-strata within sets, few of which exceed 0.5 m in thickness. Using the inequalities implicit in eq. (9.4a), Banks showed that $V_{2,x}$ could not have been less than 3.2 times $V_{1,x}$.

The type of structure shown in Fig. 9-15, with sharp, downcurrent-dipping set boundaries, occurs in several fluviatile formations (McKee, 1962; Beuf et al., 1971, fig. 165; Banks, 1973a; Martinez, 1977; Smith and Eriksson, 1979), within bars and dune-like forms in present-day rivers (Conybeare and Crook, 1968; Bluck, 1971, 1976; R.G. Jackson, 1976a), at the lips of supposed glacial-lake deltas (W.M. Davis, 1890; Helm and Roberts, 1975), within intertidal sand bodies (Klein, 1970b; Dalrymple et al., 1975), in shallow-marine, possibly tidally-influenced sandstones (H.D. Johnson, 1975; Hereford, 1977; Levell, 1980) and within aeolian sandstones (D.B. Thompson, 1969; Brookfield, 1977, 1979; Hubert and Mertz, 1980). Bluck's examples probably formed as small dunes shifted over larger and slower moving bars travelling in much the same direction. Jackson's cosets from the Wabash River seem to record the migration of many small dunes over fewer larger ones, some of which he calls transverse bars. The compound sets described by Klein and Dalrymple from the Bay of Fundy also appear to have arisen as dunes shifted over larger bed waves, though there remains the possibility that the sand transport direction was at times reversed tidally. Brookfield also proposes the migration of smaller over larger dunes, but again erosion during wind reversals cannot be entirely discounted. These forms of compound cross-stratification, together with the varieties involving small-scale sets, all testify to the diversity of relief that under appropriate circumstances may mark sedimentary surfaces. They are evidence from the past for hierarchies of relief features; in modern environments such hierarchies abound (Allen, 1966).

We may contrast these structures with cosets formed of generally sharp-based large-scale sets which visibly dip upcurrent relative to apparently horizontally deposited beds. Such are the "back-set beds" of W.M. Davis (1890), who found them in glacial-lake delta sands. Similar structures were much later noticed by Botvinkina et al. (1954) in the alluvium of the Don River, by McKee (1962) in the Nubian sandstones of the Blue Nile Canyon, by Cummins (1965) in the British Keuper Sandstone, by N.D. Smith (1972) in bar sands of the Platte River, Nebraska, and by B.R. Turner (1977) in Karoo sandstones of South Africa. Ricci-Lucchi and Valmori (1980) have even reported upcurrent-dipping cross-bedded sets from a Miocene bioclastic turbidite formation. Structures resembling geometrically the back-set beds of W.M. Davis are inferred to occur in a modern shallow-marine environment (Salsman et al., 1966), and are described from two shallow-marine sandstones (R. Anderton, 1976; Nio, 1976). However, are these upward climbing large-scale sets examples of compound cross-stratification, or merely subcritical cross-bedding? Ascription to the former is plausible for the back-set beds of Davis. Anderton's, Nio's and Cummin's climbing structures could be subcritical cross-bedding, although the general possibility is denied by Sanders (1965), for there is no direct evidence for dune migration over larger bed features. Smith's structure, closely resembled by McKee's, formed in a different way, by the lateral migration of spurs in a dune trough.

In these cases, there is explicit evidence about the shape of the larger bed features only when a modern river or tidal environment is involved. However, K.O. Stanley (1974) and Gustavson et al. (1975) have described from Pleistocene lacustrine silts a remarkable compound cross-stratification which clearly reveals both the large and small bed features. The larger forms, with a steeper lee than stoss, have wavelengths of 0.15–0.60 m and heights of 0.02–0.06 m. The smaller features, also asymmetrical, have heights less than 0.01 m and wavelengths below 0.05 m. They are generally confined to the backs of the larger forms, which they climb either subcritically or supercritically, in accordance with eq. (9.4b). Occasionally, sets with downcurrent-dipping boundaries appear on the lee sides of the larger features, in place of the more usual avalanche layers. The geometry of these compound structures does not appear inconsistent with the forced coupling of the celerities of the bed features of the two orders.

VERTICAL PATTERNS OF CROSS-STRATIFICATION

Cross-lamination in the xy-plane

Field observations show that, in cross-laminated cosets seen locally in the *xy*-plane, the attributes of angle of climb and grain size change vertically in several ways (R.G. Walker, 1963, 1969a; Jopling and Walker, 1968; Allen, 1970f, 1972b, 1973b; Leflef, 1973; Srodon, 1974). Furthermore, these cosets are often directly underlain and/or overlain by genetically related sets of parallel-laminated sand, where the genetic relationship is implied by the similarity in texture, composition and transport direction of the two kinds of

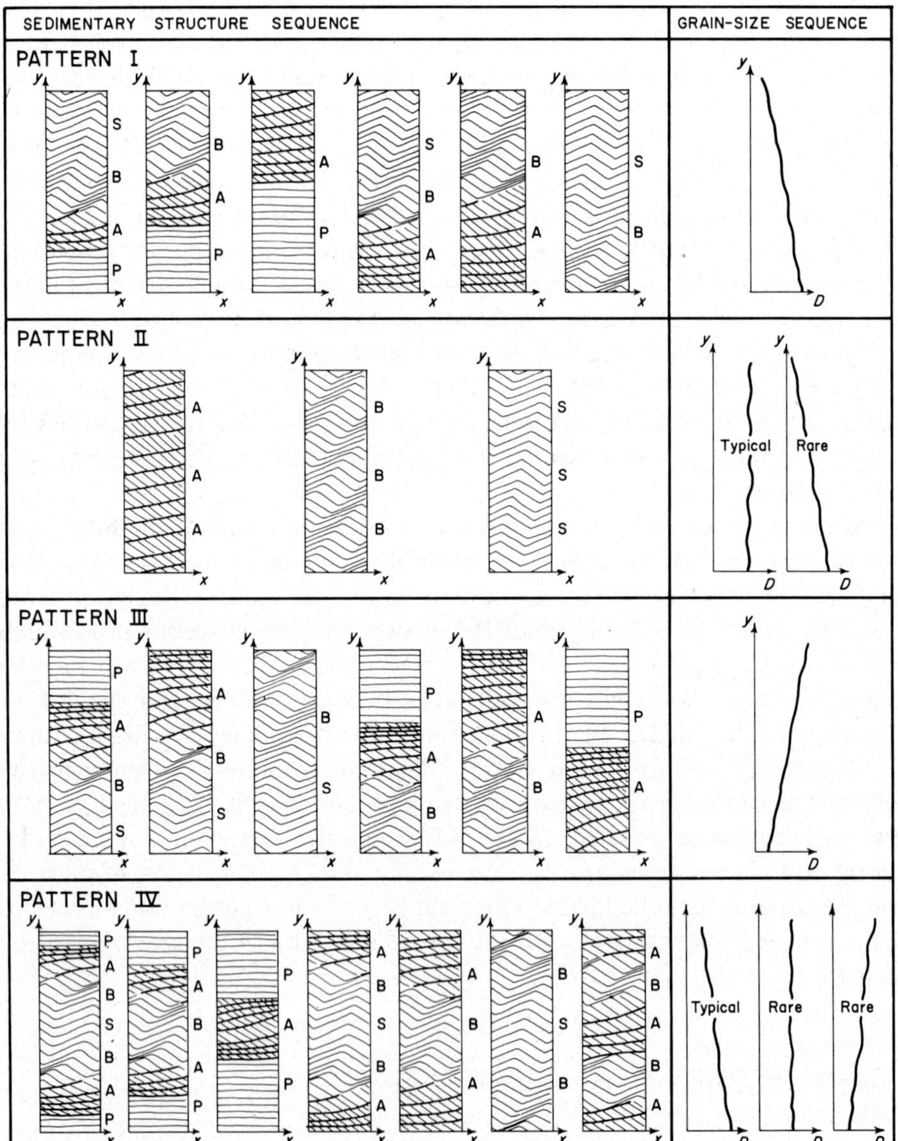

Fig. 9-16. Summary and classification of vertical patterns of grain size (D) and internal structure (particularly type of cross-lamination) observed in vertical sequences involving cross-laminated sand.

deposit, and by the fact that occasional laminae extend between the parallel-laminated and cross-laminated sediments. These vertical changes and relationships can be divided between four main patterns, summarized in Fig. 9-16 (Allen, 1973b).

Sorby (1908) gives a typical example of pattern I (Fig. 9-4). The average grain size decreases upwards within the bed, and the sequence of structures, from parallel lamination upwards to near-vertical type S cross-lamination, is complete. Many other examples are known (Illies, 1949; Kuenen, 1953; Wood and Smith, 1959; Hamblin, 1961a; De Raaf et al., 1965; McKee, 1965, 1966a; D.K. Davies, 1966; McKee et al., 1967; J.M. Coleman, 1969; Shepard et al., 1969; Gradzinski, 1970; Aario, 1972; Allen, 1972b; Singh, 1972a, 1972b; Huddart, 1973; A.F. Jacobs, 1973b; Picard and High, 1973; Grumbt, 1974; McDonnell, 1974; Anderton, 1975, 1976; Banerjee and McDonald, 1975; Helm and Roberts, 1975; Rust and Romanelli, 1975; P.K. Ray, 1976; Ruegg and Zandstra, 1977; Kumar and Singh, 1978; Nanson, 1980). The pattern in some beds is completed within a few centimetres vertically, but in others involves more than one metre of sediment. Some of these examples occur in fluviatile deposits, notably in facies accumulated high up within or just outside channels, and others are preserved in glacier melt-water deposits. Several come from turbidite facies. Pattern I cosets in which grain size is uniform or increasing upwards are rare, but Allen (1972b) describes instances from his Bed E in the Uppsala Esker.

Pattern II is arguably unsatisfactory inasmuch as it represents a special geometrical condition, namely, that the angle of climb is sensibly uniform vertically. Such cosets are fairly common, however, ranging in thickness between a few centimetres and at least one metre (McKee, 1939; Jahns, 1947; R.G. Walker, 1963; J.D. Howard, 1966; McKee et al., 1967; Conybeare and Crook, 1968; Spalletti, 1968; G.E. Williams, 1971; Singh, 1972a). There is generally no sign that the sediment either coarsens or fines vertically within these units. The facies represented by pattern II cosets include fluviatile, deltaic and turbidite. Difficulty is presented by R.G. Walker's type C cross-lamination (R.G. Walker, 1963; Jopling and Walker, 1968), in which the climb (supercritical) remains uniform vertically in the coset, formed of very fine muddy sand and silts, whereas grain size and ripple height decline upwards. Similar cosets are reported by Ricchi Lucchi (1969) and appear to occur in the Cloridorme and Pico Formations, both turbidites (Crowell et al., 1966; R.G. Walker, 1969a). In terms of currently visible attributes, these are cosets of pattern II, but the uniform climb and declining ripple height are perhaps misleading and the result of the greater compaction of the muddier sediment (see eq. 9.13). Attribution to pattern I may be more appropriate. However, Srodon (1974) takes another view, arguing that the transition from a rippled to an upper-stage plane bed is fairly fully preserved.

Cosets exemplifying pattern III are rare and chiefly restricted to glacial-lake deltas (Jopling and Walker, 1968; Aario, 1972; Srodon, 1974; Gustavson et al., 1975; Ruegg, 1975). I saw further examples in Pleistocene deltaic sands near Aspatria, north of the English Lake District, in a setting comparable with that of the Canadian studies. The cosets are relatively thick, of the order of one metre, and begin with coarse silt or very fine grained sand showing either type S cross-lamination or steeply climbing type B. Upwards the angle of climb decreases, usually to a subcritical level, while the grain size increases. Occasionally, parallel-laminated sand follows.

Pattern IV embraces the infrequent cosets which show one or more vertical "oscillations" in the cross-lamination type. Sorby (1908) gives in his plate XVI an instance from a turbidite facies, and Allen (1972b, fig. 17) records others from the Uppsala Esker. I saw further examples involving the sequence P → A → B → A → P and A → B → A, in deposits at Banc-y-Warren, Wales, thought by Helm and Roberts (1975) to represent a glacial-lake delta. Usually in pattern IV cosets, the sediment fines steadily upwards, in a manner only weakly if at all related to the oscillating climb. For example, at Banc-y-Warren, a bed 1.2 m thick with the sequence P → A → B → A → P fined steadily up from medium sand to coarse silt. A uniform or weakly increasing-upward grain size in pattern IV cosets is rare.

Cross-lamination and parallel lamination

From studies in the fluviatile Old Red Sandstone, Allen (1964a, 1971d) emphasized the common genetic association of cross-lamination with parallel lamination accompanied by parting lineation. Bouma (1962) and R.G. Walker (1965) drew attention to the same association in turbidites.

A common mode of association in fluviatile deposits appears in the Uppsala Esker (Allen, 1971d, 1972b). In places, these esker deposits present a vertical sequence of up to 30 alternations of parallel-laminated sand (erosion surface at base) merging up into a texturally related coset of cross-lamination (Fig. 9-17). An apparently similar multiple interbedding of cross-laminated with parallel-laminated sand is recorded by McKee et al. (1967), Singh (1972a, 1972b), Picard and High (1973), and Kumar and Singh (1978) from present-day fluviatile sands, and from apparently similar rocks by Hamblin (1961a), Friend (1965), and Picard and Hulen (1969). Botvinkina et al. (1954), Jopling and Walker (1968), and Aario (1971, 1972) give other records. McKee's example is particularly interesting because it is known to be related to a multi-stage flood.

The association as developed in turbidites is almost never multiply interbedded, but occurs as a part of the so-called Bouma sequence of structures and textures (Vol. II, Chapter 10). Recent turbidity current deposits show the association (Bouma, 1964, 1965; Van Straaten, 1964; Holtedahl, 1965; Von Rad, 1968; Carlson and Nelson, 1969; Shepard et al., 1969; Griggs and Kulm, 1970; Piper, 1970b), as do many older turbidites (Sorby, 1908; Wood and Smith, 1959; Bouma, 1962; McBride, 1962; Murphy and Schlanger,

Fig. 9-17. Vertical rubber-latex relief cast parallel with flow showing parts of two vertical sequences (pattern I, Fig. 9-16) of parallel laminated passing up into cross-laminated sand. Uppsala Esker, Löfstalöt, near Uppsala, Sweden. Width of cast 0.35 m. Current from left to right.

1962; Ballance, 1964b; Crowell et al., 1966; Angelucci et al., 1967; Kuenen, 1967; R.G. Walker, 1967b, 1969b; Conybeare and Crook, 1968; Ricci Lucchi, 1969; Tanaka, 1970; Jawarowski, 1971). A turbidite-like association which occurs where sands and muds are thinly interbedded in fluviatile deposits (e.g. Allen, 1964b; D.J. Stanley, 1968) is possibly of crevasse-splay and/or levee origin.

Cross-bedding set thickness

In certain fluviatile point-bar sand bodies, the thickness of the cross-bedded sandstone sets declines gradually upwards from at or near the erosional base (e.g. Allen, 1964b, 1974a; Beutner et al., 1967; Leeder, 1973). The trend is generally weak, however, and seldom involves worse than a halving of set thickness. There is no apparent thickness trend in many otherwise similar fluviatile sand bodies.

Interpretation of vertical patterns

The trivial interpretation of the patterns (Fig. 9-16) is that net deposition occurred, but it is most interesting to ask how they may have been shaped by spatio-temporal changes of flow. Equation (9.3) underpins all such interpretations of this kind, for it expresses sediment continuity and, therefore, is independent of the particular form of variation with flow ascribed to the deposition rate, R, the ripple height, H, the bedload transport rate, J_B, and, implicitly, the sediment coarseness, D.

Two concepts of interpretation rest on eq. (9.3). Allen (1970e, 1971d, 1972b) suggests that the flow may be considered to vary sufficiently gradually in space and/or time that at least the sediment transport rate and calibre of the sediment have substantially their equilibrium values. That is, R is the derivative of the local bed-material total-load transport rate, J_B depends on the local instantaneous flow and sediment properties and, from experiments (Kuenen and Sengupta, 1970) and theoretical considerations of turbulence, D is also flow-dependent (see also Middleton, 1977). The ripple height is implicitly a constant. According to the second concept, developed by Srodon (1974), grain size is independent of flow, at least on the temporal and spatial scales relevant to cross-lamination patterns. Furthermore, the influence of flow properties on the life-span, shape, size and variability of the ripples is considered significant. Finally, it is proposed that natural flows may at times hold less sediment than they can theoretically transport.

Each view finds supporting evidence. Unquestionably, the bedload transport rate and its derivatives are very strongly coupled to local flow properties. The rate varies as a large power of the mean flow velocity (eq. 2.31) but the load, being within a few particle diameters of the bed, responds quickly and over very short distances to flow changes. A less marked but still perceptible coupling should typify the suspended-load transport rate and its derivatives. This rate also seems to vary as a large power of the flow velocity (eq. 2.38), and should increase rapidly compared to the bedload rate as grain size declines (eq. 2.39). But the response to flow change should be slower and over longer distances than with the bedload, as the suspended grains are merely concentrated near the bed and can be found at any level in the fluid. Srodon (1974) is partly justified in regarding the grain size of the transported sediment as independent of flow; for however, the equilibrium turbulence intensity-flow velocity relationship is interpreted (cf. Allen, 1972b, Middleton, 1977), D varies only as the square-root of the mean flow velocity in the Stokes range, that is, for most cross-laminated sediment. Although ripple equilibrium size and shape vary in a complicated manner with grain size and flow properties (Yalin, 1972; Allen, 1977b), in real cross-laminated units they generally are constant and substantially uninfluenced by texture and climb. Early ripples in a climbing pattern visibly "force" the character of later forms. A variable compaction accounts satisfactorily for most apparent height changes.

TABLE 9-I

Patterns of erosion and climbing-ripple cross-lamination arising in aqueous flowing changing in space and/or time

Non-uniformity	Unsteadiness		
	$\frac{\partial U}{\partial t}<0$	$\frac{\partial U}{\partial t}=0$	$\frac{\partial U}{\partial t}>0$
$\frac{\partial U}{\partial x}>0$	R$\gtrless$0 $\frac{\mathrm{d}D}{\mathrm{d}y}<0,\ \frac{\mathrm{d}D}{\mathrm{d}x}>0$ $\frac{\mathrm{d}\zeta}{\mathrm{d}y}\gtrless 0,\ \frac{\mathrm{d}\zeta}{\mathrm{d}x}<0$ patterns I, IV erosion surface	R<0 erosion surface	R<0 erosion surface
$\frac{\partial U}{\partial x}=0$	R>0 $\frac{\mathrm{d}D}{\mathrm{d}y}<0,\ \frac{\mathrm{d}D}{\mathrm{d}x}=0$ $\frac{\mathrm{d}\zeta}{\mathrm{d}y}>0,\ \frac{\mathrm{d}\zeta}{\mathrm{d}x}=0$ pattern I	R=0 uniform, steady flow no pattern no erosion surface	R<0 erosion surface
$\frac{\partial U}{\partial x}<0$	R>0 $\frac{\mathrm{d}D}{\mathrm{d}y}<0,\ \frac{\mathrm{d}D}{\mathrm{d}x}<0$ $\frac{\mathrm{d}\zeta}{\mathrm{d}y}\gtrless 0,\ \frac{\mathrm{d}\zeta}{\mathrm{d}x}>0$ patterns I, IV	R>0 $\frac{\mathrm{d}D}{\mathrm{d}y}=0,\ \frac{\mathrm{d}D}{\mathrm{d}x}<0$ $\frac{\mathrm{d}\zeta}{\mathrm{d}y}=0,\ \frac{\mathrm{d}\zeta}{\mathrm{d}x}>0$ pattern II	R$\gtrless$0 $\frac{\mathrm{d}D}{\mathrm{d}y}>0,\ \frac{\mathrm{d}D}{\mathrm{d}x}<0$ $\frac{\mathrm{d}\zeta}{\mathrm{d}y}<0,\ \frac{\mathrm{d}\zeta}{\mathrm{d}x}>0$ pattern III erosion surface

Table 9-I summarizes Allen's (1970e, 1972b) interpretative model, in terms of the response of the sediment transfer rate, R, and the upward-vertical and streamwise changes in the climb ($\mathrm{d}\zeta/\mathrm{d}y$, $\mathrm{d}\zeta/\mathrm{d}x$) and sediment calibre ($\mathrm{d}D/\mathrm{d}y$, $\mathrm{d}D/\mathrm{d}x$), to variation of the mean flow velocity, U, with time t and streamwise distance x. A parallel-laminated deposit can accompany a cross-laminated coset, whenever net deposition prevails and the flow and bed conditions change appropriately. The combinations of change of climb and grain size cover all four patterns in Fig. 9-13 and most of the known examples of these patterns. For example, pattern I with grain size declining upward is typical of turbidites and river overbank deposits. According to the model, it represents a decelerating and possibly non-uniform flow, an interpretation consistent with knowledge of turbidity currents and flooding rivers (e.g. Nanson, 1980). Conversely, pattern III suggests an accelerating flow in a non-uniform environment. Jopling and Walker (1968) record it from a steep delta-front possibly affected by a seasonally fluctuating gravity current. The model collapses, however, for those occasional examples of patterns I and IV in which grain size is either uniform or gradually increasing upwards. We may then conclude with Srodon (1974) that in these cases some basic requirement of the model was lacking in the natural flow.

Where a cross-laminated deposit caps a genetically related parallel-laminated one, eq. (9.3) affords a unique opportunity for estimating the net sediment deposition rate at the instant of transition. Using Bagnold's (1966) "universal" plane-bed criterion (eq. 7.4) to define the flow, together with his bedload function (eq. 2.33) to give V_x, Allen (1971d) shows that at the bedform transition:

$$R=\frac{2\tan\zeta e_{\mathrm{b}}\sigma(\theta_{\mathrm{cr}}D)^{3/2}\{8(\sigma-\rho)g\}^{1/2}}{H\tan\alpha(f\rho)^{1/2}} \tag{9.14}$$

where e_{b} is Bagnold's (1966) bedload efficiency factor, σ and ρ the sediment and fluid densities respectively, θ_{cr} the critical Shields-Bagnold non-dimensional bed shear stress, g the acceleration due to gravity, $\tan\alpha$ the coefficient of dynamic grain friction, and f the Darcy-Weisbach bed friction coefficient. The remainder, $\tan\zeta$, D and H, are directly measurable (as is σ), and should be taken as near the transition as possible. Deposition rates in the order of 0.05 m per hour are thereby estimated for turbidites and flood deposits. Earlier claims that cross-lamination patterns record high deposition rates are not supported by the necessary independent knowledge of V_x implied in eq. (9.3) (Sorby, 1908; McKee, 1966a; Kuenen, 1967; R.G. Walker, 1969a). Using grain size to estimate the mean flow velocity in the absence of any bedform transition, Allen (1972b) suggests that cross-lamination patterns in the Uppsala Esker record flow fluctuations on a time-scale of a few hours to a few tens of hours.

The upward decline of set thickness in some cross-bedded point-bar sands is consistent with the experimental variation of dune height with flow conditions (Fig. 8-21), since across a point bar the mean flow velocity and depth generally increase in harmony (Vol. II, Chapter 2). Set thickness may increase upwards before it decreases, if the thalweg current can shape an upper-stage plane bed.

INTERNAL STRUCTURES, GRAIN SIZE, AND STREAM POWER IN WATER-LAID DEPOSITS

Although Gilbert (1914) examined bedforms experimentally, it was not until the comprehensive flume studies of Simons and Richardson (1961) and Simons et al. (1961) were published (see also Simons et al., 1965b; Guy et al.,

1966) that a satisfactory understanding of the relationship of internal structures to bedforms in sands and sandstones began to be developed.

Allen (1963a, 1963b, 1963c), Harms et al. (1963), and Gwinn (1964) drew attention to the origin of cross-lamination in current ripples, to the formation of cross-bedding by migrating dunes, and to the deposition of parallel laminae on plane beds. Each of the existence fields of Figs. 8-22 and 8-23 is therefore also defined by the occurrence of a particular internal structure, and may correspondingly be inferred from that structure.

P.G. Sheldon (1928, 1929) was amongst the first after Sorby (1859, 1908) to grasp the hydraulic significance of internal structures, especially when formed in vertical sequence. She interpreted Devonian sandstones in New York State showing the upward sequence massive bedding → parallel lamination → cross-lamination as recording a waning current. Similar models, based on extensive experimental work (Simons and Richardson, 1961; Simons et al., 1965b), were later developed for river deposits by Allen (1963a, 1963b), Harms et al. (1963), Gwinn (1964) and Harms and Fahnestock (1965), and for turbidites by Harms and Fahnestock (1965), R.G. Walker (1965, 1967b), and Allen (1968d). Hydraulically the most significant of these stress the dependence of structures on stream power (Fig. 8-22). All the models assume that flow changes are sufficiently gradual that equilibrium effectively prevails. The interpretation by Walker and by Harms and Fahnestock of massive and graded sands as deposited from supercritical flows is not now plausible.

Grain size as well as internal structure is related to stream power, at least in river and closely related deposits. Friend (1965) and Friend et al. (1976) observed from the Old Red Sandstone that cross-lamination was most prevalent in coarse siltstones and very fine sandstones, that cross-bedding typified fine to very coarse sandstones, and that parallel lamination was essentially independent of grade. Similar relationships mark certain Triassic fluviatile deposits (Grumbt, 1969, 1971; Falk et al., 1972). As Allen (1969c) noted, Friend's observations, to which Grumbt's may be added, are consistent with the experiments summarized and augmented in Figs. 8-22 and 8-23. We see no textural bound on the experimental occurrence of upper-stage plane beds and no apparent upper limit of stream power or bed shear stress. Again, whereas the ripple field has an upper grain-size limit, dunes are marked by a lower textural bound and by an existence-field that expands in range of stream power towards the coarser sediment sizes. A relationship similar to Friend's could certainly be constructed for turbidites. The experimental data, and the detailed texture–structure relationships described from the stratigraphic record, make it clear that a vertical pattern of internal sedimentary structures should never be interpreted independently of the grain size of the deposits.

SUMMARY

Cross-stratification arises in consequence of the migration of transversely oriented ripples and dunes in aeolian and aqueous environments. The attributes of cross-stratified units—the shape, arrangement, scale and attitude (cross-stratal azimuth, degree of climb) of the sets—are strictly governed by the time- and space-dependent characteristics of the parent bedforms, by the strength and direction of the driving current, and by the magnitude and sign of the net sediment transfer between bed and flow (erosion/deposition). The vertical patterns of climb and texture found in cross-laminated deposits are particularly revealing about flow properties and how these change in time and space. At least some horizontal lamination in fluviatile sands originates during the migration of bed waves of considerable uniformity and constancy, under conditions of small net deposition rate. Similarly, there are grounds for suggesting that parallel laminae formed in one-way flows record the movement across nominally flat beds of low-amplitude, damping waves created by continuous perturbations to the fluid and/or sediment flow.

Chapter 10

BEDFORMS IN SUPERCRITICAL AND RELATED FLOWS: TRANSVERSE RIBS, RHOMBOID FEATURES, AND ANTIDUNES

INTRODUCTION

The concept of free-surface subcritical and supercritical flows, introduced in Chapter 1, was developed in Chapter 7 as a short account of antidunes in terms of bed stability theory. We are now in a position to examine subcritical and supercritical flows of the same discharge as alternative states, to explore the sedimentary structures, for example, transverse ribs and rhomboid features, associated with switching from one of these states to the other, and to take a closer look at antidune surface waves and bedforms. Antidunes can occur in many different sedimentary environments, and may be useful in palaeohydraulic reconstructions.

ENERGY CONSIDERATIONS AND TRANSVERSE RIBS

Specific energy and alternate depths

The relationship which describes the partition of energy on a streamline in the steady uniform flow of an inviscid fluid is called the Bernoulli equation

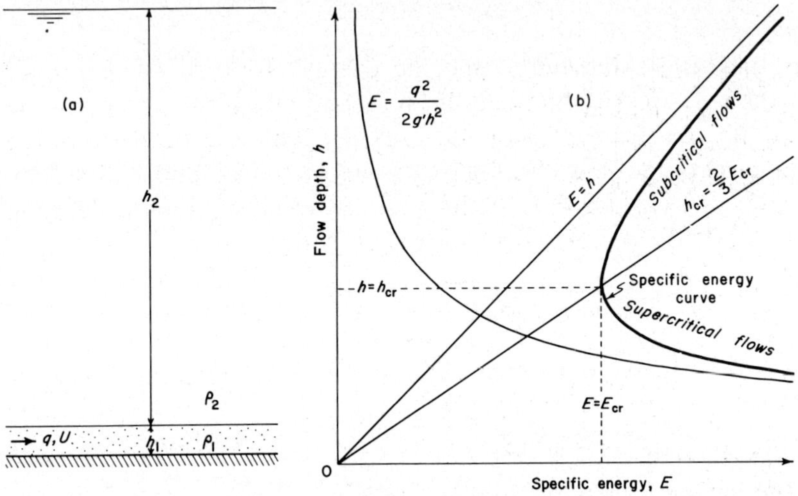

Fig. 10-1. Subcritical and supercritical flow. a. Definition diagram for the flow of a fluid beneath a second, stationary fluid. b. Specific energy curve for a fluid flowing beneath a second, stationary fluid.

(Batchelor, 1967). How for example is the energy divided when a heavy fluid in a thin layer moves beneath a much thicker layer of a lighter one, for example, a river below the air or a turbidity current beneath the ocean? If the Reynolds number is large, energy losses due to friction are negligible over short flow lengths and, if the bed slope is small, the pressure everywhere can be treated as hydrostatic.

Arguing like Yih (1965) and J.S. Turner (1973), consider the two-dimensional motion over a horizontal bed of a layer of fluid of thickness h_1, uniform velocity U, and density ρ_1, beneath a stationary fluid of thickness $h_2 \gg h_1$ and density $\rho_2 < \rho_1$ (Fig. 10-1a). Bernoulli's equation applied just above the interface gives:

$$\frac{p}{\rho_2} + gh_1 = C \tag{10.1}$$

and just below affords:

$$\frac{p}{\rho_1} + gh_1 + \frac{U^2}{2} = C \tag{10.2}$$

where p is the pressure at the interface, C is a single constant, and g is the acceleration due to gravity. Eliminating p between these equations, we obtain:

$$\frac{\rho_1 U^2}{2(\rho_1 - \rho_2)} + gh_1 = C \tag{10.3}$$

and dividing through by g find:

$$\frac{U^2}{2g'} + h_1 = E \tag{10.4}$$

where $g' = g(\rho_1 - \rho_2)/\rho_1$.

Equation (10.4) is Bakhmeteff's "specific energy" form of the Bernoulli equation for the flow of one fluid beneath a second which is stationary. The quantity E is the specific energy of the flow, written as a head measured relative to the base of the flow, $U^2/2g'$ is the velocity head, and h_1, the flow thickness, is the pressure head. Dropping the subscript on h, and defining $q = Uh$ as the discharge per unit flow width, we obtain:

$$E = h + \frac{q^2}{2g'h^2} \tag{10.5}$$

Consider eq. (10.5) in the E–h plane (Fig. 10-1b). The potential energy is given by the line:

$$E = h \tag{10.6}$$

and the kinetic energy by the hyperbola:

$$E = \frac{q^2}{2g'h^2}, \tag{10.7}$$

eq. (10.5) itself plotting between the asymptotes $E=h$ and $h=0$ for all values of the parameter q. There is a value of the flow thickness, h_{cr}, and a corresponding velocity, U_{cr}, for which E is a minimum, E_{cr}. Differentiating eq. (10.5) with respect to h:

$$\frac{dE}{dh}=1-\frac{q^2}{g'h^3} \tag{10.8}$$

or, restoring the velocity:

$$\frac{dE}{dh}=1-\frac{U^2}{g'h} \tag{10.9}$$

where the second term on the right will be recognized as the square of the Froude number, $Fr'=U/(g'h)^{1/2}$. Putting $dE/dh=0$ to obtain conditions at minimum specific energy:

$$h_{cr}=\tfrac{2}{3}E_{cr} \tag{10.10}$$

plotted in Fig. 10-1b, and:

$$\frac{U_{cr}}{(g'h_{cr})^{1/2}}=1. \tag{10.11}$$

At critical flow, the flow thickness is equal to two-thirds of the critical specific energy, and the Froude number, Fr'_{cr}, is unity. Furthermore, the flow thickness is the equal to twice the kinetic energy head $U_{cr}^2/2g'$.

The critical flow velocity, $U_{cr}=(g'h_{cr})^{1/2}$, is also the celerity of a low-amplitude long wave travelling over a fluid layer of thickness h_{cr}. Therefore in subcritical flow, $Fr'<1$, an interfacial wave can travel upstream, but in supercritical flow, $Fr'>1$, it can only travel downstream. A wave may remain stationary relative to the ground in critical flow, $Fr'=1$. The outstanding practical significance of these relationships is that, whereas a subcritical flow can be influenced by changes of bed configuration downstream, which might promote an upstream-travelling wave, a supercritical one is subject only to upstream control.

Equation (10.5) is a cubic in h with two real solutions, for each combination of E and q. The flow may either be shallow and fast (supercritical), corresponding to a point on the lower branch of the heavy curve in Fig. 10-1b, or deep and slow (subcritical), represented by a point on the upper branch. The two real flow thicknesses satisfying the relationship for each combination of E and q are called *alternate depths.* Although a steady uniform flow can be expressed in only one of these regimes, it is interesting to ask if a perturbation to the shape of the flow boundary can make the other regime locally accessible.

Figure 10-2 shows three sections in a two-dimensional flow over a bed with two changes of elevation. If the bed length over which the changes occur is large compared with the flow thickness, we can safely assume that

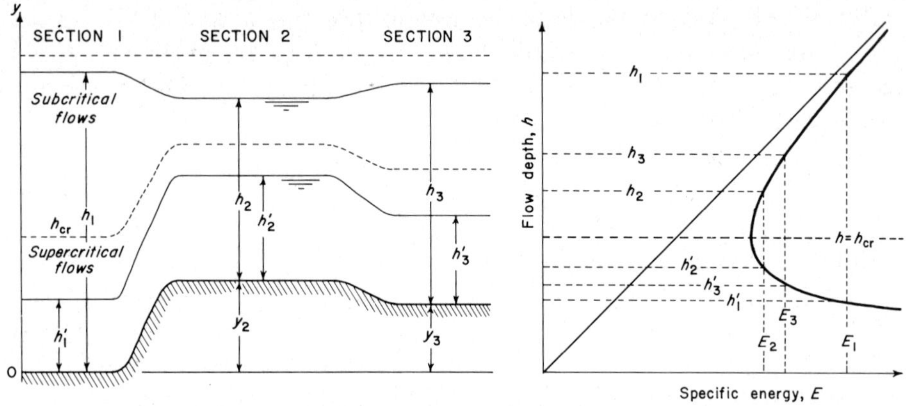

Fig. 10-2. Open-channel flow over a bed of changing elevation. a. Changes in flow depth with bed elevation for subcritical and supercritical flows. b. Flow at the stations in (a) plotted on the specific energy curve.

there is no change of total energy in the streamline at the bed, and so may apply Bernoulli's equation:

$$E_1 = E_2 + y_2 = E_3 + y_3 \tag{10.12}$$

in which E_1, E_2 and E_3 are the respective specific energies, y_2 and y_3 are the respective bed elevations referred to a datum, $y = 0$, corresponding to the bed level in section 1, and h_1, h_1', h_2, h_2', h_3, h_3' are the respective alternate flow thicknesses. Since the discharge q is the same in all three sections, and the flow width is constant, the flow surface is bowed up when a supercritical regime prevails in section 1, but is depressed when the flow there is subcritical. The flow can pass from supercritical in section 1 to subcritical in section 3 only if the change of bed level from section 1 to section 2 is sufficient that $E_2 = E_{cr}$. The change of regime will take place smoothly if the change of bed elevation is sufficiently gradual, but will otherwise occur abruptly, with some loss of energy, at the standing surge or shock wave known as a hydraulic jump (see Figs. 10-9 and 10-10). Similarly, a supercritical regime in section 3 is accessible to a subcritical flow in section 1 only if in section 2 critical flow obtains. Noting that a reduction of flow width increases the unit discharge for a given total discharge, with the effect that the specific energy curve in Fig. 10-2 is displaced to the right, it is easy to see what kinds of lateral constraint are necessary for these same effects.

We should finally note that eq. (10.5) is the most general simple form of the energy equation for free-surface flow, the quantity g' being the density-adjusted gravitational acceleration, used to calculate the densiometric Froude number Fr'. In the corresponding equation for river flow, the density of the atmosphere may be neglected compared with that of water, and g' is replaced by g (Chow, 1959; F.M. Henderson, 1966; Sellin, 1969). Chow (1959) in particular considers the forms of eq. (10.5) applicable when the bed slope cannot be regarded as small and the pressure everywhere hydrostatic.

Transverse ribs and their controls

These bedforms were but recently discovered. The title transverse ribs was given by McDonald and Banerjee (1970, 1971) to regularly spaced rows of clustered pebbles, cobbles or boulders lying transversely to flow on the bars and in the channels of braided streams. Independently, Boothroyd (1970, 1973) gave an account of these structures, also from braided streams, but his choice of name—pebble or clast stripes—has not found favour.

Figures 10-3 and 10-4 show representative transverse ribs from gravel-bed streams (McDonald and Banerjee, 1970, 1971; Gustavson, 1974, 1978; Boothroyd and Ashley, 1975; Martini, 1977), where in places they cover large areas on bar tops and channel beds. Rib streamwise spacing or wavelength is fairly uniform in each train, ranging from a mean of 0.06 m in some cases to as much as 2.26 m on the average in others. Individual ribs, formed of loosely to tightly packed clasts, are relatively straight and extend across the flow for a distance generally several times their wavelength. Ribs are seldom more than one or two clasts in height or more than a few clasts

Fig. 10-3. Transverse ribs on the Peyto Glacier outwash plain, Alberta, Canada. Trenching tool 0.5 m long. Flow toward observer. Geological Survey of Canada Photograph 157708, reproduced by permission of National Research Council of Canada (see McDonald and Banerjee, 1971).

Fig. 10-4. Transverse ribs on the Peyto Glacier outwash plain, Alberta, Canada. Trenching tool 0.5 m long. Flow from right to left. Geological Survey of Canada Photograph 157706, reproduced by permission of National Research Council of Canada (see McDonald and Banerjee, 1971).

wide. McDonald and Banerjee (1971) report that the rows of stones actually rest on a continuous layer of silt or sand, whereas Gustavson (1974) finds that clasts similar in size to those forming the ribs may commonly occur in the intervening spaces. The spreads of fine sand and silt often found between ribs that are exposed on drained beds (Fig. 10-4) may be deposited chiefly by waning flows, and an internal structure due to transverse ribs has yet to be described. It does seem clear, however, that the ribs comprise the largest clasts available to the stream. Gustavson (1978) reports that their wavelength is proportional to clast size, as also does Koster (1978).

There are other records of transverse ribs; a mosaic of photographs assembled by Klimek (1972) shows them on an active outwash plain, and Laronne and Carson (1976) figure them from a gravel-bed stream. For some years I have been familiar with the ribs as debris bars, formed where sodden pine needles or deciduous leaves are swept over steeply sloping surfaces of asphalt or closely mown grass by the run-off from summer rain-storms (Fig. 10-5a). In plan and spacing these bars are similar to the transverse ribs of gravel-bed streams, though they are generally lower in height, 0.01 m or less as compared with 0.05–0.30 m. Koster (1978) also gives accounts

The average wavelength of transverse ribs is inversely related to bed slope, but increases with ascending clast size. Figure 10-5b demonstrates that wavelength in twenty-two sets of ribs composed of pine needles or other

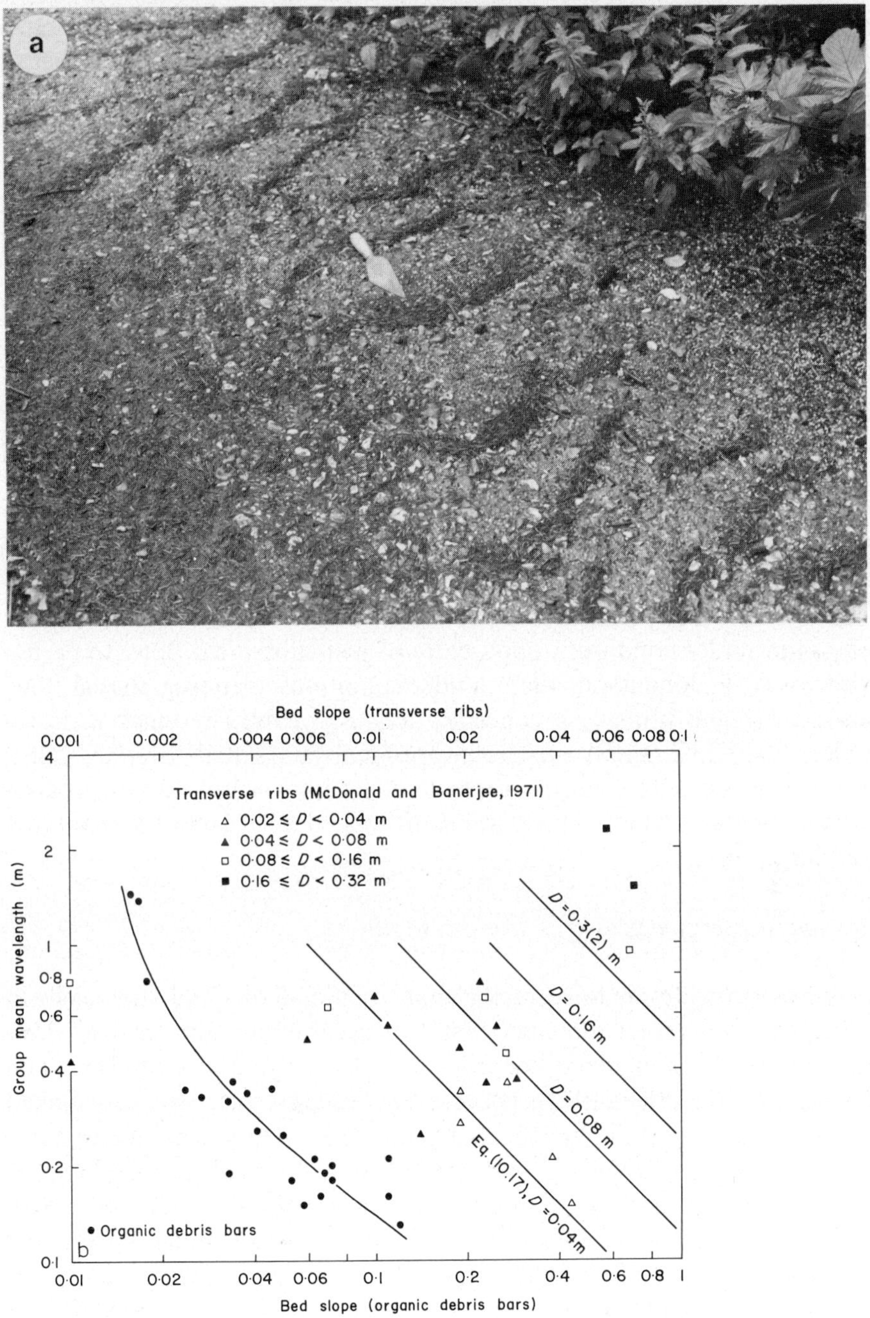

Fig. 10-5. Transverse ribs and debris bars. a. Bars of organic debris (chiefly conifer needles) on a gently sloping gravel path after summer rainstorm, University of Reading, Whiteknights Campus. Trowel 0.28 m long points in flow direction. b. Wavelength of transverse ribs (data of McDonald and Banerjee, 1971) and of organic debris bars (Whiteknights Campus), as a function of bed slope, and grain size (theoretical limiting relationship also plotted).

leaves decreases sharply with increasing bed slope. McDonald and Banerjee (1971) give a table and diagram from the Peyto outwash plain, Alberta, showing for several ranges of clasts sizes either a similar inverse relationship between rib wavelength and slope or the independence of wavelength on slope (Fig. 10-5b). They emphasize that wavelength bears a strong direct correlation with clast size (Fig. 10-5b), a relationship confirmed by Gustavson (1974, 1978) and by Boothroyd and Ashley (1975). Since the clast size available seems to increase with bar or channel slope (e.g. Boothroyd and Ashley, 1975), there may be a limiting slope for each calibre below which transverse ribs cannot exist, perhaps increasing with clast size.

From gravelly stream environments come records of other forms of stone cluster perhaps related to transverse ribs. The best defined of these are the vague cellular clast arrangements briefly reported by McDonald and Banerjee (1971), which Gustavson (1974) later figured and called stone cells. These are polygonally disposed rows of large clasts separated, as commonly with transverse ribs, by sand or silt spreads. C.E. Johansson (1965, 1976) illustrated what may be stone cells from a stream bed formed of platy gravel, and made cell-like stone clusters experimentally during the erosion of a gravelly sand under conditions apparently of just-subcritical flow. In earlier experiments (C.E. Johansson, 1963), under conditions spanning critical flow, he produced vague transverse concentrations of pebbles resembling transverse ribs. Dal Cin's (1968) clusters are probably unrelated to either transverse ribs or stone cells. Each concentration is a narrowing train of progressively smaller stones which during transport lodged upstream of a stationary isolated large clast.

Application of energy equation to transverse ribs

There exists a widespread suspicion that transverse ribs and stone cells in some way depend on phenomena that can accompany supercritical flow. Boothroyd (1970) interpreted the ribs as relict antidune bedforms, and McDonald and Banerjee (1971), because of the large clast size, also linked them with supercritical flows, observing that flow depth was of the same order as the size. Gustavson (1974) reported from the field a general spatial association between transverse ribs and supercritical flows. He also stated that Dr. B.C. McDonald had made the ribs experimentally during the upstream migration of a hydraulic jump. C.E. Johansson's (1963, 1965, 1976) experiments also tend to link ribs and cells with nearly critical if not supercritical flows. Boothroyd and Ashley (1975) in summarizing much of this evidence claim that "transverse ribs certainly are formed under upper flow regime conditions, either under near-critical to supercritical antidune breaking waves, or by supercritical flow events resulting in hydraulic jumps." However, the mechanisms involved are left unveiled.

A greater insight into the origin of transverse ribs is obtainable through a

Fig. 10-6. Cascades over transverse ribs, No-See-Um Creek, Alberta, Canada. Geological Survey of Canada Photograph 157709, reproduced by permission of National Research Council of Canada (see McDonald and Banerjee, 1971).

consideration of eq. (10.5) for specific energy. Flow over transverse ribs (Fig. 10-6) occurs as a series of cascades (McDonald and Banerjee, 1971), apparently similar to Peterson and Moharty's (1960) "tumbling" flow over

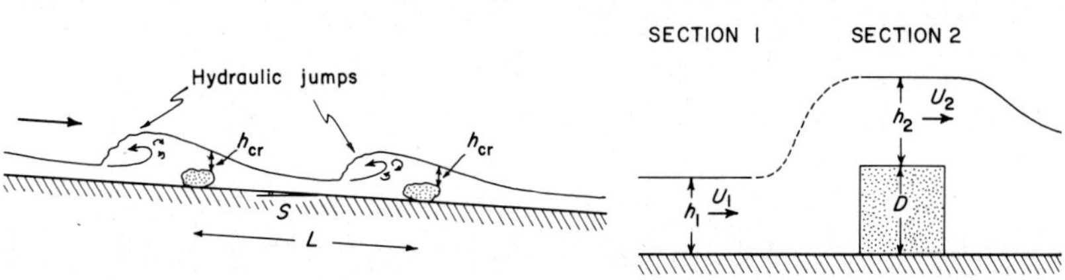

Fig. 10-7. Origin of transverse ribs. a. Suggested flow over a train of transverse ribs. b. Definition diagram for flow over a stationary clast of square section standing on a smooth bed.

regularly spaced fixed transverse obstacles acting as flow controls (Fig. 10-7a). Each obstacle (row of clasts forming a rib) generates a hydraulic jump to its upstream and a cascade downstream, the series of obstacles (train of ribs) creating an identically spaced sequence of each kind of water-surface feature. For the purposes of analysis, let the obstacle consist of a transverse row of cube-shaped clasts of side D resting face to face on a horizontal bed. Ignoring the precise form of the water surface upstream, consider the flow in a vertical streamwise plane passing through the centre of one of the clasts. Let it be further supposed that D is representative of the coarsest debris available to the flow. Equation (10.5) applied at section 1 upstream of the clast and at section 2 at the clast itself leads to:

$$\frac{q^2}{2gh_1^2} + h_1 = \frac{q^2}{2gh_2^2} + h_2 + D \quad (10.13)$$

which, on rearrangement, gives:

$$h_2^2\left(\frac{q^2}{2g} + h_1^3 - h_1^2 D\right) - h_1^2 h_2^3 - \frac{q^2 h_1^2}{2g} = 0 \quad (10.14)$$

as the relationship for the flow depth, h_2, at the clast, in terms of the upstream flow depth, h_1, the unit discharge, q, and D. For each combination

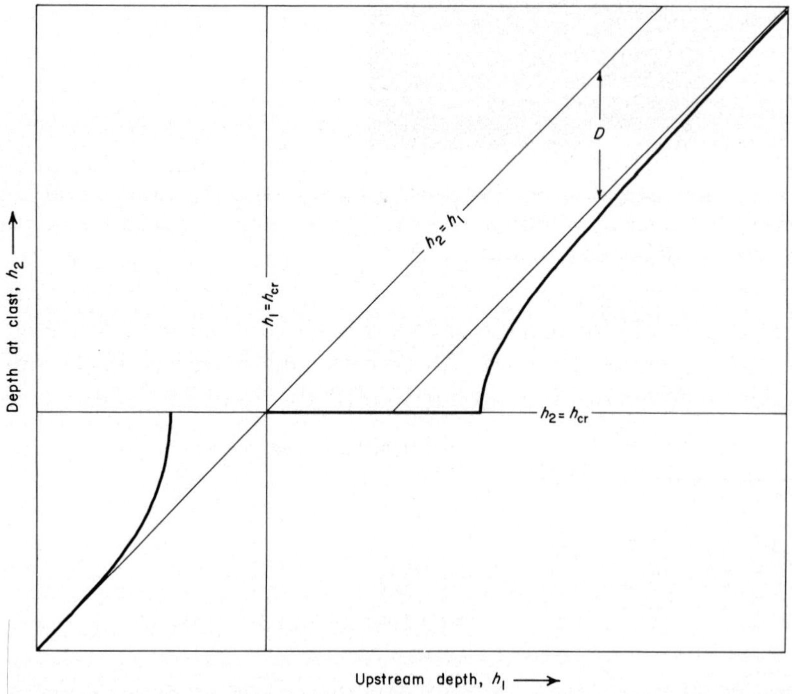

Fig. 10-8. General form of function (eq. 10.14) describing possible flows over a stationary clast of square cross-section standing on a smooth bed.

of q and D, the graph of h_2 as a function of h_1 has three portions with respect to the critical depth $h_1 = h_2 = h_{cr}$ (Fig. 10-8). One portion, to the right of the line $h_2 = h_1$, is for subcritical flow above as well as upstream of the clast, whereas a second, also to the right, is for subcritical flow upstream but critical flow at the clast. The third portion, to the left of $h_2 = h_1$, is for supercritical flow above as well as upstream of the clast. A supercritical flow occurring in the region between the two branches of the curve will promote some kind of hydraulic jump and, therefore, a subcritical depth immediately upstream of the clast.

Flows on the first portion of the graph in Fig. 10-8 cannot form transverse ribs. This conclusion hinges on the recognition that, for each clast size D, a certain constant flow velocity, U_{cr}, is necessary to initiate clast motion, and that this critical velocity can be substituted by an equivalent flow depth, h_{eq}, in a flow of constant discharge, on recalling that $q = Uh$. Now the ribs comprise stationary clasts which must have been carried from the areas between ribs. Hence in the notation of Fig. 10-7b the ribs can exist only if, in the intervening areas, $U_1 > U_{cr}$ and $h_1 < h_{eq}$, while at the ribs themselves, $U_2 < U_{cr}$ and $h_2 > h_{eq}$, that is, if $h_1 < h_{eq} < h_2$. This inequality is satisfied only by a flow occurring on either the third portion of the graph or in the region between the branches. A flow on the first portion is faster and shallower over each clast than upstream of it, perpetually urging the clast forward, and so a stationary rib could not arise. Flows on the second portion are for the same reason unable to generate ribs.

Transverse ribs are most likely to arise in flows falling on the third portion of the graph or in the gap between the two segments, where a hydraulic jump should result from the presence of clasts, for it is in these regions that the flow upstream is shallower and faster than at or near the clasts themselves, which experience arrest. The final condition restricting the occurrence of the ribs is that h_1 must exceed the depth h_{un} of the uniform flow which would otherwise exist, given by:

$$h_{un} = \left(\frac{fq^2}{8gS} \right)^{1/3} \tag{10.15}$$

where S is the bed slope and f the Darcy-Weisbach bed friction coefficient. Drawing together the several parts of the analysis:

$$h_1 < h_{cr} \tag{10.16a}$$

$$h_{un} < h_1 < h_{eq} < h_2 \tag{10.16b}$$

define the conditions for the occurrence of transverse ribs.

The analysis just sketched may be criticized on the grounds that as the depth changes are rapid rather than gradual, the friction and pressure forces cannot be neglected. It nonetheless emphasizes that transverse ribs can exist only when sediment and flow conditions fall within a precise range, and confirms as suspected the association of the ribs with hydraulic jumps and with flows which if uniform would be everywhere supercritical. Presumably ribs can be initiated downstream from any chance piling up of clasts of sufficient size in a supercritical flow. As the flow changes from subcritical to supercritical (or becomes more strongly supercritical) on passing over and downstream from the pile, the conditions which favour a further piling up are recreated but at a site still further down the flow. In this way a train of the features could arise. Overall conditions of waning flow would in particular promote their formation and preservation.

The reasons for the dependence of rib wavelength on flow and sediment conditions are less clear. However, Keutner (1929) and Peterson and Moharty (1960) developed a criterion for the occurrence of supercritical flow between regularly spaced transverse obstacles which amounts to:

$$L \geqslant \frac{0.15h_2}{S} \tag{10.17}$$

where L is the streamwise obstacle spacing. Rib wavelength should therefore vary inversely with bed slope, as is observed in some cases (Fig. 10-5). Since from eq. (10.16) the general magnitude of the permissible values of h_{eq} increases with increasing flow scale, and D is itself in the order of h_2, we should also expect wavelength to increase with ascending clast size. This dependence is particularly well documented (Fig. 10-5b), the wavelengths calculated from eq. (10.17) with $D = h_2$ being of the same general order as observed values.

McDonald and Day (1978) recently described a laboratory investigation into structures that are similar in character to transverse ribs as encountered in the field, except that the experimental forms were produced in essentially monodisperse gravels. The structures increase in height and wavelength with growing sediment calibre but, contrary to the debris bars and transverse ribs plotted in Fig. 10-5, are shown in McDonald and Day's graphs as increasing in wavelength with ascending bed slope for each bed-material and flow depth. Each rib formed as a bar beneath a temporarily stationary hydraulic jump, the train of bed features arising as the jump migrated *upstream* in short steps. Once a rib had been formed, the flow over it became subcritical, though wave-like on account of the distortion provided by the rib. As envisaged in Fig. 10-7a, however, the flow over transverse ribs formed from polydisperse sediments in the field is partly supercritical, with a train of ribs spreading *downstream* from an initial rib located where perhaps an unusually large clast had come to rest. Are the structures observed by McDonald and Day precisely the same as those reported from the field, and is there perhaps more than one kind of and origin for transverse ribs?

MOMENTUM CONSIDERATIONS AND RHOMBOID BED FEATURES

Hydraulic jumps, specific force, and conjugate depths

Perhaps the most familiar example of the abrupt transition of a supercritical into a subcritical flow at a hydraulic jump occurs where a jet of water from a tap impinges on the flat bottom of a sink (Fig. 10-9). The resulting radial jump was first remarked by Lord Rayleigh (1914) and subsequently analyzed by E.J. Watson (1964), Koloseus and Ahmad (1969), and Mehrotra (1974). Two-dimensional jumps have been most studied, however, in subatmospheric flows (e.g. Bakhmeteff and Matzke, 1936, 1938; Forster and Skrinde, 1950; Rouse et al., 1959; Rajaratnam, 1965b; Rajaratnam and Subramanya, 1968; Ali and Ridgway, 1977; Rajaratnam and Ortiz, 1977), where a flow issues as a jet into a stationary medium of the same density (Rajaratnam, 1965a), and in two-layer liquid or gaseous systems of small density contrast (e.g. Yih and Guha, 1955; Wood, 1967; Komar, 1971a; Wilkinson and Wood, 1971; J.S. Turner, 1973).

F.M. Henderson (1966) illustrates the two main forms of two-dimensional jump found in systems of fluid layers having either a density discontinuity or a thin zone of rapid density change (Fig. 10-10) and Peregrine (1966) gives numerical solutions to their equations of motion. An undular jump (Broome and Komar, 1979), occurring when the upstream Froude number is less than about 1.7, is an elevated series of unbroken standing waves decaying in

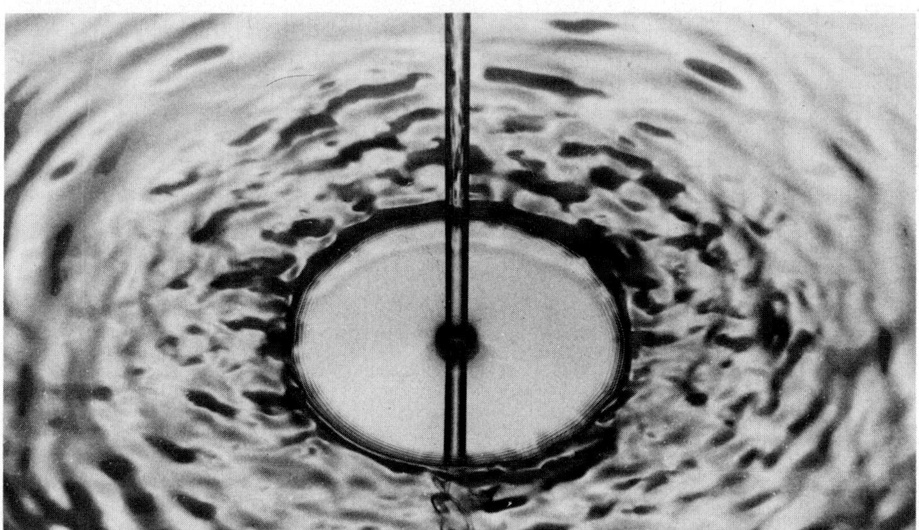

Fig. 10-9. Circular hydraulic jump formed by the impingement of a narrow jet of water vertically on to a horizontal glass plate.

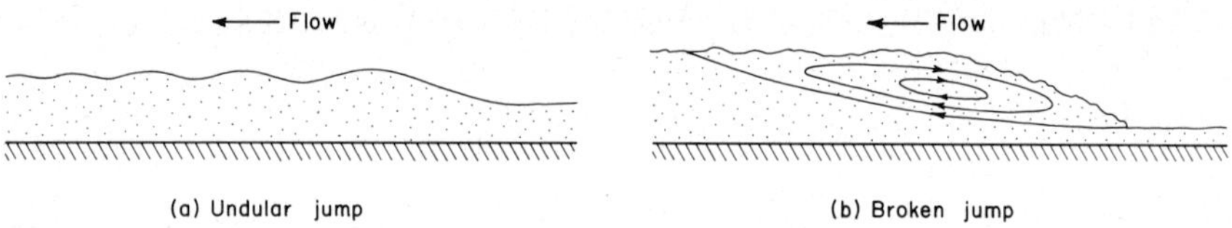

Fig. 10-10. Main forms of hydraulic jump illustrated by schematic sections in the flow plane, together with the time-averaged flow configuration within a broken jump.

height and wavelength downstream. The direct or broken jump, forming at upstream Froude numbers as low as 1.55, is a single wave of well-defined form (N.K. Gupta, 1966; Rajaratnam and Subramanya, 1968) covering a flat "roller" with reverse flow and a zone of intense vortical and turbulent mixing (Rouse et al., 1959; Longuet-Higgins, 1973b). Transitional forms of jump involve the mild breaking of one or more of the standing waves downstream of the main front. The overlap in occurrence of the main kinds of jump is probably explained by the fact that the Froude number is always based in practice on a depth-averaged velocity, the form of the velocity profile being strongly influenced by boundary shape and roughness (Lighthill, 1953). We can also classify hydraulic jumps according to their movement. Where there exists a suitable fixed upstream control, a stable stationary jump can form, as we have seen in the case of transverse ribs. Otherwise jumps travel upstream, in accordance with the implications of a Froude number in excess of unity for the celerity of a long wave.

Hydraulic jumps may be important sedimentologically for reasons other than their association with transverse ribs. At a jump there occurs a substantial increase of flow depth and corresponding decrease of flow velocity. In a river, a variety of bedforms in addition to ribs could be generated in association with jumps, for the change of flow at a jump promotes deposition, which in turn may contribute to the control for a jump still further downstream (see Jopling and Richardson, 1966). Broome and Komar (1979) provide evidence that the so-called antidunes or backwash ripples of sand beaches may form beneath the surface waves on the subcritical side of hydraulic jumps formed by wave backwash. Many an esker arises as a retrograding bar on the floor of a lake or shallow sea where a jet of meltwater issues from a deeply submerged tunnel-mouth on the front of a back-wasting glacier (R.J. Price, 1973). A jump formed where such a jet was supercritical (Rajaratnam, 1965a, 1965b) should exert a decisive control on the shape and position in relation to the tunnel-mouth of the resulting delta-like sediment body. Broken jumps in systems of miscible fluids of small density contrast are effective in mixing the ambient medium into the supercritical layer and of partially equilibrating the densities (Wilkinson and Wood, 1971). Such mixing may be significant in the production of turbidity currents (Komar, 1971a), and certainly is important in the dilution of dust-laden winds (J.S. Turner, 1973, fig. 3.11).

Hydraulic jumps may be analyzed by applying continuity with Newton's second law of motion, which states that any change in the momentum flux must be due to forces acting on the fluid (Yih and Guha, 1955; Chow, 1959; Yih, 1965; F.M. Henderson, 1966; Sellin, 1969; Komar, 1971a; Wilkinson and Wood, 1971; J.S. Turner, 1973). In the simple general case (Fig. 10-1a), the "specific force" of the flow, analogous to the "specific energy", is:

$$F = \frac{q^2}{g'h} + \frac{h^2}{2} \tag{10.18}$$

where F is the total specific force, $q^2/g'h$ measures the momentum of the moving fluid, and $h^2/2$ describes the hydrostatic pressure force. Equation (10.18) has a somewhat similar form to the corresponding eq. (10.5) but has only one asymptote, $h = 0$. Like the specific energy, the specific force is easily shown to be a minimum at critical flow. Applying eq. (10.18) to vertical sections on either side of a two-dimensional hydraulic jump beneath a stationary fluid (Fig. 10-11a), we find that:

$$\frac{h_1'}{h_1} = \tfrac{1}{2}\left\{\left(1 + 8Fr_1'^{\,2}\right)^{1/2} - 1\right\} \tag{10.19}$$

in which Fr_1' is the densiometric Froude number based on upstream conditions (Yih and Guha, 1955). Figure 10-11b shows the specific force and specific energy curves for this flow. The flow thicknesses h_1 and h_1', which occur at the same specific force, are called *conjugate depths,* but only that on the supercritical branch corresponds to an alternate depth on the energy

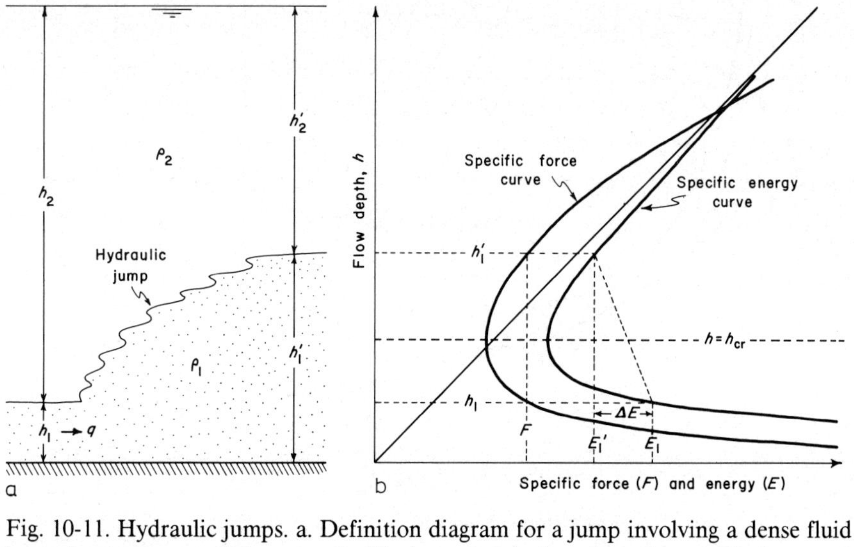

Fig. 10-11. Hydraulic jumps. a. Definition diagram for a jump involving a dense fluid flowing beneath a stationary, lighter one. b. The jump in (a) plotted on the specific force and specific energy curves.

graph. The downstream flow has a smaller specific energy, a loss of ΔE having occurred in the jump, as may have been surmized from the accompanying reversed flow and vortical mixing.

Experimental work shows that eq. (10.19) and its extensions to sloping beds satisfactorily describes the properties of jumps in subatmospheric water channels (e.g. Bakhmeteff and Matzke, 1936, 1938), as well as in layered fluids of small density difference (e.g. Yih and Guha, 1955). It may therefore be used to calculate jump characteristics, for prescribed upstream conditions.

When a supercritical flow is slightly deflected horizontally, it can be shown from momentum considerations and confirmed experimentally that an oblique hydraulic jump arises at the point of deflection (Rehbock, 1930; Ippen, 1951; Ippen and Harleman, 1956). If the deflecting object is, say, a pebble which projects through the surface of a supercritical river or wave backwash-flow, the oblique jump will form symmetrically to give a semi-angle, α, with the downstream direction:

$$\sin \alpha = \frac{(gh_1)^{1/2}}{U_1} \left\{ \frac{h_2}{h_1} \frac{1}{2} \left(1 + \frac{h_2}{h_1} \right) \right\}^{1/2} \tag{10.20}$$

where h_1 and h_2 are the flow depths upstream and downstream, respectively, and U_1 is the upstream flow velocity. Since the modifier to the inverse

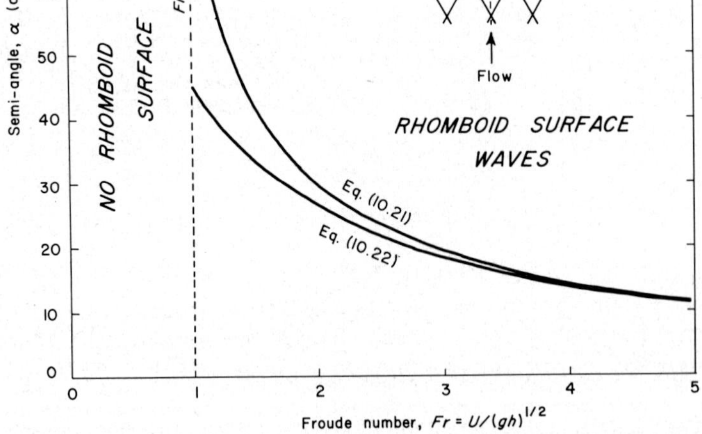

Fig. 10-12. Semi-angle between the crests of obliquely crossing hydraulic jumps, calculated as a function of Froude number and for W negligibly small (eq. 10.22).

upstream Froude number approaches unity as the jump height $(h_2 - h_1)$, grows smaller:

$$\sin \alpha = \frac{1}{Fr_1} \tag{10.21}$$

with $Fr_1 \geqslant 1$, affords the minimum obliquity of the jump for each upstream flow condition (see also Rehbock, 1930), a relation cited incorrectly by Komar (1976). Real values of the semi-angle can therefore occur only to the right of the curve for eq. (10.21) in Fig. 10-12.

It appears that symmetrically interfering oblique hydraulic jumps can also be produced in channelized flows of sufficient breadth *without* deflecting obstacles. Chang and Simons (1970) showed theoretically that the semi-angle was then given by:

$$\tan \alpha = \frac{W \pm (gh)^{1/2}}{U} \tag{10.22}$$

where U and W are the averaged velocities per unit width in the longitudinal and transverse directions and h is the local depth of flow. When W is small compared with U, their expression reduces to:

$$\tan \alpha = \pm \frac{1}{Fr} \tag{10.23}$$

for the minimum semi-angle (Fig. 10-12). Equations (10.21) and (10.23) therefore yield significantly different results for near-critical Froude numbers, but converge with increasing supercriticality.

By equations (10.21) and (10.23) the semi-angle declines with increasing flow velocity but grows with increasing flow depth. But $Fr = (8S/f)^{1/2}$ for steady uniform flow, where S is the bed slope and f the Darcy-Weisbach friction coefficient. Substituting for the Froude number, the equations then read:

$$\sin \alpha = \left(\frac{f_1}{8S_1} \right)^{1/2} \tag{10.24}$$

$$\tan \alpha = \left(\frac{f}{8S} \right)^{1/2} \tag{10.25}$$

where the subscript denotes upstream conditions, showing that the semi-angle decreases with increasing bed slope, other factors remaining constant. If there existed any bedforms generated in accordance with these equations, we should expect the semi-angle to increase with increasing sediment calibre, since an increase in grain size heightens the boundary roughness.

Rhomboid rill marks, rhomboid ripples, and rhomboid dunes

These forms, apparently restricted to sand beds, occur in trains of remarkably regular individuals, elongated with the flow and of diamond-shaped plan.

Rhomboid rill marks (Otvos, 1964, 1965), called "rhomboid ripple marks" by Woodford (1935) and Demarest (1947), and rhomboid lattice structure by Stauffer et al. (1976), are patterns of regularly criss-crossing grooves on otherwise flat, extensive sand beds (see also Lafeber and Willoughby, 1970). The grooves are straight, broad, and shallow; they are rarely deeper than a few grain-diameters. Individual rhombic elements are in the order of a few centimetres long and a centimetre or so wide. The patterns they make are accentuated where shell fragments or heavy mineral grains abound because, as James Hall (1843) observed and illustrated, "the sand of different colours is arranged in clouds and undulating stripes". Rhomboid rill marks are so far known only from relatively steep, seaward-facing beach slopes. For example, semi-angles of 17°–20° were measured by Demarest on beaches of approximately 10° slope underlain by very coarse grained quartz sand.

Rhomboid ripple marks (Fig. 10-13) in the sense of Otvos (1964, 1965)

Fig. 10-13. Rhomboid ripple marks in fine sand on the crest of an intertidal sand bar, Holm-next-the-Sea, Norfolk, England. Trowel 0.28 m long points in direction of wave overwash.

were first described by W.C. Williamson (1887). Other early accounts were published by Engels (1905), by Kindle (1917) under the name of "imbricate wave sculpture", by Bucher (1919) as "rhomboid ripple", and by D.W. Johnson (1919) under the title of "backwash mark". Many subsequent descriptions are available (Timmermans, 1935; Trusheim, 1935; W.O. Thompson, 1937; Van Straaten, 1953a, 1953b; McKee, 1957a; Hoyt and Henry, 1963; Otvos, 1965; Schwenk, 1965; Milling and Behrens, 1966; Martins, 1967; Brambati, 1968; Klein, 1970b; Rudowski, 1970b; Karcz, 1972, 1974; Picard and High, 1973; Wunderlich, 1973; Komar, 1976). Karcz and Kersey (1980) made rhomboid ripple marks experimentally, as one of a sequence of bedforms generated by very shallow, subcritical to supercritical aqueous flows. Each rhomboid ripple is a scale-like feature elongated parallel with flow and shaped in plan like a symmetrical diamond. The crest has two straight portions oblique to the current direction (shown by spurs), joined by a short curved section at the extreme downcurrent end. The lee side is tallest here, diminishing in height upcurrent. The ripple trough is also V-shaped in plan, but more sharply pointed than the crest, a deep rounded scour occasionally occupying its apical region. Commonly the stoss sides of rhomboid ripple marks carry numerous long closely placed spurs aligned nearly parallel with the mean flow direction. The semi-angle formed between a branch of the lee side and the flow is generally between 15° and 40°. Wavelengths ordinarily fall between 0.08 and 0.50 m. Rhomboid ripple

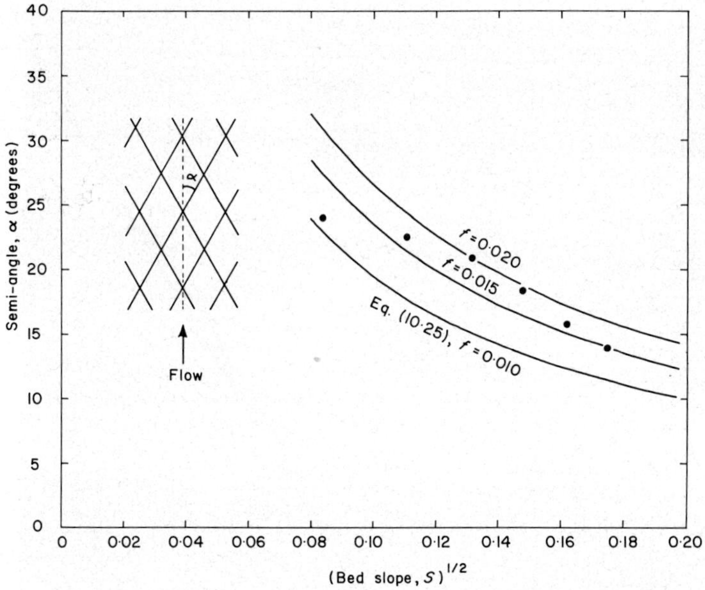

Fig. 10-14. Observed semi-angle of rhomboid ripple marks in sand as a function of beach slope (data of Hoyt and Henry, 1963), compared with values predicted by eq. (10.25) for various bed roughnesses.

marks yield values for the vertical form-index in the order of 100 and are very much flatter than current ripples.

Rhomboid ripple marks are commonest on beaches and intertidal sand flats (Timmermans, 1935; Trusheim, 1935; W.O. Thompson, 1937; Van Straaten, 1953a, 1953b; Hoyt and Henry, 1963; Otvos, 1965; Milling and Behrens, 1966; Brambati, 1968; Klein, 1970b; Rudowski, 1970b; Reineck and Singh, 1973; Wunderlich, 1973; Komar, 1976), where they are formed by backwash flows on seaward-facing slopes and by wave overwash on the crests and landward sides of bars. Hoyt and Henry found that the semi-angle of backwash-formed rhomboid ripples was inversely related to beach slope (Fig. 10-14). When present on bars, the ripples usually occur in transitional association with other bed features, for example, in the streamwise (down-slope) sequence: plane bed → rhomboid ripple marks → current ripples → wave ripples. Rhomboid ripples are also sometimes found in small channels draining sandy beaches and intertidal flats (Van Straaten, 1953b; Schwenk, 1965; Bajard, 1966). Here they occur in lateral transitional association with other structures, for example, lying between linguoid current ripples in channel deeps and plane beds near channel edges. Rhomboid ripple marks also occur in rivers, on bar crests and where flows are shallow (Karcz, 1972; Klimek, 1972; Picard and High, 1973). Young and Ross (1974) describe

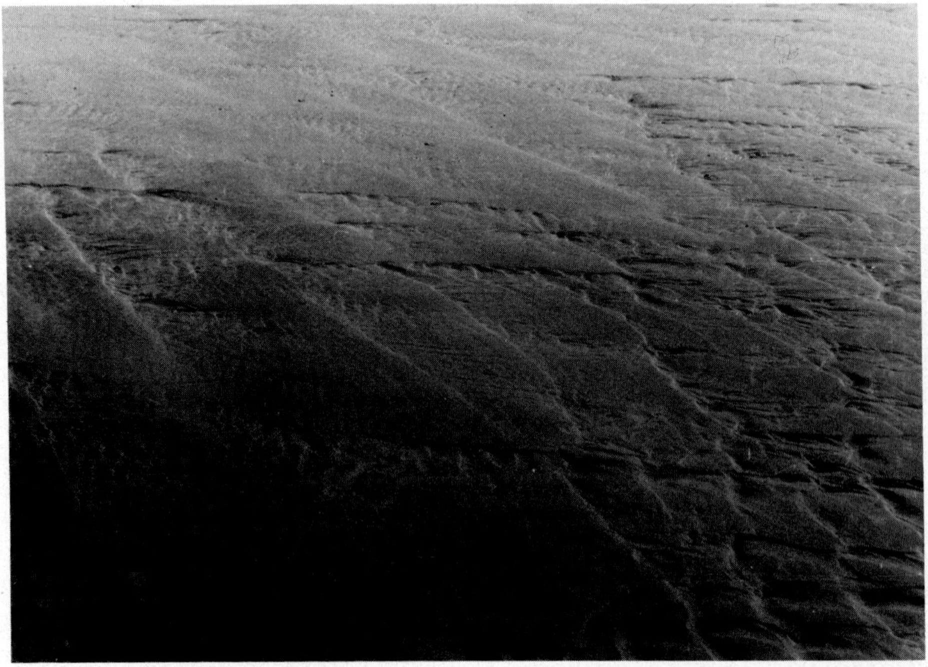

Fig. 10-15. Rhomboid dunes with decimetre-scale rhomboid ripples in fine sand, Norderney, German Friesian Islands. Photograph courtesy of H.-E. Reineck (see Reineck, 1963).

small rhomboidal features from the muddy floor beneath an area of hot brines in the Red Sea.

Rhomboid dunes (Fig. 10-15), called "rhomboid megaripples" by Van Straaten (1953a, 1953b), resemble rhomboid ripples in shape and orientation but are much larger. They are typically found on the crests and landward slopes of beach bars and on the crests of other intertidal sand bodies, where they are formed by wave overwash or other shallow flows (Van Straaten, 1953a, 1953b; Reineck, 1960a, 1963; McMullen and Swift, 1967; Rudowski, 1970b; Karcz, 1974). They occasionally arise on seaward-facing surfaces, on slopes steeper than those carrying rhomboid ripples (Hoyt and Henry, 1963). Morton (1978, 1979) has described a number of examples generated during storms on the barrier islands of the Texas coast. Some of N.D. Smith's (1971a) "lobate sand waves" are not unlike rhomboid dunes. Wavelengths range from a metre or so to a few tens of metres. The stoss sides usually carry a pattern of rhomboid ripples, often accompanied by linguoid current ripples restricted to the deeper, apical portions of the troughs. Long spurs in places extend downstream from the lee slopes. Rhomboid dunes have a close association with other bed features. They may occur on the higher parts of beach bars upstream (upslope) from fields of rhomboid ripples. McMullen and Swift (1967) reported the downcurrent (upslope) transition of transverse into rhomboid dunes on an intertidal sand ridge.

Little is known of rhomboid features from the older geological record and of their internal features. W.O. Thompson (1949) records them from a supposedly littoral sandstone, and Singh (1969) reports small rhomboid ripple marks from a late Precambrian shallow-water quartzite. Wunderlich (1973) believes that rhomboid ripples and dunes advance *against* the current, laying down laminae steeply inclined against the flow (backset bedding). In my experience the forms travel only downcurrent, and I find Wunderlich's illustration of the supposed internal structure of rhomboid ripples inconsistent with his profile through the features. Morton (1978) found the rhomboid dunes of the Texas coast to be cross-bedded internally.

Oblique jumps and rhomboid features

Rhomboid bedforms have been explained in several ways. W.C. Williamson (1887) believed that rhomboid ripples represented two sets of ordinary ripples formed subaqueously and later modified by down-beach drainage which produced the spurs. Kindle (1917) thought they were due to "the action of very small waves lapping and crossing each other from opposite sides of a miniature spit". Rudowski (1970b) proposed that "the interference of moving streamlines giving a rhomboidal effect on the water surface" was necessary for the formation of the ripples. Of these suggestions, Williamson's is not supported by field observations, Kindle's seems to imply a special circumstance, and Rudowski's refers to a consequent phenomenon. Stauffer et al. (1976) concluded that rhomboid rill marks were not due to surface flow processes, but to mechanisms acting within the already deposited sand and set in train by the drainage of the backwash into the beach. They point out that the drainage of water through the beach sediment creates a suction due to the appearance of surface-tension films, and claim that this suction gives rise to a system of conjugate shear zones within the sediment. The grains are suggested to assume a closer packing along each shear zone, with the result that a linear groove appears on the beach surface. But this explanation is unconvincing, if only because, in order to create a groove several grain-diameters deep in such a well-packed material as the sand on the seaward face of a beach, a sediment zone several tens to a few hundreds of grain-diameters deep must have been repacked under the influence of shearing. However, Stauffer et al. (1976) noticed that, at one stage in the backwash, the water surface carried a rhomboidal wave pattern, and were perhaps misled into dissociating this pattern from the rill marks by the difficulties of observing features of low relief when shallowly submerged.

There are many grounds for allying rhomboid features with supercritical flows and systems of oblique hydraulic jumps. Woodford (1935) demonstrated from the field the association of rhomboid rill marks with supercritical flow and, citing the experiments of Engels (1905) and Rehbock (1930), together with the latter's presentation of eq. (10.21), linked the marks with patterns of interfering surface waves. Demarest (1947) observed these marks to form beneath thin sheets of water which for part of the time carried roll waves (see below), a sure indicator of supercritical flow. Otvos (1965) associated rhomboid rill marks with systems of obstacle-induced oblique hydraulic jumps, but drew no definite conclusions about rhomboid ripples. Resting on his own experiments and those of Kennedy and Roubillard (1967), Kennedy and Iwasa (1968), and Chang and Simons (1970), however, Karcz (1974) associated rhomboid features generally with patterns of oblique waves developed on supercritical flows, but saw no necessity for the presence of obstacles. The rhomboidal features observed by Young and Ross (1974) from the floor of the Red Sea could record the action of oblique internal waves arising at the interface between different brine layers.

The evidence supports Karcz (1974) rather than Woodford (1935) and favours interpretation in terms of eq. (10.23) rather than eq. (10.21). Although rhomboid features of all kinds do occur on sand surfaces carrying such obstacles as stones or shells, they are typically present on beds devoid of such obstructions, and so cannot depend on Rehbock's (1930) and Ippen's (1951) mechanism, unless the sediment ridges consequent on the jumps are to be regarded also as forcing the jumps on which they depend. In my experience, comformable with the observations of Chang and Simons (1970) and of Karcz (1974), supercritical flow occurs over the downstream parts of the stoss sides of rhomboid ripples and dunes, while an undular to weakly broken hydraulic jump lies coupled to each lee and trough, somewhat as in chute-and-pool flow (Guy et al., 1966). In the apical region of the trough, the two jumps usually peak up where they cross, giving a pyramidal wave occasionally like Kennedy's (1961) rooster-tail antidunes. Rhomboid ripple marks further conform to Karcz's (1974) interpretation as expressed by eq. (10.25). Under the relatively uniform sediment and wave conditions of Sapelo Island (Hoyt and Henry, 1963), the Darcy-Weisbach friction coefficient may be supposed fairly constant, and the ripple semi-angle declines with increasing slope (Fig. 10-14), broadly in accord with eq. (10.25). Woodford (1935) and Demarest (1947) measured rhomboid rill marks with semi-angles similar in value to the angles recorded by Hoyt and Henry (1963), but from beaches an order of magnitude steeper than on Sapelo Island. Equation (10.25) is not thereby violated, however, as very coarse sand underlay the beaches and the Darcy-Weisbach coefficient must consequently have been larger.

Several workers noticed the common spatial association of (1) rhomboid ripple marks with current ripples, and (2) rhomboid dunes with normal dunes (e.g. Van Straaten, 1953a, 1953b; Bajard, 1966; McMullen and Swift, 1967). These associations suggest that two conditions must simultaneously be satisfied for the creation of these rhomboid structures. One is that the corresponding uniform flow be supercritical, a condition expressed by the Froude number and determined completely by the flow depth and velocity. The second is that the combination of stream power and grain size puts the flow in either the current ripple or the dune existence field (see Fig. 8-22). That these two conditions must simultaneously be met is also suggested by the apparent absence of rhomboid ripples from the coarser sands, although the rill marks on otherwise plane beds apparently are common in these grades (e.g. Woodford, 1935; Demarest, 1947; Stauffer et al., 1976).

SUPERCRITICAL FLOWS AND ANTIDUNES

Theoretical considerations

We saw (Chapter 7) that antidunes are trains of more or less stationary and broadly in-phase interfacial and bed waves similar in height and wavelength (Fig. 10-16). Gilbert (1914) named the features in this manner because they advanced against the flow, but Kennedy (1961, 1963), in defining antidunes, put no special emphasis on this particular property. However, Kennedy did require that bed and surface waves be in phase, a demand criticized by A.J. Reynolds (1965), on the grounds that an exact phase equivalence is seldom realized. Simons et al. (1961) distinguished stationary from non-stationary in-phase bed and water-surface waves, an interpretation of antidunes which also is unsatisfactory in practice (Henry et al., 1964; Allen, 1966), as the forms of a given train may be stationary at one

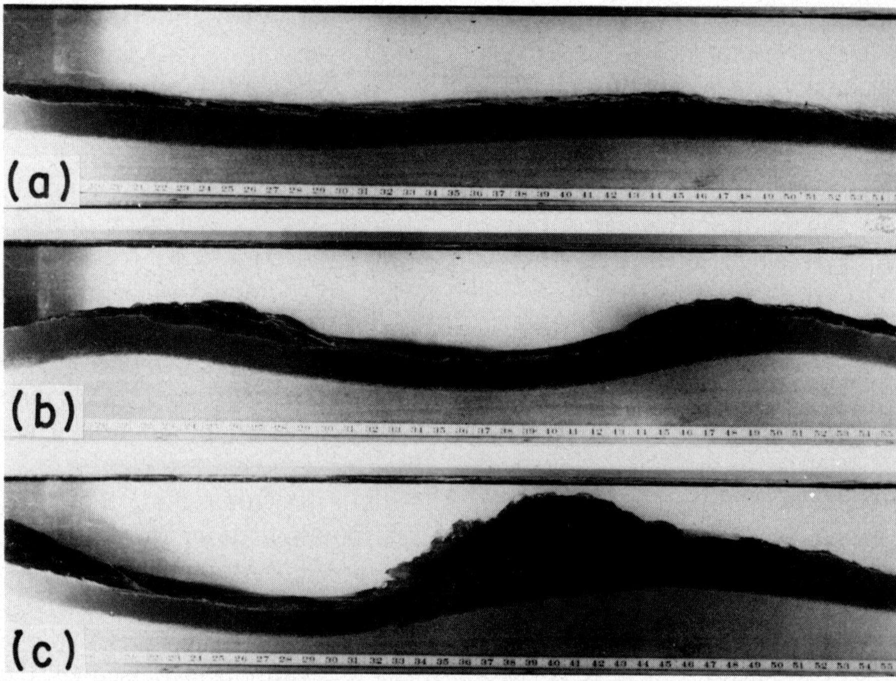

Fig. 10-16. Side views at successive times of the same portion of a laboratory flume showing (a) initiation, (b) maturation, and (c) breaking of antidune surface waves, and the related in-phase sediment waves. Current from left to right: mean depth=0.0375 m, mean flow velocity=2.42 m s^{-1}, Froude number=2.34. The bed-wave crests moved progressively upstream during this sequence, gently upstream-dipping laminae being preserved beneath the upstream faces. Reproduced from Kennedy (1961), by permission of the California Institute of Technology.

time but moving at others. Engelund and Hansen (1966) prefer to call an antidune train a sinusbed, on the grounds that it is illogical to name as an antidune a structure capable of moving either upstream or downstream with the current. Kennedy's usage, relaxed with respect to the exact phase equivalence, is preferred here.

The broad equivalence of phase and similarity of amplitude between the bed and surface waves in antidune flow means that the sedimentary boundary can be regarded as lying at an infinitely great distance below the interface between the fluids. Hence the features can be analysed as short or deep-water progressive waves (Kennedy, 1961, 1963, 1969).

In the general case of such waves advancing across the interface between two stationary fluids, of density ρ_1 below and ρ_2 above, the wave celerity, c, is given by:

$$c^2 = \frac{gL}{2\pi} \frac{(\rho_1 \pm \rho_2)}{(\rho_1 + \rho_2)} \qquad (10.26)$$

in which L is the wavelength (e.g. Kinsman, 1965; Neumann and Pierson, 1966). When the upper fluid is of negligible density compared to the lower, as with waves travelling over the sea or a river, the density term is virtually unity and eq. (10.26) reduces to:

$$c^2 = \frac{gL}{2\pi}. \tag{10.27}$$

Since antidunes are effectively stationary, Kennedy (1961) substituted for c in eq. (10.27) the mean flow velocity U, to obtain:

$$L = \frac{2\pi U^2}{g} \tag{10.28}$$

for the minimum wavelength of antidunes formed in open channels. Equation (7.10) can be rearranged as eq. (10.28).

Equation (10.28) has been thought useful in palaeohydraulic reconstruction. In discussing the relationship, Middleton (1965) proposed that "if antidune structures can be recognized in sandstone, and if the wavelength of the antidune can be estimated in the field, it will be possible to obtain a rough value for the ancient flow velocity". R.G. Walker (1967a) and Skipper (1971) used eq. (10.28) to estimate current speeds from antidune bedforms preserved in turbidite formations, and Hand et al. (1969) applied it to the features in the fluviatile Mount Toby Conglomerate. However, eq. (10.28) is invalid for antidunes as fashioned by turbidity currents, for the density of the upper fluid is then not negligibly small. Hand et al. (1972) accordingly applied eq. (10.26) with the negative root, substituting the flow speed for the celerity. On a different tack, Hand (1969) proposed that antidunes approximated to coupled sets of trochoidal waves. Introducing the maximum possible steepness for the surface waves, he deduced the maximum flow depth associated with given antidunes. Hand et al. (1969) also applied this relationship to antidunes in the Mount Toby Conglomerate. Allen (1968b) approached the prediction of flow depth from antidunes in a different way. Noting their restriction in practice to flows but little removed from critical, he substituted U_{cr} for c in eq. (10.27) to obtain an expression for the flow thickness, h_1, in terms of a measured value of L. The theoretical conclusion that antidunes can exist only for $0.844 \leqslant Fr \leqslant 1.77$ (Kennedy, 1963; Gradowczyk, 1968) further supports this derivation. In the general case:

$$U_{cr}^2 = gh_1 \frac{(\rho_1 - \rho_2)}{\rho_1} \tag{10.29}$$

for $Fr' = 1$ which, on substituting for c^2 in eq. (10.26), yields:

$$L = 2\pi h_1 \frac{(\rho_1 + \rho_2)}{\rho_1} \tag{10.30}$$

as an approximation to flow thickness, applicable to both turbidity currents and open-channel flows. Since a two-layer system is gravitationally unstable when $\rho_2 > \rho_1$, and $(\rho_1 + \rho_2)/\rho_1 \to 1$ as $\rho_2 \to 0$, the flow thickness must fall within the approximate limits $L/4\pi < h_1 \leqslant L/2\pi$. Hand et al. (1972) appear to have obtained eq. (10.30) in error by a numerical factor of two.

Shaw and Kellerhals (1977) protest that Kennedy's eq. (10.28) is unsuitable for the hydraulic interpretation of antidune bedforms from open channels, claiming that eq. (7.12) due to A.J. Reynolds (1965) is the correct limit, and that Kennedy's (1963) relationship:

$$Fr^2 = \frac{2 + kh\tanh(kh)}{(kh)^2 + 3kh\tanh(kh)} \tag{10.31}$$

for dominant antidune wavelength may also be used ($k = 2\pi/L$). These strictures are merely academic in the case of Allen's (1968b) method, since at critical flow eqs. (7.10), (7.12) and (10.31) are closely similar numerically.

One reason for the restriction of antidunes to a narrow range of Froude numbers is that the surface of a fluid stream becomes unstable at sufficiently steep slopes, with the formation of travelling surge-like waves called roll waves, found in several natural settings. As Demarest (1947) observed, small roll waves typify certain stages of wave backwash. Similar roll waves abound on films or thin sheets of water flowing over steeply inclined surfaces (e.g. Horton, 1938), for example, on asphalt roads or smooth limestone outcrops during rain, and in the broad drainage channels of sand flats. Large roll waves are frequently present in spillways and other engineered channels (Cornish, 1907; Peterson and Moharty, 1960; Brock, 1969; Road, 1977).

Roll waves have some importance in practical hydraulics and are relevant to many chemical-engineering processes involving mass transfer, a significance that extends to limestone solutional structures. For these reasons roll waves have been much studied, both theoretically (Jeffreys, 1925; Dressler, 1949, 1952b; Lighthill and Witham, 1955a; Benjamin, 1957; Tailby and Portalski, 1960; Escoffier and Boyd, 1962; Brock, 1970; Krantz and Goren, 1970, 1971a; N.O. Hansen, 1971; Javdani and Goren, 1972; Berlamont, 1976), and experimentally (Binnie, 1957, 1959; Peterson and Moharty, 1960; Tailby and Portalski, 1960, 1962a, 1962b; Mayer, 1961; Stainthorp and Allen, 1965; Brock, 1969, 1970; Krantz and Goren, 1971a, 1971b), with Fulford (1964) presenting a useful early review. A major conclusion reached first by Jeffreys is that roll waves cannot exist on an open-channel flow of uniform vertical velocity-profile unless $(8S/f) > 4$, where S is the bed slope and f the Darcy-Weisbach bed friction coefficient, that is, unless $Fr > 2$. The critical Froude number is progressively lowered as the vertical profile of velocity departs increasingly from uniformity, and is in the order of 1.7 for a turbulent flow.

Antidune surface waves

Antidunes are common naturally. The surface wave patterns were early described by Cornish (1899, 1901a, 1901b) and J.S. Owens (1908) from beach runnels, where wavelengths in the order of 0.01–0.30 m are the rule. Another early account, but of waves 4.5–6.5 m long and about one metre high, is given by Pierce (1916) from the flashy San Juan River, Colorado. Subsequently, antidunes have often been recorded from drainage channels on beaches and other inter-tidal sand flats (e.g. Van Straaten, 1953b; Bajard, 1966; Clifton et al., 1973), as well as from the steeper and flashier rivers (B.S. Young, 1960; Kennedy, 1961; Fahnestock, 1963; Nordin, 1964b; Harms and Fahnestock, 1965; Gavrilovic, 1970; Beaumont and Overlander, 1971; Klimek, 1972; Gustavson, 1974; Boothroyd and Ashley, 1975). Patterns of antidune surface waves can frequently be found on backwash flows on beaches, as recorded by Allen (1964a). Antidunes are also one of the more familiar of laboratory bedforms. Gilbert (1914) was amongst the earliest to

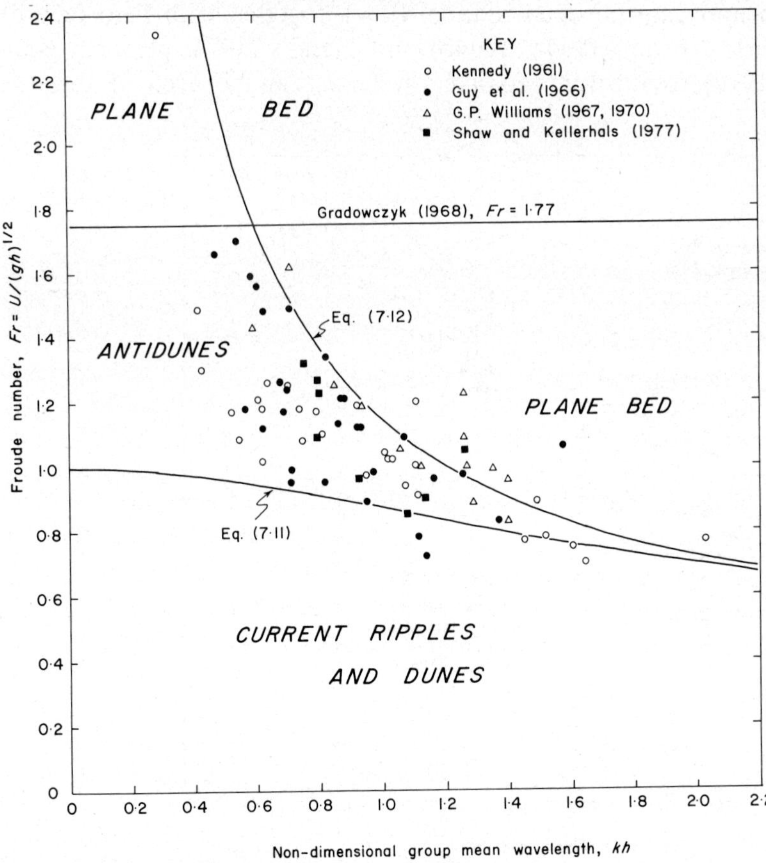

Fig. 10-17. Observed and calculated occurrence of antidunes in terms of Froude number and non-dimensional wavelength.

investigate them experimentally in open channels. A particularly important laboratory study of antidunes was made by Kennedy (1961), and others have come from Middleton (1965), Guy et al. (1966), G.P. Williams (1967, 1970), and Shaw and Kellerhals (1977). Hand (1974) produced antidune waves on the interface between a saline current separating fresh water from a charcoal bed.

It was said in Chapter 7, and also remarked later, that antidunes occur under restricted conditions, $0.844 \leqslant Fr \leqslant 1.77$, and that, according to Kennedy (1961), their wavelength is dependent on flow velocity and coupled to flow depth (eqs. 10.28 and 10.30). Figure 10-17 shows the experimental occurrence of antidunes in the *Fr–kh* plane, together with eqs. (7.11) and (7.12) limiting the development of two-dimensional forms according to Kennedy (1963) and A.J. Reynolds (1965). Also plotted is Gradowczyk's (1968) upper limit, $Fr = 1.77$. Agreement between theory and experiment is good. The discrepancies can largely be accounted for by reason of the fact that some of the antidune waves were three-dimensional, when A.J. Reynold's (1965) modified form of eq. (7.12) is applicable (cf. Figs. 7-8 and 7-13). The same experimental data on open-channel flows, together with field records (Kennedy, 1961; Nordin, 1964a, 1964b) and Hand's (1974, personal communication, 1977) results for two-layer systems, can be used to test the

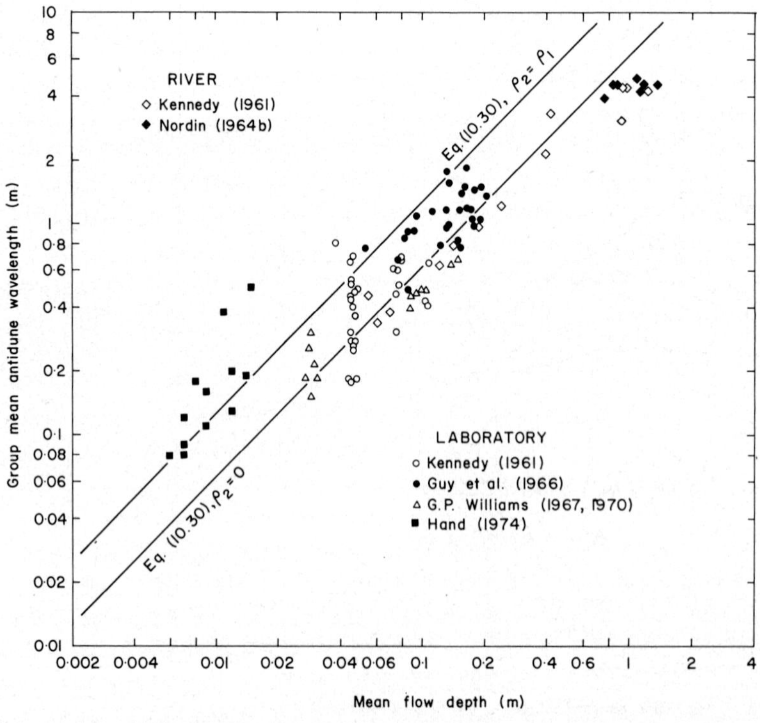

Fig. 10-18. Wavelength of antidunes in rivers and laboratory flumes (free-surface and stratified flows), as a function of flow depth.

amplified forms, eqs. (10.29) and (10.30), of Kennedy's (1961) eq. 10.28 for antidune wavelength (Fig. 10-18). For open-channel flows, the assumption that $Fr = 1$ when antidunes exist clearly permits flow depth to be estimated to the correct order. The data for river antidunes stray deeply into the subcritical range chiefly because, as Kennedy (1961) emphasizes, reliable data are difficult to obtain when bed and flow conditions vary laterally. Hand's results for a two-layer system suggests that $Fr' = 1$ is a satisfactory basis for estimating flow depth from antidune bedforms in turbidites.

These experimental and field studies (e.g. Kennedy, 1961; Guy et al., 1966) reinforce the often-graphic early accounts by Cornish (1901a), Gilbert (1914) and Pierce (1916) of the cycle of changes through which pass antidunes in nominally steady one-way flows. This cycle, visible in almost any beach runnel, presents the following temporal pattern: (1) smooth water surface, (2) rapid growth of an antidune train, (3) steepening of individual waves to a height of about $0.15L$, (4) slow to rapid upstream march of noisily breaking surface waves, (5) decay of waves and return to smooth water surface. Its period is in the order of 10 s in a typical beach runnel, but can be several to many minutes in a substantial river. It is plausible that antidune growth and breaking is due to the progressive entrainment of water into a separated region lying between the bed and the (jet-like) supercritical flow (Peregrine, 1974).

Antidune bedforms and internal structures

Beyond the laboratory, antidune bedforms are best known from beaches, where extensive trains of them are created by backwash flows on seaward-facing slopes and by wave overwash in backshore areas and beyond bar crests (Timmermans, 1935; Häntzschel, 1939; Van Straaten, 1953b; Reineck, 1960a, 1963; Martins, 1967; Panin and Panin, 1967; Davis and Fox, 1972; Hayes, 1972; Wunderlich, 1972; Reineck and Singh, 1973; Komar, 1976; Broome and Komar, 1979). In my experience they are particularly common as backwash structures (Fig. 10-19), occurring in long trains that cover areas of tens or hundreds of square metres on fine-sand beaches in the British Isles, for example, at Rhossili (South Wales), Southport (northwest England), and Brancaster (eastern England). In such environments, antidune bedforms have long and generally sinuous crests, a wavelength between 0.3 and 1.2 m, a symmetrical to weakly asymmetrical vertical streamwise profile, and a height in the order of several millimetres to a centimetre or so. On some beaches dark-coloured heavy minerals occur on their stoss (upbeach) sides while shell and coal fragments are concentrated on their leeward (down-beach) faces. Parting lineations are almost invariably superimposed on the waves, trending across their crest lines, and rhomboid rill or ripple marks are occasionally to be found (e.g. Broome and Komar, 1979). Several names are applied to beach antidune forms. Van Straaten (1953b), following Bucher

Fig. 10-19. Beach antidunes in fine sand with some shell and macerated plant debris, Sleeping Bay, near St. Ishmael's, Dyfed, Wales. Hammer 0.33 m long with head pointing down-beach.

(1919), calls them "regressive sand waves", while Reineck (1960a, 1963) applies the term "*Sandwellen*". Komar (1976) and Broome and Komar (1979) name them "backwash ripples".

Beach antidunes may have more than one mode of origin. On the Oregon coast, significant wave heights range from about 1.5 m in summer to about 4 m in winter, while wave periods average about 8 s in summer to between 10–15 s in winter. Here Broome and Komar (1979) observed that a short train of beach antidunes (they call them backwash ripples) could be formed on the *subcritical* side of an undular hydraulic generated where the wave backwash flowed into the incoming wave or swash. As we saw (Fig. 10-10a), the subcritical side of such a jump is characterized by a short train of waves that decrease in height and wavelength away from the jump. Assisted by a numerical analysis, they showed that these waves created a pattern of sediment erosion, transport and deposition expressed as a series of bed waves declining in harmony with the surface waves in height and wavelength. The short trains of bedforms observed on the Oregon beaches matched these characteristics, but in places there were extensive trains of bed features amongst which sets of forms, recognizable by a consistent seaward decline in wavelength, could not be identified. In my experience of British beaches, where waves are typically much smaller in height and period than on the Oregon coast, the antidune bedforms have a uniform wavelength value over very large areas and practically never form in sets, either isolated or juxtaposed, in the manner described by Broome and Komar. In my experience (e.g. Allen, 1964a), supporting the observations of Timmermans (1935) and Panin and Panin (1967), beach antidune bedforms arise in the supercritical part of the backwash, by the action of antidunes or very similar but solitary waves probably including some of the Peregrine's (1974, fig. 5) shear waves. The undular hydraulic jumps I have occasionally seen formed by backwash flows on British beaches have almost always been much smaller than those reported by Broome and Komar (1979) and too small generally to have formed the antidune bedforms. P. Wright (1976) has provided convincing evidence (Fig. 10-20) for the occurrence of supercritical conditions in wave backwash flows on a British beach, and his findings are confirmed circumstantially from other regions by Schiffman (1965), Dolan and Ferm

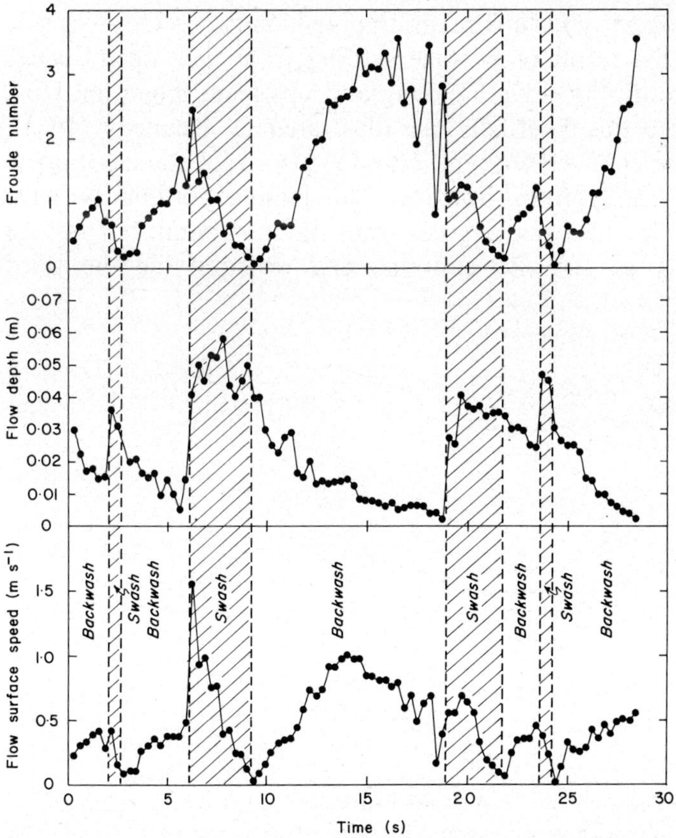

Fig. 10-20. Flow depth and speed of surface flow measured at a fixed station in a sequence of wave swashes and backwashes, beach of fine sand, Ainsdale, Merseyside. Data sequence courtesy of P. Wright (see also Wright, 1976).

(1966), Kirk (1975), and E. Waddell (1973, 1976). The origin of beach antidunes therefore calls for further study, under a much wider range of conditions than any worker has hitherto attempted.

Shaw and Kellerhals (1977) give the only record from a substantial contemporary river of what may be antidune bedforms. On parts of the drained gravel bed of the North Saskatchewan River, Alberta, they found trains of transverse gravel mounds which had wavelengths in the order of 2.5 m and heights in the order of 0.2 m. These mounds were interpreted as antidune bedforms, a link between the double-peaked ones and parallel trains of three-dimensional, "rooster-tail" surface waves being suggested. Much smaller antidune features can often be seen in dried-up beach runnels (e.g. Van Straaten, 1953b; Bajard, 1966). Foley (1977) found experimentally that a cluster of pebbles tended to accumulate beneath each antidune formed on beds of gravelly sand.

The internal structure of antidune bedforms is ill-known compared with most other transverse bed features. Referring to Kennedy's (1961) and his own experiments, Allen (1966) identified three modes of deposition on antidunes. The commonest yields upstream-dipping laminae on the stoss side of the bed wave, as the result of particle settling from the "dead" water temporarily held there during wave-breaking and upstream migration (Fig. 10-21a). Antidunes like this internally are illustrated by Kennedy (1961), Middleton (1965), Guy et al. (1966), and Hand (1974) from the laboratory, and by Wunderlich (1972) from beaches. The laminae ordinarily have maximum upcurrent dips of a few degrees from the horizontal, but in rare cases slope as steeply as 10°. By contrast, and exemplifying the third

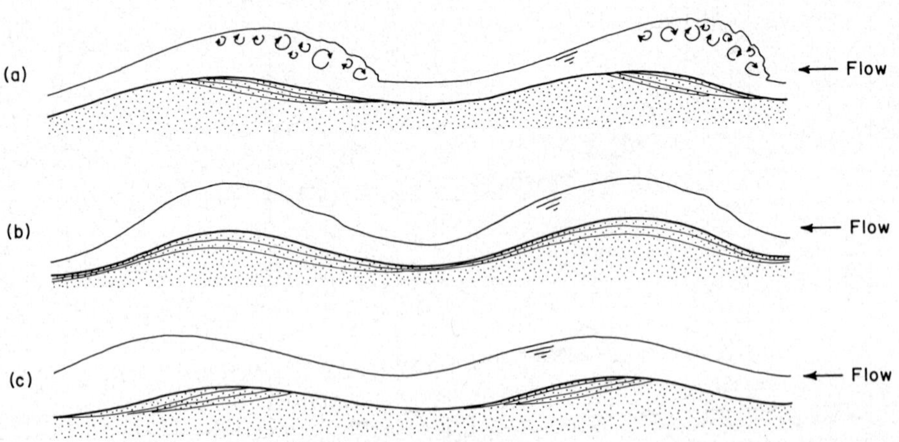

Fig. 10-21. Schematic internal structures of antidune bed waves. a. Gently upstream-dipping laminae deposited in harmony with steepening and breaking of surface wave. b. Continuous laminae draping and swelling over wave crests formed as waves grew up during prolonged period of net deposition. c. Downstream-dipping laminae (rare) formed as wave briefly migrated downstream.

depositional mode, the beach antidunes described by Panin and Panin (1967) and by Reineck (1971) have laminae dipping with the current (backwash) that formed them. In the second depositional mode (Fig. 10-21b), laminae lie draped over the whole antidune forms, as recorded by Middleton (1965) and Hand (1974). The accumulation of lee-side laminae should occur only during downstream migration (Fig. 10-21c). The repeated decay and growth in a slightly different position of antidune trains means that there is some tendency for laminae accumulated according to the first mode to be bundled into flat, gently tilted, overlapping lenses. Wunderlich's (1972) study suggests

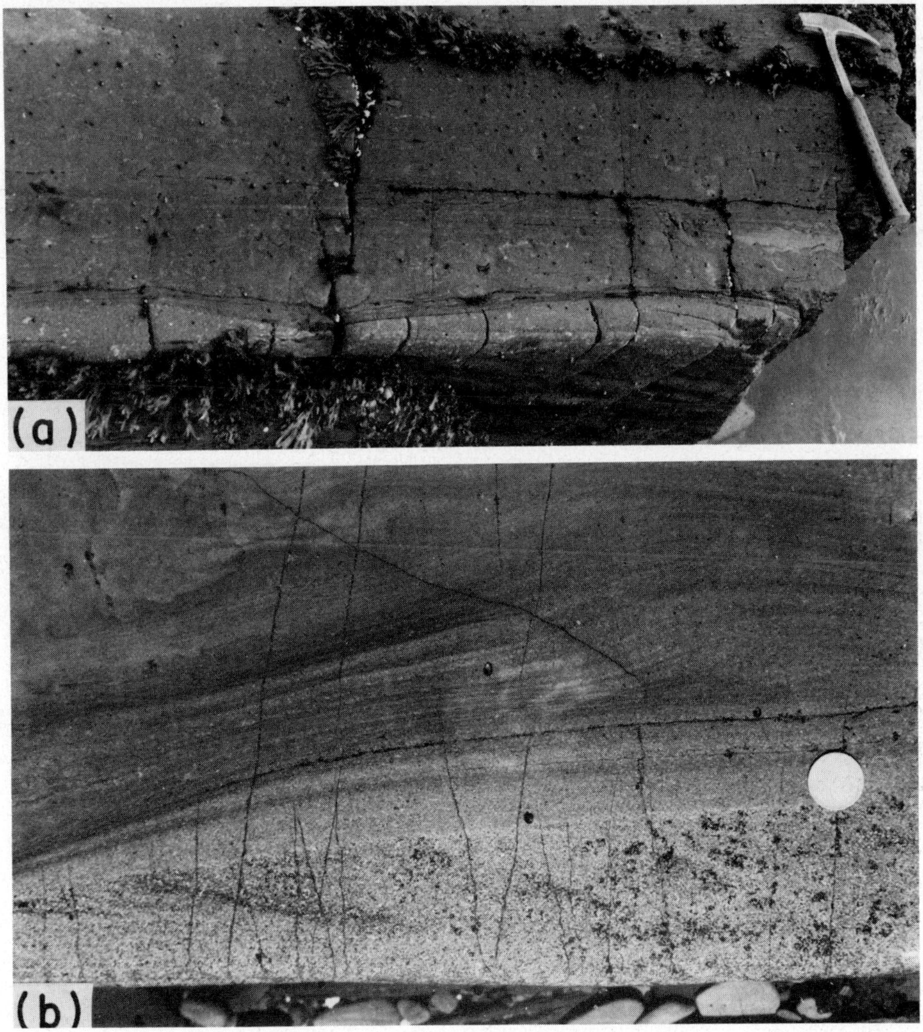

Fig. 10-22. Antidune bedding in turbidites, Cloridorme Formation (Ordovician), Gaspé, Canada. a General view showing flute marks (current from right to left) on turbidite sole and antidune cross-bedding above. b. Detail of antidune bedding, current from right to left. Photographs courtesy of K. Skipper (see also Skipper, 1971).

that such bundling in fact occurs. The laminae deposited from travelling hydraulic jumps—the so-called back-set beds—also dip against the current, but their slope is much steeper than antidune lamination, and they occur in units more extensive laterally than could be formed by antidunes (Power, 1961; Jopling and Richardson, 1966). The lamination described by Wunderlich (1973) may have been formed in this way rather than by rhomboid ripples.

Antidune bedforms have been claimed from the older geological record. R.G. Walker (1967a) interpreted as antidunes a series of gentle-sided, long-crested undulations he found in a turbidite of the Hatch Formation, New York State. This interpretation is perhaps suspect, as the crests of the undulations are subparallel with independent current-indicators. Possibly the features are related to patterns of secondary flow that prevailed during aggradation. The claim that antidune bedforms occur in fluviatile Triassic rocks is made by Hand (1969), Hand et al. (1969), and Wessel (1969). Lamination within the wave forms is inclined in the opposite direction to the local and regional palaeocurrent indicators, as McCracken (1969) also observed from antidunes in the Sespe Formation, of a similar character. An especially convincing case for antidunes was made by Skipper (1971) in respect of a thick turbidite bed in the Ordovician Cloridorme Formation of Quebec (Fig. 10-22). Its basal part is a unit of medium to coarse quartz sand with flute moulds on the base and a regularly undulating top. The unit is cross-stratified internally, the laminae dipping gently in the opposite direction to the flute moulds beneath, to cross-lamination above, and to regional current-indicators. The wavelength of the undulations is between 0.48 and 1.0 m. They can be interpreted as antidunes formed at an interface *within* the turbidity current that deposited the whole bed (Hand et al., 1972), a conclusion justified by the sharp textural change at the top of the undulose unit. A further example is discussed in detail by Skipper and Bhattacharjee (1978). There is only one record of possible antidune structures from littoral sandstones (Allen, 1974a). Collinson (1966, 1970a) assigned to antidune deposition a series of massive sandstones, in thick horizontal to lenticular units, at the bottoms of Carboniferous channel fills. The "massive" character of these rocks could have a diagenetic explanation (see Hamblin, 1962a, 1962b, 1965), and their position at the bases of channels perhaps speaks against their deposition from supercritical flows (Allen, 1971e). Lamont's (1957) "slow antidunes" are unquestionably soft-sediment deformations.

SUMMARY

The specific energy of a layer of dense fluid flowing beneath a stationary lighter one, is the sum of the flow thickness and the head equivalent to the flow velocity. For each value of the specific energy, two alternative flow depths, called alternate depths, are generally possible. For a given unit discharge, there is one value for each of the flow depth and velocity, called the critical depth and velocity, at which the specific energy is a minimum. At depths smaller than critical, the flow is supercritical and waves on the surface of the fluid layer can travel only downstream. When the depth exceeds the critical, the flow is subcritical and surface waves can be transmitted upstream as well as downstream. At the critical depth, when the flow velocity is also critical, the Froude number is unity. A flow can change from subcritical downstream to supercritical, or from supercritical to subcritical, given the presence of suitable flow controls, for example, changes in the shape and slope of the flow boundary. Transition from supercritical downstream to subcritical flow often takes place abruptly, at a stationary or travelling surge known as a hydraulic jump. Jumps can be analysed in terms of specific force, analogous to specific energy, the flow depths on either side of a jump being conjugate depths. The flow velocity decreases across a jump to match the increase of flow depth, and there is some loss of energy.

Several bedforms are associated with supercritical flows. Transverse ribs in gravel-bed rivers appear to be localized at hydraulic jumps which are stationary because the flow velocity in the subcritical part is less than the movement threshold for the coarsest sediment available. Rhomboid rill marks, ripple marks and dunes, found chiefly in coastal environments, are also coupled to supercritical-subcritical flow transitions expressed as systems of crossing waves, but in addition seem to require that the stream power-grain size properties of the flow be simultaneously appropriate to a common one-way bedform. Back-set bedding is produced when deposition occurs in a hydraulic jump retreating upcurrent. The near-stationary coupled bed and surface waves which constitute antidunes can exist in river channels, beach runnels, and wave backwash flows only within a narrow range of Froude numbers disposed about the critical value. They probably occur in some turbidity currents, perhaps chiefly on interfaces within the flow.

Chapter 11

TRANSVERSE BEDFORMS IN MULTIDIRECTIONAL FLOWS: WAVE-RELATED RIPPLE MARKS, SAND WAVES, AND EQUANT DUNES

INTRODUCTION

Several earlier chapters presented the character and relationships of bed features shaped in environments governed by currents either strictly unidirectional or effectively of this nature, as in a laboratory flume, short stretches of a river, an ebb- or flood-dominated tidal channel, or in a limited part of the desert when the wind is stable.

There are, however, bedforms present in even more complicated natural systems, in which the current, as well as being non-uniform, are also unsteady and multidirectional either in an organized, that is, periodic, or random-periodic manner. These systems involve internal and/or progressive surface waves, occasionally standing waves, the tides, and the wind acting over a cycle of seasons or longer. We shall examine wave-related ripple marks, generated mainly by wind waves and in the deep ocean by internal waves, and the generally much larger sand waves, the result of the action of tidal currents. The equant dunes found in certain deserts appear to depend on winds of a directionally uniform effectiveness.

CLASSIFICATION OF CURRENT PATTERNS

It can be inferred from the account given in Chapter 1 that a natural current acting in the horizontal plane may at any station comprise *translatory, time-periodic,* and *random* velocity components. A classification of currents based on four end-members immediately suggests itself (Fig. 11-1). Since the translatory currents of interest are unidirectional over distances comparable with the bedform spacing, one obvious end-member is a translatory-unidirectional current (unidirectional for short). Time-periodic currents can be ranged between pure oscillatory and pure rotary kinds, two further end-members identifying themselves. The fourth end-member is a random current, which lacks pattern in time and direction. This classification is less restrictive than Clifton's (1976) scheme and therefore more relevant to natural environments.

Figure 11-2 summarizes some major properties of the non-random end-member patterns and of two important intermediate cases. The ideal translatory-unidirectional current is typified by a steady velocity U_1 acting at a

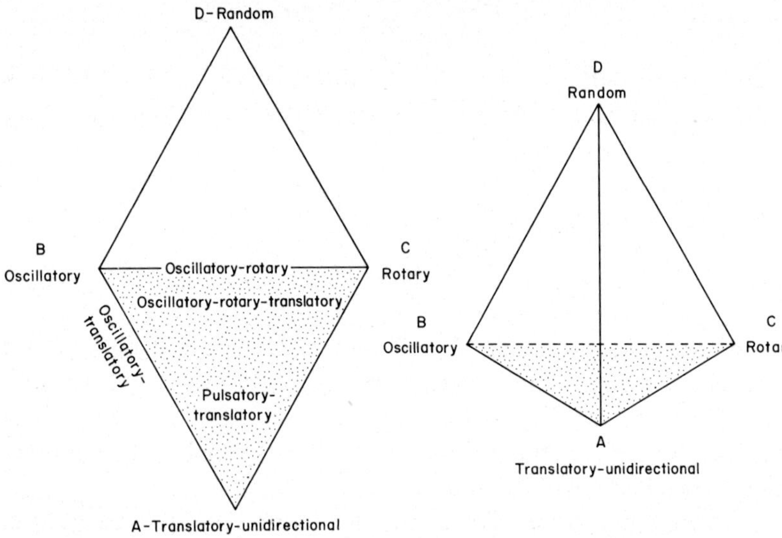

Fig. 11-1. Classification of currents in natural environments in terms of directional properties.

constant angle α_1 relative to fixed horizontal coordinates (x, z). Each fluid particle takes a straight path, and its velocity–time pattern is a straight line parallel with the time-axis (t is time), as during steady flow in a rectilinear flume channel. In a pure oscillatory flow, such as beneath an Airy wave, the velocity U_2 is a periodic (simple harmonic) function of t, and acts along a line of fixed bearing α_2 from the x-axis. Hence a fluid particle travels

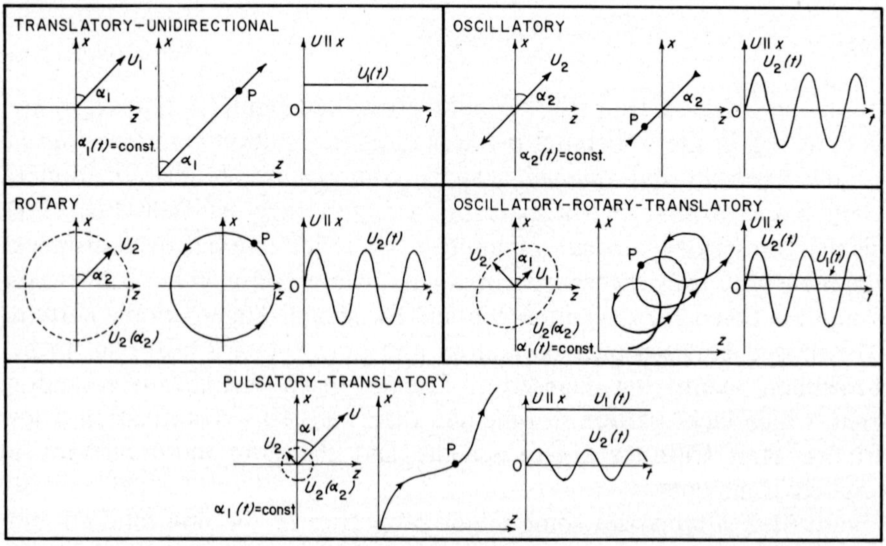

Fig. 11-2. Directional characteristics of currents in natural environments, in each case in terms of the behaviour of the flow vector in space and time, and the path of a fluid particle P.

regularly to-and-fro over the same path. Rotary currents display a constant magnitude of the velocity but a time-dependent direction, each fluid particle traversing a closed circular horizontal orbit during one flow period. An idealized wave travelling around a circular basin creates this pattern. Two velocity components in combination mark an oscillatory-rotary-translatory current. The translatory component, wholly or in part a mass-transport current, has a constant velocity U_1 and fixed direction, whereas the stronger oscillatory-rotary component U_2 varies in both magnitude and direction with time. A fluid particle now advances in a series of horizontal loops, open to an extent determined by the magnitudes and directions of U_1 and U_2. Many real tidal currents show this pattern, which as well typifies certain combinations of waves on steady currents. Two velocity components also mark a pulsatory-translatory current, the steady component U_1 dominating. Note that the particle path now lacks crossings. Some combinations of steady currents and waves can be represented by this kind of pattern. An unsteady unidirectional flow, as in a seasonally varying river, is a special case of pulsatory-translatory flow, the unidirectional and periodic (oscillatory) components being collinear.

The rate and directional pattern of sediment transport cannot be neglected from these considerations, since bedforms express the transport of debris. A translatory-unidirectional current promotes a steady sediment transport in the same direction, provided that $U_1 > U_{cr}$, where the latter is the threshold entrainment velocity (Fig. 11-3a). An oscillatory flow will afford a zero net transport (assuming U_{cr} is exceeded for some of the time) only if the velocity-time pattern is perfectly symmetrical, as in the case of Airy waves. In real waves, and in Stokes' ideal waves, this pattern is asymmetrical, on account of mass-transport, larger velocities being observed on the forward wave stroke than during the reverse (Fig. 11-3b). Hence there is a net sediment transport on the bottom beneath such waves, in the direction of wave propagation, which increases steeply with the excess of area A over area B in the diagram. But if the waves surmount an unrelated steady current opposed to their propagation, the net transport may not be in the direction of wave travel. Whereas a rotary current transports sediment

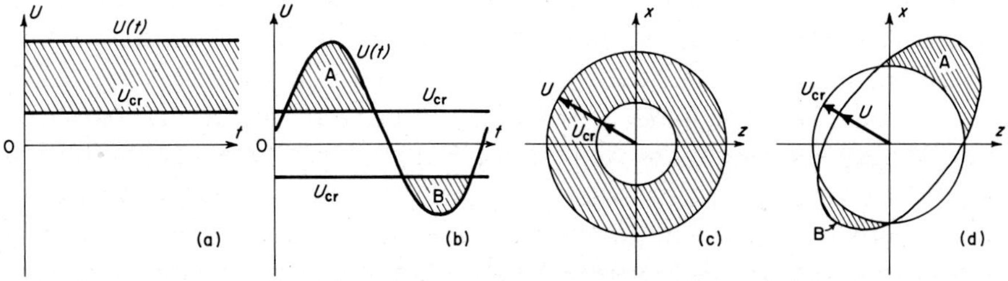

Fig. 11-3. Implication for bedload transport of oscillatory and oscillatory-rotary currents.

equally in all directions (Fig. 11-3c), an oscillatory-rotary flow moves debris only in the directions of velocities exceeding the critical, and creates a net drift only if the envelope of the vector lying outside the bounds of the critical velocity shows a suitable asymmetry (Fig. 11-3d). The net transport arising in this way can also be nullified or distorted in the presence of an unrelated translatory component of flow.

We shall see that an important family of ripple marks—wave ripples and wave-current ripples—are related to oscillatory and oscillatory-translatory flows, and that some types of three-dimensional wave-related ripples may depend on flows with a strong rotary component. Sand waves, which are much larger than wave-related ripple marks, will be found to be related to oscillatory-rotary tidal flows associated with an unrelated translatory and/or mass-transport translatory component.

The local temporal pattern of atmospheric flow is invariably disordered to some degree, including random elements which at some sites are of over-riding importance. Such organization as may be evident lies chiefly in the diurnality or seasonality of the winds of each direction, and particularly of the storm winds, capable of effecting huge sediment transports. A periodic rotational element—in the same meaning as in a tidal current—can be regarded as lacking.

WAVE RIPPLE MARKS

Character and occurrence as surface forms

These are Kindle's (1917) "symmetrical ripple-marks", Bucher's (1919) "oscillation-ripples", and *Wellenrippeln* or wave ripples (Reineck et al., 1971). A.R. Hunt (1882, 1904) clarified many aspects of their description, nomenclature and origin.

Figure 11-4 illustrates typical wave ripple marks. In plan wave ripples are remarkably regular, much more so than current ripple marks (Figs. 8-8, 8-9), having long, parallel, straight to very gently curved crests in a single set, which terminate usually in zig-zag ("tuning-fork") and less often in open junctures. The horizontal form index is seldom below 10 and is commonly in the order of 100. In cross-section the crests range from knife-sharp to well-rounded, whereas the troughs vary from gently and uniformly curved, as in Inman's (1957) "trochoidal" form, to rather flat, as in his "solitary" features. Where the crests are well-rounded, the ripple faces commonly are convex-up and occasionally plane. Whereas no wave ripple marks are *perfectly* symmetrical, which led Picard and High (1968) and Reineck et al. (1971) to categorize all the ripples as "asymmetrical", it is useful in practice to describe the features as symmetrical provided that Tanner's (1960) ripple

Fig. 11-4. Symmetrical types of wave ripple mark. a. In fine sand, coast near Burnham Overy Staithe, Norfolk, England. Trowel 0.28 m long points toward sea and source of waves. b. In pebbly very coarse sand and granules, Freshwater West, Dyfed, Wales. Hammer 0.33 m long points toward sea and source of waves. These ripples are starved and have flat, pebbly troughs.

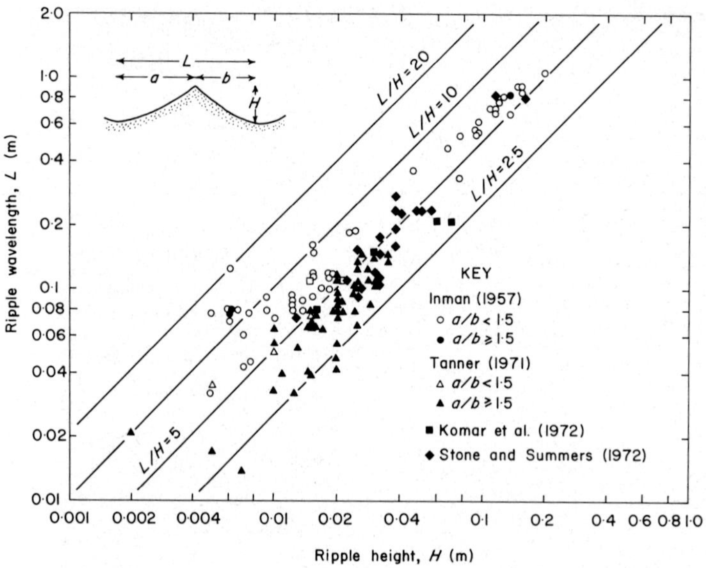

Fig. 11-5. Wavelength and height of wave-related ripple marks with the ripple symmetry index as a parameter.

symmetry-index (trough-to-crest distance divided by crest-to-trough spacing) is less than about 1.5 (see Picard and High, 1968). The wavelength of the ripple marks ranges between about 0.01 m (Singh and Wunderlich, 1978) and more than 2 m, increasing with sediment coarseness (compare Figs. 11-4a, 11-4b). Observations by T. Scott (1954), Manohar (1955), Inman (1957), Tanner (1967, 1971), Reineck and Wunderlich (1968b), Komar et al. (1972), Stone and Summers (1972), and many others show that the vertical form-index (ripple wavelength, L, divided by ripple height, H) of wave-related ripples is typically in the order of 3–8 (Fig. 11-5), substantially less than is characteristic of current ripples, but nonetheless overlapping in range with these forms. The trochoidal ripples are typically of a lower index value than those with well-rounded crests. The smaller wave and aeolian ballistic ripples are unlikely to be confused, though separation of some of the larger forms can be difficult.

Wave ripple marks occur in a wide range of modern environments: river channels and floodplains, inshore lake waters, intertidal flats, shelf seas and marine platforms, and the deep sea.

Their occurrence in river environments is apparently restricted to backwaters and flooded overbank areas, where wind-generated or other waves can over-ride any influence of the river current (Udden, 1916; Sundborg, 1956; Picard and High, 1973; Kumar and Singh, 1978). The ripple wavelength is normally small, in the order of 0.1 m, the affected sediment being very fine sand, and can be as low as 0.01 m where clean silt is available.

Wave ripples abound on sandy intertidal flats, where their character and occurrence is described by A.P. Brown (1911), Häntzschel (1938), Van Straaten (1953b, 1954a), G. Evans (1965), Bajard (1966), Land and Hoyt (1966), Mathieu (1966), B.W. Flemming (1977), and Amos and Collins (1978). Their wavelength in this environment seldom exceeds 0.20 m and commonly is substantially less. The ripples are subject to reworking by burrowing and crawling animals, and can be reshaped by the tide and modified by wind and wave during exposure.

Records of wave ripple marks from lake margins date from Forel (1883) and are supplemented by Kindle (1917), Wulf (1963), Norrman (1964), R.A. Davis (1965), and Donovan and Archer (1975). Under normal conditions the forms are restricted to water depths of a few metres only, but during storms the bottom in depths up to 10–15 m can become rippled. Wavelengths are generally in the order of 0.1 m but reach 0.4 m where the coarser sands lie. The near-perfect symmetry of the forms, and their tendency to parallel the shore, disappears in the shallows near the beach.

Many accounts of wave ripple marks come from the floors of shelf seas. The forms abound from depths of 25 m or so to the inshore, where they become confused in plan and their symmetry may be lost. The features cover very large areas and are remarkable for their long and regular crests, which not uncommonly lie parallel or sub-parallel with the coastline (Inman, 1957; Newell and Rigby, 1957; Vause, 1959; N.C. Flemming, 1965; Flemming and Stride, 1967; Newton, 1968a, 1968b; Rudowski, 1970a; Larsonneur, 1971; Cook and Gorsline, 1972; Newton and Werner, 1972; Stone and Summers, 1972; Davidson-Arnott and Greenwood, 1974; Machida et al., 1974; McKinney et al., 1974; Reineck and Dörjes, 1976; Farrow et al., 1979; K.B. Lewis, 1979; Swift et al., 1979). Wavelengths range from about 0.1 m in very fine to fine sand, up to about 1 m where either terrigeneous or biogenic coarse sediment is available. There are fewer studies of wave ripple marks on shelves 50–100 m or more deep (Siau, 1841; Conolly and Van der Borch, 1967; Conolly, 1969; Sanders et al., 1969; Komar et al., 1972; Newton et al., 1973; Sternberg and Larsen, 1975; Channon and Hamilton, 1976; J.B. Wilson, 1977; Swift and Freeland, 1978; Swift et al., 1979; Yorath et al., 1979; Hamilton et al., 1980; Scoffin et al., 1980), partly because no great attention is paid to such depths, and partly because ripples are only produced there occasionally, by storms, being prone to later destruction by the bottom fauna. The extreme depth for wave-rippling appears to be 204 m on the Oregon continental shelf, where the forms do not exceed 0.21 m in wavelength (Komar et al., 1972). Newton et al. (1973) record ripple wavelengths of 2–3 m from coarse sediments in water depths of 40–85 m on the Saharan continental shelf. On the Pacific shelf off Vancouver Island, Canada, Yorath et al. (1979) found large wave ripple marks in depths up to 105 m. Those to the southwest of Britain (Hamilton et al., 1980) occur in a depth of 140 m.

Deep-sea photography quickly brought to light wave ripple marks in depths beyond the shelf-break (Menard, 1952; Shipek, 1962; Wass et al., 1970; Heezen and Hollister, 1971; Lonsdale et al., 1972; Taylor et al., 1975; Lonsdale and Spiess, 1977; Stanley and Taylor, 1977). Occurring chiefly on the flanks and tops of sea-mounts, guyots and ridges, the forms lie in depths as great as 3300 m and range in wavelength between approximately 0.1 m and 2 m. Most seem to be constructed of biogenic sediment. Deep-water wave ripple marks are probably made by internal waves, though the operation of eddies induced by ocean-bed topographic features is also suggested (Shipek, 1962).

Wave ripple marks of moderately small wavelength have often been recorded from rocks interpreted as lacustrine (P. Allen, 1959; Prentice, 1962b; Picard, 1967; Sanders, 1968; Tanner, 1974; Martinez, 1977), and occur plentifully, often over large bedding surfaces, in rocks ranging between shallow-marine and intertidal to lagoonal (Gilbert, 1884; Hyde, 1911; Kindle, 1914; Cox and Dake, 1916; Udden, 1916; Lamar, 1927; Patton, 1933; McKee, 1954; Pepper et al., 1954; Dzulynski and Zak, 1960; Klein, 1962a, 1970a; Wulf, 1962; Otvos, 1966; Backhaus, 1967; Goldring and Curnow, 1967; Stauffer, 1967; Kemper, 1968; Picard and High, 1968, 1970b; G.M. Young, 1968; Singh, 1969; Vossmerbäumer, 1969; Wunderlich, 1970; Gall, 1971; MacKenzie, 1972; P. Hoffman, 1973; Leflef, 1973; Broekman, 1974; Hobday, 1974; Van Gelder, 1974; Bruun-Petersen and Krumbein, 1975; Andrews and Laird, 1976; H.D. Johnson, 1977; I.C. Rust, 1977; Von Bruun and Mason, 1977; Vos, 1977; Reif and Slatt, 1979). Stear (1979) reports wave ripple marks from a fluviatile succession, and Gradzinski et al. (1979) find them amongst aeolian deposits. Their presence in a continental succession is also reported by Clemmensen (1980). Large wave ripples formed from generally coarse sediments in many instances biogenic are attributed to environments ranging between inshore lacustrine to deep marine (Kindle, 1914; Prosser, 1916; Udden, 1916; Bucher, 1919; Tansey, 1953; P. Allen, 1959; Pettijohn and Potter, 1964; H.J. Hofmann, 1966; Broadhurst, 1968; McCave, 1968; Crimes, 1970; O.A. Dixon, 1970; Beuf et al., 1971; M. Lindström, 1972; Hobday, 1974; Plummer, 1978; Bergman, 1979; Birkenmajer, 1979; Singh, 1980). Some of these cases remain enigmatic. Kindle and Udden vigorously advocated a relatively deep-water, storm-related origin for the ripples they described, chiefly because of a close association with thick mudstones, whereas Hofmann and Broadhurst urged for structures with a similar association either an inshore or tidal origin. Gilbert (1899) and Campbell (1966) claimed to recognize exceptionally large wave ripples, contentions that are perhaps not supported by the fine grades of the deposits concerned.

Internal structure of wave ripple marks

Cook and Gorsline's (1972) claim that wave ripple marks are stationary is contradicted by the work of Stone and Summers (1972), also from a modern shelf, and by many other investigations into the internal structure of the features. Gilbert's (1884, 1899) work foreshadows an analysis of the internal structure of wave ripple marks (Fig. 11-5) similar to that initiated by Sorby (1859, 1908) and McKee (1939) for ripples and dunes beneath unidirectional flows (Figs. 9-1, 9-5). The difference between the two cases is that the cross-stratification in wave ripple marks is generated by currents in which an oscillatory or oscillatory-rotary element predominates over a translatory one. The horizontal component of the ripple velocity is therefore a net or time-average value.

Form sets (Fig. 11-6a) are described by P. Allen (1959) from near-shore lacustrine sediments and are fairly common amongst the large wave ripple marks of carbonate sand embedded in the Lower Palaeozoic shales of North America. Some form sets have a simple internal structure, whereas others comprise two or more bundles of oppositely-inclined laminae, showing that the ripples migrated briefly in directions differing from the dominant one.

When wave ripples in a train advance so as partly to behead their predecessors, erosively related subcritical sets are formed (Fig. 11-6b), of which Gilbert (1899) gave an early example. Further instances are described by McKee (1938) from a river deposit, and by Wulf (1962), H.J. Hofmann (1966), Harms (1969), Newton (1968b), Crimes (1970), and Broekman (1974) from shallow-marine sediments. Most sets have a simple structure, but the complex truncation and facing of laminae in other cases indicate reversals of ripple movement (e.g. Davidson-Arnott and Greenwood, 1974). De Raaf et

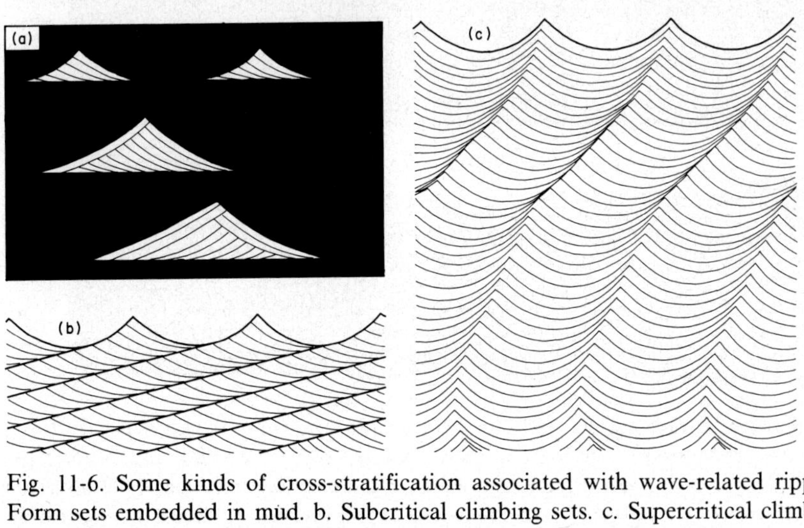

Fig. 11-6. Some kinds of cross-stratification associated with wave-related ripple marks. a. Form sets embedded in mud. b. Subcritical climbing sets. c. Supercritical climbing sets. All examples shown in vertical profile perpendicular to ripple crests.

al. (1977) describe some of the complex internal structures that can result when the ripple marks become reformed under conditions of changeable waves and a low sediment net deposition rate.

Gilbert (1884, 1899) also figured cases of supercritical cross-stratification due to wave ripple marks (Fig. 11-6c), one of which is of particular interest because of the near-vertical angle of climb. Steeply climbing sets are reported from shallow-marine sediments (Leflef, 1973; Broekman, 1973, 1974; Eriksson, 1977a), and from river backwater deposits (Gilbert, 1884; McKee, 1938, 1965). McKee's and Eriksson's sets clearly show in vertical section the trochoidal form of the parent ripples, an important feature allowing a clear distinction from climbing-ripple cross-lamination of current-ripple origin. The near-critical and subcritical sets described by McKee, however, differ from the corresponding structures of current-ripple origin only by their greater regularity.

Fluctuations in wave activity and sediment supply can create varieties of *Linsenschichten* and *Flaserschichten* (Reineck, 1960b, 1960c; Reineck and Wunderlich, 1968a; Reineck and Singh, 1973), in which layers of mud and wave-rippled sand are more or less closely interbedded. In the example

Fig. 11-7. Symmetrical wave ripple marks in plan (upper photograph) and vertical profile perpendicular to crests (lower photograph), interbedded very fine sand and mud, Coal Measures (Carboniferous), Amroth, Dyfed, Wales. Specimen 0.26 m long. Although the bedforms are symmetrical in profile, note the asymmetrical internal structure, facing mainly toward the left.

shown in Fig. 11-7 rippling occurred in four stages, during each of which, to judge from the truncations and shifts in lamina-facing, the forms changed more than once in wavelength (slightly) and sense of net movement. Related structures are reported by Häntzschel (1936), Reineck (1961), G. Evans (1965), Bajard (1966), Goldring and Curnow (1967), Reineck and Wunderlich (1968a), Singh (1969), Broekman (1974), Vos and Eriksson (1977), and Roep et al. (1979).

WAVE-CURRENT RIPPLE MARKS

There exists an important class of forms intermediate in character between wave ripple marks and long-crested current ripples. Names applied to its members include "asymmetric oscillation ripples" (O.F. Evans, 1941; Reineck and Wunderlich, 1968b), "asymmetric wave-formed ripple mark" (Tanner, 1967), "combined-flow ripples" (Harms, 1969), and "asymmetric wave ripples" and "transverse wave-current ripples" (Reineck et al., 1971; Reineck and Singh, 1973). Examples have often been lumped with current ripples, to distinguish them from the near-symmetrical wave ripple marks, but every gradational form exists.

Wave-current ripple marks are intermediate in appearance between wave ripples and long-crested current ripples (Figs. 11-8, 11-9). Forms with wavelengths between about 0.1 m and 1 m abound on modern intertidal flats (McKee, 1957a; Trefethen and Dow, 1960; Bajard, 1966; Reineck and Wunderlich, 1968b; Brambati, 1968; Davis et al., 1972; M. Lindström, 1972; Luternauer and Murray, 1973; Reineck and Singh, 1973; Rudowski and Tobolewski, 1973), as well as marking rocks of subtidal-intertidal origin (e.g. Vossmerbäumer, 1969; Wunderlich, 1970; Glen and Laing, 1975; Vos and Eriksson, 1977). The ripples occur on marine shelves and in lakes (O.F. Evans, 1941; R.A. Davis, 1965; Picard, 1967; Stanley and Swift, 1968; Rudowski, 1970a; Clifton et al., 1971; Tanner, 1971), commonly in shallower water than their symmetrical relatives, and find parallels in ancient counterparts (Prosser, 1916; Udden, 1916; McKee, 1954; Otvos, 1966; Backhaus, 1967; Picard, 1967; Sanders, 1968; Mroczkowski, 1972; McBride et al., 1975b; Martinez, 1977; Singh, 1980). Possible wave-current ripples are occasionally described from environments as different as the fluvial (Sundborg, 1956; Picard and High, 1973) and the deep-marine (Shipek, 1962; Stanley and Kelling, 1968).

Appeal must be made to several characters in order to distinguish wave-current from long-crested current ripples. As a comparison of Figs. 8-8, 11-8, and 11-9 will show, the former are more regular, with a greater parallelism and straightness of crest. Wave-current ripples either lack spurs and stoss-side ridges altogether, or have few of these features compared to current ripples, with little regularity of arrangement. Whereas the crests of two-dimensional current ripples in plan comprise linked curved elements comparable with or smaller than the ripple wavelength in scale, similar elements on wave-current

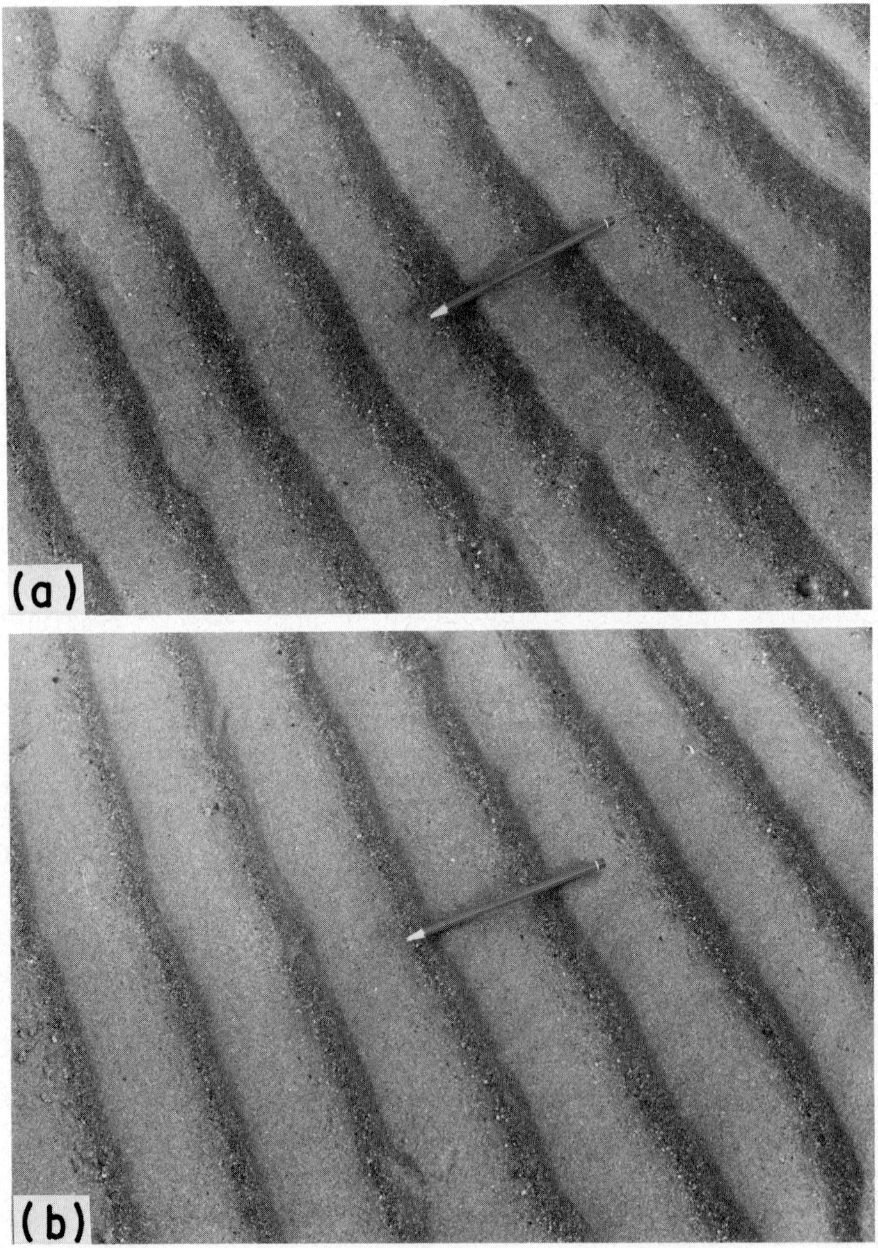

Fig. 11-8. Asymmetrical wave-related ripple marks in fine sand, Tenby North Beach, Dyfed, Wales. Pencil 0.18 m long points in direction of wave propagation. Ripples in (b) have slightly convex-up stoss sides and are more asymmetrical than the ripples in (a), which have plane to slightly concave-up backs.

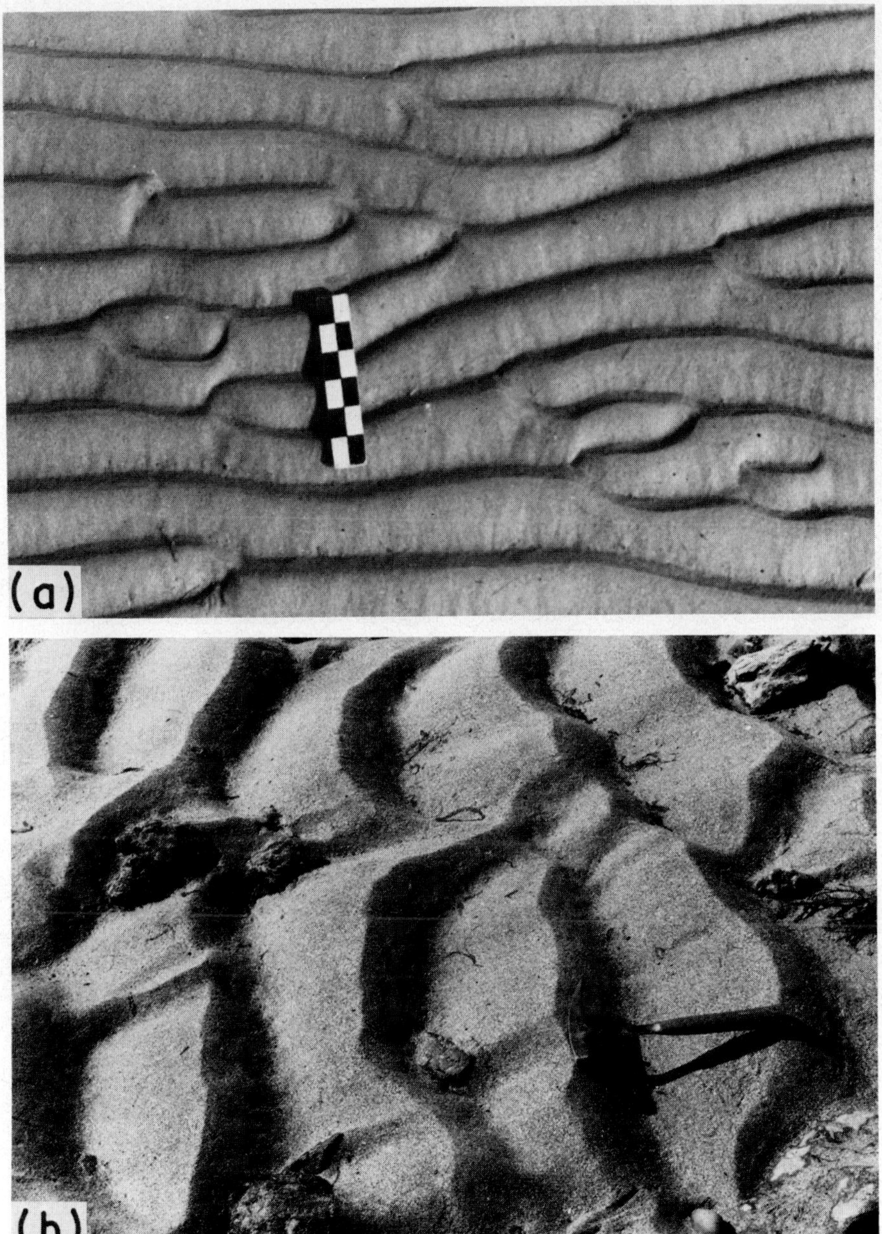

Fig. 11-9. Asymmetrical wave-related ripple marks. a. In very fine sand, Lifeboat Station, Wells-next-the-Sea, Norfolk, England. Scale 0.15 m long. Waves propagated from bottom toward top. b. In coarse to very coarse sand, Freshwater West, Dyfed, Wales. Hammer 0.33 m long points toward sea and source of waves.

ripples are generally several and not uncommonly many wavelengths in extent, if such can be recognized at all. There is a strong tendency for wave-current ripples to occur in contiguous "domains" (Fig. 11-10) each

Fig. 11-10. Flow-parallel domains in a field of asymmetrical wave-related ripple marks in very fine to fine sand, Mellum Bank, German Friesian Islands. Waves propagated from upper right toward lower left. Trowel 0.35 m long.

several metres long by several decimetres wide within a field (see Luternauer and Murray, 1973, their fig. 6d; Reineck and Singh, 1973, their fig. 24). The domains are bordered by discontinuities in the ripple pattern, for example, lines of zig-zag junctures or zones of monoclinal crestal flexuring which cut across the crestal trend, the ripples in the domain on one side of the line or zone being out of phase with forms on the other side. Tanner (1967) and Reineck and Wunderlich (1968b) explored the differences between the two kinds of ripple statistically, to find that the wave-current forms have generally a smaller vertical form-index than current ripples, and that the ripple symmetry index rarely exceeds 3, the values for current ripples ordinarily being larger. Wave-current ripples commonly have better rounded crests than current ripples, but on occasions are just as sharp.

Wave-current ripples are recognized with difficulty from internal structures. However, the frequent association of symmetrical with asymmetrical forms in *Linsenschichten* and *Flaserschichten* suggests that wave-related ripples are important in the formation of these structures. McKee's (1965) experiments suggest that cross-lamination sets due to wave-current ripples are more regular than those of current-ripple origin.

WAVE-RELATED RIPPLE MARKS WITH MULTIPLE-PARALLEL CRESTS

Observers since early times have distinguished two main kinds of wave-related ripple with multiple-parallel crests.

Forel (1883) classed amongst his *rides anormales* a form with crests composed of two and sometimes three equal to unequal smaller ridges and troughs, which lie exactly parallel with the larger features and which follow every change in their configuration (Fig. 11-11). Cornish (1901a) and Trefethen and Dow (1960) showed the ripples from intertidal flats and O.F. Evans (1943) described an example formed on a lake-bed during storm decay. The older stratigraphic yields cases (Kindle, 1917; Backhaus, 1967; Von Bruun and Hobday, 1976; Button and Vos, 1977; Tankard and Hobday, 1977).

The smaller features involved in this type of many-crested ripple are superimposed after a change of conditions on larger ripples formed earlier. The primitive forms in Fig. 11-11 are moderately asymmetrical wave-current ripples; the smaller are nearly symmetrical trochoidal wave ripples. A similar story is told by Trefethen and Dow (1960, their fig. 5), who explained the superimposition in terms of either the tide ebbing or the abatement of wind and wave, O.F. Evans having earlier favoured the latter. This type of

Fig. 11-11. Wave-related ripple marks with multiple crests, in fine sand, Wells-next-the-Sea, Norfolk, England. Trowel 0.28 m long.

many-crested ripple has strong affinities with other superimposed bedforms produced under changeable regimes (Chapter 12), but is of interest here as it demonstrates that extreme parallelism of features, indicative of a remarkably precise mechanical coupling between bedforms of different scales, does not prove a simultaneous origin.

The other kind of many-crested form, termed secondary by O.F. Evans (1943, 1949), is represented by long-crested and generally steep trochoidal wave ripples which contain centrally placed in their troughs a low trochoidal crest parallel with the main features. Occasionally the troughs of the larger forms each hold two low crests arranged symmetrically. In all cases there is strict parallelism of the various features. Van Straaten (1953b) and Bajard (1966) found secondary ripples on intertidal flats, and R.A. Davis (1965) saw many examples in the troughs between the near-shore bars of Lake Michigan. Van Hise (1896) first described the ripples from the rock record, and there are later reports from marine-influenced (W.L. Stokes, 1950; Pepper et al., 1954; McKee, 1954; Reif and Slatt, 1979) and fluvial (Stear, 1979) sequences.

WAVE RIPPLE MARKS IN BRICK AND TILE PATTERNS

Occasionally, patterns of wave ripples are found which consist of two sets of trochoidal crests arranged at a steep angle. Kindle (1917) described these as interference ripple-marks, whereas Bucher (1919) called them oscillation cross-ripples, assigning the forms to a class of complex ripple patterns (see also Tanner, 1960). Figure 11-12 shows a typical example, and another can be found in Martinez (1977).

The considerable variation shown by wave ripples in brick and tile patterns may be discussed in terms of (1) the relative development of the sets of crests, (2) the shape in plan and mutual arrangement of elementary units of the pattern, that is, the "bricks" and "tiles", and (3) the orientation of the sets of crests relative to the travel direction(s) of the parent waves. Notwithstanding the variability of these ripples, the available evidence strongly suggests that the two sets of crests are invariably formed simultaneously.

Kindle (1917) described from modern lakes interference ripples consisting of equally developed trochoidal crests arranged on the sides of squares, disposed in phase in some parts of the pattern but out of phase in others (Fig. 11-13a, b). Each set of ripples corresponded in orientation to a set of waves, the two wave sets having been generated by obstacles acting upon a single parent set as it marched toward the lake shore. Inman (1957) found a rather similar arrangement of crests on ripples which had been developed in 15 m of water where rip currents disturbed the regularity of wave action. Bagnold's (1946), Manohar's (1955), and Mogridge's (1973) experimentally

Fig. 11-12. Interference ripple marks on a bedding surface in Svartholm Sandstone Member, Porsangerfjord Group (Precambrian–Cambrian), Finnmark, Norway. The rocks are folded and the ripple marks are slightly compressed in the plane of bedding in a direction nearly parallel with hammer handle. Photograph courtesy of J.D. Roberts (see Roberts, 1974).

formed ripples of "brick pattern" are typified by one set of bold crests and by a second lower set at right angles, the elementary units being rectangular in plan and arranged mainly out of phase (Fig. 11-13d). The stronger crests are effectively at right angles to the propagation direction of but a single set of waves. The subordinate ridges in the experimental examples figured by

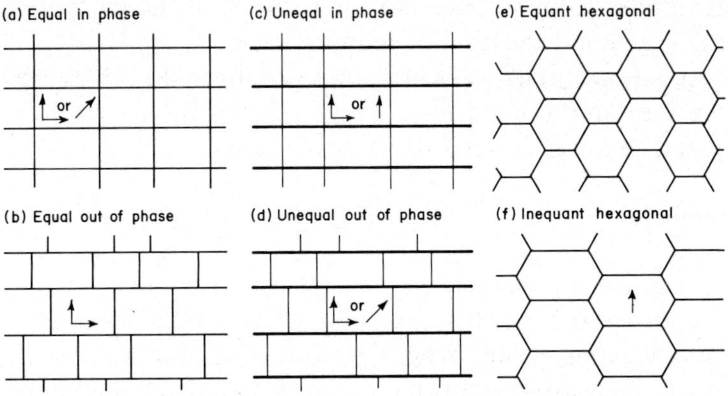

Fig. 11-13. Schematic classification of wave-related ripples of interference type, in terms of crest lines in plan. Where the crests are of unequal development, the bolder are shown by the thicker lines.

Matsunaga and Honji (1980) appear to be slightly curved. Komar (1973) claimed from the shallows of Mono Lake, California, an example of Bagnold's brick-pattern ripples in which the elementary units are in phase (Fig. 11-13c).

Three-dimensional ripples somewhat less regular than Bagnold's, Manohar's and Mogridge's were made by Carstens et al. (1969). In some experiments, crests lay diagonally across the effective direction of wave travel, the same relationship to propagation being noted by Clifton et al. (1971) and Machida et al. (1974) from inshore waters just shallow enough for ripple formation.

Kindle (1917) and Bajard (1966) illustrate interference ripples in which the elementary unit is hexagonal in plan and either equant or elongated (Fig. 11-13e, f). Equant hexagonal depressions much like interference ripples can, however, be shaped by tadpoles (Cameron and Estes, 1971).

CONTROLS ON WAVE-RELATED RIPPLE MARKS

Early work

The field observations and experimental work undertaken by A.R. Hunt (1882), De Candolle (1883), Forel (1883), Darwin (1884), Ayrton (1910) and Kindle (1917) established at an early date that wave and wave-current ripples in natural environments depended on a periodic motion of the water and that "vortices" of at least two kinds were involved in the combined motion of fluid and sediment. In addition, these studies showed qualitatively that ripple wavelength tended to (1) increase with increasing orbital diameter of near-bottom water particles, (2) grow with increasing sediment coarseness, and (3) decrease, with certain exceptions, with increasing water depth. It was also apparent that ripples, forming once the threshold of sediment motion was exceeded, could not exist beneath too vigorous waves.

Subsequent investigations, amongst which those of Bagnold (1946) and Manohar (1955) in the laboratory and of Inman (1957) in the field are pioneering, have greatly expanded upon these early results.

Theoretical considerations

Recalling the difficulties of theoretically modelling bedforms in steady one-way flows (Chapter 7), it is hardly surprising, when the fluid motion is required to be periodic, that a comprehensive analytical approach to the existence and character of wave-related ripples has not yet been attempted. The theory of Kennedy and Falcon (1965) is predicated upon a potential flow model—perhaps the least satisfactory assumption in view of the outstanding importance of viscous boundary-layer effects—and yields little beyond broad predictions about the ripple vertical form-index. Sleath's (1974b, 1975b, 1976) approach is more potent, because viscous effects are explicitly recognized.

Dimensional considerations have guided much research. The scalar attributes of wave-related ripple marks are functions of sediment properties, fluid properties, the characteristics of the near-bed wave-induced fluid motion, and g, the acceleration due to gravity. The sediment properties are the particle diameter D and density σ, whereas the fluid properties are the density ρ and kinematic viscosity ν. The fluid motion may be described in terms of either the period of oscillation T or the radian frequency ω, where $\omega = 2\pi/T$. Its specification may be completed by reference either to the horizontal length or diameter, d, of the orbit of a water particle just outside the viscous boundary layer generated by the wave motion, or to U_{max}, the maximum velocity of a water particle just outside this layer. Note that for a simple-harmonic motion, $U_{max} = (2\pi/T)(d/2)$.

Mogridge (1973) and Mogridge and Kamphuis (1973) extended Yalin and Russell's (1962) analysis to correlate ripple properties against four groups which, written using the above variables, are:

$$\frac{U_{max}D}{\nu}, \qquad \frac{\rho U_{max}^2}{g(\sigma-\rho)D}, \qquad \frac{\sigma}{\rho}, \qquad \frac{d}{D}$$

the first being a grain Reynolds number, favoured also by Komar and Miller (1975b), the second a form of the Shields-Bagnold θ, the third a density ratio, and the fourth a non-dimensional orbital diameter. Rance and Warren (1969) also use the first two of these, but in the form:

$$\frac{dD}{T\nu}, \qquad \frac{\rho d^2}{g(\sigma-\rho)T^2 D}.$$

The first group they interpret like Mogridge, but the second they treat as a grain Froude number, as Bogardi (1965) noted for bedforms in unidirectional currents (Chapter 8). Carstens et al. (1969) also use a grain Froude number in correlating sediment behaviour under wave action. Mogridge found his first two groups less convenient in practice than:

$$\frac{g(\sigma-\rho)D^3}{\rho\nu^2}, \qquad \frac{\rho D}{g(\sigma-\rho)T^2}$$

obtained by further manipulation, the first of his new pair, related to the Bagnold number, having as noted been given by Bonnefille (1965). Japanese workers, however, correlated ripple properties using Reynolds numbers based on U_{max} with d, and D combined with the sediment falling velocity (Hom-ma and Horikawa, 1963a; Hom-ma et al., 1965; Horikawa and Watanabe, 1967). Sleath's (1975b, 1976) correlations are in terms of the groups:

$$\frac{U_{max}}{(\omega\nu)^{1/2}}, \qquad \frac{g(\sigma-\rho)D}{\rho\omega\nu}, \qquad \frac{\sigma}{\rho}, \qquad D\left(\frac{\omega}{2\nu}\right)^{1/2}$$

of which, since $(2\nu/\omega)^{1/2}$ measures the viscous boundary-layer thickness, the first is a form boundary-layer Reynolds number, the second an inverse Shields-Bagnold θ, the third the density ratio, and the fourth a relative roughness. Nielsen (1977) prefers not to use grain size directly but proposes correlations partly in terms of the sediment falling velocity.

Kinds of wave ripple marks

Darwin (1884) and Ayrton (1910) found experimentally two kinds of fluid motion associated with wave ripple marks, each apparently specific to a particular ripple shape. Bagnold (1946) later distinguished what he called rolling-grain ripples and vortex ripples, partly on geometrical grounds similar to those discussed by Darwin and Ayrton, and partly in terms of fluid and sediment behaviour.

Rolling-grain ripples, which include many of Inman's (1957) solitary forms, the "ripples" of Carstens and Neilson (1967) and Carstens et al. (1969), and Clifton's (1976) anorbital ripples, are rather flat and commonly have well-rounded crests. Both Manohar (1955) and Lofquist (1978) produced them experimentally as a stable bedform. By a painstaking investigation, in which ink was used to visualize the flow, Darwin (1884) established that the flow over such ripples involved a slow but steady streaming of the fluid, in a number of distinct stationary cells or "vortices", superimposed on the oscillatory motion. Figure 11-14 shows schematically the instantaneous motion at mid-stroke of a wave. In the lowest two cells, divided symmetrically by the ripple crest, the near-bed streaming is from the trough to the crest and, a little further out in the fluid, from the crest back to the trough.

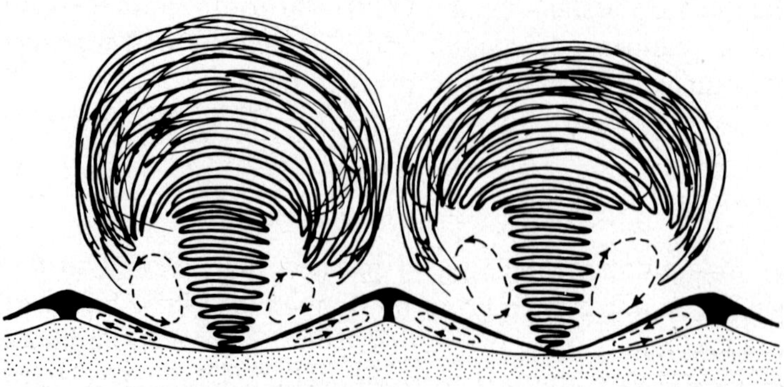

Fig. 11-14. Schematic representation of Darwin's (1884) "ink mushroom" and "ink tree" observed in the oscillatory flow over a symmetrically rippled sand bed.

The ink pattern in the crestal region made what Darwin called his "ink mushroom", the stem of which he observed to shift backwards and forwards over the crest as the waves passed by. The pattern of ink which revealed the current in the outer cells he called an "ink tree". These cells are much larger than the near-bed pair, but of a more sluggish motion. The sinuosities within the trunk of the ink tree showed clearly the coexistence of the streaming and oscillatory motions.

Darwin's important but neglected observations are consistent with the theoretical demonstration, by Lyne (1971), Hall (1974), Sleath (1974b, 1976), Uda and Hino (1975), and Kaneko and Honji (1979a), that a stationary recirculatory streaming of the fluid, in one or more vertically stacked pairs of cells depending on conditions (e.g. Fig. 1-23), is a necessary consequence of an oscillatory fluid motion above bed irregularities such as artificial rigid undulations, individual sediment grains, or wavy aggregations of grains. Darwin's observations are also supported by the experimental demonstrations of this streaming, or curvature-related mass-transport, recently given by Hino and Fujisaki (1975), Kaneko and Honji (1979a, 1979b), and Honji et al. (1980). In the light of this work, it would seem that Darwin's ink mushroom lay partly within the viscous boundary layer, while his tree, with its more complex motion, spread far into the outer potential flow. As Darwin (1884), Sleath (1974b), Uda and Hino (1975), and Kaneko and Honji (1979a) recognized, the stationary recirculatory streaming, which may be expected to arise on any real bed (never mathematically plane) affected by an oscillatory flow, provides the essential mechanism for the initiation and growth of wave ripple marks. The experiments confirm the theoretical finding that, as Darwin noticed, the streaming of fluid near the bed is from troughs on the bed toward crests, whence bed-material grains should follow a similar pattern.

Although the pattern of streaming within the body of the fluid depends crucially on flow and bed conditions (e.g. Lyne, 1971; Kaneko and Honji, 1979a), the effects of the streaming on a granular bed are much the same, provided that the amplitude of the bed waviness is not too large compared with the thickness of the boundary layer induced by the oscillatory fluid motion. Figure 11-15a, d depicts the streaming at respectively moderate and large values of the parameter $L(2\nu/\omega)^{-1/2}$, where L is the wavelength of the bed waviness and $(2\nu/\omega)^{1/2}$ is the boundary-layer thickness as before. Assuming that the oscillatory velocity-component exceeds the sediment entrainment threshold, both configurations of streaming set up at the bed a stationary pattern of sediment erosion, net transport, and deposition, $J_N(x)$ and $dJ_N/dx(x)$ where J_N is the net sediment transport rate and x is distance parallel with the oscillatory motion, that accentuates the original waviness. A mobile granular bed beneath an oscillatory current is therefore unstable on account of the initiation of a curvature-related mass-transport

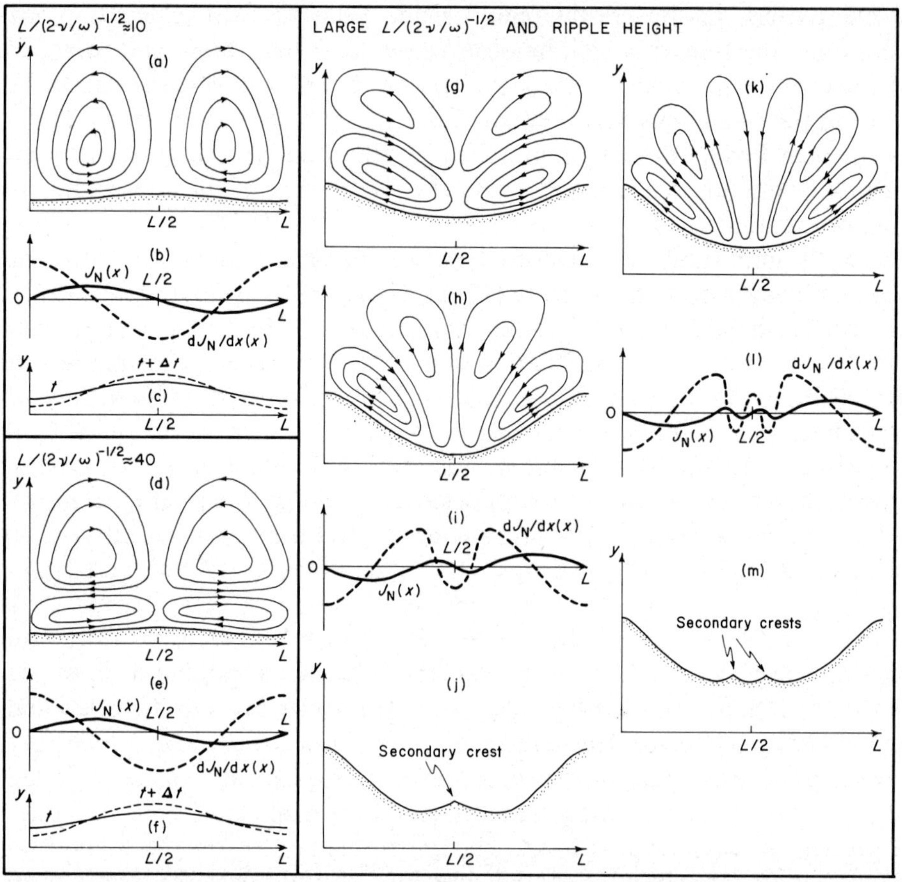

Fig. 11-15. Partly speculative relationship between the steady streaming related to bed waviness, the *local* time-averaged sediment bed load transport and transfer rates, and the cross-sectional profile of the bed, as it might apply to wave-related ripple marks on a sand bed. Flow configurations partly based on Kaneko and Honji (1979a).

by its inherent waviness. It would seem inevitable that, with some restrictions, ripple marks should grow up on the bed (Fig. 11-15b, c, e, f).

In order for ripples actually to form, not only must the oscillatory flow exceed the motion-threshold of the bed grains, but these grains must be sufficiently small in comparison with the boundary-layer thickness that they lie wholly within the lower half of the lowermost of the streaming cells. Experimentally, Sleath (1976) suggests that $1.6 < D(\omega/2\nu)^{1/2} < 3.4$ is the upper limit on grain size. Bagnold (1946) and Manohar (1955) vividly describe the motion of grains over rolling grain ripples and how the ripples grow up, though without explicitly recognizing the role of Darwin's streaming. As soon as a sufficiently strong oscillatory motion is set up, grains begin to roll over the bed and to aggregate, as the result of a series of repeated small net movements (see also Kaneko and Honji, 1979b), into parallel transverse bands which sway back and forth with the current. These bands grow in size as the streaming fluid drags ever more particles into them, until a low crest is formed. At small values of U_{max} the grains spill over the crest to form a small avalanche slope which reverses its facing with each current reversal. At larger values for the maximum oscillatory current, the grains flow *en masse* back and forth over the crest, peaking up if of sufficiently small falling velocity at the site of Darwin's ink mushroom.

Typical *vortex ripples* have sharp, steep-sided crests and rounded troughs, yielding a low value for the vertical form-index. They are called dunes by Carstens and Neilson (1967) and Carstens et al. (1969), and are the orbital and suborbital ripples of Clifton (1976). Many of Inman's (1957) trochoidal forms are vortex ripples. Ayrton (1910) observed from such ripples that the oscillatory current separated from the crest with each stroke, ultimately creating, alternately on one side and then on the other, a large and powerful vortex. Bagnold (1940, 1946), Inman and Bowen (1963), and Tunstall and Inman (1975) subsequently explored in detail the character and energy-

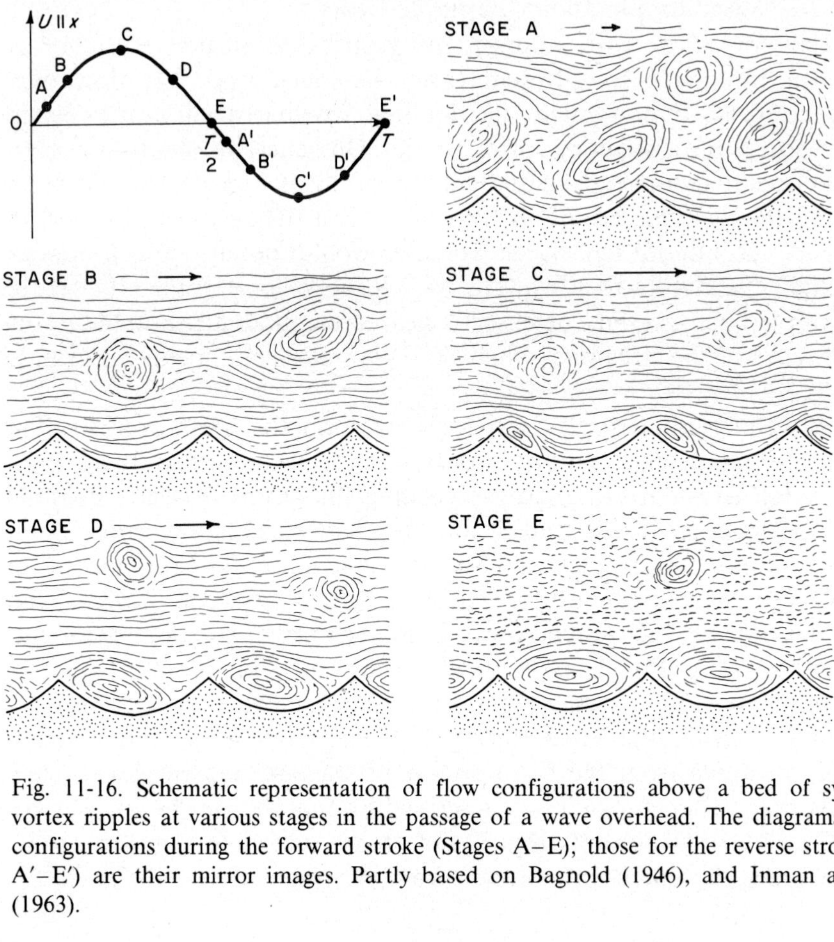

Fig. 11-16. Schematic representation of flow configurations above a bed of symmetrical vortex ripples at various stages in the passage of a wave overhead. The diagrams show the configurations during the forward stroke (Stages A–E); those for the reverse stroke (Stages A′–E′) are their mirror images. Partly based on Bagnold (1946), and Inman and Bowen (1963).

absorbing quality of these vortices (see also Keulegan, 1948; T. Scott, 1954; Tanner, 1963; Carstens et al., 1969; Cook and Gorsline, 1972). Events during essentially one stroke are summarized in Fig. 11-16. Near the start there is little or no separation to lee of the ripple crest, and the trough contains fluid largely conformable in flow pattern. The separated flow grows in size and vigour as the stroke develops until, by the time the stroke is nearing its end, the vortex fills up one-half or more of the trough and exceeds in height the ripple itself. As the flow reverses with the start of the next stroke, the vortex is abruptly swept out of the trough and upward on a steep path into main body of the fluid. Simultaneously, a new phase of separation begins to lee of the crest, but on the opposite side of the ripple. In time, as Bagnold (1946) observed, the body of the fluid fills up with the vortices swept up from the bed, the decay of each requiring a time of the order of the ripple period (Tunstall and Inman, 1975). Large amounts of relatively fine grains swept into or entrained by the vortices can thus be suspended above a rippled bed (e.g. Hom-ma and Horikawa, 1963a, 1963b; Abou-Seida, 1965; Das, 1971; Kennedy and Locher, 1972). Large particles, however, are caused to hop about the ripple crests.

The oscillatory flow with intermittent vortices associated with vortex ripples is at first sight so different from the much smoother flow over rolling-grain forms that it might be doubted whether a curvature-related mass-transport is present above vortex ripples. On general grounds, however, a stationary recirculatory (i.e. spatially periodic) mass-transport would seem essential for the maintenance of vortex ripples. But such a mass transport in the presence of intermittent large vortices would be difficult to detect experimentally, and this may explain why Kaneko and Honji (1979a), in work on rippled beds, could not observe clear streamlines attributable to the lowermost cells of the streaming when flow separation commenced at ripple crests.

Whereas rolling-grain ripples will grow even on the most expertly flattened bed, it seems that vortex ripples cannot form unless there exist sufficiently tall irregularities greatly exceeding the grains in scale. Bagnold (1946), Carstens et al. (1969), and Sleath (1976) describe how vortex ripples grow from rolling-grain ripples which have reached a critical steepness. But any sufficiently large irregularity—a pebble, shell or artificial barrier—can promote the ripples. They spread away from such irregularities in diamond-shaped patches (Fig. 11-17), as Bagnold (1946) and Shulyak (1963) saw in their experiments, and as can be found on tidal flats when the spring tides drown sands slightly hardened during the neaps (Reineck, 1961; Allen, 1968c). An example from the fossil record of wave-related ripples in such roughly diamond-shaped patches is described by MacKenzie (1972). We saw earlier (Fig. 7-1) that current ripples and transverse aeolian dunes spread away in a triangular patch downstream from an irregularity at the apex.

Fig. 11-17. Early stages in sand mobilization by waves, Wells-next-the-Sea, Norfolk, England. a. Patches of vortex ripples, many spreading from the sites of pebbles or shells. b. A triangular patch. Trowel 0.28 m long. Note ubiquitous round-crested rolling-grain ripples which preceded vortex ripples.

Existence field for wave ripple marks

Little attempt has been made to define the existence fields of bedforms in oscillatory flows. Inman (1957) plotted in the U_{max}–D plane Manohar's (1955) experimental determinations of the ripple and plane-bed modes of sediment transport. Allen (1967b, 1970g) further developed this graph and added the internal structures expected to be associated with the bedforms. Clifton (1976) also favoured Inman's form of plot. Komar and Miller (1975b), however, graphed the existence of ripples and plane beds fields in the θ–D and θ–$(U_{max}D/\nu)$ planes. Manohar's data on the replacement of ripples by a plane bed was found to agree closely with Bagnold's (1966) "universal" plane-bed criterion (eq. 7.4).

For wave-generated periodic flows, the Shields-Bagnold θ is less clear in physical meaning than for a steady unidirectional current, and is in any case

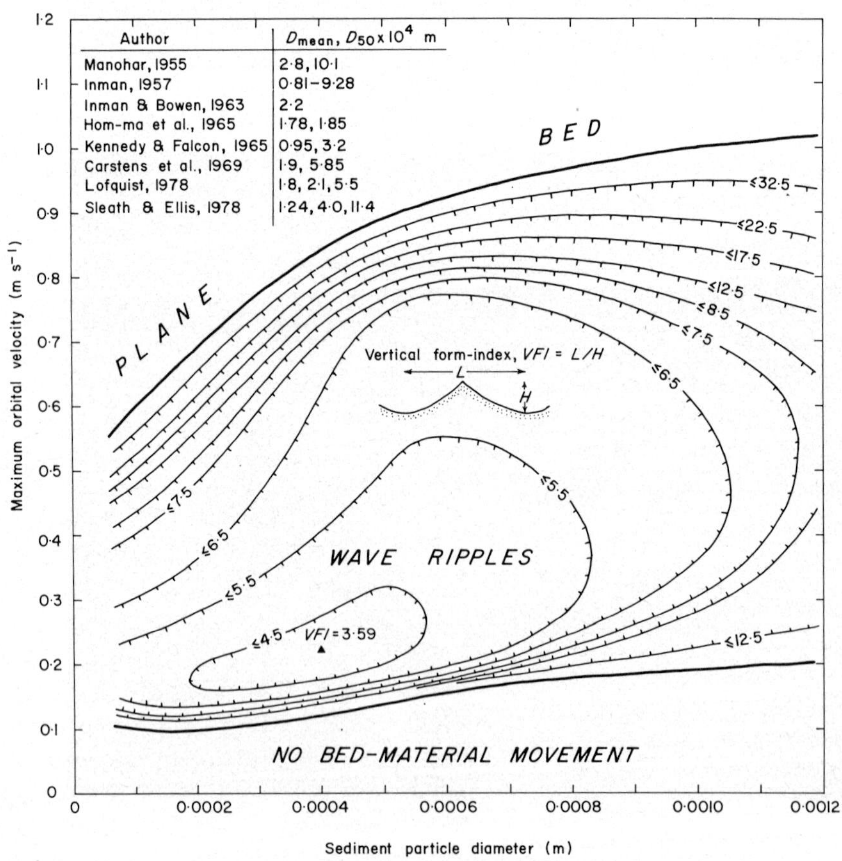

Fig. 11-18. The vertical form-index of laboratory and field examples of symmetrical wave ripple marks, as a function of the maximum orbital velocity near the bed and sediment calibre. Note that the curves for vertical form-index are *limiting* values and not isopleths. Based on a data set of 648 paired values from sources listed in the diagram.

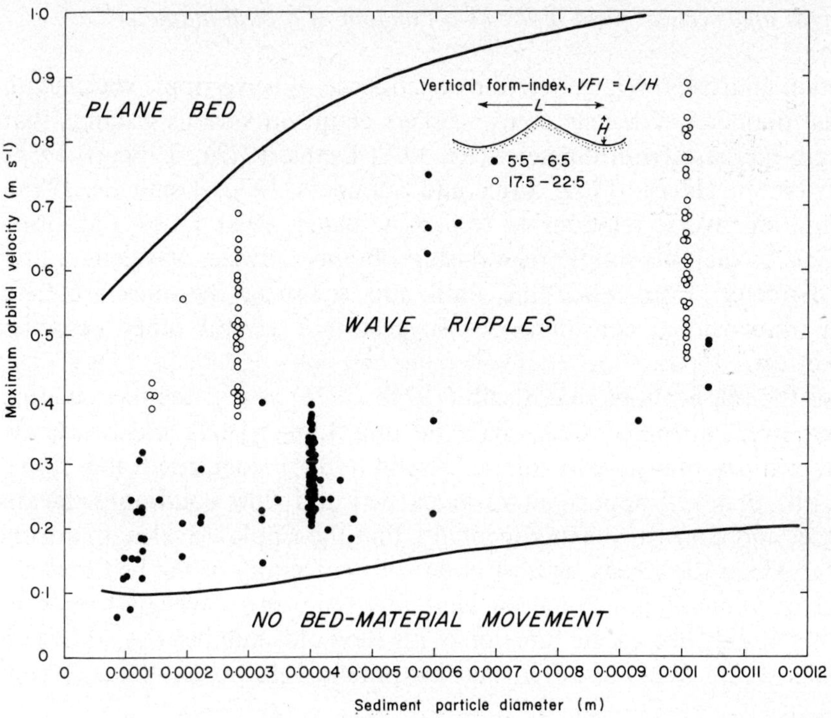

Fig. 11-19. Occurrence in terms of maximum near-bed orbital velocity and sediment calibre of symmetrical wave ripple marks of two narrow ranges in vertical form-index. Selected from data set affording Fig. 11-18.

more difficult to calculate, even using Jonsson's (1967) helpful graph. Hence a plot in the U_{max}–D is most appealing for the practical interpretation of wave ripple marks, as in Fig. 11-18 based on Allen's (1979) analysis of 648 self-consistent sets of field and experimental data provided by Manohar (1955), Inman (1957), Inman and Bowen (1963), Hom-ma et al. (1965; pers. comm., 1977), Kennedy and Falcon (1965), Carstens et al. (1969), Lofquist (1978), and Sleath and Ellis (1978). The field boundaries shown are those of Inman (1957), and a novel feature of the graph is the use of the vertical form-index as a parameter. The original data is deliberately not given in this graph, in order to conserve its value as an interpretative tool, but in Fig. 11-19 appear plots of the original data for two limiting values of the vertical form-index. The fact that this parameter appears as a limiting value (*not* an isopleth) means that ripple marks with an index say of 6.5 may occur for many combinations of orbital velocity and grain size within the bounds of the curve labelled 6.5 but not for sets of values outside it. Therefore a range of values of U_{max} may be estimated from this graph on entering it with a known sediment calibre and ripple index.

Wavelength and vertical form-index as a function of orbital diameter

A certain confusion reigns over the dependence of wave-ripple wavelength on orbital diameter. Whereas many workers claim on various grounds that wavelength increases with diameter (e.g. O.F. Evans, 1942a; T. Scott, 1954; Newton, 1968b; Harms, 1969; Stone and Summers, 1972; Trenhaile, 1973), others find an inverse relationship (e.g. R.A. Baker, 1970; Cook and Gorsline, 1972). In fact, no simple relationship obtains between wavelength and orbital diameter. Even when the fluid and sediment densities are held constant, dimensional considerations suggest that several other variables have an effect.

One of the implications that Sleath (1975b, 1976) saw in the recirculating streaming over a rippled bed which he and Lyne (1971) independently re-discovered so many years after Darwin (1884) concerned the ripple wavelengths that will appear under given bed and flow conditions. Sleath postulated, and confirmed experimentally, that the ripple-wavelength arising is that for which the steady drift of grains in the vicinity of the bed towards the crest is a maximum. For each kind of ripple, the wavelength L/d to emerge is a weakly decreasing function of his Reynolds number $U_{max}/(\omega\nu)^{1/2}$. The trends are so weak, however, that for practical purposes we may write:

$$0.65 \leqslant \frac{L}{d} \leqslant 1.0 \qquad (11.1)$$

for vortex ripples, and

$$0.036 \leqslant \frac{L}{d} \leqslant 0.059 \qquad (11.2)$$

for rolling-grain ripples, in mineral-density sands. In each case broadly two orders of magnitude of his Reynolds number are covered. Komar's (1974) arbitrarily proposed $L = 0.8d$ for vortex ripples is therefore a fair average. A more recent analysis of field and laboratory data by Miller and Komar (1980a, 1980b) has led them to prefer the lower limit of Sleath's bound as stated in eq. (11.1). Sleath's results are closely related to empirical correlations made earlier by Hom-ma and Horikawa (1963a), Hom-ma et al. (1965), and Horikawa and Watanabe (1967), who found that L/d declined gradually with the Reynolds number $U_{max}d/\nu$. This number is proportional to the square of Sleath's $U_{max}/(\omega\nu)^{1/2}$, since $d = 2U_{max}/\omega$. Kaneko (1980) has also examined the controls on ripple wavelength.

Equations (11.1) and (11.2) limit the occurrence of wave ripple marks. Figure 11-20 is based on the same data set as Fig. 11-18 and shows between these limits the non-dimensional ripple wavelength as a function of the non-dimensional water-particle orbital diameter, the vertical form-index, again a limiting value, serving as a parameter (Allen, 1979). The original data are not shown in order to give the graph maximum usefulness as an

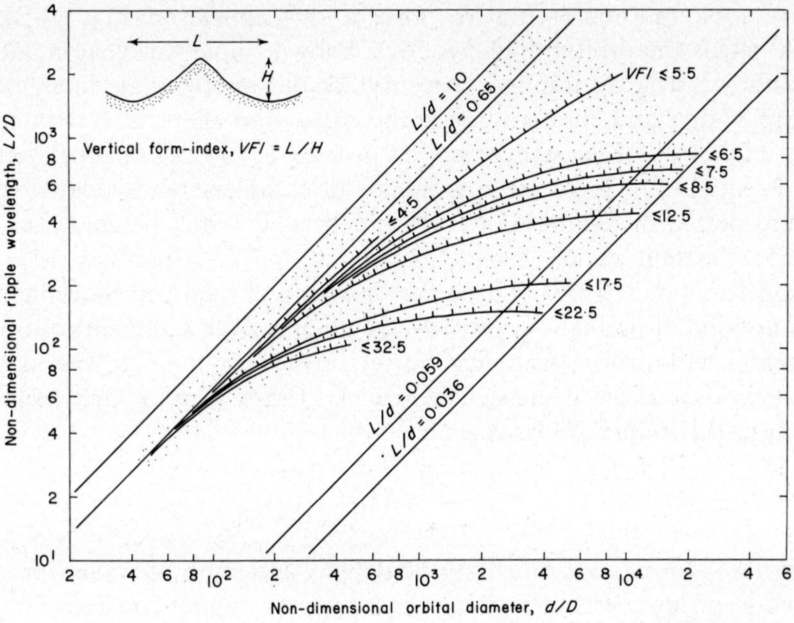

Fig. 11-20. Vertical form-index of symmetrical wave ripple marks as a function of the non-dimensional ripple wavelength and non-dimensional water-particle orbital diameter. Note that the curves for vertical form-index are *limiting* values and not isopleths. Constructed from the same data set as Fig. 11-18.

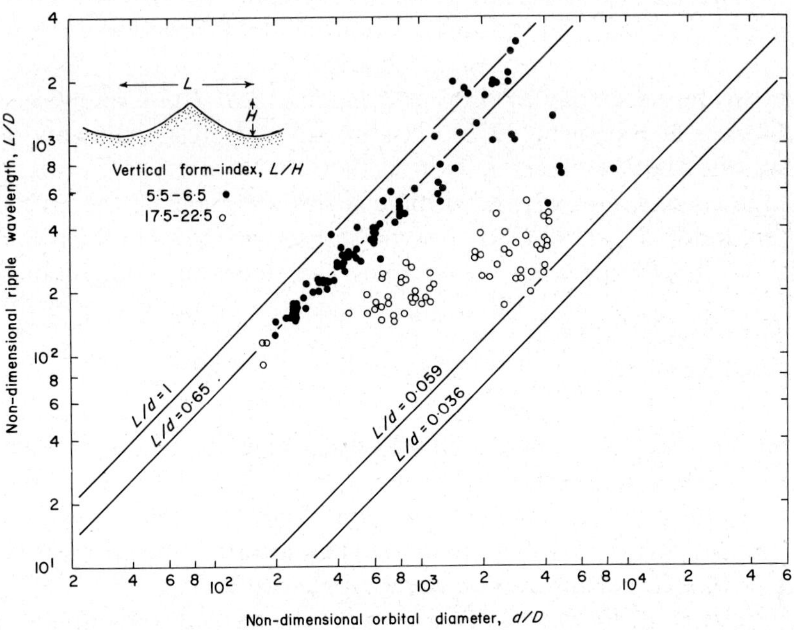

Fig. 11-21. Occurrence of symmetrical wave ripple marks in terms of non-dimensional ripple wavelength and non-dimensional water-particle orbital diameter, for two narrow ranges in the vertical form-index. Selected from data set affording Fig. 11-18.

interpretative tool, but representative observations appear in Fig. 11-21. Figure 11-20 yields the orbital diameter for a known ripple-wavelength and sediment calibre. As Sleath implied, there are ripples intermediate in shape between pure vortex and pure rolling-grain types, and there is a striking clustering in Fig. 11-20 of the steeper ripples near eq. (11.1) and of the flatter ones towards eq. (11.2). There are suggestions of complex trends related to grain size and period of oscillation (Yalin and Russell, 1962; Kennedy and Falcon, 1965; Carstens et al., 1969; Chan et al., 1972; Mogridge, 1973; Mogridge and Kamphuis, 1973). Above the threshold of sediment movement at a constant period of oscillation, the wavelength at a given sediment calibre at first increases with orbital diameter, but at sufficiently long strokes may become constant or even decrease slightly as vortex ripples are metamorphosed into the rolling-grain type.

Wavelength and grain size

Several workers found experimentally that the wavelength of wave ripple marks tended to increase with increasing grain size, even under constant flow conditions (Evans and Ingram, 1943; Bagnold, 1946; Manohar, 1955; Kennedy and Falcon, 1965; Carstens et al., 1969). The same relationship holds in the field, and is perhaps the most striking of the trends displayed by these bedforms. The largest ripples that Inman (1957), R.A. Davis (1965), and Cook and Gorsline (1972) observed generally were in the coarsest sand. Mention has already been made of many occurrences of large wave ripple marks in coarse sediments, often of biogenic origin.

Grain-size sorting accompanies rippling (Inman, 1957; Cook and Gorsline, 1972; Stone and Summers, 1972). The trough and crest sediments of ripples in very fine sands are essentially indistinguishable in grade. Forms in fine and medium sands, however, have the coarsest sediment on the crests. In still coarser grades ($D > 0.0005$ m), the coarsest sediment lies in the ripple troughs (Fig. 11-4b), as generally with ripples and dunes in unidirectional currents.

Wave-current ripples

These forms are less well known and understood than their more symmetrical relatives. Inman and Bowen's (1963), Harm's (1969), and Tietze's (1978) experiments, together with Inman's (1957) and Tanner's (1971) measurements of nearshore ripple and wave characteristics, afford a limited basis on which to assess the controls on wave-current ripple marks.

Figure 11-22 shows for Tanner's (1971) sample the ripple wavelength as a function of the orbital diameter (Airy theory) of a near-bed water particle, the vertical form-index and ripple symmetry index affording parameters. Assuming that equilibrium substantially prevailed, the forms range from

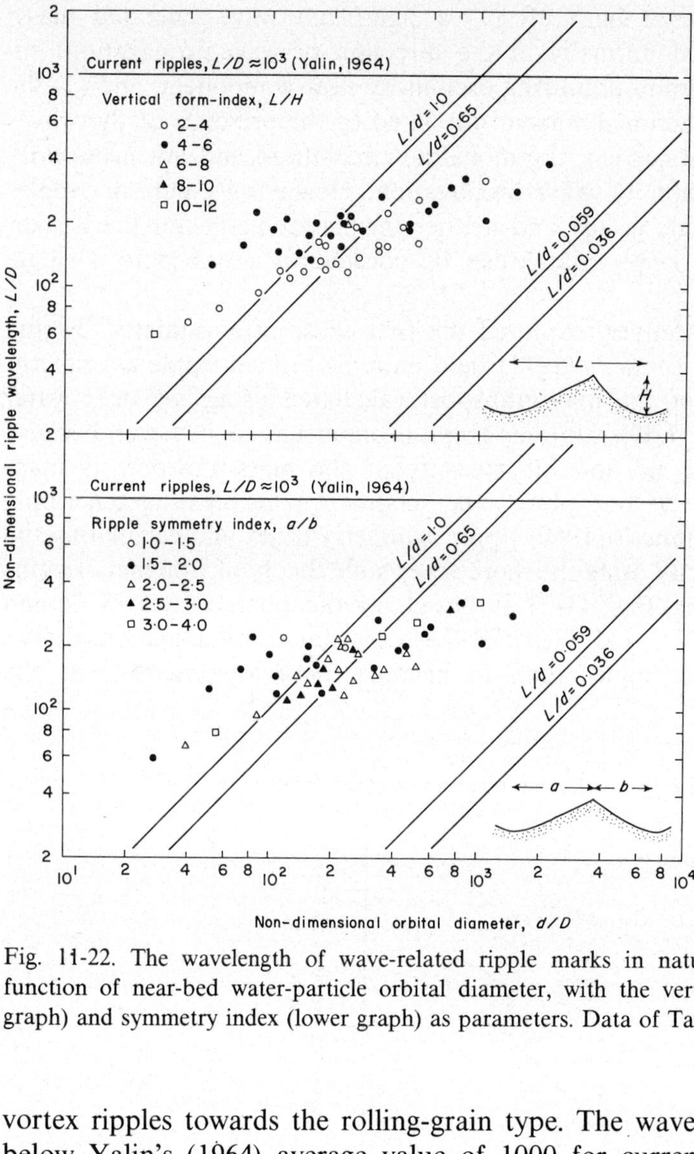

Fig. 11-22. The wavelength of wave-related ripple marks in natural environments as a function of near-bed water-particle orbital diameter, with the vertical form-index (upper graph) and symmetry index (lower graph) as parameters. Data of Tanner (1971).

vortex ripples towards the rolling-grain type. The wavelength L/D is well below Yalin's (1964) average value of 1000 for current ripples; Tanner's forms, although strongly asymmetrical, are clearly wave-dominated. That some have wavelengths exceeding the orbital diameter is consistent with Sleath's (1975b, 1976) findings, as they represent low Reynolds numbers. The vertical form-index ranges between 2.0 and 10.5, falling mostly between 3 and 5, and apparently is independent of the orbital diameter. The flatter forms, however, seem to take the larger wavelengths at each orbital diameter. The ripple symmetry index ranges to more than 3 and likewise seems uninfluenced by the orbital diameter.

Asymmetry in wave-related ripple marks depends on a net sediment transport on a spatial scale larger than the forms themselves. Such transport is promoted by one or both of: (1) a significant wave-generated mass-transport (at the bed normally in the direction of wave-propagation) superimposed on the grain-mobilizing oscillatory flow component, and (2) the presence of a unidirectional current unrelated to the presence of the waves (e.g. wind drift, tidal current, thermohaline circulation) and not necessarily acting in the direction of wave-propagation. Hence wave-current ripples comprise a polygenetic group, and neither their asymmetry nor the weaker asymmetry of wave ripple marks can be confidently ascribed to a single cause.

Clifton (1976) tentatively explored the first cause of asymmetry. Taking Inman's (1957) and Tanner's (1971) field data, he plotted ripple asymmetry against a measure of the mass-transport calculated using the deep-water theory of Stokes (eq. 1.49). Clifton's graph is unpersuasive, however, because it is partly dimensional and his measure of the mass-transport is inappropriate. Allen (1979) re-explored the relationship, expressing the ripple shape in terms of Tanner's (1960) ripple symmetry-index and calculating the mass-transport velocity using the more acceptable theory of Longuet-Higgins (1953). The resulting Fig. 11-23 is based on the observations of Inman (1957), Tanner (1971), and Tietze (1978, personal communication, 1979), and shows that the ripples tend to become more asymmetrical as the

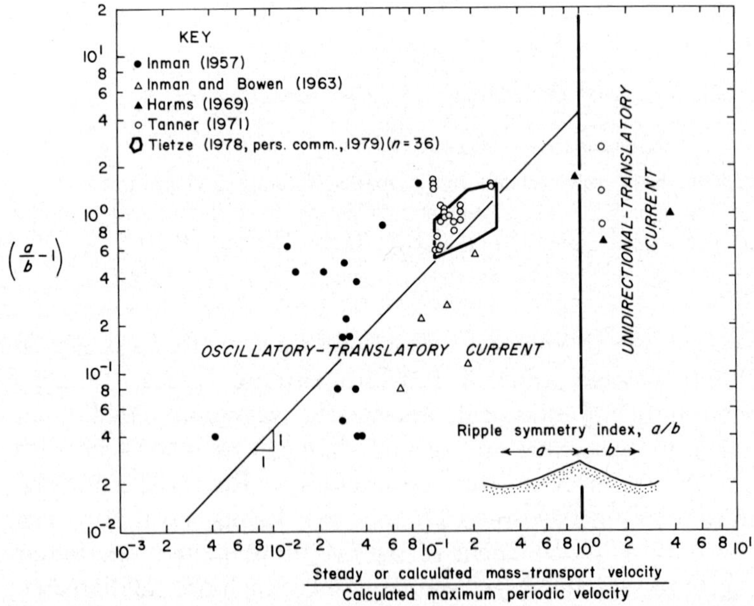

Fig. 11-23. Asymmetry of wave-related ripple marks as a function of asymmetry of governing current (steady or calculated mass-transport velocity/calculated maximum near-bed orbital velocity). In the experiments of Inman and Bowen (1963) and of Harms (1969), waves acted in combination with an applied steady current.

mass-transport velocity increases relative to the maximum near-bed orbital velocity. A comparison of eqs. (1.47) and (1.52) suggests that ripples will become increasingly asymmetrical as they are traced from deep into shallow water, for fixed wave conditions.

A unidirectional current unrelated to waves affects wave-ripple asymmetry in much the same way as a wave-induced mass-transport. Inman and Bowen (1963) and Harms (1969) measured the asymmetry of ripples created by waves superimposed on a steady current acting in the same line. Their results also appear in Fig. 11-23, in aggregate suggesting that the ripple symmetry-index is a coherent and gradually increasing function of the velocity ratio. Amos and Collins (1978) found in the field that wave-formed ripples became increasingly asymmetrical as the associated tidal current increased.

Ripples of complex pattern

Secondary ripples and ripples with crests in brick and tile patterns seem to express complex bed-fluid interactions, few of which are so far well understood.

O.F. Evans (1943) attributed secondary ripples to a reduction of wave size sufficient that the orbital diameter at the bed fell to broadly one-half of its original value, a transformation which Shulyak (1963) confirmed experimentally. They therefore regarded the minor ripple crests as superimposed on primitive larger features, and saw the combination as expressing a temporal change.

An alternative explanation relates to what happens to waves in an area of changing bottom topography. As waves propagate in shallow water (Wiegel and Fuchs, 1955; Madsen and Mei, 1970; Bryant, 1973), and particularly as they cross submerged bars which promote breaking (McNair and Sorensen, 1971; Chandler and Sorensen, 1973), the profile changes from sinusoidal to a form like the solitary wave, the crest becoming narrow and the trough broad and rather flat. Moreover, a secondary crest of a substantial but lesser height than the primary wave commonly appears in the trough during the transformation. As Morison and Crooke (1953) demonstrated, if this crest is sufficiently tall, the water particle orbits include a short subsidiary loop. It seems possible that this motion might induce ripples of two unequal wavelengths and heights simultaneously on a responsive sand bed. Evans does not comment on the location of the secondary ripples he described, but R.A. Davis (1965) and Rudowski (1970a) found them chiefly to landward of near-shore bars, in accordance with the possible influence of these structures on waves.

Perhaps the most plausible explanation of secondary ripples depends on an effect observed by Kaneko and Honji (1979a) during their experiments on waviness-induced mass-transport. As the bed-wave amplitude is increased relative to its wavelength, for large values of the parameter $L(2\nu/\omega)^{-1/2}$, the outer and weaker cells of the stationary streaming are forced down between those immediately adjacent to the bed in the troughs, themselves coming to affect the bed (Fig. 11-15g, h). The result is a doubly periodic spatial pattern of sediment erosion, net transport and deposition, which will generate a bedform containing a single subordinate crest within each trough (Fig. 11-15i, j), just as is observed. Ripples are occasionally found whose troughs hold paired minor crests. A simple extension of the model to involve at the bed the recirculatory cells squeezed down from a third level within the flow (permitted by Lyne's, 1971, analysis) suffices to account for them (Fig. 11-15k–m).

Whereas Kindle (1917) ascribed his interference ripples (Fig. 11-12) to the action of two coexisting but differently oriented trains of waves, Bucher (1919) regarded the forms as due to superimposition. That Kindle's explanation is correct follows both from his own field observations and from Silvester's (1972) experiments. The latter made interference ripples by allowing a train of progressive waves and its oblique reflection from a vertical wall to affect a sand bed. The resulting pattern of near-bed currents is identical to that Fuchs (1952) established beneath short-crested waves, which can be regarded as formed from two trains of waves propagating in directions 90° apart. Silvester found that interference ripples formed chiefly where the horizontal wave-induced bottom current was rotary, that is, in zones parallel with the reflecting wall spaced apart at one-quarter of the wavelength of the combined wave.

The strengthening of short-crestedness as waves shoal provides a satisfactory explanation of the observation by Clifton et al. (1971), Machida et al. (1974), and Clifton (1976) of interference-like ripples chiefly from the nearshore. Presumably interference ripples arise experimentally because, at ripple wavelengths sufficiently small compared to the apparatus, the sediment motion can become that which would be generated by short-crested waves, even though only one periodic motion is nominally involved. Matsunaga and Honji (1980) showed experimentally that brick-pattern ripples may be due to an instability in the stationary mass-transport currents coupled to the main transverse ridges and hollows on the bed.

Palaeohydraulic reconstructions from wave-related ripples

Wave-related ripples represent such widely ranging circumstances that their distribution and relationship to other bedforms in shallow-water environments cannot be summarized using fewer than four models, classified on the presence or absence of near-shore bars, and on wave period, the details of which are left as an exercise for the reader. Vause (1959), Clifton et al. (1971), Machida et al. (1974), and Clifton (1976) give most of the data for a non-barred large-period model. The corresponding small-period model can be derived largely from Clifton (1976). The barred large-period model depends chiefly on the work of Davidson-Arnott and Greenwood (1974, 1976), Greenwood and Davidson-Arnott (1979), and Greenwood and Mittler (1979). R.A. Davis (1965) and Rudowski (1970a) provide a satisfactory basis for a corresponding small-period model.

The deduction of ancient wave conditions in shallow-water environments is difficult, partly because the controls on ripple characteristics are ill-known, and partly because no less than four variables—wave period, wave height, water depth, and sediment calibre—combine to determine the ripple wavelength, the attribute chiefly used for reconstructions. But only grain size and ripple wavelength are readily measurable in the field; three other variables remain to be determined.

Shulyak (1963) summarized an interesting palaeohydraulic model using which, it is claimed, wave characteristics can be completely estimated from ripple and sediment attributes. For example, ripple asymmetry and the relative spacing of transverse periodic ripple elements are the basis for estimating U_{max}. Harms (1969) claimed less for his model, which he founded on wave theory and a limited range of empirical data, but failed to recognize the constraints on real waves and erroneously took ripple wavelength as invariably comparable with the near-bed water-particle orbital diameter. His model therefore has limited validity, for wavelength is a complex function of orbital diameter (Fig. 11-20), as Allen (1970g), Komar et al. (1972), and Komar (1974) emphasized. Tanner's (1971) scheme for the interpretation of wave-related ripples is empirical in basis and statistical in expression. It too has limited relevance, since a simple control on ripple wavelength is implied. The model proposed by Komar (1974), and its somewhat speculative extension by Clifton (1976), and that described by Allen (1979), making use of Figs. 11-18 and 11-20, go far towards the aim of defining wave characteristics from ripple attributes. Referring to Chapters 1 and 2, any set of estimated wave characteristics must be consistent with the sediment entrainment condition:

$$U_{max} \geqslant U_{cr} \tag{11.3}$$

where U_{cr} is the entrainment threshold velocity, and with wave stability (e.g. Miche, 1944; Iversen, 1952):

$$\frac{H}{L} \leqslant 0.142 \tanh\left(\frac{2\pi h}{L}\right) \tag{11.4}$$

$$\frac{H}{h} \leqslant 0.78 \tag{11.5}$$

where H and L are respectively the wave height and length and h is the water depth. These constraints can be used, as Komar (1974) has described, to define the physically realizable values of water depth and wave characteristics that can be inferred from the use of Fig. 11-18 for orbital velocity and Fig. 11-20 for orbital diameter. Unique solutions cannot be given, but it is possible to identify in general terms the wave-climate (e.g. oceanic, restricted sea, small lake) under which given wave-related ripples were formed.

SAND WAVES

Character and occurrence

Lüders (1929) described as *Grossrücken* trains of large submerged ridges of mixed sand and shell which he mapped transversely to the direction of reversing tidal currents in the channels of the Jade, Germany. The forms had wavelengths in the order of 100 m, heights of 5 m or so, and long and sinuous crests. Their symmetry was marked and similar inclinations, in the order of 5°, typified both the ebb-facing and flood-facing slopes. Applying the then newly discovered technique of echo-sounding, Van Veen (1935, 1936, 1938) detected similar structures, described as sand waves or *Sandwüste,* in tidal channels in the Netherlands and in the open North Sea. These bedforms are now known to occur widely in the tide-swept shallow-marine environment, usually being called either sand waves (e.g. Stride, 1963) or *Riesenrippeln* (Reineck, 1963; Reineck et al., 1971; Ulrich, 1973). Figure 11-24 shows portions of typical sand-wave fields as revealed in profile and plan by side-scan sonar.

Sand waves are sets of long-crested parallel ridges typified by a moderate to high symmetry in cross-section and by low to mild slopes. They ordinarily consist of quartz sand, but may locally be rich in biogenic material and include gravel. Although slope segments steep enough for grain avalanching occur on most sand waves, in many instances they are small in height compared with the overall height of the waves, being associated with superimposed dunes. Figure 11-25a gives the wavelength and height of representative individual sand waves. The vertical form-index is ordinarily between 20 and 100, and is greatest for the largest sand waves, which therefore are the flattest (Allen, 1963b, 1968c). Most sand waves are long-crested in plan, ranging from relatively straight or gently curved in a single direction (e.g. G.F. Jordan, 1962; Belderson and Stride, 1966; Langeraar, 1966; Belderson et al., 1972) to moderately sinuous (e.g. W.M. Gibson, 1951; G.F. Jordan, 1962; Völpel and Samu, 1966; Belderson et al., 1972). Sand wave junctures include open, zig-zag and buttress types (Allen, 1968c), as shown by Belderson et al. (1972) and Langhorne (1973).

As with ripple marks, the symmetry of sand waves can be measured by dividing the horizontal distance from trough to crest by the horizontal distance between crest and trough. The sand wave symmetry index so defined rarely exceeds 4, tending in many cases tends to unity, and size for size the more symmetrical forms prove on the whole to be the steeper (Fig. 11-26). Van Veen (1935, 1936, 1938) recognized this range in symmetry by

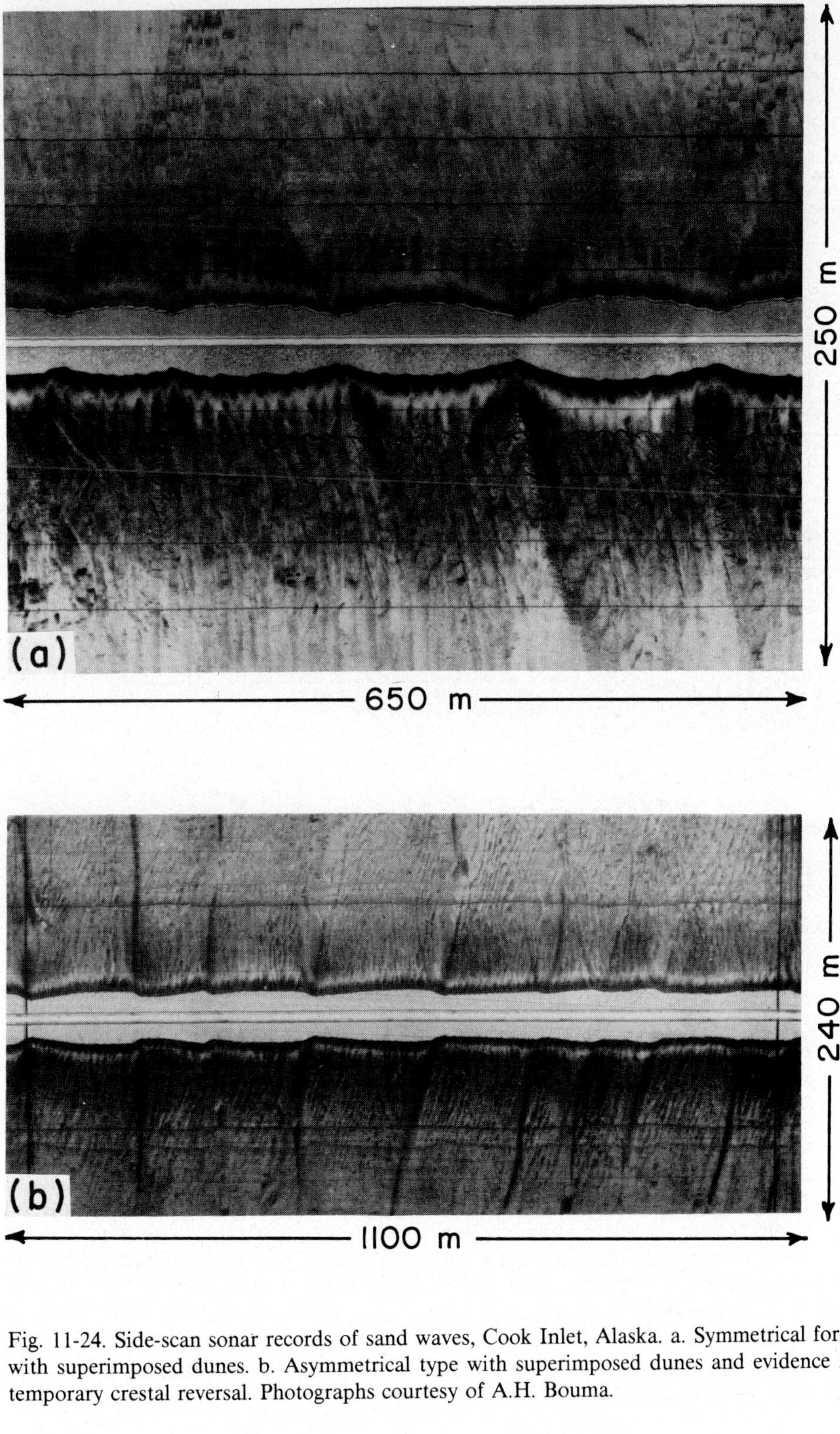

Fig. 11-24. Side-scan sonar records of sand waves, Cook Inlet, Alaska. a. Symmetrical forms with superimposed dunes. b. Asymmetrical type with superimposed dunes and evidence for temporary crestal reversal. Photographs courtesy of A.H. Bouma.

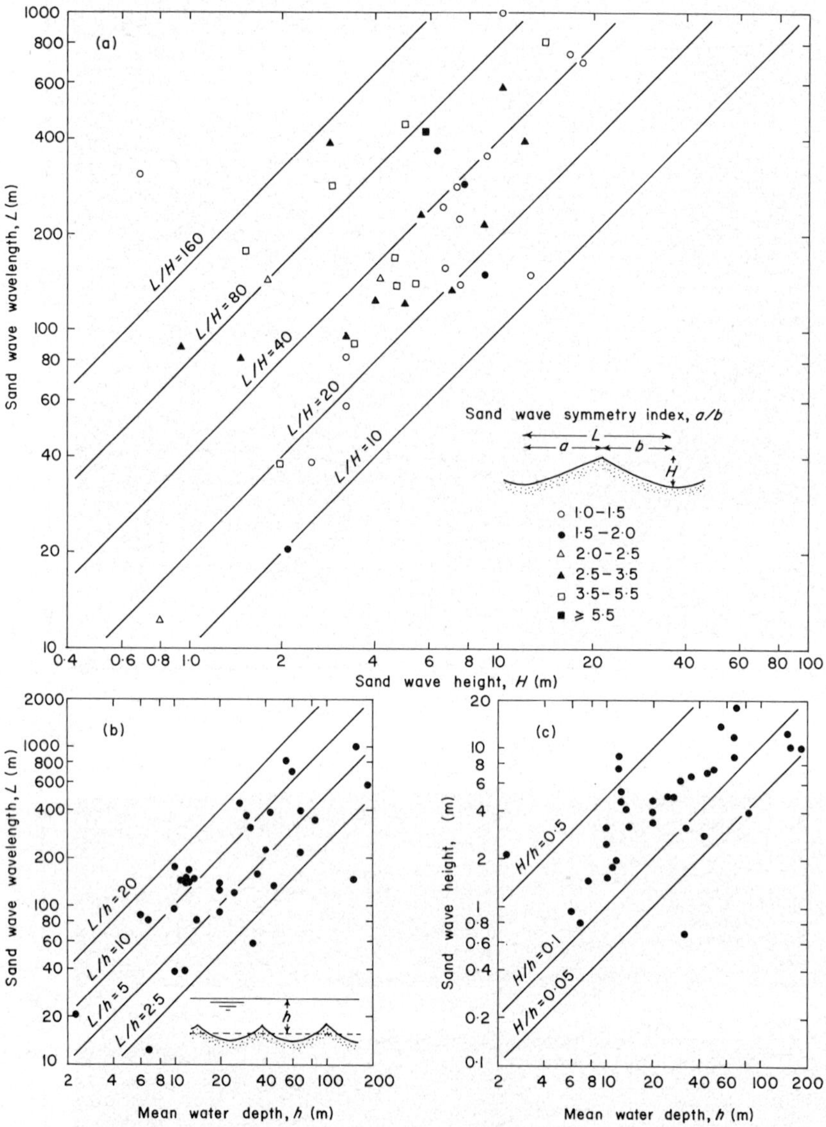

Fig. 11-25. Geometrical properties of tidal sand waves. a. Wavelength as a function of height with the vertical form-index and symmetry index as parameters. b. Wavelength as a function of water depth with the relative wavelength as a parameter. c. Height as a function of water depth with the relative height as a parameter. Based on many sources (Van Veen, 1935, 1938; Cartwright and Stride, 1958; Robinson, 1961; Hinschberger, 1963; Jones et al., 1965; Harvey, 1966; Swift et al., 1966; Volpel and Samu, 1966; Keller and Richards, 1967; Swift and Lyall, 1967; Daboll, 1969; Dyer, 1970; Ludwick, 1970c, 1971; Terwindt, 1971a; Bruun and Vollen, 1972; Boothroyd and Hubbard, 1974; Kirby and Oele, 1975; Bouysse et al., 1976; Bokuniewicz et al., 1977; Bouma et al., 1977a, 1977b, 1978; Hine, 1977).

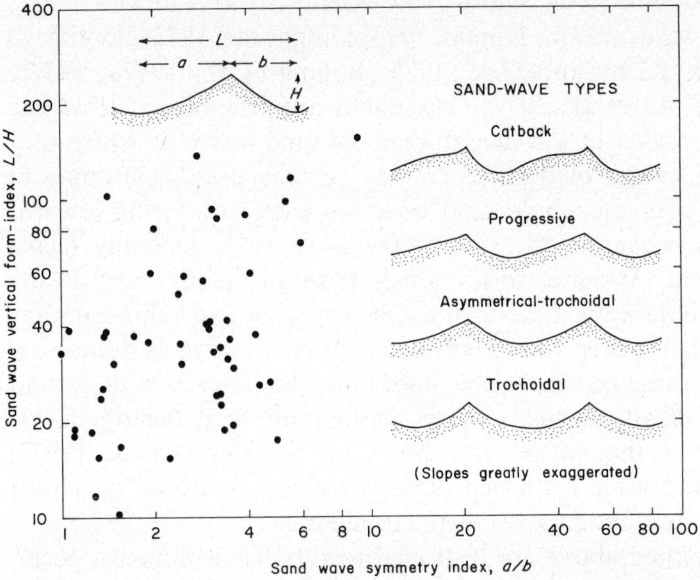

Fig. 11-26. Sand-wave vertical form-index as a function of symmetry index (data set of Fig. 11-25), together with Van Veen's (1935) types of sand wave shown in cross-section.

dividing sand wave profiles between trochoidal, asymmetrical-trochoidal, and progressive types (Fig. 11-26). Trochoidal and asymmetrical-trochoidal profiles are illustrated by Hinschberger (1963), Harvey (1966), McCave (1971a), Terwindt (1971a), Nichols (1972), D'Anglejan (1971), Bouma et al. (1978) and Beiersdorf et al. (1980), though in some cases the troughs are flatter and the crests better rounded than in Van Veen's ideal forms. Progressive sand waves are shown by G.F. Jordan (1962), V.N.D. Caston (1965), McCave (1971a), Langhorne (1973), and Bouma et al. (1977a, 1977b, 1978). Van Veen (1935, 1936, 1938) illustrated as a variant of his asymmetrical-trochoidal form a fourth type of profile, which he called "cat-back", typified by a peaked-up crest flanked on the longer slope of the wave by a flat or depressed area (Fig. 11-26). It is not yet clear if this form can be retained over a long period by a sand wave, or whether it is ephemeral and related to the reversal of the tide over a progressive or asymmetrical trochoidal type.

We cannot emphasize too strongly that sand waves possess low to mild slopes, which invariably are grossly exaggerated in size-scan sonar and echo-sounder records. From the values for the wavelength, height, and symmetry index given in Figs. 11-25 and 11-26, it is clear that the sides of the waves rarely dip more steeply than 10° overall and can slope as little as 1° (e.g. McCave, 1971a; Bouma et al., 1977b). These gentle slopes preclude any general avalanching, but not the superimposition of smaller bedforms (Fig. 11-24). Many workers find what appear to be dunes superimposed on sand waves (Reineck, 1963; J.D. Smith, 1969; Klein, 1970b; Ludwick, 1971; McCave, 1971a; Terwindt, 1971a; Kumar, 1973; Langhorne, 1973; Boothroyd and Hubbard, 1974; Kirby and Oele, 1975; Bouma et al., 1977a, 1977b, 1978; Hine, 1977; Hunt et al., 1977). The dunes are generally at least one order of magnitude smaller in wavelength than the sand waves on which they are clustered. Three modes of occurrence may be recognized: (1) dunes at any one time occur over the whole sand wave, on each side facing towards the crest, (2) dunes occur on one side of the wave only, generally facing towards the crest, and (3) dunes with a single sense of facing occur at any one time over the whole wave. Side-scan sonar surveying and echo-sounding on other sand waves, however, fails to resolve features as large as dunes (e.g. McCave, 1971a). In these instances, the slopes may be covered with current ripples, except for small segments steep enough for avalanching in the immediate vicinity of the crests (e.g. Jones et al., 1965). Hine (1977) describes a small sand wave on which dunes were superimposed only near the crest and when the tidal currents were strongest.

Sand waves as defined above are restricted to tidal environments, occurring in (1) estuarine and shoal areas typified by complexes of banks and channels, (2) restricted shelf seas, (3) shallow-marine platforms and, in rare cases, (4) open continental shelves.

The forms have been intensively studied in a number of northwest European estuarine and shoal areas (Lüders, 1929; Reineck, 1963; C.A.M. King, 1964; Völpel and Samu, 1966; Newton and Werner, 1969, 1970; Göhren, 1971c; Ulrich, 1972, 1973; Ulrich and Pasenau, 1973; C.D. Green, 1975), in the lower St. Lawrence River (D'Anglejan, 1971), and in several east Atlantic estuaries (Daboll, 1969; Ludwick, 1970a, 1970b, 1970c, 1971, 1972, 1973; Nichols, 1972; Kumar, 1973; Boothroyd and Hubbard, 1974; Kumar and Sanders, 1974; Ludwick and Wells, 1974; Visher and Howard, 1974). Ashley (1980) described what are probably sand waves from the tidal Pitt River, British Columbia.

Sand waves on shallow-marine platforms and platform-like areas are described by G.F. Jordan (1962), Stewart and Jordan (1964), Drapeau (1970) and Fillon (1976). Only Knebel and Folger (1976), Hunt et al. (1977), Swift and Freeland (1978), Swift et al. (1978) and Beiersdorf et al. (1980) report the features from open oceanic shelves.

Most accounts of sand waves come from shelf seas more or less restricted by land masses, notably the North Sea and the Celtic Sea in northwest Europe (Van Veen, 1935, 1936, 1938; Cloet, 1954a, 1954b, 1961, 1963, 1980; Cartwright and Stride, 1958; Chesterman et al., 1958; Stride and Cartwright, 1958; Cartwright, 1959; Stride, 1959, 1961, 1963, 1970, 1973; Stride and Tucker, 1960; Hinschberger, 1963; Belderson, 1964; V.N.D. Caston, 1965, 1972; Dingle, 1965; Jones et al., 1965; Vanney, 1965; Belderson and Stride, 1966, 1969; Harvey, 1966; Langeraar, 1966; Hinschberger et al., 1967; Houboult, 1968; Caston and Stride, 1970, 1973; Dyer, 1970; Kenyon and

Stride, 1970; Belderson et al., 1971, 1972; Dobson et al., 1971; McCave, 1971a; Terwindt, 1971a; Bruun and Vollen, 1972; Stride et al., 1972; Langhorne, 1973; Ulrich, 1973; Hails and Kelland, 1974; Kirby and Oele, 1975; Van Weering, 1975; Bouysse et al., 1976; G.F. Caston, 1976; Pendlebury and Dobson, 1976). Other areas of restricted shelf with sand waves are the Inland Sea, Japan (Ozasa, 1974), San Francisco Bay (W.M. Gibson, 1951; Carlson et al., 1970), the White Sea (Belderson et al., 1978), Cook Inlet in the North Pacific (Bouma et al., 1977a, 1977b, 1978, 1980), the Taiwan and Malacca Straits (Keller and Richards, 1967; Boggs, 1974), the Persian Gulf (Cloet, 1954a), the Gulf of San Matias in the South Atlantic (So et al., 1974), Long Island Sound (Bokuniewicz et al., 1977), the Cape Cod area (J.D. Smith, 1969), the Bay of Fundy (Swift et al., 1966; Swift and Lyall, 1967; Klein, 1970b; Dalrymple et al., 1975, 1978), and the Gulf of St. Lawrence (Loring et al., 1970; Loring and Nota, 1973).

Internal structure of sand waves

Few explicit models for the internal structure of sand waves have so far been proposed (Reineck, 1963; McCave, 1971a; Nio, 1976; Hine, 1977), but when attention is given to the apparent controls on these bedforms (Allen, 1980a, 1980b, 1980c), it becomes clear that a wider and more complex spectrum of internal geometries should be expected. Allen (1980c) pointed out that the degree of cross-sectional asymmetry of sand waves increases with increasing asymmetry of the associated tidal currents, just as in the case of wave-related ripple marks (Fig. 11-23), that is, as the unidirectional-translatory component (normally a mass-transport) of the observed current increases relative to the amplitude of the periodic part. An important corollary is that the amounts of bedload transport during opposed tidal semicycles become increasing disproportionate as the velocity-asymmetry grows stronger. Provided that the time-velocity pattern is not too asymmetrical, however, sediment moves as bedload to-and-fro across the sides of the sand wave. There comes a degree of asymmetry, however, when the steeper face of the bed wave becomes so precipitous (about 10° overall) that flow separation (Fig. 1-9) occurs above it, with the result that sand is deposited beneath the separated flow on a long avalanche surface. With such strongly asymmetrical sand waves, the transport of bed-material is predominantly if not wholly in one direction, that of the steeper face. The other control on the internal features of sand waves is provided by the range of velocity values through which the tidal current passes. This range is fairly restricted, however, and dunes seem to be the chief superimposed bedforms, though current ripples and plane beds can at times and in places be found. Figure 11-27 summarizes the sand-wave internal structures deduced on the basis of these controls (Allen, 1980b, 1980c).

Figure 11-27a shows the possible internal structure of a nearly symmetri-

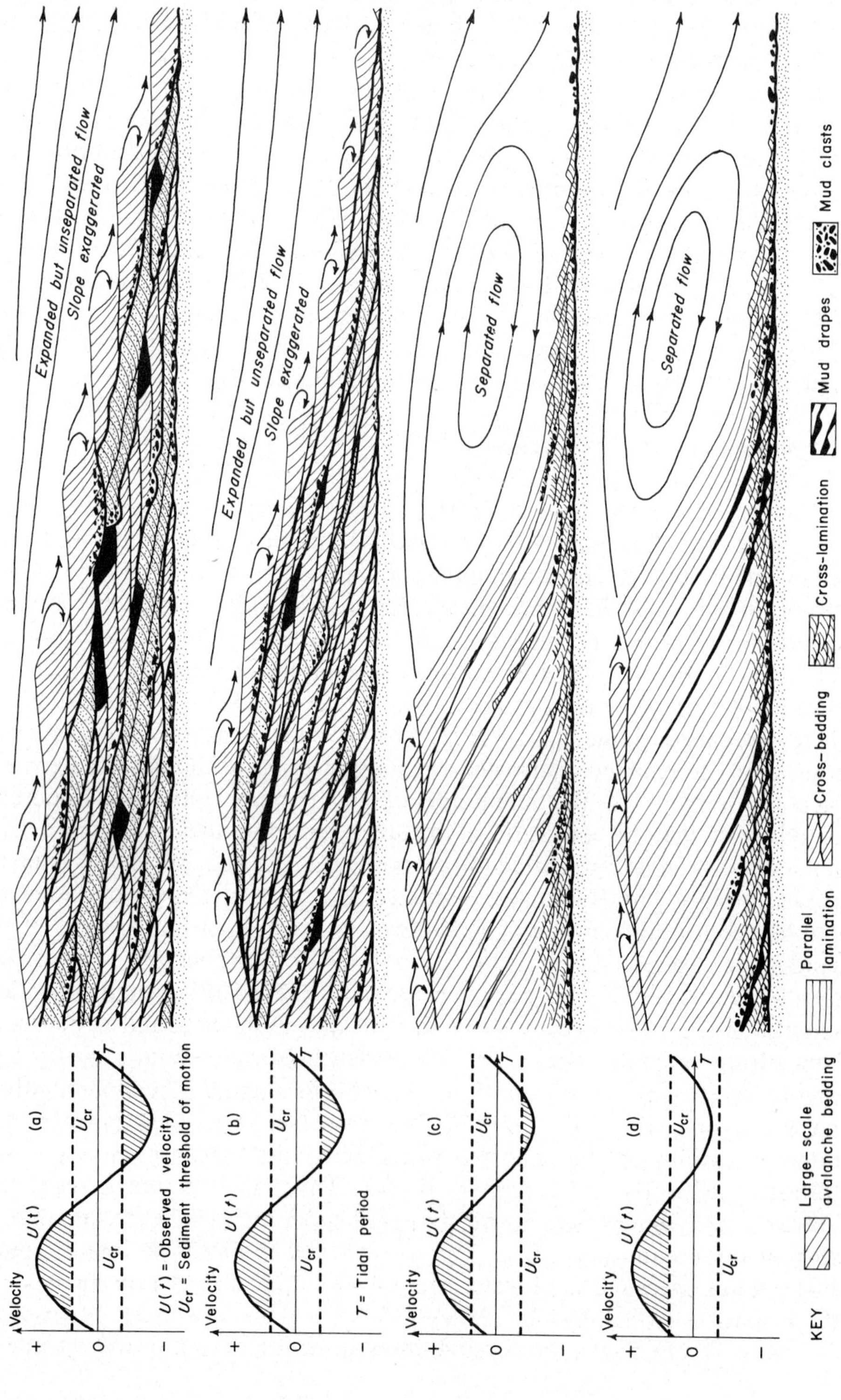

Velocity
+
0
−
U(t)
U_{cr}
(a)
(b)
(c)
(d)
T
U(t) = Observed velocity
U_{cr} = Sediment threshold of motion
T = Tidal period
Expanded but unseparated flow
Slope exaggerated
Separated flow
KEY
Large-scale avalanche bedding
Parallel lamination
Cross-bedding
Cross-lamination
Mud drapes
Mud clasts

cal sand wave shaped by an almost time-symmetrical tidal current, giving approximately equal sediment transports during each flow semi-cycle. The postulated structure somewhat resembles Reineck's (1963) and Hine's (1977) models for a similar sand wave. As with compound cross-bedding sets (Fig. 9-7), the deposit is underlain by a first-order erosion surface and marked by a master-bedding consisting of higher-order erosion surfaces—now complex and composed of intersecting scours—that dip in the direction of the steeper sand-wave face, that is, in the direction of net bed-material transport. Between the master bedding occur cross-bedding sets recording the march of the dunes superimposed on the wave. These sets form "herringbone" arrangements, those facing in the direction of net sediment transport slightly outnumbering the sets of opposite orientation. Mud drapes, formed at times of slack water or neap tides, when bed-material transport is precluded, are locally preserved, either as actual layers or as clasts due to break-up.

The structure shown in Fig. 11-27b is plausible for a slightly more asymmetrical sand wave, over which there is a moderately unequal transport of bed-material. The master bedding, recording the erosion of the steeper sand-wave face during current reversals, is clear and the cross-bedding preserved mainly unidirectional in its facing. Mud drapes and mud clasts derived from their break-up are locally evident.

Figure 11-27c depicts the possible internal structure of a sand wave that is sufficiently asymmetrical, because related to a sufficiently asymmetrical current, that flow separation takes place above the steeper face when this forms the downstream side. Deposition occurs as long foresets which lie discordantly on more gently inclined master bedding surfaces scoured during the re-orientation of the sand wave and the rounding of its crest when the flow reversed. The foresets and the master bedding dip in the direction of net sediment transport, the pattern resembling that found in the compound cross-bedding due to unidirectional flow (Fig. 9-7). Mud drapes may be preserved, either as actual layers amongst the bottomset and lower foreset deposits, or as clasts strewn over the master bedding.

It is possible for a reversing tidal current to be so asymmetrical that bed-material transport occurs during only one tidal semicycle and is therefore wholly unidirectional. A sand wave related to such a current (Fig.

Fig. 11-27. A speculative model for the internal structure of sand waves as determined by asymmetry in the governing currents, interpreted as the ratio of steady to periodic velocity components. Each diagram shows that portion of the sand wave facing in the direction of net bedload transport, that is, in terms of the model, in the direction of the dominant current. In (a) and (b) the slopes are somewhat exaggerated, together with the thicknesses of the packets of sediment likely to arise during each tidal phase. Visibly climbing cross-stratification sets in (a) and (b) are most likely to occur when the bedforms superimposed on the sand wave are current ripples or small dunes. Larger dunes would tend to yield solitary sets arranged in a herringbone pattern.

11-27d) should show internally a series of long foresets, formed when the current exceeded the motion-threshold of the sand, interspersed with mud-draped and bioturbated horizons indicating pauses related to slack water or to neap conditions. Structures indicative of reversed flow (rather than backflow within the separation bubble) are likely to be restricted to occasional discordances amongst the foresets (?exceptionally strong spring tides), though these may instead reflect dune movement under the dominant flow up to the sand-wave crest. McCave's (1971a) model for the internal structure of sand waves has some affinities with the geometries shown in Fig. 11-27c, 11-27d.

Shallow coring in modern sand waves is not yet extensive but lends support to some of these models. The superficial part of a nearly symmetrical form in the entrance to the Jade, Germany, consisted of cross-bedding sets in the order of 0.1 m thick, arrayed either in groups with a similar orientation or, less commonly, in a herring-bone pattern (Reineck, 1963). Laterally restricted mud drapes are recorded from Jade sand-waves (Wunderlich, 1978). Sand waves elsewhere have revealed a similar internal structure (Klein, 1970b; De Raaf and Boersma, 1971; Kumar, 1973; Kumar and Sanders, 1974). Newton and Werner (1969, 1970), however, found small symmetrical sand waves that consisted chiefly of current-ripple cross-laminated sand. Only Houboult (1968) and Kirby and Oele (1975) have cored large sand waves in more open waters. Houboult found mainly thin sets of cross-bedded and often bioturbated sand, consistent with either dune-movement over the waves or the development at times of short avalanche faces at the crests (see Jones et al., 1965). The occasional preservation of parallel-laminated and cross-laminated sand points to the occurrence in places of a still wider range of superimposed bedforms.

Sand-wave deposits have not yet been unequivocally recognized in the stratigraphic record. They are perhaps most likely to occur in thick sandstone sequences with herring-bone or variably oriented and complex cross-bedding (e.g. Klein, 1970a; Swett et al., 1971; Dott and Roshardt, 1972; van der Graaff, 1972; Banks, 1973b; Puigdefabrigas, 1974; R. Anderton, 1976; Eriksson, 1977b; Hereford, 1977; I.C. Rust, 1977; Hobday and Tankard, 1978; Levell, 1980). Marine cross-bedded sandstone sets individually reaching up to 5 m or more in thickness have been ascribed to sand waves (Allen and Narayan, 1964; Pryor and Amaral, 1971; R. Anderton, 1976; Nio, 1976). These units (Fig. 11-28) are relatively simple in structure, having some resemblance to McCave's (1971a) model and according well with either Fig. 11-27c or 11-27d. An intriguing feature of the cross-bedded units in the Folkstone Beds (Lower Greensand), such as that shown in Fig. 11-28, is the way the mud drapes occur in clusters along the length of the set. Allen (1980b, 1981) noticed this feature and speculated that it reflected the spring-neap variation in the strength of tidal currents, a similar observation and idea being reported independently by Visser (1980) to account for cross-bedding with clustered mud drapes in subfossil estuarine sands. It is fair to say, however, that sand-wave deposits in both the modern and stratigraphic records are poorly known, and that much remains to be learned.

Controls on sand waves

Some sand-wave controls have in broad terms been appreciated for many years. Lüders (1929) contrasted the strong asymmetry of sand bars shaped by one-way river currents with the marked symmetry of many sand waves, created by oppositely directed flood and ebb tidal currents of similar magnitude. Van Veen (1935, 1936, 1938) further developed the comparison. Firstly, he inferred that the steeper side of a sand wave faced generally in the direction of the stronger tidal flow and that the degree of asymmetry expressed the relative strength of the flood and ebb currents, the greater the asymmetry of the tidal time-velocity pattern the more asymmetrical the bed wave. Secondly, he appreciated that net sediment transport occurred in the direction faced by the steeper slope of the bed feature.

Later work has largely substantiated these inferences (W.M. Gibson, 1951; Jones et al., 1965; Langeraar, 1966; Samu, 1968; Belderson and Stride, 1969; Ludwick, 1971; Ulrich and Pasenau, 1973). Van Veen's correlation of

Fig. 11-28. Cross-bedding with mud drapes and gently sloping internal discordances, possibly of sand-wave origin, Lower Greensand, West Lavington, Sussex, England. Scale at base of section is 1 m long.

sand-wave asymmetry with velocity asymmetry can now be supported quantitatively, on similar lines to the analysis of wave-current ripples. Figure 11-29 (Allen, 1980c) shows from published sources the sand wave symmetry index as a function of the ratio, calculated from the field data, of the apparent steady current velocity-component to the amplitude of the oscillatory tidal velocity-component. The degree of correlation is surprisingly good, in view of the many evident limitations on the data. An interpretation of tidal asymmetry in terms of the presence of harmonics in the tidal wave may however be preferable (Pingree and Griffiths, 1979). Van Veen's second inference has not yet been directly tested, but it is physically plausible and, as a rule of thumb, is helpful in tracing bed-material transport paths in tidal seas (Stride, 1973).

The controls on other sand wave attributes are less well understood. On the European continental shelf, sand waves are largely restricted to places where the maximum near-surface tidal currents associated with mean spring tides range between 0.65 and 1.30 m s^{-1} (Belderson et al., 1971; Stride,

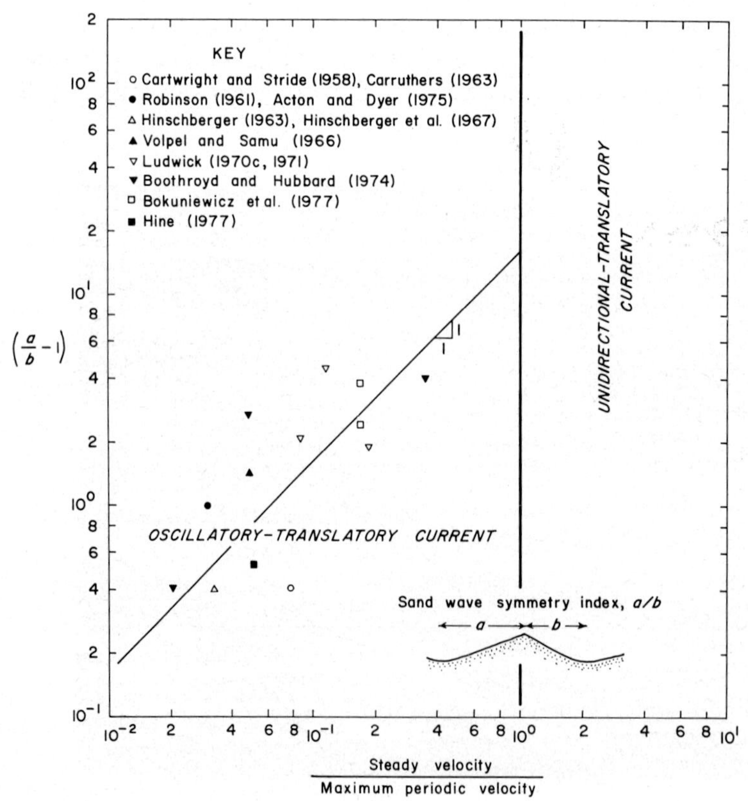

Fig. 11-29. Asymmetry of sand waves as a function of the asymmetry of the governing tidal currents, interpreted as the ratio of a steady velocity to a maximum periodic velocity. Compare with Fig. 11-23.

1973), but these limits may not be universal (Dalrymple et al., 1978). Globally, if not always locally (e.g. Stride, 1970), matured sand waves scale with water depth or with the thickness of the effective flow (Fig. 11-25b, c), the height lying for the most part between 0.05 and 0.5 times the thickness or depth. Wavelength and height may, however, be affected by grain size and current strength. For example, over a long transport path in the Southern Bight of the North Sea, in water of nearly uniform depth, wave height falls by about 40% as the bottom sediment decreases in coarseness to a similar degree and as the tidal currents weaken slightly (Stride, 1970; McCave, 1971a). A small increase of wavelength occurs in the same direction. Locally, sand-wave heights decline substantially when storm waves stir the bottom (e.g. Ludwick, 1971; Terwindt, 1971a). The coarsest debris occurs on the crests of strongly asymmetrical sand waves but in the troughs of symmetrical ones (Terwindt, 1971a; Ludwick and Wells, 1974; Wells and Ludwick, 1974).

But why should sand waves exist at all? Few have attempted this important question, the answer to which should promote a better understanding of these forms.

Cartwright (1959) analysed deep-lying sand waves in the southern Celtic Sea in terms of the action of lee waves (see R.R. Long, 1953a, 1953b, 1954; Yih, 1965), generated as the tidal flow, thermally stratified to a suitable degree, passed over the nearby shelf-edge. The analysis gave plausible results, but two general difficulties bar its wider application: (1) most sand waves occur in water too shallow and well-mixed for much thermal stratification to develop, and (2) there is no known spatial correlation between sand wave trains and the fixed mounds, hollows and steps necessary for the production of lee waves. A further difficulty arises with the sand waves in the southern Celtic Sea. The development during the summer months of the strongest thermal stratification, when lee waves seem to be present in the area (Stride and Tucker, 1960), does not match the times of year (equinoxes) when the tidal currents, assisted by wave action, are likely to have most effect on the bed. The second difficulty is not removed by Cartwright's invocation of stratification due to salt wedges and fresh water plumes, or to concentration gradients of suspended mud, even though a plausible analysis can result (Furnes, 1973, 1974). The case for lee waves as the cause of sand waves does not seem compelling. The only other attempt to account for the forms theoretically is by Yalin and Price (1975, 1976), who extended to tidal flows the Velikanov-Mikhailova bedform model (Chapter 7). This concept presents the same difficulties as before.

Sand waves may have a similar origin to wave-related ripple marks, the tidal wave substituting for wind-generated progressive waves. Although two to four orders of magnitude larger in size, sand waves found in areas of substantially continuous sediment cover are, in plan and profile, remarkably like the simple forms of wave-related ripple. For example, compare Van Veen's (1935) trochoidal sand waves with Inman's (1957) trochoidal ripples

(Figs. 11-24a, 11-4), and progressive and asymmetrical-trochoidal sand waves with Tanner's (1971) wave-current forms (Figs. 11-24b, 11-8, 11-9). Furthermore, oscillatory and oscillatory-rotary tidal currents are analogous to the oscillatory currents generated by ordinary waves, even though the periods and wavelengths involved differ by orders of magnitude. The root question is whether the stationary recirculatory streaming established by Darwin (1884), Sleath (1974b, 1975b, 1976), and Kaneko and Honji (1979a) above a granular bed acted on by comparatively small-period oscillations could also be created by the tidal wave. At present there is neither empirical nor theoretical evidence by which to resolve the question. However, given such a curvature-related streaming, it certainly seems that sand waves could grow up from a flat bed which was randomly wavy on a small amplitude-scale by virtue of the occurrence of a range of grain sizes or of small bedforms (Allen, 1979, 1980a). If this novel interpretation is correct, sand waves should be independent of the ordinary one-way bedforms (Chapter 8), whence models similar to Figs. 8-22 and 8-23 are irrelevant to them (cf. Dalrymple et al., 1978). It is perhaps telling that whereas some sand waves bear superimposed dunes, others seem to carry other bedforms (Newton and Werner, 1969; McCave, 1971a).

EQUANT DUNES

Character and distribution

Several deserts reveal examples of two kinds of large sand dune of broadly equant plan (Fig. 11-30), which seem related to winds of fairly uniform directional effectiveness.

Pyramidal dunes (Holm, 1960), shown in Figs. 11-30a, 11-31 are also called *rhourd* (Aufrère, 1934, 1935), *ghord* or *ghourd* (Capot-Rey, 1943; Capot-Rey and Capot-Rey, 1948; Clos-Arceduc, 1969), or described as peaked (H.T.U. Smith, 1963), star-shaped (McKee, 1966b), or stellate (I.G. Wilson, 1972c). Typically, these dunes rise to 150–250 m above the surrounding deflation hollows, and are regularly spaced 2–4 km apart over wide areas, locally in long chains. Each dune comprises a number of sinuous, sharp-crested, radial sand ridges which culminate in a single summit or in two or three separate peaks joined by high cols. The skirts of the dune slope gently down into the surrounding deflation hollows, where bed-rock may be exposed, but the higher slopes are steeper, with long avalanche faces flanking one or both sides of each radial ridge. Pyramidal dunes are widely present in the Sahara, notably in the Great Western and Great Eastern Ergs (Aufrère, 1934, 1935; Capot-Rey, 1943; Capot-Rey and Capot-Rey, 1948; H.T.U. Smith, 1963; Ambroggi, 1966; Clos-Arceduc, 1966, 1967a, 1967b, 1969, 1972; Hagerdorn, 1971; I.G. Wilson, 1972c; Cooke and Warren, 1973;

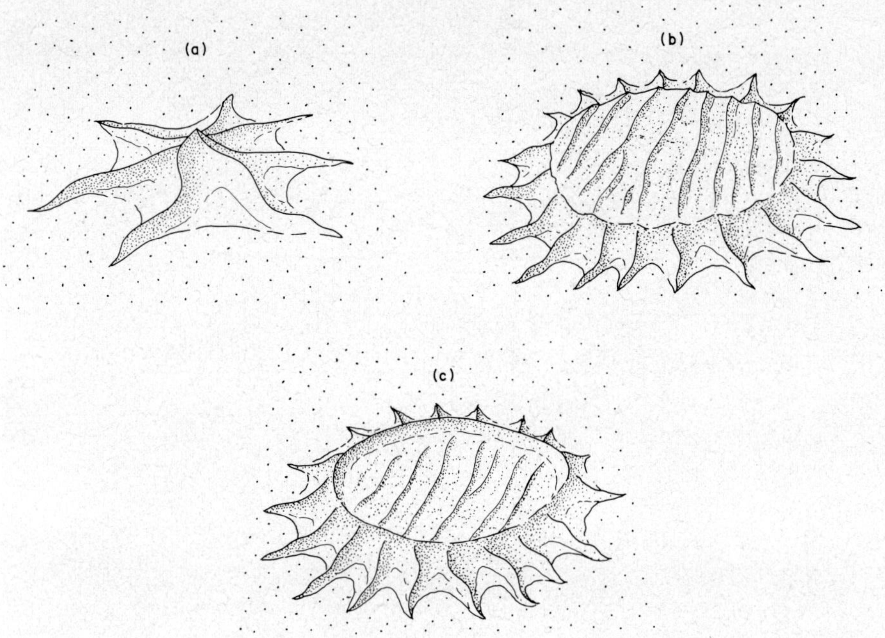

Fig. 11-30. Schematic types of equant aeolian dune. a. Pyramidal or star dune. b. Dome-shaped dune. c. Dome-shaped dune with rampart.

Mainguet and Callot, 1974), as well as in the Najd, Dahana and Rub al Khali deserts of Saudi Arabia (Holm, 1960; McKee, 1966b; Glennie, 1970), and the deserts of southern Russia (Petrov, 1962, 1976).

Transitional from pyramidal dunes are Holm's (1953) dome-shaped forms, together with the closely related "domal dunes of Medusa-head aspect" of H.T.U. Smith (1963) and I.G. Wilson's (1972c) "rounded barchanoid draas" (Fig. 11-30b, c). Dome-shaped dunes are similar in spacing to pyramidal dunes but, with a maximum recorded height of 170 m (Holm, 1953) are somewhat flatter than these forms. One extreme variety (Fig. 11-30b) has a circular, oblong or parabolic plan and a smoothly curved and rather flat-topped vertical profile. Only the forms of parabolic plan are asymmetrical in axial section, the summit lying near the focus. These forms of dome-shaped dune are usually covered by smaller dunes and sand ridges, generally of a uniform orientation but probably subject to seasonal or shorter-term changes. The second extreme variety of dome-shaped dune (Fig. 11-30c) has a relatively flat top bordered by a low rampart rising above steep flanking slopes from which radiate short, locally sinuous ridges. Their tops may show shallow depressions as well as smaller ridges and dunes. Ahlbrandt (1975) has reported small dome-shaped dunes lacking slip-faces from the edges of an inland dune-field in Wyoming.

Dome-shaped and pyramidal dunes are closely associated spatially (e.g.

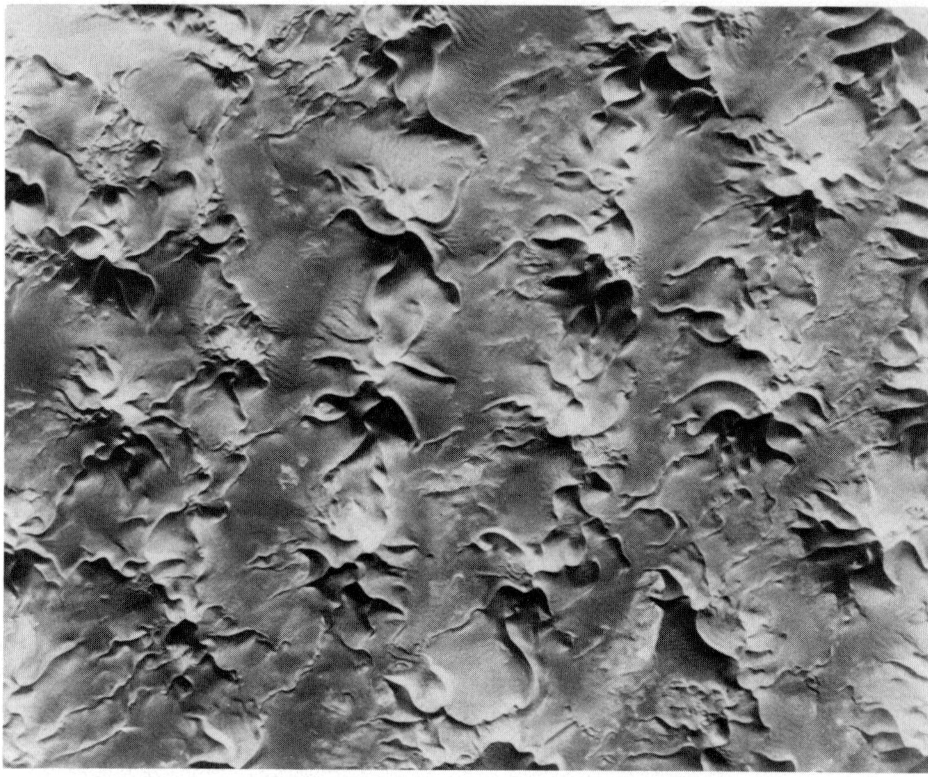

Fig. 11-31. Air photograph (8×10 km) of equant sand dunes, Grand Erg Oriental, Sahara.

Clos-Arceduc, 1967b; I.G. Wilson, 1972c). The former occur in the Sahara (H.T.U. Smith, 1963; Clos-Arceduc, 1969; I.G. Wilson, 1972c) and in many of the Saudi Arabian ergs (Holm, 1953).

Only McKee (1966b) has dissected equant dunes in hot deserts, and that on two small pyramidal forms. Avalanche-formed cross-strata dipping in three principal directions—northwest, northeast, and southeast—were found in the larger dune, which comprised four radial ridges culminating in a single summit. The smaller dune contained laminae inclined in many different directions but of lower dip.

Controls on equant dunes

It is generally believed that equant dunes in the desert are generated by effective winds which blow more or less equally from many different and opposed directions, though a lack of detailed meteorological data means that reservations must remain. Pyramidal dunes are particularly widespread and well-developed in the Great Western and Great Eastern Ergs of the Sahara, where a complex wind regime prevails, the Westerlies blowing from the west and south during the winter being replaced in summer by winds from the north, east and south (Dubief, 1943, 1952). A similarly complex regime marks the Saudi Arabian deserts where equant dunes occur (Holm, 1960). Here the Westerlies are interspersed with the *shamal* winds, which issue from the northwest, north and northeast, and the monsoonal winds blowing often hard from southerly points. Such wind regimes belong to our random class (Fig. 11-1). It is not surprising that McKee (1966b) found such complex structures in the equant dunes he studied.

The often remarkably regular spacing of the dunes may be comparable with the average excursion of a sand grain over the duration of the effective wind from each compass point, the regime of such winds being directionally uniform. Although travelling in different directions at different times, grains should in the long term under this regime tend to accumulate at fairly regularly spaced nodes (perhaps selected from amongst the larger available roughnesses), which should thereby assume a slightly higher elevation than the surrounding deflated areas. Once established, these nodes should be self-perpetuating and stationary, since any leeward surface would check the transport of sand and all slopes surrounding a node would at various times play this role. Dome-shaped dunes may be a transitional stage in the build-up of sand to form a pyramidal dune at a node. Their ramparts (Fig. 11-29c) certainly suggest a uniform inward movement of sand.

If this idea has substance, the calculated particle excursion should agree with the observed dune spacing. In the Algerian Sahara (Dubief, 1943), sand-driving winds blow from any given compass point for about 8 hours, with a typical speed of 10 m s^{-1}. Since the average speed of wind-driven grains is approximately one-ninth that of the wind (Bagnold, 1966), the characteristic grain excursion becomes 3.2 km, which is comparable with the dune spacing. Clos-Arceduc (1966, 1967b) also related equant dunes to accumulation at nodes, but in response to supposed standing waves in the lower atmosphere.

SUMMARY

Wave-related ripple marks, sand waves, and equant dunes in the desert are bedforms of high symmetry which are related, respectively, to oscillatory, oscillatory-rotary, and random current regimes. Wave-related ripples, and perhaps also sand waves, grow up because of a stationary recirculatory fluid streaming motion caused by the reversing current in the presence of bed roughnesses which impart a waviness to the bed. The wavelength of wave-related ripples is comparable with to smaller than the excursion of sediment particles driven over the bed, but is a complex function of wave characteristics and water depth, and so cannot afford a unique palaeohydraulic interpretation. Both wave-related ripple marks and sand waves increase in asymmetry of cross-sectional form as the governing currents increase in time-velocity asymmetry, that is, as the unidirectional-translatory velocity-component (attributable largely or wholly to mass-transport) increases relative to the amplitude of the periodic part. Whereas the internal structure of wave-related ripple marks is relatively simple—laminae inclined in the direction of ripple migration and net sediment movement even when the bed features are externally symmetrical—the internal facies geometry of sand waves appears to be complex. Nearly symmetrical tidal currents could lead to herringbone arrangements of cross-bedding sets subordinate to a master bedding and layering inclined in the direction of net bed-material transport. Large foresets associated with mud drapes and/or a more gently inclined master bedding may be characteristic of the more strongly asymmetrical sand waves related to relatively asymmetrical tidal currents which separate at sand-wave crests.

Equant dunes in the desert appear to depend on an essentially random wind regime, their horizontal spacing agreeing with typical particle excursions calculated for sand-driving storms.

Chapter 12

RIPPLES AND DUNES IN CHANGING FLOWS

INTRODUCTION

It was previously shown that laboratory experiments and mathematical analyses performed under quasi-steady and quasi-uniform flow conditions typically reveal unique relationships between many bedform attributes and flow properties. The insights obtained by these means have powerfully influenced the ways in which sedimentary structures have been used to deepen an understanding of sedimentary systems both of the present and the past. But are dynamical conditions in natural environments actually constant in space and/or time? Since they are patently not constant, to what extent are we justified in interpreting the sedimentary structures of real systems by the unique relationships of unchanging dynamics? Misgivings are at once prompted by the reflections of geomorphologists who, working in the context of the rapid and severe climatic fluctuations of the Quaternary, have long had a strong sense of the effects of change in the operation of real process-response systems (e.g. Chorley and Kennedy, 1971). Where a process is time-dependent, the response also changes, but usually after some delay and usually with a substantial blurring of the unique dependence that would have been ascertainable had the response been taken to equilibrium under a series of contrasted steady states. The response in fact seems "aware" of its history or endowed with "memory", for in quality and quantity it recognizably resembles what had been present at the start of the period of change. Stated non-animistically, its behavioural mode involves hysteresis or relaxation.

Hydraulic engineers have been aware for several decades of some of the effects of changing regime on river bedforms (Lane and Eden, 1940; Carey and Keller, 1957; NEDECO, 1959). Although Simons et al. (1965b) and Neill (1969) emphasized the great differences between flow in rivers and in laboratory flumes, and Simons and Richardson (1962) convincingly showed experimentally the effects on bedform development of discharge fluctuations, it is only recently that engineers have begun seriously to study bedforms in changing flows, either in the field (Stückrath, 1969; Nasner, 1973, 1974) or in the laboratory (Dillo, 1960; Bayazit, 1969; Jensen, 1973; Gee, 1973, 1975).

Sedimentologists have been just as tardy in grappling with the problems raised by the changeability of natural processes. However, dunes and cross-bedding may generally be absent from clastic turbidites either because there is insufficient time for their development beneath the rapidly changing turbidity current, or because of grain size effects (Walker, 1965; Hubert, 1966a, 1966b; Walton, 1967; Allen, 1969c, 1970f). Bedforms in tidal environments clearly display characteristics and behaviour that are peculiar to a highly changeable regime (Sternberg, 1967; Klein, 1970b; Klein and Whaley, 1972; Knight, 1972; Boothroyd and Hubbard, 1974; Allen and Friend, 1976a, 1976b), and a similar peculiarity may mark the fluvial features (R.G. Jackson, 1976a). Karcz (1972) suggested that rapid flow changes could lead to abandoned temporal sequences of bedforms being mistaken for hierarchies of simultaneously active forms. Rapid flow changes almost certainly contribute to the directional variability of sedimentary structures (Allen, 1967a; Collinson, 1971). In some natural environments, bedform attributes take double-valued relationships to flow conditions, in apparent defiance of theory (Allen, 1973c, 1974b; Allen and Collinson, 1974). Superimposition and polymodality of similarly shaped bedforms are common where hydrodynamic change is the rule. Do the modes represent hydrodynamically distinct bedform classes (Boothroyd and Hubbard, 1974; R.G. Jackson, 1976a), or do they record a series of points within a single class, each modal cluster arising at a particular time in the cycle of flow changes (Allen and Collinson, 1974)?

The study of sedimentary structures in changing flows is in its infancy. Hence we shall aim chiefly to assemble the scattered and at times contradictory observations of relevance, on the one hand at the level of the population of forms and, on the other, at the level of the supposedly representative individual feature. As yet few general principles are visible.

WHAT IS CHANGING?

The currents acting in natural sedimentary systems vary in strength with distance at a fixed time, and change in vigour and direction with time at each station. Hence they are to degrees unsteady, non-uniform, and multidirectional.

Non-uniformity in rivers is related mainly to the alternation of riffles and pools along the thalweg. In perennial streams, unsteadiness and multidirectionality depend chiefly on the annual or semi-annual flood, though diurnal effects and the shifting of large bed features may be important locally. Change in ephemeral streams is controlled by aperiodic storm-dependent floods, each generally of short duration. Tidal conditions alter on several periodicities simultaneously, those of the semi-diurnal or diurnal tide, the spring-neap cycle, and the equinoctial cycle. Tidal flows are substantially unidirectional only under rare circumstances. In restricted waters an asymmetrical oscillatory motion is the rule, but in open seas typically the current is oscillatory-rotary. Non-uniformity in tidal areas derives from the presence of channels and shoals. Environments directly affected by the wind or subject to wind-waves are under diurnal and/or annual periodic control. They are also influenced by individually aperiodic but seasonally clustered storms. The multidirectionality of such environments can be extreme when all winds are taken into account, but considering only the prevailing winds, it may not be marked. Virtually nothing is known by direct observation of the changeability of deep-water turbidity currents; gross unsteadiness may be inferred from the experimental evidence, and multidirectionality could be introduced as the result of "sloshing" between confining basin walls or the action of Coriolis force.

DUNE POPULATIONS IN UNSTEADY FLOWS

General

What happens to bedforms in these changeable regimes, and what internal structures are produced, may be analysed at two levels, that of the *individual* form, and that of the *population* of related features. Consider aqueous dunes, which have been studied at both levels.

The individual dune may be characterized in terms of its breadth, local wavelength, local height, characteristic wavelength, characteristic height, cross-sectional shape, and internal structure, as well as in other ways. The effects of changeable regimes on individual dunes is already partly known, though no investigator has so far troubled to establish the representativeness of these samples of one.

Although sedimentologists are used to the concept of population in statistics, the idea of a population of bedforms, which we shall exemplify using dunes, is perhaps less familiar. Taking a leaf from the ecologist's book (e.g., Allee et al., 1949; M. Williamson, 1972), a population of dunes comprises all those dunes existing in a defined place at a stated time. The place can be on a river bed, or a tidal sand shoal, or whatever, and it may be defined either one-dimensionally (linearly) as a sampling line (e.g. echo-sounding line) or two-dimensionally (areally) as a region marked by buoys or stakes. The time may be either an instant or some suitable interval. We may characterize the population in many ways, by the use of, for example, the mean local dune height, mean local wavelength, standard deviations of height and wavelength, mean breadth/wavelength, and mean dune age, amongst other attributes. Whilst a dune population shares certain characteristics with the individual form, it possesses others, which may be distinguished as group attributes, that are exclusive to it. For example, the population has a certain spatial density, defineable either linearly or areally, and may vary in composition and character because of emigration, immigration, creation or destruction of individual dunes. What is here said of dunes is also true of other kinds of bedform occurring in groups. How do such populations change in the flow of time?

Dunes in the River Congo near Boma, Zaire

Peters (1971) outlined the behaviour during 1968–1969 of an areally defined dune population on the sandy bed of the R. Congo, in a part of the river between the Amont Bank and the Tortues Archipelago, downstream from Boma (Fig. 12-1). The river varies in discharge over a broadly three-fold range, with a semi-annual flood, the highest flows occurring generally in December. In mid-September, shortly after the lowest water was recorded, the dunes were regular, smooth-backed, and 100–200 m in wavelength. As the stage rose, small dunes (incorrectly called antidunes by Peters) appeared on the backs of the original features, gradually by their movement and growth obliterating the large forms. At peak discharge, in December and

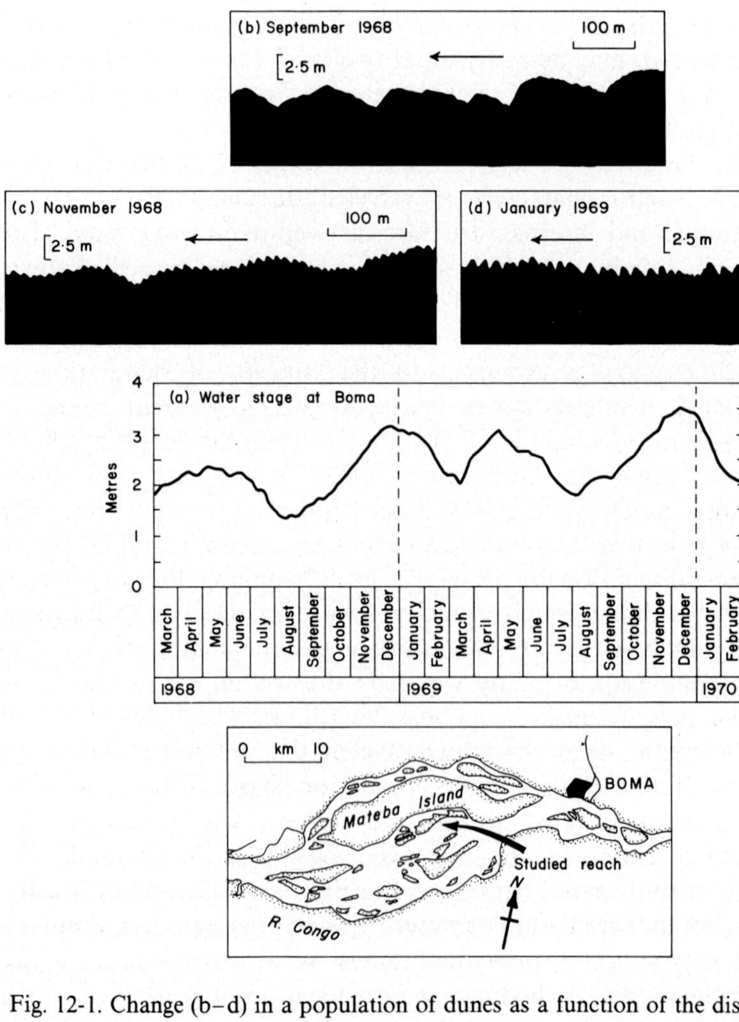

Fig. 12-1. Change (b–d) in a population of dunes as a function of the discharge (a) of the R. Congo. Data of Peters (1971).

January, the river bore little or no trace of the original large dunes and, over large areas, of the small dunes which subsequently appeared. Dunes were present, however, but they were intermediate in size and were presumably evolving towards larger dimensions still. Thus an increase of discharge is attended by dune superimposition and by the delayed evolution of the population in response to changing flow.

Dunes in the Fraser River, British Columbia, Canada

By daily echo-sounding during the flood peak of June, 1950, Pretious and Blench (1951) measured the characteristics of four linearly defined dune populations sited within 3.2 km of each other in a relatively uniform reach of the river close to Vancouver (Fig. 12-2). Each population occupied a fixed portion of the river bed approximately 608 m long. Flood stages last from May to August, and are commonly divisible between two periods of exceptionally large discharge and rapid discharge change.

The four populations are comparable in dune density and change simultaneously in density in response to the changing freshwater discharge. Initially, the dunes are relatively small, though locally there are signs of degraded larger forms. With increasing flow the dunes grow in size, reaching their least density several days after peak discharge. It will be seen (Fig. 12-2) that the average dune wavelength (inversely proportional to the dune density) is in each case a double-valued function of discharge, with an anticlockwise trajectory. As the recession advances, small dunes rather abruptly appear on the backs of the larger forms which, presumably, are then slowly degraded

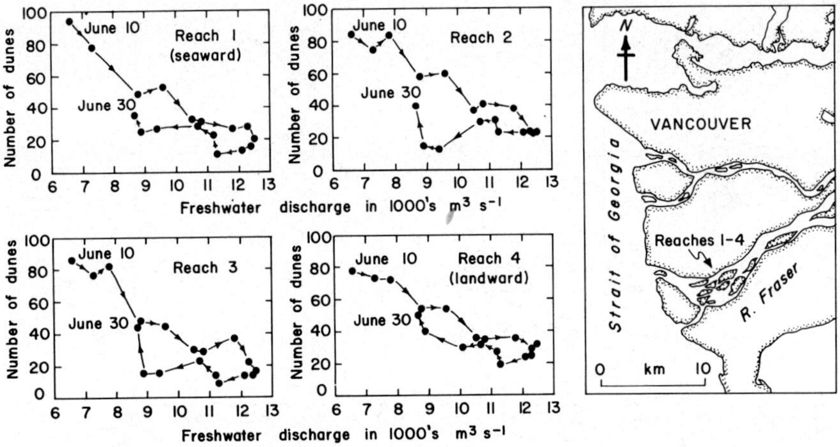

Fig. 12-2. Phase diagrams illustrating change in a population of dunes as a function of the discharge of fresh water, Fraser River, British Columbia, Canada. Data of Pretious and Blench (1951).

until the bed has an appearance similar to that at the beginning of the survey.

Intertidal dunes in the Gironde Estuary, France

The two preceding examples illustrate the delayed response of dunes to a changing regime in terms of either dune density or mean wavelength, its close relative. However, G.P. Allen et al. (1969; see also G.P. Allen et al., 1970, 1971) measured the mean height of dunes every few days during April–May, 1969, on the Banc de Plassac, an intertidal sand shoal approximately 35 km north of Bordeaux. The data were originally presented as time series for dune height and for the tidal coefficient (*coefficient de marée*) which, in French hydrological practice, is the local tidal amplitude relative to the minimum (*coefficient* = 20) and maximum (*coefficient* = 120) possible local values. G.P. Allen and his associates noted that dune size showed a delayed response to changes of flow, which was confirmed by a plot (Fig. 12-3) of mean dune height against the tidal coefficient (Allen, 1973c). Height is a multivalued function of tidal conditions, the graph consisting of loops of anticlockwise trajectory. The first spring-neap cycle (April 25–May 9) involves relatively large tidal ranges, and the loop is comparatively narrow. In the second cycle, beginning on May 9, the tidal ranges are comparatively small, and the loop is apparently more circular in form (the run of data is incomplete).

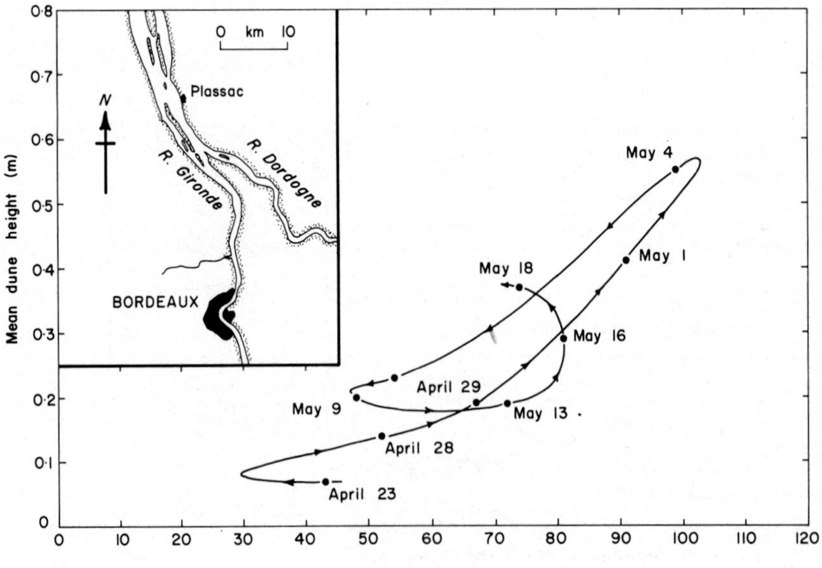

Fig. 12-3. Phase diagram illustrating change in a population of dunes as a function of tidal range (tidal coefficient), Banc de Plassac, R. Gironde, France. Data of G.P. Allen et al. (1969).

Dunes in the River Weser, Germany

An important and impressively long record of the delay in the response of dune populations to changing flows was recently assembled by Nasner (1973, 1974). He measured between 1966 and 1972 the mean height and mean wavelength of four linearly defined populations, at sites (reaches) strung out over almost 20 km in the river below Bremen (Fig. 12-4). The Weser has broadly equal flood and low-water seasons and varies in fresh-water discharge over an approximately four-fold range; the bed material is chiefly medium to coarse sand.

Nasner gave his data as time series, but by plotting mean wavelength and mean height as functions of discharge, Allen (1976d) showed that the four populations altered broadly simultaneously in response to the changing flow and differently for height than wavelength. Although the period of Nasner's observations was marked by a long-term trend of discharge, each property tended to vary on a similar pattern from one flow cycle to the next (Fig. 12-5). As in previous cases, both height and wavelength bear generally double-valued relationships to discharge. Whereas the wavelength loop has an anticlockwise trajectory, that for height can only be traced in a clockwise

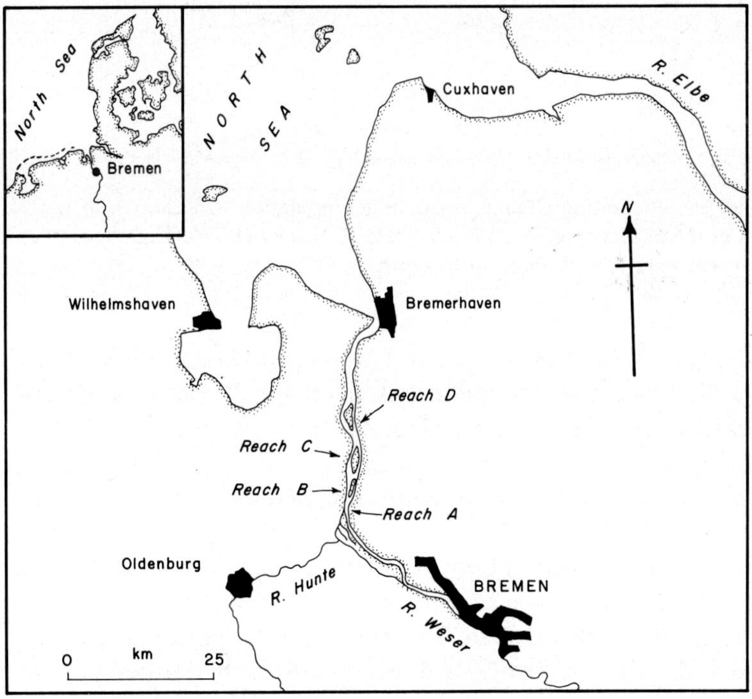

Fig. 12-4. Distribution of reaches in the tidally influenced portion of the R. Weser, Germany, at which Nasner (1973, 1974) surveyed dune populations.

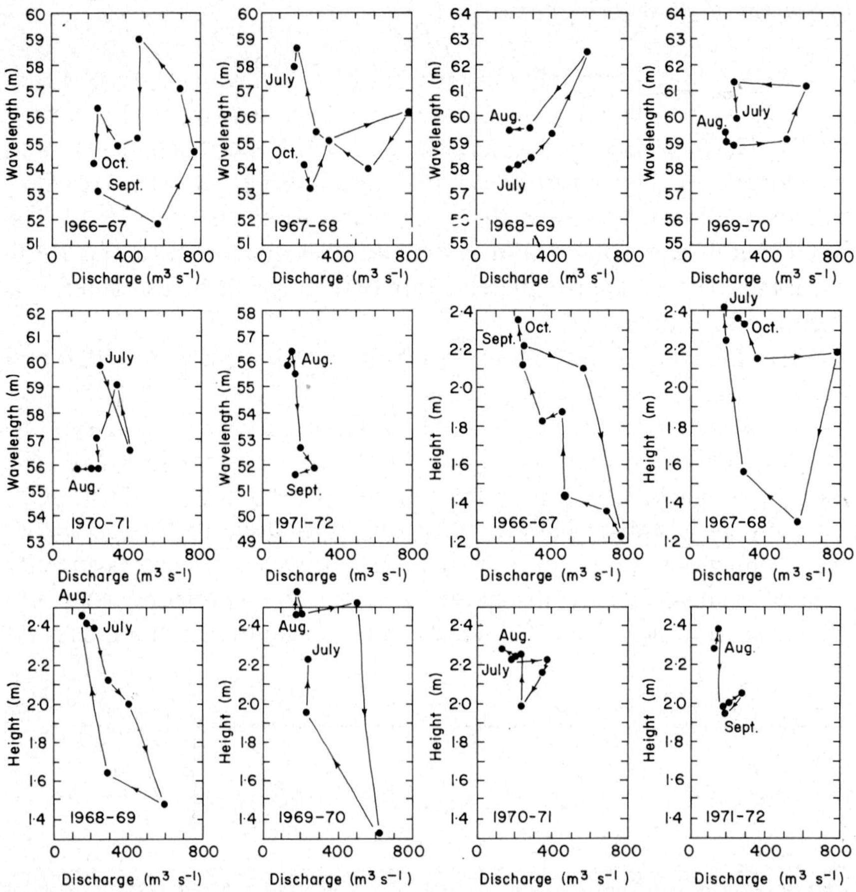

Fig. 12-5. Phase diagrams illustrating change in the dune population in Reach A of the R. Weser, as a function of the discharge of fresh water. Data of Nasner (1974). The dunes in the other reaches vary synchronously with those in Reach A.

direction, as is also true for dunes in the Rio Paraná (Stückrath, 1969; Allen, 1976a). In the latter respect, the populations from the Weser and Paraná differ from the assemblages in the Loire (Fig. 12-3).

Intertidal dunes at Wells-next-the-Sea, Norfolk, England

As an example of a complex dune population, responding in several different ways to hydraulic changes, consider Allen and Friend's (1976a, 1976b) observations at Wells-next-the-Sea (Fig. 12-6). Here, in an ebb-dominated channel, the characteristics of dunes in a linearly defined population were measured at low-tide over 68 consecutive semi-diurnal tidal cycles (August 15–September 19, 1973). Major dunes (long crests, large wavelength) and minor dunes (small wavelength, lunate crests) were recognized.

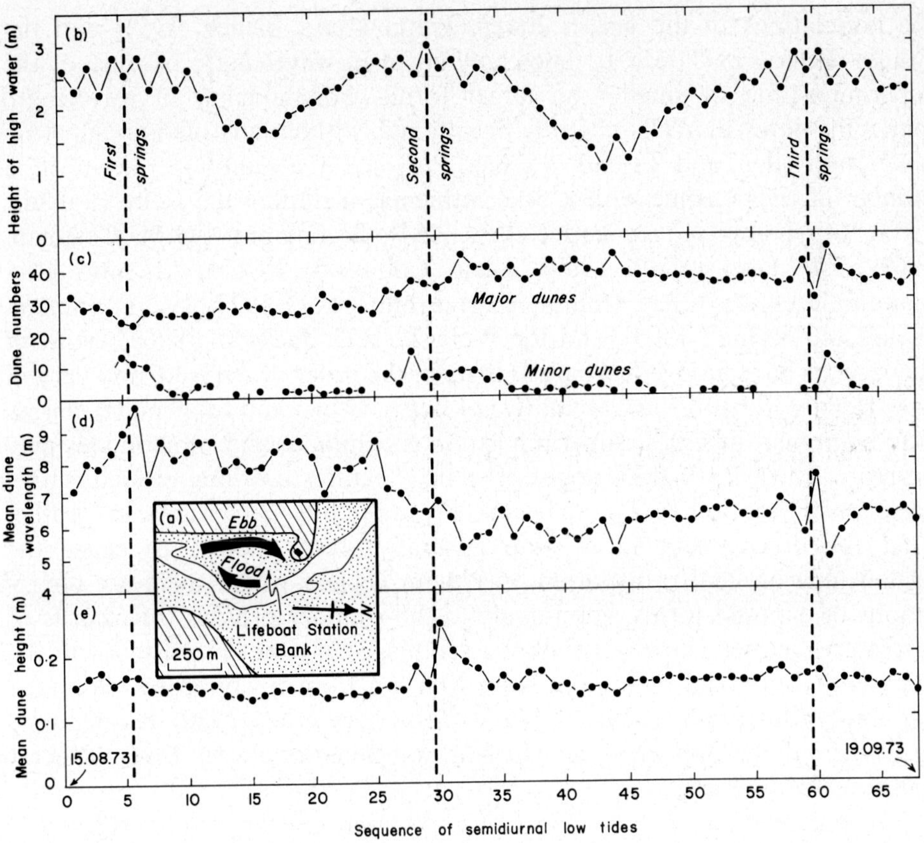

Fig. 12-6. Time-series illustrating changes in a dune population as a function of tidal height, ebb-dominated channel, Lifeboat Station Bank, Wells-next-the-Sea, Norfolk, England. Data of Allen and Friend (1976b).

The minor forms occurred scattered on the backs of the larger features, and were frequent only during spring tides. The population of major dunes was unimodal in height but generally bimodal in wavelength, making it necessary to distinguish short-wavelength (wavelength approximately 3.5 m) and long-wavelength (wavelength approximately 9 m) sub-populations. The mean wavelength of the population of major dunes was sharply changed during the middle set of spring tides, which included the largest astronomical tides of the year at the site, and showed little evidence of readjustment to subsequent lesser tides. However, the mean height, while increased significantly by the middle springs, readjusted fairly quickly to the succeeding weaker tides.

Polymodality and dune superimposition

The above emphasis on statistical population attributes should not distract attention from the structural features of dune assemblages. As noted, the populations in the Fraser River (Pretious and Blench, 1951) and the Congo (Peters, 1971) are at times polymodal in wavelength, because of the superimposition of smaller on larger forms. Polymodality of wavelength marks the dunes at Wells-next-the-Sea, though without any obvious superimpositions (Allen and Friend, 1976a). Polymodal assemblages occur in a number of rivers, some with a tidal influence, including the Loire (Ballade, 1953), Mississippi (Carey and Keller, 1957), Brahmaputra (J.M. Coleman, 1969), Red Deer (Neill, 1969), Tana (Collinson, 1970b), Missouri (Annambhotla et al., 1972), Columbia (Haushild et al., 1973, 1975), Yamuna (Singh and Kumar, 1974), and the Wabash (R.G. Jackson, 1976a). Most of these rivers have a large discharge range, in the order of ten fold, and vary in flow rapidly. Features representative of between two and four modal classes may be present at times; superimposition is common, and some modes may be represented at all flow stages. The bed features are long-crested, transverse in orientation, and unquestionably dune-like in streamwise profile. That the downstream faces were generally steep enough to cause the avalanching of sand transported over them is clear where there are photographs of exposed forms, good-quality echo-sounder profiles, or records of internal structure. These polymodal assemblages may be contrasted with the apparently unimodal dune populations known from rivers of low discharge variability (Stückrath, 1969; Nasner, 1974). Levey et al. (1980) also report a case where dunes appear to vary little in wavelength between times of flood and average discharge.

The polymodality is the subject of conflicting interpretations. Allen and Collinson (1974) argued from the echo-sounder records of Pretious and Blench (1951) that the different modal classes should all be classified as dunes hydrodynamically, mainly because assemblages of short-wavelength forms seemed able during the rise to evolve gradually into populations of long-wavelength features. In contrast, R.G. Jackson (1976a) regarded modal classes found in the River Wabash as hydrodynamically distinct, partly on slender hydraulic grounds, and partly because of the prevalence of superimposition. A similar view to this is explicit or implicit in the majority of classifications advanced for polymodal assemblages of transverse features (J.M. Coleman, 1969; Klein, 1970b; Reineck et al., 1971; Klein and Whaley, 1972; Boothroyd and Hubbard, 1974; Singh and Kumar, 1974). The available field evidence cannot resolve the issue, but the results of numerical experiments on dune populations (discussed below), suggest that polymodality may depend on a mode of self-regulation in dune assemblages formed by periodically varying flows. At large discharge ranges, dunes are created in "bursts" during particular, regularly recurrent ranges of flow stage. Since the individual dunes are long-lived, and with strongly conserved attributes, the assemblages during the whole or parts of each flow cycle become polymodal. The analysis by Allen and Collinson (1974) implies a parallel self-regulation in natural dune populations.

Water stage and bed roughness during changing flows

The preceding summary implies that in river and tidal channels where dunes abound, bed roughness may be a function of time because the discharge, interacting with the loose boundary to make the roughness elements, is itself time-dependent. Evidence supporting this comes from dune-bed rivers, in the form of loop-shaped gauge height-discharge or rating curves (e.g. Carey and Keller, 1957; NEDECO, 1959, 1961; Beckman and Furness, 1962; P.R. Jordan, 1965; Stückrath, 1969). The gauge generally reads lower on a rising river than during the recession, for the same discharge value. However, the upward shift of gauge height could represent deposition in the channel during recession, and not an increase of depth because of increased roughness.

Convincing proof that rating loops are due to a time-dependent channel roughness comes from the at-a-station variation of mean flow depth and mean velocity with discharge in the San Juan and Colorado rivers and the Rio Grande (Leopold and Maddock, 1953). The relationships are double-valued, like the curves for the dune populations, the depth-discharge loop taking an anticlockwise trajectory and the velocity–discharge curve a clockwise path. Hence for the same discharge, the flow is deeper and slower during recession than rise. Since the water-surface slope changes but little, the bed may be presumed to be roughest during recession. At least in the Mississippi, the dunes are then at their largest (Carey and Keller, 1957). Most flow properties of rivers seem to be connected together by non-unique functions (Leopold and Wolman, 1956; Beckman and Furness, 1962).

Double-valued relationships between mean depth and discharge were obtained experimentally by Simons and Richardson (1962), who admitted a time-dependent discharge simulating a flooding river to a sand bed in a straight flume. Loops of various shapes were found, depending primarily on the channel slope and discharge-range selected (hence the bedform existence-fields spanned), but all could be explained by the delayed adjustment of the bedforms to the changing flow. Dunes afforded the greatest delays. These observations may be rationalized by recalling that, since the early studies by Einstein and Barbarossa (1952), the total hydraulic resistance to flow over a mobile bed is divisible between: (1) surface resistance (grain roughness), and (2) form resistance arising from gross features, for example, dunes (Alam et al., 1966; Engelund, 1966; Simons and Richardson, 1966; Vanoni and Hwang, 1967; Bayazit, 1969). We may write:

$$f = f_1 + f_2 \qquad (12.1)$$

$$f_1 = \text{function}\left(\frac{4Uh}{\nu}, \frac{D}{4h}\right) \qquad (12.2)$$

$$f_2 = \text{function}\left(\frac{H}{4h}, \frac{H}{L}\right) \qquad (12.3)$$

where f is the total Darcy-Weisbach resistance coefficient, f_1 the grain-roughness coefficient, f_2 the form-roughness coefficient, U the mean flow velocity, h the mean depth, ν the kinematic viscosity of the fluid, D the mean sediment diameter, H the mean bedform height, and L the mean bedform wavelength. The bedform resistance laws are not well known, but it seems likely that, for a constant bedform spacing, f_2 will increase fairly strongly with increasing bedform height. By analogy with sand roughness, an increase of bedform height relative to flow depth should increase the total resistance. Hence where dunes show a delayed response to a reduction of flow, an increase of resistance and of depth may be expected. In practice, it may be necessary to add contributions to total resistance arising from flow acceleration and deceleration over the wavy bed, and the breaking of surface waves (Simons and Richardson, 1966). However, these contributions may not be great for subcritical flows.

BEDFORM POPULATIONS AND DYNAMICAL SYSTEMS THEORY

Outline of theory

We have now studied dune populations which vary in their quantitative attributes with time as the governing currents change in magnitude with time. The flows involved in our examples are comparatively simple, most occurring in relatively uniform channels and, although unsteady, are either actually or effectively unidirectional. Even so, the patterns of change are so complex, though orderly, that there is commonly little or no hint as to what might be the form of the experimental or theoretical function connecting each attribute with flow conditions given steady-state equilibrium. In particular, the field relationships are double-valued rather than unique, invariably with evidence of the delayed adjustment of the population attributes. Considering a particular population, the degree of lag seems to depend on which attribute we select (compare dune wavelength and height in the Weser) and, considering a single attribute (say, dune height, Banc de Plassac), which particular set of flows affected the population. Systems changing in time, as exemplified by dune populations which interact with tidal or river flows, are dynamical systems and their physical behaviour may be analysed or modelled using the mathematical framework and tools of dynamical systems theory. Rosen (1970) and May (1973) give lucid introductions to this theory as it relates to biology; there are several parallels between the structure and functioning of assemblages of periodic, space-filling bedforms, and the character and working of biological populations generally.

Any dynamical system may be completely characterised at a selected instant by the measurement at that time of the values of an appropriate finite number, n, of *state variables* $x_1, x_2, \ldots, x_n$. The ordered n-tuple of numbers resulting from the measurement of the n state variables define a position within the *phase space* in which the system functions. For each state variable may be regarded as specifying one of the n axes of a rectangular coordinate system in n-dimensional Euclidean space. In our examples, the state variables would include population attributes, for example, mean dune wavelength and mean dune height, together with flow properties, such as discharge. The dynamical description of the system is complete if, with t as time, the form of the functions $x_1(t)$, $x_2(t)$, ..., $x_n(t)$ is known. The behaviour of the system can then be traced as a curve, or *trajectory,* within the system phase space. The graphs of Fig. 12-5 exemplify such curves, which may be called *phase diagrams,* though these particular ones incompletely represent the phase space of the systems from which they come.

Although the form of the functions $x_1(t)$, $x_2(t)$, ..., $x_n(t)$ may be known numerically as the result of measurement of the system, it is almost never possible to write down these functions directly, and so express algebraically the physical operation of the system. Information about the system is not the same as knowledge of it. As Rosen (1970) emphasizes, one generally can only give conditions that must be satisfied by these functions, whence it may be possible to arrive at the functions themselves. The most general and fundamental of these is that the rate of change with respect to time of some state variable $x_i(t)$ must be equal to some function of the state variables alone which, letting f now stand for a function, we may call $f_i(x_1, ..., x_n)$. Making use of existence and uniqueness theorems (Rosen, 1970), so that time may be explicitly included together with any necessary system parameters $k_1, k_2, ..., k_m$, our functions $x_1(t)$, ..., $x_n(t)$ must satisfy the following set of simultaneous first-order differential equations:

$$\frac{\mathrm{d}x_i}{\mathrm{d}t} = f_i(x_i, ..., x_n, t, k_1, ..., k_m) \qquad i = 1, ..., n \tag{12.4}$$

This set comprises the *dynamical equations* of the system, and the form of the functions f_i, ..., f_n specifies, in some sense, the forces that act on the system and cause its behaviour. In systems involving bedforms, these forces are associated with the transport of sediment over the bed. The parameters k_1, ..., k_m could include such items as grain size, channel slope and, intuitively, the "changeability" of individual bedforms.

Inner controls on dynamical systems involving bedform populations

Without attempting to write down a set of equations which might represent the working of a system involving bedform populations, what insights into the inner controls and therefore functioning of such systems can be derived from general considerations? We should note that populations of bedforms in changing flows have been said to "lag behind the flow" (Carey and Keller, 1957; Simons and Richardson, 1962; Znamenskaya, 1966; Allen,

1973c), because of a "resistance to change offered by the individual components of the system" (Chorley and Kennedy, 1971) measurable in terms of a "relaxation time" or a quantity of sediment transported (Gee, 1973, 1975; Allen, 1974b; Allen and Friend, 1976a). The relaxation time (Chorley and Kennedy, 1971) should be understood as the time required for bedforms to equilibrate to a changed flow. A reminder is necessary before following these pointers. Ripples and dunes occur in communities of hydrodynamically interdependent individuals, by virtue of the kinematic structures—the pattern of separated and reattached flows—associated with them. Wherever bed and flow conditions are suitable, the entire sedimentary surface is covered by the forms, which are only exceptionally solitary. Hence in whatever way the population changes in response to a changing flow, the forms should continue to cover the surface completely.

The inner controls must include the processes by which bedform populations can respond to changing circumstances, and we should consider these in at least two general situations: (1) when the flow path or direction is relatively constant during the course of change, and (2) when the direction or path alters rapidly and substantially during the period of change.

In the first case, the shape and position of features present at one instant should exert a powerful influence on the position and shape of the forms existing a little later, because the kinematic structure of the flow (effectively the vorticity) changes little in orientation. The assemblages should be strongly coupled. Two processes of change are then plausible (Allen, 1974b, 1976a, 1976e). It is known that on current-rippled beds, there is a constant creation and destruction of individuals taking place, as in a biological population. If the period of observation is long enough, there may also be immigration into and/or emigration from the population site. By analogy, a similar behaviour should be expected of dunes and wave ripples. Hence a bedform population may change in character because of a change of *composition,* particularly if, as seems plausible, the newly created individuals are better adjusted to the changed flow conditions than the individuals they replace. The rate of population change due to this process should decline as the *lifespan* of the individuals increases, where the lifespan is the time elapsed between creation and destruction (Allen, 1974b). The lifespan itself depends on the sediment transport rate fixed by flow conditions and on the bedform *excursion,* the distance, which may be expressed non-dimensionally, over which an individual may travel. Excursion may be a fundamental property of bedforms, which may turn out to be a constant for each kind (Allen, 1976a). Attempts to measure excursions and life-spans should now be made. The second process of change in a population is through an alteration on the same general trajectory of the *individuals* that are present, independent of any compositional variation. The *coefficient of change* for each attribute, expressing the ability of an individual to change under the stress of the changing flow, may be a second little recognized intrinsic property of bedforms (Allen, 1976a). The rate of population change by this process should increase as the appropriate coefficient of change is increased. Hence each process would appear capable of moderating the effect of the other.

In the second case (current path or direction varying sufficiently quickly and markedly), a substantial or complete decoupling may be expected between the assemblages existing before and after the change, for the kinematic "grains" can only momentarily be similar and, when similar, may differ significantly in kind. Presumably the new assemblage evolves as if the flow had suddenly been imposed on an artificially smoothed bed (e.g. Bagnold, 1946; Raichlen and Kennedy, 1965; Carstens et al., 1969; Jain and Kennedy, 1974). Creation-destruction and the change of individuals may be expected to control its behaviour, but the character of the new assemblage should be unrelated to that of the earlier population.

It has been emphasized that sediment transport of an appropriate mode is a necessary condition for change by the above two processes (Allen, 1973c, 1974b; Allen and Collinson, 1974). For without transport there cannot be the local sediment transfers between bed and flow which alone can alter the shape of individual features, build up new forms, or smooth and destroy old ones. After each change of flow, there must elapse a certain amount of time, representing sediment transport to an amount controlled by the flow regime, before a new equilibrium can be reached. This concept underpins the independent attempts by Gee (1973, 1975) and by Allen and Friend (1976a) to calculate the "relaxation time" of bedforms. Unfortunately, Gee in his ultimate empirical treatment, implies that transport is merely a sufficient condition for change in population attributes. This does not seem to be correct since, when transport occurs under steady-state equilibrium conditions, the attributes are conserved in value.

Phase difference and relaxation time

The natural systems shown in Figs. 12-2 to 12-5 are dynamical systems existing in a complex phase space which may be depicted two state-variables at a time. The variables of each pair change periodically with time, on the same period but not the same phase. Consider further some of Nasner's (1974) observations (Fig. 12-7). Whereas discharge is a maximum during the winter, mean dune height attains its maximum in the following summer or autumn, and the corresponding phase diagrams are virtually closed loops (Fig. 12-5). This delay may be estimated either directly as a phase difference, because of the periodicity, or in terms of a relaxation time given enough data. Each measure is attribute-specific.

Phase difference may be estimated in two ways (Allen, 1974b). In the simplest, we measure from time series such as Fig. 12-7 the interval separating the peak value of the one state variable from the peak value of the other. This method has the grave disadvantage, however, that attention is focussed

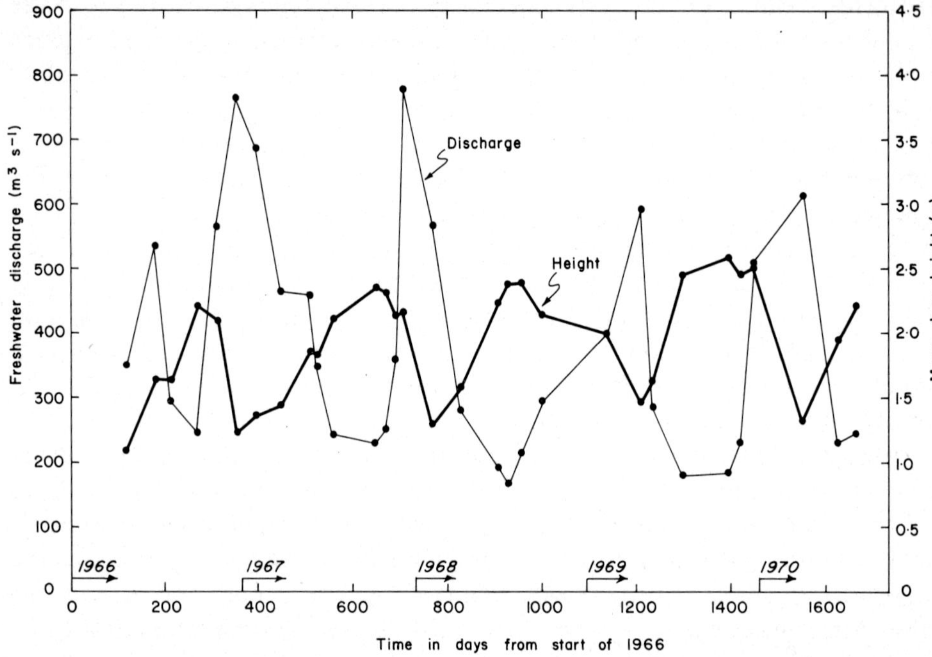

Fig. 12-7. Time-series of discharge of fresh water and mean dune height, Reach A, R. Weser, Germany. Data of Nasner (1974).

on a limited part of the range of each variable, for other values of the phase difference might emerge if instead we considered the mid-points or troughs. It would seem better to compare each complete phase diagram with a suitable theoretical model. We then obtain an "equivalent" value for the phase difference, in the same way that the size of an irregular particle can be specified by the diameter of the sphere of the same volume.

The simplest dynamical systems involve two state variables changing simple-harmonically with time. If $x(t)$ and $y(t)$ are these variables:

$$(y-y_{\mathrm{m}})=a_y \sin\left(\frac{2\pi t}{T}-\alpha_y\right) \tag{12.5}$$

$$(x-x_{\mathrm{m}})=a_x \sin\left(\frac{2\pi t}{T}-\alpha_x\right) \tag{12.6}$$

in which x_{m} and y_{m} are the respective mid-point values, a_x and a_y are the respective half-ranges, α_x and α_y are the respective phase angles, and T is the common period. We wish to eliminate t explicitly, in order to discover how x and y vary together in the phase space. Expansion yields:

$$\frac{(y-y_{\mathrm{m}})}{a_y}=\sin\frac{2\pi t}{T}\cos\alpha_y-\cos\frac{2\pi t}{T}\sin\alpha_y \tag{12.7}$$

$$\frac{(x-x_{\mathrm{m}})}{a_x}=\sin\frac{2\pi t}{T}\cos\alpha_x-\cos\frac{2\pi t}{T}\sin\alpha_x \tag{12.8}$$

Multiplying eq. (12.7) by $\sin\alpha_x$ and eq. (12.8) by $\sin\alpha_y$ and subtracting the first from the second:

$$\frac{(y-y_{\mathrm{m}})}{a_y}\sin\alpha_x+\frac{(x-x_{\mathrm{m}})}{a_x}\sin\alpha_y=\sin\frac{2\pi t}{T}(\cos\alpha_x\sin\alpha_y-\cos\alpha_y\sin\alpha_x) \tag{12.9}$$

Similarly, on multiplying eq. (12.7) by $\cos\alpha_x$ and eq. (12.8) by $\cos\alpha_y$ and subtracting the first from the second:

$$\frac{(y-y_{\mathrm{m}})}{a_y}\cos\alpha_x-\frac{(x-x_{\mathrm{m}})}{a_x}\cos\alpha_y=\cos\frac{2\pi t}{T}(\cos\alpha_x\sin\alpha_y-\cos\alpha_y\sin\alpha_x) \tag{12.10}$$

By squaring and adding eqs. (12.9) and (12.10), we obtain the implicit equation:

$$\frac{(y-y_{\mathrm{m}})^2}{a_y^2}+\frac{(x-x_{\mathrm{m}})^2}{a_x^2}-\frac{2(y-y_{\mathrm{m}})(x-x_{\mathrm{m}})\cos(\alpha_y-\alpha_x)}{a_y a_x}-\sin^2(\alpha_y-\alpha_x)=0 \tag{12.11}$$

between y and x, in which the quantity $(\alpha_y-\alpha_x)=\delta$ is the phase difference.

Equation (12.11), of quadratic form, describes an ellipse whose major and minor semi-axes are generally tilted with the respect to the coordinate axes representing the state variables. Figure 12-8 shows solutions for various phase differences, in which the state variables are normalized by the respective amplitudes and the mid-point values are set equal to zero. For special values of the phase difference we find the relationships:

$$(y-y_{\mathrm{m}})=\frac{a_y}{a_x}(x-x_{\mathrm{m}}),\qquad \delta=0,2\pi,4\pi,\text{ etc.} \tag{12.12}$$

$$(y-y_{\mathrm{m}})=-\frac{a_y}{a_x}(x-x_{\mathrm{m}}),\qquad \delta=\pi,3\pi,5\pi,\text{ etc.} \tag{12.13}$$

which are the equations of straight lines, and:

$$\frac{(y-y_{\mathrm{m}})^2}{a_y^2}+\frac{(x-x_{\mathrm{m}})^2}{a_x^2}=1,\qquad \delta=\frac{\pi}{2},\frac{3\pi}{2},\text{ etc.} \tag{12.14}$$

describing an ellipse (or circle if $a_y=a_x$). The graphs of Fig. 12-8 are phase diagrams and, like those in Fig. 12-5, have specific trajectories because of implicit time.

The relationship between the form, orientation, and trajectory of these phase diagrams and the phase difference may be expressed in a particularly

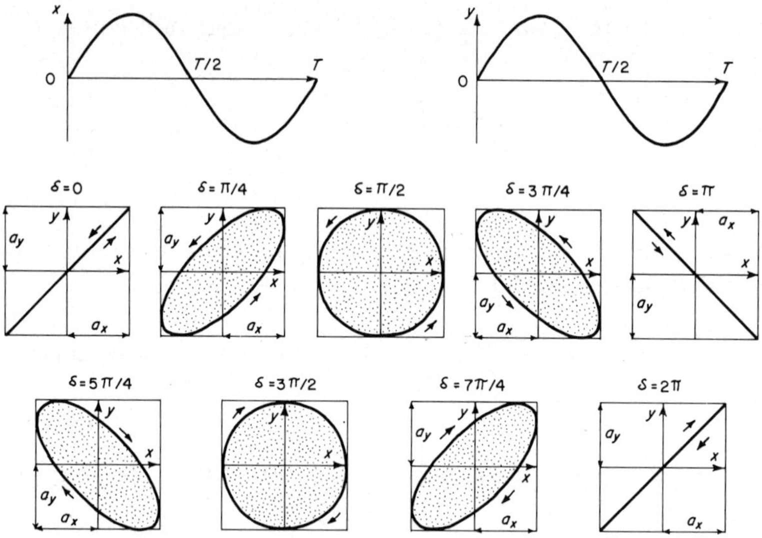

Fig. 12-8. Phase diagrams for $y(x)$ at different phase differences, where both $y(t)$ and $x(t)$ are simple-harmonic.

simple way. We calculate the area enclosed by each curve and compare this value with the area of the region defined by:

$$(x_{\mathrm{m}} - a_x) \leqslant x \leqslant (x_{\mathrm{m}} + a_x), \qquad (y_{\mathrm{m}} - a_y) \leqslant y \leqslant (y_{\mathrm{m}} + a_y)$$

a rectangle each side of which is tangent to the phase diagram (Fig. 12-9). In the case of a phase diagram of unknown phase difference, we may either estimate the difference qualitatively using Fig. 12-8, or, as Allen (1976a) suggests, enter Fig. 12-9 with the relative area and observed trajectory and orientation. For example, Nasner's (1974) data for reach A 1966–1967 (Fig. 12-5) afford a phase difference between mean dune height and discharge of $5\pi/4$ rad qualitatively and 3.57 rad by the area method.

The estimation of the relaxation time is altogether more difficult, even empirically. Consider an equilibrium steady-state bedform population which at time t is subject to an abrupt change of flow conditions, such that equilibrium is not regained until after the elapse of a time t_{r}. During the time t_{r}, which is the relaxation time, a dry mass of sediment $M_{1\to 2}$ (the subscripts refer to the initial and subsequent states) is transported through a cross-section of uniform width b normal to flow. The relaxation time and $M_{1\to 2}$, which may be called the relaxation mass, are connected by:

$$b\int_{t}^{t+t_{\mathrm{r}}} J_{\mathrm{B}}(t)\cdot \mathrm{d}t = M_{1\to 2} \tag{12.15}$$

in which $J_{\mathrm{B}}(t)$ is the dry-mass bedload transport rate prevailing during the interval t to $(t+t_{\mathrm{r}})$. Equation (12.15) epitomises the way in which Gee (1973, 1975) and Allen and Friend (1976a) have approached the problem of

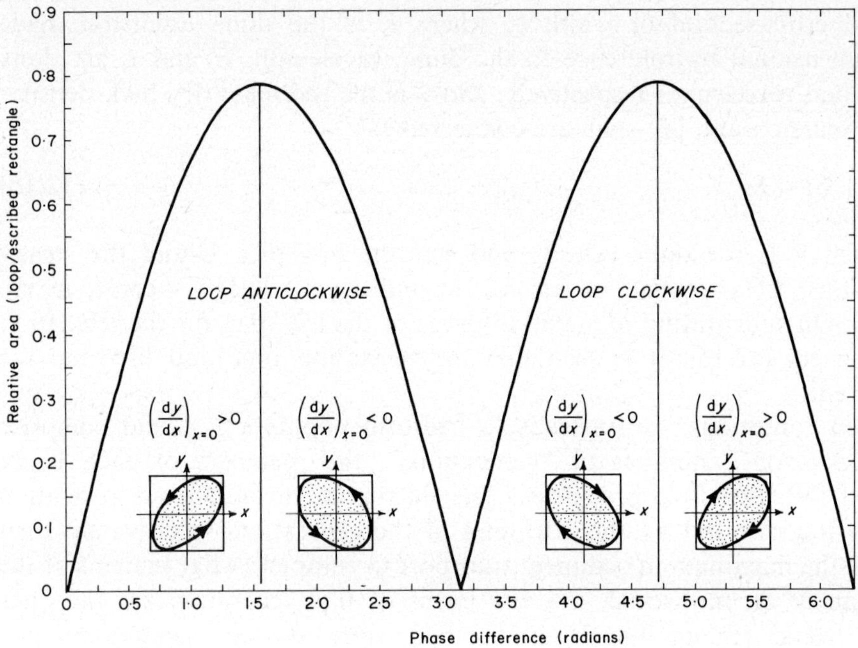

Fig. 12-9. Area of phase loop for $y(x)$ relative to that of escribed rectangle, where $y(t)$ and $x(t)$ are simple-harmonic.

relaxation time. Although the relaxation time can be measured experimentally in limited cases, the essential problem is to predict $M_{1\rightarrow 2}$ in terms of the flow conditions and the initial and subsequent states of the bed. It should be emphasized that eq. (12.15) is quite general. It refers to any two bed states, for example, the transition from dunes to current ripples, from an upper-stage plane bed to dunes, or from large to small dunes.

Bedform life-span seems to put an upper limit on the relaxation time, which can be specified even for a steady state, though under these conditions it is potential rather than effective. Consider two-dimensional dunes and remember that any population attribute, in terms of which equilibrium may be assessed, is a statistical quantity, to which each individual feature contributes. At any instant, a population of dunes uniform in shape, size and life-span in equilibrium with a steady uniform flow will present a uniform distribution of ages, with forms just created equally abundant with those mature enough to be destroyed. But the population composition will not change completely until after the elapse of one dune life-span. If there was an infinitesimally small change in flow, this is the maximum time (i.e. maximum relaxation time) which would be required for the influence of the previous conditions, as expressed by the dunes, to be erased. Under the steady conditions, and assuming dunes of triangular cross-sectional shape, a dry mass of sediment, $M=\frac{1}{2}bKHL\gamma$, is transported during this time through a fixed cross-section of width b, where K is the dune excursion made non-dimensional by reference to the dune wavelength, H and L are dune height and wavelength respectively, and γ is the sediment dry bulk density. Now excursion and life-span are connected by:

$$\int_0^{\tau} V(t)\cdot \mathrm{d}t = K \tag{12.16}$$

in which V is the dune celerity and τ is the life-span. Under the steady conditions, $V(t) = \text{const.}$ in eq. (12.16) and $J_B(t) = \frac{1}{2}HV\gamma = \text{const.}$ in eq. (12.15). On substituting M stated in terms of the life-span for the relaxation mass in eq. (12.15), it is seen that the relaxation time and life-span are equivalent.

When conditions are unsteady, a bedform population should comprise individuals of a number of "generations", the members of each being marked by distinctive properties. The maximum instantaneous relaxation time is therefore set by the attributes of the generation which would then require the maximum of sediment transport to complete what remains of the life-span of an individual. For the forms of this generation are the ones which would remain longest on the bed, since at that instant only one sediment transport rate prevails. In the case of two-dimensional dunes, the relaxation time at time t is given by:

$$\int_t^{t+t_r} J_B(t)\cdot \mathrm{d}t = \tfrac{1}{2}\gamma(aKHL)_{\max} \tag{12.17}$$

where $J_B(t)$ is the transport rate at time t and the right-hand term is the maximum transport dry mass. In this term a is the fraction of the excursion remaining at time t to the dunes in the generation affording the maximum.

Under special circumstances a minimum relaxation time is calculable. Gee (1973, 1975) and Allen and Friend (1976a) analyzed the transition from a dune to a "smooth" bed (i.e. plane or rippled), and the former extended his approach to the reverse transition. The essence of Gee's analysis is as follows. Consider a streamwise slice of uniform width b through two-dimensional dunes of uniform triangular shape, wavelength, L, and height, H (Fig. 12-10a). To change the bed into a smooth bed, we must at the

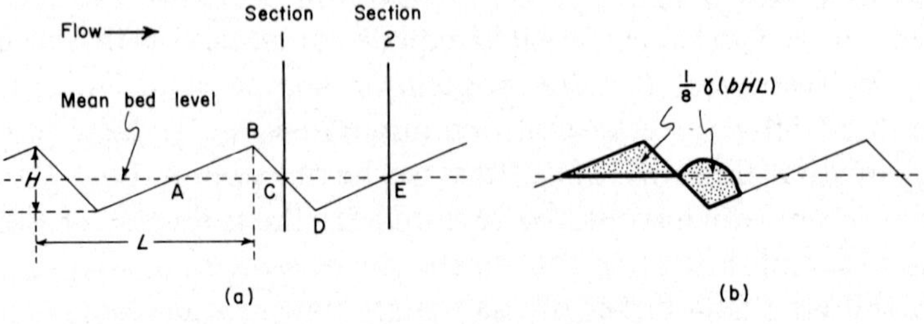

Fig. 12-10. Relaxation time of a wavy bedform.

minimum transfer the sediment in the volume represented by ABC above the mean bed level, into the space CDE below this level. Hence the minimum transport through section 1 is a dry mass equal to $\frac{1}{8}\gamma(bHL)$ and the minimum through section 2 is zero. The sediment mass equivalent to the minimum relaxation time is therefore $\frac{1}{8}\gamma(bHL)$, whence:

$$\int_{t}^{t+t_r} J_B(t) \cdot \mathrm{d}t = \tfrac{1}{8}\gamma HL \tag{12.18}$$

However, in emphasizing the average volume of sediment passing through any random section, Gee (1975) appears to accept that one-half of this minimum transported mass is equivalent to the minimum relaxation time, which does not seem correct. There is a second difficulty with Gee's analysis. We may transport the whole of the mass equal to $\frac{1}{8}\gamma(bHL)$ through section 1, yet fail to fill up completely the space CDE (Fig. 12-10b). The minimum relaxation time would therefore be underestimated by the use of this mass. Allen and Friend's (1976a) analysis does not share this weakness and, therefore, gives somewhat larger minimum relaxation times than eq. (12.18). Another minimum relaxation time is I.G. Wilson's (1971a) "reconstitution time", the time taken for a bedform to advance its own wavelength, what Sorby (1859, 1908) called the "period".

NUMERICAL MODELLING OF DUNE POPULATIONS

In an attempt to gain insights into the behaviour of dune populations in unsteady flows, Allen (1976a, 1976b, 1976c, 1978a, 1978b) developed an exploratory numerical model of dune time-lag capable of operating under simplified hydraulic conditions. The model is predicated upon the idea that the hidden, inner controls on these systems are the steady-state relationships connecting bedform and flow variables, which determine the attributes of dunes at birth, and the two processes of bedform change described above (Allen, 1976e), which determine the subsequent persistence of those attributes.

The model operates upon a prescribed, strictly periodic, unidirectional, aqueous discharge contained in a long straight channel of uniform roughness, slope, and rectangular cross-section. Lineages of sets of dunes are created in this channel in accordance with predetermined initial conditions and a boundary condition governing the ratio of new to old dunes at each creative act. Every dune is given an excursion and subsequently travels under the sequence of hydraulic conditions in accordance with a prescribed sediment-transport function, until its life-span is exhausted. Each dune when created assumes a value for wavelength and height, determined by the flow conditions then prevailing in accordance with selected equilibrium steady-state relationships. Yalin's (1964) formulae for height and wavelength were at first used, but recently a more realistic relationship for height (Fig. 8-21c) based on Stein's (1965) work has been tried (Allen, 1978b). Dune wavelength is treated as a fully conserved throughout a life-span (justified in Allen, 1976a). In contrast, dune height is permitted to respond to the changing flow conditions during a life-span, to an extent governed by a prescribed value of the coefficient of change. Hence the features as they continue in the flow of time retain a perfect "memory" of their wavelength in earlier states, when flow conditions were different, but an imperfect "memory" of their height. The dunes created at a given instant form a set which, because its components share the same wavelength and therefore life-span, is completely destroyed at a single, later instant and instantaneously replaced by another set with new characteristic attributes. Each dune lineage is a sequence of such sets.

This model yields information on the dune population over as many repetitions of the governing flow cycle as may be desired, once stability has been reached through sufficient iteration away from the initial conditions. Furthermore, analysis of the assemblage is possible at the same two levels of detail as natural populations. By averaging across a large number of lineages instant by instant, the population attributes (e.g. mean dune height, mean dune age) emerge either as time series (cf. Fig. 12-7) or as numerical solutions to functions with the form of a stable limit cycle, that is, in the system phase space as closed phase diagrams (cf. Fig. 12-5). Secondly, we may obtain for each chosen instant the density distribution representing the population structure in some individual attribute (e.g. life-span, wavelength). Analysis at both levels is required if the population behaviour is to be properly understood.

Representative experimental results are summarized in Table 12-I and Fig. 12-11 (see also Allen, 1976a, 1976b, 1976c, 1978a, 1978b). The behaviour of the population and its character at any instant are strongly related to the time-averaged dune life-span, set by the prescribed dune excursion, hydrograph, and channel geometry. Phase diagrams similar to those observed from natural populations (Figs. 12-2, 12-3, 12-5) represent the experimental assemblages (Fig. 12-11d–i). The lag between mean wavelength or height and discharge increases as the time-averaged life-span grows larger relative to the flow period. Under Yalin's (1964) relationships, phase differences for wavelength and height do not exceed approximately $\pi/2$ rad. The phase difference is greater for wavelength than height, because wavelength is fully conserved, whereas dune height can respond in a limited way to changing conditions. Using Stein's (1965) results, however, dune height lags discharge by amounts in the order of $3\pi/2$ rad, though the wavelength phase difference remains less than about $\pi/2$ rad, much as in the Weser (Fig. 12-5). Another effect of increasing life-span is to shift upward and also narrow the ranges of mean height and wavelength during a flow cycle. At life-spans comparable with to much greater than the flow period, the mean wavelength

TABLE 12-I

Summary of three numerical experiments on dune populations carried out under Yalin's (1964) relationships for dune height and wavelength, and their principal results

	Run A	Run B	Run C
Principal conditions			
Maximum discharge ($m^3 s^{-1}$)	10,000	10,000	10,000
Minimum discharge ($m^3 s^{-1}$)	1500	1000	1000
Discharge ratio	1:6.67	1:10	1:10
Flow period (years)	1	1	1
Channel slope	0.00005	0.00005	0.00005
Channel width (m)	500	500	500
Channel roughness (Darcy-Weisbach *f*)	0.08	0.08	0.08
Mean dune excursion (wavelengths)	6	18	288
Coefficient of change of dune height	0.001622	0.0045	0.0045
Principal results			
Time-averaged dune life-span (years)	0.071	0.243	3.28
Observed time-averaged dune wavelength (m)	68.83	69.69	76.92
Observed time-averaged dune height (m)	2.290	1.790	1.796
Number of modal * wavelength classes	Unimodal	2	4
Modal wavelength values in order of increasing wavelength (m)	—	20–25, 100–105	20–25, 50–55, 80–85, 100–105

* A wavelength class in the histograms of Fig. 12-11a–c is recognized as a modal class only when it is conspicuously occupied relative to neighbouring classes for a substantial portion of the flow cycle.

is virtually constant, and the time-averaged value may be much greater than when the life-span is relatively small. Increase of life-span relative to flow period also changes population structure. For large discharge ranges, assemblages are unimodal at all times when the life-span is small, bimodal for some of the time at intermediate life-spans, and permanently polymodal for large life-spans (Fig. 12-11a–c). However, provided that the discharge range is small, permanently unimodal assemblages can be formed at quite large life-spans.

The appearance of two or more wavelength modes in these experimental systems expresses their self-regulation by the concentration of dune creation and destruction into definite and regularly recurrent time intervals, each equivalent to a particular range of hydraulic conditions (Allen, 1978b). For example, the bimodality at times of the population in Fig. 12-11b is due to the fact that the large dunes created at high stages persist through the subsequent recession, to be joined during low flow by a number of smaller dunes appropriate to the now much reduced discharge. Because apparently morphologically and hydraulically distinct features are present, it might be

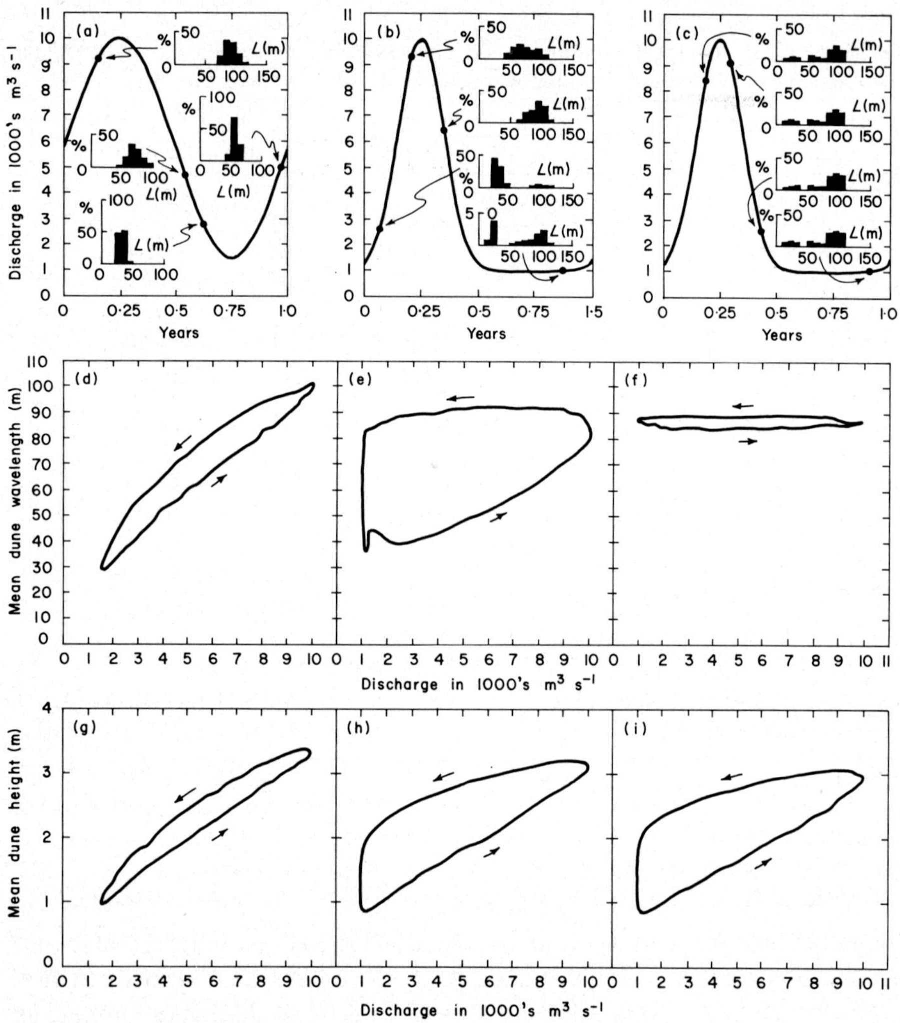

Fig. 12-11. Representative results from numerical experiments on dune populations in periodic unidirectional flows (see Table 12-I for details of experimental conditions). a–c. Hydrographs, with frequency distributions of dune wavelength at representative instants. Phase diagrams for (d–f) mean dune wavelength and (g–i) mean dune height.

supposed in the absence of knowledge of the history that the population represented two hydrodynamically distinct bedform classes. This interpretation is incorrect, for the forms were all generated under a single set of rules. These considerations put rather in doubt the usefulness of Southard's (1971b) depth–velocity–texture diagrams under unsteady conditions (e.g. Harms et al., 1974; R.G. Jackson, 1976a).

Although the above model is capable of handling discharge variations of a practical scale, it may be criticized on two main grounds: (1) the rules connecting the dune height and wavelength and the sediment transport rate to flow conditions are highly simplified and, more particularly, (2) a number of arbitrary assumptions are made concerning the response of the dunes to changing flows, such that the dunes *inevitably* lag the flow (a life-span is *assumed* and a value for it prescribed). Apparently independently, Fredsøe (1978, 1979) has also analyzed the behaviour of dunes in unsteady flows, and obtained qualitatively similar results to Allen (1976a, 1976b, 1976c, 1978a, 1978b). Fredsøe's treatment is analytical, however, and restricted to flows that vary in discharge only weakly with time. Moreover, dune creation-destruction is neglected, the only process of change considered being the response of individual dunes to changes of discharge. Although the restriction to weakly varying flows is a severe limitation, the approach is promising because a phase lag between bed and flow emerges as an explicit feature of the analysis, and is not assumed beforehand.

STRUCTURES INDICATIVE OF CHANGING RIVER FLOWS

Abandoned dunes

We shall now shift the emphasis away from ripple and dune populations, and towards more local effects, essentially at the scale of individual bed features.

Many rivers vary in discharge so rapidly and over such a large range that high-stage dunes are at low flow left either exposed without substantial modification of shape or submerged but inactive (Shantzer, 1951; Sundborg, 1956; Frazier and Osanik, 1961; Harms et al., 1963; Lundqvist, 1963; Potter and Pettijohn, 1963; Clos-Arceduc, 1967b; McKee and Murfitt, 1967; Whetten and Fulham, 1967; J.M. Coleman, 1969; Collinson, 1970b; Bluck, 1971; G.E. Williams, 1971; Karcz, 1972; Singh and Kumar, 1974; Gustavson, 1978). Usually current ripples become superimposed on the features as stage falls but, in the case of flashy ephemeral streams, a mud layer, eventually sun-cracked, may be partly or wholly draped over them (Karcz and Goldberg, 1967; Karcz, 1972). In the Brahmaputra, according to J.M. Coleman (1969), the troughs of large inactive dunes are infilled with mud during low flows. These moribund or exposed forms strikingly demonstrate the severity of lag effects in the more changeable rivers. Had bed adjustments kept pace with the waning flows, either the sedimentary surface would have been plane (gravel-bed rivers) or with features of relief no greater than current ripples (sand-bed streams). Stage in some instances fell so rapidly that drainage channels were cut across dune crests and small deltas built into troughs (Singh and Kumar, 1974).

Wind and current action during flood abatement

As floods abate river beds become exposed and may be variously modified, either directly or indirectly, through wind action.

One unquestionable action of vigorous winds is to deflate exposed sands, with the result that extensive, delicately sculptured erosion surfaces are fashioned, and internal structures are bared (Collinson, 1970b; G.E. Williams, 1971). An armouring of coarse debris may also be developed, especially under dry conditions (Glennie, 1970, McGowen and Garner, 1970). Wind ripples and blankets of drifted sand can accumulate locally on the former river beds (Collinson, 1970b), and drifts of avalanched wind-borne sand in the lee of the abandoned subaqueous dunes are even reported (Whetten and Fulham, 1967). This combination of water-laid and wind-deposited cross-strata would be of high diagnostic value if recognizable in fossilized cross-bedding sets. Sand may even be blown up on to the river banks, to form dune complexes (e.g. Sellards, 1923; Falkowski, 1972a, 1972b).

Waves where large enough act either constructively or destructively to modify shoaling or emerging dunes and sand bars. In small rivers, and in large ones of sufficiently complex bed topography, the waves seem to be due to the stream flow itself (Picard and High, 1973), probably reflecting its non-uniformity. Large rivers present large fetches, however, and for most the waves may reasonably be attributed to the wind. Smooth beaches are fashioned by wave action around the crests of large emerging bars, and vigorous waves may cut tall cliffs into the sides of the features (Collinson, 1970b; N.L. Banks, 1973c). As has often been recorded (Hjulström, 1953; Sundborg, 1956; Lundqvist, 1963; Ore, 1963; Picard and High, 1969; Williams and Rust, 1969; Collinson, 1970b; N.D. Smith, 1970, 1971a, 1971b; Klimek, 1972; Singh and Kumar, 1974), a frequent wave effect is the incision of flights of clifflets, each with its tiny beach and offshore platform, into the sides of the emerging bedforms. Such cliffs and clifflets indicative of subaerial exposure seem preservable (Collinson, 1970b; N.L. Banks, 1973c; Stear, 1979). Sufficiently vigorous waves travelling over the shoaling water can shape on the bed beneath patterns of symmetrical to near-symmetrical ripples (Picard and High, 1969, 1970a, 1973; Collinson, 1970b; Karcz, 1972; Rust, 1972a). Usually these are superimposed on the late-stage current ripples.

Internal structures

Collinson (1970b) described from his linguoid bars in the R. Tana, Norway, a variety of cross-bedding structures which he attributed to "re-activation", the renewed movement of a bed feature, possibly associated with each annual flood or with some included substantial pulse of discharge.

Fig. 12-12. Internal reactivation structure visible in vertical section through linguoid bar at low river stage, R. Tana, Norway. Current from right to left. Photograph courtesy of J.D. Collinson (see also Collinson, 1970b).

These structures (Fig. 12-12) typically comprise the streamwise sequence of normally dipping cross-beds → discordant low-angle erosion surface → normally dipping cross-beds, in some instances repeated several times. Occasionally the erosion surface is overlain by a little cross-laminated sand

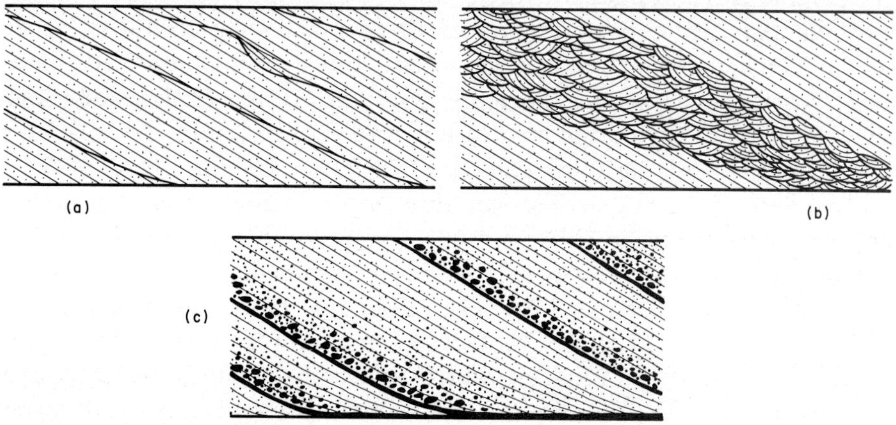

Fig. 12-13. Schematic types of reactivation structure in cross-bedded deposits depicted in vertical section. a. Erosion surfaces at a lower angle than cross-beds. b. Packet of cross-laminated sediment between groups of foresets. c. Downstream fining in groups of cross-beds.

and, in rare instances, by a substantial quantity, accreted on the earlier cross-beds by a current flowing parallel with their strike. Collinson's examples represent two of the three kinds of reactivation structure clearly recognizable in fluvial deposits (Fig. 12-13a, b). The third rests on N.D. Smith's (1974) study of cross-bedded sand wedges formed in the lee of gravel bars (Fig. 12-13c). Each wedge is composed of a number of growth increments, each increment consisting chiefly of coarse sand or granule conglomerate which grades downstream into finer sand and finally silt.

Although all reactivation structures must record a local and temporary shift of flow conditions, the change may be induced in different ways and may take a time scale from within a substantial range. Collinson (1970b) proposed that some reactivation structures belongong to the first two types were due to either wave action in the course of exposure (Fig. 12-13a) or a change of flow pattern around a bed feature as stage was lowered, but not to the point of bed emergence (Fig. 12-13b). The controlling discharge events could have been either annual or, since the R. Tana has a subordinate early summer flood following the spring snow-melt, spaced approximately 4–6 weeks apart. However, events of a much smaller time scale, of perhaps minutes or hours, and requiring no change of discharge or stage, can also explain at least the first type. G.E. Williams (1971) suggested that fluctuations of the separated flow downstream of a bar avalanche-face strongly skewed to flow might promote reactivation (see also Johansson, 1963). Allen (1973a), McCabe and Jones (1977), and C.M. Jones (1977) emphasized that the random relative movement of current ripples within a rippled bed, of dunes within a duned bed, or simply of smaller bedforms associated with larger can generate reactivation structures of the first type, even in the context of steady-state equilibrium. It is observed experimentally that, for brief periods, individual forms in some cases retreat slightly upcurrent and in others became either rounded at the crest or temporarily reduced in height due to the play of a powerful vortex, before reverting to their earlier behaviour or character. The reactivation structures within presumed dune-formed cross-bedded sets of the Knik River in Alaska (Fahnestock and Bradley, 1974) may have originated in this way. The examples described by N.D. Smith (1974) of the third type of reactivation structure (Fig. 12-13c) depend on diurnal variations of stream discharge and hence represent events on an intermediate time scale.

Reactivation structures seem to be fairly common in fluviatile deposits (Basumallick, 1966; Boersma, 1967; G.E. Williams, 1966; Gustavson, 1974; R.G. Jackson, 1976a; Miall, 1976; Worsley and Edwards, 1976; Hodgson, 1978; C.M. Jones, 1979; Blodgett and Stanley, 1980; Jones and McCabe (1980). In some of the examples the convex-up erosion surfaces within the cross-bedding sets are so closely spaced in the streamwise direction as to suggest a reactivation structure related to compound cross-stratification, that is, to the repeated march of smaller bedforms up to the crest of a larger

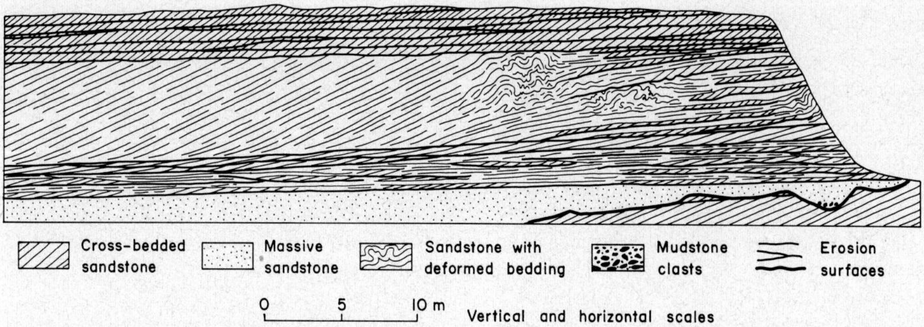

Fig. 12-14. Vertical profile showing small cross-bedding sets building up and downcurrent into a single large set, apparently recording the transformation of a low, flat sand bank into a large and relatively tall one with a steep lee side. Hawkesbury Sandstone (Triassic), Lane Cove, Sydney, Australia. After Conaghan and Jones (1975).

form, like a dune or bar. A remarkable cross-bedding pattern almost certainly indicative of a changeable river is reported by Conaghan and Jones (1975) from the Hawkesbury Sandstone of New South Wales, Australia (Fig. 12-14). The downstream change is taken by them to mean a falling river, but it is equally likely, considering the Fraser River (Fig. 12-2 and Pretious and Blench, 1951), the numerical model (Fig. 12-11b and Allen, 1976a, 1976b, 1976c) and as well the Brahmaputra (J.M. Coleman, 1969), that either an increasing flow or a peak stage may be represented.

STRUCTURES INDICATIVE OF CHANGING TIDAL AND WIND-WAVE REGIMES

Abandoned dunes

Tidal streams are unsteadier even than rivers, particularly in estuaries and near coasts where, in a matter of hours, water stage may vary vertically over several metres and flow velocity can range between zero and upward of $1-2\ m\ s^{-1}$. The total sediment transport during each interval of change is commonly adequate to modify only the smallest bedforms, and so lag effects are commonplace and usually severe (e.g., Allen and Friend, 1976a). Dunes which persist little modified after the tide has ebbed from sand banks and shoals are a widespread expression of lag. Examples abound in European waters (Cornish, 1901a, 1901b, 1914; R. Richter, 1926a; Häntzschel, 1938; Van Straaten, 1950, 1953a; Hülsemann, 1955; Reineck, 1963; G. Evans, 1965; Bajard, 1966; Allen, 1968c; Boersma, 1969; Newton and Werner, 1969, 1970; Buller et al., 1971; Göhren, 1971c; Park, 1974; Larsonneur, 1975; Allen and Friend, 1976b). Many instances are reported in North America, notably from the intricate eastern shores (Kindle, 1917; Klein, 1963b, 1970b; Land and Hoyt, 1966; Swift and McMullen, 1968; Daboll,

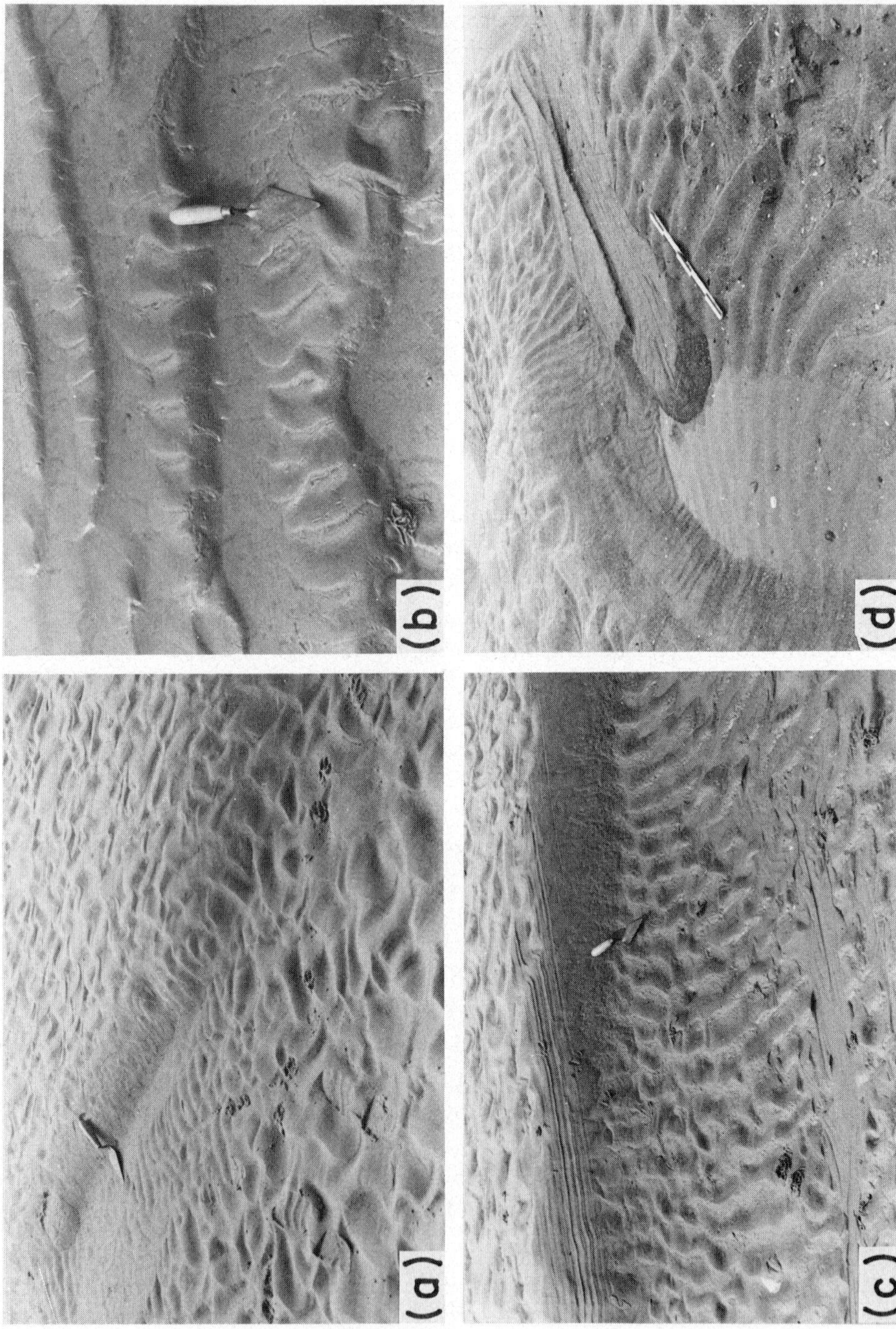
(a)
(b)
(c)
(d)

1969; Kellerhals and Murray, 1969; Farrell, 1970; Hartwell, 1970; Knight, 1972; Stone and Summers, 1972; Luternauer and Murray, 1973; Boothroyd and Hubbard, 1974, 1975; Dalrymple et al., 1975; Hine, 1975; Howard et al., 1975b; Knight and Dalrymple, 1975). Australian, South American, and African shores yield a few examples (Tricart, 1967; Gellatly, 1970; Tucker, 1973; Wright et al., 1975) and more may be expected. In areas of bioclastic sedimentation dunes abandoned at low tide are also well known (Newell and Rigby, 1957; McKee and Sterrett, 1961; Purdy, 1961; Imbrie and Buchanan, 1965; Ball, 1967; Farrow and Brander, 1971). As in ephemeral streams, intertidal dune forms may occasionally be preserved beneath mud (Wright et al., 1975).

Surface features associated with falling water level

As the tide ebbs, sand banks and shoals may be modified by (1) late-stage drainage, (2) wind-wave action at and close to the retreating strand, (3) groundwater flow out of the sand body, and (4) direct wind action. The resulting surface features generally resemble those already described from rivers; the observed differences of kind and frequency of occurrence, however, reflect the greater rates of change, and the higher intensity of wind-wave action, in tidal as compared with fluvial environments. That intertidal bedforms could be modified during falling stage was early recognized. Cornish (1914) saw that the water draining from dune-covered sand shoals was guided by the shape of the bed features. The ability of wind-waves to modify intertidal bed configurations particularly impressed Kindle (1917) and Häntzschel (1938). Interest in these modifications was recently revived, in connection with the identification of fossil strand lines and tidal ranges (Allen, 1967b; Klein, 1970a, 1970b; J.G. Ford, 1975).

The tide ebbing from dune-covered sites is usually guided during the terminal stages of drainage by the troughs of any surviving dunes. Consequently, current ripples may become superimposed at a large angle upon the earlier dunes (Fig. 12-15a), covering the troughs and occasionally the lee slopes also (Klein, 1963b, 1970a, 1970b; Imbrie and Buchanan, 1965; Bajard, 1966; Land and Hoyt, 1966; Buller et al., 1971; Knight, 1972; Luternauer and Murray, 1973; Park, 1974; Knight and Dalrymple, 1975; Larsonneur, 1975). A similar effect can occasionally be seen where a short-lived ebbing flow drains across a surface with large current ripples, the

Fig. 12-15. Superimposed bedforms. a. Ebb-tide current ripples superimposed at a steep angle across the crest of an earlier dune, Wells-next-the-Sea, Norfolk, England. b. Small current ripples superimposed across earlier current ripples due to a more powerful current, Wells-next-the-Sea. c. Braided drainage channel in dune trough and wave-cut clifflets on slip face, Wells-next-the-Sea. d. Drainage channel terminating in delta, trough of dune, Barmouth Estuary. Scale in (d) is 0.5 m long and trowel in (a–c) measures 0.28 m.

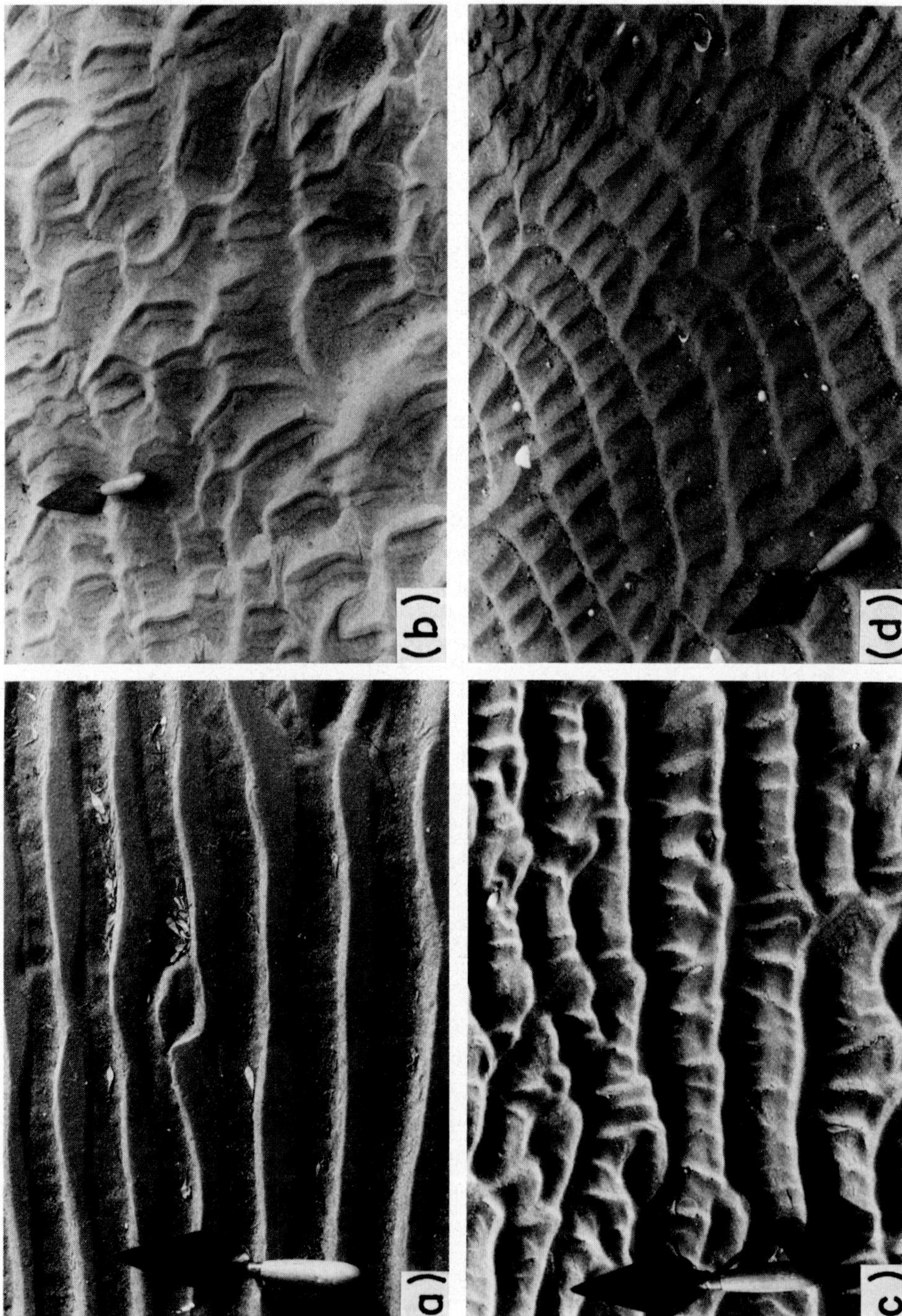
(a)
(b)
(c)
(d)

superimposed forms seldom reaching maturity (Fig. 12-15b). Where late-stage currents are sufficiently powerful, a new train of dunes can be emplaced with loss of coupling across an earlier set (Cornish, 1901a, 1914; Kellerhals and Murray, 1969). The final flows often succeed in breaching the spurs in the dune troughs. There appear short braided drainage channels (Fig. 12-15c) bounded by cliffs and terraces, and terminated by deltas (Fig. 12-15d) ranging in individual area up to several square metres (Kindle, 1917; Häntzschel, 1938; Tanner, 1959; Trefethen and Dow, 1960; R.W. Thompson, 1968; G.P. Allen et al., 1969; Klein, 1970b; Buller et al., 1971; Reyment, 1971).

Wind-waves affect shoaling bedforms in several ways. They often smooth and round the crests of emerging dunes, creating heavy mineral concentrations, and when strong may virtually obliterate these bedforms (Häntzschel, 1938; Hülsemann, 1955; Hervieu, 1968; Buller et al., 1971). Less vigorous waves superimpose wave or wave-current ripples on dunes (e.g. Häntzschel, 1938; Park, 1974) and, at the water's edge, shape flights of terraces (Fig. 12-15c) like those recorded from the river forms (Cornish, 1901a, 1914; Kindle, 1917; R. Richter, 1926a; Häntzschel, 1938; G. Evans, 1965; Land and Hoyt, 1966; Swift and McMullen, 1968; Haynes and Dobson, 1969; Wunderlich, 1969; Farrell, 1970; Gellatly, 1970; Klein, 1970b; Farrow and Brander, 1971; Boothroyd and Hubbard, 1974; Dalrymple et al., 1975). Ripple marks also may become terraced in this way (e.g. Tanner, 1962), though typically they become flat-topped and truncated (Fig. 12-16a) under wave action (Wegner, 1932; E.D. McKee, 1957a; Tanner, 1958, 1959, 1960; G. Evans, 1965; Trefethen and Dow, 1960; Bajard, 1966; Ricci-Lucchi, 1970; Rudowski and Tobolewski, 1973; Wunderlich, 1972). On barred beaches at low tide it is common to find when the wind is strong that wave ripples are superimposed at a large angle on the current ripples formed earlier in the troughs between bars during the ebb (Fig. 12-16b). The waves responsible are generated on the surface of the water ponded behind the bars, and reflect the limited fetch and local wind. In the final stages of ponding, when no more outflow can occur, mud may settle to preserve the doubly-rippled surface. The reverse temporal sequence, of current ripples on wave ripples, is also known from barred beaches (Davis et al., 1972). Not uncommonly on such beaches, wave-current ripples superimposed on wave ripples (Fig. 12-16c) and wave-current ripples superimposed on wave-current

Fig. 12-16. Superimposed bedforms. a. Flat-topped wave ripples, coast near Burnham Overy Staithe, Norfolk, England. b. Wave ripples superimposed on current ripples, barred portion of beach, Wells-next-the-Sea. c. Wave-current ripples superimposed on wave ripple marks, coast near Burnham Overy Staithe. d. Wave-current ripple marks superimposed on an earlier pattern of larger wavelength wave-current ripples, coast near Burnham Overy Staithe. Trowel measures 0.28 m long.

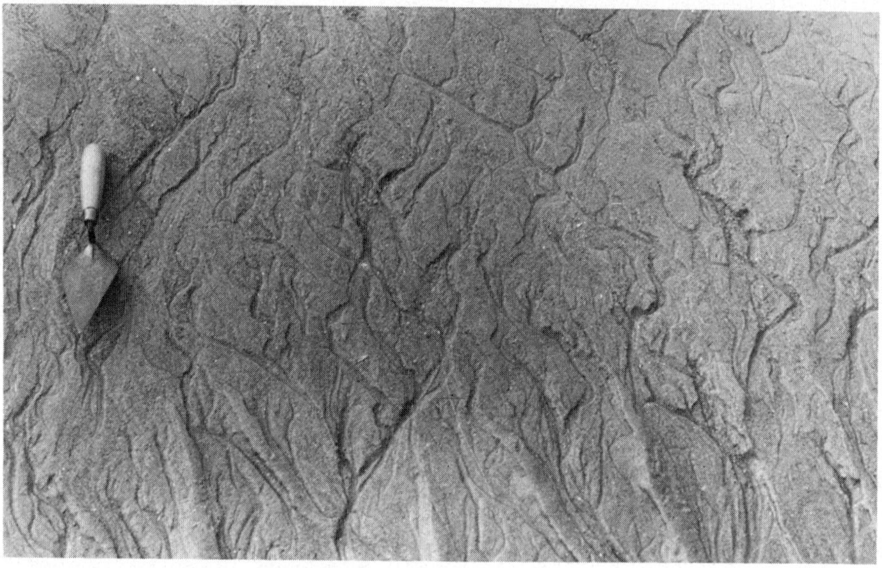

Fig. 12-17. Rill marks on steep surface underlain by fine sand, Wells-next-the-Sea, Norfolk, England. Trowel points downslope and measures 0.28 m long.

ripples (Fig. 12-16d) are also to be found. Martinez (1977) reports similar patterns from the stratigraphic record.

Small dendritic channels called rill marks (Fig. 12-17) may locally occur on an intertidal sand body, where there is a steep or abruptly changing topography (Dodge, 1894; Häntzschel, 1938, 1939; Trefethen and Dow, 1960; Pettijohn and Potter, 1964; Bajard, 1966; Martins, 1967; G.P. Allen et al., 1969; Cepek and Reineck, 1970; Reyment, 1971). These marks form where water trapped within the body oozes out over the surface, and so are attributable to the rapid fall of stage. Cepek and Reineck give an excellent description of rill marks, but incorrectly include Peabody's (1947) current crescents amongst them.

There is seldom time within one tidal cycle for the wind to have much direct effect on intertidal sands. Between sets of spring tides, however, the upper parts of a bank, shoal or bar may suffer considerable deflation, with the development of erosion surfaces, the baring of internal laminations, and the building of wind drifts and dunes (Van Straaten, 1953b; Gripp, 1961a, 1961b, 1961c, 1968; Gripp and Martens, 1963; Reineck, 1963; Allen, 1965c; Bajard, 1966; Milling and Behrens, 1966; Martins, 1967). Some of the wind-driven sand may adhere to moist surfaces as Van Straaten's anti-ripplets, now commonly termed adhesion ripples (Van Straaten, 1953b; Reineck, 1955; Gripp and Martens, 1963; Bajard, 1966; Hunter, 1969, 1973). Most authors relate the desiccation cracking of intertidally deposited mud to the prolonged exposure that accompanies the fortnightly neap tides

(Häntzschel, 1938; Van Straaten, 1954a; G. Evans, 1965; Bajard, 1966; R.W. Thompson, 1968; Haynes and Dobson, 1969).

Where waves break heavily, and the wind blows strongly onshore, clumps of foam are forced across the exposed sand. These leave the surface covered by closely packed shallow pits a few millimetres across, as described by Häntzschel (1935) and independently by Allen (1967c). The pits seemingly represent the temporary resting traces of bubbles that are jerkily over-riding each other within the moving clumps.

Structures indicative of change similar to those described above have occasionally been reported from sandstones. The significance of superimposed ripples was early appreciated by Kindle (1917), and other descriptions and interpretations have followed (E.D. McKee, 1954; Niehoff, 1958; Basumallick, 1963; Goldring, 1971; Young and Long, 1977). Allen and Kaye (1973) and Allen (1974a) described ripple marks superimposed on cross-beds attributed to tide-shaped dunes. Truncated ripples are recorded (Rücklin, 1934, 1954; Tanner, 1958; P. Allen, 1959; Dzulynski and Zak, 1960; Wulf, 1962; Crimes, 1970; Puigdefabrigas, 1974; Plummer, 1978), and there are reports of wave-terracing (e.g. P. Allen, 1959; Wunderlich, 1970), breached ripples or dunes (P. Allen, 1959; Banks et al., 1971), and of pits made by wind-driven foam (Wunderlich, 1970).

Current ripples beneath tidal currents

Tidal currents are comparatively weak away from shore and current ripples may be the predominant or only bed configuration created. Although small in scale, they may lag the changing currents (usually rotary) substantially, because of the relatively low sediment transport rates prevailing. A time-lapse study of current ripples in 8.5 m of water off Marthas Vineyard, Massachusetts, was made by Owen et al. (1967). In terms of direction of facing, the forms lagged the predicted current considerably at this site, though with no indication of significant decoupling. Kachel and Sternberg (1971) give quantitative evidence for lag, measuring the shape and motion of current ripples at a depth of 31 m in Puget Sound. Once sand movement began after slack water, the ripples migrated downcurrent and increased progressively in size throughout the 1.25 hr observation period, even though the current velocity and bed shear stress attained maxima about half-way through this interval. Additional field work, and related laboratory programmes, should cast more light on the directional and dimensional lags noted above. Under what conditions, for example, would decoupling occur?

Changes of wind-wave regime

Wind-wave induced bottom currents powerful enough to entrain sand occur from time to time offshore in water depths down to 100–200 m (e.g.

Hadley, 1964; Draper, 1967a; Drapeau, 1970; Komar et al., 1972). The changeability of these currents depends on the incoming swell and on the storm waves of more local origin. One pattern of wave ripples is occasionally superimposed on an earlier pattern (Forel, 1883; R.A. Davis, 1965; Komar et al., 1972), presumably as the result of a sufficiently rapid change of wave direction. However, angles between the crests of the successive sets of 30° and less are reported by Komar and his associates, and it would seem that quite small changes of wave direction are able to effect decoupling. We can only speculate on the possible effects of changing wave size as well. Whether superimposed or not, ripples formed well offshore seem to have little preservation potential, on account of organic reworking (Inman, 1957; Kulm et al., 1975).

Near to the shore and landward of the breakers, waves change in strength and direction partly and sometimes mainly in response to local winds and the tide which, in varying water depths and fetches, induces a constantly changing pattern of local wave generation and of wave refraction, diffraction and reflection. Changes on the scale of hours and even minutes are much in evidence, in contrast to the situation offshore. Superimposed wave-related ripples (Fig. 12-16c, d) arise frequently (e.g. R. Richter, 1926b; Häntzschel, 1939; Bajard, 1966; C.R. Harris, 1974; Reineck and Singh, 1973; Machida et al., 1974), and the comparable patterns reported from sandstones (e.g. Kindle, 1917a; Stokes, 1950; Basumallick, 1963) could record much the same conditions.

Internal structures

The relatively simple external appearance of tide-shaped dunes often belies their complex internal structure, determined by their movement pattern during the diurnal/semidiurnal and spring-neap tidal cycles. A variety of internal reactivation structures may be present, many including the bedding discontinuities first recognized by Boersma (1969), and called by him "structural diastems".

The degree of asymmetry of the tidal time-velocity pattern at a site determines how any dunes there will travel with the ebb and flood of the water mass (Fig. 12-18). As sketched in Fig. 12-18a, many forms apparently move either with the flood alone or with the ebb alone (Van Straaten, 1950; Salsman et al., 1966; Daboll, 1969; Jindrich, 1969; G.R. Davies, 1970; Terwindt, 1970; Wright et al., 1973, 1975; Hine, 1975; Allen and Friend, 1976a, 1976b). The other limiting situation (Fig. 12-18c) is when the dunes at a site completely reverse their shape during one tidal cycle, perhaps being partly or wholly flattened out during the times of maximum flow (Farrell, 1970; J.D. Smith, 1970; Terwindt, 1970; Harrison et al., 1971; Farrow and Brander, 1971; Hawkins and Sebbage, 1972; Kumar and Sanders, 1974; Boothroyd and Hubbard, 1975; C.D. Green, 1975; Boggs and Jones, 1976).

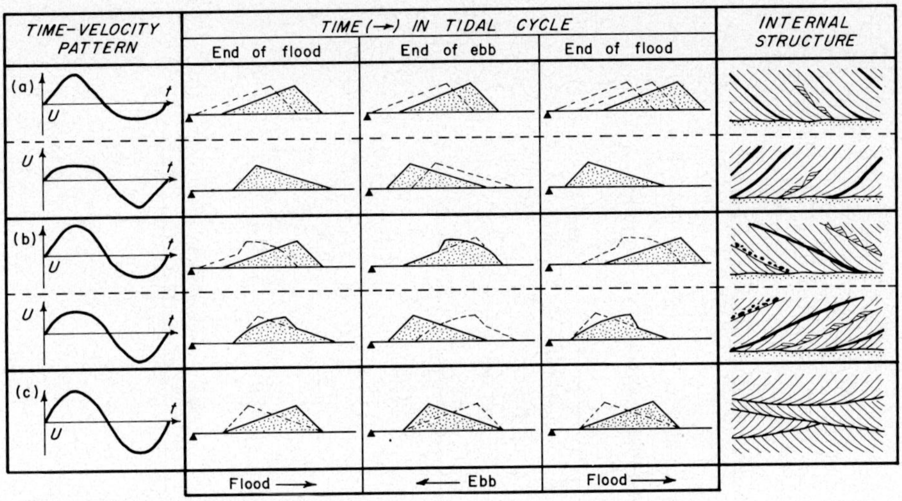

Fig. 12-18. Schematic representation of dune movement and resultant internal structures (including reactivation features) as a function of the asymmetry in time-velocity pattern of tidal currents.

In intermediate cases (Fig. 12-18b), incomplete crest reversal occurs during the part of the tidal cycle with the weaker currents, a low slip-face building up on the plinth presented by the former stoss-side of the dune (Klein, 1970b; Stone and Summers, 1972; Knight, 1972; Boothroyd and Hubbard,

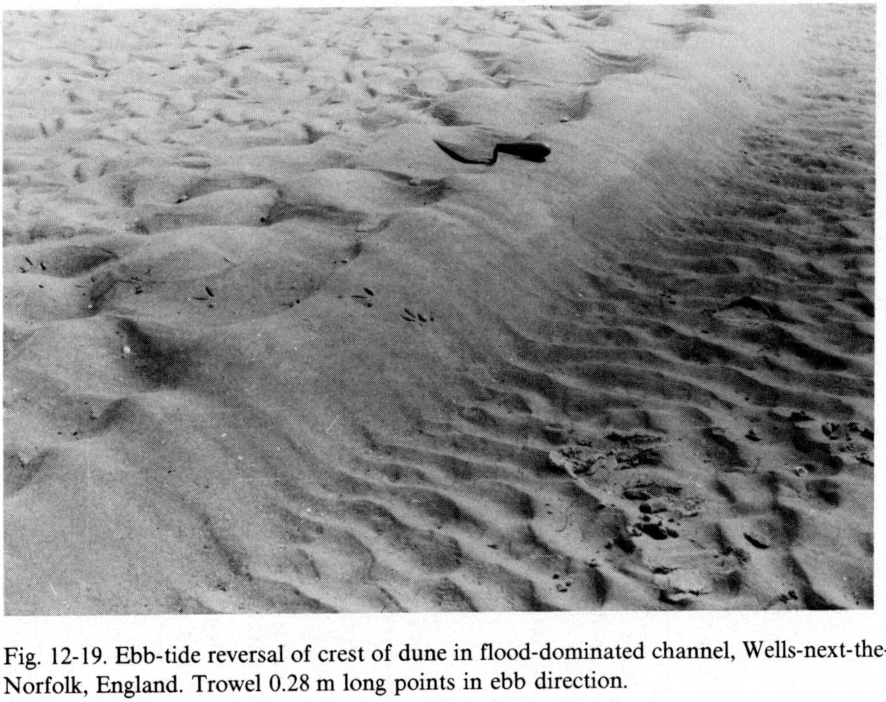

Fig. 12-19. Ebb-tide reversal of crest of dune in flood-dominated channel, Wells-next-the-Sea, Norfolk, England. Trowel 0.28 m long points in ebb direction.

TABLE 12-II

The facing (asymmetry) of dune trains in the Rhine–Meuse estuary (The Netherlands), as a function of tidal dominance and stage
Data of Terwindt (1970)

Dune orientation when train was surveyed	Ebb-dominated channels		Channels with neither ebb nor flood dominance		Flood-dominated channels		Number of dune trains
	surveyed after flood tide (%)	surveyed after ebb tide (%)	surveyed after flood tide (%)	surveyed after ebb tide (%)	surveyed after flood tide (%)	surveyed after ebb tide (%)	
Facing with ebb	9.9	39.8	22.5	19.0	2.4	6.4	374
No consistent facing	7.8	24.9	25.3	18.4	16.4	7.2	293
Facing with flood	1.2	7.0	14.8	13.3	38.7	25.0	256
Number of dune trains	63	240	196	159	156	109	923

1974; Dalrymple et al., 1975). Fig. 12-19 shows a flood-oriented intertidal dune whose crest has become rounded by an ebb-current powerful enough to make current ripples but insufficiently strong to modify the wave-related ripple marks lying in the deeper water of the dune trough. Cornish (1901a) long ago described an apparently similar superimposition of ebb-oriented current ripples on flood-facing dunes.

Because of the horizontal segregation of tidal currents, dunes representing these different modes of movement can occur in close association. Table 12-II summarizes Terwindt's (1970) extensive data on the extent to which dunes in tidal channels of the Netherlands retained their orientation from one part to another of the tidal cycle. He classified the channels as either ebb-dominated, neutral or flood-dominated, and the dunes as either ebb-oriented, "symmetrical", or flood-oriented. Most ebb-oriented dunes exist during ebb tides in ebb-dominated channels, and most flood-directed dunes occur in flood-dominated channels during floods. However, substantial numbers of ebb-oriented dunes retain that orientation during floods, as do flood-directed forms during ebbs.

Figure 12-18 also summarizes the internal structures linked to these movement patterns. The reactivation structures associated with dunes travelling during only a part of the tidal cycle (Fig. 12-18a) are defined by either a drape or infiltration of muddy sediment on certain cross-beds, or a steeply inclined erosion surface surmounted by upslope-dipping cross-laminated sets (Boersma, 1969). Analogous structures thought to be tidal in origin are reported from sandstones (Van der Linden, 1963; Allen and Narayan, 1964; De Raaf and Boersma, 1971; Sellwood, 1972, 1975; Allen and Kaye, 1973; N.L. Banks, 1973b; Bosence, 1973; Surlyk et al., 1973). Some of these structures may be of sand-wave rather than dune origin. In the intermediate case (Fig. 12-18b), the convex-up structure is marked by long, gently erosion surfaces on which new cross-strata have accumulated more steeply (J.L. Barr et al., 1970; Farrell, 1970; Klein, 1970b; Dalrymple et al., 1975). The erosion surfaces are fairly closely spaced laterally and may bear either a drape of finer sediment, an armouring, or cross-laminated ripple forms. Presumed tidal sandstones yield several examples of reactivation structures resembling this intermediate case (Reiche, 1938; Niehoff, 1958; Klein, 1970a; Swett et al., 1971; Gietelink, 1973; Chisholm and Dean, 1974; H.D. Johnson, 1975; Anderton, 1976; Levell, 1980). On a cautionary note, however, similar structures can be made by sand waves and through one-way dune migration over river shoals (N.L. Banks, 1973a). Vigorous dune movement in both parts of the tidal cycle (Fig. 12-18c) creates "herring-bone" cross-bedding internally (Farrell, 1970; Kumar and Sanders, 1974; Park, 1974), instances being known from sandstones (e.g. Niehoff, 1958; Young and Jefferson, 1975; Eriksson, 1977b).

The stage and velocity variations associated with the spring-neap cycle can also make reactivation structures. De Raaf and Boersma (1971) found at the

Ossenisse Shoals off the coast of the Netherlands that the springs caused dune movement and the deposition of steeply dipping cross-beds. The neap tides, however, rounded off the dune crests and laid down gently inclined beds, the two kinds of deposit alternating laterally within each dune. It is finally worth noting that overwash, beach, and offshore bars can include features resembling the reactivation structures seen in dunes (Hoyt, 1962; Hayes et al., 1969; Davis et al., 1972; Wunderlich, 1972; Hester and Fraser, 1973; Greenwood and Davidson-Arnott, 1975; Hine, 1979; Hobday and Jackson, 1979).

Interbedded sands and muds

An important group of structures identifiable mainly with tidal environments is classified by Reineck and Wunderlich (1968a) under the titles *Flaserschichten* (flaser bedding), *wellige Wechselschichten* (wavy bedding), and *Linsenschichten* (lenticular bedding) (Fig. 12-20). Flaser bedding, derived from a German word meaning veined or streaky, consists of superimposed and often erosively related layers of cross-laminated and ripple-marked sand associated with mainly concave-up partings of mud trapped in ripple troughs (Reineck, 1960b, 1960c). In wavy bedding, laterally continuous mud layers alternate with equally continuous and sometimes sharp-based and graded sand layers, cross-laminated internally and ripple-marked on top. In lenticu-

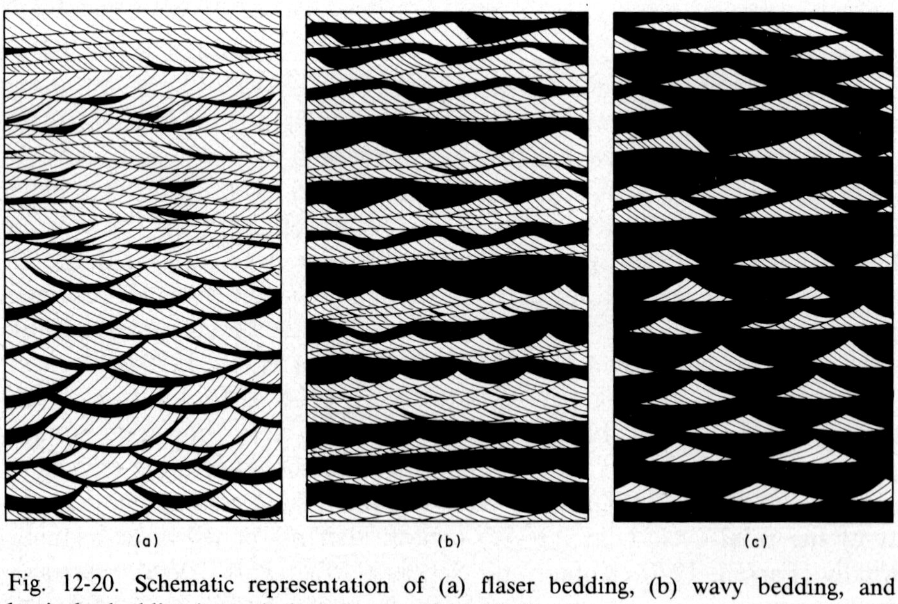

Fig. 12-20. Schematic representation of (a) flaser bedding, (b) wavy bedding, and (c) lenticular bedding in vertical section parallel with direction of wave propagation or current. The lower half of each diagram shows the structure due to wave ripples and the upper half that related to current and wave-current ripples. Adapted from Reineck and Wunderlich (1968a).

lar bedding one finds the internally cross-laminated bodies of incomplete ripples (ripple form sets) preserved in mud (Reineck, 1960b, 1960c). A progressive increase in the proportion of mud is represented by the morphological sequence from flaser bedding, through wavy bedding, to lenticular bedding. Either wave ripples, wave-current ripples, or current ripples may typify the sand layers.

These structures are known chiefly from the upper zones of intertidal flats and from subtidal depths in channels and just offshore. Most reports come from the tidal environments of northwest Germany (Lüders, 1930; Häntzschel, 1936; Reineck, 1958b, 1963; Reineck and Singh, 1967; Reineck and Wunderlich, 1969; Dörjes et al., 1970) and of the Netherlands (Van Straaten, 1954a; De Ridder, 1960; Oomkens and Terwindt, 1960; De Raaf and Boersma, 1971; Terwindt, 1971b). G. Evans (1965) found them in the intertidal sediments of the Wash, Bajard (1966) and Larsonneur (1975) figured examples off the French coast, and Howard et al. (1975a, 1975b) recorded instances from estuaries in Georgia, U.S.A. They are also reported from offshore waters as deep as 30 m (Reineck, 1963; Werner, 1968).

Reineck (1960b, 1960c) proposed a simple explanation for the structures, namely, that the rippled sand layers represented strong mid-tide currents and the mud partings or layers the deposition during slack water of otherwise suspended fines. This view was repeated by Reineck and Wunderlich (1968a) and accepted by Klein (1971). However, McCave (1970, 1971b) has vigorously challenged Reineck's interpretation, partly on theoretical grounds. According to him, it is quite exceptional that there is sufficient time and suspended material available to allow the deposition, during each period of slack water, of a mud layer of the order of thickness observed. These structures imply to McCave the action of processes on a much larger time scale, perhaps those associated with storm and calm conditions and, one may add, the spring–neap cycle. The various experiments of Reineck and Wunderlich (1967, 1969), Wunderlich (1969), and Terwindt and Breusers (1972) convincingly demonstrate that alternate sand and mud layers recording an ebb–flood cycle can be accumulated, but fail to show that flaser, wavy and lenticular bedding invariably or even predominantly record this periodicity. Each mud layer could represent many tides, or even the long calms after storms. New field and laboratory experiments seem necessary, designed to explore a much wider range of conditions.

The occurrence in sandstones of structures of the flaser-lenticular bedding family is widely taken to be evidence of a tidally-influenced depositional environment (e.g. Van Straaten, 1954b; Niehoff, 1958; Reading and Walker, 1966; Häntzschel and Reineck, 1968; Klein, 1970a; De Raaf and Boersma, 1971; Goldring, 1971; Epstein and Epstein, 1972; Kuijpers, 1972; Sellwood, 1972, 1975; Surlyk et al., 1973; R.M. Sykes, 1974; Roep et al., 1975).

BEDFORMS AND INTERNAL STRUCTURES IN CHANGING AEOLIAN ENVIRONMENTS

Ballistic ripples

These are the smallest of the aeolian bedforms and, like ripples generated by aqueous currents, often occur superimposed one set at an angle upon another (Beadnell, 1910; W.H.J. King, 1916; Matschinski, 1955; Verlaque, 1958; Sharp, 1963; Bigarella et al., 1971; Bigarella, 1972; I.G. Wilson, 1973; Ellwood et al., 1975). These superimposed patterns are observed to record changes of wind direction (Sharp, 1963; Stone and Summers, 1972). Since the two sets commonly differ in wavelength, they may also reflect variations in wind strength. Sharp found that ripple asymmetry could be reversed in 10^2 s or so by moderately strong winds, even when the trend of the new wind diverged by as much as 20° from the line of the old. A greater angular divergence, however, led to decoupling and to the superimposition of a new set of ripples.

Dunes

Transverse dunes and barkhans respond in a clear way to changes of wind direction. Hedin (1896), while travelling in the Takla Makan, observed dunes whose slip faces were reversed on a sharp change of wind direction, and others have later reported similarly (Cornish, 1897, 1914; W.H.J. King, 1916; Chudeau, 1920; Bourcart, 1928; Kádár, 1934). King attributed the reversal to seasonal changes, and Kádár to storms. Hastenrath (1967) in Peru found that regular nocturnal katabatic winds were the cause though Hörner (1957) and Verlaque (1958) in other areas preferred less frequent events. The Kelso Dunes of the Mojave Desert, developed under a particularly complex wind regime, vary in orientation on several different time scales and show little net horizontal movement (Sharp, 1966). Many workers find that seasonal changes explain the reversals and other crestal modifications they observed (e.g. Cooper, 1958; Holm, 1960; E.D. McKee, 1966b; Glennie and Evamy, 1968; Glennie, 1970; McKee and Douglass, 1971). In stable regimes, however, barkhan and transverse dunes show little tendency to crestal modification (e.g. Beadnell, 1910).

A change of wind direction often brings ballistic ripples on to the former avalanche faces of dunes (E.D. McKee, 1945; Hastenrath, 1967). As with tide-shaped dunes, the ripples commonly range steeply up and down the old slip face, or point against the former wind direction. Nevertheless, some of these ripples may reflect the action of the separated flow to lee of the dune, and not a shift of the wind (Cornish, 1897, 1900, 1914; Allen, 1968c), a possibility often discounted (Cooper, 1958; Sharp, 1966).

Several workers record dune fields showing a "cross-hatched" pattern, one set of dunes apparently lying across another unequal set (Melton, 1940; Norris and Norris, 1961; E.D. McKee, 1966b). Some of these may be due to seasonally shifting winds, and may be compared with superimposed tidal dunes (Cornish, 1901a; Kellerhals and Murray, 1969). The time and spatial scales cannot be compared in the two cases, but both record a lag in dune behaviour.

Internal structures

The response of aeolian dunes to a changeable wind regime is clearly similar to that of tide-shaped dunes to variations of tidal current strength and direction (Fig. 12-18). Hence it is unsurprising to find that the reactivation structures preserved within aeolian sands and sandstones often closely resemble their tidal counterparts.

Land (1964) first detected these structures in modern aeolian dunes. On dissecting low coastal examples in an area where the winds cluster about two direction broadly at right angles, he found that the steeply inclined cross-beds were repeatedly separated laterally into bundles by low-angle erosion surfaces (see also Bigarella et al., 1969). Perhaps E.D. McKee's (1966b) work on the gypsum dunes of White Sands, New Mexico, provides the finest record of these reactivation structures. Here the steeply dipping cross-beds and more gently inclined erosion surfaces occur on a scale almost one order of magnitude larger than those found in the coastal dunes (Fig. 12-21). Curiously, McKee offered no explanation of the reactivation structures, which are fairly clearly connected with the possible changes of dune shape which could proceed in harmony with the known wind variations. Sharp (1966) provided from the Kelso Dunes structural evidence of repeated shifts of

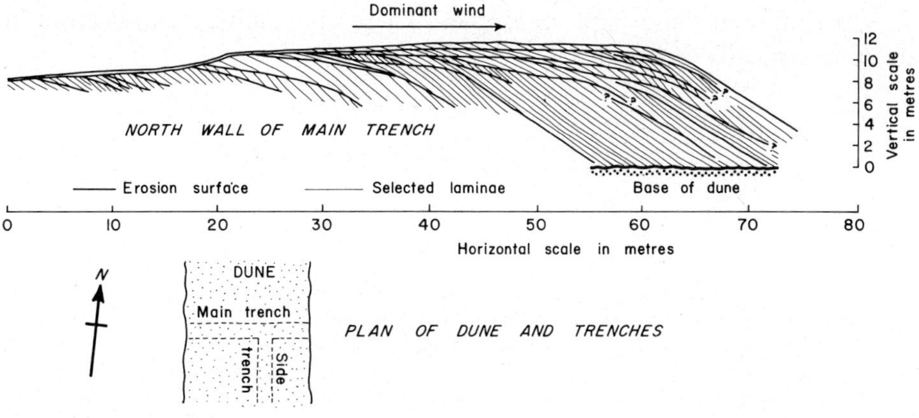

Fig. 12-21. Reactivation structures observed in a transverse dune of gypsum sand trenched vertically at right-angles to the crest, Whitesands National Monument, New Mexico, U.S.A. After McKee (1966b).

wind direction. Glennie (1970), dissecting a barkhan in the Rajasthan Desert, obtained evidence seemingly indicative of the repeated ballistic rippling of a slip face by changeable winds, a feature known from aeolian sandstones (E.D. McKee, 1945; Sanderson, 1974). Reactivation structures in some instances like those described by McKee from White Sands occur in aeolian rocks (Shotton, 1937; MacKenzie, 1964; Laming, 1966; D.B. Thompson, 1969; Horne, 1971; Vacher, 1973). Almeida (1953) and Gradzynski and Jerzykiewicz (1974a, 1974b) record presumed aeolian cross-bedding in which mud layers drape like skirts the lower foresets, plausibly the result of the partial drowning of the dunes, as in a tidal flat or fluctuating lake.

SUMMARY

The response of ripples and dunes to the changeable regimes in which they occur naturally can be considered at two levels, that of the population of forms, and that of the individual feature. Populations respond to changing environmental stresses with a degree of time-lag seemingly determined by the rate of replacement of individuals within the population and the ability of individuals to react to change during their life-spans. Change demands, and can be measured in terms of, sediment transport. At the individual level, change in the environment can lead to a wide variety of surface and internal structures. Reactivation structures, important amongst these, record changes of water stage and/or current direction. They abound in cross-bedded tidal and aeolian deposits and are also known from cross-bedded fluviatile sands. Some reactivation structures, however, could be due to the stochastic behaviour of bed features during times when conditions overall were essentially stable. Fluctuating water stage creates a wide variety of surface structures associated with bedforms, for example, planed-off ripples, terraced dunes, and superimposed dunes and/or ripples. These are significant indicators of exposure, typically at the strand.

REFERENCES

Aario, R., 1971. Bull. Geol. Soc. Finl., 43: 163–172.

Aario, R., 1972. Ann. Acad. Sci. Fenn., Ser. A3, 111: 1–55.

Aartolahti, T., 1973. Fennia, 127: 1–53.

Ab Iowerth, H., Hamilton, N. and Loveland, P.J., 1967. Geol. Mag., 104: 564–573.

Abbott, J.E. and Francis, J.R.D., 1977. Philos. Trans. R. Soc. London, Ser. A., 284: 225–254.

Abbott, M.B., 1961. Houille Blanche, 16: 622–628, 827–846.

Abou-Seida, M.M., 1965. Bed load function due to wave action. Univ. Calif. (Berkeley), Hydraul. Eng. Lab., Techn. Rep., No. Hel-2-11.

Acaroglu, E.R. and Graf, W.H., 1968. Bull. Int. Assoc. Sci. Hydrol., 18: 123–135.

Achia, B.U. and Thompson, D.W., 1977. J. Fluid Mech., 81: 439–464.

Ackerman, A.S.E., 1945. Nature (London), 155: 82.

Ackers, P., 1972. Sediment transport in channels: an alternative approach. Hydraul. Res. Stn. (Wallingford), Intern. Rep., No. 102.

Ackers, P. and White, W.R., 1973. J. Hydraul. Div. ASCE, 99 (HY11): 2041–2060.

Acton, J.R. and Dyer, C.M., 1975. J. Geol. Soc. London, 131: 63–67.

Adams, D.J. and Matheson, A.J., 1972. J. Chem. Phys., 56: 1989–1994.

Adeyemo, M.D., 1970. Proc. 12th Congr. Coastal Eng., 1: 435–460.

Ahlbrandt, T.S., 1975. Mt. Geol., 12: 61–73.

Ahlbrandt, T.S. and Andrews, S., 1978. Palaeogeogr. Palaeoclimatol. Palaeoecol., 25: 327–351.

Aksoy, S., 1973. Proc. 15th Congr. Inst. Assoc. Hydraul. Res., 1: 217–224.

Alam, A.M.Z., Cheyer, T.F. and Kennedy, J.R., 1966. Friction factors for flow in sand bed channels. Mass. Inst. Technol., Dep. Civ. Eng., Res. Rep., No. 78.

Albertson, M.L., 1953. Proc. 5th Hydraul. Conf., Univ. Iowa, pp. 243–261.

Albertson, M.L., Simons, D.B. and Richardson, E.V., 1958. J. Hydraul. Div. ASCE, 84 (HY1): 23–31 (Pap. 1558).

Alder, B.J., 1955. J. Chem. Phys., 23: 263–271.

Alder, B.J. and Wainwright, T.E., 1957. J. Chem. Phys., 27: 1208–1209.

Alder, B.J., Frankel, S.P. and Lewison, V.A., 1955. J. Chem. Phys., 23: 417–422.

Alexander, H.S., 1932. J. Geol. 40: 305–337.

Alger, G.R. and Simons, D.B., 1968. J. Hydraul. Div. ASCE, 94 (HY3): 721–737.

Ali, K.H.M. and Ridgway, A., 1977. Proc. Inst. Civ. Eng., 63 (2): 203–214.

Allee, W.C., Emerson, A.E., Park, O., Park, T. and Schmidt, K.P., 1949. Principles of Animal Ecology. Saunders, Philadelphia, Penn., 837 pp.

Allen, G.P., Deresseguier, A. and Klingebiel, A., 1969. C.R. Acad. Sci., Sér. D, 269: 2167–2169.

Allen, G.P., Castaing, P., Feral, A., Klingebiel, A. and Vigneaux, M., 1970. Bull. Inst. Géol. Bassin Aquitaine, 8: 99–154.

Allen, G.P., Courtois, G., Jeanneau, B. and Klingebiel, A., 1971. Bull. Inst. Géol. Bassin Aquitaine, 11: 287–315.

Allen, J., 1942. J. Inst. Civ. Eng., 18: 1–34.

Allen, J. and Gibson, D.H., 1959. Proc. Inst. Civ. Eng., 13: 363–386.

Allen, J.R.L., 1962a. J. Sediment. Petrol., 32: 657–697.

Allen, J.R.L., 1962b. Nature (London), 194: 167–169.

Allen, J.R.L., 1963a. Geol. Mijnbouw, 42: 223–238.

Allen, J.R.L., 1963b. Liverpool Manchester Geol. J., 3: 187–236.

Allen, J.R.L., 1963c. Sedimentology, 2: 93–114.

Allen, J.R.L., 1963d. Nature (London), 198: 847–849.

Allen, J.R.L., 1964a. Sedimentology, 3: 89–108.

Allen, J.R.L., 1964b. Sedimentology, 3: 163–198.
Allen, J.R.L., 1965a. Bull. Am. Assoc. Pet. Geol., 49: 547–600.
Allen, J.R.L., 1965b. J. Geol., 73: 95–116.
Allen, J.R.L., 1965c. J. Sediment. Petrol., 35: 331–338.
Allen, J.R.L., 1966. Sedimentology, 6: 153–190.
Allen, J.R.L., 1967a. Sedimentology, 9: 75–88.
Allen, J.R.L., 1967b. Mar. Geol., 5: 429–446.
Allen, J.R.L., 1967c. J. Sediment. Petrol., 37: 691–692.
Allen, J.R.L., 1968a. Geol. Mijnbouw, 47: 173–185.
Allen, J.R.L., 1968b. Sedimentology, 10: 161–182.
Allen, J.R.L., 1968c. Current Ripples. North-Holland, Amsterdam.
Allen, J.R.L., 1968d. Geol. Mijnbouw, 47: 3–16.
Allen, J.R.L., 1969a. J. Sediment. Petrol., 39: 607–623.
Allen, J.R.L., 1969b. Geogr. Ann., A51: 61–96.
Allen, J.R.L., 1969c. Proc. Geol. Assoc., 80: 1–42.
Allen, J.R.L., 1969d. Geol. Mag., 106: 309–321.
Allen, J.R.L., 1969e. Geol. Soc. Am. Bull., 80: 1923–1930.
Allen, J.R.L., 1969f. Geol. Mijnbouw, 48: 3–8.
Allen, J.R.L., 1970a. Geol. Mijnbouw, 42: 211–220.
Allen, J.R.L., 1970b. Geol. Mijnbouw, 49: 13–22.
Allen, J.R.L., 1970c. J. Geol., 78: 326–351.
Allen, J.R.L., 1970d. J. Sediment. Petrol., 40: 298–323.
Allen, J.R.L., 1970e. Sedimentology, 14: 5–26.
Allen, J.R.L., 1970f. Scott. J. Geol., 6: 146–161.
Allen, J.R.L., 1970g. Physical Processes of Sedimentation. Allen and Unwin, London.
Allen, J.R.L., 1971a. J. Sediment. Petrol., 41: 695–702.
Allen, J.R.L., 1971b. J. Sediment. Petrol., 41: 97–113.
Allen, J.R.L., 1971c. Sediment. Geol., 5: 167–385.
Allen, J.R.L., 1971d. J. Geol. Soc., London, 127: 553–561.
Allen, J.R.L., 1971e. Proc. Yorks Geol. Soc., 38: 293–294.
Allen, J.R.L., 1972a. Sedimentology, 18: 105–111.
Allen, J.R.L., 1972b. Geogr. Ann., A53: 157–187.
Allen, J.R.L., 1972c. Sediment. Geol., 8: 309–316.
Allen, J.R.L., 1973a. Sedimentology, 20: 189–202.
Allen, J.R.L., 1973b. J. Geol. Soc., London, 129: 537–541.
Allen, J.R.L., 1973c. Sedimentology, 20: 323–329.
Allen, J.R.L., 1974a. Sediment. Geol., 12: 73–167.
Allen, J.R.L., 1974b. Earth-Sci. Rev., 10: 263–242.
Allen, J.R.L., 1974c. Sedimentology, 21: 71–86.
Allen, J.R.L., 1976a. Sediment. Geol., 15: 1–53.
Allen, J.R.L., 1976b. Sediment. Geol., 16: 99–130.
Allen, J.R.L., 1976c. Sediment. Geol., 16: 255–279.
Allen, J.R.L., 1976d. Sediment. Geol., 15: 309–321.
Allen, J.R.L., 1976e. Earth Surf. Processes, 1: 361–374.
Allen, J.R.L., 1977a. In: K.J. Gregory (Editor), River Channel Changes. Wiley, London, pp. 15–45.
Allen, J.R.L., 1977b. Sedimentology, 24: 53–62.
Allen, J.R.L., 1978a. Sediment. Geol., 20: 17–28.
Allen, J.R.L., 1978b. Sediment. Geol., 20: 165–216.
Allen, J.R.L., 1979. J. Geol. Soc., London, 136: 672–683.
Allen, J.R.L., 1980a. Sedimentology, 27: 317–323.
Allen, J.R.L., 1980b. Geol. Mag., 117: 437–446.

Allen, J.R.L., 1980c. Sediment. Geol., 26: 281–328.
Allen, J.R.L., 1981. Nature (London), 289: 579–581.
Allen, J.R.L. and Collinson, J.D., 1974. Sediment. Geol., 12: 169–178.
Allen, J.R.L. and Friend, P.F., 1968. Geol. Soc. Am., Spec. Pap., 106: 21–74.
Allen, J.R.L. and Friend, P.F., 1976a. J. Geol. Soc., London, 132: 17–26.
Allen, J.R.L. and Friend, P.F., 1976b. Sedimentology, 23: 329–346.
Allen, J.R.L. and Kaye, P., 1973. Geol. Mag., 110: 153–163.
Allen, J.R.L. and Leeder, M.R., 1980. Sedimentology, 27: 209–217.
Allen, J.R.L. and Narayan, J., 1964. Geol. Mijnbouw, 43: 451–461.
Allen, J.R.L. and Williams, B.P.J., 1978. Geol. J., 13: 113–136.
Allen, J.R.L. and Williams, B.P.J., 1979. J. Geol. Soc., London, 136: 361–366.
Allen, P., 1959. Philos. Trans. R. Soc. London, Ser. B, 242: 282–346.
Allen, P., Allen, J.R.L., Goldring, R. and Maycock, I.D., 1960. Geol. Mag., 97: 261–263.
Allen, P.B. and Welch, N.H., 1967. Publ. Int. Assoc. Sci. Hydrol., 75: 355–366.
Allen, W.A., Mayfield, E.B. and Morrison, H.L., 1957. J. Appl. Phys., 28: 370–376.
Allix, A., 1924. Geogr. Rev., 14: 519–560.
Almagor, G., 1967. In: A.F. Richards (Editor), Marine Geotechnique. University of Chicago Press, Urbana, Ill., p. 131–147.
Almagor, G. and Garfunkel, Z., 1979. Bull. Am. Assoc. Pet. Geol., 63: 324–340.
Almeida, F.F.M., 1953. Proc. 19th Int. Geol. Congr., pp. 9–24.
Al-Zanaidi, M.A. and Dore, B.D., 1976. Geofis Pura Appl., 114: 403–414.
Ambroggi, R.P., 1966. Sci. Am., 214: 21–29.
Amein, M., 1966. J. Geophys. Res., 71: 401–410.
Amos, A.F., Langseth, M.G. and Markl, R.G., 1972. Stud. Phys. Oceanogr., 1: 49–62.
Amos, C.L. and Collins, M.B., 1978. J. Sediment. Petrol., 48: 849–856.
Amstutz, G.C. and Chico, R., 1958. Bull. Ver. Schweiz. Pet.-Geol.-Ing., 24: 47–52.
Anczurowski, E. and Mason, S.G., 1967. J. Colloid Interface Sci., 23: 522–546.
Anczurowski, E. and Mason, S.G., 1968. Trans. Soc. Rheol., 12: 209–215.
Anczurowski, E., Cox, R.G. and Mason, S.G., 1967. J. Colloid Interface Sci., 23: 547–562.
Anderegg, F.O., 1931. Ind. Eng. Chem., 23: 1058–1064.
Andersen, S.A., 1931. Dan. Geol. Unders., Raeke II, No. 54: 201 pp.
Andersen, S.A., 1937. Dan. Geol. Unders., Raeke II, No. 59: 52 pp.
Anderson, A.G., 1942. Trans. Am. Geophys. Union, 23: 678–683.
Anderson, A.G., 1953. Proc. 3rd Mid-Western Conf. Fluid Mech., pp. 379–395.
Anderson, F.E., 1968. Mar. Geol., 6: 419–438.
Anderson, F.E., 1973. Mar. Geol., 14: 101–116.
Anderson, T. and Flett, J.S., 1903. Philos. Trans. R. Soc., Ser. A., 200: 353–553.
Andersson, J.G., 1906. J. Geol., 14: 91–112.
Anderton, R., 1975. Geol. Mag., 112: 337–348.
Anderton, R., 1976. Sedimentology, 23: 429–458.
Anding, M.G., 1970. Hydraulic characteristics of Mississippi River channels. U.S. Army Corps Eng., Vicksburg, Potamol. Invest. Rep. No. 19-3.
Andreasen, A.H.M. and Andersen, J., 1930. Kolloidzeitschrift, 50: 217–228.
Andrée, K., 1915. Geol. Rundsch., 6: 351–397.
Andrews, J.T., 1961. Cah. Géogr. Québec, No. 10: 219–230.
Andrews, J.T., 1963. Geogr. Bull., No. 20: 82–129.
Andrews, J.T., 1965. J. Sediment. Petrol., 35: 753–758.
Andrews, J.T. and King, C.A.M., 1968. Proc. Yorks. Geol. Soc., 36: 435–461.
Andrews, J.T. and Smith, D.I., 1970. Q. J. Geol. Soc. London, 125: 503–542.
Andrews, J.T. and Smithson, B.B., 1966. Geol. Soc. Am. Bull., 77: 271–290.
Andrews, P.B., 1970. Rep. Invest. Bur. Econ. Geol. Univ. Texas, No. 67: 147 pp.
Andrews, P.B. and Laird, M.G., 1976. Sediment. Geol., 16: 21–44.

Angelucci, A., De Rosa, E., Fierro, G., Gnaccolini, M., La Monica, G.B., Martinis, B., Parea, G.C., Pescatore, T., Rizzini, A. and Wezel, F.C., 1967. Geol. Rom., 6: 345–420.
Annambhotla, V.S.S., Sayre, W.W. and Livesey, R.H., 1972. ASCE J. Waterw., Harbors Coastal Eng. Div., 98 (WW4): 489–510.
Ansley, R.W. and Smith, T.N., 1967. AIChE J., 13: 1193–1196.
Antevs, E., 1928. Geogr. Rev., 18: 675–676.
Antonia, R.A., Danh, H.Q. and Prabhu, A., 1976. Phys. Fluids, 19: 1680–1686.
Antsyferov, S.M., 1969. Meteorol. Gidrol., 1969 (3): 79–86 (NLL Transl. RTS 5976).
Aoi, T., 1955. J. Phys. Soc. Jpn., 10: 119–129.
Ariathurai, R. and Arulanandan, K., 1978. J. Hydraul. Div. ASCE, 104 (HY2): 279–283.
Armstrong, J.E. and Brown, W.C., 1954. Bull. Geol. Soc. Am., 65: 349–364.
Arnborg, L., Walker, H.J. and Peippo, J., 1966. Geogr. Ann., A48: 195–210.
Arons, A.B., Wetheim, G. and Krumholz, M., 1951. U.S. Navy Dep., Bur. Ordnance, Rep. No. 485.
Ashida, K. and Tanaka, Y., 1967. Proc. 12th Congr. Int. Assoc. Hydraul. Res., 2: 103–110.
Ashley, G.M., 1980. Earth Surf. Processes, 5: 347–368.
Ashwell, I.Y., 1966. Ann. Assoc. Am. Geogr., 56: 529–540.
Ashwell, I.Y., 1972. Geogr. Mag., 44: 322–327.
Ashwell, I.Y., 1973. Rep. Br. Schools Explor. Soc., 1971–72, pp. 39–51.
Asquith, S.M. and Cramer, S.L., 1975. J. Geol., 83: 657–661.
Athy, L.F., 1930. Bull. Am. Assoc. Pet. Geol., 14: 1–35.
Atlas, D., 1960. J. Meteorol., 17: 244–258.
Audley-Charles, M.G., Curray, J.R. and Evans, G., 1977. Geology, 5: 341–344.
Auffret, G.A., Berthois, L., Cabioch, L., Douville, J.L. and Kenyon, N.L., 1972. Ann. Inst. Océanogr., Monaco, 48: 157–172.
Aufrère, L., 1934. Bull. Assoc. Géogr. Fr., No. 83: 130–142.
Aufrère, L., 1935. Geogr. Ann., 17: 481–498.
Augustinus, P.G.E.F. and Riezbos, H.T., 1971. Geol. Mijnbouw, 50: 341–348.
Axelsson, V., 1967. Geogr. Ann., A49: 3–127.
Ayer, J.E. and Soppett, F.E., 1965. J. Am. Ceram. Soc., 48: 180–183.
Ayer, J.E. and Soppett, F.E., 1966. J. Am. Ceram. Soc., 49: 207–210.
Aylmore, L.A.G. and Quirk, J.P., 1959. Nature (London), 183: 1752–1753.
Aylmore, L.A.G. and Quirk, J.P., 1960. Nature (London), 187: 1046–1048.
Aylmore, L.A.G. and Quirk, J.P., 1962. Clays Clay Miner., 9: 104–130.
Ayrton, H., 1910. Proc. R. Soc. London, Ser. A., 84: 285–310.
Aziz-ur-Rahman, Gough, D.I. and Evans, M.E., 1975. Can. J. Earth Sci., 12: 1465–1473.
Bache, D.H., 1976. J. Hydraul. Res., 14: 1–16.
Backhaus, E., 1967. Jahresber. Mitt. Oberrhein. Geol. Ver., NF49: 157–171.
Bagnold, R.A., 1931. Geogr. J., 78: 13–33.
Bagnold, R.A., 1933. Geogr. J., 82: 106–126.
Bagnold, R.A., 1935. Geogr. J., 85: 342–365.
Bagnold, R.A., 1936. Proc. R. Soc. London, Ser. A., 157: 594–620.
Bagnold, R.A., 1937a. Geogr. J., 89: 409–438.
Bagnold, R.A., 1937b. Proc. R. Soc. London, Ser. A., 163: 250–265.
Bagnold, R.A., 1940. J. Inst. Civ. Eng., 15: 27–52.
Bagnold, R.A., 1946. Proc. R. Soc. London, Ser. A., 187: 1–16.
Bagnold, R.A., 1947. J. Inst. Civ. Eng., 27: 447–469.
Bagnold, R.A., 1951. Geogr. J., 117: 78–86.
Bagnold, R.A., 1954a. Proc. R. Soc. London, Ser. A., 225: 49–63.
Bagnold, R.A., 1954b. The Physics of Blown Sand and Desert Dunes. Methuen, London, 2nd ed., 265 pp.
Bagnold, R.A., 1955. Proc. Inst. Civ. Eng., 4: 174–205.

Bagnold, R.A., 1956. Philos. Trans. R. Soc. London, Ser. A., 259: 235–297.
Bagnold, R.A., 1960. U.S. Geol. Surv., Prof. Pap., 282-E: 135–144.
Bagnold, R.A., 1963. In: M.N. Hill (Editor), The Sea, vol. 3. Interscience, New York, N.Y., pp. 507–528.
Bagnold, R.A., 1966. U.S. Geol. Surv., Prof. Pap., 422-I.
Bagnold, R.A., 1967. Proc. R. Soc. London, Ser. A., 295: 219–232.
Bagnold, R.A., 1968. Sedimentology, 10: 45–56.
Bagnold, R.A., 1973. Proc. R. Soc. London, Ser. A., 332: 473–504.
Bagnold, R.A., 1974. Proc. R. Soc. London, Ser. A., 340: 147–171.
Bagnold, R.A., 1977. Water Resour. Res., 13: 303–312.
Bagnold, R.A., 1980. Proc. R. Soc. London, Ser. A., 372: 453–473.
Bailey, A.B., 1974. J. Fluid Mech., 65: 401–410.
Bailey, A.B. and Hiatt, J., 1972. AIAA J., 10: 1436–1440.
Bairstow, L., Cave, B.M. and Lang, E.D., 1923. Philos. Trans. R. Soc. London, Ser. A., 223: 383–432.
Bajard, J., 1966. Rev. Géogr. Phys. Géol., Dyn., (2) 8: 39–111.
Baker, C.A., 1976. Proc. Geol. Assoc., 87: 285–306.
Baker, R.A., 1970. Geol. Soc. Am. Bull., 81: 1589–1590.
Baker, V.R., 1973. Geol. Soc. Am., Spec. Pap., No. 144.
Baker, V.R. and Ritter, D.F., 1975. Geol. Soc. Am. Bull., 86: 975–978.
Bakewell, H.P. and Lumley, J.L., 1967. Phys. Fluids, 10: 1880–1889.
Bakhmeteff, B.A. and Matzke, A.E., 1936. Trans. ASCE, 101: 630–647.
Bakhmeteff, B.A. and Matzke, A.E., 1938. Trans. ASME, 60: 111–118.
Ball, D.F., 1976. Biul. Perygl., 1976 (25): 5–7.
Ball, F.K., 1967. Deep-Sea Res., 14: 79–88.
Ball, M.M., 1967. J. Sediment. Petrol., 37: 556–591.
Ballade, P., 1953. Bull. Inf. Comm. Cent. Océanogr. Étude Côtes, No. 4: 163–173.
Ballance, P.F., 1964a. J. Sediment. Petrol., 34: 91–101.
Ballance, P.F., 1964b. N.Z. J. Geol. Geophys., 7: 466–499.
Bandyopadhyay, S., 1971. J. Sediment. Petrol., 41: 585–587.
Banerjee, I., 1973. Bull. Geol. Surv. Can., No. 226: 1–44.
Banerjee, I., 1977. J. Sediment. Petrol., 47: 771–783.
Banerjee, I. and McDonald, B.C., 1975. In: A.V. Jopling and B.C. McDonald (Editors), Glaciofluvial and Glaciolacustrine Sedimentation. Soc. Econ. Paleontol. Mineral., Spec. Publ., 23: 132–154.
Banham, P.H., 1966. Proc. Geol. Assoc., 77: 469–474.
Banks, N.L., 1973a. J. Sediment. Petrol., 43: 423–427.
Banks, N.L., 1973b. Sedimentology, 20: 213–228.
Banks, N.L., 1973c. Sediment. Geol., 10: 147–154.
Banks, N.L. and Collinson, J.D., 1975. Sedimentology, 22: 583–599.
Banks, N.L., Edwards, M.B., Geddes, W.P., Hobday, D.A. and Reading, H.G., 1971. Norg. Geol. Unders., No. 269: 197–236.
Banner, M.L. and Phillips, O.M., 1974. J. Fluid Mech., 65: 647–656.
Barclay, W.S., 1917. Geogr. J., 49: 53–56.
Barden, L., 1972a. Geotechnique, 22: 159–163.
Barden, L., 1972b. Q. J. Eng. Geol., 5: 85–102.
Barden, L. and Sides, G.R., 1970a. J. Soil Mech. Found., ASCE, 96 (SM4): 1171–1200.
Barden, L. and Sides, G., 1970b. Q. J. Eng. Geol., 3: 259–260.
Barlow, W., 1883. Nature (London), 29: 186–188, 205–207.
Barlow, W. and Pope, W.J., 1907. J. Chem. Soc., 91: 1150–1214.
Barnes, H.L., 1956. Am. J. Sci., 254: 493–505.
Barnes, R.S.K. and Green, J., 1972. The Estuarine Environment, Applied Science Publishers, London, 133 pp.

Barnett, T.P. and Kenyon, K.E., 1975. Rep. Prog. Phys., 38: 667–729.
Barr, D.I.H., 1963. Houille Blanche, 18: 739–756.
Barr, D.I.H., 1967. Houille Blanche, 22: 619–632.
Barr, D.I.H. and Hassan, A.M.M., 1963. Houille Blanche, 18: 757–766.
Barr, D.I.H. and Herbertson, J.G., 1968. Wat. Resour. Res., 4: 307–315.
Barr, J.L., Dinkelman, M.G. and Sandusky, C.L., 1970. J. Sediment. Petrol., 40: 445–449.
Barrett, P.J., 1970. J. Sediment. Petrol., 40: 395–411.
Barrett, P.J. and Kohn, B.P., 1975. In: K.S.W. Campbell (Editor), Gondwana Geology, Australian National University Press, Canberra, pp. 15–35.
Bart, H.A., 1977. Sediment. Geol., 19: 165–184.
Barth, H.K., 1978. Z. Geomorphol., S30: 39–54.
Bartok, W. and Mason, S.G., 1957. J. Colloid Sci., 12: 243–262.
Baschin, O., 1899. Z. Ges. Erdk. Berlin, 34: 408–421.
Bassett, A.B., 1888. A Treatise on Hydrodynamics, vol. 2. Deighton Bell, Cambridge.
Bassett, A.B., 1910. Q. J. Math., 41 (1): 369–381.
Bassett, D.A. and Walton, E.K., 1960. Q. J. Geol. Soc. London, 116: 85–105.
Basumallick, S., 1963. J. Sediment. Petrol., 33: 948–950.
Basumallick, S., 1966. Sedimentology, 6: 35–68.
Bata, G.L., 1959. Proc. 8th Congr. Int. Assoc. Hydraul. Res., pp. 13-C-1–13-C-18.
Bata, G.L. and Bogich, K., 1953. Proc. Minn. Int. Hydraul. Conv., Univ. Minneapolis, Minn., pp. 387–400.
Batchelor, G.K., 1967. An Introduction to Fluid Dynamics. Cambridge University Press, London.
Batchelor, G.K., 1970. J. Fluid Mech., 44: 419–440.
Batchelor, G.K., 1971. J. Fluid Mech., 46: 813–829.
Bates, C.C., 1953. Bull. Am. Assoc. Pet. Geol., 37: 2119–2162.
Bayazit, M., 1969. J. Hydraul. Div. ASCE, 95 (HY4): 1109–1127.
Beadnell, H.J.L., 1910. Geogr. J., 35: 379–392.
Bearman, P.W. and Zdravkovich, M.M., 1978. J. Fluid Mech., 89: 33–47.
Beaty, C.B., 1963. Ann. Assoc. Am. Geogr., 53: 516–535.
Beaumont, P., 1971. Proc. Yorks Geol. Soc., 38: 343–360.
Beaumont, P. and Overlander, T.M., 1971. Bull. Geol. Soc. Am., 82: 1695–1698.
Beavers, G.S., Sparrow, E.M. and Rodenz, D.E., 1973. Trans. ASME, E40: 655–660.
Becker, H.A., 1959. Can. J. Chem. Eng., 37: 85–91.
Beckinsale, R.P., 1969. In: R.J. Chorley (Editor), Water, Earth and Man. Methuen, London, pp. 455–472.
Beckman, E.W. and Furness, L.W., 1962. Water-Supply Irrig. Pap., Wash., No. 1498-B.
Beiersdorf, H., Kudrass, H.-U. and Von Stackelberg, U., 1980. Geol. Jahrb., D36: 5–85.
Belderson, R.H., 1964. Mar. Geol., 2: 147–163.
Belderson, R.H. and Kenyon, N.H., 1969. J. Sediment. Petrol., 39: 1249–1250.
Belderson, R.H. and Stride, A.H., 1966. Mar. Geol., 4: 237–257.
Belderson, R.H. and Stride, A.H., 1969. Nature (London), 222: 74–75.
Belderson, R.H., Johnson, M.A. and Stride, A.H., 1978. Mar. Geol., 28: 65–75.
Belderson, R.H., Kenyon, N.H. and Stride, A.H., 1971. Rep. Inst. Geol. Sci., No. 70/14: 151–170.
Belderson, R.H., Kenyon, N.H., Stride, A.H. and Stubbs, A.R., 1972. Sonographs of the Sea Floor. Elsevier, Amsterdam.
Bell, H.S., 1942. J. Geol., 50: 512–547.
Belly, P.-Y., 1964. Tech. Mem. U.S. Coastal Eng. Res. Cent., No. 1.
Ben Aïm, R. and Le Goff, P., 1967. Powder Technol., 1: 281–290.
Ben Aïm, R. and Le Goff, P., 1968a. Powder Technol., 2: 169–174.
Ben Aïm, R. and Le Goff, P., 1968b. Powder Technol., 2: 1–12.

Benedict, B.A. and Christensen, B.A., 1972. In: H.W. Shen (Editor), Sedimentation. Colorado State University, Fort Collins, Colo., pp. 5.1–5.17.
Benedict, J.B., 1969. Arct. Alp. Res., 1: 45–48.
Benedict, J.B., 1970. Arct. Alp. Res., 2: 165–226.
Benenati, R.F. and Brosilow, C.B., 1962. AIChE J., 8: 359–361.
Benjamin, T.B., 1957. J. Fluid Mech., 2: 554–574.
Benjamin, T.B., 1959. J. Fluid Mech., 6: 161–205.
Benjamin, T.B., 1968. J. Fluid Mech., 31: 209–248.
Benjamin, T.B. and Ellis, A.T., 1966. Philos. Trans. R. Soc., Ser. A., 260: 221–240.
Bennett, H.W. and Lee, C.A., 1955. Proc. ASCE, Pap. No. 796: 34 pp.
Bennett, J.G. and Brown, R.L., 1940. J. Inst. Fuel, 13: 232–246.
Benney, D.J., 1961. J. Fluid Mech., 10: 209–236.
Benney, D.J., 1964. Phys. Fluids, 7: 319–326.
Benney, D.J. and Lin, C.C., 1960. Phys. Fluids, 3: 656–657.
Benson, C.S., 1962. U.S. Army Snow Ice and Permafrost Research Establishment, Res. Rep., No. 70.
Beresford, R.H., 1969. Nature (London), 224: 550–553.
Berg, T.G.O., McDonald, R.L. and Trainer, R.J., 1969. Powder Technol., 3: 183–188.
Berger, W.H. and Piper, D.J.W., 1972. Limnol. Oceanogr., 17: 275–287.
Bergh, H., 1958. In: H. Görtler (Editor), Boundary Layer Research. Springer, Berlin, pp. 173–178.
Bergh, H. and Van den Bergh, B., 1958. Z. Angew. Math. Phys., Festschr. Jacob Ackeret, 9b: 97–104.
Bergman, C.F., 1979. Geol. För. Stockholm Förh., 101: 217–222.
Berlamont, J., 1976. Proc. Int. Symp. Unsteady Flow in Open Channels, BHRA Fluid Eng., Cranfield, U.K., Pap. A2, pp. 13–26.
Bernal, J.D., 1959. Nature (London), 183: 141–147.
Bernal, J.D., 1960a. Sci. Am., August 1960: 9 pp.
Bernal, J.D., 1960b. Nature (London), 185: 68–70.
Bernal, J.D., 1964. Proc. R. Soc. London, Ser. A., 280: 299–322.
Bernal, J.D., 1965. In: T.J. Hughel (Editor), Liquids: Structure, Properties, Solid Interactions. Elsevier, Amsterdam, pp. 25–47.
Bernal, J.D., Cherry, I.A., Finney, J.L. and Knight, K.R., 1970. J. Phys., E3: 388–390.
Bernal, J.D. and Finney, J.L., 1967. Discuss. Faraday Soc., No. 43: 62–69.
Bernal, J.D. and Mason, J., 1960. Nature (London), 188: 910–911.
Bernard, H.C. and LeBlanc, R.J., 1965. In: H.E. Wright and D.G. Frey (Editors), The Quaternary of the United States. Princeton University Press, Princeton, N.J., pp. 137–185.
Berson, F.A., 1958. Q. J. R. Meteorol. Soc., 84: 1–16.
Bertelsen, A.F., 1974. J. Fluid Mech., 64: 589–597.
Berthois, L., 1962. Sedimentology, 1: 40–49.
Berthois, L. and Le Calvez, Y., 1966. Bull. Inst. Pêch. Mar., 24: 293–301.
Betchov, R. and Criminale, W.O., 1967. Stability of Parallel Flows. Academic Press, New York, N.Y.
Bettany, E., 1962. J. Soil Sci., 13: 10–17.
Betzer, P.R., Richardson, P.L. and Zimmerman, H.B., 1974. Mar. Geol., 16: 21–29.
Beuf, S., Biju-Duval, B., De Charpal, O., Rognon, P., Gariel, O. and Bennacef, A., 1971. Les Grès du Paléozoïque Inférieure au Sahara. Editions Technip, Paris.
Beutelspacher, H. and Van der Marel, H.W., 1968. Atlas of Electron Microscopy of Clay Minerals and their Admixtures. Elsevier, Amsterdam.
Beutner, E.C., Flueckinger, L.A. and Gard, T.M., 1967. Geol. Soc. Am. Bull., 78: 911–916.
Bhattacharya, P.K. and Kennedy, J.F., 1971. Proc. 14th Congr. Int. Assoc. Hydraul. Res., 4: 137–144.

Bhattacharya, S.K., 1973. ASCE, J. Waterw., Harbors Coastal Eng. Div., 99 (WW1): 69–87.
Bhattacharyya, D.S., 1966. Tectonophysics, 3: 29–33.
Bierwagen, G.P. and Saunders, T.E., 1974. Powder Technol., 10: 111–119.
Biesel, F., 1950. Houille Blanche, 5: 279–284.
Biesel, F., 1952. Circ. U.S. Nat. Mus., No. 521: 243–253.
Bigarella, J.J., 1965. Mar. Geol., 3: 269–278.
Bigarella, J.J., 1972. In: J.K. Rigby and W.K. Hamblin (Editors), Recognition of Ancient Sedimentary Environments. Soc. Econ. Paleontol. Mineral., Spec. Publ., 16: 12–62.
Bigarella, J.J., 1975. Bol. Paraná. Geocienc., No. 33: 133–167.
Bigarella, J.J. and Mousinho, M.R., 1965. Bol. Paraná. Geogr., Nos. 16/17: 153–197.
Bigarella, J.J., Mousinho, M.R. and Silva, J.X., 1965. Processes and environments of the Brazilian Quaternary. Imprensa Univ. Fed. Paraná, Curitiba.
Bigarella, J.J., Becker, R.D. and Duarte, G.M., 1969. Mar. Geol., 7: 5–55.
Bigarella, J.J., Duarte, G.M. and Becker, R.D., 1971. Bol. Paraná. Geocienc., Nos. 28/29: 9–72.
Bijker, E.W., 1967a. Proc. 10th Conf. Coastal Eng., 1: 746–765.
Bijker, E.W., 1967b. Publ. Delft Hydraul. Lab., No. 46.
Bijker, E.W., 1971. ASCE J. Waterw., Harbors Coastal Eng. Div., 97 (WW4): 687–701.
Bijker, E.W., Kalkwijk, J.P.T. and Pieters, T., 1975. Proc. 14th Conf. Coastal Eng., 1: 447–465.
Binder, R.C., 1939. J. Appl. Phys., 10: 711–713.
Binnie, A.M., 1957. J. Fluid Mech., 2: 551–553.
Binnie, A.M., 1959. J. Fluid Mech., 5: 561–576.
Birkeland, P.W., 1968. Geol. Soc. Am. Bull., 79: 137–141.
Birkenmajer, K., 1959. Stud. Geol. Pol., 3: 1–133.
Birkenmajer, K., 1979. Stud. Geol. Pol., 60: 45–55.
Bisal, F., 1973. Can. J. Soil Sci., 53: 481–482.
Bisal, F. and Nielsen, K., 1962. Can. J. Soil Sci., 42: 81–86.
Biscaye, P.E. and Eittreim, S.L., 1977. Mar. Geol., 23: 155–172.
Biscoe, J. and Warren, B.E., 1942. J. Appl. Phys., 13: 364–371.
Bishop, A.A., Simons, D.B. and Richardson, E.V., 1965. J. Hydraul. Div. ASCE, 91 (HY2): 175–191.
Bitter, J.G.A., 1963a. Wear, 6: 5–21.
Bitter, J.G.A., 1963b. Wear, 6: 169–190.
Bjørlykke, K., 1968. Norg. Geol. Unders., No. 251: 18–44.
Blackwelder, E., 1928. Bull. Geol. Soc. Am., 39: 465–484.
Blackwelder, R.F., 1979. Phys. Fluids, 22: 583–584.
Blackwelder, R.F. and Eckelmann, H., 1979. J. Fluid Mech., 94: 577–594.
Blackwelder, R.F. and Kaplan, R.E., 1976. J. Fluid Mech., 76: 89–112.
Blake, W.K., 1970. J. Fluid Mech., 44: 637–660.
Blake, W.P., 1855. Am. J. Sci., (2) 20: 178–181.
Blanchet, C. and Villatte, H., 1954. U.N. Economic Commission for Asian and Far East Water Resources, Flood Control Ser., U.N. Ser. No. 9, ST/ECAFE/Ser F/9, pp. 270–277.
Blasius, H., 1910. Z. Bauw., 60: 466–472.
Blench, T., 1957. Regime Behaviour of Canals and Rivers. Butterworths, London.
Blench, T., 1969a. Mobile-bed Fluviology. University of Alberta Press. Edmonton, Alta.
Blench, T., 1969b. J. Hydraul. Div. ASCE, 95 (HY6): 1871–1898.
Blench, T., 1970. J. Hydraul. Res., 8: 159–187.
Blinco, P.H. and Simons, D.B., 1975. J. Eng. Mech. Div. ASCE, 101 (EM3): 241–255.
Blissenbach, E., 1954. Bull. Geol. Soc. Am., 65: 175–190.
Blodgett, R.H. and Stanley, K.O., 1980. J. Sediment. Petrol., 50: 139–148.

Blong, R.J. and Dunkerley, D.L., 1976. Geogr. Ann., A58: 139–147.
Bloomfield, K. and Valastro, S., 1977. Overseas Geol. Miner. Resour., No. 46: 15 pp.
Bloomfield, K., Rubio, G.S. and Wilson, L., 1977. Geol. Rundsch., 66: 120–146.
Bluck, B.J., 1967. J. Sediment. Petrol., 37: 128–156.
Bluck, B.J., 1971. Scott. J. Geol., 7: 93–138.
Bluck, B.J., 1976. Trans. R. Soc. Edinburgh, 69: 425–456.
Bo, M.K., Freshwater, D.C. and Scarlett, B., 1965. Trans. Inst. Chem. Eng., 43: T228–232.
Boeder, P., 1932. Z. Phys., 75: 258–281.
Boerdijk, A.H., 1952. Philips Res. Rep., 7: 303–313.
Boersma, J.R., 1967. Geol. Mijnbouw, 46: 217–235.
Boersma, J.R., 1969. Geol. Mijnbouw, 48: 409–414.
Bogardi, J.L., 1961. J. Geophys. Res., 66: 3337–3346.
Bogardi, J.L., 1965. J. Hydraul. Div. ASCE, 91 (HY1): 29–54.
Bogardi, J.L., 1974. Sediment Transport in Alluvial Streams. Akademiai Kiado, Budapest.
Bogardi, J.L. and Szucs, E., 1970. Acta Techn. Acad. Sci. Hung., 69: 3–27.
Boggs, A., 1974. Geology, 2: 251–253.
Boggs, S. and Jones, C.A., 1976. Geol. Soc. Am. Bull., 87: 419–426.
Bohlen, W.F., 1976. In: M. Wiley (Editor), Estuarine Processes, vol. 2. Academic Press, New York, N.Y., pp. 109–123.
Bokuniewicz, H.J., Gordon, R.B. and Kastens, K.A., 1977. Mar. Geol., 24: 185–199.
Bolt, G.H., 1956. Geotechnique, 6: 86–93.
Bond, A. and Sparks, R.S.J., 1976. J. Geol. Soc. London, 132: 1–16.
Bonington, C., 1976. Everest the Hard Way. Hodder and Stoughton, London.
Bonnefille, R., 1965. Bull. Cent. Rech. Essais Chatou, No. 11: 17–22.
Bonnefille, R. and Goddet, J., 1959. Proc. 8th Congr. Int. Assoc. Hydraul. Res., pp. 14-C-1–14-C-29.
Booth, B., 1973. Proc. Geol. Assoc., 84: 353–370.
Booth, B. and Walker, G.P.L., 1973. Philos. Trans. R. Soc. London, Ser. A., 274: 147–151.
Booth, B., Croasdale, R. and Walker, G.P.L., 1978. Philos. Trans. R. Soc. London, Ser. A., 288: 271–319.
Boothroyd, J.C., 1970. Bull. Am. Assoc. Pet. Geol., 54: 836 (Abstr.).
Boothroyd, J.C., 1973. Coarse-grained sedimentation on a braided outwash fan, northwest Gulf of Alaska. Univ. South Carolina, Dep. Geol., Coastal Res. Div., Tech. Rep., No. 6-CRD.
Boothroyd, J.C. and Ashley, G.M., 1975. In: A.V. Jopling and B.C. McDonald (Editors), Glaciofluvial and Glaciolacustrine Sedimentation. Soc. Econ. Paleontol. Mineral., Spec. Publ., 23: 193–222.
Boothroyd, J.C. and Hubbard, D.K., 1974. Misc. Pap. U.S. Coastal Eng. Res. Cent., No. 1–74.
Boothroyd, J.C. and Hubbard, D.K., 1975. In: L.E. Cronin (Editor), Estuarine Research, vol. 2. Academic Press, New York, N.Y., pp. 217–234.
Bordner, G.L., 1978. Phys. Fluids, 21: 1471–1474.
Borowka, K.R., 1980. Poznan. Tow. Przyj. Nauk, Pr. Kom. Geogr.-Geol., Wydz. Mat.-Przyr., 20: 1–126.
Borradaile, C., 1973. J. Sediment. Petrol., 43: 1160.
Borradaile, G.J. and Johnson, H.D., 1973. Tectonophysics, 18: 249–259.
Borszy, Z., 1973. Földrajzi Ert., 22: 109–115.
Bosence, D.W.J., 1973. Geol. Mijnbouw, 52: 63–67.
Bosence, D.W.J., 1976. Palaeontology, 19: 365–395.
Boswell, P.G.H., 1961. Muddy Sediments. Heffer, Cambridge.
Bottinga, Y. and Weill, D.F., 1972. Am. J. Sci., 272: 438–475.
Botvinkina, L.N., 1959. Izv. Akad. Nauk SSSR, Geol. Ser., 1959 (6): 13–30 (in Russian).

Botvinkina, L.N., Feofilova, A.P. and Jablokov, V.C., 1954. Trud. Inst. Geol. Nauk Mosk., 1954 (151): 30–89 (in Russian).
Boulaine, J., 1954. Rev. Géomorph. Dyn., 5: 102–123.
Boulton, G.S., 1967. J. Glaciol., 6: 717–735.
Boulton, G.S., 1968. J. Glaciol., 7: 391–412.
Boulton, G.S., 1970a. J. Glaciol., 9: 213–229.
Boulton, G.S., 1970b. J. Glaciol., 9: 231–245.
Boulton, G.S., 1971. In: R.P. Goldthwait (Editor), Till, a Symposium. Ohio State University Press, Columbus, Ohio, pp. 41–72.
Bouma, A.H., 1962. Sedimentology of some Flysch Deposits. Elsevier, Amsterdam.
Bouma, A.H., 1964. Geol. Mijnbouw, 43: 375–379.
Bouma, A.H., 1965. Mar. Geol., 3: 291–320.
Bouma, A.H., Hampton, M.A. and Orlando, R.C., 1977a. Mar. Geotechn., 2: 291–308.
Bouma, A.H., Hampton, M.A., Wennekens, M.P. and Dygas, J.A., 1977b. Proc. 9th Offshore Technol. Conf., pp. 79–84.
Bouma, A.H., Hampton, M.A., Rappeport, M.L., Whitney, J.L., Teleki, P.G., Orlando, R.C. and Torresan, M.E., 1978. Proc. Offshore Technol. Conf. 1978, Pap. OTC 3311, pp. 2271–2284.
Bouma, A.H., Rappeport, M.L., Orlando, R.C. and Hampton, M.A., 1980. Sediment. Geol., 26: 157–177.
Bourcart, J., 1928. Revue Géogr. Phys. Géol. Dyn., 1: 26–54, 195–265.
Bouysse, P., Horn, R., Lapierre, F. and Le Lann, F., 1976. Mar. Geol., 20: 251–275.
Bowden, K.F., 1962. J. Geophys. Res., 67: 3181–3186.
Bowden, K.F., 1964. Annu. Rev. Mar. Biol. Oceanogr., 2: 11–30.
Bowden, K.F., 1970. Annu. Rev. Mar. Biol. Oceanogr., 8: 11–32.
Bowden, K.F. and Fairbairn, L.S., 1956. Proc. R. Soc. London, Ser. A., 237: 422–438.
Bowden, K.F., Fairbairn, L.A. and Hughes, P., 1959. Geophys. J. R. Astron. Soc., 2: 288–305.
Bowen, A.J., Inman, D.L. and Simmons, V.P., 1968. J. Geophys. Res., 73: 2569–2577.
Bowen, B.D. and Masliyah, J.H., 1973. Can. J. Chem. Eng., 51: 8–15.
Bowler, J.M., 1971. In: D.J. Mulvaney and J. Gobson (Editors), Aboriginal Man and Environment in Australia. Australian National University Press, Canberra, pp. 47–65.
Bowler, J.M., 1973. Earth-Sci. Rev., 9: 315–338.
Bowler, J.M., 1976. Earth-Sci. Rev., 12: 279–310.
Bowles, F.A., 1968. Electron microscope investigation of the microstructure in sediment samples from the Gulf of Mexico. Texas A & M Univ., Dep. Oceanogr., Tech. Rep., No. 68-24T.
Bowles, F.A., Bryant, W.R. and Wallin, C., 1969. J. Sediment. Petrol., 39: 1546–1551.
Bradley, W.H., 1965. Science, 150: 1423–1428.
Bradshaw, P., 1964. Experimental Fluid Mechanics. Pergamon Press, Oxford.
Bradshaw, P., 1971. An Introduction to Turbulence and its Measurement. Pergamon, Oxford.
Braithwaite, C.J.R., 1973. Sedimentology, 20: 251–262.
Brambati, A., 1968. Stud. Trentini Sci. Nat., Ser. A., 45: 188–223.
Brasier, M.D., Hewitt, R.A. and Brasier, C.J., 1978. Geol. Mag., 115: 21–36.
Breach, D.R., 1961. J. Fluid Mech., 10: 306–314.
Brebner, A. and Collins, J.I., 1961. Trans. Eng. Inst. Can., 5: 55–62.
Brebner, A., Askew, J.A. and Law, S.W., 1967. Proc. 10th Conf. Coastal Eng., 1: 175–184.
Bremhurst, K. and Walker, T.B., 1973. J. Fluid Mech., 61: 173–186.
Brenchley, P.J., 1972. Proc. Yorks Geol. Soc., 39: 199–224.
Brenchley, P.J. and Newall, G., 1970. Palaeogeogr. Palaeoclimatol. Palaeoecol., 7: 185–220.
Brenner, H., 1961. Chem. Eng. Sci., 16: 242–251.
Brenner, H., 1964. Chem. Eng. Sci., 19: 631–651.

Brenner, K., 1976. Neues Jahrb. Geol. Paläontol. Abh., 151: 101–118.
Bretherton, F.P., 1962a. J. Fluid Mech., 12: 591–613.
Bretherton, F.P., 1962b. J. Fluid Mech., 14: 284–304.
Bretherton, F.P. and Lord Rothschild, 1961. Proc. R. Soc. London, Ser. B., 153: 490–502.
Bretschneider, C.L., 1952a. Trans. Am. Geophys. Union, 33: 381–389.
Bretschneider, C.L., 1952b. Proc. 2nd Conf. Coastal Eng., pp. 1–5.
Bretschneider, C.L., 1966. In: A.T. Ippen (Editor), Estuary and Coastline Hydrodynamics. McGraw-Hill, New York, N.Y., pp. 133–198.
Bretz, J.H., 1959. Wash., Div. Mines Geol., Bull., No. 45.
Bretz, J.H., Smith, H.T.U. and Neff, G.E., 1956. Bull. Geol. Soc. Am., 67: 957–1049.
Brewer, R., 1964. Fabric and Mineral Analysis of Soils. Wiley, New York, N.Y.
Brewer, R. and Haldane, A.D., 1957. Soil Sci., 84: 301–309.
Brezina, J., 1963. J. Sediment. Petrol., 33: 931–937.
Brice, J.C., 1964. U.S. Geol. Surv., Prof. Pap., 422-D.
Bridge, J.S., 1978. Sediment. Geol., 20: 1–16.
Bridge, J.S. and Jarvis, J., 1976. Earth Surface Processes, 1: 303–306.
Briggs, L.I., McCulloch, D.S. and Moser, F., 1962. J. Sediment. Petrol., 32: 645–656.
Brinkmann, R., 1933. Nachr. Ges. Wiss. Göttingen, Kl. IV, Fachgr. IV, No. 32, pp. 1–12.
Briscoe, M.G., 1975. Rev. Geophys., 13: 591–598.
Britter, R.E. and Linden, P.F., 1980. J. Fluid Mech., 99: 531–543.
Britter, R.E. and Simpson, J.E., 1978. J. Fluid Mech., 88: 223–240.
Broadhurst, F.M., 1968. Lethaia, 1: 28–38.
Broch, E.K., 1932. Z. Phys., 78: 257–270.
Brock, R.H., 1969. J. Hydraul. Div. ASCE, 95 (HY4): 1401–1427.
Brock, R.R., 1970. J. Hydraul. Div. ASCE, 96 (HY2): 2565–2580.
Brodkey, R.S., Wallace, J.M. and Eckelman, H., 1974. J. Fluid Mech., 63: 209–224.
Broekman, J.A., 1973. Proc. K. Ned. Akad. Wet., B76: 423–445.
Broekman, J.A., 1974. Proc. K. Ned. Akad. Wet., B77: 303–318.
Broggi, J.A., 1952. Bol. Soc. Geol. Peru, 24: 3–29.
Brookes, G.F. and Whitmore, R.L., 1968. Rheol. Acta, 7: 188–193.
Brookfield, M.E., 1977. Sedimentology, 24: 303–332.
Brookfield, M.E., 1979. Scott. J. Geol., 15: 81–96.
Brooks, F.A. and Berggren, W.P., 1943. Trans. Am. Geophys. Union, 24: 189–204.
Brooks, R.L., Jasper, N.H. and James, R.W., 1958. Trans. Am. Geophys. Union, 39: 1064–1075.
Broome, R. and Komar, P.D., 1979. Sedimentology, 26: 543–559.
Brosset, D., 1939. Bull. Inst. Fr. Afr. Noire, 1: 657–690.
Broussard, M.L., 1975. Deltas. Models for Exploration. Houston Geological Society, Houston, Texas.
Brown, A.P., 1911. Proc. Acad. Nat. Sci. Philadelphia, 63: 536–547.
Brown, F.N.M., 1959. Proc. 6th Midwestern Conf. Fluid Mechanics, pp. 331–349.
Brown, F.N.M., 1965. Develop. Mech., 3: 421–431.
Brown, F.R., 1963. J. Hydraul. Div. ASCE, 90 (HY1): 99–115.
Brown, G.L. and Thomas, A.S.W., 1977. Phys. Fluids, 20 (10, II): S243–S252.
Brown, R.L. and Hawksley, P.G.W., 1945. Nature (London), 156: 421–422.
Brown, R.L. and Hawksley, P.G.W., 1946. Nature (London), 157: 585.
Brown, R.L. and Hawksley, P.G.W., 1947. Fuel, 26: 159–173.
Browning, K.A., 1972. Weather, 27: 2–13.
Browning, K.A. and Ludlam, F.H., 1962. Q. J. R. Meteorol. Soc., 88: 117–135.
Brückner, W.D. and Anderson, M.M., 1971. Proc. Geol. Assoc. Can., 24: 95–102.
Brunton, J.H., 1966. Philos. Trans. R. Soc. London, Ser. A., 260: 79–85.
Brush, L.M., Ho, H.W. and Yen, B.C., 1964. J. Hydraul. Div. ASCE, 90 (HY1): 149–160.

Bruun, P. and Vollen, Ø., 1972. Schiff Hafen, 3: 139–142.
Bruun-Petersen, J. and Krumbein, W.E., 1975. Geol. Rundsch., 64: 126–143.
Bryant, P.J., 1973. J. Fluid Mech., 59: 625–644.
Bryant, W.R., Cernock, P. and Morelock, J., 1967. In: A.F. Richards (Editor), Marine Geotechnique. University of Chicago Press, Urbana, Ill., pp. 41–62.
Brynhi, I., 1978. Norsk Geol. Tidsskr., 58: 273–300.
Bucher, W.H., 1919. Am. J. Sci., 47: 149–210, 241–269.
Buckmaster, J., 1973. J. Fluid Mech., 59: 481–491.
Buckmaster, J., 1977. J. Fluid Mech., 81: 735–756.
Bull, M.K., 1967. J. Fluid Mech., 28: 719–754.
Bull, W.B., 1964. U.S. Geol. Surv., Prof. Pap., 437-A.
Buller, A.T., McManus, J. and Williams, D.J.A., 1971. Investigations in the estuarine environments of the Tay. Univ. Dundee, Tay Res. Cent., Res. Rep., No. 1.
Bumpus, D.F., 1973. Progr. Oceanogr., 6: 111–157.
Burke, M.J., 1969. Scott. J. Geol., 5: 286–292.
Burnham, C.P., 1970. In: D.A. Osmond and P. Bullock (Editors), Micromorphological Techniques and Applications. Soil Surv. Tech. Monogr., No. 2: 83–96.
Busby, J.S.J., 1950. J. Soc. Glass Technol., 34: 10–16.
Busse, F.H., 1970. J. Fluid Mech., 41: 219–240.
Butterfield, R. and Andrawes, K.Z., 1970. Geotechnique, 20: 97–100.
Button, A., 1976. Trans. Geol. Soc. S. Afr., 78: 173–183.
Button, A. and Vos, R.G., 1977. Sediment. Geol., 18: 175–200.
Bye, J.A.T., 1967. J. Mar. Res., 25: 95–102.
Cacchione, D.A., 1970. Experimental study of internal gravity waves over a slope. Mass. Inst. Technol., Dep. Earth Planet. Sci., Exp. Sedimentol. Lab., Rep., No. 70-6.
Cacchione, D.A. and Southard, J.B., 1974. J. Geophys. Res., 79: 2237–2242.
Cacchione, D.A. and Wunsch, C., 1974. J. Fluid Mech., 66: 223–239.
Cailleux, A., 1938. Rev. Géogr. Phys. Géol. Dyn., 11: 171–198.
Cailleux, A., 1945. Bull. Soc. Géol. Fr., (5) 15: 375–404.
Cain, J.D.B., 1968. Scott. J. Geol., 4: 191–208.
Caine, N., 1967. J. Sediment. Petrol., 37: 796–803.
Caine, N., 1969. Arct. Alp. Res., 1: 127–134.
Caligny, A.F.H., 1878. C.R. Acad. Sci., 87: 19–23.
Calkin, P.E. and Rutford, R.H., 1974. Geogr. Rev., 64: 189–216.
Callander, R.A., 1969. J. Fluid Mech., 36: 465–480.
Cameron, B. and Estes, R., 1971. J. Sediment. Petrol., 41: 171–178.
Campbell, C.V., 1966. J. Sediment. Petrol., 36: 825–828.
Campbell, E.M., 1968. Trans. R. Soc. S. Aust., 92: 85–109.
Campbell, R.H., 1975. U.S. Geol. Surv., Prof. Pap., 851.
Cant, D.J., 1978. J. Sediment. Petrol., 48: 1321–1330.
Cantwell, B., Coles, D. and Dimotakis, P., 1978. J. Fluid Mech., 87: 641–672.
Capot-Rey, R., 1943. Trav. Inst. Rech. Sahariennes, 2: 69–102.
Capot-Rey, R., 1947. Trav. Inst. Rech. Sahariennes, 4: 67–109.
Capot-Rey, R., 1957. Trav. Inst. Rech. Sahariennes, 15: 149–157.
Capot-Rey, R. and Capot-Rey, F., 1948. Trav. Inst. Rech. Sahariennes, 5: 47–80.
Capot-Rey, R. and Grémion, M., 1965. Trav. Inst. Rech. Sahariennes, 23: 153–163.
Carey, S.W. and Ahmed, N., 1961. In: G.O. Raasch (Editor), Geology of the Arctic, University of Toronto Press, Toronto, Ont., 2: 865–895.
Carey, W.C. and Keller, M.D., 1957. J. Hydraul. Div. ASCE, 83 (Pap. 1331): 1–24.
Carlson, P.R., 1974. Proc. 9th Int. Symp. Remote Sensing of the Environment, pp. 1279–1288.
Carlson, P.R. and McCulloch, D.S., 1974. J. Res. U.S. Geol. Surv., 2: 519–526.
Carlson, P.R. and Nelson, H.C., 1969. J. Sediment. Petrol., 39: 1269–1282.

Carlson, P.R., Alpha, T.R. and McCulloch, D.S., 1970. Calif. Div. Mines Geol. Inf. Serv., 23: 97–107.
Carman, P.C., 1937. Trans. Inst. Chem. Eng., 15: 150–166.
Carman, P.C., 1938. Trans. Inst. Chem. Eng., 16: 168–188.
Carr, A.P., 1969. J. Sediment. Petrol., 38: 297–311.
Carr, A.P., 1975. Proc. 14th Conf. Coastal Eng., 1: 851–870.
Carr, A.P., Gleason, R. and King, A., 1970. Sediment. Geol., 4: 89–101.
Carr, D.D., 1973. Bull. Indiana Div. Geol., No. 48.
Carrigy, M.A., 1967. Sediment. Geol., 1: 327–352.
Carrigy, M.A., 1970. Sedimentology, 14: 147–158.
Carruthers, J.N., 1963. Nature (London), 197: 942–946.
Carson, M.A., 1977. Earth Surf. Processes, 2: 363–380.
Carson, M.A. and Kirkby, M.J., 1972. Hillslope Form and Process. Cambridge University Press, Cambridge.
Carstens, M.R. and Neilson, F.M., 1967. J. Geophys. Res., 72: 3053–3059.
Carstens, M.R., Neilson, F.M. and Altinbilek, H.D., 1969. Tech. Mem. U.S. Coastal Eng. Res. Cent., No. 28.
Carstens, M.R. and Altinbilek, H.D., 1972. J. Hydraul. Div. ASCE, 98 (HY5): 787–794.
Carter, L. and Heath, R.A., 1975. N.Z. J. Mar. Freshwater Res., 9: 423–448.
Carter, R.M., 1975. Earth-Sci. Rev., 11: 145–177.
Cartwright, D.E., 1959. Proc. R. Soc. London, Ser. A., 253: 218–241.
Cartwright, D.E. and Stride, A.H., 1958. Nature (London), 181: 41.
Casagrande, A., 1932. Boston Soc. Civ. Eng., 19: 168–221.
Casshyap, S.M., 1968. J. Sediment. Petrol., 38: 920–942.
Casshyap, S.M. and Qidwai, H.A., 1974. Bull. Geol. Soc. Am., 85: 749–760.
Caston, G.F., 1976. Rep. Inst. Geol. Sci., No. 76/7, 16 pp.
Caston, V.N.D., 1965. Mar. Geol., 3: 401–410.
Caston, V.N.D., 1972. Sedimentology, 18: 63–78.
Caston, V.N.D. and Stride, A.H., 1970. Mar. Geol., 9: M38–M42.
Caston, V.N.D. and Stride, A.H., 1973. Estuarine Coastal Mar. Sci., 1: 379–386.
Cegla, J. and Dzulynski, S., 1967. Annu. Soc. Géol. Pol., 37: 489–497.
Cepek, P. and Reineck, H.-E., 1970. Senckenbergiana Marit., 2: 3–30.
Chabert, J. and Chauvin, J.L., 1963. Bull. Cent. Rech. Essais Chatou, No. 4: 31–51.
Chaffey, C.E., Takano, M. and Mason, S.G., 1965. Can. J. Phys., 43: 1269–1287.
Chan, K.W., Baird, M.H.I. and Round, G.F., 1972. Proc. R. Soc. London, Ser. A., 330: 537–559.
Chanda, S.K. and Bhhattacharyya, A., 1974. Geol. Soc. Am. Bull., 85: 1117–1122.
Chandler, P.L. and Sorensen, R.M., 1973. Proc. 10th Conf. Coastal Eng., 1: 385–404.
Chang, H.Y. and Simons, D.B., 1970. J. Fluid Mech., 42: 491–495.
Chang, J.H., 1972. Atmospheric Circulation Systems and Climates. Oriental Publishing Co., Honolulu, Hawaii.
Chang, M.S., 1969. J. Geophys. Res., 74: 1515–1536.
Chang, P.K., 1970. Separation of Flow. Pergamon, Oxford.
Chang, Y.L., 1939. Trans. ASCE, 104: 1246–1313.
Channon, R.D., 1971. J. Sediment. Petrol., 41: 867–870.
Channon, R.D. and Hamilton, D., 1976. Sedimentology, 23: 17–42.
Charba, J., 1974. Mon. Weather Rev., 102: 140–156.
Chassefière, B. and Leenhardt, O., 1967. Vie Milieu, B18: 1–32.
Chen, C.N. and Carstens, M.R., 1973. Proc. 15th Congr. Int. Assoc. Hydraul. Res., 1: 225–233.
Chen, C.S. and Goodell, H.G., 1964. J. Sediment. Petrol., 34: 46–72.
Cheng, E.D.H. and Clyde, G.G., 1972. In: H.W. Shen (Editor), Sedimentation. Colorado

State University, Fort Collins, Colo., pp. 3.1–3.20.
Chepil, W.S., 1945a. Soil Sci., 60: 305–320.
Chepil, W.S., 1945b. Soil Sci., 60: 397–411.
Chepil, W.S., 1946c. Soil Sci., 60: 475–480.
Chepil, W.S., 1957a. Am. J. Sci., 255: 12–22.
Chepil, W.S., 1957b. Am. J. Sci., 255: 206–213.
Chepil, W.S., 1958. Trans. Am. Geophys. Union, 39: 397–404.
Chepil, W.S., 1961. Proc. Soil Sci. Soc. Am., 25: 343–345.
Chepil, W.S. and Milne, R.A., 1939. Sci. Agric., 19: 249–257.
Chepil, W.S. and Woodruff, N.P., 1957. Am. J. Sci., 255: 104–114.
Chester, R., 1972. Nobel Symposium, No. 20, pp. 291–305.
Chester, R. and Johnson, L.R., 1971a. Nature (London), 229: 105–107.
Chester, R. and Johnson, L.R., 1971b. Mar. Geol., 11: 251–260.
Chester, R., Elderfield, H., Griffin, J.J., Johnson, L.R. and Padgham, R.C., 1972. Mar. Geol., 13: 91–105.
Chester, W., Breach, D.R. and Proudman, I., 1969. J. Fluid Mech., 37: 751–760.
Chesterman, W.D., Clynick, P.R. and Stride, A.H., 1958. Acoustica, 8: 285–290.
Chisholm, J.I. and Dean, J.M., 1974. Scott. J. Geol., 10: 1–30.
Chiu, C.L., 1967. J. Eng. Mech. Div. ASCE, 93 (EM4): 69–77.
Chorley, R.G. and Kennedy, B.A., 1971. Physical Geography: A Systems Approach. Prentice-Hall International, London.
Chow, V.T., 1959. Open Channel Hydraulics. McGraw-Hill, New York.
Chowdhury, K.C.R. and Fritz, W., 1959. Chem. Eng. Sci., 11: 92–98.
Christiansen, E.D. and Barker, D.H., 1965. AIChE J., 11: 145–151.
Christiansen, J.E., 1935. Trans. Am. Geophys. Union, 16: 478–485.
Christman, T. and Shewman, P.G., 1979. Wear, 52: 57–70.
Chudeau, R., 1920. Ann. Géogr., 29: 334–351.
Church, M., 1972. Bull. Geol. Surv. Can., No. 216.
Chwang, A.T. and Wu, T.Y., 1976. J. Fluid Mech., 75: 677–689.
Clague, J.J., 1974. Can. J. Earth Sci., 11: 916–938.
Clague, J.J., 1976. Can. J. Earth Sci., 13: 803–815.
Clapperton, C.M. and Hamilton, P., 1971. Geogr. Mag., 43: 632–639.
Clark, B.R., 1970. Am. J. Sci., 269: 250–266.
Clark, J.A., 1968. Trans. ASME, D90: 455–468.
Clark, J.A. and Markland, E., 1971. J. Hydraul. Div. ASCE, 97 (HY10): 1653–1664.
Clarke, B.S., 1964. Proc. Geol. Assoc., 75: 345–355.
Clarke, R.H., 1961. J. Meteorol., 18: 715–735.
Clarke, R.H., 1965. Aust. Meteorol. Mag., 50: 1–25.
Clarke, R.H., 1973. Proc. 1st Australian Conf. Heat and Mass Transfer, Sect. 1, pp. 41–48.
Clauser, F.H., 1956. Adv. Appl. Mech., 4: 1–51.
Clemmensen, L.B., 1978. Palaeogeogr. Palaeoclimatol. Palaeoecol., 24: 111–135.
Clemmensen, L.B., 1980. Medd. Grønl., No. 136.
Clift, R., Grace, J.R. and Weber, M.E., 1978. Bubbles, Drops and Particles. Academic Press, New York, N.Y.
Clifton, H.E., 1971. J. Sediment. Petrol., 41: 671–682.
Clifton, H.E., 1976. In: R.A. Davis and R.L. Ethington (Editors), Beach and Nearshore Sedimentation. Soc. Econ. Paleontol. Mineral., Spec. Publ., 24: 126–148.
Clifton, H.E. and Boggs, S., 1970. J. Sediment. Petrol., 40: 888–897.
Clifton, H.E., Hunter, R.E. and Phillips, R.L., 1971. J. Sediment. Petrol., 41: 651–670.
Clifton, H.E., Philips, R.L. and Hunter, R.E., 1973. In: D.R. Coates (Editor), Coastal Geomorphology. State University of New York, Binghamton, N.Y., pp. 115–140.
Clifton, R.L., 1944. Bull. Am. Assoc. Pet. Geol., 28: 1012–1031.

Cloet, R.L., 1954a. J. Inst. Navig., 7: 272–279.
Cloet, R.L., 1954b. Geogr. J., 120: 203–215.
Cloet, R.L., 1961. Geogr. J., 127: 335–339.
Cloet, R.L., 1963. Admiralty Mar. Sci. Publ., No. 6.
Cloet, R.L., 1980. Hydrogr. J., No. 17 (April 1980): 19–30.
Cloos, H., 1938. Geol. Rundsch., 29: 357–367.
Clos-Arceduc, A., 1966. C.R. Acad. Sci., D262: 2673–2676.
Clos-Arceduc, A., 1967a. C.R. Acad. Sci., D264: 1393–1396.
Clos-Arceduc, A., 1967b. Trans. 2nd Int. Symp. on Photo-Interpretation, Paris, pp. IV-1-17–IV-1-34.
Clos-Arceduc, A., 1969. Essai d'explication des formes dunaires Sahariennes. Inst. Geogr. Nat., Paris, Étud. Photo-interpretation, No. 4.
Clos-Arceduc, A., 1972. Trav. Inst. Géogr. Reims, 6: 63–72.
Cloud, P.E., 1962. U.S. Geol. Surv., Prof. Pap., 350.
Clyde, C.G. and Einstein, H.A., 1966. J. Eng. Mech. Div. ASCE, 92 (EM2): 251–275.
Coantic, M., 1967. Proc. 4th Euromech. Colloq. Southampton.
Coffey, G.N., 1909. J. Geol., 17: 754–755.
Cokelet, E.D., 1977. Nature (London), 267: 769–774.
Colburn, I.P., 1968. J. Sediment. Petrol., 38: 146–158.
Colby, B.R., 1964. U.S. Geol. Surv., Prof. Pap., 462-A.
Colby, B.R. and Hembree, C.H., 1955. Water-Supply Irrig. Pap., Wash., No. 1357.
Coleman, J.M., 1969. Sediment. Geol., 3: 129–239.
Coleman, J.M. and Gagliano, S.M., 1964. Trans. Gulf-Coast Assoc. Geol. Soc., 14: 67–80.
Coleman, J.M. and Gagliano, S.M., 1965. In: G.V. Middleton (Editor), Primary Sedimentary Structures and their Hydrodynamic Interpretation. Soc. Econ. Paleontol. Mineral., Spec. Publ., 12: 133–248.
Coleman, J.M. and Wright, L.D., 1975. In: M.L. Broussard (Editor), Deltas. Models for Exploration. Houston Geological Society, Houston, Tex., pp. 99–149.
Coleman, J.M., Gagliano, S.M. and Webb, J.E., 1964. Mar. Geol., 1: 240–258.
Coleman, J.M., Suhayoa, J.N., Whelan, T. and Wright, C.D., 1974. Trans. Gulf-Coast Assoc. Geol. Soc., 24: 49–68.
Coleman, N.L., 1967. Proc. 12th Congr. Int. Assoc. Hydraul. Res., 3: 185–192.
Coleman, N.L., 1969. J. Hydraul. Res., 67: 67–82.
Coleman, N.L., 1970. Water Resour. Res., 6: 801–809.
Coles, E.E., 1938. Prof. Not. Meteorol. Off., London, No. 84.
Collins, J.I., 1963. J. Geophys. Res., 68: 6007–6014.
Collins, J.I., 1964. J. Geophys. Res., 69: 1051–1056.
Collinson, J.D., 1966. Geol. Mijnbouw, 45: 262–264.
Collinson, J.D., 1970a. Proc. Yorks. Geol. Soc., 37: 495–515.
Collinson, J.D., 1970b. Geogr. Ann., A52: 31–56.
Collinson, J.D., 1971. Geol. Mijnbouw, 50: 671–678.
Colton, G.W. and De Witt, W., 1959. Bull. Geol. Soc. Am., 70: 1759–1760.
Compton, R.R., 1962. Manual of Field Geology. Wiley, New York, N.Y.
Conaghan, P.J. and Jones, J.G., 1975. J. Geol. Soc. Aust., 22: 275–283.
Connary, S.D. and Ewing, M., 1972. In: A.L. Gordon (Editor), Studies in Physical Oceanography. Gordon and Breach, New York, N.Y., pp. 169–184.
Conolly, J.R., 1964. Aust. J. Sci., 27: 113–115.
Conolly, J.R., 1965. J. Sediment. Petrol., 35: 765–768.
Conolly, J.R., 1969. N.Z. J. Geol. Geophys., 12: 310–343.
Conolly, J.R. and Van der Borch, C.C., 1967. Sediment. Geol., 1: 181–200.
Conybeare, C.E.B. and Crook, K.A.W., 1968. Manual of Sedimentary Structures. Bull. Bur. Miner. Resour. Geol. Geophys. Aust., No. 102.

Coogan, A.H. and Manus, R.W., 1975. In: G.V. Chilingarian and K.H. Wolf (Editors), Compaction of Coarse Grained Sediments, vol. 1. Elsevier, Amsterdam, pp. 79–166.
Cook, D.O. and Gorsline, D.S., 1972. Mar. Geol., 13: 31–55.
Cooke, R.U. and Warren, A., 1973. Geomorphology in Deserts. Batsford, London.
Cooper, R.H., Peterson, A.W. and Blench, T., 1972. J. Hydraul. Div. ASCE, 98 (HY5): 827–843.
Cooper, W.S., 1958. Mem. Geol. Soc. Am., No. 72.
Cooper, W.S., 1967. Mem. Geol. Soc. Am., No. 104.
Corino, E.R. and Brodkey, R.S., 1969. J. Fluid Mech., 37: 1–30.
Cornish, V., 1897. Geogr. J., 9: 278–302.
Cornish, V., 1899. Geogr. J., 13: 624–626.
Cornish, V., 1900. Geogr. J., 15: 1–30.
Cornish, V., 1901a. Geogr. J., 18: 170–200.
Cornish, V., 1901b. Scott. Geogr. Mag., 17: 1–11.
Cornish, V., 1902. Geogr. J., 20: 137–175.
Cornish, V., 1904. Geogr. J., 23: 623–645.
Cornish, V., 1907. Geogr. J., 29: 23–31.
Cornish, V., 1908. Geogr. J., 31: 400–402.
Cornish, V., 1914. Waves of Sand and Snow. Fisher Unwin, London.
Cornish, V., 1934. Ocean Waves. Cambridge University Press, Cambridge.
Costello, W.R., 1974. Development of bed configurations in coarse sands. Mass. Inst. Technol., Dep. Earth Planet. Sci., Rep. No. 74-1.
Costello, W.R. and Walker, R.G., 1972. J. Sediment. Petrol., 42: 389–400.
Coulson, J.M., 1949. Trans. Inst. Chem. Eng., 27: 237–253.
Coulson, J.M. and Richardson, J.F., 1965. Chemical Engineering, vol. 1. Pergamon, Oxford, 2nd ed.
Coursin, A., 1964. Bull. Inst. Fr. Afr. Noire, A26: 989–1022.
Coutenceau-Clarke, J.A.R., 1972. Nature (London), 240: 408–410.
Cowan, W.R., 1968. Can. J. Earth Sci., 5: 1145–1159.
Cowie, J.D., 1963. N.Z. J. Geol. Geophys., 6: 268–280.
Cowperthwaite, I.A., Fitch, F.J., Miller, R.G. and Robertson, R.H.S., 1972. Clay Miner., 9: 309–327.
Cox, G.H. and Dake, C.L., 1916. Bull. School Mines Missouri Univ., 2 (4): 1–59.
Cox, R.G., 1970. J. Fluid Mech., 44: 791–810.
Cox, R.G., 1971. J. Fluid Mech., 45: 625–657.
Coxeter, H.S.M., 1961. Introduction to Geometry. Wiley, New York, N.Y.
Craig, G.Y. and Walton, W.K., 1962. Trans. Edinburgh Geol. Soc., 19: 100–119.
Craig, R.E., 1973. Marine Physics. Academic Press, New York, N.Y.
Craik, A.D.D., 1971. J. Fluid Mech., 50: 393–413.
Crandell, D.R., 1971. U.S. Geol. Surv., Prof. Pap., 677.
Crandell, D.R., 1980. Bull. U.S. Geol. Surv., No. 1492.
Crandell, D.R. and Fahnestock, R.K., 1965. Bull. U.S. Geol. Surv., 1221-A.
Crandell, D.R. and Waldron, H.H., 1956. Am. J. Sci., 254: 349–362.
Crawford, C.B., 1968. Eng. Geol., 2: 239–265.
Cressey, G.B., 1928. Bull. Geogr. Soc. Chicago, No. 8.
Crickmore, M.J., 1970. J. Hydraul. Div. ASCE, 96 (HY2): 473–496.
Crimes, T.P., 1970. Palaeogeogr. Palaeoclimatol. Palaeoecol., 7: 113–170.
Criminale, W.O. and Kovasznay, L.G., 1962. J. Fluid Mech., 14: 59–80.
Cronin, L.E., 1975. Estuarine Research, vol. 2. Academic Press, New York, N.Y.
Crook, K.A.W., 1957. Proc. Linn. Soc. N.S.W., 82: 157–166.
Crook, K.A.W., 1965. Sedimentology, 5: 249–254.
Crosby, E.J., 1960. Am. Perfumer, 75: 43–48.

Crowell, J.C., 1955. Bull. Geol. Soc. Am., 66: 1351–1384.
Crowell, J.C. and Frakes, L.A., 1971. Geol. Soc. Am. Bull., 82: 2515–2540.
Crowell, J.C., Hope, R.A., Kahle, J.E., Ovenshine, A.T. and Sams, R.H., 1966. Spec. Rep. Calif. Div. Mines Geol., No. 89.
Csanady, G.T., 1973. Turbulent Diffusion in the Environment. Reidel, Dordrecht, The Netherlands.
Culbertson, J.K. and Dawdy, D.R., 1964. Water-Supply Irrig. Pap., Wash, 1498-F.
Culbertson, J.K. and Scott, C.H., 1970. U.S. Geol. Surv., Prof. Pap., 700-B, pp. 237–241.
Culbertson, J.K., Scott, C.H. and Bennett, J.P., 1971. Open File Rep., U.S. Geol. Surv. Water Resour. Div.
Culbertson, J.K., Scott, C.H. and Bennett, J.P., 1972. U.S. Geol. Surv., Prof. Pap., 562-J.
Cummins, W.A., 1965. Mercian Geol., 1: 153–160.
Cunge, J.A. and Simons, D.B., 1975. Proc. 16th Congr. Int. Assoc. Hydraul. Res., 2: 48–56.
Curray, J.R., 1956. Bull. Am. Assoc. Pet. Geol., 40: 2440–2456.
Curry, R.R., 1966. Bull. Geol. Soc. Am., 77: 771–776.
Cutts, J.A. and Smith, R.S.U., 1973. J. Geophys. Res., 78: 4139–4154.
Daboll, J.M., 1969. Holocene sediments of the Parker River Estuary, Massachusetts. Univ. Mass., Dep. Geol., Contrib. No. 3-CRG.
Dahl, R., 1965. Geogr. Ann., A47: 83–140.
Daily, J.W. and Stephan, S.C., 1953. Proc. 3rd Conf. Coastal Eng., pp. 13–30.
Dal Cin, R., 1968. Sediment. Geol., 2: 233–241.
Dal Cin, R. and Sperandio, A., 1966. Stud. Trentini Sci. Nat., Ser. A., 43: 334–358.
Dallavalle, J.M., 1948. Micromeritics. Pittman, London, 2nd ed.
Dalrymple, R.A., 1974. J. Geophys. Res., 78: 4498–4504.
Dalrymple, R.W., Knight, R.J. and Middleon, G.V., 1975. In: L.E. Cronin (Editor), Estuarine Research, vol. 2. Academic Press, New York, N.Y., pp. 293–307.
Dalrymple, R.W., Knight, R.J. and Lambiase, J.J., 1978. Nature (London), 275: 100–104.
Daly, B.J. and Pracht, W.E., 1968. Phys. Fluids, 11: 15–30.
Danel, P., Durand, R. and Condiolis, E., 1953. Houille Blanche, 8: 815–829.
D'Anglejan, B.F., 1971. Can. J. Earth Sci., 8: 1480–1486.
D'Anglejan, B.F. and Ingram, R.G., 1976. Estuarine Coastal Mar. Sci., 4: 401–416.
D'Anglejan, B.F. and Smith, E.C., 1973. Can. J. Earth Sci., 10: 1380–1396.
Dapples, W.C. and Rominger, J.F., 1945. J. Geol., 53: 246–261.
Darabaner, C.L., Raasch, J.K. and Mason, S.G., 1967. Can. J. Chem. Eng., 45: 3–12.
Darbyshire, M., 1962. Dtsch. Hydrog. Z., 15: 256–264.
Darbyshire, M. and Draper, L., 1963. Engineering, 5 April: 482–484.
Darwin, G.H., 1884. Proc. R. Soc. London, 36: 18–43.
Das, M.M., 1971. Mechanics of sediment suspension due to oscillatory water waves. Univ. Calif. Coll. Eng. Hydraul. Eng. Lab. Tech. Rep., No. HEL-2-32.
Dattatri, J., 1973. ASCE J. Waterw., Harbours Coastal Eng. Div., 99(WW1): 39–58.
Davidson, B.J. and Riley, N., 1972. J. Fluid Mech., 53: 287–303.
Davidson-Arnott, R.G.D. and Greenwood, B., 1974. J. Sediment. Petrol., 44: 698–704.
Davidson-Arnott, R.G.D. and Greenwood, B., 1976. In: R.A. Davis and R.L. Ethington (Editors), Beach and Nearshore Sedimentation. Soc. Econ. Paleontol. Mineral., Spec. Publ., 24: 149–168.
Davies, A.G., 1977. Geophys. J.R. Astron. Soc., 51: 503–529.
Davies, D.K., 1966. J. Geol., 74: 234–239.
Davies, G.R., 1970. In: B.W. Logan (Editor), Carbonate Sedimentation and Environments, Shark Bay, Western Australia. Mem. Am. Assoc. Pet. Geol., 13: 85–168.
Davies, T.R.H. and Samad, M.F.A., 1978. J. Hydraul, Div. ASCE Eng., 104(HY8): 1171–1182.
Davis, R.A., 1965. J. Sediment. Petrol., 35: 857–866.
Davis, R.A., 1978. Coastal Sedimentary Environments. Springer, New York, N.Y.

Davis, R.A. and Fox, W.T., 1972. Coastal dynamics along Mustang Island, Texas. Western Mich. Univ., Off. Nav. Res. Tech. Rep., No. 9.
Davis, R.A., Fox, W.T., Hayes, M.O. and Boothroyd, J.C., 1972. J. Sediment. Petrol., 42: 413–421.
Davis, W.M., 1890. Bull. Geol. Soc. Am., 1: 195–202.
Dawson, J.B. and Hawthorne, J.B., 1973. J. Geol. Soc. London, 129: 61–84.
De Candolle, C., 1883. Archs. Sci. Phys. Nat., (3)9: 241–278.
De Geer, S., 1911. Årsb. Sver. Geol. Unders., 4: 198 pp.
De Mestre, N.J., 1973. J. Fluid Mech., 58: 641–656.
De Raaf, J.F.M. and Boersma, J.R., 1971. Geol. Mijnbouw, 50: 479–504.
De Raaf, J.F.M., Reading, H.G. and Walker, R.G., 1965. Sedimentology, 4: 1–52.
De Raaf, J.F.M., Boersma, J.R. and Van Gelder, A., 1977. Sedimentology, 24: 451–483.
De Ridder, N.A., 1960. Geol. Mijnbouw, 39: 679–685.
De Rop, W., 1971. Tellus, 23: 261–262.
De Vries, M., 1965. Delft Hydraul. Lab. Publ., No. 36.
Dean, R.G., 1966. In: A.T. Ippen (Editor), Estuary and Coastline Hydrodynamics. McGraw-Hill, New York, N.Y., pp. 197–230.
Dean, R.G. and Eagleson, P.S., 1966. In: A.T. Ippen (Editor), Estuary and Coastline Hydrodynamics. McGraw-Hill, New York, N.Y., pp. 93–132.
Debbas, S. and Rumpf, H., 1966. Chem. Eng. Sci., 21: 583–607.
Deelman, J.C., 1974a. Neues Jahrb. Geol. Paläontol., Abh., 147: 1–29.
Deelman, J.C., 1974b. Neues Jahrb. Geol. Paläontol., Abh., 147: 237–268.
Defant, A., 1921. Beitr. Phys. Frei. Atmos., 9: 99–113.
Defant, A., 1958. Ebb and Flow. University of Michigan Press, Ann Arbor, Mich.
Deigaard, R. and Fredsøe, J., 1978. Nordic Hydrol., 9: 7–16.
Delft Hydraulics Laboratory, 1962. Report on siltation of Demerara bar channel and coastal erosion in British Guiana, Demerara Coastal Investigation. Hydraulics Laboratory, Delft, The Netherlands.
Demarest, D.F., 1947. J. Sediment. Petrol., 17: 18–22.
Dennis, S.C.R. and Chang, G.Z., 1970. J. Fluid Mech., 42: 471–489.
Denton, W.H., 1957. Rep. U.K. Atom. Energy Auth., E/R 1095, 13 pp.
Depuydt, F., 1972. Verh. K. Vlaam. Akad. Wet., 34: 1–229.
Deresiewicz, H., 1958. Adv. Appl. Mech., 5: 233–306.
Detterman, R.L. and Reed, B.L., 1973. Bull. U.S. Geol. Surv., 1368-A.
Dexter, A.R. and Tanner, D.W., 1971. Nature (London), Phys. Sci., 230: 177–179.
Dhawan, S. and Narasimha, R., 1958. J. Fluid Mech., 3: 418–436.
Deutsches Hydrographisches Institut (DHIH), 1963. Atlas der Gezeitenströme für die Nordsee, den Kanal und die Britischen Gewässer. DHIH, Hamburg.
Dillo, H.G., 1960. Mitt. Franzius Inst., 17: 135–253.
Dingle, R.V., 1965. Mar. Geol., 3: 391–400.
Dingler, J.R., 1979. J. Sediment. Petrol., 49: 287–294.
Dionne, J.C., 1971. Can. J. Earth Sci., 8: 1585–1591.
Divoky, D., Le Mehauté, B. and Lin, A., 1970. J. Geophys. Res., 75: 1681–1692.
Dixon, O.A., 1970. J. Sediment. Petrol., 40: 682–687.
Dobrowolski, L., 1924. Geogr. Ann., 6: 300–367.
Dobson, M.R., Evans, W.E. and James, K.H., 1971. Mar. Geol., 11: 27–69.
Dodds, J.A., 1975. Nature (London), 256: 187–189.
Dodge, R.E., 1894. Science, 23: 38–39.
Doeglas, D.J., 1962. Sedimentology, 1: 167–190.
Dolan, R. and Ferm, J.C., 1966. Prof. Geogr., 18: 210–213.
Donat, J., 1929. Wasserwirtschaft, 26: 27–36.
Donner, J.J. and West, R.G., 1957. Skr. Nor. Polarinst., No. 109.

Donovan, R.N. and Archer, R., 1975. Proc. Yorks. Geol. Soc., 40: 547–562.
Dore, B.D., 1970. J. Fluid Mech., 40: 113–126.
Dore, B.D., 1973. Proc. Cambridge Philos. Soc., Math. Phys. Sci., 74: 333–347.
Dore, B.D., 1974. Mecanica, 9: 172–178.
Dore, B.D., 1975. J. Fluid Mech., 70: 257–266.
Dore, B.D., 1976a. Pure Appl. Geophys., 114: 629–637.
Dore, B.D., 1976b. J. Fluid Mech., 76: 819–828.
Dore, B.D., 1977. Q. J. Mech. Appl. Math., 30: 157–173.
Dore, B.D., 1978a. Geophys. Astrophys. Fluid Dyn., 10: 215–230.
Dore, B.D., 1978b. J. Eng. Math., 12: 289–301.
Dörjes, J. and Reineck, R.-E., 1977. Senckenbergiana Marit., 9: 19–45.
Dörjes, J., Gadow, S., Reineck, H.-E. and Singh, I.B., 1970. Senckenbergiana Marit., 2: 31–59.
Dorrepaal, J.M., 1976. Z. Angew. Math. Phys., 27: 739–748.
Dorrepaal, J.M., O'Neill, M.E. and Ranger, K.B., 1976. J. Fluid Mech., 75: 273–286.
Dott, R.H., 1963. Bull. Am. Assoc. Pet. Geol., 47: 104–128.
Dott, R.H. and Roshardt, M.A., 1972. Geol. Soc. Am. Bull., 83: 2589–2596.
Doubiansky, V., 1928. Bull. Appl. Bot., Genet. Plant Breed., 19: 225–285.
Douglas, I., 1967. Publ. Int. Assoc. Sci. Hydrol., No. 75: 17–29.
Drake, L.D., 1974. Geol. Soc. Am. Bull., 85: 247–250.
Drake, L.D., 1977. Earth Surf. Processes, 2: 309–317.
Drapeau, G., 1970. Marit. Sediments, 6: 90–101.
Draper, L., 1967a. Mar. Geol., 5: 133–140.
Draper, L., 1967b. Proc. 10th Conf. Coastal Eng., 1: 12–17.
Draper, L., 1971. J. Inst. Navig., 24: 273–277.
Draper, L. and Driver, J.S., 1971. Proc. 1st Int. Congr. Port and Ocean Engineering under Arctic Conditions, 2: 966–978.
Draper, L. and Whitaker, M.A.B., 1965. Dtsch. Hydrogr. Z., 18: 25–30.
Drazin, P.G. and Howard, L.N., 1966. Adv. Appl. Mech., 9: 1–89.
Dresch, J., 1975. Bull. Assoc. Géogr. Fr., No. 430: 337–351.
Dressler, R.F., 1949. Commun. Pure Appl. Math., 2: 149–194.
Dressler, R.F., 1952a. J. Res. Natl. Bur. Stand., 49: 217–225.
Dressler, R.F., 1952b. Circ. U.S. Natl. Bur. Stand., No. 521: 237–241.
Dressler, R.F., 1954. Publ. Inst. Assoc. Sci. Hydrol., No. 38: 319–328.
Drew, D.A., 1975. Trans. ASME, E42: 38–44.
Drew, D.A. and Kogelman, S., 1975. Appl. Sci. Res., 30: 279–290.
Dryden, H.L., 1959. In: C.C. Lin (Editor), Turbulent Flows and Heat Transfer. Oxford University Press, London, 549 pp.
Dubief, J., 1943. Trav. Inst. Rech. Sahariennes, 2: 11–35.
Dubief, J., 1952. Trav. Inst. Rech. Sahariennes, 8: 123–162.
Duck, P.W., 1978. Proc. R. Soc. London, Ser. A., 363: 335–355.
Dunagan, W.M., 1940. J. Am. Concr. Inst., 36: 649–684.
Dunn, I.S., 1959. J. Soil Mech. Found. Div. ASCE, 85(SM3): 1–24.
Duquennois, H., 1951. Trans. 4th Congr. Large Dams, 4: 119–135.
Durand, R., 1951. Houille Blanche, 6: 609–619.
Durand, R. and Cohen de Lara, G., 1953. Houille Blanche, 8: 254–259.
Duxbury, A.C. and McGary, N.B., 1967. Int. J. Limnol. Oceanogr., 1: 71–84.
Dyer, K.R., 1970. Geophys. J. R. Astron. Soc., 22: 153–161.
Dyer, K.R., 1973. Estuaries. Wiley, London.
Dyer, K.R., 1976. In: M.L. Wiley (Editor), Estuarine Processes, vol. 2. Academic Press, New York, N.Y., pp. 124–135.
Dziedzic, K., 1971. Geol. Sudetica, 5: 7–75.

Dzulynski, S., 1963. Stud. Geol. Pol., 12: 1–136.
Dzulynski, S., 1965. J. Sediment. Petrol., 35: 196–212.
Dzulynski, S. and Sanders, J.E., 1962a. Ann. Soc. Géol. Pol., 32: 143–146.
Dzulynski, S. and Sanders, J.E., 1962b. Trans. Conn. Acad. Arts Sci., 42: 57–96.
Dzulynski, S. and Slaczka, A., 1958. Ann. Soc. Géol. Pol., 28: 205–259.
Dzulynski, S. and Walton, E.K., 1963. Trans. Edinburgh Geol. Soc., 19: 279–305.
Dzulynski, S. and Walton, E.K., 1965. Sedimentary Features of Flysch and Graywackes. Elsevier, Amsterdam.
Dzulynski, S. and Zak, C., 1960. Ann. Soc. Géol. Pol., 30: 213–241.
Eager, R.M.C., 1971. Proc. Geol. Assoc., 82: 71–85.
Eagleson, P.S., 1956. Trans. Am. Geophys. Union, 37: 565–572.
Eagleson, P.S. and Dean, R.G., 1961. Trans. Am. Soc. Civ. Eng., 126: 1162–1186.
Eagleson, P.S. and Dean, R.G., 1966. In: A.T. Ippen (Editor), Estuary and Coastline Hydrodynamics. McGraw-Hill, New York, pp. 1–10.
Eagleson, P.S., Dean, R.G. and Peralta, L.A., 1957. The mechanics of the motion of discrete spherical bottom sediment particles due to shoaling waves. Mass. Inst. Technol., Dep. Civ. Eng., Hydrodyn. Lab., Rep. No. 26.
Eagleson, P.S., Dean, R.G. and Peralta, L.A., 1958. Tech. Mem. U.S. Beach Erosion Board, No. 104.
Earle, M.D., 1975. J. Geophys. Res., 80: 377–379.
Earle, M.D. and Malahoff, A., 1979. Ocean Wave Climate. Plenum, New York, N.Y.
Eastwood, J., Matzen, E.J.P., Young, M.J. and Epstein, N., 1969. Br. Chem. Eng., 14: 1542–1545.
Eaton, G.P., 1963. J. Geophys. Res., 68: 521–528.
Eaton, G.P., 1964. J. Geol., 72: 1–30.
Eckart, C., 1952. Circ. Natl. Bur. Stand., No. 521: 165–173.
Einsele, G., 1967. In: A.F. Richards (Editor), Marine Geotechnique, University of Chicago Press, Urbana, Ill., pp. 154–169.
Einsele, G., Overbeck, R., Schwarz, H.U. and Unsöld, G., 1974. Sedimentology, 21: 339–372.
Einstein, H.A., 1942. Trans. ASCE, 107: 561–597.
Einstein, H.A., 1950. Tech. Bull. U.S. Dep. Agric., No. 1026: 70 pp.
Einstein, H.A., 1968. J. Hydraul. Div. ASCE, 94(HY5): 1197–1205.
Einstein, H.A. and Barbarossa, N.L., 1952. Trans. ASCE, 117: 1121–1146.
Einstein, H.A. and Chien, N., 1953. Transport of sediment mixtures with large ranges of grain sizes. U.S. Army Corps Eng., Missouri River Div., Sediment Ser., No. 2.
Einstein, H.A. and Chien, N., 1955. Effects of heavy sediment concentration near the bed on velocity and sediment distribution. University of California, Berkeley, California, Institute of Engineering Research.
Einstein, H.A. and El-Samni, E.A., 1949. Rev. Mod. Phys., 21: 520–524.
Einstein, H.A. and Krone, R.B., 1962. J. Geophys. Res., 67: 1451–1461.
Einstein, H.A. and Li, H., 1958. Trans. ASCE, 123: 293–317.
Einstein, H.A., Anderson, A.G. and Johnson, J.W., 1940. Trans. Am. Geophys. Union, 21: 628–632.
Eirich, F.R., Margaretha, H. and Bunzl, M., 1936. Kolloidzeitschrift, 75: 20–37.
Eisenberg, P., 1961. In: V.L. Streeter (Editor), Handbook of Fluid Dynamics. McGraw-Hill, New York, N.Y., pp. 12-1–12-24.
Eisenschitz, R., 1932. Z. Phys. Chem., A158: 78–90.
Eittreim, S., Gordan, A.L., Ewing, M., Thorndike, E.M. and Bruchhausen, P., 1972. In: A.L. Gordon (Editor), Studies in Physical Oceanography, vol. 2. Gordon and Breach, New York, pp. 19–35.
Ekman, V.W., 1905. Ark. Mat., Astron. Fys., 2(11): 1–53.
Elder, J.W., 1960. J. Fluid Mech., 9: 235–246.

Elliott, R.E., 1964. Nature (London), 202: 451–452.
Elliott, R.E., 1965. Sedimentology, 5: 193–209.
Ellwood, J.M., Evans, P.D. and Wilson, I.G., 1975. J. Sediment. Petrol., 45: 554–561.
Embleton, C. and King, C.A.M., 1975. Glacial and Periglacial Geomorphology. 1. Glacial Geomorphology. Arnold, London, 2nd ed.
Embley, R.W., 1976. Geology, 4: 371–374.
Embley, R.W., 1980. Mar. Geol., 38: 23–50.
Emerson, W.W., 1959. J. Soil Sci., 10: 235–244.
Emerson, W.W., 1962. J. Soil Sci., 13: 40–45.
Emerson, W.W., 1963. Aust. J. Soil Res., 1: 129–143.
Emery, K.O., 1963. In: M.N. Hill (Editor), The Sea, vol. 3. Interscience, New York, N.Y., pp. 776–793.
Emery, K.O., 1968. J. Sediment. Petrol., 38: 1264–1269.
Emery, K.O. and Gunnerson, C.G., 1973. Proc. Natl. Acad. Sci. U.S.A., 70: 2379–2380.
Emmons, H.W., 1951. J. Aeronaut. Sci., 18: 490–498.
Engels, H., 1905. Z. Bauw., 55: 663–680.
Engelund, F., 1966. J. Hydraul. Div. ASCE, 92(HY2): 315–326.
Engelund, F., 1969. Tech. Univ. Den. Hydraul. Lab. Prog. Rep., No. 18: 1–4.
Engelund, F., 1970. J. Fluid Mech., 42: 225–244.
Engelund, F., 1971. Tech. Univ. Den., Inst. Hydrodyn. Hydraul. Eng., Prog. Rep., No. 24: 51–54.
Engelund, F., 1973. Tech. Univ. Den., Inst. Hydrodyn. Hydraul. Eng., Prog. Rep., No. 31: 25–30.
Engelund, F., 1974. Tech. Univ. Den., Inst. Hydrodyn. Hydraul. Eng., Prog. Rep., No. 32: 37–40.
Engelund, F., 1975. J. Fluid Mech., 72: 145–160.
Engelund, F. and Fredsøe, J., 1970. Dan. Cent. Appl. Math. Mech., Rep. No. 6.
Engelund, F. and Fredsøe, J., 1974. Tech. Univ. Den., Inst. Hydrodyn. Hydraul. Eng., Ser. Pap., No. 4.
Engelund, F. and Fredsøe, J., 1977. Nordic Hydrol., 7: 293–306.
Engelund, F. and Hansen, E., 1966. Acta Polytech. Scand., Civ. Eng. Build. Constr. Ser., No. 35.
Engelund, F. and Hansen, E., 1967. A Monograph on Sediment Transport in Alluvial Streams. Technisk Forlag, Copenhagen.
Enos, P., 1969a. J. Sediment. Petrol., 39: 680–723.
Enos, P., 1969b. Geol. Soc. Am., Spec. Pap., 117.
Enquist, F., 1932. Geol. För. Stockholm Förh., 54: 19–59.
Epstein, J.B. and Epstein, A.G., 1972. U.S. Geol. Surv., Prof. Pap., 744.
Epstein, N. and Young, M.J., 1962. Nature (London), 196: 885–886.
Ergun, S. and Ornung, A.A., 1949. Ind. Eng. Chem., 41: 1179–1184.
Erickson, J.M., 1971. J. Sediment. Petrol., 41: 589–593.
Eriksson, K.A., 1977a. Sediment. Geol., 18: 223–244.
Eriksson, K.A., 1977b. Sediment. Geol., 18: 257–281.
Eriksson, K.A. and Vos, R.G., 1979. Precambrian Res., 9: 169–188.
Erkek, C., 1973. Proc. 15th Congr. Inst. Assoc. Hydraul. Res., 1: 411–416.
Ertel, H., 1968. Beitr. Geophys., 77: 337–341.
Escoffier, F.F. and Boyd, M.B., 1962. J. Hydraul. Div. ASCE 88(HY6): 145–166.
Etheridge, D.W. and Kemp, P.H., 1979. J. Hydraul. Res., 17: 107–119.
Euler, F., 1957. J. Appl. Phys., 28: 1342–1345.
Evans, G., 1965. Q. J. Geol. Soc. London, 121: 209–241.
Evans, O.F., 1938a. J. Sediment. Petrol., 8: 71.
Evans, O.F., 1938b. J. Sediment. Petrol., 8: 33–35.

Evans, O.F., 1941. J. Sediment. Petrol., 11: 37–41.
Evans, O.F., 1942a. J. Sediment. Petrol., 12: 31–35.
Evans, O.F., 1942b. J. Geol., 50: 846–865.
Evans, O.F., 1943. J. Sediment. Petrol., 13: 35–39.
Evans, O.F., 1949. J. Sediment. Petrol., 19: 82–86.
Evans, O.F. and Ingram, R.L., 1943. J. Sediment. Petrol., 13: 117–120.
Evans, P.E. and Millman, R.S., 1964. Powder Metall., 7(13): 50–63.
Everts, C.H., 1973. ASCE J. Waterw. Harbors Coastal Eng. Div., 99(WW4): 425–438.
Ewing, J.A., 1973. Mar. Geol., 15: M31–M35.
Exner, F.M., 1920. Sitzungsber. Akad. Wiss. Wien, Math.-Naturw. Kl., Abt. 2a, 129: 929–952.
Exner, F.M., 1925. Sitzungsber. Akad. Wiss. Wien, Math.-Naturw. Kl. Abt. 2a, 134: 165–203.
Fage, A. and Townend, H.C.H., 1932. Proc. R. Soc. London, Ser. A, 135: 656–677.
Fahnestock, R.K., 1963. U.S. Geol. Surv., Prof. Pap., 422-A.
Fahnestock, R.K. and Bradley, W.C., 1974. In: M. Morisawa (Editor), Fluvial Geomorphology. State University of New York, Binghamton, N.Y., pp. 221–250.
Fahnestock, R.K. and Haushild, W.L., 1962. Geol. Soc. Am., Bull., 72: 1431–1436.
Fahnestock, R.K., Bradley, W.C. and Leveen, L.S., 1969. Geol. Soc. Am., Abstr. Progr., 7: 61–62.
Fahrig, W.F., 1961. Bull. Geol. Surv. Can., No. 68.
Fairchild, H.L., 1901. Am. Geol., 28: 9–14.
Fairchild, I.J., 1980. J. Sediment. Petrol., 50: 423–446.
Fairchild, J.C., 1959. Tech. Mem. U.S. Beach Erosion Board, No. 115.
Falco, R.E., 1977. Phys. Fluids, 20(10,II): S124–132.
Falcon, M., 1969. Theoretical description of free surface two-phase flow over a wavy bed. Cent. Univ. Venezuela, Caracas, Hydraul. Lab., Bull., No. 2: 87–102.
Fales, E.N., 1955. J. Franklin Inst., 259: 491–515.
Falk, F., Lützner, H., Grumbt, E. and Ellenberg, J., 1972. Geologie, 21: 585–608.
Falkowski, E., 1972a. In: E. Falkowski (Editor), Change in the Palaeogeography of Valley Floors of the Vistula Drainage System during the Holocene. Polish Academy of Sciences, Committee for Quaternary Research, pp. 94–100.
Falkowski, E., 1972b. In: E. Falkowski (Editor), Changes in the Palaeogeography of Valley Floors of the Vistula Drainage System during the Holocene. Polish Academy of Sciences, Committee for Quaternary Research, pp. 3–30.
Fan, J.H., 1960. Scientia Sin., 9: 275–303.
Fannelop, T.K. and Waldman, G.D., 1972. AIAA J., 10: 506–510.
Farrell, S.C., 1970. Sediment distributions and hydrodynamics: Sacco River and Scarboro Estuaries. Univ. Mass., Dep. Geol., Contrib. No. 6-CRG.
Farrow, G.E., 1971. Symp. Zool. Soc. London, No. 28: 455–500.
Farrow, G.E. and Brander, K.M., 1971. Philos. Trans. R. Soc. London, Ser. B., 260: 93–121.
Farrow, G.E., Scoffin, T., Brown, B. and Cucci, M., 1979. Scott. J. Geol., 15: 13–29.
Fasel, H., 1976. J. Fluid Mech., 78: 355–383.
Fatt, I., 1956. Trans. Am. Inst. Min. Metall. Eng., 207: 144–159, 160–163, 164–177.
Fay, J.A., 1969. In: D.P. Hoult (Editor), Oil on the Sea. Plenum Press, New York, N.Y., pp. 53–63.
Fejes Toth, L., 1953. Lagerungen in der Ebene, auf der Kugel und im Raum. Springer, Berlin.
Fenton, J.D. and Abbott, J.E., 1977. Proc. R. Soc. London, Ser. A., 352: 523–537.
Ferguson, C.C., 1979. Tectonophysics, 60: 247–262.
Fidleris, V. and Whitmore, R.L., 1961. Br. J. Appl. Phys., 12: 490–494.
Fiedler, H. and Head, M.R., 1966. J. Fluid Mech., 25: 719–735.
Fietz, T.R., 1966. The measurement of characteristics of a three-dimensional density current. Univ. N. S. W., Manley Vale, Water Res. Lab., Rep. No. 85.

Fietz, T.R. and Wood, I.R., 1967. J. Hydraul. Div. ASCE, 93(HY6): 1–23.
Fillon, R.H., 1976. Mar. Geol., 20: 7–25.
Finkel, H.J., 1959. J. Geol., 67: 614–647.
Finkel, H.J., 1961. Photogramm. Eng., June 1961: 439–444.
Finney, J.L., 1970a. Proc. R. Soc. London, Ser. A., 319: 479–493.
Finney, J.L., 1970b. Proc. R. Soc. London, Ser. A., 319: 495–507.
Finnie, I., 1960a. Wear, 3: 87–103.
Finnie, I., 1960b. Proc. Soc. Exp. Stress Anal., 17: 65–75.
Finnie, I., 1962. In: Symposium on Erosion and Cavitation, American Society for Testing and Materials, Spec. Publ. No. 307: 70–82.
Finnie, I., 1972. Wear, 19: 81–90.
Finnie, I. and McFadden, D.H., 1978. Wear, 48: 181–190.
Finnie, I., Wolak, J. and Kabil, Y., 1967. J. Mater., 2: 682–700.
Fischer, M.C., 1972. AIAA J., 10: 1387–1389.
Fisher, R.V., 1964. J. Geophys. Res., 69: 341–355.
Fisher, R.V., 1965. Deep-Sea Res., 12: 345–348.
Fisk, H.N., 1944. Geological investigation of the alluvial valley of the Lower Mississippi River. Mississippi River Commission, Vicksburg, Miss.
Fisk, H.N., 1947. Fine grained alluvial deposits and their effects on Mississippi River activity. Mississippi River Commission, Vicksburg, Miss.
Fisk, H.N., 1959. Proc. 2nd Coastal Geogr. Conf., Natl. Acad. Sci., Wash., pp. 103–151.
Flaxman, E.M., 1963. J. Hydraul. Div. Am. Soc. Civ. Eng., 89(HY2): 87–96.
Fleming, C.A. and Hunt, J.N., 1976. Proc. Inst. Civ. Eng., (2)61: 297–310.
Fleming, R.H., 1938. J. Mar. Res., 1: 192–206.
Flemming, B.W., 1977. Sediment. Geol., 18: 61–95.
Flemming, B.W., 1978. Mar. Geol., 26: 177–198.
Flemming, B.W., 1980. Mar. Geol., 26: 179–205.
Flemming, N.C., 1964. J. Sediment. Petrol., 34: 112–122.
Flemming, N.C., 1965. J. Sediment. Petrol., 35: 381–390.
Flemming, N.C. and Stride, A.H., 1967. J. Mar. Biol. Assoc. U.K., 47: 433–444.
Fliegel, M. and Nowroozi, A.A., 1970. Limnol. Oceanogr., 15: 615–624.
Flint, R.F., 1961. Geol. Soc. Am., Bull., 72: 1687–1691.
Flood, P.G., Orme, G.R. and Scoffin, T.P., 1978. Philos. Trans. R. Soc. London, Ser. A, 291: 78–83.
Fluendy, M.A. and Smith, E.B., 1962. Quart. Rev., Chem. Soc., 16: 241–266.
Foley, M.G., 1977. J. Sediment. Petrol., 47: 738–746.
Folger, D.W., 1970. Deep-Sea Res., 17: 337–352.
Folk, R.L., 1971. Sedimentology, 16: 5–54.
Folk, R.L., 1976. Sedimentology, 23: 649–669.
Folk, R.L., 1977. Sedimentology, 24: 864–874.
Folk, R.L. and Robles, R., 1964. J. Geol., 72: 255–292.
Foord, S.G., 1945. Nature (London), 155: 427.
Ford, E.F., 1957. Trans. Am. Geophys. Union, 38: 171–174.
Ford, J.G., 1975. Mercian Geol., 5: 205–239.
Forel, F.A., 1883. Arch. Sci. Phys. Nat., (3)10: 39–72.
Forel, F.A., 1885. C. R. Acad. Sci., 101: 725–728.
Forgacs, O.L. and Mason, S.G., 1959. J. Colloid Sci., 14: 457–491.
Forster, J.W. and Skrinde, R.A., 1950. Trans. Am. Soc. Civ. Eng., 115: 973–987.
Förstner, U., Müller, G. and Reineck, H.-E., 1968. Neues Jahrb. Miner., Abh., 109: 33–62.
Fowler, R.T. and Chodziesner, W.B., 1959. Chem. Eng. Sci., 10: 157–162.
Frakes, L.A. and Crowell, J.C., 1969. Geol. Soc. Am. Bull., 80: 1007–1042.
Frakes, L.A., Matthews, J.L. and Crowell, J.C., 1971. Geol. Soc. Am., Bull., 82: 1581–1604.

Francis, J.R.D., 1970. S. Afr. Mech. Eng., 20: 72–75.
Francis, J.R.D., 1973. Proc. R. Soc. London, Ser. A., 332: 443–471.
Francis, J.R.D., 1975. Fluid Mechanics for Engineering Students. Arnold, London, 4th ed.
Francis, J.R.D. and Dudgeon, C.R., 1967. Q. J. R. Meteorol. Soc., 93: 247–253.
Francis, J.R.D. and Vickers, J.A., 1968. 3rd Australas. Conf. Hydraul. Fluid Mech., Sydney, Pap. No. 2577: 38–40.
Francis, P.J., 1976. Volcanoes. Penguin Books, Harmondsworth.
Fraser, C., 1966. The Avalanche Enigma. Murray, London.
Fraser, H.J., 1935. J. Geol., 43: 910–1010.
Frazier, D.E. and Osanik, A., 1961. Trans. Gulf-Coast Assoc. Geol. Soc., 11: 121–137.
Fredsøe, J., 1972. Tech. Univ. Den., Inst. Hydrodyn. Hydraul. Eng., Prog. Rep., No. 25: 9–16.
Fredsøe, J., 1974a. Tech. Univ. Den., Inst. Hydrodyn. Hydraul. Eng., Prog. Rep., No. 32: 29–36.
Fredsøe, J., 1974b. J. Fluid Mech., 64: 1–16.
Fredsøe, J., 1974c. Tech. Univ. Den., Inst. Hydrodyn. Hydraul. Eng., Prog. Rep., No. 33: 15–22.
Fredsøe, J., 1974d. Tech. Univ. Den., Inst. Hydrodyn. Hydraul. Eng., Prog. Rep., No. 34: 25–29.
Fredsøe, J., 1975. Tech. Univ. Den., Inst. Hydrodyn. Hydraul. Eng., Prog. Rep., No. 37: 31–36.
Fredsøe, J., 1978. Tech. Univ. Den., Inst. Hydrodyn. Hydraul. Eng., Prog. Rep., No. 45: 23–28.
Fredsøe, J., 1979. J. Fluid Mech., 91: 497–512.
Fredsøe, J. and Engelund, F., 1975. Tech. Univ. Den., Inst. Hydrodyn. Hydraul. Eng., Ser. Pap., No. 8.
Friedman, G.M. and Sanders, J.E., 1978. Principles of Sedimentology. Wiley, New York, N.Y., 792 pp.
Friend, P.F., 1965. Sedimentology, 5: 39–68.
Friend, P.F., Alexander-Marrack, P.D., Nicholson, J. and Yeats, A.K., 1976. Medd. Grønl., No. 206: 91 pp.
Fristrup, N., 1952. Proc. 19th Int. Geol. Congr., Algiers, Sect. 7, pp. 91–99.
Frost, B.A., 1966. Sci. Pap. Meteorol. Off., London, No. 25.
Frostick, L.E. and Reid, I., 1977. Sedimentology, 24: 1–9.
Fryberger, S.A., Ahlbrandt, T.S. and Andrews, S., 1979. J. Sediment. Petrol., 49: 733–746.
Fuchs, O.P., 1963. AIAA J., 1: 2124–2126.
Fuchs, R.A., 1952. Circ. Natl. Bur. Stand., No. 521: 187–200.
Fudali, R.F. and Melson, W.G., 1972. Bull. Volcanol., 35: 383–401.
Führböter, A., 1967. Mitt. Franzius Inst., 29: 1–35.
Fukuoka, S., 1968. Generation, development and spectrum of sand waves. Tokyo Inst. Technol., Dep. Civ. Eng. Tech. Rep., No. 4: 45–55.
Fulford, G.D., 1964. Adv. Chem. Eng., 5: 151–236.
Furnas, C.C., 1929. Bull. Bur. Mines U.S. Dep. Commerce, No. 307.
Furnas, C.C., 1931. Ind. Eng. Chem., 23: 1052–1058.
Furnes, G.K., 1973. On the formation of waves on an erodible bed. University of Bergen, Geophysical Institute, Rep. No. 32 (Div. A, Phys. Oceanogr.).
Furnes, G.K., 1974. Mar. Geol., 16: 145–160.
Furnes, G.K., 1976a. Danish Centre for Applied Mathematics and Mechanics, Rep. No. 105.
Furnes, G.K., 1976b. Danish Centre for Applied Mathematics and Mechanics, Rep. No. 105: 25–34.
Futterer, E., 1974. Neues Jahrb. Geol. Paläontol. Abh., 144: 342–372.
Futterer, E., 1977. Einregelung, Transport und Ablagerung biozener Hartteile im Strömungs-

kanal. Geol. Inst. Ebehard-Karls-University, Tübingen.
Futterer, E., 1978. Neues Jahrb. Geol. Paläontol. Abh., 155: 318–357.
Gabriel, A., 1938. Geogr. J., 92: 193–208.
Gagny, C., 1964. Bull. Soc. Géol. Fr., (7)6: 43–54.
Gaither, A., 1953. J. Sediment. Petrol., 23: 180–195.
Galay, V.J., 1967. Proc. 12th Congr. Int. Assoc. Hydraul. Res., 1: 85–94.
Galay, V.J., Kellerhals, R. and Bray, D.I., 1973. In: Fluvial Processes and Sedimentation. Department of Environment, Canada, Inland Waterways Directorate, pp. 217–250.
Gall, J.C., 1971. Mem. Serv. Carte Géol. Alsace Lorraine, No. 34: 318 pp.
Gallenne, B. and Salomon, J.-C., 1975. Rev. Géogr. Phys. Géol. Dyn., (2)17: 147–162.
Galloway, R.W., 1956. J. Glaciol., 2: 730–733.
Galvin, C.J., 1968. J. Geophys. Res., 73: 3651–3659.
Garde, R.J. and Albertson, M.L., 1959. Proc. Congr. Int. Assoc. Hydraul. Res., Montreal, 4: 28-SII-1–28-SII-5.
Garde, R.J. and Ranga Raju, K.G., 1963. J. Hydraul. Div. ASCE, 89(HY6): 153–164.
Garde, R.J. and Ranga Raju, 1978. Mechanics of Sediment Transportation and Alluvial Stream Problems. Wiley Eastern Limited, New Delhi, 483 pp.
Gardner, J.S., 1971. Can. Geogr., 15: 318–322.
Garrett, C. and Munk, W., 1979. Annu. Rev. Fluid Mech., 11: 339–369.
Garvine, R.W., 1974a. J. Geophys. Res., 79: 831–846.
Garvine, R.W., 1974b. J. Phys. Oceanogr., 4: 557–569.
Garvine, R.W., 1977. J. Geophys. Res., 82: 441–454.
Garvine, R.W. and Monk, J.D., 1974. J. Geophys. Res., 79: 2251–2259.
Gaughan, M.K., 1975. J. Geophys. Res., 80: 2991–2996.
Gauss, G.A. and House, M.R., 1972. J. Geol. Soc. London, 128: 151–172.
Gavrilovic, D., 1970. Z. Geomorphol., 14: 202–218.
Gay, N.C., 1966. Tectonophysics, 3: 29–33.
Gay, N.C., 1968a. Tectonophysics, 5: 211–234.
Gay, N.C., 1968b. Tectonophysics, 5: 295–302.
Gay, N.C., 1968c. Tectonophysics, 5: 81–88.
Gay, P., 1962. Bol. Soc. Geol. Peru, 37: 37–58.
Gee, D.M., 1973. Sediment transport in non-steady flow. Univ. Calif. Berkeley, Coll. Eng., Hydraul. Eng. Lab., Rep. No. HEL 22-3.
Gee, D.M., 1975. J. Hydraul. Div. ASCE, 101(HY3): 437–449.
Gellatly, D.C., 1970. Sediment. Geol., 4: 185–191.
Gent, P.R. and Taylor, P.A., 1976. J. Fluid Mech., 77: 105–128.
Gessler, J., 1971. In: H.W. Shen (Editor), River Mechanics, vol. 1. Colorado State University, Fort Collins, Colo., pp. 8-1–8-24.
Gevers, T.W., 1936. S. Afr. Geogr. J., 19: 61–79.
Ghatage, V.M., 1936. Modelversuche über die gegenzeitige Bewegung von Luftmassen verschiedener Temperaturen. University of Göttingen.
Ghibaudo, G., Mutti, E. and Rosell, J., 1974. Mem. Soc. Geol. Ital., 13: 497–537.
Ghosh, S.N. and Jena, S.B., 1971. Proc. Inst. Civ. Eng., 49: 417–430.
Ghosh, S.N. and Kar, S.K., 1975. Proc. Inst. Civ. Eng., (2)59: 805–811.
Ghosh, S.N. and Mehta, P.J., 1974. Proc. Inst. Civ. Eng., 57: 159–163.
Gibbings, J.C., 1971. J. Mech. Eng. Sci., 13: 234–236.
Gibbs, R.J., Matthews, M.D. and Link, D.A., 1971. J. Sediment. Petrol., 41: 7–18.
Gibson, W.M., 1951. J. Coast Geod. Surv., No. 4: 54–58.
Gierszewski, P.L. and Chaffey, C.E., 1978. Can. J. Phys., 56: 6–11.
Giesekus, H., 1962a. Rheol. Acta, 2: 101–112.
Giesekus, H., 1962b. Rheol. Acta, 2: 112–122.
Gietelink, G., 1973. Leid. Geol. Meded., 49: 125–144.

Gilbert, G.K., 1884. Science, 3: 375–376.
Gilbert, G.K., 1885. Rep. U.S. Geol. Surv., pp. 75–123.
Gilbert, G.K., 1899. Bull. Geol. Soc. Am., 10: 135–140.
Gilbert, G.K., 1914. U.S. Geol. Surv., Prof. Pap., 86.
Gill, M.A., 1971. J. Hydraul. Div. ASCE, 97 (HY12): 2067–2074.
Gillberg, G., 1961. Geol. För. Stockh. Förh., 83: 335–369.
Gillette, D. and Goodwin, P.A., 1974. J. Geophys. Res., 79: 4080–4084.
Gillott, J.E., 1968. Clay in Engineering Geology. Elsevier, Amsterdam, 296 pp.
Gillott, J.E., 1969. J. Sediment. Petrol., 39: 90–105.
Gillott, J.E., 1970. Eng. Geol., 4: 133–153.
Gilmore, J., 1874. Storm Warriers; or Life-boat Work on the Goodwin Sands. Macmillan, London.
Gipson, M., 1965. J. Sediment. Petrol., 35: 408–414.
Gipson, M., 1966. J. Sediment. Petrol., 36: 888–903.
Gladfelter, W.I., Hawkinson, E.E., Neumann, A.P. and Nichols, V.W., 1953. J. Soc. Automot. Eng., October 1953: 63–68.
Glastonbury, J.R. and Bratel, P.G., 1966. Trans. Inst. Chem. Eng., 44: T28–T135.
Glen, J.W., 1958. Adv. Phys., 7: 254–265.
Glen, J.W., Donner, J.J. and West, R.G., 1957. Am. J. Sci., 255: 194–205.
Glen, R.A. and Laing, W.P., 1975. Proc. Australas. Inst. Min. metall., No. 256: 15–20.
Glennie, K.W., 1970. Desert Sedimentary Environments. Elsevier, Amsterdam, 222 pp.
Glennie, K.W., 1972. Bull. Am. Soc. Pet. Geol., 56: 1048–1071.
Glennie, K.W. and Evamy, B.D., 1968. Palaeogeogr. Palaeoclimatol. Palaeoecol., 4: 77–87.
Gluckman, M.J., Weinbaum, S. and Pfeffer, R., 1972. J. Fluid Mech., 55: 677–709.
Goddet, J., 1960. Houille Blanche, 15: 122–133.
Godin, G., 1972. The Analysis of Tides. Liverpool University Press, Liverpool.
Goff, R.C., 1976. Mon. Weath. Rev., 104: 1429–1440.
Göhren, H., 1971a. Die Küste, 21: 6–16.
Göhren, H., 1971b. Proc. 12th Conf. Coastal Eng., 3: 2147–2162.
Göhren, H., 1971c. Hamburger Küstenforsch., No. 19.
Golden, J.H., 1974. J. Appl. Meteorol., 13: 676–692.
Goldring, R., 1971. Mem. Geol. Soc. London, No. 5.
Goldring, R. and Curnow, C.N., 1967. J. Geol. Soc. Aust., 14: 195–214.
Goldsmith, H.L., 1966. Science, 153: 1406–1407.
Goldsmith, H.L. and Mason, S.G., 1962a. J. Colloid. Sci., 17: 448–476.
Goldsmith, H.L. and Mason, S.G., 1962b. J. Fluid Mech., 12: 88–96.
Goldsmith, H.L. and Mason, S.G., 1967. In: F.R. Eirich (Editor), Rheology: Theory and Applications, vol. 4. Academic Press, New York, N.Y., pp. 85–250.
Goldsmith, V., 1973. J. Sediment. Petrol., 43: 1128–1142.
Goldstein, S., 1929. Proc. R. Soc. London, Ser. A., 123: 225–235.
Goldstein, S., 1965. Modern Developments in Fluid Dynamics, vol. 1. Dover Publ., New York, N.Y.
Goldthwait, R.P., 1951. J. Geol., 59: 567–577.
Goodarz-Nia, I. and Sutherland, D.N., 1975. Chem. Eng. Sci., 30: 407–412.
Goodman, M.A. and Cowin, S.C., 1971. J. Fluid Mech., 45: 321–339.
Goodman, M.A. and Cowin, S.C., 1972. Archs Ration. Mech. Anal., 44: 249–266.
Goodwin, J.E., Sage, W. and Tilly, G.P., 1969. Proc. Inst. Civ. Eng., 184: 279–289.
Gordon, C.M., 1974. Nature (London), 248: 392–394.
Gordon, C.M., 1975a. Mar. Geol., 18: M57–M64.
Gordon, C.M., 1975b. Phys. Fluids, 18: 141–143.
Gordon, C.M. and Witting, J., 1977. In: J.C.J. Nihoul (Editor), Bottom Turbulence. Elsevier, Amsterdam, pp. 59–81.

Gordon, J.E., Birnie, R.V. and Timmis, R., 1978. Arct. Alp. Res., 10: 49–60.
Gordon, R., Carmichael, J.B. and Isackson, F.J., 1972. Water Resour. Res., 8: 444–459.
Gorshkov, G.S., 1959. Bull. Volcanol., (2) 20: 77–109.
Görtler, H. and Witting, H., 1958. In: H. Görtler (Editor), Boundary Layer Research. Springer, Berlin, pp. 110–126.
Gotoh, K., 1971. Ind. Eng. Chem., Fundam., 10: 161–164.
Gotoh, K. and Finney, J.L., 1974. Nature (London), 252: 202–205.
Gotoh, K., Jodrey, W.S. and Tory, E.M., 1978a. Powder Technol., 20: 233–242.
Gotoh, K., Jodrey, W.S. and Tory, E.M., 1978b. Powder Technol., 20: 257–260.
Goudie, A.S., Allchin, B. and Hegde, K.T.M., 1973. Geogr. J., 139: 243–257.
Gough, D.I., Aziz-ur-Rahman and Evans, M.E., 1977. Geophys. J. R. Astron. Soc., 50: 685–697.
Gould, H.R., 1951. In: J.L. Hough (Editor), Turbidity Currents. Soc. Econ. Paleontol. Mineral., Spec. Publ., 2: 34–52.
Gould, W.J., 1978. Philos. Trans. R. Soc., Ser. A., 290: 87–98.
Gouleau, D., 1968. Étude Hydrologique et Sédimentologique de la Baie de Bourgneuf, University of Caen, Faculty of Science.
Gradowczyk, M.H., 1968. J. Fluid Mech., 33: 93–112.
Gradowczyk, M.H., 1971. In: H. Liepholz (Editor), Instability of Continuous Systems. Springer, Berlin, pp. 143–150.
Gradzinski, R., 1970. Palaeontol. Pol., No. 21: 147–229.
Gradzinski, R. and Jerzykiewicz, T., 1974a. Sediment. Geol., 12: 249–278.
Gradzinski, R. and Jerzykiewicz, T., 1974b. Palaeontol. Pol., No. 30: 111–146.
Gradzinski, R., Gagol, J. and Slaczka, A., 1979. Acta Geol. Pol., 29: 151–175.
Graf, W.H., 1965. J. Geol., 73: 550–552.
Graf, W.H., 1971. Hydraulics of Sediment Transport. McGraw-Hill, New York, N.Y.
Graf, W.H. and Acaroglu, E.R., 1966. Bull. Int. Assoc. Sci. Hydrol., 16: 27–43.
Graf, W.H. and Acaroglu, E.R., 1968. Bull. Int. Assoc. Sci. Hydrol., 18: 20–39.
Graf, W.H. and Mansour, F.F., 1975. J. Hydraul. Res., 13: 127–147.
Graham, R.W. and Deissler, R.G., 1967. Trans. ASME, C89: 371–372.
Grancini, G. and Cescon, B., 1973. Limnol. Oceanogr., 18: 705–710.
Grant, H.L., 1958. J. Fluid Mech., 4: 149–190.
Grass, A.J., 1970. J. Hydraul. Div. ASCE, 95(HY3): 619–632.
Grass, A.J., 1971. J. Fluid Mech., 50: 233–255.
Graton, L.C. and Fraser, H.J., 1935. J. Geol., 43: 785–909.
Gravenor, C.P. and Meneley, W.A., 1958. Am. J. Sci., 256: 715–728.
Gray, W.A., 1968. The Packing of Solid Particles. Chapman and Hall, London.
Greely, R., Iversen, J.D., Pollack, J.D., Udovich, N. and White, B., 1974. Proc. R. Soc. London, Ser. A., 341: 331–360.
Green, C.D., 1975. In: L.E. Cronin (Editor), Estuarine Research, vol. 2. Academic Press, New York, N.Y., pp. 323–344.
Greene-Kelly, R. and Mackney, D., 1970. In: D.A. Osmond and P. Bullock (Editors), Micromorphological Techniques and Applications. Soil Surv. Tech. Monogr., No. 2: 43–52.
Greenspan, H.P., 1968. The Theory of Rotating Fluids. Cambridge University Press, Cambridge.
Greensmith, J.T. and Tucker, E.V., 1968. Nature (London), 220: 1115–1116.
Greenwood, B. and Davidson-Arnott, R.G.D., 1975. In: J. Hails and A. Carr (Editors), Nearshore Sediment Dynamics and Sedimentation. Wiley, London, pp. 123–150.
Greenwood, B. and Davidson-Arnott, R.G.D., 1979. Can. J. Earth Sci., 16: 312–332.
Greenwood, B. and Mittler, P.R., 1979. Mar. Geol., 32: 191–203.
Griffith, G.A. and Sutherland, A.J., 1977. J. Hydraul. Div. ASCE, 103(HY11): 1279–1291.

Griffith, J.S., 1962. Nature (London), 196: 764–765.
Griffiths, J.C., 1950. Prod. Mont., 14: 26–32.
Griffiths, J.C. and Rosenfeld, M.A., 1953. Am. J. Sci., 251: 192–214.
Grigg, N.S., 1970. J. Hydraul. Div. ASCE, 96(HY2): 2501–2518.
Griggs, G.B. and Kulm, L.D., 1970. Geol. Soc. Am. Bull., 81: 1361–1384.
Grim, R.E., 1962. Applied Clay Mineralogy. McGraw-Hill, New York, N.Y.
Grim, R.E., 1968. Clay Mineralogy. McGraw-Hill, New York, N.Y., 2nd ed.
Grim, R.E., Bradley, W.F. and White, W.A., 1957. Rep. Invest. Ill. St. Geol. Surv., No. 203.
Grimshaw, R., 1974. J. Fluid Mech., 62: 775–791.
Grinnell, R.S., 1974. J. Sediment. Petrol., 44: 116–122.
Gripp, K., 1961a. Neues Jahrb. Geol. Paläontol. Monatsh., No. 9: 474–477.
Gripp, K., 1961b. Z. Geomorphol., 5: 24–36.
Gripp, K., 1961c. Neues Jahrb. Geol. Paläontol. Monatsh., No. 3: 156–160.
Gripp, K., 1968. Die Küste, 16: 76–117.
Gripp, K. and Martens, L., 1963. Wenn die Natur im Sande Spielt. Verlag der Gesellschaft der Freunde des vaterländischen Schul- und Erziehungswesens, Hamburg
Grissinger, E.H., 1966. Water Resour. Res., 2: 131–138.
Groen, P., 1967. Neth. J. Sea Res., 3: 564–574.
Groth, K., 1961. Geologie, 10: 169–184.
Grove, A.T., 1969. Geogr. J., 135: 191–212.
Grove, A.T. and Warren, A., 1968. Geogr. J., 134: 194–208.
Grover, N.C. and Howard, C.S., 1938. Trans. Am. Soc. Civ. Eng., 103: 720–790.
Grumbt, E., 1966. Ber. Dtsch. Ges. Geol. Wiss., Reihe A, 11: 217–234.
Grumbt, E., 1969. Geologie, 18: 151–168.
Grumbt, E., 1971. Ber. Dtsch. Ges. Geol. Wiss., Reihe A, 16: 297–314.
Grumbt, E., 1974. Schriftenr. Geol. Ges., No. 1: 205 pp.
Gry, H., 1942. Folia Geogr. Danica, 2: 1–138.
Guilcher, A., 1958a. Coastal and Submarine Morphology. Methuen, London.
Guilcher, A., 1958b. Scott. Geogr. Mag., 74: 137–149.
Gunaratnam, D.J. and Perkins, F.E., 1970. Numerical solution of unsteady flows in open channels. Mass. Inst. Technol., Dep. Civ. Eng., Hydrodyn. Lab., Rep. No. 127.
Gupta, A.K., Laufer, J. and Kaplan, R.E., 1971. J. Fluid Mech., 50: 493–512.
Gupta, N.K., 1966. J. Hydraul. Div. Am. Soc. Civ. Eng., 92(HY3): 115–117.
Gupta, S.C., 1957. Z. Angew. Math. Mech., 8: 257–261.
Gürich, G., 1933. Z. Dtsch. Geol. Ges., 85: 652–664.
Gurka, J.J., 1976. Mon. Weather Rev., 104: 1484–1493.
Gust, G., 1976. J. Fluid Mech., 75: 29–47.
Gust, G. and Walger, E., 1976. Mar. Geol., 22: 189–206.
Gustavson, T.C., 1974. J. Sediment. Petrol., 44: 374–389.
Gustavson, T.C., 1978. Sedimentology, 25: 401–426.
Gustavson, T.C., Ashley, G.M. and Boothroyd, J.C., 1975. In: A.V. Jopling and B.C. McDonald (Editors). Soc. Econ. Paleontol. Mineral., Spec. Publ., 23: 264–280.
Guy, H.P., Simons, D.B. and Richardson, E.V., 1966. U.S. Geol. Surv., Prof. Pap., 462-I.
Guza, R.T. and Bowen, A.J., 1976. J. Mar. Res., 34: 269–293.
Gwinn, V.E., 1964. J. Sediment. Petrol., 34: 656–658.
Gyr, A., 1976. Z. Angew. Math. Mech., 27: 717–725.
Hack, J.T., 1941. Geogr. Rev., 31: 240–263.
Hadley, M.L., 1964. Mar. Geol., 2: 164–167.
Hagemeyer, R.W., 1960. Tappi, 43: 277–288.
Hagerdorn, H., 1971. Z. Geomorphol., S11: 1–251.
Hails, J.R. and Kelland, N.C., 1974. J. Geol. Soc. London, 131: 7–35.
Hälbich, I.W., 1964. Trans. Proc. Geol. Soc. S. Afr., 67: 95–109.

Hall, F.F., Neff, W.D. and Frazier, T.V., 1976. Nature (London), 264: 408–411.
Hall, J., 1843. Geology of New York. IV. Geology of the Fourth District. Caroll and Cook, Albany, N.Y.
Hall, P., 1974. J. Fluid Mech., 64: 209–226.
Hallier, U.W., 1976. Z. Geomorphol., 20: 108–120.
Halliwell, A.R. and O'Connor, B.A., 1965. J. Liverpool Eng. Soc., 11: 21–45.
Halliwell, A.R. and O'Connor, B.A., 1966. Proc. 10th Conf. Coastal Eng., 1: 687–706.
Halliwell, A.R. and O'Dell, M., 1971. Proc. 12th Conf. Coastal Eng., 3: 1959–1977.
Hama, F.R., 1962. Phys. Fluids, 5: 644–650.
Hama, F.R., Long, J.D. and Hegarty, J.C., 1957. J. Appl. Phys., 28: 388–394.
Hamblin, W.K., 1958. Mich. Geol. Surv. Div., Publ., No. 51.
Hamblin, W.K., 1961a. J. Sediment. Petrol., 31: 390–401.
Hamblin, W.K., 1961b. Geol. Soc. Am. Bull., 72: 1–18.
Hamblin, W.K., 1962a. J. Sediment. Petrol., 32: 201–210.
Hamblin, W.K., 1962b. J. Sediment. Petrol., 32: 530–533.
Hamblin, W.K., 1965. Kans. State Geol. Surv., Bull., No. 175-1.
Hamelin, H.E., 1958. Rev. Géogr. Alp., 46: 429–439.
Hamilton, D., Sommerville, J.H. and Stanford, P.N., 1980. Sediment. Geol., 26: 115–138.
Hamilton, N.L. and Rees, A.I., 1971. Geol. Rundsch., 60: 1103–1124.
Hammond, T.M. and Collins, M.B., 1979. Sedimentology, 26: 795–812.
Hampton, M.A., 1972. J. Sediment. Petrol., 42: 775–793.
Hand, B.M., 1961. Compass Sigma Gamma Epsilon, 38: 133–144.
Hand, B.M., 1969. J. Sediment. Petrol., 39: 1302–1309.
Hand, B.M., 1974. J. Sediment. Petrol., 44: 637–648.
Hand, B.M., Wessel, J.M. and Hayes, M.O., 1969. J. Sediment. Petrol., 39: 1310–1316.
Hand, B.M., Middleton, G.V. and Skipper, K., 1972. Sedimentology, 18: 135–138.
Hanratty, T.J., 1956. AIChE J., 2: 359–362.
Hansen, N.O., 1971. Dan. Cent. Appl. Math. Mech. Rep., No. 16.
Häntzschel, W., 1935. Natur Volk, 65: 461–465.
Häntzschel, W., 1936. Senckenbergiana, 18: 316–356.
Häntzschel, W., 1938. Senckenbergiana, 20: 1–42.
Häntzschel, W., 1939. Natur Volk, 69: 40–48.
Häntzschel, W. and Reineck, H.-E., 1968. Mitt. Geol. Staatsinst. Hamburg, 37: 5–39.
Happel, J., 1949. Ind. Eng. Chem., 41: 1161.
Happel, J. and Brenner, H., 1965. Low Reynolds Number Hydrodynamics. Prentice-Hall, Englewood Cliffs, N.J.
Hardy, R.M. and Legget, R.F., 1960. Bull. Geol. Soc. Am., 71: 93–94.
Harlett, J.C. and Kulm, L.D., 1973. Geol. Soc. Am. Bull., 84: 3815–3826.
Harms, J.C., 1969. Geol. Soc. Am. Bull., 80: 363–396.
Harms, J.C. and Fahnestock, R.K., 1965. In: G.V. Middleton (Editor), Primary Sedimentary Structures and their Hydrodynamic Interpretation. Soc. Econ. Paleontol. Mineral., Spec. Publ., 12: 84–115.
Harms, J.C., MacKenzie, D.B. and McCubbin, D.G., 1963. J. Geol., 71: 566–580.
Harms, J.C., Choquette, P.W. and Brady, M.J., 1974. In: A.E. Weidie (Editor), Field Seminar on Water and Carbonate Rocks of the Yucatan Peninsula, Mexico. New Orleans Geological Society, New Orleans, La., pp. 123–147.
Harris, C., 1975. Area, 6: 243–249.
Harris, C.R., 1974. Sci. Proc. R. Dublin Soc., Ser. A., 5: 237–252.
Harris, D.L., 1971. Proc. 12th Conf. Coastal Eng., 1: 85–100.
Harris, D.L., 1972a. In: D.J.P. Swift and O.H. Pilkey (Editors), Shelf Sediment Transport: Processes and Pattern. Dowden, Hutchinson and Ross, Stroudsburg, Penn., pp. 99–125.
Harris, D.L., 1972b. In: R.E. Meyer (Editor), Waves on Beaches and Resulting Sediment

Transport. Academic Press, New York, N.Y., pp. 1–51.
Harris, J.B., Nawaz, M. and Pittman, J.F.T., 1979. J. Fluid Mech., 95: 415–429.
Harris, S.A., 1967. Can. Geogr., 11: 16–34.
Harris, S.A., 1968. Prof. Geogr., 20: 195–198.
Harrison, P.W., 1956. Description and origin of some orientation properties of englacial debris near Thule, Greenland. U.S. Army Snow Ice Permafrost Res. Establishment, Rep. S-4835.
Harrison, P.W., 1957. J. Glaciol., 65: 275–308.
Harrison, S.S. and Clayton, L., 1968. Proc. North Dakota Acad. Sci., 22: 93–96.
Harrison, S.S. and Clayton, L., 1970. Geol. Soc. Am. Bull., 81: 1217–1226.
Harrison, W., Byrne, R.J., Boon, J.D. and Moncure, R.W., 1971. Proc. 12th Conf. Coastal Eng., 2: 1201–1222.
Harrison, W.J., 1908. Proc. London Math. Soc., (2)7: 107–121.
Hartwell, A.D., 1970. Hydrography and Holocene sedimentation of the Merrimack River Estuary, Massachusetts. Univ. Mass. Dep. Geol., Contrib. No. 5-CRG.
Harvey, J.G., 1966. Mar. Geol., 4: 49–55.
Hastenrath, S.L., 1967. Z. Geomorphol., 11: 300–331.
Hattori, M., 1969. Proc. 13th Congr. Int. Assoc. Hydraul. Res., 3: 399–406.
Haughey, D.P. and Beveridge, G.S.G., 1966. Chem. Eng. Sci., 21: 905–916.
Haughey, D.P. and Beveridge, G.S.G., 1967. Chem. Eng. Sci., 22: 715–718.
Haughey, D.P. and Beveridge, G.S.G., 1969. Can. J. Chem. Eng., 47: 130–140.
Haushild, W.L., Stevens, H.H., Nelson, J.L. and Dempster, G.R., 1973. U.S. Geol. Surv., Prof. Pap., 433-N.
Haushild, W.L., Dempster, G.R. and Stevens, H.H., 1975. U.S. Geol. Surv., Prof. Pap., 433-0.
Hausner, H.H., 1962. In: W.A. Knepper (Editor), Agglomeration. Interscience, New York, N.Y., pp. 55–88.
Havelock, T.H., 1908. Proc. R. Soc. London, Ser. A., 81: 398–430.
Havelock, T.H., 1918. Proc. R. Soc. London, Ser. A., 95: 38–51.
Hawkins, A.B. and Sebbage, M.J., 1972. Mar. Geol., 12: M7–M9.
Hay, R.L., 1976. Geology of the Olduvai Gorge: a Study of Sedimentation in a Semiarid Basin. University of California Press, Berkeley and Los Angeles, Calif.
Hayashi, T., 1970. J. Hydraul. Div. ASCE, 96(HY2): 357–366.
Hayes, M.O., 1967. Univ. Tex., Bur. Econ. Geol., Rep. Invest., No. 61.
Hayes, M.O., 1972. In: R.E. Meyer (Editor), Waves on Beaches and Resulting Sediment Transport. Academic Press, New York, N.Y., pp. 297–356.
Hayes, M.O., Anan, F.S. and Bozeman, R.N., 1969. Sediment dispersal trends in the littoral zone: a problem in palaeogeographic reconstruction. In: Coastal Environments of North-eastern Massachusetts and New Hampshire. Univ. Mass., Dep. Geol., Contrib. No. 1-CRG, pp. 290–315.
Haynes, J. and Dobson, M., 1969. Geol. J., 6: 217–256.
Head, W.J. and Harr, M.E., 1970. Wear, 15: 1–46.
Heathershaw, A.D., 1974. Nature (London), 248: 394–395.
Heathershaw, A.D., 1979. Geophys. J. R. Astron. Soc., 58: 395–430.
Hedberg, H.D., 1936. Am. J. Sci., 224: 241–297.
Hedin, S., 1896. Geogr. J., 8: 264–278, 356–372.
Hedin, S., 1904. The scientific results of a journey in Central Asia, 1899–1902. 1. The Tarim Basin. Lithographic Institute of the General Staff of the Swedish Army, Stockholm.
Heesch, H. and Laves, F., 1933. Z. Kristallogr., Kristallgeom., Kristallphys., Kristallchem., 85: 443–453.
Heezen, B.C., 1956. Sci. Am., 195: 36–41.
Heezen, B.C. and Drake, C.L., 1964. Bull. Am. Assoc. Pet. Geol., 48: 221–233.
Heezen, B.C. and Ewing, M., 1952. Am. J. Sci., 250: 849–873.

Heezen, B.C. and Ewing, M., 1955. Bull. Am. Assoc. Pet. Geol., 39: 2506–2514.
Heezen, B.C. and Hollister, C., 1971. The Face of the Deep. Oxford University Press, London–New York–Oxford.
Heezen, B.C. and Johnson, G.L., 1969. Bull. Inst. Océanogr., Monaco, No. 1382.
Heezen, B.C., Ericson, D.B. and Ewing, M., 1954. Deep-Sea Res., 1: 193–202.
Hefley, H.M. and Sidwell, R., 1945. Am. J. Sci., 243: 361–376.
Heidel, S.G., 1956. Trans. Am. Geophys. Union, 37: 56–66.
Heidrick, T.R., Banerjee, S. and Azad, R.S., 1977. J. Fluid Mech., 82: 705–723.
Heiken, G.H., 1972. Geol. Soc. Am. Bull., 83: 1961–1988.
Heiken, G., 1974. Smithson. Contrib. Earth Sci., No. 12.
Heiss, J.F. and Coull, J., 1952. Chem. Eng. Prog., 48: 133–140.
Hele-Shaw, H.D., 1898. Trans. Inst. Nav. Archit., 40: 21–40.
Heling, D., 1970. Sedimentology, 15: 247–260.
Helley, E.J., 1969. U.S. Geol. Surv., Prof. Pap., 562-G.
Helm, D.G. and Roberts, B., 1975. Geol. J., 10: 131–146.
Hemingway, J.E. and Clarke, A.M., 1963a. Nature (London), 197: 550–551.
Hemingway, J.E. and Clarke, A.M., 1963b. Nature (London), 198: 849–850.
Hendershott, M. and Munk, W., 1970. Annu. Rev. Fluid Mech., 2: 205–224.
Henderson, F.M., 1964. Proc. 1st Australas. Conf. Hydraul. Fluid Mech., pp. 51–67.
Henderson, F.M., 1966. Open Channel Flow. Macmillan, New York, N.Y.
Hendry, H.E., 1976. J. Sediment. Petrol., 46: 48–55.
Henningsen, D., 1968. Neues Jahrb. Geol. Paläontol., Monatsh., No. 3: 153–163.
Henry, H.R., Laursen, E.M., Adams, J.R., Hwang, L.J. and Chen, C.L., 1964. Pressure and shear distribution on schematic dunes. Nat. Sci. Found., Wash., Final Rep., NSF G-7409.
Herbertson, J.G., 1968. Bull. Int. Assoc. Sci. Hydrol., 18: 5–19.
Herbertson, J.G., 1969. J. Hydrol., 8: 1–26.
Herczynski, R., 1975. Nature (London), 255: 540–541.
Herdervari, P., 1968. Bull. Volcanol., (2)32: 439–461.
Hereford, R., 1977. Geol. Soc. Am. Bull., 88: 199–211.
Hervieu, J., 1968. Contribution a l'étude de l'alluvionnement en milieux tropical. Office Rech. Sci. Tech. Outre-Mer, Mem., No. 24.
Herzer, R.H., 1975. N.Z. J. Geol. Geophys., 18: 183–188.
Hess, S., 1959. Introduction to Theoretical Meteorology. Henry Holt, New York, N.Y.
Hester, N.C. and Fraser, G.S., 1973. Environ. Geol. Notes, Ill. State Geol. Surv., No. 63.
Heywood, H., 1938. Proc., Inst. Mech. Eng., London, 140: 257–308.
Heywood, H., 1946. J. Chem. Eng. Soc. Imp. Coll., 2: 9–26.
Heywood, H., 1962. In: Proceedings Symposium on Interaction between Fluids and Particles. Institution of Chemical Engineers, London, pp. 1–8.
Higgins, G.M., Baig, S. and Brinkman, R., 1974. Z. Geomorphol., 18: 272–290.
Higuti, I., 1961. Ann. Inst. Statist. Math., Tokyo, 12: 257–271.
Hill, A.R., 1971. Geogr. Ann., A53: 14–31.
Hill, H.M., 1966. J. Hydraul. Div. ASCE, 92(HY2): 127–143.
Hill, H.M., Srinivasan, V.S. and Unny, T.E., 1969. J. Hydraul. Div. ASCE, 95(HY5): 1545–1558.
Hill, M.N., 1963. The Sea, vol. 3. Interscience, New York, N.Y.
Hill, R.D., 1966. J. Trop. Geogr., 23: 19–27.
Hills, E.S., 1939. Proc. R. Soc. Victoria, 51: 297–323.
Hills, E.S., 1940. Aust. Geogr., 3: 15–21.
Hinch, E.J. and Leal, L.G., 1979. J. Fluid Mech., 92: 591–608.
Hine, A.C., 1975. In: L.E. Cronin (Editor), Estuarine Sediments. 2. Geology and Engineering. Academic Press, New York, N.Y., pp. 235–252.
Hine, A.C., 1977. J. Sediment. Petrol., 47: 1554–1581.

Hine, A.C., 1979. Sedimentology, 26: 333–351.
Hine, A.C. and Boothroyd, J.C., 1978. J. Sediment. Petrol., 48: 901–920.
Hino, M., 1963. J. Hydraul. Div. ASCE, 89(HY4): 161–185.
Hino, M., 1968. J. Fluid Mech., 34: 565–573.
Hino, M. and Fujisaki, H., 1975. Proc. 3rd Symposium on Flow Visualization. University of Tokyo, Tokyo, pp. 143–146.
Hinschberger, F., 1963. Bull. Sect. Géogr. Comm. Trav. Hist. Sci., 75: 53–80.
Hinschberger, F., Saint-Requier, A. and Toulemont, A., 1967. Rev. Trav. Inst. (Sci. Tech.) Pêches Marit., 31: 425–445.
Hinze, J.O., 1960. Geol. Mijnbouw, 39: 18–25.
Hiscott, R.N. and Middleton, G.V., 1980. J. Sediment. Petrol., 50: 703–722.
Hjelmfelt, A.T. and Lenau, C.W., 1970. J. Hydraul. Div. ASCE, 96(HY7): 1567–1586.
Hjelmfelt, A.T. and Mockros, L.F., 1967. J. Eng. Mech. Div. ASCE, 93(EM6): 87–102.
Hjulström, F., 1935. Bull. Geol. Inst. Univ. Upsala, 25: 221–527.
Hjulström, F., 1953. Ymer, 3: 161–191.
Hobbs, B.E., Means, W.D. and Williams, P.F., 1976. An Outline of Structural Geology. Wiley, New York, N.Y.
Hobbs, J.M., 1966. Philos. Trans. R. Soc. London, Ser. A., 260: 267–275.
Hobbs, W.H., 1906. Bull. Geol. Soc. Am., 17: 285–294.
Hobday, D.K., 1974. Norsk. Geol. Unders., No. 303: 39–56.
Hobday, D.K. and Jackson, M.P.A., 1979. J. Sediment. Petrol., 49: 145–158.
Hobday, D.K. and Tankard, A.J., 1978. Geol. Soc. Am. Bull., 89: 1733–1744.
Hobday, D.K., Brauteseth, S.V. and Mathew, D., 1971. Petros, 3: 51–56.
Hodgson, A.V., 1978. Proc. Yorks Geol. Soc., 41: 509–532.
Hoffman, P., 1973. Philos. Trans. R. Soc., London, Ser. A., 273: 547–581.
Hofmann, H.J., 1966. J. Geol., 74: 868–885.
Hofmann, U., 1942. Angew. Chem., 55: 282–289.
Hofmann, U., 1952. Kolloidzeitschrift, 125: 86–97.
Högbom, I., 1923. Geogr. Ann., 5: 113–243.
Hogendijk, M.J., 1963. Philips Res. Rep., 18: 109–126.
Holland-Batt, A.B., 1972a. Trans. Inst. Chem. Eng., 50: 12–20.
Holland-Batt, A.B., 1972b. Trans. Inst. Chem. Eng., 50: 156–167.
Holm, D.A., 1953. Proc. 19th Int. Geol. Congr., part 7, pp. 107–112.
Holm, D.A., 1960. Science, 132: 1369–1379.
Holmes, C.D., 1941. Bull. Geol. Soc. Am., 52: 1299–1354.
Holtedahl, H., 1965. In: W.F. Whittard and R. Bradshaw (Editors), Submarine Geology and Geophysics. Butterworths, London, pp. 107–140.
Holtorff, G., 1972. Wasserwirtschaft, 62(10): 314–317.
Holyer, J.Y. and Huppert, H.E., 1980. J. Fluid Mech., 100: 739–767.
Hom-Ma, M. and Horikawa, K., 1963a. Proc. 7th Conf. Coastal Eng., pp. 168–193.
Hom-Ma, M. and Horikawa, K., 1963b. A laboratory study on suspended sediment due to wave action. Univ. Tokyo, Dep. Civ. Eng., Rep. No. 6405.
Hom-Ma, M., Horikawa, K. and Kajima, R., 1965. Coastal Eng. Jpn., 8: 85–103.
Honji, H., Kaneko, A. and Matsunaga, N., 1980. Rep. Res. Inst. Appl. Mech. Kyushu Univ., 27: 225–229.
Hooke, R. LeB., 1967. J. Geol., 75: 438–460.
Hopfinger, E.K. and Tochon-Danguy, J.C., 1977. J. Glaciol., 19: 343–356.
Hoppe, G., 1952. Geogr. Ann., 34: 1–72.
Hoppe, G., 1957. Geogr. Ann., 39: 1–18.
Horikawa, K. and Shen, H.W., 1960. Tech. Mem. U.S. Beach Erosion Board, No. 119.
Horikawa, K. and Watanabe, A., 1967. Coastal Eng. Jpn., 10: 39–57.
Horikawa, K. and Watanabe, A., 1968. Coastal Eng. Jpn., 11: 13–28.

Horikawa, K. and Watanabe, A., 1970. Coastal Eng. Jpn., 13: 15–24.
Horne, R.R., 1971. Geol. Mag. 108: 151–158.
Horne, R.R., 1975. J. Sediment. Petrol., 45: 535–540.
Hörner, N.G., 1927. Sver. Geol. Unders., Ser. C, No. 342.
Hörner, N.G., 1957. Some notes and data concerning dunes and sand drift in the Gobi Desert. Reports from the Scientific Expedition to the North-western Provinces of China under the leadership of Dr. Sven Hedin. Vol. III (Geol.), No. 5.
Horsfield, H.T., 1934. J. Soc. Chem. Ind., 53: T107–T115.
Horton, R.E., 1938. Trans. Am. Geophys. Union, 19: 367–374.
Hoskin, C.M. and Nelson, R.V., 1969. J. Sediment. Petrol., 39: 581–590.
Hoskins, C.W., 1967. Bull. Am. Assoc. Pet. Geol., 51: 470 (abstr.).
Houbolt, J.C., 1973. AIAA J., 11: 421–437.
Houboult, J.J.H.C., 1968. Geol. Mijnbouw, 47: 245–273.
Houghton, J.T., 1977. The Physics of Atmospheres. Cambridge University Press, Cambridge.
Hoult, D.P., 1972. Annu. Rev. Fluid Mech., 4: 341–368.
Houtz, R.E. and Wellman, H.W., 1962. Geol. Mag., 99: 57–62.
Howard, A.D., 1977. Geol. Soc. Am. Bull., 88: 853–856.
Howard, A.D., Morton, J.B., Gad-el-Hak, M. and Pierce, D.B., 1978. Sedimentology, 25: 307–328.
Howard, J.D., 1966. Bull. Utah Geol. Miner. Surv., 80: 23–33.
Howard, J.D., Elders, C.A. and Heinbokel, J.F., 1975a. Senckenbergiana Marit., 7: 181–203.
Howard, J.D., Remmer, G.H. and Jewitt, J.L., 1975b. Senckenbergiana marit., 7: 237–256.
Howarth, R.J., 1971. Proc. R. Ir. Acad., Sect. B, 71: 1–35.
Hoyer, M.C., 1971. Geol. Soc. Am. Bull., 82: 1267–1284.
Hoyt, J.H., 1962. J. Sediment. Petrol., 32: 309–311.
Hoyt, J.H., 1966. Sedimentology, 7: 137–147.
Hoyt, J.H., 1967. Sedimentology, 8: 229–238.
Hoyt, J.H. and Henry, V.J., 1963. J. Sediment. Petrol., 33: 604–608.
Hrabar, S.V., Cressman, E.R. and Potter, P.E., 1971. Brigham Young Univ. Geol. Stud., 18: 99–114.
Hrubisek, J., 1941. Kolloidbeihefte, 53: 385–452.
Hsu, J.R.C., Tsuchiya, Y. and Silvester, R., 1979. J. Fluid Mech., 90: 179–196.
Hsu, J.R.C., Silvester, R. and Tsuchiya, Y., 1980. J. Fluid Mech., 99: 321–342.
Hsu, K.J., 1959. Am. J. Sci., 257: 529–536.
Hsu, K.J., 1975. Geol. Soc. Am. Bull., 86: 129–140.
Hsu, S.A., 1971. J. Geophys. Res., 76: 8684–8686.
Hsu, S.A., 1974. J. Geol., 81: 739–743.
Hsu, S.T. and Kennedy, J.F., 1971. J. Fluid Mech., 47: 481–502.
Huang, E.Y., Squier, L.R. and Triffo, R.P., 1963. Highw. Res. Rec. No. 22: 38–47.
Huang, N.E., 1971. Deep-Sea Res., 18: 255–259.
Huang, N.E., 1972. J. Phys. Oceanogr., 2: 420–431.
Huang, N.E., 1979. J. Fluid Mech., 91: 191–208.
Huang, T.C., Watkins, N.D. and Wilson, L., 1979. Geol. Soc. Am. Bull., (II)90: 235–288.
Hubbard, D.K., 1975. In: L.E. Cronin (Editor), Estuarine Research, 2. Geology and Engineering. Academic Press, New York, N.Y., pp. 253–266.
Hubbell, D.W. and Sayre, W.W., 1964. J. Hydraul. Div. ASCE, 90(HY3): 39–68.
Hubert, J.F., 1966a. J. Sediment. Petrol., 36: 677–699.
Hubert, J.F., 1966b. Nature (London), 211: 614–615.
Hubert, J.F. and Mertz, K.A., 1980. Geology, 8: 516–519.
Hübner, H., 1965. Stockholm Contrib. Geol., 13(5): 41–61.
Huddart, D., 1973. Proc. Cumberland Geol. Soc., 3(2): 59–69.
Huffman, G.F. and Price, W.A., 1949. J. Sediment. Petrol., 19: 118–127.

Hughes, B.P., 1960. Proc. Inst. Civ. Eng. (London), 17: 315–332.
Hughes, B.P., 1962. Proc. Inst. Civ. Eng. (London), 21: 927–952.
Hülsemann, J., 1955. Senckenbergiana Lethaea, 36: 359–388.
Hume, W.F., 1925. Geology of Egypt, vol. 1. Government Press, Cairo.
Hume, T.M., Sherwood, A.M. and Nelson, C.S., 1975. J. R. Soc. N.Z., 5: 421–462.
Hung, C.S. and Shen, H.W., 1976. J. Hydraul. Div. ASCE, 102(HY12): 1745–1759.
Hung, C.S. and Shen, H.W., 1979. J. Hydraul. Div. ASCE, 105(HY3): 213–227.
Hunkins, K., Mathieu, G., Teeter, S. and Gill, A., 1970. Arctic, 23: 175–189.
Hunt, A.R., 1882. Proc. R. Soc. London, 34: 1–18.
Hunt, A.R., 1904. Geol. Mag., N. S., 1: 410–418.
Hunt, J.N., 1954. Proc. R. Soc. London, Ser. A., 24: 322–335.
Hunt, J.N., 1969. Q. J. Mech. Appl. Math., 22: 235–246.
Hunt, J.N. and Johns, B., 1963. Tellus, 15: 343–351.
Hunt, R.E., Swift, D.J.P. and Palmer, H., 1977. Geol. Soc. Am. Bull., 88: 299–311.
Hunter, R.E., 1969. J. Sediment. Petrol., 39: 1573–1578.
Hunter, R.E., 1973. J. Sediment. Petrol., 43: 1125–1127.
Hunter, R.E., 1977a. Sedimentology, 24: 361–387.
Hunter, R.E., 1977b. J. Sediment. Petrol., 47: 697–706.
Hunter, R.E., Watson, R.L., Hill, G.W. and Dickinson, K.A., 1972. In: R.F. Travis, R.N. Tench and W.D. Hodgson (Editors), Padre Island National Sea Shore Field Guide. Gulf Coast Association of Geological Societies, pp. 1–27.
Huntley, D.A., 1976. J. Geophys. Res., 81: 6441–6449.
Huppert, H.E. and Simpson, J.E., 1980. J. Fluid Mech. 99: 785–799.
Hurault, J., 1966. Rev. Inst. Fr. Pét., 21: 1837–1896.
Hurley, R.J., 1964. In: R.L. Miller (Editor), Papers in Marine Geology. Macmillan, New York, N.Y., pp. 117–132.
Hutchings, I.M., Winter, R.E. and Field, J.E., 1976. Proc. R. Soc. London, Ser. A., 348: 379–392.
Hutchinson, J.N., 1967. Proc. Geotechnical Conference Oslo, pp. 113–118.
Hutchison, H.P. and Sutherland, D.N., 1965. Nature (London), 206: 1036–1037.
Huthnance, J.M., 1973. Estuarine Coastal Mar. Sci., 1: 89–99.
Hyde, J.E., 1911. J. Geol., 19: 257–269.
Hyde, J.H., 1975. Bull. U.S. Geol. Surv., 1383-B.
Hyyppä, E., 1948. Bull. Comm. Géol. Finl., No. 142: 97–122.
Iczkowski, R.P., 1966. Ind. Eng. Chem. Fund, 5: 516–519.
Idso, S.B., 1973. Weather, 28: 154–155.
Idso, S.B., 1974. Am. Sci., 62: 530–541.
Idso, S.B., 1976. Weather, 31: 224–226.
Idso, S.B., Ingram, R.S. and Pritchard, J.M., 1972. Bull. Am. Meteorol. Soc., 53: 930–935.
Iida, K., 1938. Bull. Earthquake Res. Inst. Tokyo Univ., 16: 658–681.
Illies, H., 1949. Mitt. Geol. Staatsinr. Hamburg, 19: 89–109.
Illing, L.V., 1954. Bull. Am. Soc. Pet. Geol., 38: 1–95.
Imbrie, J. and Buchanan, H., 1965. In: G.V. Middleton (Editor), Primary Sedimentary Structures and their Hydrodynamic Interpretation. Soc. Econ. Paleontol. Mineral., Spec. Publ., 12: 149–172.
Ingles, O.G., 1968. In: I.K. Lee (Editor), Soil Mechanics. Butterworths, London, pp. 1–57.
Inglis, C.C., 1949. Res. Publ. Cent. Water Power Irrig. Navig. Res. Stn., Poona, 13(II): 459–467.
Inglis, C.C. and Allen, F.H., 1957. Proc. Inst. Civ. Eng., 7: 827–868.
Ingram, R.G., 1976. J. Geophys. Res., 81: 1951–1959.
Inman, D.L., 1949. J. Sediment. Petrol., 19: 51–70.
Inman, D.L., 1957. Tech. Mem. U.S. Beach Erosion Board, No. 100.

Inman, D.L. and Bagnold, R.A., 1963. In: M.N. Hill (Editor), The Sea, vol. 3. Interscience, New York, N.Y., pp. 529–553.
Inman, D.L. and Bowen, A.J., 1963. Proc. 8th Conf. Coastal Eng., pp. 137–150.
Inman, D.L. and Nasu, N., 1956. Tech. Mem. U.S. Beach Erosion Board, No. 79.
Inman, D.L., Ewing, G.C. and Corliss, J.B., 1966. Geol. Soc. Am. Bull., 77: 787–802.
Inokuchi, M. and Takayama, S., 1973. Sci. Rep. Tokyo Kyoiku Daigaku, C11: 157–169.
Ippen, A.T., 1951. Trans. Am. Soc. Civ. Eng., 116: 268–295.
Ippen, A.T., 1966. Estuary and Coastline Hydrodynamics. McGraw-Hill, New York, N.Y.
Ippen, A.T., 1971. J. Boston Soc. Civ. Eng., 58: 131–163.
Ippen, A.T. and Harleman, D.R.F., 1956. Trans. Am. Soc. Civ. Eng., 121: 678–694.
Ippen, A.T. and Kulin, G., 1955. Proc. 5th Conf. Coastal Eng., pp. 27–47.
Ippen, A.T. and Verma, R.P., 1955. Trans. Am. Soc. Civ. Eng., 120: 921–938.
Irmay, S., 1960. Trans. ASME, D82: 961–968.
Irvine, T.N., 1965. In: G.V. Middleton (Editor), Primary Sedimentary Structures and their Hydrodynamic Interpretation. Soc. Econ. Paleontol. Mineral., Spec. Publ., 12: 220–232.
Irvine, T.N., 1980. Am. J. Sci., 280(A): 1–58.
Isaacson, M. de St. Q., 1976a. J. Fluid Mech., 74: 401–413.
Isaacson, M. de St. Q., 1976b. J. Fluid Mech., 78: 445–457.
Isaacson, M. de St. Q., 1978. ASCE J. Waterw., Harbors Coastal Eng. Div., 104(WW2): 215–225.
Ismail, H.M., 1952. Trans. Am. Soc. Civ. Eng., 117: 409–434.
Ito, S. and Kajiuchi, T., 1969. J. Chem. Eng. Soc. Jpn., 2: 19–24.
Iversen, H.W., 1952. Circ. U.S. Natl. Bur. Stand., No. 521: 9–32.
Iversen, J.D., Greely, R., White, B.R. and Pollack, J.B., 1976. J. Geophys. Res., 81: 4846–4856.
Ives, J.D., Mears, A.I., Carrara, P.E. and Bovis, M.J., 1976. Ann. Assoc. Am. Geogr., 66: 129–144.
Iwasa, Y. and Kennedy, J.F., 1968. J. Hydraul. Div. ASCE, 94(HY2): 321–454.
Jackson, R.G., 1975. Geol. Soc. Am. Bull., 86: 1523–1533.
Jackson, R.G., 1976a. Sedimentology, 23: 593–623.
Jackson, R.G., 1976b. J. Sediment. Petrol., 46: 579–594.
Jackson, R.G., 1976c. J. Fluid Mech., 77: 531–560.
Jackson, W.H., 1964. Dock Harb. Auth., 45: 120–122.
Jacobi, R.D., 1976. Mar. Geol., 22: 157–173.
Jacobs, A.F., 1973a. Geology, 1: 103–105.
Jacobs, A.F., 1973b. Bull. Am. Assoc. Petr. Geol., 57: 1038–1052.
Jago, C.F., 1980. Sediment. Geol., 26: 21–49.
Jahns, R.H., 1947. Water-Supply Irrig. Pap., Wash., 996.
Jain, S.C. and Kennedy, J.F., 1971. Proc. Int. Symp. Stochastic Hydraul., pp. 449–471.
Jain, S.C. and Kennedy, J.F., 1974. J. Fluid Mech., 63: 301–314.
Jäkel, D., 1980. Z. Geomorphol., 24: 141–159.
Jamieson, T.F., 1860. Q. J. Geol. Soc. London, 16: 347–371.
Jauzein, A., 1958. C. R. Acad. Sci., 247: 2396–2399.
Javdani, K. and Goren, S.L., 1972. Progress in Heat and Mass Transfer, 6: 253–262.
Jawarowski, K., 1971. Acta Geol. Pol., 21: 519–571.
Jayaweera, K.O.L.F. and Mason, B.J., 1965. J. Fluid Mech., 22: 709–720.
Jayaweera, K.O.L.F. and Mason, B.J., 1966. Q. J. R. Meteorol. Soc., 92: 151–156.
Jeffery, G.B., 1922. Proc. R. Soc. London, Ser. A., 102: 161–179.
Jeffery, A. and Tin, S., 1973. Proc. R. Soc. Edinburgh, Ser. A, 71: 181–192.
Jeffreys, H., 1924. Philos. Mag., (6)48: 44–48.
Jeffreys, H., 1925. Philos. Mag., (6)49: 793–807.
Jeffreys, H., 1929. Proc. Cambridge Philos. Soc., Math. Phys. Sci., 25: 272–276.

Jenkins, B.S., 1970. Density currents and turbidity currents in waste disposal in the ocean. Univ. N. S. W., Water Res. Lab., Rep. No. 119.
Jennings, J.N., 1967. In: J.N. Jennings and J.A. Mabbutt (Editors), Land Form Studies in Australia and New Guinea. Cambridge University Press, Cambridge, pp. 256–292.
Jensen, P.D., 1973. Proc. 15th Congr. Int. Assoc. Hydraul. Res., 1: 173–179.
Jeschar, R., 1964. Arch. Eisenhüttenw., 35: 91–108.
Jewtuchowicz, S., 1954. Bull. Soc. Sci. Lett. Lodz, (III)4(4): 1–23.
Jindrich, V., 1969. J. Sediment. Petrol., 39: 531–553.
Jipa, D., 1965. Rep. 7th Congr. Carpatho-Balkan Geol. Assoc., (II)1: 291–299.
Jipa, D., 1967. Rep. 8th Congr. Carpatho-Balkan Geol. Assoc., Sedimentol., pp. 349–356.
Jipa, D., 1968. Proc. 23rd Int. Geol. Congr., Part 8, pp. 203–210.
Johansson, C.-E., 1960. Svensk Geogr. Årsb., 36: 13–144.
Johansson, C.-E., 1963. Geogr. Ann., 45: 85–112.
Johansson, C.-E., 1965. Geol. För. Stockh. Förh., 87: 3–61.
Johansson, C.-E., 1976. Geogr. Ann., A58: 201–300.
Johansson, H.G., 1968. Geol. För. Stockh. Förh., 90: 205–212.
Johns, B., 1966. Geophys. J. R. Astron. Soc., 12: 103–110.
Johns, B., 1967. Geophys. J. R. Astron. Soc., 13: 377–386.
Johns, B., 1968. Geophys. J. R. Astron. Soc., 15: 501–507.
Johns, B., 1969. Geophys. J. R. Astron. Soc., 17: 39–44.
Johns, B., 1970a. J. Fluid Mech., 43: 177–185.
Johns, B., 1970b. Geophys. J. R. Astron. Soc., 20: 159–175.
Johns, B., 1973. Pure Appl. Geophys., 104: 594–607.
Johns, B., 1977. J. Phys. Oceanogr., 7: 733–738.
Johns, B. and Dyke, P., 1972. J. Phys. Oceanogr., 2: 73–79.
Johnson, A.M., 1970. Physical Processes in Geology. Freeman and Cooper, San Francisco, Calif.
Johnson, D.A. and Damuth, J.E., 1979. Mar. Geol., 33: 1–44.
Johnson, D.W., 1919. Shore Processes and Shoreline Development. Wiley, New York, N.Y.
Johnson, H.D., 1975. Sedimentology, 22: 45–74.
Johnson, H.D., 1977. Sedimentology, 24: 245–270.
Johnson, J.W., 1943. Trans. ASCE, 69: 941–956.
Johnson, J.W., 1947. Trans. Am. Geophys. Union, 28: 867–874.
Johnson, J.W., 1948. Trans. Am. Geophys. Union, 29: 671–681.
Johnson, M.A., 1962. Sci. Prog., Oxford, 50: 257–273.
Johnson, M.A., 1964. Ann. Rev. Océanogr. Mar. Biol., 2: 31–43.
Johnson, P.G., 1978. Can. J. Earth Sci., 15: 1496–1507.
Johnson, R.B., 1967. U.S. Geol. Surv., Prof. Pap., 575-C, pp. 177–183.
Johnson, R.G., 1957. J. Geol., 65: 527–535.
Johnson, R.R., Elkins, R.E., Lindgren, E.R. and Yoo, J.K., 1976. Phys. Fluids, 19: 1422–1423.
Johnson, V.E., 1963. J. Hydraul. Div. ASCE, 89(HY3): 251–275.
Joliffe, I.P., 1964. Proc. Geol. Assoc., 75: 67–84.
Joliffe, K.H., 1968. Proc. R. Soc. London, Ser. A., 303: 193–205.
Jones, A.M. and Knudsen, J.G., 1961. AIChE J, 7: 20–25.
Jones, C.M., 1977. Geology, 5: 567–570.
Jones, C.M., 1979. Sediment. Geol., 24: 85–104.
Jones, C.M. and McCabe, P.J., 1980. J. Sediment. Petrol., 50: 613–620.
Jones, G.P., 1961. Rec. Geol. Surv. Nigeria 1959, pp. 5–14.
Jones, I.P., 1973. Proc. Cambridge Philos. Soc., Math. Phys. Sci., 73: 231–238.
Jones, M.E. and Dennison, J.M., 1970. J. Sediment. Petrol., 40: 642–649.
Jones, N.S., Kain, J.M. and Stride, A.H., 1965. Mar. Geol., 3: 329–336.
Jonsson, I.G., 1967. Proc. 10th Conf. Coastal Eng., 1: 127–148.

Jonsson, I.G. and Carlsen, N.A., 1976. J. Hydraul. Res., 14: 45–60.
Jopling, A.V., 1961. U.S. Geol. Surv., Prof. Pap., 424-D, pp. 15–17.
Jopling, A.V., 1963. Sedimentology, 2: 115–121.
Jopling, A.V., 1964a. J. Geophys. Res., 69: 3403–3418.
Jopling, A.V., 1964b. J. Sediment. Petrol., 34: 165–172.
Jopling, A.V., 1965. J. Sediment. Petrol., 35: 777–791.
Jopling, A.V., 1966. Sedimentology, 7: 71–102.
Jopling, A.V., 1967. J. Geol., 75: 287–305.
Jopling, A.V. and Forbes, D.L., 1979. Geogr. Ann., A61: 67–85.
Jopling, A.V. and Richardson, E.V., 1966. J. Sediment. Petrol., 36: 821–825.
Jopling, A.V. and Walker, R.G., 1968. J. Sediment. Petrol., 38: 971–984.
Jordan, G.F., 1962. Science, 136: 839–848.
Jordan, P.R., 1965. Water-Supply Irrig. Pap., Wash., 1802.
Jordinson, R., 1970. J. Fluid Mech., 43: 801–811.
Joseph, D.D. and Tao, L.N., 1964. Z. Angew. Math. Phys., 44: 361–364.
Juignet, P., Dangeard, L. and Le Guyader, M.-T., 1965. Rev. Géogr. Phys. Géol. Dyn., (2)7: 97–112.
Jukes, J.B., 1872. The Student's Manual of Geology. Black, Edinburgh.
Jüngst, H., 1938. Geologie Meer Binnengewäss., 2: 229–277.
Juvigné, E., 1977. Z. Geomorphol., 21: 323–342.
Kaarsberg, E.A., 1959. J. Geol., 67: 446–472.
Kachel, N.B. and Sternberg, R.W., 1971. Mar. Geol., 10: 229–244.
Kádár, L., 1934. Geogr. J., 83: 470–478.
Kádár, L., 1966. Földrajzi Ert., 15: 413–448.
Kahn, J.S., 1956a. J. Geol., 64: 385–395.
Kahn, J.S., 1956b. J. Geol., 64: 578–606.
Kajikawa, M., 1976. J. Meteorol. Soc. Jpn., 54: 276–283.
Kajiura, K., 1968. Bull. Earthquake Res. Inst. Tokyo Univ., 46: 75–123.
Kalkanis, G., 1964. Tech. Mem. U.S. Coastal Eng. Res. Cent., No. 2.
Kalinske, A.A., 1942. Trans. Am. Geophys. Union, 23: 639–643.
Kalinske, A.A., 1947. Trans. Am. Geophys. Union, 28: 615–620.
Kalinske, A.A. and Hsia, C.H., 1945. Bull. State Univ. Iowa, No. 29.
Kalterherberg, J., 1956. Neues Jahrb. Geol. Paläontol., Abh., 104: 3–57.
Kamphuis, J.W., 1967. Proc. 10th Conf. Coastal Eng., 1: 766–789.
Kamphuis, J.W., 1975. ASCE J. Waterw. Harbors Coastal Eng. Div., 101(WW2): 135–144.
Kamphuis, J.W. and Mohamed, N., 1978. ASCE J. Waterw. Harbors Coastal Eng. Div., 104(WW1): 135–146.
Kaneko, A., 1980. Rep. Res. Inst. Appl. Mech. Kyushu Univ., 28: 57–71.
Kaneko, A. and Honji, H., 1979a. J. Fluid Mech., 93: 727–736.
Kaneko, A. and Honji, H., 1979b. Sedimentology, 26: 101–113.
Kao, T.W., 1977. J. Hydraul. Div. ASCE, 103(HY5): 543–555.
Kao, T.W., Park, V. and Pao, H.P., 1978. Phys. Fluids, 21: 1912–1922.
Karcz, I., 1970. J. Geophys. Res., 75: 2869–2873.
Karcz, I., 1972. Sediment. Geol., 7: 161–182.
Karcz, I., 1974. In: M. Morisawa (Editor), Fluvial Geomorphology. State University of New York, Binghamton, N.Y., pp. 149–173.
Karcz, I. and Goldberg, M., 1967. J. Sediment. Petrol., 37: 1244–1245.
Karcz, I. and Kersey, D., 1980. Sediment. Geol., 27: 263–300.
Karnis, A. and Mason, S.G., 1966. Trans. Soc. Rheol., 10: 571–592.
Karnis, A., Goldsmith, H.L. and Mason, S.G., 1963. Nature (London), 200: 159–160.
Karnis, A., Goldsmith, H.L. and Mason, S.G., 1966. Can. J. Chem. Eng., 44: 181–193.
Katz, E.J., 1964. J. Mar. Res., 22: 175–189.

Kaufmann, H., 1929. Rhythmische Phänomene der Erdoberfläche. Vieweg, Braunschweig.
Kauranne, L.K., 1960. Bull. Comm. Géol. Finl., No. 188: 87–99.
Kausch, H.H., Fesko, D.G. and Tschoegl, N.W., 1971. J. Colloid Interface Sci., 37: 603–611.
Kawamura, R., 1964. Study of sand movement by wind. Univ. Calif., Berkeley, Hydraul. Eng. Lab., Tech. Rep., HEL-2-8.
Kayser, K., 1973. Z. Geomorphol., S17: 156–167.
Keller, G.H., 1969. J. Soil Mech. Found. Div. ASCE, 95(SM6): 1379–1392.
Keller, G.H. and Bennett, R.H., 1970. Mar. Geol., 9: 215–223.
Keller, G.H. and Richards, A.F., 1967. J. Sediment. Petrol., 37: 102–127.
Keller, G.H., Anderson, S.H. and Lavelle, J.W., 1975. Can. J. Earth Sci., 12: 703–710.
Keller, W.D., 1946. J. Sediment. Petrol., 16: 63–71.
Kellerhals, P. and Murray, J.W., 1969. Bull. Can. Pet. Geol., 17: 67–91.
Kelling, G. and Moshrif, M.A., 1977. J. Sediment. Petrol., 47: 1342–1346.
Kelling, G. and Stanley, D.J., 1972. In: D.J. Stanley (Editor), The Mediterranean Sea. Dowden, Hutchinson and Ross, Stroudsburg, Penn., pp. 489–519.
Kelling, G. and Williams, P.F., 1967. J. Geol., 75: 243–267.
Kemper, E., 1968. Geol. Jahrb., 86: 49–106.
Kendall, C.G.St.C. and Skipwith, Sir P.A. d'E., 1969. Geol. Soc. Am. Bull., 80: 865–892.
Kendall, J.M., 1970. J. Fluid Mech., 41: 259–281.
Kendrick, M.P., 1972. Philos. Trans. R. Soc. London, Ser. A., 272: 223–243.
Kenn, M.J., 1966. Civ. Eng. Publ. Works Rev., November 1966.
Kenn, M.J., 1968. Tech. Note Construct. Ind. Res. Inf. Serv., No. 1.
Kenn, M.J. and Minton, P., 1968. Nature (London), 217: 633–634.
Kennedy, J.F., 1961. Calif. Inst. Technol., W.M. Keck Lab. Hydraul. Water Resour., Rep. KH-R-2.
Kennedy, J.F., 1963. J. Fluid Mech., 16: 521–544.
Kennedy, J.F., 1964. J. Geophys. Res., 69: 1517–1524.
Kennedy, J.F., 1969. Annu. Rev. Fluid Mech., 1: 147–168.
Kennedy, J.F. and Brooks, N.H., 1965. U.S. Dept. Agric., Misc. Publ., No. 970: 320–330.
Kennedy, J.F. and Falcon, M., 1965. Mass. Inst. Technol. Hydrodyn. Lab., Rep. No. 86.
Kennedy, J.F. and Iwasa, Y., 1968. J. Hydraul. Div. Am. Soc. Civ. Eng., 94(HY2): 431–454.
Kennedy, J.F. and Locher, F.A., 1972. In: R.E. Meyer (Editor), Waves on Beaches and Resulting Sediment Transport. Academic Press, New York, N.Y., pp. 249–295.
Kennedy, J.F. and Roubillard, L., 1967. Proc. 12th Congr. Int. Assoc. Hydraul. Res., 1: 41–48.
Kenyon, K.E., 1969. J. Geophys. Res., 74: 6991–6994.
Kenyon, K.E., 1970. J. Geophys. Res., 75: 1133–1135.
Kenyon, N.H. and Belderson, R.H., 1973. Sediment. Geol., 9: 77–99.
Kenyon, N.H. and Stride, A.H., 1968. J. Sediment. Petrol., 38: 255–259.
Kenyon, N.H. and Stride, A.H., 1970. Sedimentology, 14: 159–173.
Kerr, R.C. and Nigra, J.O., 1952. Bull. Am. Assoc. Pet. Geol., 36: 1541–1573.
Kersey, D.G. and Hsu, K.J., 1976. Sedimentology, 23: 761–789.
Kessler, L.G., 1971. Earth Sci. Bull. Western Geol. Assoc., 4: 25–35.
Keulegan, G.H., 1948. Tech. Rep. U.S. Beach Erosion Board, No. 3: 42 pp.
Keulegan, G.H., 1950. In: H. Rouse (Editor), Engineering Hydraulics. Wiley, New York, N.Y., pp. 711–768.
Keulegan, G.H., 1957. Rep. Natl. Bur. Stand., No. 5168.
Keulegan, G.H., 1958. Rep. Natl. Bur. Stand., No. 5831.
Keutner, C., 1929. Bautechnik, 7: 575–582.
Kezdi, A., 1966. In: Kravtochenko and P.M. Sirieys (Editors), Rheology and Soil Mechanics. Springer, Berlin, pp. 164–178.
Khanna, S.D., 1970. J. Hydraul. Div. ASCE, 96(HY10): 2029–2040.

Kiersch, G.S., 1950. Bull. Am. Assoc. Pet. Geol., 34: 923–942.
Kim, H.T., Kline, S.J. and Reynolds, W.C., 1971. J. Fluid Mech., 50: 133–160.
Kindle, E.M., 1914. J. Geol., 22: 703–713.
Kindle, E.M., 1917. Bull. Geol. Surv. Can., No. 25.
Kindle, E.M., 1938. J. Paleontol., 12: 515–516.
King, C.A.M., 1964. East Midl. Geogr., 3: 230–243.
King, C.A.M., 1972. Beaches and Coasts. Arnold, London, 2nd ed.
King, F.H., 1899. U.S. Geol. Surv., Annu. Rep., Part II, pp. 59–294.
King, W.H.J., 1916. Geogr. J., 47: 189–207.
King, W.H.J., 1918. Geogr. J., 51: 16–33.
Kingma, J.T., 1958. N. Z. J. Geol. Geophys., 1: 1–30.
Kinsman, B., 1965. Wind Waves. Prentice Hall, Englewood Cliffs, N.J.
Kirby, R. and Oele, E., 1975. Philos. Trans. R. Soc. London, Ser. A., 279: 257–267.
Kirby, R.P., 1968. Scott. J. Geol., 4: 209–220.
Kirby, R.P., 1969a. Geogr. Ann., A51: 48–60.
Kirby, R.P., 1969b. Scott. J. Geol., 5: 49–53.
Kirk, R.M., 1975. Geogr. Ann., A57: 117–133.
Kirkby, M.J., 1967. J. Geol., 75: 359–378.
Kittleman, L.R., 1973. Geol. Soc. Am. Bull., 84: 2957–2980.
Klebanoff, P.S., 1955. NASA Rep. No. 1247.
Klebanoff, P.S. and Tidstrom, K.D., 1972. Phys. Fluids, 15: 1173–1188.
Klebanoff, P.S., Tidstrom, K.D. and Sargent, L.M., 1962. J. Fluid Mech., 12: 1–34.
Klein, G. deV., 1962a. Geol. Mag., 99: 137–144.
Klein, G. deV., 1962b. Geol. Soc. Am. Bull., 73: 1127–1146.
Klein, G. deV., 1963a. J. Sediment. Petrol., 33: 49–52.
Klein, G. deV., 1963b. J. Sediment. Petrol., 33: 844–854.
Klein, G. deV., 1970a. J. Sediment. Petrol., 40: 973–985.
Klein, G. deV., 1970b. J. Sediment. Petrol., 40: 1095–1127.
Klein, G. deV., 1971. Geol. Soc. Am. Bull., 82: 2585–2592.
Klein, G. deV. and Whaley, M.L., 1972. Geol. Soc. Am. Bull., 83: 3465–3470.
Klemm, W. and Smekal, A., 1941. Naturwissenschaften, 29: 688–689.
Klimek, K., 1972. Pol. Acad. Sci., Geogr. Ser., No. 94.
Kline, S.J., 1967. In: G. Sovran (Editor), Fluid Mechanics of Internal Flow. Elsevier, Amsterdam, pp. 27–68.
Kline, S.J., Reynolds, W.C., Schraub, F.A. and Runstadler, P.W., 1967. J. Fluid Mech., 30: 741–773.
Kluwick, A., 1977. Acta Mech., 26: 15–46.
Knapp, R.T., 1938. Trans. Am. Geophys. Union, 19: 501–505.
Knapp, R.T., 1943. Proc. 2nd Iowa Conf. Hydraul., pp. 289–306.
Knapp, R.T., Daily, J.W. and Hammitt, F.G., 1970. Cavitation. McGraw-Hill, New York, N.Y.
Knauss, J.A., 1957. Tellus, 9: 235–237.
Knebel, H.J. and Folger, D.W., 1976. Mar. Geol., 22: M7–M15.
Knewtson, S.L. and Hubert, J.F., 1969. J. Sediment. Petrol., 39: 954–968.
Knight, D.W., 1975. Proc. 16th Congr. Int. Assoc. Hydraul. Res., 1: 66–72.
Knight, D.W., 1978. J. Hydraul. Div. ASCE, 104(HY6): 839–855.
Knight, R.J., 1972. Marit. Sediments, 8: 45–60.
Knight, R.J. and Dalrymple, R.W., 1975. In: R.N. Ginsburg (Editor), Tidal Deposits. Springer, Berlin–New York, pp. 47–55.
Knight, S.H., 1929. Univ. Wyo., Publ. Sci., Geol., 1: 1–82.
Knox, G.J. and Short, N.M., 1964. Bull. Volcanol., (2)27: 5–24.
Kobayashi, S., 1978. J. Meteorol. Soc. Jpn., 56: 130–139.

Koch, A.J. and McLean, H., 1975. Geol. Soc. Am. Bull., 86: 529–541.
Koerner, R.M., 1969. Powder Technol., 3: 208–212.
Koh, R.C.Y. and Le Mehauté, B., 1966. J. Geophys. Res., 71: 2005–2012.
Kohn, H. and Gonell, H.W., 1950. Staub, 22: 293–322, 23: 420–456.
Kojan, E., 1967. Rep. U.S. Forest Serv. Exp. Stn., Berkeley, Calif., pp. 233–253.
Kolbuszewski, J., 1948a. Proc. 2nd Int. Conf. Soil Mech., 1: 158–165.
Kolbuszewski, J., 1948b. Proc. 2nd Int. Conf. Soil Mech., 7: 47–49.
Kolbuszewski, J., 1950a. Research, 3: 478–483.
Kolbuszewski, J., 1950b. Geol. Mag., 87: 433–435.
Kolbuszewski, J., 1953. Geol. Mag., 90: 48–56.
Koloseus, H.J. and Ahmad, D., 1969. J. Hydraul. Div. ASCE, 95(HY1): 409–422.
Komar, P.D., 1969. J. Geophys. Res., 74: 4544–4558.
Komar, P.D., 1971a. Geol. Soc. Am. Bull., 82: 1477–1488.
Komar, P.D., 1971b. J. Geophys. Res., 76: 713–721.
Komar, P.D., 1972. Geol. Soc. Am. Bull., 83: 1151–1156.
Komar, P.D., 1973a. Geol. Soc. Am. Bull., 84: 3329–3338.
Komar, P.D., 1973b. J. Sediment. Petrol., 43: 1111–1113.
Komar, P.D., 1974. J. Sediment. Petrol., 44: 169–180.
Komar, P.D., 1976. Beach Processes and Sedimentation. Prentice-Hall, Englewood Cliffs, N.J.
Komar, P.D., 1977a. In: E.D. Goldberg, I.N. McCave and J.J. O'Brien (Editors), The Sea, vol. 6. Wiley, New York, N.Y., pp. 603–621.
Komar, P.D., 1977b. In: E.D. Goldberg, I.N. McCave and J.J. O'Brien (Editors), The Sea, vol. 6. Wiley, New York, N.Y., pp. 499–513.
Komar, P.D., 1977c. J. Waterw. Port Coastal Ocean Div. Am. Soc. Civ. Eng., 103(WW2): 225–239.
Komar, P.D., 1977d. J. Sediment. Petrol., 47: 1444–1453.
Komar, P.D., 1978. J. Sediment. Petrol., 48: 921–932.
Komar, P.D. and Inman, D.L., 1970. J. Geophys. Res., 75: 5914–5927.
Komar, P.D. and Miller, M.C., 1973. J. Sediment. Petrol., 43: 1101–1110.
Komar, P.D. and Miller, M.C., 1974. Summaries 14th Int. Conf. Coastal Eng., pp. 353–356.
Komar, P.D. and Miller, M.C., 1975a. J. Sediment. Petrol., 45: 362–367.
Komar, P.D. and Miller, M.C., 1975b. J. Sediment. Petrol., 45: 697–703.
Komar, P.D. and Reimers, C.E., 1978. J. Geol., 86: 193–209.
Komar, P.D., Newdeck, R.H. and Kulm, L.D., 1972. In: D.J.P. Swift, D.B. Duane and O.H. Pilkey (Editors), Shelf Sediment Transport: Process and Pattern. Dowden, Hutchinson and Ross, Stroudsburg, Penn., pp. 601–619.
Komar, P.D., Kulm, L.D. and Harlett, J.C., 1974. J. Geol., 82: 104–111.
Kondelik, P., Horak, J. and Tesarova, J., 1968. Ind. Eng. Chem., Proc. Res. Div., 7: 250–252.
Kondrat'yev, N.E., 1962. River Flow and River Channel Formation. Israel Program for Scientific Translations, Jerusalem.
Kopstein, F.P.H.W., 1954. Graded Bedding in the Harlech Dome. Geological Institute, University of Groningen.
Korchokha, Y.M., 1969. Sov. Hydrol. Selected Pap., 1968(6): 541–559.
Korchokha, Y.M., 1972. Sov. Hydrol. Selected Pap., 1972(4): 339–348.
Kornicker, L.S. and Armstrong, N., 1959. Publ. Inst. Mar. Sci., Univ. Tex., No. 6: 171–185.
Kornicker, L.S. and Purdy, E.G., 1957. J. Sediment. Petrol., 27: 126–128.
Kornicker, L.S., Wise, C.D. and Wise, J.M., 1963. J. Sediment. Petrol., 33: 703–712.
Koster, E.H., 1978. In: A.D. Miall (Editor), Fluvial Sedimentology. Mem. Can. Soc. Pet. Geol., No. 5: 161–186.
Koutaniemi, L., 1979. Fennia, 157(1): 13–73.
Kovasznay, L.S.G., 1967. Phys. Fluids, S10: 25–30.
Kovasznay, L.S.G., 1970. Annu. Rev. Fluid Mech., 2: 95–112.

Kovasznay, L.S.G., Kibbens, V. and Blackwelder, R.F., 1970. J. Fluid Mech., 41: 283–325.
Kranck, K., 1972. J. Sediment. Petrol., 42: 596–601.
Krantz, W.B. and Goren, S.L., 1970. Ind. Eng. Chem. Fund., 9: 107–113.
Krantz, W.B. and Goren, S.L., 1971a. Ind. Eng. Chem. Fund., 10: 91–101.
Krantz, W.B. and Goren, S.L., 1971b. AIChE J., 17: 494–496.
Krause, D.C., White, W.C., Piper, D.J.W. and Heezen, B.C., 1970. Geol. Soc. Am. Bull., 81: 2153–2160.
Krauss, W., 1966. Annu. Rev. Oceanogr. Mar. Biol., 4: 11–32.
Krinsley, D., 1960. J. Sediment. Petrol., 30: 321–323.
Krinsley, D., 1970. A geomorphological and paleoclimatological study of the playas of Iran. Rep. U.S. Geol. Surv. No. AFCRL-70-0503.
Krinsley, D.H. and Doornkamp, J.C., 1973. Atlas of Quartz Sand Surface Textures. Cambridge University Press, Cambridge.
Krüger, J., 1970. Geogr. Tidsskr., 69: 139–169.
Krumbein, W.C., 1939. J. Geol., 47: 673–706.
Krumbein, W.C., 1940. Bull. Geol. Soc. Am., 51: 636–676.
Krumbein, W.C., 1941a. J. Geol., 49: 482–520.
Krumbein, W.C., 1941b. J. Sediment. Petrol., 11: 64–72.
Krumbein, W.C., 1942a. Trans. Am. Geophys. Union, 23: 621–633.
Krumbein, W.C., 1942b. Bull. Geol. Soc. Am., 53: 1355–1402.
Krumbein, W.C. and Pettijohn, F.J., 1938. Manual of Sedimentary Petrography. Appleton-Century-Crofts, New York, N.Y.
Kuenen, P.H., 1928. Leidse Geol. Meded., 3: 17–38.
Kuenen, P.H., 1950. Proc. 18th Int. Geol. Congr., part 8, pp. 44–51.
Kuenen, P.H., 1951. In: J.L. Hough (Editor), Turbidity Currents. Soc. Econ. Paleontol. Mineral., Spec. Publ., 2: 14–33.
Kuenen, P.H., 1952. Am. J. Sci., 250: 874–884.
Kuenen, P.H., 1953. Verh. K. Ned. Akad. Wet., (I)20(3): 1–47.
Kuenen, P.H., 1957. J. Geol., 65: 231–258.
Kuenen, P.H., 1960. J. Geol., 68: 427–449.
Kuenen, P.H., 1965. In: W.F. Whittard and R. Bradshaw (Editors), Submarine Geology and Geophysics. Butterworths, London, pp. 47–74.
Kuenen, P.H., 1966. J. Geol., 74: 523–545.
Kuenen, P.H., 1967. Sedimentology, 9: 203–243.
Kuenen, P.H. and Humbert, F.L., 1969. Sedimentology, 13: 253–261.
Kuenen, P.H. and Menard, H.W., 1952. J. Sediment. Petrol., 22: 83–96.
Kuenen, P.H. and Migliorini, C.I., 1950. J. Geol., 58: 91–127.
Kuenen, P.H. and Sengupta, S., 1970. Geol. Mijnbouw, 49: 89–118.
Kuettner, J.P. and Soules, S.D., 1966. Bull. Am. Meteorol. Soc., 47: 364–370.
Kuhn, W., 1933. Kolloidzeitschrift, 62: 269–285.
Kuhn, W. and Kuhn, H., 1945. Helv. Chim. Acta, 28: 97–127.
Kuijpers, E.P., 1972. Upper Devonian tidal deposits and associated sediments south and southwest of Kinsale (southern Ireland). Geological Institute, University of Utrecht.
Kulm, L.D., Roush, R.C., Harlett, J.C., Neudeck, R.H., Chambers, D.M. and Runge, E.J., 1975. J. Geol., 83: 145–175.
Kumar, N., 1973. Ann. N.Y. Acad. Sci., 220: 245–340.
Kumar, N. and Sanders, J.E., 1974. Sedimentology, 21: 491–532.
Kumar, N. and Singh, I.B., 1978. Senckenbergiana Marit., 10: 145–211.
Kunii, D. and Smith, J.M., 1960. AIChE J., 6: 71–78.
Kürsten, M., 1960. Geol. Rundsch., 49: 498–501.
Lacombe, H., 1965. Houille Blanche, 20: 38–43.
Lafeber, D., 1964. In: A. Jongerius (Editor), Soil Micromorphology. Elsevier, Amsterdam, pp. 351–360.

Lafeber, D. and Willoughby, D.R., 1970. In: Foundations on Interbedded Sands. Division of Applied Mechanics C.S.I.R.O. and Western Australia Group of Australian Geomechanics Society, Perth, pp. 80–86.
Lai, R.Y.S., 1974. Int. J. Eng. Sci., 12: 645–655.
Laitone, E.V., 1963. Tech. Mem. U.S. Beach Erosion Board, No. 133.
Laitone, J.A., 1979. Wear, 56: 239–246.
Lamar, J.E., 1927. Bull. Geol. Soc. Am., 38: 459–468.
Lamb, H., 1911. Philos. Mag., (6)21: 112–121.
Lamb, H., 1932. Hydrodynamics. Cambridge University Press, Cambridge, 6th ed.
Lambe, T.W., 1953. Proc. Am. Soc. Civ. Eng., Pap. No. 315, 49 pp.
Lambe, T.W., 1960a. Trans. Am. Soc. Civ. Eng., 125: 682–706.
Lambe, T.W., 1960b. Trans. Am. Soc. Civ. Eng., 125: 718–741.
Lambermont, J. and LeBon, G., 1978. J. Hydraul. Res., 16: 27–44.
Lambert, A.M., Kelts, K.R. and Marshall, N.F., 1976. Sedimentology, 23: 87–105.
Laming, D.J.C., 1966. J. Sediment. Petrol., 36: 940–959.
Lamont, A., 1957. Geol. Mag., 94: 472–480.
Lancaster, I.N., 1978a. Geogr. J., 144: 81–98.
Lancaster, I.N., 1978b. Z. Geomorphol., 22: 148–169.
Land, L.S., 1964. J. Sediment. Petrol., 34: 389–394.
Land, L.S. and Hoyt, J.H., 1966. Sedimentology, 6: 191–207.
Landsberg, S.Y., 1956. Geogr. J., 122: 176–189.
Lane, D.W., 1963. J. Sediment. Petrol., 33: 350–354.
Lane, E.W. and Eden, E.W., 1940. J. Western Soc. Eng., 45: 281–291.
Lane, E.W. and Kalinske, A.A., 1939. Trans. Am. Geophys. Union, 20: 637–641.
Lane, E.W. and Kalinske, A.A., 1941. Trans. Am. Geophys. Union, 22: 603–607.
Langbein, W.B., 1942. Trans. Am. Geophys. Union, 23: 615–621.
Langbein, W.B. and Leopold, L.B., 1968. U.S. Geol. Surv., Prof. Pap., 422-L.
Langeraar, W., 1966. Hydrogr. Newslett., 1: 243–246.
Langhorne, D.N., 1973. Mar. Geol., 14: 129–143.
Laronne, J.B. and Carson, M.A., 1976. Sedimentology, 23: 67–85.
Larsonneur, C., 1971. Manche centrale et Baie de Seine: géologie du substratum et des dépôts meubles. University of Caen, Caen.
Larsonneur, C., 1975. In: R.N. Ginsburg (Editor), Tidal Deposits. Springer, Berlin, pp. 21–30.
Larsson, W., 1937. Bull. Geol. Inst. Uppsala, 26: 27–52.
Laufer, J., 1951. NACA Rep. 1053.
Laufer, J., 1954. NACA Rep. 1174.
Laufer, J. and Narayanan, M.A.B., 1971. Phys. Fluids, 14: 182–183.
Laursen, E.M., 1958. J. Hydraul. Div. ASCE, 84(HY1), Pap. 1530.
Lawler, M.T. and Lu, P.C., 1971. In: I. Zandi (Editor), Advances in Solid-Liquid Flow in Pipes and its Applications. Pergamon, Oxford, pp. 39–57.
Lawson, T.J., 1971. Weather, 26: 105–112.
Le Clair, B.P., Hamielec, A.E. and Pruppacher, H.R., 1970. J. Atmos. Sci., 27: 308–315.
Le Feuvre, A.R., Altinbilek, H.D. and Carstens, M.R., 1970. J. Hydraul. Div. ASCE, 96(HY10): 2051–2063.
Le Lacheur, E.A., 1924. Geogr. Rev., 14: 282–286.
Le Mehauté, B., 1968. J. Geophys. Res., 73: 5973–5979.
Le Mehauté, B., 1976. An Introduction to Hydrodynamics and Water Waves. Springer, New York, N.Y.
Le Pichon, X., Hekinian, R., Francheteau, J. and Carré, D., 1975. Deep-Sea Res., 22: 667–670.
Leach, S.J. and Walker, G.L., 1966. Philos. Trans. R. Soc. London, Ser. A., 260: 295–308.

LeBlond, P.H. and Mysak, L.A., 1978. Waves in the Ocean. Elsevier, Amsterdam.
LeBlond, P.H. and Mysak, L.A., 1979. SIAM Rev., 21: 289–328.
Ledbetter, M.T. and Sparks, R.S.J., 1979. Geology, 7: 240–244.
Lee, B.K. and Jobson, H.E., 1977. U.S. Geol. Surv., Prof. Pap., 1040: 1–72.
Lee, C.Y. and Masch, F.D., 1971. Proc. 12th Conf. Coastal Eng., 1: 223–242.
Lee, K.L. and Morrison, R.A., 1970. J. Soil Mech. Found. Div. ASCE, 96(SM6): 2025–2043.
Lee, M.K., Eckelman, L.D. and Hanratty, T.J., 1974. J. Fluid Mech., 66: 17–33.
Leeder, M.R., 1973. Scott. J. Geol., 9: 117–144.
Leeder, M.R., 1977a. Sedimentology, 24: 863–874.
Leeder, M.R., 1977b. Earth Surf. Processes, 2: 3–12.
Leeder, M.R., 1979a. Earth Surf. Processes, 4: 229–240.
Leeder, M.R., 1979b. Earth Surf. Processes, 4: 291–295.
Leeder, M.R., 1980. J. Geol. Soc. London, 137: 423–429.
Lees, G., 1969. Q. J. Eng. Geol., 2: 129–147.
Leflef, D., 1973. Geol. Mijnbouw, 52: 257–276.
Leith, W.C. and McIlquham, W.S., 1962. In: Symposium on Erosion and Cavitation. Am. Soc. Test. Mater., Spec. Tech. Publ., 307: 46–63.
Lené, G.W. and Owed, D.E., 1969. J. Sediment. Petrol., 39: 737–743.
Leopold, L.B., 1962. Am. Sci., 50: 511–537.
Leopold, L.B. and Maddock, J.T., 1953. U.S. Geol. Surv., Prof. Pap., 252.
Leopold, L.B. and Wolman, M.G., 1956. Int. Assoc. Hydraul., Res. Publ., No. 42: 85–98.
Leopold, L.B. and Wolman, M.G., 1960. Bull. Geol. Soc. Am., 71: 769–794.
Leopold, L.B., Wolman, M.G. and Miller, J.P., 1964. Fluvial Processes in Geomorphology. Freeman, San Francisco, Calif.
Lerbekmo, J.F. and Campbell, F.A., 1969. Can. J. Earth Sci., 6: 109–116.
Lerbekmo, J.F., Westgate, J.A., Smith, D.G.W. and Denton, G.H., 1975. Bull. R. Soc. N. Z., No. 13: 203–209.
Lettau, K. and Lettau, H., 1969. Z. Geomorphol., 13: 182–195.
Leva, M. and Grummer, M., 1947. Chem. Eng. Prog., 43: 713–718.
Levashov, A.A., 1976. Soviet Meteorology and Hydrology for 1976. No. 3, pp. 85–88.
Levell, B.K., 1980. Sedimentology, 27: 539–557.
Lever, J. and Thijssen, R., 1968. Symp. Zool. Soc. London, 22: 259–271.
Levey, R.A., 1976. In: M.O. Hayes and T.W. Kana (Editors), Terrigenous Clastic Depositional Environments. Univ. S. C., Dep. Geol., Rep. No. 11-CRD: II-38–II-51.
Levey, R.A., Kjerve, B. and Getzen, R.T., 1980. J. Sediment. Petrol., 50: 149–155.
Levine, M.M. and Chernik, J., 1965. Nature (London), 208: 68–69.
Lewis, A.D., 1921. Minutes Proc. Inst. Civ. Eng., 212: 393–399.
Lewis, A.D., 1936a. S. Afr. Geogr. J., 19: 22–32.
Lewis, A.D., 1936b. S. Afr. Geogr. J., 19: 33–49.
Lewis, K.B., 1971. Sedimentology, 16: 97–110.
Lewis, K.B., 1979. Mar. Geol., 31: 32–43.
Lewis, M.S., 1969. Mar. Geol., 7: 95–127.
Li, H., 1954. Tech. Mem. U.S. Beach Erosion Board, No. 47.
Liang, S.S. and Wang, H., 1973. Sediment transport in random waves. Univ. Del., Coll. Mar. Stud., Tech. Rep., No. 26.
Liboriussen, J., 1975. Sediment. Geol., 14: 235–251.
Lighthill, M.J., 1953. Proc. Cambridge Philos. Soc., Math. Phys. Sci., 49: 704–706.
Lighthill, M.J., 1978. Waves in Fluids. Cambridge University Press, Cambridge.
Lighthill, M.J. and Whitham, G.B., 1955a. Proc. R. Soc. London, Ser. A., 229: 281–316.
Lighthill, M.J. and Whitham, G.B., 1955b. Proc. R. Soc. London, Ser. A., 229: 317–345.
Lin, C.C., 1945. Q. Appl. Math., 3: 117–142, 218–234, 277–301.
Lin, C.C., 1955. The Theory of Hydrodynamic Stability. Cambridge University Press, Cambridge.

Lin, C.C., 1958. In: H. Görtler (Editor), Boundary Layer Research, Springer, Berlin, pp. 144–160.
Lin, C.C. and Benney, D.J., 1964. Proc. 11th Int. Congr. Appl. Mech., pp. 797–802.
Lindsay, J.F., 1966. J. Sediment. Petrol., 36: 719–732.
Lindsay, J.F., 1968. J. Sediment. Petrol., 38: 1242–1253.
Lindsay, J.F., 1970a. J. Sediment. Petrol., 40: 629–641.
Lindsay, J.F., 1970b. Geol. Soc. Am. Bull., 81: 1149–1172.
Lindsay, J.F., 1970c. J. Geol., 78: 597–603.
Lindsay, J.F., 1973. Geol. Soc. Am. Bull., 84: 1799–1806.
Lindsey, D.A., 1966. Science, 154: 1442–1443.
Lindsey, D.A., 1969. Geol. Soc. Am. Bull., 80: 1685–1702.
Lindsey, D.A., 1971. Palaeogeogr. Palaeoclimatol. Palaeoecol., 9: 7–25.
Lindström, E., 1973. Deglaciation, sediment och högsta kustlinje i nordvästre Ångermanland. Univ. Uppsala, UNGI Rep., 26.
Lindström, M., 1972. Geol. Palaeontol., 6: 9–23.
Lirer, L. and Pescatore, T., 1968. Atti Acad. Sci. Fis. Mat., Napoli, (3a) 7: 139–187.
Lirer, L., Pescatore, T., Booth, B. and Walker, G.P.L., 1973. Geol. Soc. Am. Bull., 84: 759–772.
Liu, A.K. and Davis, S.H., 1977. J. Fluid Mech., 81: 63–84.
Liu, H.K., 1957. Proc. Soc. Civ. Eng., No. HY3, pp. 1–23.
Liu, H.K. and Hwang, S.Y., 1961. Trans. ASCE, 126: 1787–1822.
Ljungner, E., 1930. Bull. Geol. Inst. Univ. Uppsala, 21: 255–478.
Lliboutry, L., 1964. Traité de Glaciologie, vol. 1. Masson, Paris.
Lliboutry, L., 1965. Traité de Glaciologie, vol. 2. Masson, Paris.
Lloyd, E.F., 1972. N. Z. Geol. Surv. Bull., 85.
Lofquist, K.E.B., 1978. U.S. Coastal Eng. Res. Cent., Tech. Pap., 78-5.
Lombard, A., 1963. J. Sediment. Petrol., 33: 14–22.
Long, J.T. and Sharp, R.T., 1964. Geol. Soc. Am. Bull., 75: 149–156.
Long, R.R., 1953a. Bull. Am. Meteorol. Soc., 34: 205–211.
Long, R.R., 1953b. Tellus, 5: 42–58.
Long, R.R., 1954. Tellus, 6: 97–115.
Longuet-Higgins, M.S., 1952. J. Mar. Res., 2: 245–266.
Longuet-Higgins, M.S., 1953. Philos. Trans. R. Soc. London, Ser. A., 245: 535–581.
Longuet-Higgins, M.S., 1958. Proc. 6th Conf. Coastal Eng., pp. 184–193.
Longuet-Higgins, M.S., 1960. J. Fluid Mech., 8: 293–306.
Longuet-Higgins, M.S., 1970. J. Geophys. Res., 75: 6778–6789, 6790–6801.
Longuet-Higgins, M.S., 1973a. Proc. R. Soc. London, Ser. A, 331: 445–456.
Longuet-Higgins, M.S., 1973b. J. Fluid Mech., 57: 129–148.
Longuet-Higgins, M.S. and Stewart, R.W., 1963. J. Mar. Res., 21: 4–10.
Longuet-Higgins, M.S. and Stewart, M.S., 1964. Deep-Sea Res., 11: 529–562.
Lonsdale, P. and Malfait, B., 1974. Geol. Soc. Am. Bull., 85: 1697–1712.
Lonsdale, P. and Southard, J.B., 1974. Mar. Geol., 17: M51–M60.
Lonsdale, P. and Spiess, F.N., 1977. Mar. Geol., 23: 57–75.
Lonsdale, P., Normark, W.R. and Newman, W.A., 1972. Geol. Soc. Am. Bull., 83: 289–316.
Lopatin, G.V., 1952. Vses. Geogr. Obschch. Zap., 14: 1–366.
Loring, D.H. and Nota, D.J.G., 1973. Bull. Fish. Res. Board Can., No. 182.
Loring, D.H., Nota, D.J.G., Chesterman, W.D. and Wong, K.H., 1970. Mar. Geol., 8: 337–354.
Loubere, P., 1977. J. Sediment. Petrol., 47: 419–427.
Lowe, D.R., 1976a. Sedimentology, 23: 285–308.
Lowe, D.R., 1976b. J. Sediment. Petrol., 46: 188–199.
Lowe, D.R. and Knauth, L.P., 1978. J. Sediment. Petrol., 48: 709–722.

Lu, S.S. and Willmarth, W.W., 1973. J. Fluid Mech., 60: 481–511.
Luckman, B.H., 1976. Earth Surf. Processes, 1: 287–298.
Lüders, K., 1929. Senckenbergiana, 11: 123–142.
Lüders, K., 1930. Senckenbergiana, 12: 229–254.
Ludwick, J.C., 1970a. Chesapeake Sci., 11: 98–110.
Ludwick, J.C., 1970b. Va. J. Sci., 21: 178–184.
Ludwick, J.C., 1970c. Sand waves and tidal channels in the entrance to Chesapeake Bay. Inst. Oceanogr., Old Domin. Univ., Norfolk, Va., Tech. Rep., No. 1.
Ludwick, J.C., 1971. Migration of tidal sand waves in Chesapeake Bay entrance. Inst. Oceanogr., Old Domin. Univ., Norfolk, Va., Tech. Rep., No. 2.
Ludwick, J.C., 1972. In: D.J.P. Swift, D.B. Duane and H.O. Pilkey (Editors), Shelf Sediment Transport: Process and Pattern. Dowden, Hutchinson and Ross, Stroudsburg, Penn., pp. 377–410.
Ludwick, J.C., 1973. Tidal currents and zig-zag sand shoals in a wide estuary entrance. Inst. Oceanogr., Old Domin. Univ., Norfolk, Va., Tech. Rep., No. 7.
Ludwick, J.C. and Wells, J.T., 1974. Particle size distribution and small-scale bed-forms on sand waves, Chesapeake Bay entrance. Inst. Oceanogr., Old Domin. Univ., Norfolk, Va., Tech. Rep., No. 12.
Lundqvist, G., 1949. Geogr. Ann., 31 (Glaciers and Climate): 335–347.
Lundqvist, G., 1951. Sver. Geol. Unders., Ser. C., No. 21.
Lundqvist, G., 1963. Sver. Geol. Unders., Ser. C., No. 42.
Lunson, E.A., 1950. Prof. Note Meteorol. Off., London, No. 102: 1–12.
Luque, R.F. and Van Beek, R., 1976. J. Hydraul. Res., 14: 127–144.
Lustig, L.K., 1965. U.S. Geol. Surv., Prof. Pap., 352-F.
Luternauer, J.L. and Murray, J.W., 1973. Can. J. Earth Sci., 10: 1642–1663.
Lützner, H., 1966. Ber. Dtsch. Ges. Geol. Wiss., A11: 137–160.
Lyles, L. and Woodruff, N.P., 1972. In: H.W. Shen (Editor), Sedimentation. Colorado State University, Fort Collins, Colo.
Lyne, W.H., 1971. J. Fluid Mech., 50: 33–48.
Mabbutt, J.A., 1977. Desert Landforms. Australian National University Press, Canbera.
MacClintock, P. and Dreimanis, A., 1964. Am. J. Sci., 262: 133–142.
MacClintock, P. and Terasome, J., 1960. J. Geol., 68: 232–241.
MacDonald, G.A., 1972. Volcanoes. Prentice-Hall, Englewood Cliffs, N.J.
MacDonald, T.C., 1977. U.S. Coastal Eng. Res. Cent., Tech. Pap., TP 77-4.
MacDougal, D.T., 1912. Geogr. J., 39: 105–120.
Machida, T., Inokuchi, M., Matsumoto, E., Ishii, T. and Ikeda, H., 1974. Sci. Rep. Tokyo Kyoiku Daigaku, Sect. C, No. 112.
MacKenzie, D.B., 1972. Mt. Geol., 9: 269–277.
MacKenzie, F.T., 1964. Science, 144: 1449–1450.
Macrae, J.C. and Gray, W.A., 1961. Br. J. Appl. Phys., 12: 164–172.
Macumber, P.G., 1970. Min. Geol. J., 6: 16–18.
Maddock, T., 1969. U.S. Geol. Surv., Prof. Pap., 622-A.
Maddock, T., 1976. Water Resour. Res., 12: 11–21.
Madsen, O.S., 1978. J. Phys. Oceanogr., 8: 1009–1015.
Madsen, O.S. and Grant, W.D., 1975. J. Sediment. Petrol., 45: 360–361.
Madsen, O.S. and Mei, C.C., 1970. J. Fluid Mech., 44: 195–208.
Maggiolo, O.J. and Borghi, J., 1965. Rev. Ing. (Montevideo), 59: 149–160.
Maiklem, W.R., 1968. Sedimentology 10: 101–109.
Mainguet, M., 1968. Ann. Géogr., 77: 296–322.
Mainguet, M., 1976. Z. Geomorphol., 20: 275–296.
Mainguet, M. and Callot, Y., 1974. Z. Geomorphol., S20: 62–68.
Malmsheimer, W.K., 1968. Forschber. Wirtsch.-Verkehrminist. Nordrhein-Westfalen, No. 2000.

Manegold, E. and Von Engelhardt, W., 1933. Kolloidzeitschrift, 62: 285–294, 63: 12–24, 149–154.
Manegold, E., Hofman, R. and Solf, K., 1931. Kolloidzeitschrift, 56: 142–159.
Mangin, J.P., 1962. C. R. Soc. Géol. Fr., 2: 34–36.
Manglesdorf, P.C. and Washington, E.L., 1960. Nature (London), 187–930.
Manley, R.St.J. and Mason, S.G., 1952. J. Colloid Sci., 7: 354–369.
Manley, R.St.J., Arlov, A.P. and Mason, S.G., 1955. Nature (London), 175: 682–683.
Manohar, M., 1955. Tech. Mem. U.S. Beach Erosion Board, No. 75.
Mantz, P.A., 1973. Nature (London), Phys. Sci., 246: 14–16.
Mantz, P.A., 1977a. Sedimentology, 24: 819–832.
Mantz, P.A., 1977b. J. Hydraul. Div. ASCE, 103(HY6): 601–615.
Mantz, P.A., 1978. Sedimentology, 25: 83–103.
Marchildon, E.K., Clamen, A. and Gauvin, W.H., 1964a. Phys. Fluids, 7: 2018–2019.
Marchildon, E.K., Clamen, A. and Gauvin, W.H., 1964b. Can. J. Chem. Eng., 42: 178–182.
Marmer, H.A., 1923. Geogr. Rev., 13: 413–444.
Marsal, D., 1950. Neues Jahrb. Geol. Paläontol. Monatsh., No. 10: 295–304.
Martin, C.S., 1970. Water Resour. Res., 6: 1162–1174.
Martin, J.C. and Moyse, W.J., 1952a. Philos. Trans. R. Soc. London, Ser. A., 244: 312–324.
Martin, J.C. and Moyse, W.J., 1952b. Philos. Trans. R. Soc. London, Ser. A., 244: 324–334.
Martin, R.C., 1965. N.Z. J. Geol. Geophys., 8: 680–701.
Martin, R.T., 1965. Qualitative fabric of consolidated kaolinite. Mass. Inst. Technol., Dep. Civ. Eng., Soils Publ., No. 179.
Martinec, J., 1967. Int. Assoc. Sci. Hydrol., Publ. No. 75: 243–249.
Martinez, J.F., 1977. Sedimentation Triasica en el borde sureste de la Meseta. University of Granada.
Martini, I.P., 1971. Sedimentology, 17: 265–275.
Martini, I.P., 1977. Sedimentology, 24: 603–622.
Martins, L.R., 1967. Aspectos texturais e deposicionais dos sedimentos praiais e eolicos da planicie costeira do Rio Grande do Sul. Esc. Geol. Porto Alegre, Publ. No. 13.
Marvin, J.W., 1937. Science, 86: 493–494.
Marvin, J.W., 1939. Am. J. Bot., 26: 280–288.
Masch, F.D., Epsey, W.H. and Moore, W.L., 1965. U.S. Dept. Agric., Agric. Res. Serv., Misc. Publ., No. 970: 151–155.
Masliyah, J. and Epstein, N., 1970. J. Fluid Mech., 44: 493–512.
Mason, G., 1968. Nature (London), 217: 733–735.
Mason, G., 1971. J. Colloid Interface Sci., 35: 279–287.
Mason, G. and Clark, W., 1965. Nature (London), 207: 512.
Mason, G. and Clark, W., 1966. Nature (London), 211: 957.
Mason, J.S. and Smith, B.V., 1972. Powder Technol., 6: 323–335.
Mason, M.A., 1941. Tech. Mem. U.S. Beach Erosion Board, No. 1.
Mason, S.G., 1954. TAPPI, 37: 494–501.
Mason, S.G. and Bartok, W., 1959. In: C.C. Mills (Editor), Rheology of Disperse Systems. Pergamon, London, pp. 16–47.
Mason, S.G. and Manley, R.St.J., 1956. Proc. R. Soc. London, Ser. A., 238: 117–131.
Matalucci, R.V., Shelton, J.W. and Abdel-Hady, M., 1969. J. Sediment. Petrol., 39: 969–979.
Mathieu, R., 1966. Rev. Géogr. Phys. Géol. Dyn., (2)8: 113–122.
Matschinski, M., 1952. C. R. Soc. Géol. Fr., 9/10: 171–174.
Matschinski, M., 1954. Boll. Serv. Geol. Ital., 75: 579–592.
Matschinski, M., 1955. Nature, Paris, No. 3241: 169–175.
Matsunaga, N. and Honji, H., 1980. Rep. Res. Inst. Appl., Mech., Kyushu Univ., 28: 27–38.
Matthes, G.H., 1947. Trans. Am. Geophys. Union, 28: 255–262.
Matthews, R.K., 1966. J. Sediment. Petrol., 36: 428–454.

Matzke, E.B., 1939. Am. J. Bot., 26: 288–295.
Matzke, E.B., 1946. Am. J. Bot., 33: 58–80.
Maude, A.D. and Whitmore, R.L., 1958. Brit. J. Appl. Phys., 9: 477–482.
Mavis, F.T. and Laushey, L.M., 1948. Rep. 2nd Meet., Int. Assoc. Hydraul. Res., pp. 213–218.
Maxwell, W.G.H., 1970. Deep-Sea Res., 17: 1005–1018.
Maxwell, W.G.H., Day, R.W. and Fleming, P.J.G., 1961. J. Sediment. Petrol., 31: 215–230.
Maxworthy, T., 1965. J. Fluid Mech., 23: 369–372.
May, R.M., 1973. Stability and Complexity in Model Ecosystems. Princeton University Press, Princeton, N.J.
Mayer, P.G., 1961. Trans. ASCE, 126: 505–535.
McBride, E.F., 1962. J. Sediment. Petrol., 32: 39–91.
McBride, E.F., 1966. Rep. Invest. Bur. Econ. Geol. Univ. Tex., No. 57.
McBride, E.F. and Kimberley, J.E., 1963. Bull. Am. Assoc. Pet. Geol., 47: 1840–1854.
McBride, E.F. and Yeakel, L.S., 1963. J. Sediment. Petrol., 33: 779–782.
McBride, E.F., Shepherd, R.G. and Crawley, R.A., 1975a. J. Sediment. Petrol., 45: 132–139.
McBride, E.F., Weidie, A.E. and Wolleben, J.A., 1975b. In: M.L. Broussard (Editor), Deltas. Models for Exploration. Houston Geological Society, Houston, Tex., pp. 485–522.
McCabe, P.J. and Jones, C.M., 1977. J. Sediment. Petrol., 47: 707–715.
McCann, S.B., 1961. Geol. Mag., 98: 131–142.
McCave, I.N., 1968. Geol. Soc. Am., Spec. Pap., No. 106: 75–107.
McCave, I.N., 1969. Nature (London), 224: 1288–1289.
McCave, I.N., 1970. J. Geophys. Res., 75: 4151–4159.
McCave, I.N., 1971a. Mar. Geol., 10: 199–225.
McCave, I.N., 1971b. J. Sediment. Petrol., 41: 89–96.
McCave, I.N., 1973a. Mém. Soc. R. Sci. Liège, (6)6: 187–206.
McCave, I.N., 1973b. In: E.D. Goldberg (Editor), North Sea Science. Massachusetts Institute of Technology Press, Cambridge, Mass., pp. 75–100.
McCave, I.N., 1975. Deep-Sea Res., 22: 491–502.
McCave, I.N. and Geiser, A.C., 1979. Sedimentology, 26: 353–369.
McCave, I.N. and Swift, D.J.P., 1976. Geol. Soc. Am. Bull., 87: 541–546.
McCormick, M.E., 1968. Trans. ASME, E35: 216–219.
McCracken, W.A., 1969. Bull. Am. Soc. Pet. Geol., 53: 463.
McCulloch, D.S. and Janda, R.J., 1964. J. Sediment. Petrol., 34: 694.
McDonald, B.C. and Banerjee, I., 1970. Geol. Surv. Can., Pap. 70-1A: 199.
McDonald, B.C. and Banerjee, I., 1971. Can. J. Earth Sci., 8: 1282–1301.
McDonald, B.C. and Day, T.J., 1978. Geol. Surv. Can., Pap. 78-1A: 441–451.
McDonald, B.C. and Vincent, J.S., 1972. Geol. Surv. Can., Pap. 72-27.
McDonnell, K.L., 1974. J. Geol. Soc. Aust., 21: 107–132.
McDowell, J.P., 1957. Circ. Ont. Dep. Mines, No. 6.
McDowell, J.P., 1960. Bull. Geol. Soc. Am., 71: 1925 (abstract).
McGeary, R.K., 1961. J. Am. Ceram. Soc., 44: 513–522.
McGowen, J.H., 1971. Rep. Invest. Bur. Econ. Geol. Univ. Tex., No. 71.
McGowen, J.H. and Garner, L.E., 1970. Sedimentology, 14: 77–111.
McGowen, J.H. and Scott, A.J., 1975. In: L.E. Cronin (Editor), Estuarine Research, vol. 2. Academic Press, New York, N.Y., pp. 23–46.
McGregor, B.A. and Bennett, R.H., 1977. Bull. Am. Assoc. Pet. Geol., 61: 918–928.
McGregor, B.A. and Bennett, R.H., 1979. Mar. Geol., 33: 163–174.
McGregor, R.C., 1972. Geophys. J. R. Astron. Soc., 29: 103–108.
McKee, E.D., 1933. Publ. Carnegie Inst. Wash., No. 440: 79–115.
McKee, E.D., 1938. J. Sediment. Petrol., 8: 77–83.
McKee, E.D., 1939. J. Geol., 47: 64–81.

McKee, E.D., 1945. J. Geol., 53: 313–325.
McKee, E.D., 1954. Mem. Geol. Soc. Am., No. 61.
McKee, E.D., 1957a. Bull. Am. Assoc. Pet. Geol., 41: 1704–1747.
McKee, E.D., 1957b. J. Sediment. Petrol., 27: 129–134.
McKee, E.D., 1959. J. Sediment. Petrol., 29: 354–364.
McKee, E.D., 1962. Geol. Rundsch., 52: 551–587.
McKee, E.D., 1965. In: G.V. Middleton (Editor), Primary Sedimentary Structures and their Hydrodynamic Interpretation. Soc. Econ. Paleontol. Mineral., Spec. Publ., 12: 66–83.
McKee, E.D., 1966a. U.S. Geol. Surv., Prof. Pap., 550-D: 94–103.
McKee, E.D., 1966b. Sedimentology, 7: 1–69.
McKee, E.D. and Douglass, J.R., 1971. U.S. Geol. Surv., Prof. Pap., 750-D: 108–114.
McKee, E.D. and Moiola, R.J., 1975. J. Res. U.S. Geol. Surv., 3: 59–66.
McKee, E.D. and Sterrett, T.S., 1961. In: J.A. Peterson and J.C. Osmond (Editors), Geometry of Sandstone Bodies. Am. Assoc. Pet. Geol., Tulsa, Okla., pp. 13–28.
McKee, E.D. and Tibbitts, G.C., 1964. J. Sediment. Petrol., 34: 5–17.
McKee, E.D. and Weir, G.W., 1953. Bull. Geol. Soc. Am., 64: 381–390.
McKee, E.D., Crosby, E.J. and Berryhill, H.L., 1967. J. Sediment. Petrol., 37: 829–851.
McKee, R.F. and Murfitt, R.F.A., 1967. Sierra Leone Geogr. J., No. 11: 4–10.
McKenzie, G.D., 1970. Glacial geology of Adams Inlet, southeastern Alaska. Ohio State Univ., Inst. Polar Stud., Rep. No. 25.
McKinney, T.F., Stubblefield, W.C. and Swift, D.J.P., 1974. Mar. Geol., 17: 79–102.
McLean, R.F. and Stoddart, D.R., 1978. Philos. Trans. R. Soc. London, Ser. A., 291: 101–117.
McMullen, R.M. and Swift, D.J.P., 1967. J. Sediment. Petrol., 37: 705–706.
McNair, E. and Sorensen, R.M., 1971. Proc. 12th Conf. Coastal Eng., 1: 415–434.
McNown, J.S., 1951. Houille Blanche, 6: 701–710.
McNown, J.S. and Malaika, J., 1950. Trans. Am. Geophys. Union, 31: 74–82.
Meade, R.H., 1964. U.S. Geol. Surv., Prof. Pap., 497-B.
Meade, R.H., 1966. J. Sediment. Petrol., 36: 1085–1101.
Meade, R.H., 1968. U.S. Geol. Surv., Prof. Pap., 497-D.
Meckel, L.D., 1967. J. Sediment. Petrol., 37: 80–86.
Mehrotra, S.C., 1974. J. Hydraul. Div. ASCE, 100(HY8): 1133–1140.
Mehta, A.J., Lee, J. and Christensen, B.A., 1980. J. Hydraul. Div. ASCE, 106(HY11): 1724–1744.
Mei, C.C., Liu, P.L.F. and Carter, T.G., 1972. Mass transport in water waves. Mass. Inst. Technol., Dep. Civ. Eng., Rep. No. 146.
Meinster, B. and Tickell, S.J., 1975. Trans. Geol. Soc. S. Afr., 78: 191–199.
Meissner, H.P., Michaelis, A.S. and Kaiser, R., 1964. Ind. Eng. Chem., Process Des. Dev., 3: 202–205.
Meland, N. and Norrman, J.O., 1966. Geogr. Ann., A48: 165–182.
Meland, N. and Norrman, J.O., 1969. Geogr. Ann., A51: 127–144.
Meldau, R. and Stach, E., 1934. J. Inst. Fuel, 7: 336–354.
Melières, F., Nesteroff, W.D. and Lancelot, Y., 1970. Cah. Océanogr., 22: 63–72.
Mellor, M., 1965. Blowing snow. U.S. Army Cold Reg. Res. Eng. Lab., Rep. III-A3c.
Mellor, M., 1978. In: B. Voight (Editor), Rockslides and Avalanches, vol. 1. Elsevier, Amsterdam.
Melmore, S., 1942a. Nature (London), 149: 412.
Melmore, S., 1942b. Nature (London), 149: 669.
Melmore, S., 1947. Nature (London), 159: 817.
Melmore, S., 1949. Mineral. Mag., 28: 479–483.
Melton, F.A., 1936. Geogr. Rev., 26: 593–609.
Melton, F.A., 1940. J. Geol., 48: 113–145.

Menard, H.W., 1950a. J. Sediment. Petrol., 20: 148–160.
Menard, H.W., 1950b. J. Geol., 58: 152–153.
Menard, H.W., 1952. J. Sediment. Petrol., 22: 3–9.
Menard, H.W. and Boucot, A.J., 1951. Am. J. Sci., 249: 131–151.
Menard, H.W. and Ludwick, J.C., 1951. In: J.L. Hough (Editor), Turbidity Currents. Soc. Econ. Paleontol. Mineral., Spec. Publ., 2: 2–13.
Mercer, A.G., 1971a. In: H.W. Shen (Editor), River Mechanics, vol. 1. Colo. State Univ., Fort Collins, Colo., pp. 10-1–10-26.
Mercer, A.G., 1971b. J. Eng. Mech. Div. ASCE, 97(EM1): 175–180.
Mercer, A.G. and Haque, M.I., 1973. J. Hydraul. Div. ASCE, 99(HY3): 441–459.
Merkli, P. and Thomann, H., 1975. J. Fluid Mech., 68: 567–575.
Merzkirch, W., 1974. Flow Visualization. Academic Press, New York, N.Y.
Metcalf, J.R., 1966. Int. J. Rock Mech. Min. Sci., 3: 151–161.
Metropolis, N., Rosenbluth, A.W., Rosenbluth, M.N. and Teller, A.H., 1953. J. Chem. Phys., 21: 1087–1092.
Meyer-Peter, E. and Müller, R., 1948. Int. Assoc. Hydraul. Struct. Res., 2nd Meet., Stockholm, pp. 39–64.
Miall, A.D., 1976. Can. J. Earth Sci., 13: 1422–1432.
Miall, A.D., 1979. Mem. Geol. Surv. Can., No. 387.
Michael, P., 1966. Phys. Fluids, 9: 466–471.
Miche, R., 1944. Ann. Ponts Chauss., pp. 25–78, 131–164, 270–292, 369–406.
Miche, R., 1958. Houille Blanche, 13: 40–74.
Michell, J.H., 1893. Philos. Mag., (5)36: 430–437.
Michon, S., Godet, J. and Bonnefille, R., 1955. Étude Theorique et Experimentale des Courants de Densité. Laboratoire Nationale d'Hydraulique, Chatou, 2 vols.
Mickelson, D.M., 1971. Glacial geology of the Burroughs Glacier area, southeastern Alaska. Ohio State Univ., Inst. Polar Stud., Rep. No. 40.
Mickelson, D.M., 1973. Arct. Alp. Res., 5: 17–27.
Middleton, G.V., 1965. J. Sediment. Petrol., 35: 922–927.
Middleton, G.V., 1966a. Can. J. Earth Sci., 3: 523–546.
Middleton, G.V., 1966b. Can. J. Earth Sci., 3: 627–637.
Middleton, G.V., 1967a. Can. J. Earth Sci., 4: 475–504.
Middleton, G.V., 1967b. J. Sediment. Petrol., 37: 229–232.
Middleton, G.V., 1970. Geol. Assoc. Can., Spec. Pap., 7: 253–272.
Middleton, G.V., 1977. J. Geol., 84: 405–426.
Middleton, G.V. and Hampton, M.A., 1973. In: G.V. Middleton and A.H. Bouma (Editors), Short Course, Turbidites and Deep Water Sedimentation. SEPM Pacific Section, Anaheim, pp. 1–38.
Middleton, G.V. and Hampton, M.A., 1976. In: D.J. Stanley and D.J.P. Swift (Editors), Marine Sediment Transport and Environmental Management. Wiley, New York, N.Y., pp. 197–218.
Migniot, C., 1968. Houille Blanche, 23: 519–620.
Mikhailova, N.A., 1952. Izv. Akad. Nauk SSSR, Ser. Geofyz., 1952(1): 47–56 (in Russian).
Mikhailova, N.A. and Naidenova, I.B., 1953. Izv. Akad. Nauk SSSR, Ser. Geofyz., 1953(5): 445–450 (in Russian).
Miller, D.G. and Bailey, A.B., 1979. J. Fluid Mech., 93: 449–464.
Miller, D.J., 1953. J. Glaciol., 61: 17–40.
Miller, H., 1884. Proc. R. Phys. Soc. Edinburgh, 8: 156–189.
Miller, M.C. and Komar, P.D., 1977. Sedimentology, 24: 709–721.
Miller, M.C., McCave, I.N. and Komar, P.D., 1977. Sedimentology, 24: 507–527.
Miller, M.C. and Komar, P.D., 1980a. J. Sediment. Petrol., 50: 173–182.
Miller, M.C. and Komar, P.D., 1980b. J. Sediment. Petrol., 50: 183–191.

Miller, R.L., 1968. J. Geophys. Res., 73: 4497–4510.
Miller, R.L. and Byrne, R.J., 1966. Sedimentology, 6: 303–314.
Miller, R.L. and Zeigler, J.M., 1965. Proc. 9th Conf. Coastal Eng., pp. 103–122.
Miller, W. and M'Inally, T.W., 1936. J. R. Tech. College Glasgow, 3: 682–689.
Milling, M.E. and Behrens, E.W., 1966. Publ. Inst. Mar. Sci., Univ. Tex., 11: 135–148.
Mills, D. and Mason, J.S., 1977a. Wear, 44: 311–328.
Mills, D. and Mason, J.S., 1977b. Powder Technol., 17: 37–53.
Mills, F.S., 1903. Am. Geol., 32: 162–170.
Mills, H.H., 1977a. J. Sediment. Petrol., 47: 728–737.
Mills, H.H., 1977b. Geol. Soc. Am. Bull., 88: 824–828.
Milne-Thompson, L.M., 1955. Theoretical Hydrodynamics. Macmillan, London, 3rd ed.
Minakami, T., 1942a. Bull. Earthquake Res. Inst. Tokyo Univ., 20: 65–92.
Minakami, T., 1942b. Bull. Earthquake Res. Inst. Tokyo Univ., 20: 94–105.
Minakami, T., 1950. Bull. Volcanol., (2)10: 59–87.
Minkowski, H., 1904. Nachr. Ges. Wiss. Göttingen, Math. Phys. Kl., No. 4: 311–355.
Minzoni, A.A. and Whitham, G.B., 1977. J. Fluid Mech., 79: 273–287.
Mishra, S.K., 1971. Proc. Indiana Acad. Sci., A37: 308–314.
Mitchell, J.E. and Hanratty, T.J., 1966. J. Fluid Mech., 26: 199–221.
Mitchell, J.K., 1956. Proc. Highw. Res. Board, 35: 693–713.
Mitchell, J.K., 1960. In: R.G.H. Parry (Editor), Interparticle Forces in Clay-Water-Electrolyte Systems. C.S.I.R.O., Melbourne, Aust., pp. 292–297.
Mitchell, J.K. and Houston, W.N., 1969. J. Soil Mech. Found. Div. Am. Soc. Civ. Eng., 95(SM3): 845–871.
Mitchell, K.E. and Hovermale, J.B., 1977. Mon. Weather Rev., 105: 657–675.
Mitchim, C.F., 1940. Mil. Eng., March–April 1940: 107–109.
Mitchner, M., 1954. J. Aeronaut. Sci., 21: 350–351.
Mogridge, G.R., 1973. Bed forms generated by wave action. Natl. Res. Counc. Can., Rep. No. DME/NAE 1973(2), pp. 21–41.
Mogridge, G.R. and Kamphuis, J.W., 1973. Proc. 13th Conf. Coastal Eng., 2: 1123–1142.
Molder, K., 1948. Bull. Comm. Géol. Finl., No. 142: 45–52.
Monod, T., 1958. Mem. Inst. Fr. Afr. Noire, No. 52.
Moon, C.F., 1972. Earth-Sci. Rev., 8: 303–321.
Moore, D.G., 1961. J. Sediment. Petrol., 31: 343–357.
Moore, D.G., 1978. In: B. Voight (Editor), Rockslides and Avalanches, vol. 1. Elsevier, Amsterdam, pp. 563–604.
Moore, D.G., Curray, J.R. and Emmel, F.J., 1966. Mar. Geol., 21: 211–226.
Moore, J.G., 1967. Bull. Volcanol., (2)30: 337–363.
Moore, J.G. and Melson, W.G., 1969. Bull. Volcanol., (2)33: 600–620.
Moore, J.G., Nakamura, K. and Alcarez, A., 1966. Science, 151: 955–960.
Moore, W.L. and Masch, F.D., 1962. J. Geophys. Res., 67: 1437–1446.
Moors, H.T., 1969. Sediment. Geol., 3: 241–261.
Morell, W.E. and Hildebrand, J.H., 1936. J. Chem. Phys., 4: 224–227.
Morgan, J.P., Nichols, L.G. and Wright, M., 1958. Morphological effects of Hurricane Audrey on the Louisiana coast. La. State Univ., Coastal Stud. Inst., Tech. Rep., No. 10.
Morgenstern, N.R., 1967. In: A.F. Richards (Editor), Marine Geotechnique. University of Illinois Press, Urbana, Ill., pp. 189–220.
Morgenstern, N.R. and Tchalenko, J.S., 1967a. Proc. R. Soc. London, Ser. A., 300: 218–234.
Morgenstern, N.R. and Tchalenko, J.S., 1967b. Proc. R. Soc. London, Ser. A., 300: 235–250.
Morgenstern, N.R. and Tchalenko, J.S., 1967c. Geotechnique, 17: 308–328.
Morgenstern, N.R. and Tchalenko, J.S., 1967d. Proc. Geotech. Conf., Oslo, 1: 147–152.
Morisawa, M., 1968. Streams. Their Dynamics and Morphology. McGraw-Hill, New York, N.Y.

Morison, J.R. and Crooke, R.C., 1953. Tech. Mem. U.S. Beach Erosion Board, No. 40.
Morrison, W.R.B., Bullock, K.J. and Kronauer, R.E., 1971. J. Fluid Mech., 47: 639–656.
Morrow, N.R. and Graves, J.R., 1969. Soil Sci., 108: 102–107.
Morton, D.M. and Campbell, R.H., 1974. Q. J. Eng. Geol., 7: 377–384.
Morton, R.A., 1978. Sedimentology, 25: 183–204.
Morton, R.A., 1979. Sediment. Geol., 24: 105–122.
Moss, A.J., 1962. Am. J. Sci., 260: 337–373.
Moss, A.J., 1963. Am. J. Sci., 261: 297–343.
Moss, A.J., 1966. J. Geol. Soc. Aust., 13: 97–136.
Moss, A.J., 1975. J. Geol. Soc. Aust., 22: 79–90.
Moss, J.H. and Ritter, D.F., 1962. Am. J. Sci., 260: 81–106.
Mottershead, D.N., 1971. Field Stud., 3: 433–453.
Mroczkowski, J., 1972. Acta Geol. Pol., 22: 351–377.
Mukherjee, S., 1968. Q. J. Geol., Min. Metall. Soc. India, 40: 189–190.
Mukherjee, S. and Haldar, D., 1975. Mineralia Deposita, 10: 109–119.
Müller, S.H., 1951. Abh. Dtsch. Akad. Wiss., Berlin, 3: 95–96.
Müller, G., 1967. In: G. Larsen and G.V. Chillingar (Editors), Diagenesis in Sediments. Elsevier, Amsterdam, pp. 127–177.
Müller, G. and Förstner, U., 1968. Nature (London), 217: 244–245.
Müller, J., 1969. Bull. Acad. Sci. Pol., Ser. Sci. Géol. Géogr., 17: 49–56.
Murphy, M.A. and Schlanger, S.O., 1962. Bull. Am. Assoc. Pet. Geol., 46: 457–477.
Myers, R.C. and Elsaway, E.M., 1975. J. Hydraul. Div. ASCE, 101(HY7): 933–946.
Naar, J., Wygal, R.J. and Henderson, J.H., 1963. Rev. Inst. Fr. Pét., 18: 369–386.
Nachtigall, K.H., 1962. Meyniana, 12: 9–24.
Nagle, J.S., 1967. J. Sediment. Petrol., 37: 1124–1138.
Nagtegaal, P.J.C., 1965. Geol. Mijnbouw, 44: 347–352.
Nairn, A.E.M., 1958. J. Sediment. Petrol., 28: 57–64.
Nakagawa, H. and Nezu, I., 1977. J. Fluid Mech., 80: 99–128.
Nakato, T., Locher, F.A., Glover, J.R. and Kennedy, J.F., 1977. ASCE J. Waterw. Harbors Coastal Eng. Div., 103(WW1): 83–99.
Nanson, G.C., 1980. Sedimentology, 27: 3–29.
Narasimha, R., 1957. J. Aeronaut. Sci., 24: 711–712.
Nasner, H., 1973. Proc. 15th Congr. Int. Assoc. Hydraul. Res., 1: 181–188.
Nasner, H., 1974. Mitt. Franzius Inst., 40: 1–149.
Natland, M.L., 1957. Mem. Geol. Soc. Am., No. 67: 543–571.
Navntoft, E., 1970. A theory of the velocity and suspended load distribution in a two-dimensional steady and uniform open-channel flow. Dan. Cent. Appl. Math. Mech., Rep. No. 3.
Neale, G., Epstein, N. and Nader, W., 1973. Chem. Eng. Sci., 28: 1865–1874.
Neall, V.E., 1972. N.Z. J. Geol. Geophys., 15: 507–557.
NEDECO (Netherlands Engineering Consultants), 1959. River Studies and Recommendations on Improvement of Niger and Benue. North-Holland, Amsterdam.
NEDECO (Netherlands Engineering Consultants), 1961. The Waters of the Niger Delta. NEDECO, The Hague.
Neill, C.R., 1967. Proc. 12th Congr. Int. Assoc. Hydraul. Res., 3: 46–64.
Neill, C.R., 1968a. Hydraul. Res. Stn. Rep. INT-68.
Neill, C.R., 1968b. J. Hydraul. Res., 6: 173–176.
Neill, C.R., 1969. J. Hydrol., 7: 58–85.
Neill, C.R. and Yalin, M.S., 1969. J. Hydraul. Div. ASCE, 95(HY1): 585–588.
Neilson, J.H. and Gilchrist, A., 1968a. Wear, 11: 111–122.
Neilson, J.H. and Gilchrist, A., 1968b. Wear, 11: 123–143.
Nesbitt, R.W. and Talbot, J.L., 1966. Contrib. Mineral. Petrol., 13: 1–11.

Neumann, B.S., 1953. In: J.J. Hermans (Editor), Flow Properties of Disperse Systems. North-Holland, Amsterdam.
Neumann, G., 1968. Ocean Currents. Elsevier, Amsterdam.
Neumann, G. and Pierson, W.J., 1966. Principles of Physical Oceanography. Prentice-Hall, Englewood Cliffs, N.J.
Nevin, C.M. and Trainer, D.W., 1927. Bull. Geol. Soc. Am., 38: 451–458.
Newell, N.D. and Boyd, D.W., 1955. J. Sediment. Petrol., 25: 226–228.
Newell, N.D. and Rigby, J., 1957. In: R.J. LeBlanc and J.G. Breeding (Editors), Regional Aspects of Carbonate Deposition. Soc. Econ. Paleontol. Mineral., Spec. Publ., 5: 15–72.
Newell, N.D., Rigby, J.K., Whiteman, A.J. and Bradley, J.S., 1951. Bull. Am. Mus. Nat. Hist., 97: 1–29.
Newell, N.D., Imbrie, J., Purdy, E. and Thurber, D.L., 1959. Bull. Am. Mus. Nat. Hist., 117: 183–228.
Newton, R.S., 1968a. Mar. Geol., 6: 73–75.
Newton, R.S., 1968b. Sedimentology, 11: 275–292.
Newton, R.S. and Werner, F., 1969. Hamb. Küstenforsch., No. 8.
Newton, R.S. and Werner, F., 1970. Geol. Rundsch., 60: 321–330.
Newton, R.S. and Werner, F., 1972. Meyniana, 22: 89–94.
Newton, R.S., Seibold, E. and Werner, F., 1973. "Meteor" Forsch. Ergeb., C15: 55–77.
Nichols, M.M., 1972. Mem. Geol. Soc. Am., No. 133: 169–212.
Nickling, W.G., 1978. Can. J. Earth Sci., 15: 1069–1084.
Niddrie, D.L., 1964. Geogr. Mag., 37: 228–235.
Niehoff, W., 1958. Geol. Rundsch., 47: 252–321.
Nielsen, P., 1977. Tech. Univ. Den., Inst. Hydrodyn. Hydraul. Eng., Progr. Rep., No. 43: 17–22.
Niemeyer, H.D., 1974. Mitt. Franzius Inst., 41: 124–256.
Nikuradse, J., 1933. Forschh. Ver. Dtsch. Ing., No. 361.
Nilsen, T.H., 1968. Norg. Geol. Unders., No. 259.
Nilsen, T.H., 1969. Sediment. Geol., 3: 35–57.
Nilsson, G. and Martvall, S., 1972. The Öre River and its meanders. University of Uppsala, UNGI Rep., No. 19.
Nio, S-D., 1976. Geol. Mijnbouw, 55: 18–40.
Nir, A., 1976. Appl. Sci. Res., 32: 313–325.
Nizery, A. and Bonnin, J., 1953. Proc. Minn. Int. Hydraul. Conv., Univ. Minneapolis, Minn., pp. 369–386.
Noda, H., 1969. Proc. 11th Conf. Coastal Eng., 1: 227–247.
Nordin, C.F., 1963. U.S. Geol. Surv., Prof. Pap., 462-C.
Nordin, C.F., 1964a. J. Hydraul. Div. ASCE, 90(HY4): 173–192.
Nordin, C.F., 1964b. Water Supply Irrig. Pap., Wash., 1498-H.
Nordin, C.F., 1971a. In: H.W. Shen (Editor), River Mechanics. Colorado State University, Fort Collins, Colo., pp. 12-1–12-31.
Nordin, C.F., 1971b. U.S. Geol. Surv., Prof. Pap., 562-F.
Nordin, C.F. and Algert, J.H., 1966. J. Hydraul. Div. ASCE, 92(HY5): 95–114.
Nordin, C.F. and Beverage, J.P., 1965. U.S. Geol. Surv., Prof. Pap., 462-F.
Nordin, C.F. and Dempster, G.R., 1963. U.S. Geol. Surv., Prof. Pap., 462-B.
Nordseth, K., 1973. Norsk Geogr. Tidsskr., 27: 77–108.
Norin, R., 1940. Geol. För. Stockh. Förh., 62: 31–44.
Norman, L.D., Maust, E.E. and Skolnick, L.P., 1971a. Bull. U.S. Bur. Min, No. 658: 18–25.
Norman, L.D., Maust, E.E. and Skolnick, L.P., 1971b. Bull. U.S. Bur. Mines, No. 658: 48–55.
Normark, W.R., 1970a. Bull. Am. Assoc. Pet. Geol., 54: 2170–2195.
Normark, W.R., 1970b. Deep-Sea Res., 17: 837–846.
Normark, W.R., 1978. Bull. Am. Assoc. Pet. Geol., 62: 912–931.

Normark, W.R. and Dickson, F.H., 1976. Sedimentology, 23: 815–831.
Norris, R.M., 1956. Bull. Am. Assoc. Pet. Geol., 40: 1681–1686.
Norris, R.M., 1966. J. Geol., 74: 292–306.
Norris, R.M. and Norris, K.S., 1961. Geol. Soc. Am. Bull., 72: 605–620.
Norrman, J.O., 1964. Geogr. Ann., 46: 3–238.
Nutting, P.G., 1921. J. Franklin Inst., 191: 679–685.
Nychas, S.G., Hershey, H.G. and Brodky, R.S., 1973. J. Fluid Mech., 61: 513–540.
Nye, J.F., 1952. J. Glaciol., 2: 82–93.
Nye, J.F., 1957. Proc. R. Soc. London, Ser. A., 239: 113–133.
Nystuen, J.P., 1976. Norg. Geol. Unders., No. 329.
Oberbeck, A., 1876. J. Reine Angew. Math., 81: 62–80.
Obremski, H.J. and Fejer, A.A., 1967. J. Fluid Mech., 29: 93–111.
O'Brien, M.P., 1933. Trans. Am. Geophys. Union, 14: 487–491.
O'Brien, M.P. and Cherno, J., 1934. Trans. ASCE, 99: 576–594.
O'Brien, M.P. and Rindlaub, B.D., 1934. Trans. Am. Geophys. Union, 15: 593–603.
O'Brien, M.P. and Rindlaub, B.D., 1936. Civil Eng., 6: 325–327.
O'Brien, N.R., 1968a. Naturwissenschaften, 55: 490–491.
O'Brien, N.R., 1968b. Marit. Sediments, 4: 104–105.
O'Brien, N.R., 1970a. Sedimentology, 15: 229–246.
O'Brien, N.R., 1970b. Marit. Sediments, 6: 79–80.
O'Brien, N.R. and Harrison, W., 1969. Naturwissenschaften, 56: 135–136.
O'Brien, N.R. and Suito, E., 1969. Marit. Sediments, 5: 58–61.
Ockendon, J.R., 1968. J. Fluid Mech., 34: 229–239.
Odar, F., 1966. J. Fluid Mech., 25: 591–592.
Odar, F. and Hamilton, W.S., 1964. J. Fluid Mech., 18: 302–314.
Odd, N.V.M. and Owen, M.V., 1972. Proc. Inst. Civ. Eng., Suppl., No. 9: 175–205.
Odom, I.E., 1967. J. Sediment. Petrol., 37: 610–623.
Oettli, M., 1917. Naturforsch. Wochenschr., 16: 593–594.
Offen, G.R. and Kline, S.J., 1974. J. Fluid Mech., 62: 223–239.
Offen, G.R. and Kline, S.J., 1975. J. Fluid Mech., 70: 209–228.
Officer, C.B., 1976. Physical Oceanography of Estuaries. Wiley, New York, N.Y.
Ojakangas, R.W. and Matsch, C.L., 1980. Geol. Soc. Am. Bull., 91: 495–501.
Oldaker, D.K. and Tiederman, W.G., 1977. Phys. Fluids, 20(10, II): S133–S144.
Oldenziel, D.M. and Brink, W.E., 1974. J. Hydraul. Div. ASCE, 100(HY7): 935–949.
Oliver, D.R., 1962. Nature (London), 194: 1269–1271.
Ollier, C.D., 1969. Volcanoes. Australian National University Press, Canberra.
O'Loughlin, E.M. and Squarer, D., 1967. Proc. 12th Congr. Int. Assoc. Hydraul. Res., 2: 118–127.
Olson, R.E., 1962. Geotechnique, 12: 23–43.
Olson, R.E. and Mesri, G., 1970. J. Soil Mech. Found. Div. ASCE, 96(SM6): 1863–1878.
Olsson-Seffer, P., 1910. Publ. Augustana Libr., No. 7: 10–41.
Olszewski, A. and Szupryczynski, J., 1975. Bull. Acad. Sci. Pol., Sér. Sci. Terre, 23: 59–67.
Oman, A.O. and Watson, K.M., 1944. Nat. Pet. News, 36: R795–R802.
Onions, D. and Middleton, G.V., 1968. J. Sediment. Petrol., 38: 164–174.
Oomkens, E. and Terwindt, J.H.J., 1960. Geol. Mijnbouw, 30: 701–710.
Ooms, G., Mijnlieff, P.F. and Becker, H.L., 1970. J. Chem. Phys., 53: 4123–4130.
Ore, H.T., 1963. Wyo. Contrib. Geol., 3: 1–14.
Orme, A.R., 1973. Coastal dune systems of northwest Baja California. Univ. Calif., Riverside, Dep. Earth Sci., Tech. Rep., No. 73-1.
Oseen, C.W., 1927. Hydrodynamik. Akademische Verlagsgesellschaft, Leipzig.
Østrem, G., 1975. In: A.V. Jopling and B.C. McDonald (Editors), Glaciofluvial and Glaciolacustrine Environments. Soc. Econ. Paleontol. Mineral., Spec. Publ., 23: 101–122.

Ostry, R.C. and Deane, R.E., 1963. Geol. Soc. Am. Bull., 74: 165–168.
Otvos, E.G., 1964. J. Sediment. Petrol., 34: 683–687.
Otvos, G., 1965. Am. J. Sci., 263: 271–276.
Otvos, G., 1966. Bull. Am. Assoc. Pet. Geol., 50: 159–165.
Owen, D.M., Emery, K.O. and Hoadley, L.D., 1967. In: J.B. Hersey (Editor), Deep-Sea Photography. Johns Hopkins Press, Baltimore, Md., pp. 159–166.
Owen, F.K. and Horstman, C.C., 1972. J. Fluid Mech., 53: 611–616.
Owen, M.W., 1977. In: E.D. Goldberg, I.N. McCave and J.J. O'Brien (Editors), The Sea, vol. 6. Wiley, New York, N.Y., pp. 515–537.
Owen, P.R., 1964. J. Fluid Mech., 20: 225–242.
Owens, J.S., 1908. Geogr. J., 31: 415–425.
Owens, J.S., 1911. Geogr. J., 37: 59–77.
Owens, J.S., 1912. Geogr. J., 39: 247–257.
Owens, J.S., 1927. Engineer, 8 April, 1927: 377.
Owens, W.H., 1974. Tectonophysics, 24: 115–131.
Ozasa, H., 1974. Coastal Eng. Jpn., 17: 155–184.
Pain, C.F. and Blong, R.J., 1976. In: R.W. Johnson (Editor), Volcanism in Australia. Elsevier, Amsterdam, pp. 239–251.
Paintal, A.S., 1971. J. Hydraul. Res., 9: 91–113.
Panin, N. and Panin, S., 1967. Mar. Geol., 5: 221–226.
Pantin, H.M., 1979. Mar. Geol., 31: 59–99.
Pardé, M., 1947. Fleuves et Rivières. Collection Armand Colin, Paris, 2nd ed.
Pardee, J.T., 1942. Bull. Geol. Soc. Am., 53: 1569–1600.
Parea, G.C., 1970. Natura, 61: 297–307.
Park, Y.A., 1974. Meyniana, 24: 73–89.
Parkash, B. and Middleton, G.V., 1970. Sedimentology, 14: 259–293.
Parker, G., 1975. J. Hydraul. Div. ASCE, 101(HY2): 211–221.
Parker, W.R., 1973. Sedimentology, 20: 615–623.
Parker, W.R., 1975. In: J. Hails and A. Carr (Editors), Nearshore Sediment Dynamics and Sedimentation. Wiley, London, pp. 150–178.
Parsick, R.J., Jones, S.C. and Hatch, L.P., 1966. Nucl. Appl., 2: 221–225.
Parsons, D.A., 1972. In: H.W. Shen (Editor), Sedimentation. Colorado State University, Fort Collins, Colo., pp. 1.1–1.22.
Partheniades, E., 1965. J. Hydraul. Div. ASCE, 91(HY1): 105–139.
Partheniades, E., 1971. In: H.W. Shen (Editor), River Mechanics. Colorado State University, Fort Collins, Colo., pp. 25.1–25.91.
Partheniades, E., 1972. In: H.W. Shen (Editor), Sedimentation. Colorado State University, Fort Collins, Colo., pp. 20.1–20.39.
Partheniades, E. and Paaswell, R.E., 1970. J. Hydraul. Div. ASCE, 96(HY3): 755–771.
Pasquill, F., 1962. Atmospheric Diffusion. Van Nostrand, London.
Paterson, W.S.B., 1969. The Physics of Glaciers. Pergamon, Oxford.
Patton, H.B., 1910. Bull. Geol. Soc. Am., 21: 663–676.
Patton, L.T., 1933. J. Sediment. Petrol., 3: 77–82.
Paulissen, E., 1973. Verh. K. Vlaam. Acad. Wet., No. 35.
Payne, L.E. and Pell, W.H., 1960. J. Fluid Mech., 7: 529–549.
Pazwash, H. and Robertson, J.M., 1975. J. Hydraul. Res., 13: 35–55.
Peabody, F.E., 1947. J. Sediment. Petrol., 17: 73–76.
Pedersen, G.K. and Surlyk, F., 1977. Sedimentology, 24: 581–590.
Peirce, J.J., Jarman, R.T. and DeTurville, C.M., 1970. Proc. Inst. Civ. Eng., 45: 231–243.
Peleg, K., 1972. Israel J. Agric. Res., 22: 225–236.
Pelletier, B.R., 1958. Bull. Geol. Soc. Am., 69: 1033–1064.
Pendlebury, D.C. and Dobson, M.R., 1976. Scott. J. Geol., 11: 315–332.

Penner, E., 1963. In: I.T. Rosenqvist and P. Graf-Peterson (Editors), Proc. Int. Clay Conf., London, 1: 365–376.

Penny, L.F. and Catt, J.A., 1967. Geol. Mag., 104: 344–360.

Penny, W.G. and Thornhill, C.K., 1951. Philos. Trans. R. Soc. London, Ser. A., 244: 285–311.

Pepper, J.F., De Witt, W. and Demarest, D.F., 1954. U.S. Geol. Surv., Prof. Pap., 259.

Peregrine, D.H., 1966. J. Fluid Mech., 25: 321–330.

Peregrine, D.H., 1974. J. Hydraul. Div. ASCE, 100(HY9): 1215–1227.

Perkins, R.D. and Enos, P., 1968. J. Geol., 76: 710–717.

Perret, F.A., 1935. Publ. Carnegie Inst. Wash., No. 458.

Perthuisot, J.P. and Jauzein, A., 1975. Rev. Géogr. Phys. Géol. Dyn., (2)17: 295–306.

Pestrong, R., 1965. Publ. Earth Sci. Stanford Univ., 10(2): 1–87.

Pestrong, R., 1972a. Sediment. Geol., 8: 251–288.

Pestrong, R., 1972b. Calif. Geol., 25(2): 27–40.

Peterlin, A., 1938. Z. Phys., 111: 232–263.

Peters, J.J., 1971. La dynamique de la sédimentation de la région divagante du bief maritime du flueve Congo. Laboratoire de Recherche Hydraulique à Borgerhout (Belgium).

Peters, J.J., 1977. In: J.C.J. Nihoul (Editor), Bottom Turbulence. Elsevier, Amsterdam, pp. 221–236.

Peterson, D.F. and Moharty, P.K., 1960. J. Hydraul. Div. ASCE, 85(HY9): 55–76.

Petrov, M.P., 1962. Ann. Géogr., 71: 131–155.

Petrov, M.P., 1976. Deserts of the World. Wiley, New York, N.Y.

Petryk, S. and Grant, E.U., 1978. J. Hydraul. Div. ASCE, 104(HY5): 583–594.

Pettijohn, F.J., 1957. Sedimentary Rocks. Harper, New York, N.Y., 2nd ed.

Pettijohn, F.J., 1962. Bull. Am. Assoc. Pet. Geol., 46: 1468–1493.

Pettijohn, F.J. and Potter, P.E., 1964. Atlas and Glossary of Primary Sedimentary Structures. Springer, Berlin.

Pettijohn, F.J., Potter, P.E. and Siever, R., 1972. Sand and Sandstone. Springer, Berlin.

Pettyjohn, E.A. and Christiansen, E.B., 1948. Chem. Eng. Prog., 44: 157–172.

Péwé, T.L., 1975. U.S. Geol. Surv., Prof. Pap., 835.

Phillips, O.M., 1977. The Dynamics of the Upper Ocean. Cambridge University Press, Cambridge.

Picard, M.D., 1967. Bull. Am. Assoc. Pet. Geol., 51: 383–392.

Picard, M.D. and High, L.R., 1964. Wyo. Contrib. Geology, 3: 27–31.

Picard, M.D. and High, L.R., 1968. J. Sediment. Petrol., 38: 411–423.

Picard, M.D. and High, L.R., 1969. Bull. Geol. Surv. Utah, 82: 175–190.

Picard, M.D. and High, L.R., 1970a. J. Sediment. Petrol., 40: 708–711.

Picard, M.D. and High, L.R., 1970b. Spec. Stud. Utah Geol. Miner. Surv., No. 33.

Picard, M.D. and High, L.R., 1973. Sedimentary Structures of Ephemeral Streams. Elsevier, Amsterdam.

Picard, M.D. and Hulen, J.B., 1969. Geol. Soc. Am. Bull., 80: 2631–2636.

Pick, M.C., 1964. Proc. Geol. Assoc., 75: 199–221.

Pick, W.H., 1962. A Course in Elementary Meteorology. Meteorological Office, H.M.S.O., London.

Pierce, R.C., 1916. Water Supply Irrig. Pap., Wash., No. 400.

Pilgrim, A.T. and Conacher, A.J., 1974. Aust. Geogr. Stud., 12: 38–56.

Pillai, K.K., 1977. Chem. Eng. Sci., 32: 59–61.

Pilpel, N., 1969. Endeavour, No. 204: 73–76.

Pingree, R.D. and Griffiths, D.K., 1979. J. Mar. Biol. Assoc. U.K., 59: 497–513.

Piper, D.J.W., 1970a. Scott. J. Geol., 6: 295–308.

Piper, D.J.W., 1970b. Mar. Geol., 8: 211–227.

Piper, D.J.W., 1972. Geol. Mag., 108: 517–523.

Plafker, G. and Eriksen, G.E., 1978. In: B. Voight (Editor), Rockslides and Avalanches, vol. 1. Elsevier, Amsterdam.
Plapp, J.E. and Mitchell, J.P., 1960. J. Geophys. Res., 65: 983–992.
Plate, E.J., 1971. Aerodynamic characteristics of atmospheric boundary layers. U.S. Atomic Energy Commission, Oak Ridge, Tenn.
Platzman, G.W., 1971. In: W.H. Reid (Editor), Mathematical Problems in the Geophysical Sciences. American Mathematical Society, Providence, R.I., pp. 239–291.
Plessman, W., 1961. Geol. Jahrb., 78: 503–566.
Plummer, R.S., 1978. Trans. R. Soc. S. Aust., 102(2): 25–38.
Polya, G., 1937. Zur Kinematik der Geschiebebewegung. Verlag Rascher, Zurich.
Porter, D. and Dore, B.D., 1974. Proc. Cambridge Philos. Soc., Math. Phys. Sci., 75: 283–294.
Postma, H., 1954. Archs Néerl. Zool., 10: 405–511.
Postma, H., 1961. Neth. J. Sea Res., 1: 148–190.
Postma, H., 1967. In: G.H. Lauff (Editor), Estuaries. American Association for the Advancement of Science, Washington, D.C., pp. 158–180.
Potter, P.E., 1963. Rep. Invest. Geol. Surv. Ill., No. 217.
Potter, P.E., 1978. J. Geol., 86: 13–33.
Potter, P.E. and Glass, H.D., 1958. Rep. Invest. Geol. Surv. Ill., No. 204.
Potter, P.E. and Mast, R.F., 1963. J. Geol., 71: 441–471.
Potter, P.E. and Pettijohn, F.J., 1963. Palaeocurrents and Basin Analysis. Springer, Berlin.
Pounder, E.R., 1965. The Physics of Ice. Pergamon, Oxford.
Power, W.R., 1961. J. Sediment. Petrol., 31: 603–607.
Powers, W.H., Draper, L. and Briggs, P.M., 1969. Proc. 11th Conf. Coastal Eng., 1: 1–8.
Prakash, S. and Gupta, M.K., 1971. In: D.A. Howells, I.P. Haigh and C. Taylor (Editors), Dynamic Waves in Civil Engineers. Wiley–Interscience, London, pp. 228–244.
Prandtl, L., 1933. Z. Ver. Dtsch. Ing., 77: 105–114.
Prandtl, L., 1952. Essentials of Fluid Dynamics. Blackie, London.
Prandtl, L. and Tietjens, O.G., 1957. Fundamentals of Hydro- and Aeromechanics. Dover Publ., New York.
Pratt, C.J. and Smith, K.V.H., 1972. J. Hydraul. Div. ASCE, 98(HY5): 859–874.
Praturi, A.K. and Brodkey, R.S., 1978. J. Fluid Mech., 89: 251–272.
Prentice, J.E., 1956. Geol. Mag., 93: 393–400.
Prentice, J.E., 1962a. In: K. Coe (Editor), Some Aspects of the Variscan Fold Belt. Manchester University Press, Manchester, pp. 93–108.
Prentice, J.E., 1962b. Proc. Geol. Assoc., 73: 171–185.
Pretious, E.S. and Blench, T., 1951. Final report on special observations of bed movement in Lower Fraser River at Ladner Reach during 1950 freshet. Rep. Nat. Res. Counc. Can., Fraser River Model, Vancouver.
Price, N.J., 1966. Fault and Joint Development. Pergamon, New York, N.Y.
Price, R.J., 1969. Trans. Inst. Br. Geogr., No. 46: 17–43.
Price, R.J., 1970. Arct. Alp. Res., 2: 27–42.
Price, R.J., 1973. Glacial and Fluvioglacial Landforms. Oliver and Boyd, Edinburgh.
Price, W.A., 1933. Bull. Am. Assoc. Pet. Geol., 17: 907–962.
Price, W.A., 1958. Trans. Gulf-Coast Assoc. Geol. Soc., 8: 41–75.
Price, W.A., 1963. J. Sediment. Petrol., 33: 765–778.
Price, W.A. and Kornicker, L.S., 1961. J. Sediment. Petrol., 31: 245–255.
Price. W.H., 1947. J. Am. Concr. Inst., 19: 1009–1023.
Prior, D.B. and Coleman, J.M., 1980a. Mar. Geol., 36: 227–239.
Prior, D.B. and Coleman, J.M., 1980b. Mar. Geotech., 3: 37–60.
Prior, D.B. and Stephens, N., 1972. Geol. Soc. Am. Bull., 83: 2533–2544.
Prospero, J.M. and Carlson, T.N., 1972. J. Geophys. Res., 77: 5255–5265.
Prospero, J.M., Bonatti, E., Schubert, C. and Carlson, T.N., 1970. Earth Planet. Sci. Lett., 9: 287–293.

Prosser, C.S., 1916. J. Geol., 24: 456–475.
Proudman, I. and Pearson, J.R.A., 1957. J. Fluid Mech., 2: 237–262.
Proudman, J., 1925. Philos. Mag., (6)49: 465–475.
Pruppacher, H.R. and Steinberger, E.H., 1968. J. Appl. Phys., 39: 4129–4132.
Pruppacher, H.R., Le Clair, B.P. and Hamielec, A.E., 1970. J. Fluid Mech., 44: 781–790.
Pryor, W.A., 1971a. Reservoir inhomogeneities of some Recent sand bodies. Soc. Pet. Eng., Am. Inst. Min. Eng., Pap. SPE 3607.
Pryor, W.A., 1971b. Sediment. Geol., 6: 221–254.
Pryor, W.A. and Amaral, E.J., 1971. Geol. Soc. Am. Bull., 82: 239–244.
Pryor, W.A., Cottrill, N.E. and Klein, H.M., 1972. Geol. Soc. Am., Abstr. Progr., 4: 345.
Psuty, N.P., 1966. The geomorphology of beach ridges in Tabasco, Mexico. La. State Univ., Coastal Stud. Inst., Tech. Rep., No. 30.
Puchegger, F., 1952. Naturwissenschaften, 39: 428–429.
Puigdefabrigas, C., 1974. Bull. Cent. Rech. Pau, 8: 305–325.
Purdy, E.G., 1961. Bahamian oolite shoals. In: J.A. Peterson and J.C. Osmond (Editors), Geometry of Sandstone Bodies. American Association of Petroleum Geologists, Tulsa, Okla., pp. 53–62.
Purdy, E.G., 1963a. J. Geol., 71: 334–355.
Purdy, E.G., 1963b. J. Geol., 71: 472–497.
Pusch, R., 1966. Eng. Geol., 1: 433–443.
Pusch, R., 1969. Eng. Geol., 3: 135–148.
Pusch, R., 1970. Can. Geotech. J., 7: 1–7.
Pusch, R. and Arnold, M., 1969. Eng. Geol., 3: 135–148.
Py, B., 1973. Int. J. Heat Mass Transfer, 16: 129–144.
Quigley, R.M. and Thompson, C.D., 1966. Can. Geotech. J., 3: 61–73.
Raask, E., 1969. Wear, 13: 301–315.
Rabien, A., 1956. Abh. Hess. Landesamt. Bodenforsch., 16: 1–83.
Radok, U., 1977. J. Glaciol., 19: 123–138.
Raichlen, F. and Kennedy, J.F., 1965. Proc. 11th Int. Congr. Int. Assoc. Hydraul. Res., 8 pp.
Rajar, R., 1978. J. Hydraul. Div. ASCE, 104(HY9): 1011–1026.
Rajaratnam, N., 1965a. J. Hydraul. Div. ASCE, 91(HY4): 71–96.
Rajaratnam, N., 1965b. J. Hydraul. Div. ASCE, 91(HY5): 107–132.
Rajaratnam, N. and Ahmadi, R.M., 1979. J. Hydraul. Div. ASCE, 105(HY5): 573–588.
Rajaratnam, N. and Ortiz, N.V., 1977. J. Hydraul. Div. ASCE, 103(HY4): 381–394.
Rajaratnam, N. and Subramanya, K., 1968. J. Hydraul. Div. ASCE Eng., 94(HY3): 663–673.
Ramsay, J.G., 1967. Folding and Fracturing of Rocks. McGraw-Hill, New York, N.Y.
Rana, S.A., Simons, D.B. and Mahmood, K., 1973. J. Hydraul. Div. ASCE, 90(HY11): 1967–1980.
Rance, P.J. and Warren, N.F., 1969. Proc. 11th Conf. Coastal Eng., 1: 487–491.
Rao, G.N.V., 1974. Z. Angew. Math. Phys., 25: 63–75.
Rao, K.N., Narasimha, R. and Narayanan, M.A.B., 1971. J. Fluid Mech., 48: 339–352.
Rao, N.S.G. and Thiruvengadam, A., 1961. J. Hydraul. Div. ASCE, 87(HY5): 37–62.
Rapp, A., 1959. Geogr. Ann., 41: 34–48.
Rapp, A., 1960a. Geogr. Ann., 42: 65–200.
Rapp, A., 1960b. Skr. Norsk Polarinst., No. 119.
Rathburn, R.E. and Guy, H.P., 1967. Water Resour. Res., 3: 107–122.
Rattigan, J.H., 1967. J. Geol. Soc. Aust., 14: 5–18.
Raudkivi, A.J., 1963. J. Hydraul. Div. ASCE, 89(HY6): 15–36.
Raudkivi, A.J., 1966a. J. Fluid Mech., 26: 507–514.
Raudkivi, A.J., 1966b. Proc. 2nd Australas. Conf. Hydraul. Fluid Mech., pp. A135–A142.
Raudkivi, A.J., 1976. Loose Boundary Hydraulics. Pergamon, Oxford, 2nd ed.
Raudkivi, A.J. and Hutchinson, D.L., 1974. Proc. R. Soc. London, Ser. A., 337: 537–554.

Ray, P.K., 1976. J. Sediment. Petrol., 46: 788–801.
Rayleigh, Lord, 1883. Philos. Trans. R. Soc. London, Ser. A., 175: 1–21.
Rayleigh, Lord, 1914. Proc. R. Soc. London, Ser. A., 90: 324–328.
Raynaud, J.P., 1951. Trans. 4th Congr. Large Dams, 4: 137–161.
Read, W.A. and Johnson, S.R.H., 1967. Scott. J. Geol., 3: 242–267.
Reade, T.M., 1884. Q. J. Geol. Soc. London, 40: 267–269.
Reading, H.G., 1978. Sedimentary Environments and Facies. Blackwell, Oxford.
Reading, H.G. and Walker, R.G., 1966. Palaeogeogr. Palaeoclimatol. Palaeoecol., 2: 177–212.
Reed, L.J. and Tryggvason, E., 1974. Tectonophysics, 24: 85–98.
Reed, W.E., Le Fever, R. and Moir, G.J., 1975. Geol. Soc. Am. Bull., 86: 1321–1328.
Rees, A.I., 1961. Geophys. J.R. Astron. Soc., 5: 235–251.
Rees, A.I., 1964. Geophys. J. R. Astron. Soc., 8: 356–369.
Rees, A.I., 1966a. J. Geol., 74: 856–867.
Rees, A.I., 1966b. Sedimentology, 6: 209–240.
Rees, A.I., 1968. J. Geol., 76: 457–465.
Rees, A.I., 1971. J. Sediment. Petrol., 41: 307–309.
Rees, A.I., 1979. Tectonophysics, 55: 275–287.
Rees, A.I. and Woodall, W.A., 1975. Earth Planet. Sci. Lett., 25: 121–130.
Rees, A.I., Von Rad, U. and Shepard, F.P., 1968. Mar. Geol., 6: 145–178.
Rehbock, T., 1930. Z. Dtsch. Geol. Ges., 81: 497–534.
Reiche, P., 1938. J. Geol., 46: 905–932.
Reid, J.C. and Nowlin, W.D., 1971. Deep-Sea Res., 18: 51–64.
Reif, D.M. and Slatt, R.M., 1979. J. Sediment. Petrol., 49: 869–890.
Reineck, H.-E., 1955. Senckenbergiana Lethaea, 36: 347–357.
Reineck, H.-E., 1958a. Senckenbergiana Lethaea, 39: 29–41.
Reineck, H.-E., 1958b. Natur Volk, 88: 81–84.
Reineck, H.-E., 1960a. Jahresber. Forschungsanst. Norderney, 11: 21–38.
Reineck, H.-E., 1960b. Abh. Dtsch. Akad. Wiss. Berlin, (III)1: 369–374.
Reineck, H.-E., 1960c. Geol. Rundsch., 49: 149–161.
Reineck, H.-E., 1960d. Natur Volk, 90: 334–337.
Reineck, H.-E., 1961. Senckenbergiana Lethaea, 42: 51–67.
Reineck, H.-E., 1963. Abh. Senckenb. Naturforsch. Ges., No. 505.
Reineck, H.-E., 1971. Natur Mus., 101: 45–60.
Reineck, H.-E., 1974. Senckenbergiana Marit., 6: 135–143.
Reineck, H.-E. and Dörjes, J., 1976. Senckenbergiana Marit., 8: 111–153.
Reineck, H.-E. and Singh, I.B., 1967. Mar. Geol., 5: 227–235.
Reineck, H.-E. and Singh, I.B., 1973. Depositional Sedimentary Environments. Springer, Berlin.
Reineck, H.-E. and Wunderlich, F., 1967. Natur Mus., 97: 193–197.
Reineck, H.-E. and Wunderlich, F., 1968a. Sedimentology, 11: 99–104.
Reineck, H.-E. and Wunderlich, F., 1968b. Senckenbergiana Lethaea, 49: 321–345.
Reineck, H.-E. and Wunderlich, F., 1969. Senckenbergiana Marit., 50: 79–84.
Reineck, H.-E., Singh, I.B. and Wunderlich, F., 1971. Senckenbergiana Marit., 3: 93–101.
Reiner, M., 1959. The Flow of Matter. Sci. Am., December 1959: 11 pp.
Reiner, M., 1969. Deformation, Strain and Flow. Lewis, London.
Reinson, G.E., 1977. Geol. Surv. Can., Pap. 77-1B: 99–105.
Reinson, G.E., 1979. Mar. Geol., 31: 279–296.
Rempel, P.J., 1936. Ecology, 17: 347–358.
Reshotko, E., 1976. Annu. Rev. Fluid Mech., 8: 311–349.
Reyment, R.A., 1968. J. Sediment. Petrol., 38: 1387–1389.
Reyment, R.A., 1971. Palaeogeogr. Palaeoclimatol. Palaeoecol., 9: 265–275.
Reynolds, A.J., 1965. J. Fluid Mech., 22: 113–133.

Reynolds, A.J., 1976. Nordic Hydrol., 7: 161–180.
Reynolds, O., 1883. Philos. Trans. R. Soc. London, Ser. A., 174: 935–982.
Reynolds, O., 1885. Philos. Mag., (5)20: 469–489.
Reynolds, O., 1886. Proc. R. Inst. Gt. Br., 11: 354–363.
Ricci-Lucchi, F., 1969. J. Geol., (2a)36: 203–282.
Ricci-Lucchi, F., 1970. Sedimentografia. Zanichelli, Bologna.
Ricci-Lucchi, F. and Valmori, E., 1980. Sedimentology, 27: 241–270.
Rice, O.K., 1944. J. Chem. Phys., 12: 1–23.
Richards, K.J., 1980. J. Fluid Mech., 99: 597–618.
Richards, K.S., 1978. Earth Surf. Processes, 3: 345–354.
Richardson, E.G., 1934. Philos. Mag., (7)17: 769–783.
Richardson, F.M. and Beatty, K.O., 1959. Phys. Fluids, 2: 718–719.
Richardson, J.F. and Zaki, W.N., 1954. Trans. Inst. Chem. Eng., 32: 35–53.
Richter, K., 1932. Z. Geschiebeforsch. Flachlandgeol., 8: 62–66.
Richter, K., 1933. Abh. Geol.-Palaeontol. Inst. Greifswald, 11: 1–63.
Richter, K., 1936. Geol. Rundsch., 27: 196–206.
Richter, R., 1926a. Senckenbergiana, 8: 297–305.
Richter, R., 1926b. Natur Mus., 56: 389–407.
Ricketts, B.D. and Donaldson, J.A., 1979. Can. J. Earth Sci., 16: 1887–1891.
Riddell, J.F., 1969. Can. J. Earth Sci., 6: 231–246.
Riddell, J.F., 1970. Houille Blanche, 25: 317–329.
Ridgway, K. and Tarbuck, K.J., 1966. J. Pharm. Pharmac., 18: S168–S175.
Ridgway, K. and Tarbuck, K.J., 1967. Br. Chem. Eng., 12: 384–388.
Ridgway, K. and Tarbuck, K.J., 1968a. Chem. Eng. Sci., 23: 1147–1155.
Ridgway, K. and Tarbuck, K.J., 1968b. Chem. Process Eng., February 1968: 103–105.
Riley, N., 1965. Mathematica, 12: 161–175.
Riley, N., 1967. J. Inst. Math. Appl., 3: 419–434.
Riley, N., 1975. SIAM Rev., 17: 274–297.
Rim, M., 1953. Centre National de la Recherche Scientifique, Paris, Colloq. Int., No. 35: 261–276.
Ritchie, W., 1972. Scott. Geogr. Mag., 88: 19–35.
Ritter, A., 1892. Z. Ver. Dtsch. Ing., 36: 947–954.
Road, A., 1977. Valley of the dammed. The Observer, Newspaper, Colour Magazine Suppl., 25 November 1973.
Roberts, J., 1975. Internal Gravity Waves in the Ocean. Dekker, New York.
Roberts, J.D., 1974. Norg. Geol. Unders., No. 303: 57–118.
Robertson, J.M. and Rouse, H., 1941. Civ. Eng., Easton, Pa., 11(3): 169–171.
Robillard, L. and Kennedy, J.F., 1967. Proc. 12th Congr. Int. Assoc. Hydraul. Res., 1: 41–48.
Robinson, A.H.W., 1961. Geogr. J., 127: 63–77.
Roblee, L.H.S., Baird, T.M. and Tierney, J.W., 1958. AIChE J., 4: 460–464.
Rocke, F.A., 1971. Powder Technol., 4: 180–186.
Rodda, J.C., 1969. In: R.J. Chorley (Editor), Water, Earth and Man. Methuen, London.
Roep, T.B., Beets, D.J., Dronkert, H. and Pagnier, H., 1979. Sediment. Geol., 22: 135–163.
Roep, T.B., Holst, H., Vissers, R.L.M., Pagnier, H. and Postma, D., 1975. Palaeogeogr. Palaeoclimatol. Palaeoecol., 17: 289–308.
Rogers, C.A., 1958. Proc. Math. Soc. London, (3): 609–620.
Rogers, J.J.W. and Head, W.B., 1961. J. Sediment. Petrol., 31: 467–470.
Rohrer, E., 1954. Die Alpen, 30: 62–68.
Rohrer, E., 1955. Die Alpen, 31: 68–72.
Roll, H.U., 1965. Physics of the Marine Atmosphere. Academic Press, New York, N.Y.
Rose, H.E., 1945. Proc. Inst. Mech. Eng., 153: 141–148.
Rose, J., 1974. Proc. Geol. Assoc., 85: 223–237.

Rosen, R., 1970. Dynamical Systems Theory in Biology. Wiley, New York, N.Y.
Rosenbluth, M.N. and Rosenbluth, A.W., 1954. J. Chem. Phys., 22: 881–884.
Rosenhead, L., 1963. Laminar Boundary Layers. Clarendon Press, Oxford.
Rosenqvist, I.T., 1958. Geol. För. Stockh. Förh., 80: 433–457.
Rosenqvist, I.T., 1962. Clays Clay Miner., 9: 12–27.
Roshko, A., 1954a. NACA Rep. No. 1191.
Roshko, A., 1954b. NACA, Tech. Note, No. 3169.
Roshko, A., 1955. J. Aeronaut. Sci., 22: 124–132.
Roshko, A., 1961. J. Fluid Mech., 10: 345–356.
Rosin, P.O., 1937. Trans. Inst. Chem. Eng., 15: 167–184.
Ross, J.A., Barnes, F.H., Burns, J.G. and Ross, M.A.S., 1970. J. Fluid Mech., 43: 819–832.
Rossinsky, K.I. and Lyubomirova, K.S., 1970. Sov. Hydrol. Selected Pap., 1969(1): 38–49.
Rotenburg, A., 1965. J. Chem. Phys., 43: 4377–4381.
Round, G.F. and Newton, R., 1963. Nature (London), 198: 747–749.
Rouse, H., 1937. Trans. ASCE, 102: 463–543.
Rouse, H. 1961. Fluid Mechanics for Hydraulic Engineers. Dover Publ., New York.
Rouse, H., Siao, T.T. and Nagaratnam, S., 1959. Trans. ASCE, 124: 926–966.
Rowland, R.A., 1946. Bull. Geol. Soc. Am., 59: 547–564.
Rowley, K., 1978. Geol. Soc. am. Bull., 89: 825–835.
Rozovskii, I.L., 1961. Flow of Water in Bends of Open Channels. Israel Program for Scientific Translations, Jerusalem.
Rubatta, A., 1971. Meccanica, 6: 218–222.
Rubey, W.W., 1930. U.S. Geol. Surv., Prof. Pap., 165-A.
Rubey, W.W., 1933. Am. J. Sci., 225: 325–338.
Rubey, W.W., 1938. U.S. Geol. Surv., Prof. Pap., 189-E: 121–141.
Rubinow, S.I. and Keller, J.B., 1961. J. Fluid Mech., 11: 447–459.
Rücklin, H., 1934. Bad. Geol. Abh., 6: 81–100.
Rücklin, H., 1954. Jahresber. Mitt. Oberrhein. Geol. Ver., 35: 26–42.
Rudowski, S., 1970a. Acta Geol. Pol., 20: 451–483.
Rudowski, S., 1970b. Bull. Acad. Sci. Pol., Série Sci. Géol. Géogr., 18: 179–187.
Rudowski, S. and Tobolewski, J., 1973. Acta Geol. Pol., 23: 519–528.
Ruedemann, R., 1897. Am. Geol., 19: 367–391.
Ruegg, G.H.J., 1975. Meded. Rijks Geol. Dienst, N. S., 26(1): 17–37.
Ruegg, G.H.J. and Zandstra, J.G., 1977. Meded. Rijks Geol. Dienst, N. S., 28(4): 65–99.
Rumer, R.R., 1973. J. Hydraul. Div. ASCE, 99(HY3): 509–524.
Rukhin, L.B., 1958. Grundzüge der Lithologie. Akademie-Verlag, Berlin.
Rusnak, G.A., 1957a. J. Geol., 65: 384–409.
Rusnak, G.A., 1957b. J. Sediment. Petrol., 27: 41–55.
Russell, R.C.H. and Macmillan, D.H., 1952. Waves and Tides. Hutchinson's, London.
Russell, R.C.H. and Osorio, J.D.C., 1958. Proc. 6th Conf. Coastal Eng., pp. 171–183.
Rust, B.R., 1972a. Sedimentology, 18: 221–245.
Rust, B.R., 1972b. J. Sediment. Petrol., 42: 384–388.
Rust, B.R., 1975. In: A.V. Jopling and B.C. McDonald (Editors), Glaciofluvial and Glaciolacustrine Sedimentation. Soc. Econ. Paleontol. Mineral., Spec. Publ., 23: 238–248.
Rust, B.R. and Romanelli, R., 1975. In: A.V. Jopling and B.C. McDonald (Editors), Glaciofluvial and Glaciolacustrine Sedimentation. Soc. Econ. Paleontol. Mineral., Spec. Publ., 23: 177–192.
Rust, I.C., 1977. Sediment. Geol., 18: 123–133.
Rutgers, R., 1962. Nature (London), 193: 465–466.
Ryan, W.B.F. and Heezen, B.C., 1965. Geol. Soc. am. Bull., 76: 915–932.
Sabot, J., Saleh, I. and Comte-Bellot, G., 1977. Phys. Fluids, 20(10, II): S150–155.
Sachdev, P.S. and Seshadri, V.S., 1976. J. Fluid Mech., 78: 481–487.

Saffman, P.G., 1956. J. Fluid Mech., 1: 540–553.
Saffman, P.G., 1965. J. Fluid Mech., 22: 385–400.
Sage, W. and Tilly, G.P., 1969. J. R. Aeronaut. Soc., 73: 427–428.
Sager, G. and Sammler, R., 1975. Atlas der Gezeitenströme für die Nordsee, den Kanal und die Irische See. Seehydrographischer Dienst D.D.R., Rostock, 3rd ed.
Sahgal, P.P. and Singh, B., 1974. Irrig. Power, 30: 387–396.
Sakkas, J.G. and Strelkoff, T., 1973. J. Hydraul. Div. ASCE, 99(HY12): 2195–2216.
Salsburg, Z.W. and Wood, W.W.J., 1962. J. Chem. Phys., 37: 798–804.
Salsman, G.G., Tolbert, W.H. and Villars, R.G., 1966. Mar. Geol., 4: 11–19.
Samu, G., 1968. Mitteilungsbl. Bundesanst. Wasserbau, 26: 13–61.
Sander, B., 1930. Gefügekunde der Gesteine. Springer, Berlin.
Sanders, J.E., 1960. Geol. Mag., 97: 409–421.
Sanders, J.E., 1963. J. Sediment. Petrol., 33: 173–179.
Sanders, J.E., 1965. In: G.V. Middleton (Editor), Primary Sedimentary Structures and their Hydrodynamic Interpretation. Soc. Econ. Paleontol. Mineral., Spec. Publ., 12: 192–217.
Sanders, J.E., 1968. Geol. Soc. Am., Spec. Pap., 106: 265–305.
Sanders, J.E., Emery, K.O. and Uchupi, E., 1969. Geol. Soc. Am. Bull., 80: 561–572.
Sanderson, D.J. and Donovan, R.N., 1974. J. Sediment. Petrol., 44: 680–688.
Sanderson, I.D., 1974. Geol. Stud. Brigham Young Univ., 21: 215–246.
Sarkar, S.K. and Basumallick, S., 1968. J. Sediment. Petrol., 38: 747–754.
Sarnthein, M. and Walger, E., 1974. Geol. Rundsch., 63: 1065–1087.
Saunders, G.E., 1968. Geol. J., 6: 105–118.
Saunders, P.M., 1962. Weather, 17: 390–400.
Saunders, P.M., 1977. Deep-Sea Res., 24: 381–384.
Saunderson, H.C., 1976. J. Sediment. Petrol., 46: 761–769.
Saunderson, H.C., 1977. Z. Geomorphol., 21: 44–56.
Savage, S.B., 1978. Z. Angew. Math. Mech., 58: T313–T315.
Savage, S.B., 1979. J. Fluid Mech., 92: 53–96.
Sarfe, C.M., 1973. Nature (London), Phys. Sci., 241: 101–102.
Scheidegger, A.E., 1960. The Physics of Flow through Porous Media. University of Toronto Press, Toronto, Ont.
Scheidegger, A.E., 1975. Physical Aspects of Natural Catastrophes. Elsevier, Amsterdam.
Scheidegger, A.E. and Potter, P.E., 1965. Sedimentology, 5: 289–304.
Scheidegger, A.E. and Potter, P.E., 1968. Sedimentology, 11: 163–170.
Schenk, P.E., 1970. Geol. Assoc. Can., Spec. Pap., 7: 127–153.
Schiemenz, S., 1960. Geol. Jahrb., Beih., No. 38.
Schiffers, H., 1957. Trav. Inst. Rech. Sahariennes, 15: 65–88.
Schiffman, A., 1965. Limnol. Oceanogr., 10: 255–260.
Schijf, J.B. and Schönfeld, J.C., 1953. Proc. Minn. Int. Hydraul. Conv., pp. 321–333.
Schiller, L. and Naumann, A., 1933. Z. Dtsch. Ver. Ing., 77: 318–320.
Schindewolf, O.H. and Seilacher, A., 1955. Abh. Math.-Naturw. Kl. Akad. Wiss. Mainz, No. 10: 259–446.
Schlee, J., 1957. J. Sediment. Petrol., 27: 162–176.
Schleiger, N.W., 1968. J. Sediment. Petrol., 38: 262–272.
Schlichting, H., 1932. Phys. Z., 33: 327–335.
Schlichting, H., 1933. Nachr. Ges. Wiss. Göttingen, Math.-Phys. Kl., pp. 181–208.
Schlichting, H., 1935. Nachr. Ges. Wiss. Göttingen, Math.-Phys. Kl., 1: 47–68.
Schlichting, H., 1960. Boundary Layer Theory. McGraw-Hill, New York, N.Y., 4th ed.
Schlichting, H., 1961. Trans. ASME, E38: 289–300.
Schmidt, W., 1910. Akad. Wiss. Wien, (IIa): 119: 1101–1213.
Schmidt, W., 1911. Meteorol. Z., 8: 355–362.
Schmiedel, J., 1928. Phys. Z., 29: 593–610.

Schmincke, H.-U., 1967. J. Sediment. Petrol., 37: 438–448.
Schoewe, W.H., 1932. Am. J. Sci., (5)24: 111–134.
Schofield, R.K. and Samson, H.R., 1953. Clay Miner. Bull., 2: 45–51.
Schofield, R.K. and Samson, H.R., 1954. Discuss. Faraday Soc., 18: 135–145.
Schoklitsch, A., 1914. Über Schleppkraft und Geschiebebewegung. Engelman, Leipzig.
Schoklitsch, A., 1930. Handbuch des Wasserbaues. Springer, Vienna.
Schraub, F.A., Kline, S.J., Henry, V., Runstadler, P.W. and Littell, A., 1964. Trans. ASME, D87: 429–444.
Schreiber, J.F., Pine, G.L., Pipkin, B.W., Robinson, R.C. and Wilt, J.C., 1972. In: C.C. Reeves (Editor), Playa Lake Symposium, Lubbock, Texas. ICASALS Publ., No. 4: 133–184.
Schröder, L., 1965. Geol. Jahrb., 82: 655–704.
Schröter, H., 1933. Z. Ver. Dtsch. Ing., 77: 865–869.
Schubauer, G.B., 1958. In: H. Görtler (Editor), Boundary Layer Research. Springer, Berlin, pp. 85–107.
Schubauer, G.B. and Klebanoff, P.S., 1955. In: Boundary Layer Effects in Aerodynamics. National Physical Laboratory of Great Britain, H.M.S.O., pp. 41–42.
Schubauer, G.B. and Klebanoff, P.S., 1956. NACA Rep. No. 1289.
Schubauer, G.B. and Skramstad, H.K., 1947. J. Res. Natl. Bur. Stand., 38: 251–292.
Schubel, J.R., 1968. Science, 161: 1013–1015.
Schubel, J.R., 1969. Neth. J. Sea Res., 4: 283–309.
Schubert, G. and Corcos, G.M., 1967. J. Fluid Mech., 29: 113–135.
Schulz, W., 1961. Geologie, 10: 30–49.
Schumm, S.A., 1977. The Fluvial System. Wiley, New York, N.Y.
Schwarzacher, W., 1951. J. Sediment. Petrol., 21: 162–172.
Schwarzacher, W., 1963. J. Sediment. Petrol., 33: 580–586.
Schwenk, T., 1965. Sensitive Chaos. Steiner, London.
Schytt, V., 1962. Geogr. Ann., 44: 411–412.
Scoffin, T.P., Alexandersson, E.T., Bowes, G.E., Clokie, J.J. and Farrow, G.E., 1980. J. Sediment. Petrol., 50: 331–356.
Scorer, R.S., 1958. Natural Aerodynamics. Pergamon, London.
Scorer, R.S., 1978. Environmental Aerodynamics. Horwood, Chicester.
Scott, G.D., 1960. Nature (London), 188: 908–909.
Scott, G.D., 1962. Nature (London), 194: 956–957.
Scott, G.D. and Mader, D.L., 1964. Nature (London), 201: 382–383.
Scott, G.D. and Kilgour, D.M., 1969. Br. J. Appl. Phys., 2: 863–866.
Scott, G.D., Charlesworth, A.M. and Mak, M.K., 1964. J. Chem. Phys., 40: 611–612.
Scott, J.T. and Csanady, G.T., 1976. J. Geophys. Res., 81: 5401–5409.
Scott, K.M., 1967a. Scott. J. Geol., 3: 268–281.
Scott, K.M., 1967b. Int. Assoc. Sci. Hydrol., Publ. No. 75, pp. 309–318.
Scott, T., 1954. Tech. Mem. U.S. Beach Erosion Board, No. 48.
Scott Blair, G.W., 1965. Rheol. Acta, 4: 53–55.
Scrivenor, J.B., 1929. Geol. Mag., 66: 433–434.
Scruton, P.C. and Moore, D.G., 1953. Bull. Am. Assoc. Pet. Geol., 37: 1067–1074.
Searle, A.B. and Grimshaw, R.W., 1959. The Chemistry and Physics of Clays, Benn, London, 3rd ed.
Sedimentary Petrology Seminar, 1965. Sedimentology, 4: 273–283.
Seed, H.B. and Chan, C.K., 1959. J. Soil Mech. Found. Div. ASCE, 85(SM5): 87–128.
Seed, H.B. and Silver, M.L., 1972. J. Soil Mech. Found. Div. ASCE, 98(SM4): 381–397.
Segerstrom, K., 1950. Bull. U.S. Geol. Surv., 965-A.
Segré, G. and Silberberg, A., 1962. J. Fluid Mech., 14: 115–157.
Seilacher, A., 1953. Natur Volk, 83: 263–272.

Seilacher, A., 1959. Aus der Heimat, 67: 170–177.
Seilacher, A., 1960. Notizbl. Hess. Landesamt. Bodenforsch., Wiesbaden, 88: 88–106.
Seilacher, A., 1961. Natur Volk, 91: 257–264.
Seitz, R.C., 1971. J. Mar. Res., 29: 140–150.
Selby, M.J., Rains, R.B. and Palmer, R.W.P., 1974. N. Z. J. Geol. Geophys., 17: 543–562.
Self, S., 1976. J. Geol. Soc. London, 132: 645–666.
Self, S. and Booth, B., 1975. Geogr. Mag., 47: 447–450.
Self, S., Sparks, R.S.J., Booth, B. and Walker, G.P.L., 1974. Geol. Mag., 111: 539–548.
Seligman, G., 1936. Snow Structure and Ski Fields. Macmillan, London.
Sellard, H.C. and Draper, L., 1975. Estuarine Coastal Mar. Sci., 3: 219–228.
Sellards, E.H., 1923. Tex. Univ. Bull., No. 2327: 27–87.
Sellin, R.H.J., 1964. Houille Blanche, 19: 793–801.
Sellin, R.H.J., 1969. Flow in Channels. Macmillan, London.
Sellwood, B.W., 1972. Palaeogeogr. Palaeoclimatol. Palaeoecol., 11: 93–106.
Sellwood, B.W., 1975. In: R.N. Ginsburg (Editor), Tidal Deposits. Springer, New York, N.Y., pp. 93–101.
Sempere, R., 1969. Powder Technol., 3: 1–8.
Sengupta, S., 1966. J. Sediment. Petrol., 36: 362–369.
Seppälä, M., 1971. Fennia, No. 104: 1–88.
Seppälä, M. 1972. Geogr. Ann., A54: 85–104.
Seppälä, M. and Lindé, K., 1978. Geogr. Ann., A60: 29–42.
Sestini, G. and Pranzini, G., 1965. J. Sediment. Petrol., 35: 100–108.
Sevenet, Lieut., 1943. Bull. Inst. Fr. Afr. Noire, 5: 1–26.
Shackleton, J.S., 1962. Liverpool Manchester Geol. J., 3: 109–118.
Shal'nev, K.K., Varga, I.I. and Sebestyn, D., 1966. Philos. Trans. R. Soc., London, Ser. A., 260: 256–266.
Shamov, G.I., 1959. Alluvial Rivers. Gidrometeorizdat, Leningrad, 2nd ed. (in Russian).
Shantzer, E.V., 1951. Tr. Inst. Geol. Nauk, Akad. Nauk SSSR, Geol. Ser., 1951(135): 1–271 (in Russian).
Sharp, J.J., 1971. J. Hydraul. Div. ASCE, 97(HY9): 1471–1492.
Sharp, R.P., 1963. J. Geol., 71: 617–636.
Sharp, R.P., 1964. Geol. Soc. Am. Bull., 75: 785–804.
Sharp, R.P., 1966. Geol. Soc. Am. Bull., 77: 1045–1074.
Sharp, R.P. and Nobles, L.H., 1953. Bull. Geol. Soc. Am., 64: 547–560.
Sharp, W.E., 1960. J. Geol., 68: 567–572.
Sharpe, C.F.S., 1938. Landslides and Related Phenomena. Columbia University Press, New York, N.Y.
Shaw, D.M., Watkins, N.D. and Huang, T.C., 1974. J. Geophys. Res., 79: 3087–3094.
Shaw, H.R., 1963. J. Geophys. Res., 68: 3637–3643.
Shaw, H.R., 1972. Am. J. Sci., 272: 870–893.
Shaw, J., 1975. In: A.V. Jopling and B.C. McDonald (Editors), Glaciofluvial and Glaciolacustrine Sedimentation. Soc. Econ. Paleontol. Mineral., Spec. Publ., 23: 281–303.
Shaw, J., 1977. Geogr. Ann., A59: 221–240.
Shaw, J. and Freschauf, R.C., 1973. Can. Geogr., 17: 19–35.
Shaw, J. and Kellerhals, R., 1977. J. Sediment. Petrol., 47: 257–266.
Shaw, M.C., 1954. Proc. Natl. Acad. Sci., 40: 394–401.
Shaw, M.S., Hopkins, J.S. and Caton, P.G.F., 1976. Weather, 31: 172–183.
Shaw, W.B.K., 1936. Geogr. J., 87: 193–217.
Shearer, J.M., MacNab, R.F., Pelletier, B.R. and Smith, T.B., 1971. Science, 174: 816–818.
Sheldon, G.L., 1970. Trans. ASME, D92: 619–626.
Sheldon, G.L. and Finnie, I., 1966a. Trans ASME, B88: 387–392.
Sheldon, G.L. and Finnie, I., 1966b. Trans. ASME, B88: 393–399.

Sheldon, G.L. and Kanhere, A., 1972. Wear, 21: 195–209.
Sheldon, P.G., 1928. Am. J. Sci., (5)15: 243–252.
Sheldon, P.G., 1929. Am. J. Sci., 5(17): 525–533.
Sheldon, R.W., 1968. Limnol. Oceanogr., 13: 72–83.
Shelton, J.W. and Mack, D.E., 1970. Bull. Am. Assoc. Pet. Geol., 54: 1108–1119.
Shen, H.W. and Cheong, H.F., 1977. J. Hydraul. Div. ASCE, 103(HY11): 1303–1321.
Shen, H.W. and Roper, A.T., 1970. Bull. Inst. Assoc. Sci. Hydrol., 15(1): 7–26.
Shen, S.F., 1954. J. Aeronaut. Sci., 21: 62–64.
Shepard, F.P., 1951. Trans. Am. Geophys. Union, 32: 405–418.
Shepard, F.P., 1963. Submarine Geology. Harper and Row, New York, 2nd ed.
Shepard, F.P., Dill, R.F. and Von Rad, U., 1969. Bull. Am. Assoc. Pet. Geol., 53: 390–420.
Shergold, F.A., 1953. Mag. Concr. Res., 5(13): 3–10.
Sherman, P., 1970. Industrial Rheology. Academic Press, London.
Shields, A., 1936. Mitt. Preuss. Versuchsanst. Wasserbau Schiffbau, No. 26.
Shinn, E.A., 1973. In: B.H. Purser (Editor), The Persian Gulf. Springer, Berlin, pp. 199–209.
Shinohara, K. and Tsubaki, T., 1959. Rep. Res. Inst. Appl. Mech., Kyushu Univ., 7(25): 15–45.
Shipek, C.J., 1962. Geol. Soc. Am. Bull., 73: 805–812.
Shirasuna, T., 1973. Proc. 15th Conf. Int. Assoc. Hydraul. Res., 1: 107–114.
Shizgal, B., Goldsmith, H.L. and Mason, S.G., 1965. Can. J. Chem. Eng., 43: 97–101.
Shonting, D.H., 1968. J. Mar. Res., 26: 43–65.
Shotton, F.W., 1937. Geol. Mag., 74: 534–553.
Shotton, F.W., 1956. Liverpool Manchester Geol. J., 1: 450–465.
Shrock, R.R., 1948. Sequence in Layered Rocks. McGraw-Hill, New York, N.Y.
Shulyak, B.A., 1963. Deep-Sea Res., 10: 488–497.
Shutler, N.D. and Mesler, R.B., 1965. Trans ASME, D87: 511–517.
Siau, M., 1841. C. R. Acad. Sci., 12: 774–775.
Sides, G. and Barden, L., 1971. Can. Geotech. J., 8: 391–399.
Sidwell, R. and Tanner, W.P., 1938. Bull. Geol. Soc. Am., 49: 139 (abstr.).
Siever, R., 1966. Bull. Am. Assoc. Pet. Geol., 50: 635 (abstr.).
Silva, A.J. and Hollister, C.D., 1973. J. Geophys. Res., 78: 3597–3616.
Silvester, R., 1970. Civ. Eng. Trans., CE12(1): 63–71.
Silvester, R., 1972. Proc. Inst. Civ. Eng., 51: 123–131.
Silvester, R. and Mogridge, G.R., 1971. Proc. 12th Conf. Coastal Eng., 2: 651–667.
Silvester, R. and Vongvisessomjai, S., 1971. Proc. Inst. Civ. Eng., 48: 259–283.
Simon, J.A. and Hopkins, M.C., 1966. Guideb. Ser. Ill. State Geol. Surv., No. 7.
Simonett, D.S., 1960. Ann. Am. Assoc. Geogr., 50: 216–241.
Simons, D.B., 1971. In: H.W. Shen (Editor), River Mechanics, vol. 2. Colorado State University, Fort Collins, Colo., pp. 20-1–20-60.
Simons, D.B. and Richardson, E.V., 1961. J. Hydraul. Div. ASCE, 87(HY3): 87–105.
Simons, D.B. and Richardson, E.V., 1962. Water Supply Irrig. Pap., Wash., 1498-E.
Simons, D.B. and Richardson, E.V., 1965. Misc. Publ. U.S. Dep. Agric., No. 970: 193–207.
Simons, D.B. and Richardson, E.V., 1966. U.S. Geol. Surv., Prof. Pap., 422-J.
Simons, D.B. and Richardson, E.V., 1971. In: H.W. Shen (Editor), River Mechanics, vol. 1. Colorado State University, Fort Collins, Colo., pp. 9-1–9-89.
Simons, D.B., Richardson, E.V. and Albertson, M.C., 1961. Water-Supply Irrig. Pap., Wash., 1498-A.
Simons, D.B., Richardson, E.V. and Nordin, C.F., 1965a. U.S. Geol. Surv., Prof. Pap., 462-H.
Simons, D.B., Richardson, E.V. and Nordin, C.F., 1965b. In: G.V. Middleton (Editor), Primary Sedimentary Structures and their Hydrodynamic Interpretation. Soc. Econ. Paleontol. Mineral., Spec. Publ., 12: 34–52.
Simons, F.S., 1956. J. Geol., 64: 517–521.

Simons, F.S. and Ericksen, G.E., 1953. Bol. Soc. Geol. Peru, 26: 229–246.
Simpson, J.E., 1964. Weather, 19: 208–220.
Simpson, J.E., 1967. Weather, 22: 306–327.
Simpson, J.E., 1969. Q. J. R. Meteorol. Soc., 95: 758–765.
Simpson, J.E., 1972. J. Fluid Mech., 53: 759–768.
Simpson, J.E. and Britter, R.E., 1980. Q. J. R. Meteorol. Soc., 106: 485–500.
Simpson, J.E., Mansfield, D.A. and Milford, J.R., 1977. Q. J. R. Meteorol. Soc., 103: 47–76.
Simpson, R.L., Strickland, J.H. and Barr, P.W., 1977. J. Fluid Mech., 79: 553–594.
Singh, B. and Shah, C.R., 1971. Houille Blanche, 26: 59–64.
Singh, I.B., 1969. Norsk Geol. Tiddskr., 49: 1–33.
Singh, I.B., 1972a. Sediment. Geol., 7: 309–317.
Singh, I.B., 1972b. Natur Mus., 102: 50–62.
Singh, I.B., 1980. Sediment. Geol., 25: 83–103.
Singh, I.B. and Kumar, S., 1974. Sediment. Geol., 12: 53–66.
Singh, I.B. and Wunderlich, F., 1978. Senckenbergiana Marit., 10: 75–83.
Singh, M.P. and Gupta, J.L., 1971. Z. Angew. Math. Mech., 51: 27–32.
Sioli, H., 1965. Naturwissenschaften, 53: 104.
Sitler, R.F., 1968. Proc. 23rd Int. Geol. Congr., Part 8, pp. 283–295.
Sitler, R.F. and Chapman, C.A., 1955. J. Sediment. Petrol., 25: 262–269.
Skelland, A.H.P., 1967. Non-Newtonian Flow and Heat Transfer. Wiley, New York, N.Y.
Skipper, K., 1971. Sedimentology, 17: 51–68.
Skipper, K. and Bhattacharjee, S.B., 1978. J. Sediment. Petrol., 48: 193–202.
Slattery, J.C. and Bird, R.B., 1961. Chem. Eng. Sci., 16: 231–241.
Slaughter, M. and Earley, J.W., 1965. Geol. Soc. Am., Spec. Pap., No. 83.
Slaughter, M. and Hamil, M., 1970. Geol. Soc. Am. Bull., 81: 961–968.
Sleath, J.F.A., 1970. J. Fluid Mech., 42: 111–123.
Sleath, J.F.A., 1973. J. Hydraul. Res., 11: 369–383.
Sleath, J.F.A., 1974a. ASCE J. Waterw. Harbors Coastal Eng. Div., 100(WW2): 105–122.
Sleath, J.F.A., 1974b. J. Mar. Res., 32: 13–24.
Sleath, J.F.A., 1975a. Proc. Inst. Civ. Eng., 59: 309–322.
Sleath, J.F.A., 1975b. J. Hydraul. Res., 13: 315–328.
Sleath, J.F.A., 1976. J. Hydraul. Res., 14: 69–81.
Sleath, J.F.A., 1978. ASCE J. Waterw. Harbors Coastal Eng. Div., 104(WW4): 291–307.
Sleath, J.F.A. and Ellis, A.C., 1978. Ripple geometry in oscillatory flow. Univ. Cambridge, Dep. Eng., Rep. A-Hydraulics/TR2.
Slichter, C.S., 1899. Ann. Rep. U.S. Geol. Surv., Part II, pp. 295–384.
Sloane, R.L. and Kell, T.R., 1966. Clays Clay Miner., 26: 289–296.
Smalley, I.J., 1962. Nature (London), 194: 1271.
Smalley, I.J., 1964. Am. J. Sci., 262: 242–248.
Smalley, I.J., 1970. Powder Technol., 4: 97–101.
Smalley, I.J. and Cabrera, J.G., 1969. Nature (London), 222: 80–81.
Smart, P., 1967. Clays Clay Miner., 27: 241–254.
Smekal, A. and Klem, W., 1951. Monatsh. Chem., 82: 411–421.
Smeltzer, C.E., Gulden, M.E. and Compton, W.A., 1970. Trans. ASME, D92: 639–652.
Smerdon, E.T. and Beasley, R.P., 1959. Bull. Mo. Agric. Exp. Stn., No. 715.
Smith, A.J., 1973. Nature (London), Phys. Sci., 246: 10–11.
Smith, A.L., 1909. Bull. Am. Geogr. Soc., 41: 729–742.
Smith, E.B. and Lea, K.R., 1960. Nature (London), 186: 714.
Smith, G. LeB. and Eriksson, K.A., 1979. Palaeogeogr. Palaeoclimatol. Palaeoecol., 27: 67–84.
Smith, H.T.U., 1946. Trans. N.Y. Acad. Sci., (2)8: 197–199.
Smith, H.T.U., 1963. Eolian geomorphology, wind direction, and climatic change in North

Africa. U.S. Air Force, Cambridge Res. Lab., Rep. 63-443.
Smith, H.T.U., 1965a. Geol. Soc. Am., Spec. Pap., No. 87: 160 (abstr.).
Smith, H.T.U., 1965b. J. Geol., 73: 557–578.
Smith, H.T.U., 1968. In: C.B. Schultz and J.C. Frye (Editors), Loess and Related Eolian Deposits of the World. Proc. 7th INQUA Congr., 12: 29–47.
Smith, H.T.U., 1969. Photo-interpretation studies of desert basins in northern Africa. U.S. Air Force, Cambridge Res. Lab., Rep. 68-0590.
Smith, J.D., 1969. J. Geol., 77: 39–55.
Smith, J.D., 1970. J. Geophys. Res., 75: 5928–5940.
Smith, L.S., 1976. Stud. Geol. Brigham Young Univ., 23: 113–171.
Smith, N.D., 1970. Geol. Soc. Am. Bull., 81: 2993–3014.
Smith, N.D., 1971a. J. Sediment. Petrol., 41: 69–73.
Smith, N.D., 1971b. Geol. Soc. Am. Bull., 82: 3407–3420.
Smith, N.D., 1972. J. Sediment. Petrol., 42: 624–634.
Smith, N.D., 1974. J. Geol., 82: 205–223.
Smith, W.O., Foote, P.D. and Busang, R.F., 1929. Phys. Rev., 34: 1271–1274.
Sneed, E.D. and Folk, R.L., 1958. J. Geol., 66: 114–150.
Snider, R.H. and Chakrabarti, S.K., 1973. J. Geophys. Res., 78: 8793–8807.
Snischenko, B.F., 1968. Soviet Hydrology Selected Papers for 1966, pp. 486–493.
So, C.L., Pierce, J.W. and Siegel, F.R., 1974. Geogr. Ann., A56: 227–235.
Sohn, H.Y. and Moreland, C., 1968. Can. J. Chem. Eng., 46: 162–167.
Solle, G., 1966. Notizbl. Hess. Land. Bodenforsch. Wiesbaden, 95: 54–121.
Sommer, F. and Soeder, H., 1963. Arch. Eisenhüttenw., 34: 1–8.
Sommerfield, R. and Businger, J.A., 1965. J. Geophys. Res., 70: 3303–3306.
Sonntag, G., 1960. Chem. Ing. Tech., 32: 317–329.
Soo, S.L., 1967. Fluid Dynamics of Multiphase Systems. Blaisdell, Waltham, Mass.
Soo, S.L., 1977. Powder Technol., 17: 259–263.
Sorby, H.C., 1859. Geologist, 2: 137–147.
Sorby, H.C., 1908. Q. J. Geol. Soc. London, 64: 171–233.
Southard, J.B., 1967. Science, 156: 1725–1727.
Southard, J.B., 1971a. J. Sediment. Petrol., 41: 320–324.
Southard, J.B., 1971b. J. Sediment. Petrol., 41: 903–915.
Southard, J.B. and Boguchwal, L.A., 1973. J. Sediment. Petrol., 43: 1114–1121.
Southard, J.B. and Cacchione, D.A., 1972. In: D.J.P. Swift, D.B. Duane and O.H. Pilkey (Editors), Shelf Sediment Transport: Process and Pattern. Dowden, Hutchinson and Ross, Stroudsburg, Penn., pp. 83–97.
Southard, J.B. and Dingler, J.R., 1971. Sedimentology, 16: 251–263.
Southard, J.B., Young, R.A. and Hollister, C.D., 1971. J. Geophys. Res., 87: 5903–5909.
Spalletti, L.A., 1968. Rev. Mus. La Plata, 6: 235–272.
Sparks, R.S.J. and Walker, G.P.L., 1973. Nature (London), Phys. Sci., 241: 62–64.
Sparks, R.S.J. and Wilson, L., 1976. J. Geol. Soc. London, 132: 441–451.
Sparks, R.S.J., Wilson, L. and Hulme, G., 1978. J. Geophys. Res., 83: 1727–1739.
Sparks, R.S.J., Wilson, L. and Sigurdsson, H., 1981. Philos. Trans. R. Soc. London, Ser. A., 229: 241–273.
Spencer, A.M., 1971. Mem. Geol. Soc. London, No. 6.
Spielvogel, E.R. and Spielvogel, L.Q., 1974. J. Geophys. Res., 79: 2752–2754.
Spjeldnaes, N., 1965. Geol. Rundsch., 54: 24–45.
Spotts, J.H., 1964. J. Sediment. Petrol., 34: 229–253.
Spotts, J.H. and Weser, O.E., 1964. In: A.H. Bouma and A. Brouwer (Editors), Turbidites. Elsevier, Amsterdam, pp. 199–221.
Sprigg, R.G., 1978. Sediment. Geol., 22: 53–96.
Spurr, J.E., 1894a. Am. Geol., 13: 43–47.

Spurr, J.E., 1894b. Am. Geol., 13: 201–206.
Squarer, D., 1970. J. Hydraul. Div. ASCE, 96 (HY4): 995–1017.
Srodon, J., 1974. Ann. Soc. Géol. Pol., 44: 449–473.
Stainthorp, F.P. and Allen, J.M., 1965. Trans. Ins. Chem. Eng., 43: T85–T91.
Stander, G.H., Shannon, L.V. and Campbell, J.A., 1969. J. Mar. Res., 27: 293–300.
Standish, N. and Burger, D.E., 1979. Powder Technol., 22: 121–125.
Stanley, D.J., 1968. Geol. Soc. Am., Spec. Pap., No. 106: 211–239.
Stanley, D.J., 1974. Bull. Cent. Rech. Pau, 8: 351–371.
Stanley, D.J. and Kelling, G., 1968. Oceanogr. Rep. U.S. Coast Guard, No. 22.
Stanley, D.J. and Silverberg, N., 1969. Earth Planet. Sci. Lett., 6: 123–133.
Stanley, D.J. and Swift, D.J.P., 1968. Mar. Geol., 6: 479–500.
Stanley, D.J. and Swift, D.J.P., 1976. Marine Sediment Transport and Environmental Management. Wiley, New York, N.Y.
Stanley, D.J. and Taylor, P.T., 1977. Mar. Geol., 23: 77–88.
Stanley, K.O., 1974. J. Sediment. Petrol., 44: 472–483.
Stanley, S.M., 1970. Mem. Geol. Soc. Am., No. 125.
Statham, I., 1974. Sedimentology, 21: 149–162.
Statham, I., 1976. Earth Surf. Processes, 1: 173–180.
Statham, I., 1977. Earth Surf. Processes, 2: 437–440.
Stauffer, M.R., Hajnal, Z. and Gendzwill, D.J., 1976. Can. J. Earth Sci., 13: 1667–1677.
Stauffer, P.H., 1967. J. Sediment. Petrol., 37: 487–508.
Stear, W.M., 1979. Trans. Geol. Soc. S. Afr., 81: 393–399.
Steedman, R.A. and Ashour, Y., 1976. Tellus, 28: 299–306.
Steenstrup, K.J.V., 1894. Medd. Dan. Geol. Foren., 1: 1–14.
Steidtmann, J.R., 1974. Geol. Soc. Am. Bull., 85: 1835–1842.
Stein, R.A., 1965. J. Geophys. Res., 70: 1831–1842.
Steinour, H.H., 1944. Ind. Eng. Chem., 36: 618–624, 840–847, 901–907.
Stephens, C.G. and Crocker, R.L., 1946. Trans. R. Soc. S. Aust., 70: 303–312.
Sternberg, J., 1962. J. Fluid Mech., 13: 241–271.
Sternberg, R.W., 1967. Mar. Geol., 5: 195–205.
Sternberg, R.W., 1971. Mar. Geol., 10: 113–119.
Sternberg, R.W., 1972. In: D.J.P. Swift, D.B. Duane and O.H. Pilkey (Editors), Shelf Sediment Transport: Process and Pattern. Dowden, Hutchinson and Ross, Stroudsburg, Penn., pp. 61–82.
Sternberg, R.W. and Larsen, L.H., 1975. Deep-Sea Res., 22: 299–309.
Stetson, H.C. and Smith, J.F., 1937. Am. J. Sci., 35: 1–13.
Stevens, M.A., Simons, D.B. and Richardson, E.V., 1975. J. Hydraul. Div. ASCE, 101 (HY5): 557–566.
Stewart, D.A., 1951. The Design and Placing of High Quality Concrete. Spon, London.
Stewart, H.G. and Jordan, G.F., 1964. In: R.L. Miller (Editor), Papers in Marine Geology. Macmillan, New York, N.Y., pp. 102–114.
Stewart, J.H., 1961. U.S. Geol. Surv., Prof. Pap., 424-B: 127–129.
Stewart, J.H., Poole, F.G. and Wilson, R.F., 1972. U.S. Geol. Surv., Prof. Pap., 690.
Stoa, P.N., 1978. Tech. Pap. U.S. Coastal Eng. Res. Cent., 78-2.
Stoddart, D.R., 1962. Nature (London), 196: 512–515.
Stoker, J.J., 1957. Water Waves. Interscience, New York, N.Y.
Stoker, R.L., 1949. Ind. Eng. Chem., 41 (II): 1196–199.
Stokes, G.G., 1847. Trans. Cambridge Philos. Soc., Math. Phys. Sci., 8: 441–473.
Stokes, G.G., 1851. Trans. Cambridge Philos. Soc., Math. Phys. Sci., 9: 8–27.
Stokes, W.L., 1947. J. Geol., 55: 52–54.
Stokes, W.L., 1950. J. Geol., 58: 153–154.
Stokes, W.L., 1953. U.S. At. Energy Comm. Rep. RME 3043.

Stokes, W.L., 1961. In: J.A. Peterson and J.C. Osmond (Editors), Geometry of Sandstone Bodies. American Association of Petroleum Geologists, Tulsa, Okla., pp. 151–178.
Stokes, W.L., 1968. J. Sediment. Petrol., 38: 510–515.
Stone, R. and Vondra, C.F., 1972. J. Sediment. Petrol., 32: 227–229.
Stone, R.O. and Summers, H.J., 1972. Study of subaqueous and subaerial sand ripples. Univ. S. Calif., Dep. Geol. Sci., Final Rep., 72-1.
Straub, L.G., 1935. Missouri River. Appendix XV, House Document No. 238, 73rd Congr. U.S., 2nd Sess., pp. 1032–1245.
Straub, L.G., 1936. Civ. Eng., Eaton, Pa., 6: 321–323.
Straub, L.G., 1939. Trans. Am. Soc. Civ. Eng., 105: 1298–1303.
Street, W.B. and Tildesley, D.J., 1976. Proc. R. Soc. London, Ser. A., 348: 485–510.
Strickland, J.H. and Simpson, R.L., 1975. Phys. Fluids, 18: 306–308.
Stricklin, F.L. and Smith, C.I., 1973. Geol. Soc. Am. Bull., 84: 1349–1368.
Stride, A.H., 1959. Dock Harbor Auth., 40: 145–147.
Stride, A.H., 1961. New Sci., 10: 304–306.
Stride, A.H., 1963. Q. J. Geol. Soc. London, 119: 175–197.
Stride, A.H., 1970. Geol. Mag., 107: 469–477.
Stride, A.H., 1973. In: E.D. Goldberg (Editor), North Sea Science. Massachusetts Institute of Technology Press, Cambridge, Mass., pp. 101–130.
Stride, A.H. and Cartwright, D.E., 1958. Dock Harbor Auth., 38: 323–324.
Stride, A.H. and Tucker, M.J., 1960. Nature (London), 188: 933.
Stride, A.H., Belderson, R.H. and Kenyon, N.H., 1972. Mem. Bur. Rech. Géol. Min., No. 79: 233–240.
Stringham, G.E., Simons, D.B. and Guy, H.P., 1969. U.S. Geol. Surv., Prof. Pap., 562-C.
Stuart, J.T., 1962. Adv. Aeronaut. Sci., 3: 121–142.
Stuart, J.T., 1963. In: L. Rosenhead (Editor), Laminar Boundary Layers. Clarendon Press, Oxford, pp. 349–488.
Stuart, J.T., 1965. Appl. Mech. Rev., 18: 523–531.
Stuart, J.T., 1966. J. Fluid Mech., 24: 673–687.
Stückrath, T., 1969. Mitt. Franzius Inst., 32: 267–293.
Sugden, D.E. and John, B.S., 1976. Glaciers and Landscape. Arnold, London.
Sullwold, H.H., 1960. Bull. Am. Assoc. Pet. Geol., 44: 433–457.
Sumer, B.M., 1977. J. Hydraul. Div. ASCE, 103 (HY11): 1323–1337.
Sundborg, Å., 1955. Geogr. Ann., 37: 94–111.
Sundborg, Å., 1956. Geogr. Ann., 38: 127–316.
Suquet, F., 1950. Houille Blanche, 5: 342–361.
Surlyk, F., Callomon, J.H., Bromley, R.G. and Birkelund, T., 1973. Bull. Grønlands Geol. Unders., No. 105.
Susskind, H. and Becker, W., 1966. Nature (London), 212: 1564–1565.
Sutherland, A.J., 1967. J. Geophys. Res., 72: 6183–6194.
Sutherland, D.N. and Goodarz-Nia, I., 1971. Chem. Eng. Sci., 26: 2071–2085.
Sutherland, D.N. and Tan, C.T., 1970. Chem. Eng. Sci., 25: 1948–1950.
Sutherland, W., 1893. Philos. Mag., (5) 36: 507–531.
Sutton, J. and Watson, J., 1960. Geol. Mag., 97: 102–122.
Svasek, J.N. and Terwindt, J.H.J., 1974. Sedimentology, 21: 211–322.
Swett, K., Klein, G. deV. and Smit, D.E., 1971. J. Geol., 79: 400–415.
Swett, K. and Smit, D.E., 1972. Geol. Soc. Am. Bull., 83: 3223–3248.
Swift, D.J.P. and Freeland, G.L., 1978. J. Sediment. Petrol., 48: 1257–1266.
Swift, D.J.P. and Lyall, A.K., 1967. Marit. Sediments, 3: 67–70.
Swift, D.J.P. and McMullen, R.M., 1968. Can. J. Earth Sci., 5: 175–183.
Swift, D.J.P., Cok, A.E. and Lyall, A.K., 1966. Marit. Sediments, 2: 175–180.
Swift, D.J.P., Duane, D.B. and Pilkey, O.H., 1972. Shelf Sediment Transport: Process and

Pattern. Dowden, Hutchinson and Ross, Stroudsburg, Penn.
Swift, D.J.P., Sears, P.C., Bohlke, B. and Hunt, R., 1978. Mar. Geol., 27: 19—42.
Swift, D.J.P., Freeland, G.L. and Young, R.A., 1979. Sedimentology, 26: 389–406.
Swift, E., 1962. Tech. Rep. U.S. Naval Ordnance Lab., 62-191.
Sykes, R.M., 1974. Bull. Geol. Soc. Den., 23: 213–224.
Tabakoff, W., Kotwal, R. and Hamed, A., 1979. Wear, 52: 161–173.
Tailby, S.R. and Portalski, S., 1960. Trans. Inst. Chem. Eng., 38: 324–334.
Tailby, S.R. and Portalski, S., 1962a. Chem. Eng. Sci., 17: 283–290.
Tailby, S.R. and Portalski, S., 1962b. Trans. Inst. Chem. Eng., 40: 114–122.
Tan, T.-K., 1959. Sci. Sin., 8: 93–97.
Tanaka, K., 1970. Rep. Geol. Surv. Jpn., No. 236.
Tani, I., 1967. Phys. Fluids, 10: S11–S16.
Tani, I., 1969. Annu. Rev. Fluid Mech., 1: 169–196.
Tani, I. and Hama, F.R., 1953. J. Aeronaut. Sci., 29: 289–290.
Tankard, A.J. and Hobday, D.K., 1977. Sediment. Geol., 18: 135–159.
Tanner, W.F., 1958. J. Sediment. Petrol., 28: 95–96.
Tanner, W.F., 1959. J. Sediment. Petrol., 29: 564–574.
Tanner, W.F., 1960. J. Sediment. Petrol., 30: 481–485.
Tanner, W.F., 1962. Trans. Gulf-Coast Assoc. Geol. Soc., 12: 295–301.
Tanner, W.F., 1963. Sedimentology, 2: 307–311.
Tanner, W.F., 1967. Sedimentology, 9: 89–104.
Tanner, W.F., 1971. Sedimentology, 16: 71–88.
Tanner, W.F., 1974. Guideb. 25th Field Conf., N.M. Geol. Soc., pp. 219–223.
Tansey, V.O., 1953. J. Sediment. Petrol., 23: 62–64.
Taylor, B.D., 1971. Temperature effects in alluvial streams. Calif. Inst. Technol., W.M. Keck Lab. Hydraul. Water Resour., Rep. KH-R-27.
Taylor, B.D. and Vanoni, V.A., 1972a. J. Hydraul. Div. ASCE, 98 (HY8): 1427–1445.
Taylor, B.D. and Vanoni, V.A., 1972b. J. Hydraul. Div. ASCE, 98 (HY12): 2191–2206.
Taylor, G.A., 1958. Bull. Aust. Bur. Min. Res. Geol. Geophys., No. 38.
Taylor, G.I., 1917. Proc. R. Soc. London, Ser. A., 93: 99–113.
Taylor, G.I., 1923. Proc. R. Soc. London, Ser. A, 103: 58–61.
Taylor, G.I., 1946. Proc. R. Soc. London, Ser. A., 187: 16–18.
Taylor, J.M., 1950. Bull. Am. Assoc. Pet. Geol., 34: 701–716.
Taylor, P.A., 1977. Boundary-Layer Meteorol., 11: 439–465.
Taylor, P.A. and Dyer, K.R., 1977. In: E.D. Goldberg, I.N. McCave and J.J. O'Brien (Editors), The Sea, vol. 6. Wiley, New York, N.Y., pp. 579–601.
Taylor, P.A., Gent, P.R. and Keen, J.M., 1976. Geophys. J. R. Astron. Soc., 44: 177–201.
Taylor, P.T., Stanley, D.J. and Jahn, W., 1975. Mar. Geol., 19: 139–157.
Teisseyre, A.K., 1975. Geol. Sudetica, 10: 7–56.
Teisseyre, A.K., 1978. Geol. Sudetica, 13: 89–92.
Temple, P.H. and Sundborg, Å., 1973. Geogr. Ann., A54: 345–368.
Ten Haaf, E., 1959. Graded Beds of the Northern Apennines. University of Groningen, The Netherlands.
Terwindt, J.H.J., 1970. Geol. Mijnbouw, 49: 489–501.
Terwindt, J.H.J., 1971a. Mar. Geol., 10: 51–67.
Terwindt, J.H.J., 1971b. Geol. Mijnbouw, 50: 515–526.
Terwindt, J.H.J. and Breusers, H.N.C., 1972. Sedimentology, 19: 85–98.
Terwindt, J.H.J., De Jong, J.D. and Van der Wilk, E., 1963. Verh. K. Ned. Geol.-Mijnbouwk. Genoot., Geol. Ser., 21/22: 243–258.
Terzaghi, K., 1925. Erdbaumechanik auf Bodenphysikalischer Grundlage. Deuticke, Wien.
Tesaker, E., 1969a. Proc. 12th Congr. Int. Assoc. Hydraul. Res., 2: 1–8.
Tesaker, E., 1969b. Uniform turbidity current experiments. Technical University of Norway.

Thadani, M.C. and Peebles, F.N., 1966. Ind. Eng. Chem. Proc. Res. Dev., 5: 265–268.
Theodorsen, T., 1955. In: H. Görtler and W. Tollmien (Editors), 50 Years of Boundary Layer Research. Vieweg, Braunschweig, pp. 55–62.
Thiel, G.A., 1932. J. Geol., 40: 452–458.
Thom, B.G., 1967. Coastal and fluvial landforms: Horry and Marion Counties, South Carolina. La. State Univ., Coastal Stud. Inst., Tech. Rep., No. 44.
Thom, H.S.C., 1973. ASCE J. Waterw., Harbor Coastal Eng. Div., 99 (WW3): 355–374.
Thompson, D.B., 1969. Sediment. Geol., 3: 263–289.
Thompson, E.F., 1977. Tech. Rep. U.S. Coastal Eng. Res. Cent., 77-1.
Thompson, E.F. and Harris, D.L., 1972. Reprint U.S. Coastal Eng. Res. Cent., No. 1-72.
Thompson, R.N. and Patrick, D.J., 1968. Geol. J., 6: 139–146.
Thompson, R.W., 1968. Geol. Soc. Am. Mem., No. 107.
Thompson, W.O., 1937. Bull. Geol. Soc. Am., 48: 723–752.
Thompson, W.O., 1949. Bull. Am. Assoc. Pet. Geol., 33: 52–72.
Thomson, W., 1889. Proc. R. Soc. Edinburgh, 16: 693–724.
Thorn, M.F.C., 1975a. HRS Notes, Hydraul. Res. Stn., Wallingford, June 1975: 2–3.
Thorn, M.F.C., 1975b. Proc. 16th Congr. Int. Assoc. Hydraul. Res., pp. 217–223.
Thornton, R.B., 1973. Proc. 13th Cong. Coastal Eng., 1: 1049–1068.
Thornton, E.B. and Krapohl, R.F., 1974. J. Geophys. Res., 79: 847–852.
Thoulet, M., 1887. Ann. Mines, (8) 11: 199–224.
Thwaites, B., 1960. Incompressible Aerodynamics. Clarendon Press, Oxford.
Tickell, F.G., Mechem, O.E. and McCurdy, R.C., 1933. Trans. Am. Inst. Min., Metall. Pet. Eng., 103: 250–260.
Tietjens, O.G., 1925. Z. Angew. Math. Mech., 5: 200–217.
Tietze, K.W., 1975. Geol. Rundsch., 64: 100–118.
Tietze, K.W., 1978. Geol. Rundsch., 67: 1016–1033.
Tilly, G.P., 1969. Wear, 14: 63–74.
Tilly, G.P., 1973. Wear, 23: 87–97.
Tilly, G.P. and Sage, W., 1970. Wear, 16: 447–465.
Timmermans, P.D., 1935. Leid. Geol. Meded., 6: 231–386.
Tinney, E.R. and Bassett, D.L., 1961. J. Hydraul. Div. ASCE, 87 (HY5): 117–133.
Todd, D.K. and Wiegel, R.L., 1952. Trans. Am. Geophys. Union, 33: 217–225.
Toebes, G.H. and Sooky, A.A., 1967. ASCE J. Waterw., Harbors Coastal Eng. Div., 93 (WW2): 213–236.
Toffaleti, F.B., 1965. Misc. Publ. U.S. Dept. Agric., No. 970: 207–228.
Toffaleti, F.N., 1969. J. Hydraul. Div. ASCE, 95 (HY1): 225–248.
Tollmien, W., 1929. Nachr. Ges. Wiss. Göttingen, No. 28: 21–44.
Tollmien, W. and Grohne, P., 1961. In: G.V. Lachmann (Editor), Boundary Layer and Flow Control, vol. 2. Pergamon, Oxford, pp. 602–636.
Tomotika, S. and Aoi, T., 1953. Q. J. Mech. Appl. Math., 6: 290–312.
Topping, W.W., 1972. N.Z. J. Geol. Geophys., 15: 476–490.
Topping, W.W., 1973. N.Z. J. Geol. Geophys., 16: 497–523.
Topping, W.W. and Kohn, B.P., 1973. N.Z. J. Geol. Geophys., 16: 375–395.
Torobin, L.B. and Gauvin, W.H., 1959a. Can. J. Chem. Eng., 37: 129–141.
Torobin, L.B. and Gauvin, W.H., 1959b. Can. J. Chem. Eng., 37: 167–176.
Torobin, L.B. and Gauvin, W.H., 1959c. Can. J. Chem. Eng., 37: 224–236.
Torobin, L.B. and Gauvin, W.H., 1960a. Can. J. Chem. Eng., 38: 142–153.
Torobin, L.B. and Gauvin, W.H., 1960b. Can. J. Chem. Eng., 38: 189–200.
Tory, E.M., Corchrane, N.A. and Waddell, S.R., 1968. Nature (London), 220: 1023–1024.
Tory, E.M., Church, B.H., Tam, R.K. and Ratner, M., 1973. Can. J. Chem. Eng., 51: 484–493.

Townsend, A.A., 1956. The Structure of Turbulent Shear Flow. Cambridge University Press, Cambridge.
Townsend, A.A., 1958. In: H. Görtler (Editor), Boundary Layer Research. Springer, Berlin, pp. 1–15.
Townsend, A.A., 1970. J. Fluid Mech., 41: 13–46.
Townsend, A.A., 1972. J. Fluid Mech., 55: 719–735.
Townsend, A.A., 1976. The Structure of Turbulent Shear Flow. Cambridge University Press, Cambridge, 2nd ed.
Trefethen, J.M. and Dow, R.L., 1960. J. Sediment. Petrol., 30: 589–602.
Trendall, A.F., 1965. Annu. Rep. Geol. Surv. W. Aust. 1964, pp. 51–55.
Trenhaile, A.S., 1973. J. Sediment. Petrol., 43: 558–568.
Trevelyan, B.J. and Mason, S.G., 1951. J. Colloid Sci., 6: 354–367.
Tricart, J., 1954a. Rev. Géomorphol. Dyn., 5: 97–101.
Tricart, J., 1954b. Rev. Géomorphol. Dyn., 5: 124–132.
Tricart, J., 1957. Rev. Géomorphol. Dyn., 8: 11–12.
Tricart, J., 1967. Formes littoral tropicaux. Photo Interpretation No. 2, Fasc. 5, pp. 29–47. Editions Technip, Paris.
Tricart, J. and Mainguet, M., 1965. Rev. Géomorphol. Dyn., 15: 110–121.
Tricart, J.L.F., 1977. Ann. Géogr., 86: 1–54.
Trichet, J., 1963. Bull. Soc. Géol., Fr., (7) 5: 617–621.
Tritton, D.J., 1967. J. Fluid Mech., 28: 439–462.
Trusheim, F., 1931. Senckenbergiana, 13: 124–139.
Trusheim, F., 1935. Natur Volk, 65: 521–524.
Tschirwinsky, P.N., 1908. Z. Gletscherk. Eiszeitforsch. Gesch. Klimas, 2: 103–113.
Tsien, H.S., 1943. Q. Appl. Math., 1: 130–148.
Tsoar, H., 1974. Z. Geomorphol., S20: 41–61.
Tsuchiya, A. and Ishizaki, K., 1967. Proc. 12th Congr. Int. Assoc. Hydraul. Res., 1: 479–486.
Tsuchiya, Y., 1969a. Bull. Disaster Prev. Res. Inst., Kyoto Univ., 19: 31–44.
Tsuchiya, Y., 1969b. Proc. 13th Congr. Int. Assoc. Hydraul. Res., 2: 191–198.
Tsuchiya, Y., 1971. Proc. 12th Conf. Coastal Eng., 2: 1417–1427.
Tsuchiya, Y. and Aoyama, T., 1970. Annu. Disaster Prev. Res. Inst., Kyoto Univ., No. 13B: 199–216.
Tsuchiya, Y. and Kawata, Y., 1970. Annu. Disaster Prev. Res. Inst., Kyoto Univ., No. 13B: 217–232.
Tsuchiya, Y. and Kawata, Y., 1971. Annu. Disaster Prev. Res. Inst., Kyoto Univ., No. 14B: 311–325.
Tsuchiya, Y. and Kawata, Y., 1973. Proc. 13th Conf. Coastal Eng., 2: 1617–1625.
Tsuchiya, Y., Watado, K. and Aoyama, T., 1969. Annu. Disaster Prev. Res. Inst., Kyoto Univ., No. 12B: 475–490.
Tsutsumi, Y., 1973. Powder Technol., 7: 181–188.
Tucholke, B.E. and Eittreim, S., 1974. J. Geophys. Res., 79: 4115–4118.
Tucker, M.E., 1973. Geol. Mijnbouw, 52: 203–215.
Tucker, M.J., 1963. Proc. Inst. Civ. Eng., 26: 305–316.
Tullett, M.T., 1980. J. Earth Sci. R. Dublin Soc., 3: 35–39.
Tunstall, E.B. and Inman, D.L., 1975. J. Geophys. Res., 80: 3475–3484.
Turner, B.R., 1977. Trans. Proc. Geol. Soc. S. Afr., 80: 241–252.
Turner, F.J. and Weiss, L.E., 1963. Structural Analysis of Metamorphic Tectonites. McGraw-Hill, New York, N.Y.
Turner, J.S., 1973. Buoyancy Effects in Fluids. Cambridge University Press, Cambridge.
Twenhofel, W.H., 1932. Treatise on Sedimentation. Williams and Wilkins, Baltimore, Md., 2nd ed.
Uchupi, E., 1967. Deep-Sea Res., 14: 635–639.

Uda, T. and Hino, M., 1975. Proc. Jpn. Soc. Civ. Eng., No. 237: 27–36.
Udden, J.A., 1916. J. Geol., 24: 123–129.
Ueda, H. and Hinze, J.O., 1975. J. Fluid Mech., 67: 125–143.
Ulate, C.A. and Corrales, M.F., 1966. J. Hydraul. Div. ASCE, 96 (HY6): 117–129.
Ulrich, J., 1972. Die Küste, 23: 112–121.
Ulrich, J., 1973. Dtsch. Hydrogr. Z. Ergänzungsh., B14: 1–31.
Ulrich, J. and Pasenau, H., 1973. Die Küste, 22: 95–112.
Umeji, A.C., 1975. Geol. J., 10: 107–130.
Ünlüata, Ü. and Mei, C.C., 1970. J. Geophys. Res., 75: 7611–7618.
Unrug, R., 1957. Acta Geol. Pol., 7: 217–257.
Unrug, R., 1959. Ann. Soc. Géol. Pol., 29: 197–225.
Urbain, P., 1937. Bull. Soc. Géol. Fr., (5) 7: 341–346.
Ursell, F., 1951. Proc. Cambridge Philos. Soc. Math. Phys. Sci., 47: 348–358.
Ursell, F., 1952. Proc. R. Soc. London, Ser. A., 214: 79–97.
Ursell, F., 1953. Proc. Cambridge Philos. Soc. Math. Phys. Sci., 49: 145–150.
USACERC (U.S. Coastal Engineering Research Center), 1975. Shore Protection Manual, vols. I–III. Coastal Engineering Research Center, Fort Belvoir, Va.
USWES (U.S. Waterways Experiment Station), 1935. Studies of river bed materials and their movement with special reference to the lower Mississippi River. U.S. Army Corps Eng., Waterw. Exp. Stn., Pap. No. 17.
USWES (U.S. Waterways Experiment Station), 1939. Study of materials in suspension, Mississippi River. Tech. Mem. U.S. Army Corps Eng., Waterw. Exp. Stn., No. 122-1.
Vaché-Grandet, C., 1959. Trav. Inst. Rech. Sahariennes, 18: 161–171.
Vacher, L., 1973. In: D.R. Coates (Editor), Coastal Geomorphology. State University of New York, Binghamton, N.Y., pp. 355–391.
Valentik, L. and Whitmore, R.C., 1965. Br. J. Appl. Phys., 16: 1197–1203.
Van Andel, T.H. and Komar, P.D., 1969. Geol. Soc. Am. Bull., 80: 1163–1190.
Van Beek, J.L. and Koster, E.A., 1972. Sedimentology, 19: 237–256.
Van Bemmelen, R.W., 1949. The Geology of Indonesia. 1. General Geology. Government Printing Office, The Hague.
Van Burkalow, A., 1945. Bull. Geol. Soc. Am., 56: 669–708.
Van der Graaff, W.J.E., 1972. Annu. Rep. W. Aust. Geol. Surv., 1971, pp. 56–88.
Van der Knaap, W. and Eijpe, R., 1968. Sedimentology, 11: 115–124.
Van der Linden, W.J.M., 1963. Geol. Ultraiectina, No. 12.
Van Gelder, A., 1974. Sedimentation in the marine margin of the Old Red continent, south of Cork, Ireland. University of Utrecht, Utrecht.
Van Hise, C.R., 1896. Rep. U.S. Geol. Surv., Part I, No. 16: 571–843.
Van Olphen, H., 1963. An Introduction to Clay Colloid Chemistry. Interscience, New York, N.Y.
Van Straaten, L.M.J.U., 1950. Tijdschr. K. Ned. Aardrijksk. Genoot., 67: 76–81.
Van Straaten, L.M.J.U., 1953a. Geol. Mijnbouw, 15: 1–11.
Van Straaten, L.M.J.U., 1953b. Geol. Mijnbouw, 15: 31–43.
Van Straaten, L.M.J.U., 1954a. Leid. Geol. Meded., 19: 1–110.
Van Straaten, L.M.J.U., 1954b. Geol. Mijnbouw, 16: 25–47.
Van Straaten, L.M.J.U., 1961. Geol. Mijnbouw, 40: 333–346, 363–391.
Van Straaten, L.M.J.U., 1964. In: A.H. Bouma and A. Brouwer (Editors), Turbidites. Elsevier, Amsterdam, pp. 142–147.
Van Straaten, L.M.J.U. and Kuenen, P.H., 1957. Geol. Mijnbouw, 19: 329–354.
Van Straaten, L.M.J.U. and Kuenen, P.H., 1958. J. Sediment. Petrol., 28: 406–413.
Van Veen, J., 1935. Hydrogr. Rev., 12: 21–29.
Van Veen, J., 1936. Onderzoekingen in De Hoofden. Algemeene Landsdrukkerij, The Hague.
Van Veen, J., 1938. Geol. Meere Binnengewäss., 2: 62–86.

Van Weering, J.C.E., 1975. Geol. Mijnbouw, 54: 130–145.

Van Wezer, J.R., Lyons, J.W., Kim, K.Y. and Colwell, R.E., 1963. Viscosity and Flow Measurement. Interscience, New York, N.Y.

Vanney, J.R., 1965. Mar. Geol., 3: 195–222.

Vanoni, V.A., 1946. Trans. ASCE, 111: 67–102.

Vanoni, V.A., 1953. Proc. 5th Hydraul. Conf. Univ. Iowa, pp. 137–158.

Vanoni, V.A., 1964. Measurements of the critical shear stress for entraining five sediments in a boundary layer. Calif. Inst. Technol., W.M. Keck Lab. Hydraul. Water Resour., Rep. KH-R-7.

Vanoni, V.A., 1974. J. Hydraul. Div. ASCE, 100 (HY3): 363–377.

Vanoni, V.A. and Brooks, N.H., 1957. Laboratory Studies of the Roughness and Suspended Load of Alluvial Streams. U.S. Army Corps Eng., Mo. River Div., Sediment Ser., No. 11.

Vanoni, V.A. and Hwang, L.S., 1967. J. Hydraul. Div. ASCE, 93 (HY3): 121–144.

Varnes, D.J., 1958. In: E.B. Eckell (Editor), Landslides and Engineering Practice. U.S. Highw. Res. Board, Spec. Rep., No. 29: 20–47.

Vasseur, P. and Cox, R.G., 1976. J. Fluid Mech., 78: 385–413.

Vasseur, P. and Cox, R.G., 1977. J. Fluid Mech., 80: 561–591.

Vause, J.E., 1959. J. Sediment. Petrol., 29: 555–563.

Velinkanova, M. and Yarnykh, N.A., 1972. Soviet Hydrology Selected Papers for 1970, No. 5: 426–440.

Velthuizen, H.G.M. and Van Wijngaarden, L., 1969. J. Fluid Mech., 39: 817–829.

Verlaque, C., 1958. Trav. Inst. Rech. Sahariennes, 17: 13–57.

Verma, P.D. and Bhatt, B.S., 1974. Proc. Indian Acad. Sci., A80: 174–183.

Verma, P.D. and Bhatt, B.S., 1976. Appl. Sci. Res., 32: 61–72.

Verman, L.C. and Banerjee, S., 1946. Nature (London), 157: 584–585.

Verstappen, H.T., 1970. Z. Geomorphol., S10: 104–120.

Vickers, J.A., Francis, J.R.D. and Grant, A.W., 1968. Erosion of sewers and drains. CIRIA Res. Rep. No. 14.

Vincent, C.E. and Smith, D.J., 1976. Estuarine Coastal Mar. Sci., 5: 641–652.

Vincent, G.E., 1958. Proc. 6th Conf. Coastal Eng., pp. 326–355.

Vincent, G.E. and Ruellan, F., 1957. Houille Blanche, 12 (B): 693–706.

Vinopal, R.J. and Coogan, A.H., 1978. J. Sediment. Petrol., 48: 7–24.

Virkkala, K., 1951. Bull. Comm. Géol. Finl., No. 155: 1–66.

Virkkala, K., 1960. Bull. Comm. Géol. Finl., No. 188: 161–176.

Virkkala, K., 1961. Bull. Comm. Géol., Finl., No. 196: 214–241.

Visher, G.S. and Howard, J.D., 1974. J. Sediment. Petrol., 44: 502–521.

Visscher, W.M. and Bolsterli, M., 1972a. Nature (London), 239: 504–507.

Visscher, W.M. and Bolsterli, M., 1972b. Random packing of equal and unequal spheres in two and three dimensions. Univ. Calif., Los Alamos Sci. Lab., Rep. LA-D-C-72-445.

Visser, M.J., 1980. Geology, 8: 543–546.

Vitale, P., 1979. ASCE J. Waterw., Harbors Coastal Eng. Div., 105 (WW3): 229–245.

Voight, B., 1978. Rockslides and Avalanches. 1. Natural Phenomena. Elsevier, Amsterdam.

Voight, B. and Pariseau, W.G., 1978. In: B. Voight (Editor), Rockslides and Avalanches. 1. Natural Phenomena. Elsevier, Amsterdam, pp. 1–67.

Vold, M.J., 1959a. J. Colloid Sci., 14: 168–174.

Vold, M.J., 1959b. J. Phys. Chem., 63: 1608–1612.

Vold, M.J., 1963. J. Colloid Sci., 18: 684–695.

Volkov, I.A., 1957. Meteorol. Abstr., 8: 334.

Vollbrecht, K., 1953. Geologie, 2: 268–276.

Vollmers, H. and Wolf, G., 1969. Wasserwirtschaft, 59: 292–297.

Völpel, A. and Samu, G., 1966. Mitteilungsbl. Bundesanst. Wasserbau, 24: 1–20.

Von Bertsbergh, J.W.B., 1940. Geol. Rundsch., 31: 328–364.

Von Bruun, V. and Hobday, D.K., 1976. J. Sediment. Petrol., 46: 670–679.
Von Bruun, V. and Mason, T.R., 1977. Sediment. Geol., 18: 245–255.
Von Engelhardt, W., 1960. Der Porenraum der Sedimente. Springer, Berlin.
Von Engelhardt, W. and Gaida, K.H., 1963. J. Sediment. Petrol., 33: 919–930.
Von Engelhardt, W. and Pitter, H., 1951. Heidelb. Beitr. Mineral. Petrogr., 2: 477–491.
Von Kármán, T., 1940. Bull. Am. Math. Soc., 46: 615–683.
Von Kármán, T., 1947. Collected Works, 4: 352–356.
Von Kármán, T., 1953. Colloq. Int. C.N.R.S., No. 35: 103–108.
Von Rad, U., 1968. J. Sediment. Petrol., 38: 1120–1154.
Von Rad, U., 1970. Geol. Rundsch., 60: 331–354.
Von Seifert, G., 1954. Meyniana, 2: 126–189.
Vos, R.G., 1977. J. Sediment. Petrol., 47: 1242–1260.
Vos, R.G. and Eriksson, K.A., 1977. Sediment. Geol., 18: 161–173.
Vossmerbäumer, H., 1969. Neues Jahrb. Geol. Paläontol. Monatsh., No. 9: 559–576.
Vossmerbäumer, H., 1970. Neues Jahrb. Geol. Paläontol. Monatsh., No. 7: 427–432.
Vreugdenhil, C.B. and De Vries, M., 1967. Computations on non-steady bedload transport by a pseudo-viscosity method. Delft Hydraul. Lab. Publ., No. 45.
Vucetich, C.G. and Pullar, W.A., 1973. N.Z. J. Geol. Geophys., 16: 745–780.
Waddell, E., 1973. Coastal Eng. Jpn., 16: 61–71.
Waddell, E., 1976. In: R.A. Davis and R.L. Ethrington (Editors), Beach and Nearshore Sedimentation. Soc. Econ. Paleontol. Mineral., Spec. Publ., 24: 115–125.
Wadell, H., 1934. J. Franklin Inst., 217: 459–490.
Wadell, H., 1935. J. Geol., 43: 250–280.
Wadell, H., 1936. Geogr. Ann., 18: 74–92.
Wadsworth, J., 1960. Experimental examination of local processes in packed beds of homogeneous spheres. Nat. Res. Coun. Can., Mech. Eng. Rep., No. MT-41.
Wadsworth, J., 1963. In: International Developments in Heat Transfer. American Society of Mechanical Engineers, New York, N.Y., pp. 760–769.
Wadsworth, W.J., 1973. Miner. Sci. Eng., 5: 25–35.
Wakeman, R.J., 1975. Powder Technol., 11: 297–299.
Wakiya, S., 1956. Res. Rep. Fac. Eng. Niigata Univ., No. 5.
Wakiya, S., 1957. J. Phys. Soc. Jpn., 12: 1130–1141.
Walden, H., 1964. Dtsch. Hydrogr. Z., 17: 232–235.
Waldron, H.H., 1967. Bull. U.S. Geol. Surv., 1241-I.
Walker, B.P. and Whitaker, T., 1967. Geotechnique, 17: 161–167.
Walker, G.P.L., 1971. J. Geol., 79: 696–714.
Walker, G.P.L., 1973. Geol. Rundsch., 62: 431–446.
Walker, G.P.L. and Croasdale, R., 1971. J. Geol. Soc. London, 127: 17–55.
Walker, G.P.L. and Croasdale, R., 1972. Bull. Volcanol., (2) 35: 303–317.
Walker, G.P.L., Wilson, L. and Bowell, E.L.G., 1971. Geophys. J. R. Astron. Soc., 22: 377–383.
Walker, H.J., 1973. In: M.E. Britton (Editor), Alaskan Arctic Tundra. Arct. Inst. North America, Tech. Pap., No. 25: 49–92.
Walker, R.G., 1963. Sedimentology, 2: 173–188.
Walker, R.G., 1965. Proc. Yorks. Geol. Soc., 35: 1–32.
Walker, R.G., 1966. J. Sediment. Petrol., 36: 90–114.
Walker, R.G., 1967a. J. Sediment. Petrol., 37: 1052–1058.
Walker, R.G., 1967b. J. Sediment. Petrol., 37: 25–43.
Walker, R.G., 1969a. Can. J. Earth Sci., 6: 383–391.
Walker, R.G., 1969b. J. Geol., 77: 125–143.
Walker, R.G. and Middleton, G.V., 1977. Geosci. Can., 4: 182–190.
Wallace, J.M., Eckelmann, H. and Brodkey, R.S., 1972. J. Fluid Mech., 54: 39–48.

Wallet, A. and Ruellan, F., 1950. Houille Blanche, 5: 483–489.
Walling, D.E., 1974. Spec. Publ. Inst. Br. Geogr., No. 6: 169–192.
Walling, D.E. and Gregory, K.J., 1970. J. Hydrol., 11: 129–144.
Wallington, C.E., 1959. Weather, 14: 263–270.
Walton, E.K., 1967. Scott. J. Geol., 3: 306–317.
Walton, F.D. and Goodell, H.G., 1972. Mar. Geol., 13: 1–29.
Wang, H. and Liang, S.S., 1975. J. Geophys. Res., 80: 3488–3494.
Wang, W.L., 1971. J. Soil Mech. Found. Div. ASCE, 97 (SM12): 1635–1655.
Ward, B.D., 1969. Water Resour. Res., 5: 1090–1096.
Ward, W.H., 1945. Geogr. J., 105: 170–191.
Warren, A., 1970. Z. Geomorphol., S10: 154–180.
Warren, A., 1971. Geogr. J., 137: 458–461.
Warren, A., 1972. Sedimentology, 19: 37–44.
Warren, A., 1976. J. Geol., 84: 685–700.
Washburn, A.L., 1967. Medd. Grønl., 166 (4): 1–297.
Washburn, A.L., 1973. Periglacial Processes and Environments. Arnold, London.
Wass, R.E., Conolly, J.E. and MacIntyre, R.J., 1970. Mar. Geol., 9: 63–73.
Watkins, N.D., Sparks, R.S.J., Sigurdson, H., Huang, T.C., Federman, A. and Carey, S., 1978. Nature (London), 271: 122–126.
Watson, D.F., 1975. Math. Geol., 7: 349–354.
Watson, E., 1969. Biul. Periglac., No. 18: 95–112.
Watson, E. and Watson, S., 1967. Geol. J., 5: 419–440.
Watson, E.J., 1964. J. Fluid Mech., 20: 481–499.
Watson, R.L., 1969. Water Resour. Res., 5: 1147–1150.
Watters, G.Z. and Rao, M.V.P., 1971. J. Hydraul. Div. ASCE, 97 (HY3): 421–439.
Way, J.H., 1968. J. Sediment. Petrol., 38: 424–433.
Webber, N.B. and Bullock, G.N., 1971. In: D.A. Howells, I.P. Haigh and C. Taylor (Editors), Dynamic Waves in Civil Engineering. Wiley-Interscience, London.
Webster, F., 1971. Deep-Sea Res., 18: 885–893.
Weertman, J., 1961. J. Glaciol., 3: 965–978.
Wegner, T., 1932. Zentralbl. Mineral. Geol. Paläontol., Abt. B: 31–34.
Wehausen, J.V. and Laitone, E.V., 1960. In: Handbuch der Physik, vol. 9. Springer, Berlin, pp. 446–778.
Weir, J.E., 1962. Geol. Soc. Am., Spec. Pap., No. 73: 72.
Weller, J.M., 1959. Bull. Am. Assoc. Pet. Geol., 43: 273–310.
Wells, J.T. and Ludwick, J.C., 1974. J. Sediment. Petrol., 44: 1029–1036.
Wells, M.K., 1962. Overseas Geol. Miner. Resour. Suppl. Ser. No. 4.
Wentworth, C.K., 1922. J. Geol., 30: 377–392.
Wermund, E.G., 1965. Sedimentology, 5: 69–79.
Werner, F., 1968. Meyniana, 18: 79–105.
Werner, F. and Newton, R.S., 1970. Meyniana, 20: 83–90.
Werner, F. and Newton, R.A., 1975a. In: L.E. Cronin (Editor), Estuarine Research, vol. 2. Academic Press, New York, N.Y.
Werner, F. and Newton, R.S., 1975b. Mar. Geol., 19: 29–59.
Werner, F., Arntz, W.E. and Tauchgruppe Kiel, 1974. Meyniana, 26: 39–62.
Wessel, J.M., 1969. Sedimentary history of Upper Triassic alluvial fan complexes in north-central Massachusetts. Univ. Mass., Dep. Geol., Contrib., No. 2.
Wessels, W.P.J. and Strelkov, T., 1968. J. Irrig. Drainage Div. ASCE, 94 (IR1): 1–22.
West, I.M., 1964. Clays Clay Miner., 19: 209–222.
West, R.C. and Donner, J.J., 1956. Q. J. Geol. Soc. London, 112: 69–91.
Westman, A.E.R. and Hugill, H.R., 1930. J. Am. Ceram. Soc., 9: 767–779.
Whetten, J.T., 1965. Geol. Soc. Am. Bull., 76: 43–56.

Whetten, J.T. and Fulham, J.J., 1967. Proc. 12th Congr. Int. Assoc. Hydraul. Res., 1: 107–114.
Whetten, J.T., Kelley, J.C. and Hanson, L.G., 1969. J. Sediment. Petrol., 39: 1149–1166.
White, B.R. and Schulz, J.C., 1977. J. Fluid Mech., 81: 497–512.
White, C.M., 1940. Proc. R. Soc. London, Ser. A., 174: 322–338.
White, H.E. and Walton, S.F., 1937. J. Am. Ceram. Soc., 20: 155–166.
White, L.P., 1971. Geogr. J., 137: 69–73.
White, S.J., 1970. Nature (London), 228: 152–153.
White, W.A., 1961. J. Sediment. Petrol., 31: 560–570.
White, W.R., Milli, H. and Crabbe, A.D., 1973a. Sediment transport: an appraisal of available methods. 1. Survey of existing theories. Hydraul. Res. Stn., Rep. INT 119.
White, W.R., Milli, H. and Crabbe, A.D., 1973b. Sediment transport: an appraisal of existing methods. 2. Performance of theoretical methods when applied to flume and field data. Hydraul. Res. Stn., Rep. INT 119.
White, W.R., Milli, H. and Crabbe, A.D., 1975. Proc. Inst. Civ. Eng., 59: 265–292.
White, W.S., 1952. J. Sediment. Petrol., 22: 189–199.
Whitmore, R.L., 1969. Proc. 5th Int. Congr. Rheol., 1: 353–360.
Wiegel, R.L., 1950. Trans. Am. Geophys. Union, 31: 377–385.
Wiegel, R.L., 1964. Oceanographical Engineering. Prentice-Hall, Englewood Cliffs, N.J.
Wiegel, R.L. and Fuchs, R.A., 1955. Trans. Am. Geophys. Union, 36: 975–984.
Wilcox, R.E., 1959. Bull. U.S. Geol. Surv., 1028-N.
Wilcoxson, J.A., 1962. Compass Sigma Gamma Epsilon, 39: 65–76.
Wilkinson, D.L., 1970. Studies in density stratified flows. Univ. N.S.W., Water Res. Lab., Rep. No. 118.
Wilkinson, D.L. and Wood, I.R., 1971. J. Fluid Mech., 47: 241–256.
Wilkinson, D.L. and Wood, I.R., 1972. J. Hydraul. Res., 10: 305–324.
Wilkinson, W.L., 1960. Non-Newtonian Fluids. Pergamon, New York, N.Y.
Willett, H.C. and Sanders, F., 1959. Descriptive Meteorology. Academic Press, New York, N.Y., 2nd ed.
Willetts, B.B., 1970. Proc. 1st Conf. Hydraul. Transp. Solids Pipes, Pap. H2, pp. H2-17–H2-35.
Willetts, B.B. and Drossos, M.E., 1975. J. Hydraul. Div. ASCE, 101 (HY12): 1477–1488.
Williams, A.T. and Gulbrandsen, L.F., 1977. Cambria, 4: 174–186.
Williams, G.E., 1966. J. Sediment. Petrol., 36: 742–746.
Williams, G.E., 1968. J. Sediment. Petrol., 38: 136–140.
Williams, G.E., 1970a. Z. Geomorphol., S10: 40–63.
Williams, G.E., 1970b. Scott. J. Geol., 6: 409–411.
Williams, G.E., 1971. Sedimentology, 17: 1–40.
Williams, G.P., 1964. Sedimentology, 3: 257–287.
Williams, G.P., 1966. J. Sediment. Petrol., 36: 255–259.
Williams, G.P., 1967. U.S. Geol. Surv., Prof. Pap., 562-B.
Williams, G.P., 1970. U.S. Geol. Surv., Prof. Pap., 562-H.
Williams, P.B. and Kemp, P.H., 1971. J. Hydraul. Div. ASCE, 97 (HY4): 505–522.
Williams, P.B. amd Kemp, P.H., 1972. J. Hydraul. Div. ASCE, 98 (HY6): 1057–1070.
Williams, P.F., 1968. J. Sediment. Petrol., 38: 332–362.
Williams, P.F. and Rust, B.R., 1969. J. Sediment. Petrol., 39: 349–379.
Williams, R.E., 1968. Science, 161: 276–277.
Williamson, M., 1972. The Analysis of Biological Populations. Edward Arnold, London.
Williamson, W.C., 1887. Mem. Proc. Manchester Lit. Philos. Soc., (3) 10: 19–29.
Willis, D.G., 1977. Geol. Soc. Am. Bull., 88: 883–894.
Willis, J.C., 1969. Water Resour. Res., 5: 1322–1329.
Willis, J.C., 1972. J. Hydraul. Res., 10: 205–225.
Willis, J.C., 1979. J. Hydraul. Div. ASCE, 105 (HY7): 801–816.

Willis, J.C. and Coleman, 1969. Water Resour. Res., 5: 1330–1336.
Willis, J.C. and Kennedy, J.F., 1977. Sediment discharge of alluvial streams calculated from bed-form statistics. Iowa Inst. Hydraul., Res. Rep. No. 202.
Willis, J.C., Coleman, N.L. and Ellis, W.M., 1972. J. Hydraul. Div. ASCE, 98 (HY3): 489–501.
Willmarth, W.W., 1964. Phys. Fluids, 7: 219–220.
Willmarth, W.W., 1975. Annu. Rev. Fluid Mech., 7: 13–38.
Willmarth, W.W. and Bogar, T.J., 1977. Phys. Fluids, 20 (10, II): 9–21.
Willmarth, W.W. and Lu, S.S., 1972. J. Fluid Mech., 55: 65–92.
Willmarth, W.W. and Tu, B.J., 1967. Phys. Fluids, 10: S134–137.
Willmarth, W.W. and Woolridge, C.E., 1962. J. Fluid Mech., 14: 187–210.
Willmarth, W.W. and Yang, C.S., 1970. J. Fluid Mech., 41: 47–80.
Willmarth, W.W., Hawk, N.E. and Harvey, R.L., 1964. Phys. Fluids, 7: 197–208.
Wills, J.A.B., 1971. J. Fluid Mech., 45: 65–90.
Wilson, I.G., 1971a. Geogr. J., 137: 180–199.
Wilson, I.G., 1971b. Geogr. Mag., January 1971: 264–270.
Wilson, I.G., 1972a. New Sci.; 53: 634–637.
Wilson, I.G., 1972b. J. Sediment. Petrol., 42: 667–669.
Wilson, I.G., 1972c. Sedimentology, 19: 173–210.
Wilson, I.G., 1973. Sediment. Geol., 10: 77–106.
Wilson, J.B., 1967. Scott. J. Geol., 3: 329–371.
Wilson, J.B., 1977. In: R.A. Geyer (Editor), Submersibles and their Uses in Oceanography and Engineering. Elsevier, Amsterdam, pp. 151–167.
Wilson, Lee, 1972. Water Resour. Res., 8: 1470–1479.
Wilson, Lionel, 1972. Geophys. J. R. Astron. Soc., 30: 381–392.
Windom, H.L. and Chamberlain, C.F., 1978. J. Sediment. Petrol., 48: 385–388.
Wing-Fatt, L. and Stacey, F.D., 1966. Pure Appl. Geophys., 64: 78–80.
Winkelmolen, A.M., 1969. Experimental rollability and natural shape sorting of sand. University of Groningen, Groningen.
Winn, R.D. and Dott, R.H., 1977. Geology, 5: 41–44.
Wise, M.E., 1952. Philips Res. Rep., 7: 321–343.
Wise, M.E., 1960. Philips Res. Rep., 15: 101–106.
Wobber, F.J., 1966. Proc. Geol. Assoc., 77: 127–137.
Wobber, F.J., 1967. J. Sediment. Petrol., 37: 1233–1235.
Wollaston, W.H., 1813. Philos. Trans. R. Soc. London, 103: 51–63.
Wood, A. and Smith, A.J., 1959. Q. J. Geol. Soc. London, 114: 163–190.
Wood, I.R., 1965. Studies in unsteady self-preserving turbulent flows. Univ. N.S.W., Water Res. Lab., Rep. No. 81.
Wood, I.R., 1967. J. Hydraul. Div. ASCE, 93 (HY2): 34–42.
Wood, I.R., 1970. J. Fluid Mech., 42: 671–687.
Wood, P.A., 1977a. Sedimentology, 24: 437–445.
Wood, P.A., 1977b. J. Trop. Geogr., 45: 65–69.
Wood, W.H., 1970. J. Sediment. Petrol., 40: 7–28.
Wood, W.W. and Jacobson, J.D., 1957. J. Chem. Phys., 27: 1207–1208.
Wood, W.W. and Parker, F.R., 1957. J. Chem. Phys., 27: 720–733.
Woodcock, N.H., 1976. Proc. Geol. Assoc., 87: 169–182.
Woodford, A.O., 1935. Am. J. Sci., 223: 518–525.
Woodworth, J.N., 1901. Am. Geol., 27: 281–283.
Wopfner, H., 1970. Q. Geol. Notes, Geol. Surv. S. Aust., No. 34: 5–10.
Wopfner, H., Freytag, I.B. and Heath, G.R., 1970. Bull. Am. Assoc. Pet. Geol., 54: 383–416.
Worrall, G.A., 1974. Z. Geomorphol., 18: 291–302.
Worsley, D. and Edwards, M.B., 1976. Årb. Nor. Polarinst., 1974: 17–34.

Wortmann, F.X., 1953. Z. Angew. Phys., 5: 201–206.
Wright, C.A., 1936. J. Res. U.S. Natl. Bur. Stand., 17: 193–206.
Wright, H.E., 1957. Geogr. Ann., 39: 19–31.
Wright, H.E., 1962. Geol. Soc. Am. Bull., 73: 73–100.
Wright, L.D. and Coleman, J.M., 1973. Bull. Am. Assoc. Pet. Geol., 57: 370–398.
Wright, L.D. and Coleman, J.M., 1974. J. Geol., 82: 751–778.
Wright, L.D., Sonu, C.J. and Kielhorn, W.V., 1972. Mar. Geol., 12: 43–58.
Wright, L.D., Coleman, J.M. and Thom, B.G., 1973. J. Geol., 81: 15–41.
Wright, L.D., Coleman, J.M. and Thom, B.G., 1975. In: L.E. Cronin (Editor), Estuarine Research, vol. 2. Academic Press, New York, N.Y., pp. 309–321.
Wright, P., 1976. Sedimentology, 23: 705–712.
Wulf, G.R., 1962. J. Sediment. Petrol., 32: 327–329.
Wulf, G.R., 1963. Bull. Am. Assoc. Pet. Geol., 47: 691–695.
Wunderlich, F., 1967. Senckenbergiana Lethaea, 48: 337–343.
Wunderlich, F., 1969. Senckenbergiana Marit., 50: 107–146.
Wunderlich, F., 1970. J. Sediment. Petrol., 40: 102–130.
Wunderlich, F., 1972. Senckenbergiana Marit., 4: 47–79.
Wunderlich, F., 1973. Senckenbergiana Marit., 5: 161–164.
Wunderlich, F., 1978. Secnkenbergiana Marit., 10: 257–267.
Wurster, P., 1958a. Geol. Rundsch., 47: 322–359.
Wurster, P., 1958b. Neues Jahrb. Geol. Paläontol. Monatsh., pp. 479–489.
Wurster, P., 1964. Mitt. Geol. Staatsanst. Hamb., 33: 1–140.
Wygnanski, I.J. and Champagne, F.H., 1973. J. Fluid Mech., 59: 281–335.
Wygnanski, I., Sokolov, M. and Friedman, D., 1976. J. Fluid Mech., 78: 785–819.
Xanthopoulos, T. and Koutitas, C., 1976. J. Hydraul. Res., 14: 321–331.
Yagashita, K., 1973. J. Geol. Soc. Jpn., 79: 381–390.
Yagashita, K. and Morris, R.C., 1979. Geol. Mag., 116: 105–116.
Yalin, M.S., 1963. J. Hydraul. Div. ASCE, 89 (HY3): 221–250.
Yalin, M.S., 1964. J. Hydraul. Div. ASCE, 90 (HY5): 105–119.
Yalin, M.S., 1971. Proc. 14th Congr. Int. Assoc. Hydraul. Res., 3: 101–108.
Yalin, M.S., 1972. Mechanics of Sediment Transport. Pergamon, Oxford.
Yalin, M.S., 1975. Proc. 16th Congr. Int. Assoc. Hydraul. Res., 2: 212–219.
Yalin, M.S. and Karahan, E., 1979. J. Hydraul. Div. ASCE, 105 (HY4): 381–392.
Yalin, M.S. and Price, W.A., 1975. Proc. 13th Conf. Coastal Eng., 2: 991–1008.
Yalin, M.S. and Price, W.A., 1976. In: Symposium on Modelling Techniques. American Society of Civil Engineers, New York, N.Y., pp. 936–961.
Yalin, M.S. and Russell, R.C.H., 1962. Proc. 8th Conf. Coastal Eng., pp. 151–167.
Yamagata, T., 1978. J. Oceanogr. Soc. Jpn., 34: 97–104.
Yamamoto, K., 1971. J. Phys. Soc. Jpn., 31: 1572–1575.
Yang, C.T., 1972. J. Hydraul. Div. ASCE, 98 (HY10): 1805–1826.
Yang, C.T., 1973. J. Hydraul. Div. ASCE, 99 (HY10): 1679–1704.
Yang, C.T. and Stall, J.B., 1974. Unit stream power for sediment transport in natural rivers. Univ. Ill. Water Resour. Cent., Res. Rep., No. 88.
Yang, T.D., 1979. J. Hydrol., 40: 123–138.
Yen, B.C., 1971. Proc. 14th Congr. Int. Assoc. Hydraul. Res., 3: 83–90.
Yen, B.C. and Liou, Y.C., 1969. Hydraulic resistance in alluvial channels. Univ. Ill. Water Resour. Cent., Res. Rep., No. 22.
Yerazunis, S., Bartlett, J.W. and Nissan, A.H., 1962. Nature (London), 195: 33–35.
Yerazunis, S., Cornell, S.W. and Wintner, B., 1965. Nature (London), 207: 835–837.
Yih, C.S., 1965. Dynamics of Nonhomogeneous Fluids. Macmillan, New York, N.Y.
Yih, C.S. and Guha, C.R., 1955. Tellus, 7: 358–366.
Yorath, C.J., Bornhold, B.D. and Thomson, R.W., 1979. Mar. Geol., 31: 45–58.

Young, A., 1972. Slopes. Oliver and Boyd, Edinburgh.
Young, B.S., 1960. Weather, 15: 92–97.
Young, G.A., 1965. The physics of the base surge. U.S. Nav. Ordnance Lab. Tech. Rep., NOL TR 64-103.
Young, G.M., 1968. Palaeogeogr. Palaeoclimatol. Palaeoecol., 4: 125–153.
Young, G.M. and Jefferson, C.W., 1975. Can. J. Earth Sci., 12: 1734–1748.
Young, G.M. and Long, D.G.F., 1977. Can. J. Earth Sci., 14: 2246–2261.
Young, J.A.T., 1969. Geol. Soc. Am. Bull., 80: 2343–2352.
Young, R.A. and Ross, D.A., 1974. Deep-Sea Res., 21: 289–297.
Young, R.N. and Southard, J.B., 1978. Geol. Soc. Am. Bull., 89: 663–672.
Yu, Y.Y., 1952. Trans. Am. Geophys. Union, 33: 39–41.
Yuen, A.F.H. and Kennedy, J.F., 1971. A laboratory investigation of free surface flows over wavy beds. Univ. Iowa, Inst. Hydraul. Res., Res. Rep., No. 121.
Zabawa, C.F., 1978. Science, 202: 49–51.
Zagustin, K., 1968. J. Hydraul. Res., 6: 163–171.
Zanke, U., 1976a. Mitt. Franzius Inst., 44: 1–111.
Zanke, U., 1976b. Mitt. Franzius Inst., 44: 289–324.
Zeller, J., 1963. Schweiz. Bauztg., 81: 597–602, 620–626, 627–634.
Zenkovitch, V.P., 1967. Processes of Coastal Development. Wiley, New York, N.Y.
Zhemchuzhnikov, Y.A., 1926. Ann. Inst. Mines (Leningrad), 7: 35–69.
Zia, I.Y., Cox, R.G. and Mason, S.G., 1966. Science, 153: 1405–1406.
Zia, I.Y., Cox, R.G. and Mason, S.G., 1967. Proc. R. Soc. (London), Ser. A., 300: 421–444.
Zilberman, M., Wygnanski, I. and Kaplan, R.E., 1977. Phys. Fluids, 20 (10, II): 258–271.
Zilker, D.P., Cook, G.W. and Hanratty, T.J., 1977. J. Fluid Mech., 82: 29–51.
Zimmerman, C. and Kennedy, J.F., 1973. Proc. 15th Congr. Int. Assoc. Hydraul. Res., 1: 1–18.
Zimmerman, H.B., 1971. J. Geophys. Res., 76: 5865–5876.
Zingg, A.W., 1953. Proc. 5th Hydraul. Conf. Iowa City, pp. 111–135.
Znamenskaya, N.S., 1962. Soviet Hydrology Selected Papers 1962, No. 2: 111–116.
Znamenskaya, N.S., 1963. Soviet Hydrology Selected Papers 1963, No. 3: 253–275.
Znamenskaya, N.S., 1966. Soviet Hydrology Selected Papers 1965, No. 5: 415–432.